Edited by
Philippe Serp and
Karine Philippot

Nanomaterials in Catalysis

Related Titles

Tao, F., Schneider, W. A.

Heterogeneous Catalysis at Nanoscale and Energy Applications

2012
ISBN: 978-0-470-95260-3

Lyon, L. A., Serpe, M. J. (eds.)

Hydrogel Micro and Nanoparticles

2012
ISBN: 978-3-527-33033-1

Su, B.-L., Sanchez, C., Yang, X.-Y. (eds.)

Hierarchically Structured Porous Materials

From Nanoscience to Catalysis, Separation, Optics, Energy, and Life Science

2012
ISBN: 978-3-527-32788-1

Farrusseng, D. (ed.)

Metal-Organic Frameworks

Applications from Catalysis to Gas Storage

2011
ISBN: 978-3-527-32870-3

Zecchina, A., Bordiga, S., Groppo, E. (eds.)

Selective Nanocatalysts and Nanoscience

Concepts for Heterogeneous and Homogeneous Catalysis

2011
ISBN: 978-3-527-32271-8

Schmid, G. (ed.)

Nanoparticles

From Theory to Application

2010
ISBN: 978-3-527-32589-4

Cejka, J., Corma, A., Zones, S. (eds.)

Zeolites and Catalysis

Synthesis, Reactions and Applications

2010
ISBN: 978-3-527-32514-6

Astruc, D. (ed.)

Nanoparticles and Catalysis

2008
ISBN: 978-3-527-31572-7

Edited by Philippe Serp and Karine Philippot

Nanomaterials in Catalysis

With a Foreword by Gabor A. Somorjai and Bruno Chaudret

WILEY-VCH

WILEY-VCH Verlag GmbH & Co. KGaA

The Editors

Prof. Dr. Philippe Serp
Toulouse University
Laboratoire de Chimie de Coordination -
UPR8241 CNRS, composante ENSIACET
4 allée Emile Monso
31030 Toulouse Cedex 4
France

Dr. Karine Philippot
Laboratoire de Chimie de Coordination
du CNRS
205 route de Narbonne
BP44099
31077 Toulouse Cedex 4
France

Library of Congress Card No.: applied for

British Library Cataloguing-in-Publication Data
A catalogue record for this book is available from the British Library.

Bibliographic information published by the Deutsche Nationalbibliothek
The Deutsche Nationalbibliothek lists this publication in the Deutsche Nationalbibliografie; detailed bibliographic data are available on the Internet at http://dnb.d-nb.de.

Print ISBN: 978-3-527-33124-6
ePDF ISBN: 978-3-527-65690-5
ePub ISBN: 978-3-527-65689-9
mobi ISBN: 978-3-527-65688-2
oBook ISBN: 978-3-527-65687-5

Typesetting Thomson Digital, Noida, India

Printing and Binding Markono Print Media Pte Ltd, Singapore

Cover Design Grafik-Design Schulz, Fußgönheim

Printed on acid-free paper

Contents

Foreword

Catalysis has been the leading science and technology in the 60s, associated with the raise of the use of oil. Refineries and petrochemistry have then known their golden age. The rarefaction of energy sources and the need for a more rational use of the available energy provides a new opportunity for catalysis to play a leading role in society. If catalysis is traditionally divided into three main fields: heterogeneous, homogeneous and enzymatic, the present tendency is the convergence of these fields towards a molecular approach.

Thus, heterogeneous catalysis that usually is carried out at the solid-gas interface and is perhaps the technologically most utilized form as it provides high turnovers to yield products and low deactivation, which provides their long life. Homogeneous catalysis is usually carried out on transition metal ions surrounded by ligands and dissolved in organic solvents. These catalysts show very high selectivity and they play a major role in the pharmaceutical industry. Enzyme catalysts, of which there are about 3000 in the human body, usually operating in water solution under physiological conditions at room temperature and neutral pH. Catalysis is life! It accounts for most of the processes on this planet, whether operating under planetary conditions of near room temperature or making products with high selectivity by the chemical technologies. Revolution in the synthesis of nanomaterials that provides the opportunity to produce catalysts with controlled size and shape has led to the discovery that the catalytic selectivity and turnover correlates with the size and shape of nanoparticles. This observation, along with characterization techniques, allow us to carry out molecular or atomic scale studies of catalyst particles under reaction conditions and provides revolutionary developments in the field of catalysis science, which is named "nanocatalysis". The result is a rapidly increasing interest for the molecular aspect of catalysis aiming at a new understanding of how catalysts work on molecular and atomic scales. The challenge for the future will be to develop technologies that are very (totally?) selective while little energy demanding and respecting the environment. This process in our contemporary world is commonly called "green chemistry" and can rely on nanocatalysis.

This book entitled *Nanomaterials in Catalysis* by K. Phillippot and P. Serp is a very timely exposure of the new science and technologies of nanocatalysis. Following an excellent introduction of the concepts in nanocatalysis, the fabrication of nanocatalysts in various media is discussed in several important papers ranging from nanoparticles in aqueous phase, nanoclusters and colloids as catalysts precursors, nanoparticles in ionic liquid and supercritical fluids, dendrimers that serve as excellent polymeric supports for nanoparticles and finally nanocatalysts recovery, which addresses the problem of deactivation and regeneration. A section is dedicated to nanoparticle supports like carbon nanotubes and nano oxides which are discussed. The last chapter reviews modeling of nanocatalysts to show the foundation of theoretical treatment of nanocatalysis and nanomaterials that are used as catalysts.

This is a timely book and it will be a very useful addition to those interested in the field of catalysis and its most important extension by the use of nanomaterials to carry out heterogeneous, homogeneous and enzyme catalysis. Studies of nano-catalysts and characterization of these nanomaterial systems lead to an atomic and molecular level understanding of how catalyst materials work. The book describes the frontiers of catalysis on a broad front, and we believe it adds to our knowledge and perhaps rapid evolution of the field of catalysis for the near future.

Berkeley and Toulouse
October 2012

G. A. Somorjai
B. Chaudret

Preface

Process catalysts, for petroleum products, chemicals, pharmaceuticals, synthetic rubber and plastics, among others, represent a \$13 billion-per-year business worldwide, and as such play a vital role in the economy. Thus, the importance of catalysis in the chemical industry is reflected by the following points:

– $>$95% of all products (volume) and $>$70% of all products (processes) are synthesized by means of catalysis,
– $>$80% of the added value in chemical industry is based on catalysis,
– $\sim$20% of the world economy depends directly or indirectly on catalysis.

Catalysts can be classified into two main groups: heterogeneous catalysts (solid-state catalysts) and homogeneous catalysts (transition metal complexes). Of increasing importance are the biocatalysts, enzymes that are protein molecules of colloidal size, which can be classified somewhere between molecular homogeneous catalysts and macroscopic heterogeneous catalysts. There are also intermediate forms, such as homogeneous catalysts attached to solids, also known as immobilized catalysts.

The suitability of these catalysts for an industrial process is governed mainly by the following properties:

– recovery (heterogeneous catalysts $\gg$ homogeneous catalysts),
– activity (heterogeneous catalysts $<$ homogeneous catalysts),
– selectivity (heterogeneous catalysts $\ll$ homogeneous catalysts),
– stability (heterogeneous catalysts $>$ homogeneous catalysts).

By far, and mainly because of the recovery aspect, the most applied catalysts at an industrial level are the heterogeneous ones. Indeed, approximately 80% of all catalytic processes involve heterogeneous catalysts against 15% homogeneous catalysts and 5% biocatalysts. The homogeneous systems are currently limited to reactions for which i) heterogeneous catalysts are not stable with a dissolution of the active phase (methanol carbonylation, hydroformylation), ii) the catalyst can be lost (polymerization), or iii) heterogeneous catalysts are inefficient (asymmetric catalysis).

In the past century, catalysis became the basis of large-scale processes in bulk chemistry and petrochemistry. In the XXI century, shifting demands, energy and new environmental challenges require new catalytic solutions. The two major issues are related to the depletion of raw materials and to the environment/health.

On one hand the depletion of raw materials drives researches:

–on catalysis for energy aiming at: i) improvements in the catalytic processes across the chemical and petroleum industries that will increase resource and energy utilization efficiencies; ii) cost-effective, environmentally sound utilization of energy resources such as coal and biomass; iii) efficient photocatalysts for water splitting and electrocatalysts for fuel cells, and iv) hydrogen storage and carbon dioxide capture and sequestration could benefit from new abilities to catalyze solid-state reactions.

–on new catalytic systems which: i) show high selectivity (atom economy), ii) are able to transform biomass in various chemicals, and iii) are based on cheap and abundant metals.

In parallel, the growing concern over environmental issues and the successful implementation of legislation drives researches in catalysis on two ways to improve the quality of our environment by:

–cleaning up emissions from current processes (catalytic converters, VOC, ozone, pesticides, dioxin . . . , indoor air quality, water pollution, organic particulate), and

–developing alternative cleaner catalytic processes (safer reagents, limited by-products, reduced mass and energy inputs . . .).

Although incremental improvements to catalytic processes will result in a better control of desired chemical transformations, in order to fully realize the needed advancements real breakthroughs must also be achieved:

- in photocatalysis to successfully and economically use solar energy to convert water into oxygen and hydrogen;
- in interfacial electrocatalysis to realize, at a realistically low cost, the efficiency potentially attainable by fuel cells;
- in the gasification and conversion of coal/biomass into transportation fuels;
- in the catalytic conversion of biomass to more useful products;
- in the development of catalysts that operate in aqueous media to fully enable the promise of "green chemistry";
- in the construction of multi-component (nanoscale) catalysts.

In such a challenging context, it is obvious that any new branches of catalysis science should be considered as a potential spur to reach the objectives. A promising approach consists in bridging the gap between homogeneous and heterogeneous catalysis, in order to combine the advantages of each ones. The first efforts date from the 70's with the first International Symposium on Relations between Homogeneous and Heterogeneous Catalysis in 1973. In the middle of the 90's, catalysis joined the nanosciences and nanotechnologies wave, which significantly contributed to reinforce the connections between molecular and solid state catalysis communities.

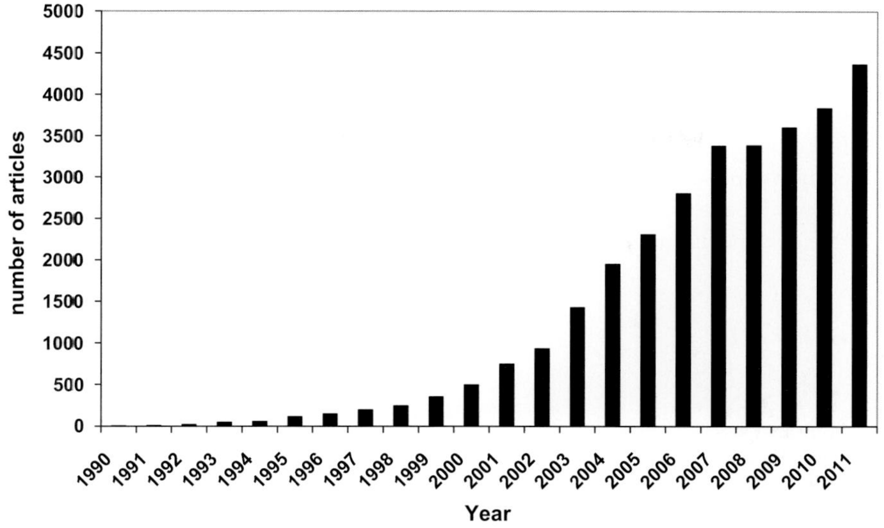

Figure 1 Number of articles including the prefix nano- in their title (from Web of Knowledge).

At that time, we both started our scientific carrier, during a period in which the prefix nano was poorly used (Figure 1). One of us (Karine) as a PhD student and after as a postdoc at Rhodia worked on the homogeneous rhodium catalyzed hydroamination reaction and the synthesis of carbonates with tin-supported catalysts, respectively, before integrating the *Centre National de la Recherche Scientifique* where she develops ligand-functionalized colloidal nanoparticles for application in catalysis. The other (Philippe) as a PhD and then a post-doc student, investigated the potential of chemical vapor deposition methods to prepare supported catalysts or carbon nanomaterials, and is now the team leader of a group that develops new catalytic systems for both homogeneous and supported catalysis. We are both very excited by the catalysis with nanomaterials, due to our double scientific background that allows to conciliate the molecular and solid state point of view, and to work on the bridge in between homogeneous and heterogeneous catalysis.

Thus, the terms "nanocatalyst" and "nanocatalysis" have appeared some years ago as a continuation of the development of nanotechnologies. Today, few books are dedicated to the subject,[1–4] and a recent article [5] raises this interesting question: "Nanocatalysis: Mature science revisited or something really new?" In the present book, the concepts of nanocatalysis are defined, to give to the reader a comprehensive overview of what is a nanocatalyst, and to rationalize the advantages of nanocatalysts related to their activity, selectivity and stability. Thus, each chapter will provide a critical overview of a specific domain of nanocatalysis through the most relevant examples of the literature.

In that sense, this book is the first one that introduces concepts and main achievements, and covers the main aspects of nanocatalysis in general, considering both the active phase, and the support as well as their modeling and characterization.

Since this is a multi-authored book, significant differences in style from chapter to chapter are inevitable, but we have tried to avoid overlaps as much as possible. We thank all the authors for their efforts to meet deadlines, and to follow the format defined for the book. We would also like to acknowledge the assistance of Anne Brennfueher and Lesley Belfit at Wiley-VCH, whose advice has been most helpful at the various stages of preparation of the manuscript. Finally, we hope that the book will be useful to fellow scientists and practitioners and will stimulate further research and discussion on the development of nanomaterials for catalysis.

Toulouse
October 2012

Philippe Serp
Karine Philippot

References

1 Scott, S.L., Crudden, C.M., Jones, C.W. (Eds.) (2003) Kluwer Academics/Plenum Publishers, *Nanostructured catalysts*, New-York.
2 Zhou, B., Hermans, S., Somorjai, G.A. (Eds.) (2004) Kluwer Academics/Plenum Publishers, *Nanotechnology in catalysis*, New-York.
3 Heiz, U., Landman, U. (Eds.) (2008) Springer-Verlag, Berlin, *Nanocatalysis*, Heidelberg, 2007.
4 Astruc, D. (Ed.) (2008) Wiley-VCH Verlag GmbH & Co KGaA, *Nanoparticles and catalysis*, Weinheim.
5 Schlögl, R., Abd Hamid S.B. (2004) Nanocatalysis: Mature Science Revisited or Something Really New? *Angew. Chem. Int. Ed.* **43** (13), 1628–1637.

List of Contributors

Didier Astruc
Université Bordeaux 1
ISM, UMR CNRS 5255
351 Cours de la Libération
33405 Talence Cedex
France

Cyril Aymonier
ICMCB-CNRS-Université
de Bordeaux
87 avenue du docteur Albert
Schweitzer
33608 Pessac Cedex
France

Xinhe Bao
State Key Laboratory of Catalysis
Dalian Institute of Chemical Physics
Zhongshan Road 457
Dalian 116023
China

Helmut Bönnemann
Max-Planck-Institut für
Kohlenforschung
Kaiser-Wilhelm-Platz 1
45470 Mülheim an der Ruhr
Germany

Dirk E. De Vos
K. U. Leuven
Dept. M^2S - Faculteit
Bio-ingenieurswetenschappen
Postbus 2461, Kasteelpark
Arenberg 23
3001 Heverlee
Belgium

Audrey Denicourt-Nowicki
Ecole Nationale Supérieure
de Chimie de Rennes
CNRS, UMR 6226
Avenue du Général Leclerc,
CS 50837
35708 Rennes Cedex 7
France

Abdou Diallo
Université Bordeaux 1
ISM, UMR CNRS 5255
351 Cours de la Libération
33405 Talence Cedex
France

Emil Dumitriu
Technical University of Iasi
Faculty of Chemical Engineering and
Environment Protection
71 D. Mangeron Ave
700050 Iasi
Romania

Isabelle Favier
Université Paul Sabatier, Laboratoire Hétérochimie Fondamentale et Appliquée, UMR CNRS 5069
118 route de Narbonne
31062 Toulouse Cedex 9
France

and

CNRS, LHFA UMR 5069
31062 Toulouse Cedex 9
France

Iann C. Gerber
Université de Toulouse
Laboratoire de Physique et Chimie des Nano-Objets (LPCNO-IRSAMC, INSA, UPS, CNRS-UMR 5215)
Équipe Modélisation Physique et Chimique
135 avenue de Rangueil
31077 Toulouse Cedex
France

Inge Geukens
K. U. Leuven
Dept. M^2S - Faculteit Bio-ingenieurswetenschappen
Postbus 2461, Kasteelpark Arenberg 23
3001 Heverlee
Belgium

Montserrat Gómez
Université Paul Sabatier, Laboratoire Hétérochimie Fondamentale et Appliquée, UMR CNRS 5069
118 route de Narbonne
31062 Toulouse Cedex 9
France

and

CNRS, LHFA UMR 5069
31062 Toulouse Cedex 9
France

Marco Haumann
FAU Erlangen-Nuremberg
Chemical Reaction Engineering
Egerlandstr. 3
91058 Erlangen
Germany

Josef Hormes
University of Saskatchewan
Canadian Light Source, Inc.
101 Perimeter Road
Saskatoon, SK S7N 0X4
Canada

Timna-Joshua Kühn
University of Saskatchewan
Canadian Light Source, Inc.
101 Perimeter Road
Saskatoon, SK S7N 0X4
Canada

Vasile Hulea
Ecole Nationale Supérieure de Chimie de Montpellier
Institut Charles Gerhard
Equipe MACS, UMR 5253
8, rue de l'Ecole Normale
34296 Montpellier Cedex 5
France

Guram Khelashvili
Strem Chemicals, Inc.
7 Mulliken Way
Newburyport, MA 01950-4098
USA

David Madec
Université Paul Sabatier, Laboratoire
Hétérochimie Fondamentale et
Appliquée, UMR CNRS 5069
118 route de Narbonne
31062 Toulouse Cedex 9
France

and

CNRS, LHFA UMR 5069
31062 Toulouse Cedex 9
France

Samuel Marre
ICMCB-CNRS-Université
de Bordeaux
87 avenue du docteur Albert
Schweitzer
33608 Pessac Cedex
France

Catia Ornelas
Université Bordeaux 1
ISM, UMR CNRS 5255
351 Cours de la Libération
33405 Talence Cedex
France

Xiulian Pan
State Key Laboratory of Catalysis
Dalian Institute of Chemical Physics
Zhongshan Road 457
Dalian 116023
China

Karine Philippot
Laboratoire de Chimie de
Coordination du CNRS
205 route de Narbonne
BP44099
31077 Toulouse Cedex 4
France

Romuald Poteau
Université de Toulouse
Laboratoire de Physique et Chimie
des Nano-Objets (LPCNO-UMR5215,
IRSAMC)
Équipe Modélisation Physique et
Chimique
135 avenue de Rangueil
31077 Toulouse Cedex
France

Wolf-Jürgen Richter
Max-Planck-Institut für
Kohlenforschung
Kaiser-Wilhelm-Platz 1
45470 Mülheim an der Ruhr
Germany

Alain Roucoux
Ecole Nationale Supérieure
de Chimie de Rennes
CNRS, UMR 6226
Avenue du Général Leclerc,
CS 50837
35708 Rennes Cedex 7
France

Judith Scholz
FAU Erlangen-Nuremberg
Chemical Reaction Engineering
Egerlandstr. 3
91058 Erlangen
Germany

Philippe Serp
Laboratoire de Chimie de
Coordination - UPR8241 CNRS
composante ENSIACET
4 allée Emile Monso
Toulouse University
31030 Toulouse Cedex 4
France

Dang Sheng Su
Chinese Academy of Science
Institute of Metal Research
Shenyang National Laboratory for Materials Science
72 Wenhua Road
Shenyang 110016
China

and

Fritz Haber Institute of the Max Planck Society
Department of Inorganic Chemistry
Faradayweg 4–6
14195 Berlin
Germany

1
Concepts in Nanocatalysis

Karine Philippot and Philippe Serp

1.1
Introduction

Catalysis occupies an important place in chemistry, where it develops in three directions, which still present very few overlaps: heterogeneous, homogeneous and enzymatic. Thus, homogeneous and heterogeneous catalysis are well-known as being two different domains defended by two scientific communities (molecular chemistry and solid state), although both are looking for the same objective, the discovery of better catalytic performance. This difference between homogeneous and heterogeneous catalysis is mainly due to the materials used as catalysts (molecular complexes in solution versus solid particles, often grafted onto a support), as well as to the catalytic reaction conditions applied (for example liquid-phase reactions versus gas-phase ones). Considering the advantages of these two catalytic approaches, on the one hand heterogeneous catalysts are easy to recover but present some drawbacks, such as the drastic conditions they require to be efficient and the mass transport problems; on the other hand, homogeneous catalysts are known for their higher activity and selectivity, but the separation of expensive transition metal catalysts from substrates and products remains a key issue for industrial applications [1]. The first attempts to bridge the gap between these two communities date from the 1970s to the early 1980s. From one side chemists working in the molecular field, such as J.M. Basset, M. Che, B.C. Gates, Y. Iwasawa and R. Ugo, among others, initiated pioneering works on surface molecular chemistry to develop single-site catalysts, and/or reach a better understanding of conventional supported catalyst preparation through a molecular approach; from the other side, chemists of the solid state, such as G. Ertl and G. Somorjai, were interested in the molecular understanding of surface chemical catalytic processes. For the latter, the revolutionary development of surface science at the molecular level was possible thanks to the development of techniques of preparation of clean single crystal surfaces and characterization of structure and chemical composition under ultrahigh vacuum [(X-ray photoelectron spectroscopy (XPS), atomic emission spectroscopy (AES), low energy electron diffraction (LEED) etc]. Once again, although these scientists aimed at a common objective, little interaction or cross-fertilization action

Nanomaterials in Catalysis, First Edition. Edited by Philippe Serp and Karine Philippot.

has appeared during the last 20 years. One should however cite the first International Symposium on Relations between Homogeneous and Heterogeneous Catalysis, organized on Prof. Delmon's initiative in Brussels (Belgium) in 1973. Interestingly, this event appeared 17 years after the first International Congress on Catalysis (Philadelphia, 1956) and 5 years before the first International Symposium on Homogeneous Catalysis (Corpus Christi, 1978). In parallel, although colloidal metals of Group 8 were among the first catalysts employed in the hydrogenation of organic compounds, the advent of high pressure hydrogenation and the development of supported and skeletal catalysts meant that colloidal catalysis has hardly been explored for many years [2–4].

Since the end of the 1990s, and with the development of nanosciences, nanocatalysis has clearly emerged as a domain at the interface between homogeneous and heterogeneous catalysis, which offer unique solutions to answer the demanding conditions for catalyst improvement [5, 6]. The main focus is to develop well-defined catalysts, which may include both metal nanoparticles and a nanomaterial as support. These nanocatalysts should be able to display the ensuing benefits of both homogenous and heterogeneous catalysts, namely high efficiency and selectivity, stability and easy recovery/recycling. Specific reactivity can be anticipated due to the nanodimension that can afford specific properties which cannot be achieved with regular, non-nano materials (Figure 1.1).

In this approach, the environmental problems are also considered. Definitions can be given: the term 'colloids' is generally used for nanoparticles (NPs) in liquid-phase catalysis, giving rise to 'colloidal catalysis,' while 'nanoparticle' is more often attributed to NPs in the solid state, thus related to the heterogeneous catalysis domain. The terms 'nanostructured' or 'nanoscale' materials (and by extension 'nanomaterials') are any solid that has a nanometer dimension. Despite these differences in nomenclature, NPs are always implicated and 'nanocatalysts' or 'nanocatalysis' summarize well all the different cases.

In the nanoscale regime, neither quantum chemistry nor the classical laws of physics hold. In materials where strong chemical bonding is present, delocalization of electrons can be extensive, and the extent of delocalization can vary with the size

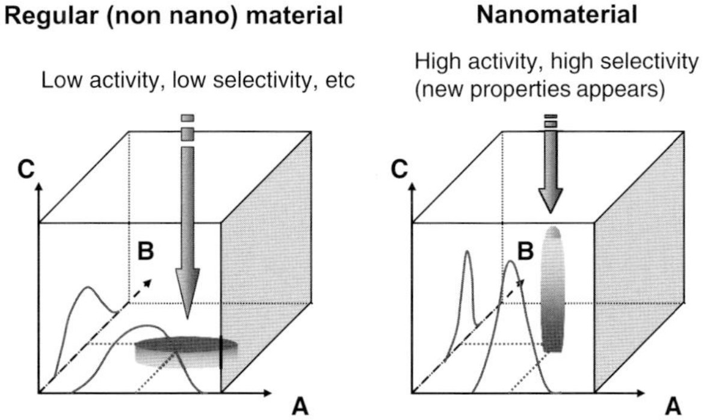

Figure 1.1 Nanoarchitecture: an avenue to superior precision. Axes are: A: composition of functional sites; B: ordering level of sites; C: functional properties of material.

of the system. This effect, coupled with structural changes, can lead to different chemical and physical properties, depending on size. As for other properties, surface reactivity of nanoscale particles is thus highly size-dependent. Of particular importance for chemistry, surface energies and surface morphologies are also size-dependent, and this can translate to enhanced intrinsic surface reactivity. Added to this are large surface areas for nanocrystalline powders and this can also affect their chemistry in substantial ways [7]. Size reduction to the nanometer scale thus leads to particular intrinsic properties (quantum size effect) for the materials that render them very promising candidates for various applications, including catalysis. Such interest is well established in heterogeneous catalysis, but colloids are currently experiencing renewed interest to get well-defined nanocatalysts to increase selectivity.

Much work in the field has focused on the elucidation of the effects of nanoparticle size on catalytic behavior. As early as 1966, Boudart asked fundamental questions about the underlying relationship between particle size and catalysis, such as how catalyst activity is affected by size in the regime between atoms and bulk, whether some minimum bulk-like lattice is required for normal catalytic behavior, and whether an intermediate ideal size exists for which catalytic activity is maximized [8]. Somorjai's group has studied this issue extensively. Although there is tremendous variation in the relationships between size and activity depending on the choice of catalyst and choice of reaction, these relationships are often broken into three primary groups: positive size-sensitivity reactions, negative size-sensitivity reactions, and size-insensitive reactions. There is also a fourth category composed of reactions for which a local minima or maxima in activity exists at a particular NP size (see Figure 1.2) [9, 10]. Positive size-sensitivity reactions are those for which turnover frequency increases with decreasing particle size. The prototypical reaction demonstrating positive size-sensitivity is methane activation. Dissociative bond cleavage via σ-bond activation as the rate-limiting step is a

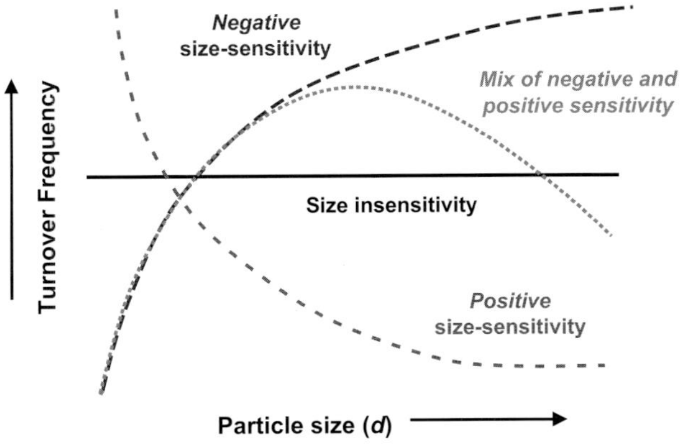

Figure 1.2 Major classes of size-sensitivity, which describe the relationships between NP size and turnover frequency for a given combination of reaction and NP catalyst. (------) negative size-sensitivity; (- - - -) positive size-sensitivity; (·······) Mix of negative and positive sensitivity.

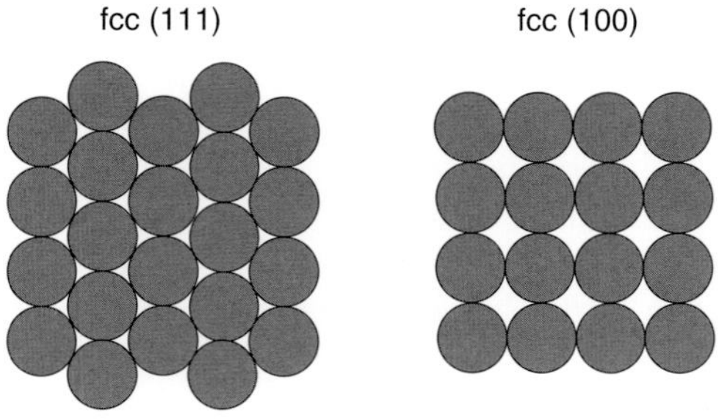

Figure 1.3 Two of the most common fcc crystal faces, (111) (left) and (100) (right).

common feature in reactions with positive size-sensitivity. Negative size-sensitivity reactions are those for which turnover frequency decreases with decreasing particle size. In this case, formation or dissociation of a π-bond is often the rate-limiting step. The prototypical reactions for this group are dissociation of CO and N_2 molecules, which each require step-edge sites and contact with multiple atoms. These sites do not always exist on very small NPs, in which step-edges approximate adatom sites. These reactions also sometimes fall into the fourth category of those with a local maximum in turnover frequency versus particle size because certain particle sizes geometrically favor the formation of these sorts of sites. The third type of reaction is the size-insensitive reaction, for which there is no significant dependence of turnover frequency on nanoparticle diameter. The prototypical size-insensitive reaction is hydrocarbon hydrogenation on transition metal catalysts, for which the rate-limiting step is complementary associative σ-bond formation. Although these effects are often referred to as structure-sensitivity effects, they are referred to as size-sensitivity effects here in order to further distinguish them from another type of structure-sensitivity, which is derived from differences in crystal face and which is discussed below.

Aside from considerations of NP size, a second major area of inquiry is that of the effect of nanoparticle shape on reaction rate, selectivity, and deactivation. This work is derived from the abundance of research done on single crystal surfaces, which has demonstrated what is known as structure sensitivity in catalysis. Experiments on a wide variety of catalysts have determined that the atomic arrangement of atoms on a surface has a significant effect on catalyst behavior. As demonstrated in Figures 1.3 and 1.4, the type of crystal face dramatically affects the coordination,

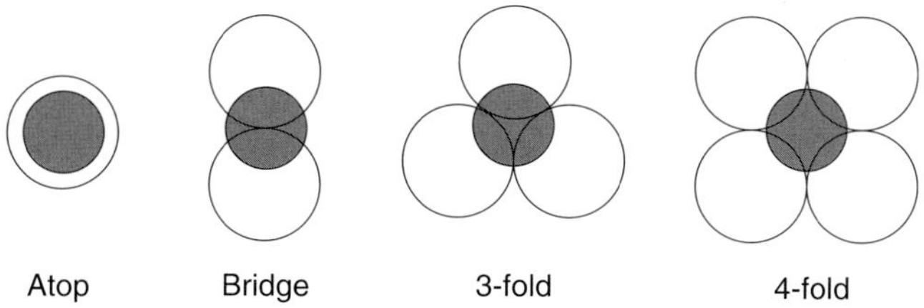

Figure 1.4 Four of the most common adsorption sites found on single crystal terraces.

number of nearest neighbors, and both two- and three-dimensional geometry of the catalytically active surface atoms. The availability of particular types of adsorption sites can have a large effect on catalysis, as it is common for adsorbates to differ in their affinity for each type of adsorption site. Consequently, the presence or absence of a particular type of site can affect not only reaction rates, but also selectivity. However, not all reactions are structure sensitive and some reactions are known to be structure sensitive only within a range of specific conditions. In the case of nanoparticle catalysts, structure-sensitivity is manifested in terms of NP shape. When little attention is given to shape, most NPs adopt roughly spherical shapes, often referred to as polyhedra or octahedra, in order to minimize surface energy.

These NPs predominately feature (111)-oriented surface atoms, which is the lowest energy crystal face. Under certain conditions, however, nanoparticle catalysts can be synthesized such that the shape, and consequently the surface atom orientation, is kinetically trapped into a nonequilibrium shape, such as a cube, triangle, platelet, or rod [11]. Nanoparticles of different shape have been shown to have different activity and selectivity [12, 13], as well as stability [14] in catalytic reactions. Shape-controlled NPs play an important role as model catalysts in furthering the large-scale effort to bridge the 'materials gap' between the real-world systems and scientific understanding in catalysis.

Since nanocatalysts are made of nanoparticles or/and nanomaterials, as a metal or metal oxide active phase or as a support or a combination of both, nanoparticles and nanomaterials have been the object of an ever increasing interest during recent decades. The common goal is the development of well-defined nanoparticles/nanomaterials displaying well-controlled properties to get efficient and selective nanocatalysts for numerous relevant catalytic reactions (as examples arene hydrogenation, carbon–carbon coupling, CO oxidation. . .).

1.2
The Impact of the Intrinsic Properties of Nanomaterials on Catalysis

Involvement of interatomic interaction causes the performance of a solid, or a cluster of atoms, to vary from that of an isolated atom. Adjustment of the relative number of the under-coordinated surface atoms provides an additional freedom that allows one to tune the properties of a nanosolid with respect to that of its bulk counterpart. Hence, contribution from the under-coordinated atoms and the involvement of interatomic interaction can be the starting point of consideration to bridge the gap between an isolated atom and a bulk solid in chemical and physical performances. The impact of atomic coordination reduction (deviation of bond order, length, and angle) is tremendous. It unifies the performance of a surface, a nanosolid, and a solid in amorphous state consistently in terms of bond relaxation and its consequences on bond energy [15, 16]. The unusual behavior of a surface and a nanosolid has been consistently understood and systematically formulated as functions of atomic coordination reduction and its derivatives (size dependence) on the atomic trapping potential, crystal binding intensity, and electron–phonon coupling. If one could establish the functional dependence of a detectable quantity, Q. on atomic separation or its derivatives, the size dependency of the quantity Q is then certain. One can hence design a nanomaterial with desired functions based on such prediction. The physical quantities of a solid can be normally categorized as follows:

- Quantities that are directly related to bond length, such as the mean lattice constant, atomic density, and binding energy. Lattice contraction in a nanosolid induces densification and surface relaxation.
- Quantities that depend on the cohesive energy per discrete atom, such as self-organization growth; thermal stability; Coulomb blockade; critical temperature for phase transitions, and evaporation in a nanosolid; and the activation energy for atomic dislocation, diffusion, and chemical reactions.
- Properties that vary with the binding energy density in the relaxed continuum region such as the Hamiltonian that determine the entire band structure and related properties such as band gap, core level energy, photoabsorption, and photoemission.
- Properties from the joint effect of the binding energy density and atomic cohesive energy such as the mechanical strength Young's modulus, surface energy, surface stress, extensibility and compressibility of a nanosolid, as well as the magnetic performance of a ferromagnetic nanosolid.

Structural miniaturization has indeed given a new freedom that allows us to tune the physical properties that are initially nonvariable for the bulk chunks by simply changing the shape and size to make use of the effect of atomic coordination reduction.

The intrinsic properties of nanomaterials and their size dependency will induce, directly or not, several effects on catalysis (Figure 1.5), that will be discussed in detail throughout this book. Some relevant examples are given below, which will be developed in more details in the following chapters.

1.2.1
Metallic Nanoparticles

About two-thirds of chemical elements are metals. Using the molecular orbital description, as is usual for covalently bonded atoms in molecules, the generation of a metallic material can simply be understood as the formation of an infinitely extended molecular orbital, leading to energy bands. The development of a metallic band structure requires a minimum number of electronic levels, which have to be very similar in energy so that electrons can move by only thermal activation. All the properties that we know for a bulk metal derive from the existence of such a band. The most important property of a metal is its ability to transport electrons, namely the property of conductivity. To understand what conductivity is based on, it is necessary to consider the relation between occupied and unoccupied electronic bands, as electrons can become mobile only if the energy band of which they are part is

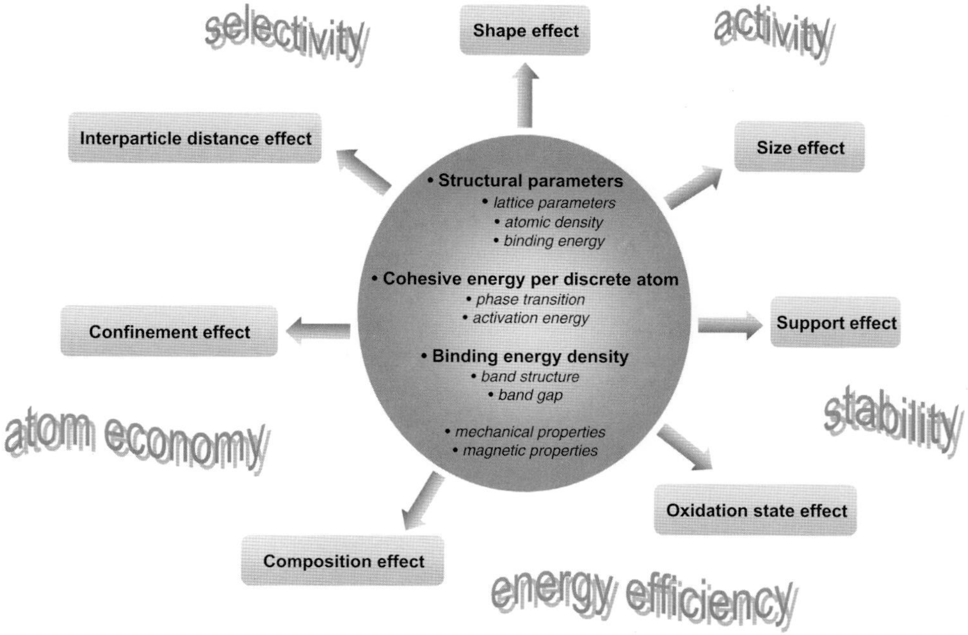

Figure 1.5 The impact of the intrinsic properties of nanomaterials on catalysis.

not fully occupied. Most of the *d*-type transition metals are characterized by only partially filled *d*-orbitals so that incompletely filled bands result in any case. d^{10} elements such as palladium, platinum or gold have nearby *s*-bands that can be used for electron transport. Another important property of metals, at least for some of them, is magnetism, for example the well-known ferromagnetism of iron, cobalt and nickel. The existence of unpaired electrons is a condition for magnetism; however, only the uniform orientation of free spins over a large area results in ferromagnetism while non-oriented free spins produce paramagnetic materials. Copper and gold are the only colored metals, the others looking 'silvery' when they have smooth surfaces. Finely dispersed metals are all dark brown or black. The silvery luster and the dark appearance are caused by the total reflection of light in the first case, and by the total absorption of light in the latter. Color is caused by the partial absorption of light by electrons in matter, resulting in the visibility of the complementary part of the light. On smooth metal surfaces, light is totally reflected by the high density of electrons and no color results; instead a mirror-like effect is observed [17].

The description of bulk materials is made by means of the laws of classical physics. A metal particle will present properties different from those of metal bulks, because of the reduction in the size (quantum size effect). Indeed, if a size range is attained where the band structure begins to disappear and discrete energy levels become dominant, quantum mechanical rules, which are well-established for describing electronic situations in molecules and atoms have to replace those of

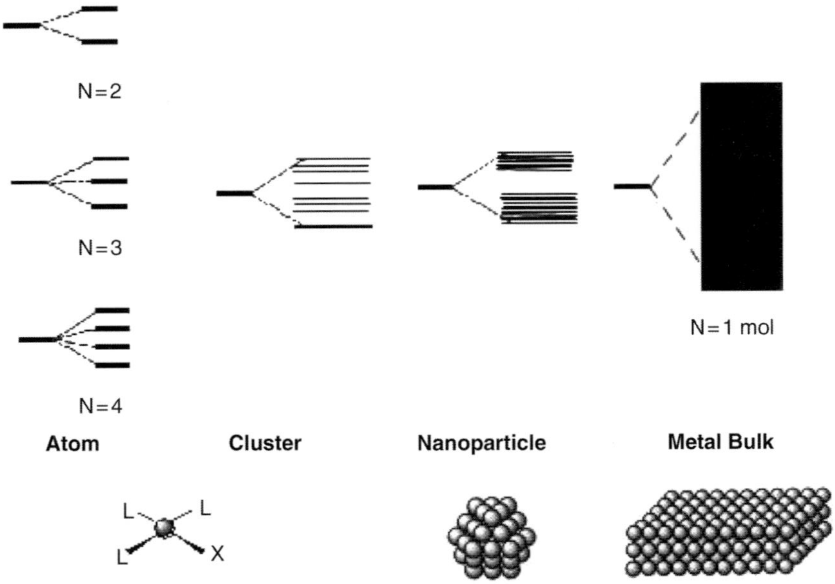

Figure 1.6 Electron levels of atom, cluster, nanoparticle and bulk metal.

classical physics suitable for bulk materials (Figure 1.6). Nevertheless, small particles are parts of a material and not atoms nor molecules, and are thus considered as intermediate species.

Metallic NPs, also called nanoclusters, are pieces of metal at the nanometer scale, of one to a few nanometers in size. They can be noncrystalline, aggregates of crystallites or single crystallites (nanocrystals). Due to the number of bound metal atoms they contain, metallic nanoparticles display intermediate electronic energy levels in comparison with molecules and metal bulks [18]. As a result, particular physical and chemical properties are expected for metallic nanoparticles that can lead to applications in various areas such as in catalysis [19]. In this latter domain, metallic nanoparticles are generally considered as intermediate species between metal complexes and metal surfaces, and the term 'nanocatalysts' is now commonly used to describe them.

In heterogeneous catalysis, the use of metallic nanoparticles is well established, mainly based on their high reactivity. One of their properties is their high number of surface atoms that increases with decreasing particle size (Table 1.1), these surface atoms being the active sites for catalysis. In addition, surface atoms which are at the edges or in the corners are more active than those in planes, and their number also increases with decreasing particle size. Since the number of surface atoms present in NPs will govern their catalytic reactivity, control of the size of NPs is thus of high importance. For industrial processes, the particles are generated on supports such as alumina, silica or charcoal by impregnation from solutions of the corresponding metal salts followed by a reduction procedure giving rise to

Table 1.1 Number of surface atoms in relation with the total number of atoms in full shell clusters.

Full-shell clusters	Total number of atoms	Number of surface atoms	Surface atoms (%)
1 shell	13	12	92
2 shells	55	42	76
3 shells	147	92	63
4 shells	309	162	52
5 shells	561	252	45
6 shells	923	362	39

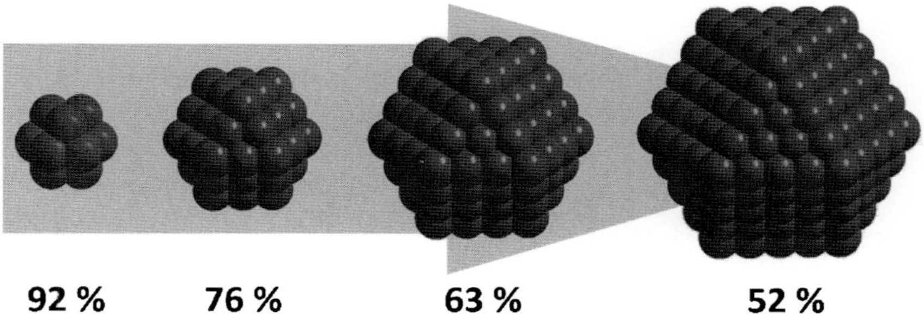

Decrease of surface to volume ratio for clusters of 13, 55, 147 to 309 atoms

zero-valent nanoparticles. Since the larger particles are less active than the smaller ones, only a portion of the metal particles contribute to the catalytic process [17]. To overcome this problem, other ways of synthesis have been explored like solution procedures, but in that case, the addition of a stabilizing agent (polymer, surfactant, ligand. . .) is necessary to keep the NPs stable. Besides their protecting role, and although they occupy some active sites at the surface of nanoparticles, stabilizers can tune their reactivity by influencing their morphology or/and their surface chemistry. When the nanoparticles are deposited onto a support, the organic ligands can be eliminated from the nanomaterial by washings with appropriate solvents or by calcination under air at high temperature or under plasma conditions, to obtain naked NPs.

A modern approach of colloid chemistry is presently being developed to increase the reactivity of NPs in a limited size up to 10 nm, using several types of stabilizer as well as several types of support for their heterogenization. It appears that a fundamental understanding of the surface properties of such metallic nanoparticles is needed to get more efficient and selective nanocatalysts in the future.

1.2.2 Metal Oxide Nanoparticles

The metal elements can form a large diversity of oxide compounds, which can adopt structural geometry with an electronic structure that can exhibit metallic, semiconductor, or insulator character [20]. Most of the catalysts used in industrial applications involve an oxide as active phase, promoter, or support. At the nanoscale, these compounds can exhibit unique physical and chemical properties due to their limited size and a high density of defect sites such as edges, corners and point defects [21]. As for other materials, the process of size reduction is expected to dictate structural, transport and chemical properties, which themselves will influence the final catalytic performance. In this section we will concentrate on pure oxide nanoparticles, but one should keep in mind the important place of mixed nano-oxides, such as nanozeolites [22] or others [23–26] in catalysis.

Bulk oxides are usually robust and stable systems with well-defined crystallographic structures. However, the growing importance of surface free energy and stress with decreasing particle size must be considered, since changes in thermodynamic stability associated with size can induce modification of cell parameters and/or structural transformations [27], and in extreme cases, the NP can disappear because of interactions with its surrounding environment and of its high surface free energy [28]. To display mechanical or structural stability, a NP must have a low surface free energy. As a consequence of this requirement, phases that have a low stability in bulk materials can become very stable in nanostructures. This structural phenomenon has been detected in TiO_2, VO_x, $A1_2O_3$, or MoO_x [29–33]. Thus, in the case of alumina the stable structure for micro-sized samples is the α phase while γ appears more stable for nanostructured materials [14]. Size-induced structural distortions have been observed in NPs of $A1_2O_3$ [28], Fe_2O_3 [34], ZrO_2 [35] or CeO_2 [36].

The NP size is also related to the transport properties of the oxide, since, as already stated, the nanostructure produces the so-called quantum size or confinement effects, which essentially arise from the presence of discrete, molecular-like electronic states. Additional general electronic effects of quantum confinement experimentally probed on oxides are related to the energy shift of exciton levels and optical bandgap [37]. Oxide materials can present ionic or mixed ionic/electronic conductivity and it is experimentally well established that both can be influenced by the nanostructure of the solid. The number of electronic charge carriers in a metal oxide is a function of the band gap energy according to the Boltzmann statistics. The electronic conduction is referred to as n- or p-hopping-type depending on whether the principal charge carrier are, respectively, electrons or holes. In an analogous manner to hopping-type conduction, ionic conduction takes place when ions can hop from site to site within a crystal lattice as a result of thermal activation. As a result of the nanoscale derived effects, it is well known that CeO_2 exhibits an improved n-type conductivity, which may be four orders of magnitude greater than that corresponding to bulk/micro-crystalline ceria, and is ascribed to a significant enhancement of the electronic contribution [38]. The strong size-dependence observed for the electrical conductance in the context of gas-sensing devices has been recently reviewed [39–41]. Some of the most dramatic effects of the nanostructure on ionic transport in oxides are observed in the field of Li-ion batteries. An outstanding enhancement of Li-ion vacancy conductivity has been achieved using Li-infiltrated nanoporous Al_2O_3 [42].

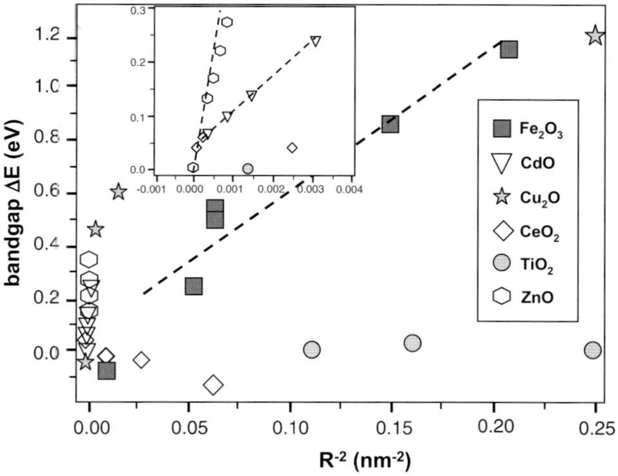

Figure 1.7 Optical band gap energy as a function of the inverse squared primary particle size for common metal oxides. (Adapted from [56]).

Structural and electronic properties obviously drive the physical and chemical properties of the solid, and this last group of properties is influenced by size in a simple classification. In their bulk state, many oxides have wide band gaps and a low reactivity [43]. A decrease in the average size of an oxide particle does in fact change the magnitude of the band gap (Figure 1.7), with strong influence on the conductivity and (photo)chemical reactivity [41, 44–46]. Surface properties of oxides nanomaterials are of central importance in catalysis. Solid–gas or solid–liquid chemical reactions can be mostly confined to the surface and/or subsurface regions of the solid. As mentioned, the two-dimensional (2D) nature of surfaces has notable structural consequences, typically a rearrangement or reconstruction of bulk geometries, and electronic consequences, such as the presence of mid-gap states, which may act as trapping centers in photocatalysis, whose behavior depend on the relative position of their energy with respect to the valence and conduction band edge position [46]. In the case of nanostructured oxides, surface properties are strongly modified with respect to 2D-infinite surfaces, producing solids with unprecedented sorption [47, 48] and acid/base characteristics [49], or metal–support interaction/epitaxy [50, 51]. Finally, the presence of under-coordinated atoms or O vacancies in an oxide NP should produce specific geometrical arrangements as well as occupied electronic states located above the valence band of the corresponding bulk material, enhancing in this way the chemical reactivity of the system [35, 45, 52–54]. This latter remark also concerns the cytotoxicity of these materials [55].

We feel that it is important to stress the need for a fundamental understanding of the properties of nanostructured oxides, particularly for sizes in which the atoms directly affected in their properties are a significant percentage of the total number of atoms present in the solid particle; this usually implies a dimension limited to about or below 10 nm. When this fact occurs exclusively in one dimension, we are dealing with a surface or film, whereas in two dimensions, nanotubes, nanowires, and other interesting morphologies are obtained. Finally, when the three dimensions are limited to the nanoscale, nanoparticles are formed.

1.2.3 Carbon Nanoparticles

Carbon is unique in the number and the variety of its polymorphs. Figure 1.8 illustrates how the inorganic (nano)carbons can result from the extension of organic materials through large molecules [57]. These inorganic (nano)materials are very different in structure and properties, and their structural as well as surface chemistry is extremely complex.

In heterogeneous catalysis, carbon materials are unique catalyst supports, allowing the anchoring of the active phase, and can also be catalysts or catalyst poisons (carbon deposits) by themselves [58]. Although activated carbon and carbon blacks (CBs) are the most commonly used carbon supports, there is an increasing interest in the application of new carbon nanoparticles [fullerenes [59], carbon nanotubes (CNTs), carbon nanofibers (CNFs) [60], and graphene [61]] as supports for catalysis

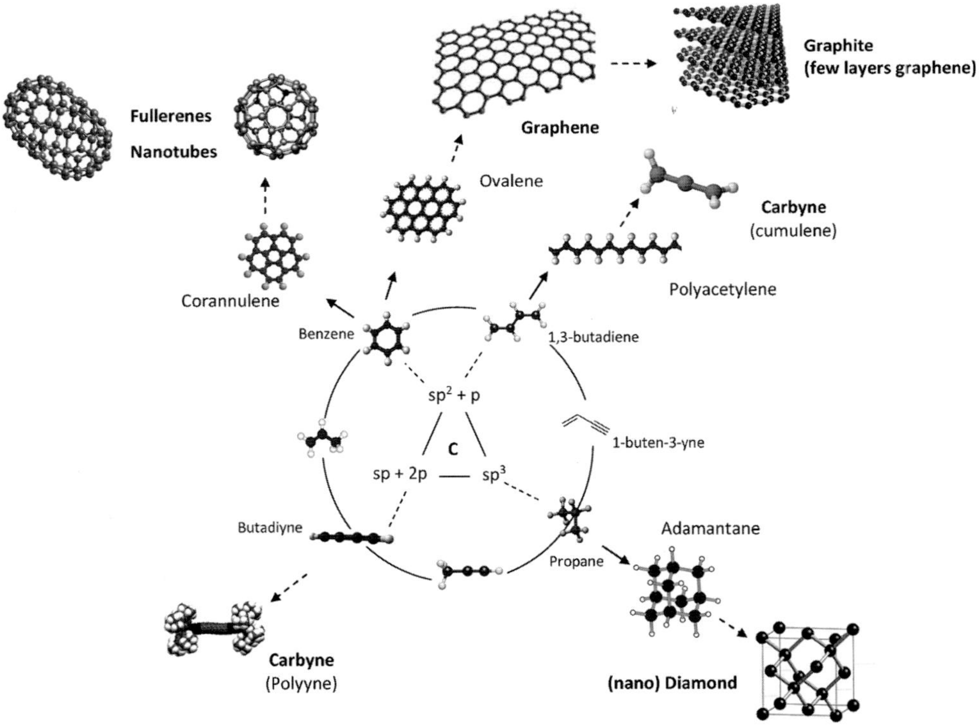

Figure 1.8 C—C bonds to form a large number of hydrocarbons and their extension to carbon families. (Adapted from [58]).

or catalysts since the nanostructure of these materials can offer a unique combination of properties. The catalytic behavior of solid carbons depends of course on their surface properties, but these surface properties are to a large extent a direct consequence of their bulk properties.

The controlled curvature or the orientation of the graphene layers in carbon nanoparticles dictated important properties. The curvature in fullerenes, CBs or CNTs is introduced by including pentagons and heptagons, together with hexagons, as starting fragments. This curvature of the graphene sheets induces strong modifications of the electronic properties; and comparison with graphite shows modification of the π-electron cloud [62]. The rolling-up of the graphene sheet to form a CNT causes a rehybridization of carbon orbital's (nonplanar sp^2) configuration, thus leading to modification of the π density in the graphene sheet, which will depends on CNT diameter. It is worth noting that the theoretically predicted electronic properties are often modified by the presence of defects such as pentagons, heptagons, vacancies or impurities [63]. Similarly, in order to account for the bonding of the carbon atoms of a fullerene molecule, the hybridization must be a modification of the sp^3 hybridization of diamond and sp^2 hybridization of graphite. It is such that the σ orbital no longer contain all of the s-orbital character and the π orbital is no longer of the purely p-orbital character, as they are in graphite. Unlike the sp^3 or sp^2 hybridizations, the fullerene hybridization is not fixed but has variable characteristics depending on the number of carbon atoms in the molecule and consequently of its diameter. The number of carbon atoms, the pyramidization angle ($\theta - 90°$), and the nature of the hybridization are related and this relationship (in this case the s character in the π-orbital) is given in Figure 1.9 [64].

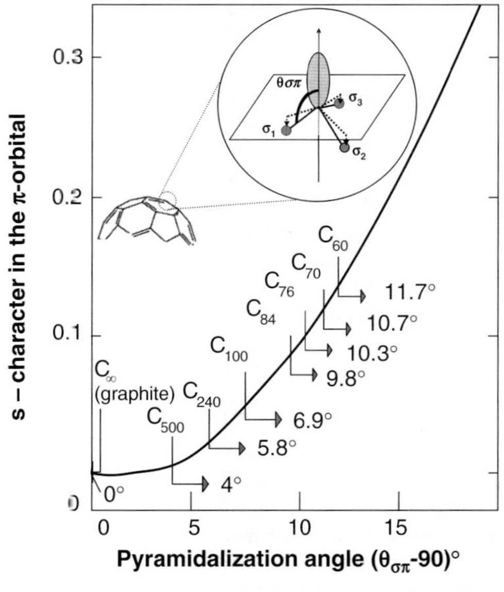

Figure 1.9 Hybridization of fullerene molecules as a function of pyramidization angle ($\theta_{\sigma\pi} - 90°$). $\theta_{\sigma\pi}$ is the common angle of the three σ bonds. (Adapted from [64]).

The rehybridization plays an important role in determining the electronic structure of the fullerene's family and it is the combination of topology and rehybridization that together account for the possible specific reactivity of all the curved sp^2 nanostructures. The influence of carbon curvature on molecular adsorption of hydrogen has been reported [65, 66]. The hydrogen adsorption energy barrier is found to strongly depend on the local curvature of the carbon network whereby the barrier is lowered with increasing curvature. Whereas in the case of C_{60} and CNTs, hydrogen chemisorption can be achieved by exposure to atomic hydrogen, the chemisorption on graphite (0001) requires hydrogen ions of low kinetic energy ($\sim$1 eV).

For CNTs, the presence of relatively well-defined and nanometric inner hollow cavities can also induce differences of reactivity between the convex (external) and concave (internal) surfaces. Thus, it has been experimentally proven that hematite NPs located inside the CNT inner cavity are more easily reduced (873 K) by the support than those on the outer surface (1073 K) [67]. Beside CNTs, other carbon nanostructures with the negative curvature analog of fullerenes have been proposed as materials with interesting structural and functional properties [68]. Another property of CNTs is the possibility to perform reactivity in a well-defined confined space (see Chapter 11), and to take advantage of plausible confinement effects [69, 70]. The confinement effects that influence chemical reactions can be classify into three groups: (i) shape-catalytic effects, that is, the effect of the shape of the confining material and/or the reduced dimensionality of the porous space; (ii) physical (or 'soft') effects including the influence of dispersion and electrostatic interactions with the confining material; and (iii) chemical (or 'hard') effects that involve significant electron rearrangement, including the formation and breaking of chemical bonds with the confining material [71]. The latter is usually considered to be the actual catalytic effect, and it is the one that has the most obvious influence on the reaction rates, as it alters the reaction mechanism. However, the first and second types of effect can also have a strong influence on both the rates and equilibrium yields, as has been shown in several recent theoretical calculations [72] and experimental studies [73].

By careful manipulation of various synthesis parameters, it is possible to generate filamentous carbon nanostructures in assorted conformations and also to control their crystalline order (Figure 1.10). The tunable orientation of the graphene layers can directly affect catalytic activity and selectivity, for example by specific metal catalyst crystallographic face exposure according to the support [60].

Finally, the thermal control on nanocatalysts becomes increasingly important as the size of the system diminishes. Therefore, for exothermic reactions the thermal conductivity of CNTs or graphene should play a critical role in controlling the performance of the catalyst.

Besides CNTs, graphene and fullerenes, catalytic applications of other carbon nanomaterials such as carbon nano-onions [74], or recently nanodiamonds [75, 76], and carbon nanohorns [77] have been much less studied.

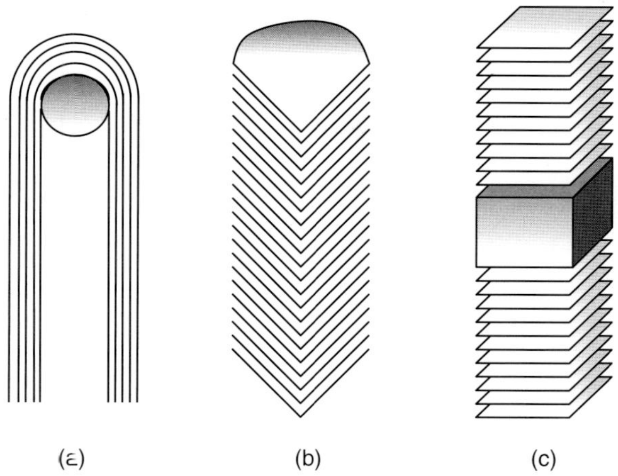

Figure 1.10 Different carbon nanostructures produced by catalytic chemical vapor deposition: (a) multiwalled carbon nanotubes; (b) fishbone-carbon nanofibers (f-CNFs); and (c) platelet-carbon nanofibers (p-CNFs).

1.3 How can Nanocatalyst Properties be Tailored?

The use of metallic/oxide NPs in catalysis is crucial as they mimic metal surface activation and catalysis at the nanoscale and thereby bring selectivity and efficiency to heterogeneous catalysis [78]. But, to be of interest, NPs should at least: (i) have a specific size (1–10 nm); (ii) have a well-defined surface composition; (iii) have reproducible syntheses and properties; and (iv) be able to be isolated and redissolved [79]. If the nanoparticle is supported, the question of precise control of its location, and thus of its spatial and chemical environment should also be addressed. Tailoring nanocatalysts properties thus necessitates being able to control these NP characteristics as well as their morphology, crystalline structure and composition (intrinsic composition and surface state).

1.3.1 Size, Shape and Surface Chemistry of Nanoparticles

Nowadays, to develop efficient catalytic systems, two important concepts are considered in nanocatalysis, namely the bottom-up strategy for the synthesis of well-controlled in size/shape NPs and the molecular approach to obtain more selective nanocatalysts. The bottom-up strategy allows the building of metallic nanoparticles from monometallic species. The control of NPs size is made possible by addition of a stabilizing agent, also called a capping agent (ligand, surfactant, polymer, dendrimer. . .). The introduction of ligands as nanoparticles stabilizers is of special interest because it focuses on the precise molecular definition of the catalytic materials. This strategy potentially allows optimization of the parameters that govern the efficiency in catalytic reactions, including enantioselectivity [80, 81]. As for homogeneous catalysts, an appropriate choice of the protecting agent means that the surface properties of the NPs can be tuned, as it can modify the nature of active sites (morphology) and the surface chemical environment (steric and/or electronic effect). As a result, NPs are very soluble in water or classic solvents, depending on the way of preparation and the stabilizer. The surface of nanoparticles can also be modified to render them more soluble in specific media (for example ionic liquids (IL)s or *sc*CO_2). They can also be handled and even characterized as molecular compounds by spectroscopic techniques [nuclear magnetic resonance (NMR), infrared (IR), ultraviolet-visible (UV-vis) spectroscopy, electrochemistry] in addition to solid-state techniques [transmission electron microscopy (TEM), wide-angle X-ray scattering (WAXS), powder-X-ray diffraction (XRD), X-ray excited photoelectron spectroscopy (XPS), extended X-ray absorption fine structure (EXAFS). . .].

Concerning the control of nanoparticle size on which the number of surface active sites will be dependent, chemists have developed several methods, and very small NPs are already produced by different procedures [82–84]. The most well-known method is the reduction of a metal salt which gives rise to nanoparticles in an aqueous phase followed by the decomposition of metal-organic precursors also called the organometallic approach, which is more appropriate for obtaining NPs in organic media [85]. Size control is attained by the use of a large variety of capping agents, which limit the growth of nanoparticles. One can cite as examples of stabilizing agents, ions, polyoxoanions, surfactants (ammonium salts), polymers (polyvinylalcohol, polyvinylpyrrolidone, block-copolymers. . .), dendrimers such as polyamidoamine (PAMAM) and ligands (thiols, phosphines, amines. . .).

Considering reactivity and selectivity, the control of the surface state of the NPs is of critical importance as it can influence the course of a reaction. Since nanocrystals in their native form are dominated by the surface species [86], the protective agents used during the synthesis of metallic nanoparticles in solution play an important role. Two points are concerned, the control of the morphology and crystal structure [87] and the control of the surface composition [88].

The catalytic activity of metal nanocrystals is highly dependent on the nature of their surface structure [89, 90], exposure of different crystallographic facets, together with the increased number of edges, corners and faces, being key parameters. Therefore, NPs of different shapes are highly desirable as catalysts. While the effect of metallic nanoparticles size on the catalytic activity is well documented, knowledge about the influence of metallic nanoparticle shapes has started to develop only recently [91, 92]. Thus, the efficient control of the morphology is an on-going project all over the world and very interesting works have appeared in recent years [93, 94]. Solution-phase based methods have been shown to have great capability and flexibility to produce metal nanocrystals with well-defined morphologies with crystallographic control [87]. Capping agents are chosen for their influence on the shape of the particles, and consequently on the nature of surface active sites (edges, corners, faces, kinks, terraces, defects. . .) [95–98]. Controlling precisely the kinetics of the reaction appears as a key point to control NP shape [99, 100]. This is illustrated on Figure 1.11 that depicts different Co

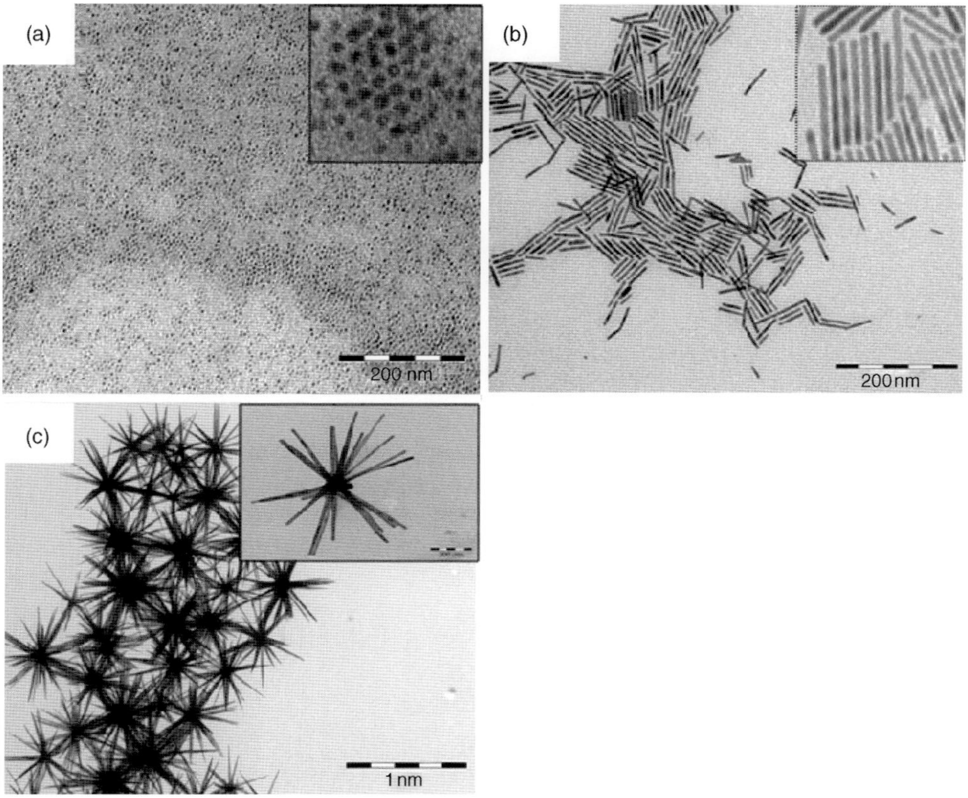

Figure 1.11 Various cobalt nano-objects synthesized from $[Co_2\{N(SiMe_3)_2\}_2(THF)]$/ lauric acid/hexadecyl amine mixtures by careful control of the reaction conditions, (a) spherical Co NPs (scale bar = 200 nm); (b) Co nanorods (scale bar = 200 nm); (c) Co nano-urchins (scale bar = 1 mm). With courtesy of K. Soulantica (LPCNO, France).

nano-objects synthesized from the same precursor and using the same stabilizing agents by careful adjustment of kinetic parameters.

Second, the intrinsic composition of metallic nanoparticles has also to be controlled to tune their reactivity and selectivity. For example, the synthesis of alloyed versus core-shell bimetallic systems or the synthesis of NPs with a well-controlled surface state (meaning that the influence of stabilizing agents and/or the eventual poisoning of active sites are perfectly known and directed) are key points of current interest. Bimetallic nanocrystals with core-shell, heterostructure, or intermetallic and alloyed structures are emerging as a new class of nanocatalysts. They are expected to display not only a combination of the properties associated with two distinct metals, but also new properties and capabilities due to a synergy between the two metals [101]. More importantly, bimetallic nanocrystals usually show composition-dependent surface structure and atomic segregation behavior, and therefore more interesting potential applications. Compared with monometallic nanocrystals, preparation of bimetallic ones is much more complicated and difficult to achieve. In recent years, many research groups have made great efforts in this area, and bimetallics with controllable structures could be obtained, following different ways of synthesis [102]. For example, the preparation of uniform bimetallic Rh/Fe NPs in a phenylazomethine dendrimer, which provides improved catalytic reactivity for the hydrogenation of olefins and nitroarenes compared with monometallic RhNPs in a dendrimer cage has been reported recently [101]. Nevertheless, one of the most challenging problems is the comprehension of nucleation and growth mechanism of nanocrystals in solution, which would make the synthesis more efficient and better control the catalytic properties.

The properties of nanocrystals are also dependent on the surface chemistry. Chemical modifications of nanoparticles, such as by the use of ligands or adatoms to decorate their surface can thus provide new catalytic properties [87]. Thus, one can expect to be able to modify the chemical properties of nanoparticles by an appropriate choice of the capping agents, due to their own electronic or/and steric properties. At least, the coordination of ligands on surface atoms can block some metallic active sites and further orient catalytic reactions. Such studies are presently emerging, comparing for example the influence of a polymer and simple ligands in the dynamics and reactivity of carbon monoxide at the surface of small ruthenium NPs [103] or the influence of more sophisticated ligands as carbenes [104] or alkyl/arylphosphines [105] in the hydrogenation of aromatic derivatives. In asymmetric catalysis, the chiral capping agent used for the stabilization of metallic nanoparticles is of fundamental importance as it is expected to induce enantioselectivity. This area of nanocatalysis should be developed as only a very few examples of enantioselective nanocatalysts are known, mainly for the hydrogenation of ethyl pyruvate with cinchonidine-stabilized NPs [106]. This aspect will be treated in a specific section below.

The use of colloidally synthesized nanoparticles for the preparation of supported catalysts offers several advantages (e.g., precise control of particle size and morphology) when compared with traditional preparation techniques. Although such NPs have already been successfully used for catalytic applications in the liquid phase, applications in heterogeneous gas-phase catalysis are still scarce [107]. This is mainly due to the fact that in heterogeneous gas-phase catalysis organic stabilizers are often considered to have a detrimental effect on catalytic activity since, due to their presence, the active centers on the NPs are partly blocked. But recent studies have been published, in which the influence of ligands on heterogeneously catalyzed reactions in the gas phase was investigated with positive impact on catalytic reactivity and selectivity.

The development of new reaction media and recovery of the nanocatalysts are also crucial aspects that are widely studied to solve environmental problems. There are presently many investigations to develop metallic nanoparticles for catalysis in green solvents [108, 109]. For example, ILs [110], in particular imidazolium-based ILs, have proven to be suitable media for the generation and stabilization of soluble metallic nanoparticles. Such metallic nanoparticles immobilized in ILs appeared as efficient green catalysts for several reactions in multiphase conditions [111]. This aspect of nanocatalysis will be presented in Chapter 5 by M. Gomez *et al.* as well as in Chapter 6 by M. Haumann *et al.* dealing with supported IL-phase catalysis. Fluorous solvents, perfluorinated alkanes being the most representative, are also green solvents which have been used as reaction media for nanocatalysts [112]. The first work was reported by R. Crooks and co-workers who described dendrimer-encapsulated PdNPs for alkene hydrogenation in a mixture THF/perfluoro butyltetrahydrofuran with successful recycling tests up to 12 runs, with no loss of activity nor detectable leaking of the catalyst into the organic phase [113]. Other groups stabilized metallic nanoparticles with heavily fluorinated ligands or polymers to increase their solubility in fluorous solvents for catalysis [114–117]. Nevertheless, applications of fluorous solvents in nanocatalysis are practiced to a much smaller extent compared with other green solvents. In another way, some groups are developing the synthesis of metallic nanoparticles and nanomaterials soluble in supercritical fluids [118, 119]. Chapter 7 by C. Aymonier *et al.* is dedicated to this subject.

In parallel with progress in the efficiency of metallic nanoparticles, progress has also been made in their recovery and recyclability [80]. This point is of high importance as it is probable that recyclability will be the bottleneck for the industrial application of NPs in solution phase, the good dispersion of NPs within a solvent making the separation of the catalyst from the products more complicated. To solve this problem, a good strategy for easy and efficient recycling has to be applied [108]. Thus, different groups are working to obtain nanoparticles as recyclable catalysts. Most often, the experiences gained from the recycling of homogeneous catalysts is a good source of ideas, as some methods can be easily adapted to nanoparticle catalysts to get recyclable nanocatalysts, such as the use of polymers [120] or dendrimers [121]. In addition, the deposition of pre-formed NPs onto supports to get heterogeneous systems easy to recover is being explored. Traditional supports such as silica, alumina, titania, ceria or other oxides as well as carbon supports in various forms (mesostructured silica [122], alumina membranes [123], carbon nanotubes [124]. . .) are often used, but a new trend appeared in the few past years which consists in the use of magnetic supports to recover the catalyst from the catalytic media by simply applying a magnet. Indeed, magnetic separation has emerged as a robust, highly efficient and fast separation tool with many advantages compared with product/catalyst isolation by means of other chemical or physical procedures, such as liquid–liquid extraction, chromatography, distillation, filtration or centrifugation [125]. The catalyst immobilized on superparamagnetic nanomaterials can be easily separated from the products due to a strong interaction between the magnetic nanoparticles and an external applied magnetic field, and it can be easily redispersed in the absence of the magnetic field due to the absence of 'magnetic memory.' For example, silica-coated magnetite NPs functionalized at their surface with a pendant phosphine ligand were used as a support for the deposition of PdNPs giving rise to novel nanocatalysts active in Suzuki cross-coupling reactions and that could be reused up to ten recycles [126]. More details about the recovery of nanocatalysts will be given by D. De Vos *et al.* in Chapter 8. Today, much research work concerns the recovery of the catalysts for economical and environmental reasons.

In summary, different ways can be followed to tailor the surface properties of nanoparticles and further to optimize their catalytic performance. It appears clearly that the capping agents used for NP stabilization have to be chosen properly, as they play a key role, by influencing the growth of the particles in terms of size and crystal structure as well as in terms of surface composition. Besides development of synthetic methods to get well-defined nanocatalysts, it is also necessary to develop a toolbox of techniques for a better understanding of surface coordination chemistry at a molecular level, and even better directly under catalytic reaction conditions [127].

1.3.2
Assembling Strategies to Control Active Site Location

Well-controlled, large-scale synthesis of nanomaterials, and their nanoscale (*in situ*) characterization should enable unambiguous correlation of the structural properties, with the physical and chemical properties of the nanocatalyst. For supported nanocatalysts, the synthesis should integrate strategies to control (i) active NP structure; (ii) nano support structure; and (iii) their assembly. This also constitutes one of the major challenges in many nanotechnology applications, that is, the development of synthesis and assembly approaches for manufacturing complex and durable nanostructures. Different strategies have been investigated to reach an optimal control of active phase location in a supported nanocatalyst (Figure 1.12).

The one-pot synthesis should provide a well-defined active site in a porous structure. The metal–organic frameworks (MOFs), a synthetic, crystalline, microporous metal oxide structure linked together by organic 'struts', constitutes a representative example. It has been shown that, under mild conditions, MOFs successfully compete with the classical molecular sieves zeolites in selective catalysis. Their primary fields of application are fine organic synthesis and enantioselective catalysis [128]. Of course, zeolite catalysts, for which the active site (most often an acidic site) is integrated in a perfectly controlled environment, should be included in this class of catalysts. More sophisticated synthetic routes have also been developed for the one-pot preparation of PtNPs into the meso-channels of mesoporous materials such as SBA-15 [129].

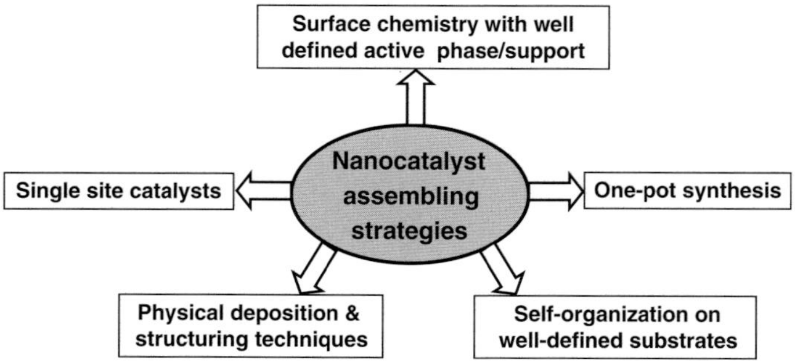

Figure 1.12 Nanocatalysts assembling strategies.

Another elegant strategy is to reduce the size of the active site to a very small number of atoms to prepare 'single-site catalysts,' which present an opportunity to exactly control the active site and its environment on the molecular scale. For this approach knowledge and control of the surface chemistry is an absolute prerequisite [130]. The strong structural response of such species to subtle changes in their electronic structure, and the absence of a 'flexible' support structure comparable to that provided by the protein structure of enzymes, limit the durability of such bio-inspired structures if they are active, or prevent their catalytic activity if they are fixed too strongly to the support. The fixation of the active species into nanostructured compartments, such as the 'ship in the bottle' approach in zeolites [131], constitutes an efficient way to circumvent stability issues, but introduces the problem of active site accessibility and of leaching, limiting the application of these systems to reactions performed under mild conditions. It is also worth mentioning that the rational building of single chains of a specially designed polymer that folds up in water to form an encapsulated catalytic chamber, that mimics the one used by enzymes, has recently been proposed [132]. In water, intramolecular interactions cause single chains of the polymer to adopt an enzyme-like structure (Figure 1.13): the self-assembling groups form helices, and the ruthenium ions bound to ligands become surrounded by a hydrophobic shell (black), which in turn is surrounded by hydrophilic groups (blue). These folded structures act as ruthenium catalysts for the reaction in which cyclohexanone reacts with hydrogen to form cyclohexanol.

Another solution to avoid heterogeneity is to prepare solid catalysts via self-organization. In self-assembly, interactions between atoms, nanostructures, and their environment produce kinetically or thermodynamically preferred structures and patterns. Self-assembly is inherently parallel and thus is able to produce macroscopic amounts of nanostructures. Two types of spontaneous processes are important for creating nanostructures: *self-assembly* and *self-organization*.

- **Self-assembly** of nanostructures can be defined as the *spontaneous* formation of structures smaller than 10 nm (in at least one dimension) and with a *well-defined size distribution*.

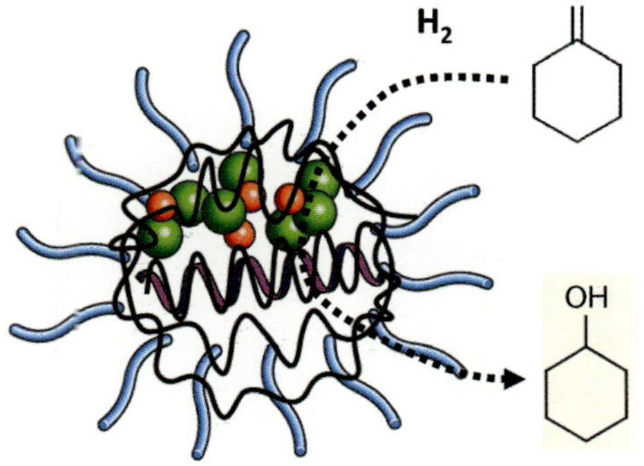

Figure 1.13 Enzyme-like polymer folding. (Adapted from [132]).

- **Self-organization** in nanoscale systems involves their natural tendency to form structures with long-range order, that is, order at the scale much larger than the size of an individual nanostructure.

In general, self-assembly is required to produce large numbers of individual nanoparticles (or nanostructures) having unique and/or tailored properties. Self-organization provides a pathway for combining nanostructures into novel materials or for establishing connections between nanostructures and the macroscopic world. Self-assembly and self-organization need not coexist in the same system, that is, self-assembled nanostructures do not always self-organize, and self-organization is not exclusive to self-assembled nanostructures. But systems that exhibit *both* self-assembly and self-organization are desirable.

Preparation of NPs by the self-organization of polymers constitutes a representative example of this approach [133]. Colloidal dispersions of polymer-protected Pd/Ag/Rh trimetallic nanoparticles were synthesized via self-organization by mixing colloidal dispersions of polymer-protected RhNPs with those of polymer-protected core–shell (Pd-core/Ag-shell) Pd/Ag bimetallic NPs. The Pd/Ag/Rh trimetallic NPs appeared to have a triple core–shell (Pd-core/Ag-interlayer/Rh-shell) [134]. A solid-phase self-organized catalyst of palladium nanoparticles was prepared from $PdCl_2$ with main-chain viologen polymers via complexation and reduction [135]. Convincing results on the precise spatial distribution control of metallic nanoparticles into domains of microphase-separated block copolymers have been reported [136]. The selective incorporation into a specific microphase of a block copolymer is the key issue there.

A rigorous approach to heterogeneity issues is the application of a combination of physical deposition and structuring techniques, such as physical vapor deposition (PVD), that are commonly used in nanoelectronics. This powerful and well-mastered methodology allows structuring with geometric precision and exact reproducibility at the 10-nm level. These techniques, however are extremely limited in producing large surface area materials and thus are only suitable for producing model catalysts [137]. Additionally, PVD processes are also limited in generating moderate chemical complexity.

The concept of molecular recognition through the control of surface chemistry can also be exploited to drive metallic nanoparticles on or in specific sites of the support, or to prepare bimetallic catalysts. Thus metallic nanoparticles have been selectively confined in the inner cavity of carbon nanotubes thanks to a careful choice of nanoparticle stabilizer and to the passivation of carbon nanotube surface (Figure 1.14) [124]. The specific association of metallic nanoparticles to dendrimers also represents a nice example of the surface chemistry approach [138–140].

To conclude this section, we have seen that metallic nanoparticles have been studied extensively due to their superior catalytic properties over their bulk counterparts. One of the challenging issues facing the community is to synthesize well-controlled (in size and shape) nanoparticles, and to address the issue of their precise location on a support. Although significant amounts of experimental work has

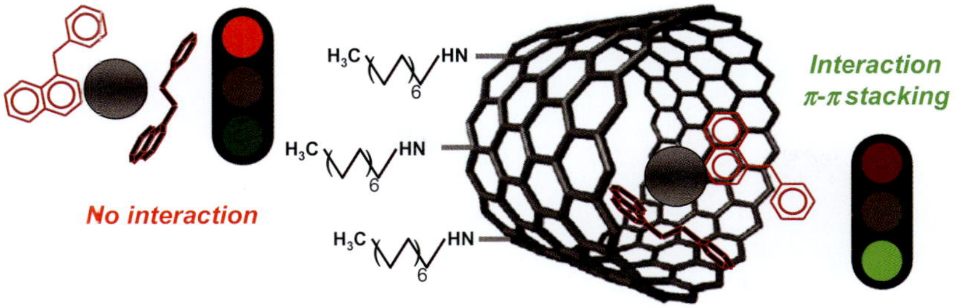

Figure 1.14 Molecular recognition: specific binding between the active phase and the support.

been devoted to these challenges, it is worth mentioning that there is an increasing need of theoretical studies to characterize and understand these complex but well-defined systems. This topic will be discussed in detail in Chapter 12 of this book.

1.4
Nanocatalysis: Applications in Chemical Industry

Nanocatalysis is a rapidly growing field, which involves the use of nanomaterials as catalysts for a variety of catalytic reactions. Heterogeneous catalysis represents one of the oldest commercial applications of nanoscience; nanoparticles of metals, semiconductors, oxides, and other compounds have been widely used for important chemical reactions. Although surface science studies have contributed significantly to our fundamental understanding of catalysis, most commercial catalysts are still produced by 'mixing, shaking and baking' mixtures of multicomponents; their nanoscale structures are not well controlled and the synthesis–structure–performance relationships are poorly understood. Due to their complex physicochemical properties at the nanometer scale, even characterization of the various active sites of most commercial catalysts proves to be elusive. Key objectives of nanocatalysis research should include:

- Increased selectivity and activity, low energy consumption, and long lifetime of catalysts by controlling pore size and particle characteristics.
- Replacement of precious metal catalysts by catalysts tailored at the nanoscale and use of base metals, thus improving chemical reactivity and reducing process costs.
- Catalytic membranes by design that can remove unwanted molecules from gases or liquids by controlling the pore size and membrane characteristics.

As discussed in the previous section, this can be achieved only by precisely controlling the size, shape, spatial distribution, surface composition and electronic structure, and thermal and chemical stability of the individual nano components. In view of the numerous potential benefits (Figure 1.15a) that can accrue through their use, nanostructured catalysts have been the subject of considerable academic and industrial research attention in recent times.

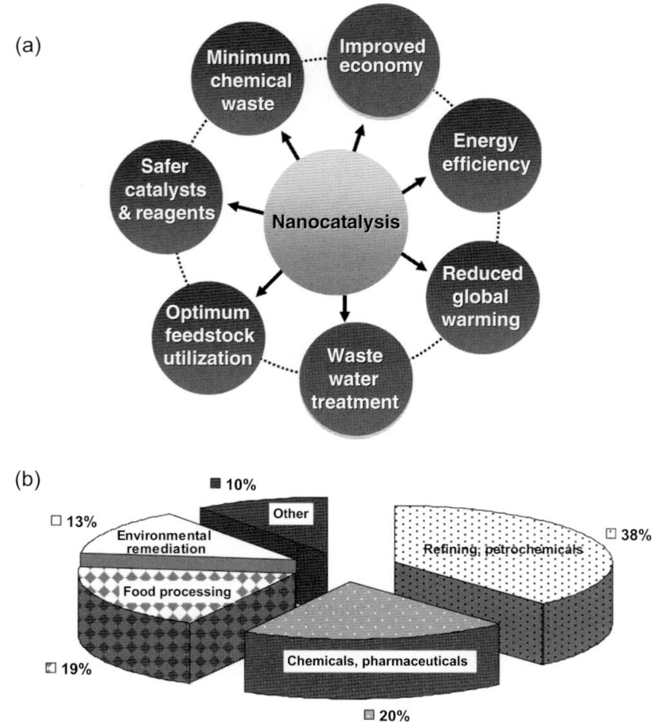

Figure 1.15 (a) Expected benefits of nanocatalysis; and (b) catalyst market by end-use.

Catalysts find application in a variety of industry segments spanning refinery, petrochemical, pharmaceuticals, chemical, food processing sectors and others. An assessment of the global catalysts' market reveals the following:

- The global market for NPs used in catalytic applications increased from US $193.74 million in 2006 to an estimated US $200.695 million in 2007. It is expected to reach US $325 million by 2012, corresponding to a growth of 9.5% in the preceding five years, from 2007 onwards.
- Commercially well-established catalysts such as industrial enzymes, zeolites and transition metal catalysts, till recently, accounted for about 98% of global sales.
- However, nanocatalysts, such as transition metal oxides, gold catalysts, carbon nanotubes and others are expected to more than triple their combined market share from 2009 onwards.
- It is the smaller end-user segments whose consumption is growing the fastest.

Not surprisingly, nanocatalysis is a growing business. The list of companies that have already patented and/or commercialized technologies relating to nanocatalysts is already impressive. The dominant global players include Argonide Corporation, BASF Catalyst LLC, BASF SE, Bayer AG, Catalytic Solution, Inc., Evonik Degussa GmbH, Genencor International, Inc., Headwaters Nanokinetix, Inc.,

Hyperion Catalysis International, Johnson Matthey PLC, Mach I, Inc., Nanophase Technologies Corporation, NanoScale Corporation, NexTech Materials Ltd., Oxonica, PQ Corporation, Sachtleben Chemie Gmbh, Süd-Chemie AG, Umicore NV and Zeolyst International, among others.

In this section, the exciting opportunities of nanocatalysis in selected chemical processes, as well as the challenges in developing nanostructured catalysts for industrial applications, are discussed.

1.4.1 Fuel Cells

One of the most important challenges for the ultimate commercialization of fuel cells is the preparation of active, robust, and low-cost catalysts. Recently, the number of publications related to the fundamentals of fuel cell (nano)electrocatalysts has grown significantly [141–145]. Catalysts composed of small, metallic particles, such as platinum or PtRu, supported on carbons or oxides are widely used in fuel cells.

The basic function of the catalyst layer present in a fuel cell is to provide a conductive environment for electrochemical reactions. The main processes that occur in the catalyst layer include mass transport, interfacial reactions at electrochemically active sites, proton transport in the electrolyte phase, and electron conduction in the electronic phase. The oxidation reaction occurs in the anode catalyst layer, while oxygen reduction occurs in the cathode catalyst layer. Anodic and cathodic reactions require metallic active sites to break molecular bonds of gaseous diatomic reactants. The oxidation reaction possesses a lower over-potential and higher rate than oxygen reduction; thus, oxygen reduction is a significant source of voltage loss.

Platinum catalysts suffer from several drawbacks including slow kinetics, low efficiency, high cost, and limited lifetime. Thus, current research is focused on the development of catalyst materials with low cost, high performance, high stability, and durability. In general, literature data show that Pt-based binary catalysts perform better as catalysts for oxygen reduction than pure Pt. However, the electrochemical activities of nanosized Pt-based materials are strongly influenced by their physical and chemical properties which are, in many cases, mutually dependent and determined by the synthetic conditions. So, it comes as no surprise to find that the electrocatalytic activities reported in the literature are dependent on the preparation methodology adopted. Different reports have interpreted activities as associated to various physical parameters, such as particle size, Pt enrichment of the surface, distance between Pt atoms, and occupancy of the 5d Pt band.

Since the activity of the Pt and Pt alloy catalysts increases as the reaction surface area of the catalyst increases, catalyst particles should be reduced in diameter to increase the active surface. So the catalysts are supported on a high surface area substrate. The use of carbon materials as catalyst supports for precious metals has increased rapidly in recent years, due to the continuous advancing development of fuel cells. The main requirements of suitable supports for fuel cell catalysts are:

high surface area, good electrical conductivity, suitable porosity to allow good reactant flux, and high stability in fuel cell environment. Carbon blacks are commonly used as low-temperature fuel cell catalysts. They are usually submitted to chemical activation to increase anchoring centers for metal catalysts (to increase metal loading and dispersion). The high availability and low cost make carbon blacks the most used support for fuel cell catalysts. The disadvantage of these carbons is the presence of a high amount of micropores, which can hinder the reactant flow. Moreover, these materials present low stability at temperatures higher than 373 K.

In that context, fuel cell electrocatalysis constitutes one of the best examples of the contribution that nanocatalysis can bring [141–143]. Nanoscience has stimulated extensive interest in nanostructured catalysts to significantly improve the energy density, power density and operation reliability while greatly reducing the manufacturing expense. In particular, nanostructured materials play a critical role in the catalysis of various innovative fuel cells, which not only possess high specific surface area and good conductivity for low polarization, but also provide unique nanoporous structures and functional chemical properties for highly intrinsic electroactivity and excellent utilization. In particular, the use of nanomaterials in fuel cells can significantly improve the electrocatalytic performance for high energy density and high power density, while reducing the manufacturing cost. The prominent electrocatalytic behavior of the nanomaterials mainly derives from their unique physico-chemical properties such as size, shape, pore structure/distribution, surface defects and chemical properties. The great challenges to synthesize and further use the various nanostructured catalysts are not only from chemistry but also from nanoengineering approaches.

In general, core-shell nanostructures provide an economic and effective way to prepare precious metal catalysts while remarkably reducing the usage of platinum. To enhance the mass activity of Pt, core-shell nano-alloy catalysts with a Pt shell have been developed to remarkably reduce Pt loading while enhancing the catalytic performance [141]. Recent work has shown that the activity and durability of the oxygen reduction reaction catalysts can be further enhanced by controlling the core particle composition and tailoring the shape of the Pt shell structure [146]. As far as pure Pt is concerned, it has been shown that the specific activity of the oxygen reduction reaction on Pt nanoparticles decreases with decreasing the particle size, with a maximum in mass activity for particles with a diameter of 3 nm [147]. Additionally, unique nanostructures such as nanowires and nanotubes provide high specific active surface area, superior conductivity, better mass transport and high intrinsic catalytic activity [148]. Sun *et al.* have developed a promising electrocatalyst based on starlike Pt nanowires presenting both good catalytic activity and durability (Figure 1.16) [149]. The progress in nanoscience should provide good opportunities to explore alternatives to replace Pt and other noble metal catalysts. Cr-decorated NiNPs [150] or cobalt porphyrin-templated nanostructured polyaniline and polypyrrole array nanorods [151, 152] have already shown interesting performances for hydrogen oxidation and oxygen reduction reactions, respectively.

The replacement of carbon black by carbon nanostructures has led to significant changes in the catalytic layer of fuel cells. The arrangement of carbon nanoparticles

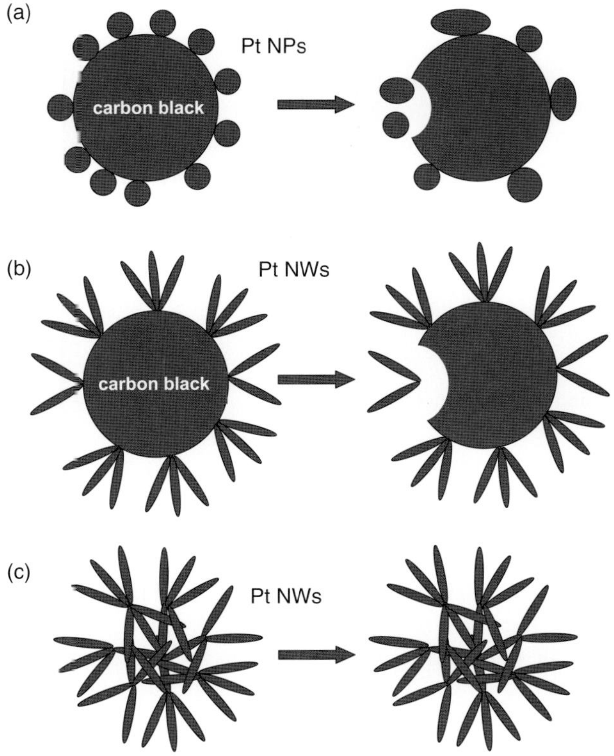

Figure 1.16 Scheme of morphology changes that occur for Pt during accelerated electrochemical cycling. (a) Pt nanoparticles on carbon; (b) Pt nanowires (NWs) on carbon; (c) unsupported Pt NWs. (Adapted from [149]).

at the anode and cathode differs significantly, which prevents the formation of small or closed pores (aggregates) in the catalytic layer of fuel cells. The high surface area and high number of mesopores in ordered mesoporous carbons and carbon gels allow high diffusivity and good reactant flux. Thus, catalysts supported on these materials show higher catalytic activity than the same catalysts supported on carbon black. Among new carbon-based materials, nanotubes are the most investigated catalyst support for low-temperature fuel cells [145]. The highly crystalline structure of CNTs provides high conductivity, surface area and porosity, resulting in exceptional diffusivity. Moreover, CNTs and CNFs have a positive effect on catalyst structure, yielding higher catalytic activity and stability than carbon black [153]. Few studies have been conducted on carbon nanohorns, nanocoils and fibers as Pt catalyst supports. Graphene could be a promising alternative for CNTs and other carbon-based supports due to its superior electronic and chemical properties [154]. If the graphene nanosheets can be fabricated onto a 3D network or vertical array structure, the surface area can be dramatically increased over that of 2D graphene. Mixtures of graphene and CNTs as support have already shown interesting synergetic effects [155].

In brief, the nanostructured catalysts for fuel cell applications currently have been fueled up and some have been used in practical devices. However, many more efforts are needed to investigate the mass production methods, and the interaction of these nanostructured materials with practical electrodes such as MEA and/or large-scale fuel cells for optimization and performance evaluations. Finally, and as for conventional nanocatalysis, for electro-nanocatalysis, computational approaches for theoretical modeling of nanomaterials such as carbon nanotubes and metallic nanoparticles through molecular dynamic techniques are needed for a better understanding of these complex systems under operating conditions. These issues will be discussed in more detail in Chapter 4, Section 4.2.1.2.

1.4.2
Nanostructured Exhaust Catalysts

Catalytic abatement of toxic automotive exhaust represents one of the major applications of heterogeneous catalysts all over the world. Nowadays, more than 95% of vehicles produced in the world (more than 700 million automobiles) are equipped with a catalytic converter, which, for gasoline-fueled engines, is almost exclusively based on the so-called three-way catalyst (TWC). TWCs are capable of simultaneously and efficiently converting CO, hydrocarbon and NO_x into CO_2, H_2O and N_2, provided that the air-to-fuel ratio is constantly kept at the stoichiometry, that is, under conditions where the amount of oxidants is equal to that of reducing agents. Catalytic converters comprise a high surface area mixed oxide monolith (often incorporating oxides such as CeO_2, ZrO_2, and Al_2O_3) with noble metal catalysts dispersed into the monolith. The noble metals most often include Rh, Pt, and Pd, which are used to perform seven basic reactions listed in Table 1.2. The main use of these metals is in automobile catalytic converters, where over 60% of the procured metals are used. Two of the most important reactions in the catalytic converter are the oxidation of CO by O_2 to produce CO_2 and the oxidation of CO by NO to produce CO_2 and N_2. Rhodium, in particular, is recognized as critical to efficient reduction of NO. Besides the noble metals (Pd and Pt to promote oxidation reaction and Rh to promote NO_x reduction), the current TWCs contain CeO_2-ZrO_2 mixed oxide promoters as the so-called oxygen storage/release component. The oxygen storage capacity is the ability of CeO_2-containing oxides to adsorb and release

Table 1.2 The seven basic reactions used in the catalytic converter. Several of these reactions, including all of the hydrocarbon (HC) reactions, are selective or have variable stoichiometry.

Oxidation	$2\,CO + O_2 \rightarrow 2\,CO_2$
	$HC + O_2 \rightarrow CO_2 + H_2O$
Reduction/three-way	$2\,CO + 2\,NO \rightarrow 2\,CO_2 + N_2$
	$HC + NO \rightarrow CO_2 + H_2O + N_2$
	$2\,H_2 + 2\,NO \rightarrow 2\,H_2O + N_2$
Water Gas Shift	$CO + H_2O \rightarrow CO_2 + H_2$
Steam Reforming	$HC + H_2O \rightarrow CO_2 + H_2$

oxygen under, respectively, fuel-lean and fuel-rich conditions, according to the reactions:

Lean $$\overset{\text{III}}{\text{Ce}}_2\text{O}_3 + \tfrac{1}{2}\,\text{O}_2 \longrightarrow 2\,\overset{\text{IV}}{\text{Ce}}\text{O}_2$$

Rich $$2\,\overset{\text{IV}}{\text{Ce}}\text{O}_2 + \text{CO} \longrightarrow \text{CO}_2 + \overset{\text{III}}{\text{Ce}}_2\text{O}_3$$

Ce_2O_3 captures excess oxygen that would escape the tailpipe and saves it for CO oxidation when in short supply. The act of O_2 storage enhances NO reduction. The oxygen storage/release is a crucial property of the TWCs since it helps to cope with the air-to-fuel oscillations and maintains a stoichiometric composition of the exhaust at the noble metal catalyst, necessary to attain the highest exhaust conversions.

Since the advent of the TWCs, in the early 1980s, there has been a progressive tightening of the environmental legislation aimed at minimizing the amount of harmful pollutants emitted during the vehicle use. Besides the quite restrictive limits on the emissions (>97% compared with uncontrolled emissions), new materials with enhanced thermal stability and high durability are also being developed. This is due to the fact that TWCs feature the so-called light-off type of conversion *vs.* temperature behavior, where the conversion steadily increases from 0 to 100%. The light-off temperature, conventionally taken as corresponding to 50% of conversion, is typically 513–623 K. This means that to achieve the improvement of efficiency as required by the EU and US legislation, the catalyst heating time, that is, time to reach light-off temperature, must be decreased to 10–20 s. A cost-effective solution is to mount a secondary converter directly on the exhaust manifold. This, however, exposes the catalysts to extremely harsh conditions, where temperatures as high as 1373 K are reached.

Thus, today the development of new and efficient catalytic converters constitutes a real challenge for the chemists, because the use of nanostructured materials is needed both for the oxide support/promoters and the metallic particles [156, 157]. Additionally, due to the technological importance of this catalyst, much effort has been devoted to the study by sophisticated *in situ* surface science tools [X-ray absorption near edge structure (XANES), environmental TEM, high-energy-resolution fluorescence-detected, resonant inelastic X-ray scattering, time-resolved fluorescence X-ray absorption spectroscopy (XAS)] of the nanoscale compositional and structural evolution of these systems [158–161]. The development of specific high-temperature reactors for *in situ* XAS measurements should permit the study of commercial three-way catalysts under realistic gas composition and temporal conditions [162]. The in-depth and *in situ* characterization of these nanocatalysts permits the identification of the active sites, analyzes the genesis of the catalyst, ascertains its dynamic structure, and can be used to evaluate synthetic strategies. Representative examples of the input of nanomaterials in understanding and designing active sites at the surface of these catalytic materials are given below.

The importance of ceria originates from its remarkable redox and oxygen storage capability. It can undergo repeatable Ce^{IV}/Ce^{III} redox cycles depending on the conditions in the exhaust stream. Using organic molecules to precisely control CeO_2 crystal growth, Adschiri *et al.* have prepared 10 nm, 61 $m^2\,g^{-1}$ CeO_2 nanocubes [163]. These authors have demonstrated the extra-low-temperature oxygen storage capacity of these CeO_2 nanocrystals with cubic (100) facets. A considerable oxygen storage capacity occurs at 423 K without active species loading. This temperature is 250 K lower than that of irregularly shaped cerium oxide (10 nm, 100 $m^2\,g^{-1}$). This result indicates that cubic (100) facets of cerium oxide have the characteristics to be a superior low-temperature catalyst.

In fact, the actual importance of oxygen uptake/release is determined by not only the oxygen storage capability but also the rates of the redox cycles. However, while the rate of cerium oxidation is very fast, reduction of ceria is generally sluggish. It has been proposed that the reduction of ceria is controlled by the nature of the oxygen vacancies, since oxygen diffusion, the rate-controlling step, depends on the type, size, and concentration of oxygen vacancies [164]. For example, Esch *et al.* showed that small size surface oxygen vacancies on CeO_2 (111) were immobile at room temperature, but linear clusters of these vacancies formed at higher temperatures [165]. This indicates that any processing condition that favors the formation of more desired oxygen vacancies will result in enhanced reducibility.

Two ceria nanorod samples with different types and distributions of oxygen vacancies were synthesized by Li *et al.* [162]. Two types of oxygen vacancies, small neutral Ce^{III}-oxygen vacancy associates and larger size clusters of these defects, were discovered in ceria nanorods from XRD and XPS analyses. The larger size oxygen vacancy clusters (i.e., dimers, trimers, or larger) result from interaction between the small neutral Ce^{III}-oxygen vacancy associates. The synthesis method was crucial to determine their distribution. These results show direct evidence of promoting the reducibility and activity of ceria nanorods with a high concentration of larger size oxygen vacancy clusters.

An inherent problem for nanocatalysts is their well-documented lack of stability under realistic technical conditions. Supported and unsupported metallic nanoparticles show rapid growth under reactive conditions, especially if the reaction temperature is very high as for automobile-exhaust control, where reaction temperatures are typically in excess of 873 K, and thermal stability of the catalyst becomes a crucial issue. Organic capping agents, which are widely used in colloidal chemistry to stabilize NPs under mild conditions, decompose at temperatures above 573 K, leaving the uncapped NPs unprotected against sintering and deactivation. Veser *et al.* prepared bimetallic PtRh-BHA nanocomposites (BHA: barium hexa-aluminate, a high temperature-stabilized alumina with a crystallization temperature around 1473 K) in a straightforward and efficient one-pot synthesis, which results in PtRhNPs with a narrow size distribution around 4 nm, homogeneously dispersed throughout the BHA support [166]. They discover that supported PtRhNPs with sufficiently high Rh content survive extended calcination at temperatures up to 1123 K without significant sintering [167]. The extreme thermal stability of these particles depends critically on the Pt/Rh ratio, with higher Rh contents leading to increased thermal stability. Phase separation is observed when the PtRhNPs start to sinter: Pt bleeds out from PtRhNPs and the Rh content in the remaining nanoparticle increases accordingly. This 'distillation' of the low-melting-point Pt from the PtRhNP results in a sacrificial self-stabilization of the bimetallic NPs.

Finally, it is worth noting that this type of catalytic reactor also raises the critical issue of nanoparticle possible (eco)toxicity. For example, palladium nanoparticles are recognized as components of airborne automotive pollution produced by abrasion of catalyst materials in the car exhaust system [168, 169]. The possible (eco)toxicity of cerium oxide nanoparticles, which are also used in emission control systems in automobile engines as a diesel fuel-borne catalyst to reduce particulate matter emissions, could also be envisaged [170].

1.4.3 Gas Sensors

Gas sensors, also called gas detectors, can be used to detect combustible, flammable and toxic gases, and oxygen depletion. A gas sensor is a device that detects the presence of various gases within an area, usually as part of a safety system. Such equipment can be used to detect a gas leak for example. The gas detector will thus interface with a control system so a process can be automatically shut down, or it will sound an alarm to operators in the area where the leak is occurring, giving them the opportunity to leave the area. The development of devices to be used as gas detectors is thus of high importance because there are many gases that can be harmful to organic life, such as humans or animals.

Gas detectors are of two main types: portable devices and fixed gas detectors. Portable devices are often used to monitor the atmosphere around personnel and are worn on clothing or on a belt/harness. The fixed gas detectors are often used for detection of one or more gas types. Fixed type detectors are generally mounted near the process area of a plant or control room. Generally, they are installed on fixed type mild steel structures and a cable connects the detectors to a system for continuous monitoring and where a tripping interlock can be activated for an emergency situation.

Concerning the mode of control, gas detectors can be classified according to the operation mechanism (semiconductors, oxidation, catalytic, infrared, etc.). Several types of system have thus been developed to act as gas sensors able to detect target species. Among them, electrochemical or chemical sensors are well-known.

Electrochemical gas sensors [171] are like fuel cells, but instead of being designed to produce power, they are designed to produce an electric current. Thus, electrochemical gas sensors measure the concentration of a target gas by oxidizing or reducing the target gas at an electrode and measuring the resulting current, this latter being thus precisely related to the amount of the target gas. Commercial devices based on this principle are currently used in industries or laboratories to detect carbon monoxide in the atmosphere for example. Carbon monoxide is oxidized at one electrode to carbon dioxide while oxygen is consumed at the other electrode. For carbon monoxide detection, the electrochemical cell has advantages over other technologies in that it has a highly accurate and linear output to carbon monoxide concentration, requires minimal power as it is operated at room temperature, and has a long lifetime (typically commercial available cells now have lifetimes of five years or greater). Until recently, the cost of these cells and concerns about their long term reliability had limited uptake of this technology in the marketplace, although these concerns are now largely overcome. This technology is now the dominant technology in USA and Europe for CO detection.

The detection of various species in gas mixtures, in order to evaluate the impact of these species on the quality of a product or on the environment (toxic gases, offensive odors, volatile organic compounds (VOCs) and other air pollutants) in low concentrations is a challenging subject (Figure 1.17). A practical approach can be achieved by using 'chemical sensors.'

A chemical sensor consists of a system, including a sensitive layer (the receptor) and a device (the transducer) that transforms the atomic scale interaction between the gas and the material of the sensitive layer into a mechanical or electrical response. When the sensed molecules are physically or chemically adsorbed on the

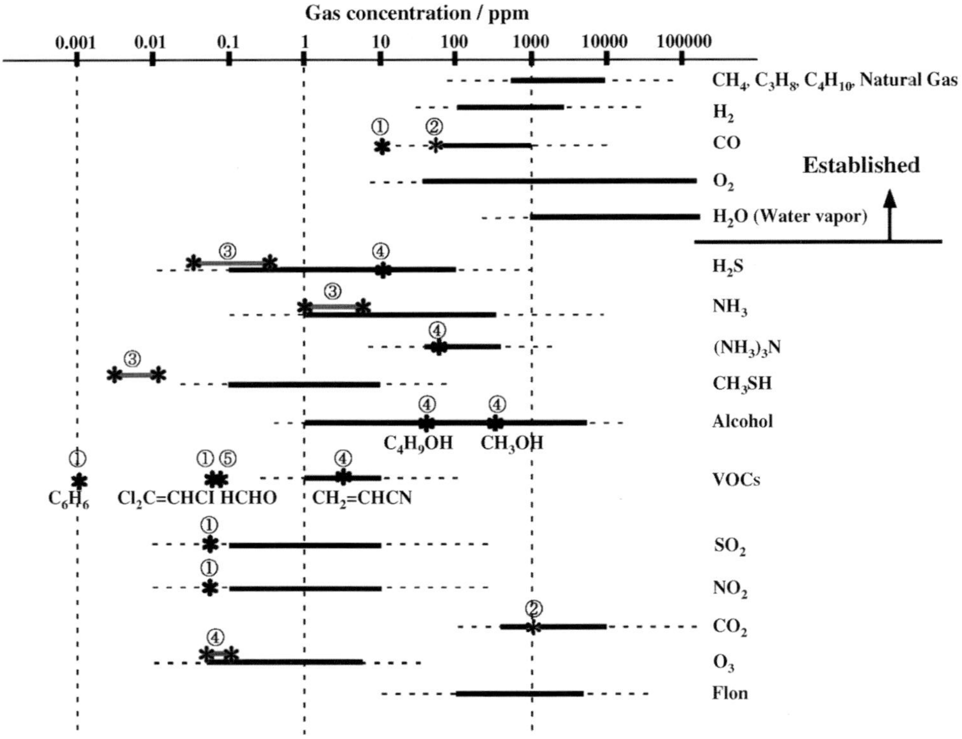

Figure 1.17 Concentration levels of typical gas components concerned. Star marks indicate the standards of the gases legislated in Japan by (1) Environmental Standard, (2) Ordinance on Health Standards in the Office, (3) Offensive Odor Control Law, (4) Working Environment Measurement Law, and (5) Ordinance by Ministry of Health, Labor and Welfare. (Reproduced with permission from [188]. Copyright © (2005) Elsevier).

sensitive layer, chemical reactions like combustion often happen. These phenomena occur at the molecular or atomic level as for a nanocatalyst. The interaction between molecules of the fluid (gas or liquid) phase induces a variation in the physical properties of the sensitive layer, such as electron density, optical properties, mass or temperature [172]. The most common sensors systems contain a solid material as sensitive layer deposited on an electronic device, and deal with the changes in the electronic charge density and charge carrier mobility when a gas interacts with the solid surface. Thus, the mode of detection of a target gas is based on the variation of the electrical properties of the sensitive layer, such as resistance or capacitance, which is detected by the device. In the research of chemical sensors, one of the main problems is the high number of parameters that interfere during the measurement:

- the *lack of specificity* for a given chemical species;
- the *cross-sensitivity*, meaning that the signal of one target species can change if this is in a mixture of species;
- the *temperature dependence*, as adsorption and chemical reactions are widely influences by the temperature;
- the *'memory effect' of the sensor*, meaning modification of the signal with use and thus the need for recalibration;
- *drift problems* after long-term use, in other words, *stability problems*.

Gas sensors based on semiconducting materials have become of great interest. Metal oxides are widely employed as sensitive elements in gas sensors [173] and commercial devices produced by classical ceramic processes have been available for a long time. SnO_2 sensors are the best-understood prototype of oxide-based gas sensors [174]. Nevertheless, highly specific and sensitive SnO_2 sensors are not yet available [175].

If such devices are able to detect low concentrations of flammable or toxic gases at a ppm level, their lack of selectivity and long term stability remains their major drawback, and improving the cross sensitivities between gases such as CO, hydrocarbons, and NO_2 is still a major issue. Enhancement of sensor performance can be achieved through (i) an accurate control of the morphology in the nanometer range of the composition and of the surface state of the sensing materials [176, 177]; (ii) the integration of the sensor components on micromachined silicon platforms [178]; (iii) the optimization of the working temperature of the sensitive layer [179], the operating mode, and the signal processing [180]; and (iv) the use of filters exhibiting selective gas retention or conversion properties [181]. The use of filters combining catalyst-loaded mesoporous materials represents a very attractive way to improve the selectivity of a sensor toward a target gas and to overcome the classical problem of saturation exhibited by passive filtering membranes [182, 183]. The filtering efficiency of such a material, in terms of selectivity, reproducibility, and long term stability, nevertheless demands requirements of crucial importance regarding composition and morphology. As for any nanocatalyst, these requirements include the homogeneous dispersion and high surface area of the active sites on the support, the homogeneous distribution of size-controllable pore channels in the support, and the immobilization of the active sites inside the pore channels to prevent any migration and coalescence into the host structure.

Besides tin oxide based nanomaterials, other nanomaterials such as zinc oxide nanostructures [184, 185] metal nanoparticle-functionalized oxide nanostructures [186] or carbon nanotubes decorated with metallic nanoparticles [187, 188] to cite a few examples, are also used as sensitive layers. Because of the small size of nanotubes, nanowires, or NPs, a few gas molecules are sufficient to change the electrical properties of the sensing elements. This allows the detection of a very low concentration of chemical species.

Whatever the nature of the nanomaterials used, the on-going research on gas sensors is mainly directed toward the increase in sensitivity and selectivity of the devices. This is performed through the development of novel sensitive layers and catalytic filters made of well-controlled nanomaterials in terms of structure and composition, to lead to efficient and reproducible devices able to detect very low quantities of target molecules in mixtures of species like in urban environments. Numerous works concern for example the shape control of the NPs acting as detecting layer or the integration of nanoparticles of a doping metal for synergy effect. Another current trend of research concerns the assembly of several individual sensors having adequate detection properties, to get multiple systems able to detect simultaneously several gases for example.

Although gas sensors are well-established in some application fields (like combustible gas monitoring, oxygen sensing for combustion exhaust control and humidity sensing for amenity or living space), there are a variety of newly emerging markets (CO_2 monitoring, air quality, odors and humidity are in increasing demand for various purposes; Figure 1.18), which will be substantiated when gas sensors are sufficiently developed [189]. The main commercial applications are nowadays environment monitoring and food quality analysis. Materials design is an important point in innovating gas sensors but attempts to make the sensor devices more intelligent and more quantitative are also important for the further advancement of gas sensor technology.

1.4.4
Photocatalysis

Photoinduced processes are studied in several industrial-oriented applications. Despite the difference in character and use, all photoinduced processes are based on the excitation of a semi-conductor by a light energy higher than the band gap thus inducing the formation of energy-rich electron-hole pairs. All photoinduced chemical processes have in common the so-called 'initial' steps, which include the absorption of light, diffusion, and trapping of charge and fate (recombination or phase transfer to the gas/liquid media) of charge carriers. By photocatalysis, it is commonly understood to be any chemical process catalyzed by a solid where the external energy source is an electromagnetic field with wavenumber in the UV-vis range. Photocatalysis is mainly involved in three areas namely organic synthesis,

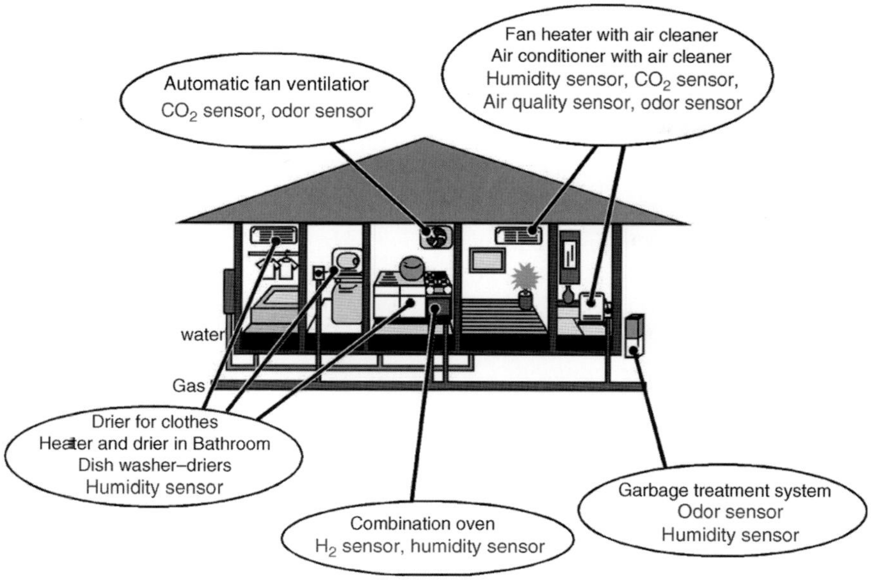

Figure 1.18 Various kinds of gas sensors equipped at various sites in a house (Reproduced with permission from [188]. Copyright © (2005) Elsevier).

degradation of pollutants (in both gas and liquid phases) and special reactions like H_2O reduction or N_2 fixation [190].

Photocatalysts are most of the time solid oxide semiconductors, among which TiO_2 in its anatase form is by far the most known and investigated photocatalyst, due to its chemical stability, nontoxicity and well positioned valence and conduction bands. Besides the research concerning metal oxide materials, recent progress has expanded the chemical nature of photoactive systems by including (oxy)sulfides and (oxy)nitrides, doped-zeolites, and molecular entities embodied in zeolites or other composite materials.

The interest of metal oxide nanomaterials either as active phases, unsupported/supported nanoparticles or as supports for metallic nanoparticles, in photocatalysis is discussed by V. Hulea *et al.* in Chapter 10, Section 10.3.2 devoted to nano-oxides. In particular, the influence of nanomaterials characteristics (size, morphology) on their photocatalytic behavior is described. For instance, the major effect of morphology and crystal structure on the photocatalytic activity has been reported for TiO_2 and ZnO, two materials usually used in photocatalytic reactions.

Photocatalysts resulting from the combination of metallic nanoparticles and metal oxides are the subject of intensive work because of the benefits induced by such association in terms of activity and selectivity. The deposition of transition metals co-catalysts (Pt, Rh, Au, Cu and Ag) on TiO_2 catalyst usually improves the photocatalytic activity, which strongly depends on the extent, nature and the work function of the metals [191]. Several strategies have been followed to improve the photocatalytic activity by tailoring size and shape of titania and metal nanostructures. For instance, the photoactivity of TiO_2 has been drastically enhanced with the decreasing size of gold NPs loading even at very low metal content (0.01 wt%).

Carbon materials have some beneficial effects on the photocatalytic activity of TiO_2 [192, 193]. Different preparation methods have been used to introduce carbon into the TiO_2 and activated carbon system such as: mixing TiO_2 with activated carbon, precipitating TiO_2 onto the surface of activated carbon, polymerizing a carbon source with a titanium precursor, inclusion of carbon nanotubes into an inorganic film of TiO_2 by a sol-gel method, or preparation of titania composite materials based on multiwalled carbon nanotubes (MWCNTs) by an impregnation method, and so on. MWCNTs are incorporated to optimize the surface and catalytic properties of TiO_2 since the composites containing carbon nanotubes are believed to exhibit cooperative or synergetic effects between the metal oxide and carbon phase. For example, MWCNT-TiO_2 composite catalysts prepared by a modified sol-gel method gave rise to synergetic effects on the photocatalytic degradation of phenol (Figure 1.19). The increase of MWCNT/TiO_2 ratio from 1 to 20% favored the enhancement of the synergetic effect on phenol disappearance [194]. The results were explained in terms of the intimate contact between the MWCNT and TiO_2 phases, suggesting that a strong interphase interaction may be triggered between these two phases and MWCNTs might behave as photosensitizer in the composite catalysts.

The future of energy supply depends on innovative breakthroughs regarding the design of cheap, sustainable, and efficient systems for the conversion and the storage of renewable energy sources, such as solar energy. In that context, the production of hydrogen through sunlight-driven water splitting appears to be a promising and appealing solution (see also Chapter 4, Section 4.3.3.1). This solution, inspired by the biological process photosynthesis, has attracted the interest of numerous bioinorganic chemists in the past years. Designing original synthetic molecular systems to achieve a similar function is presently the base of a large field of research called 'artificial photosynthesis.' Recent developments regarding the

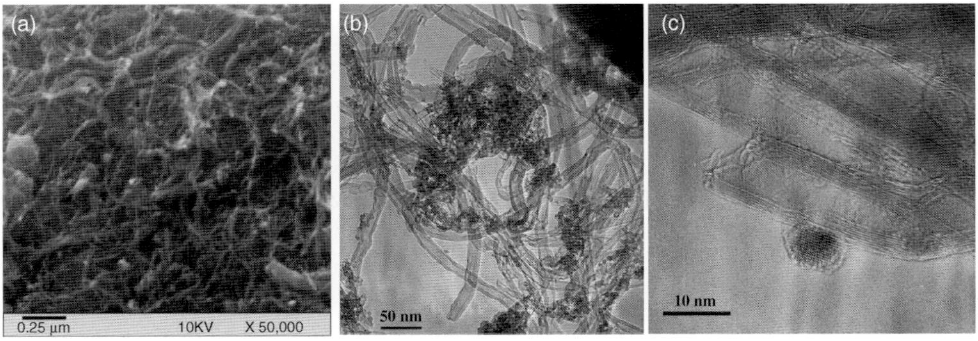

Figure 1.19 SEM micrograph (a) and TEM images (b, c) of MWCNT-TiO_2 composite material (20% CNT w/w).

design, characterization, and evaluation of cobalt-based molecular catalysts and their combination with photosensitizers to generate light-driven systems for H_2 or O_2 evolution were reported in 2011 by Artero *et al.* [195]. In their review, the authors showed how cobalt-based molecular catalysts can be immobilized onto conducting materials so as to form electrodes or photoelectrodes to be integrated in a photoelectrochemical cell for light-driven hydrogen generation from water. They also mention that besides the development of multi-component photocatalytic systems based on a molecular complex, a renewed interest regarding solid metal oxide/hydroxide materials and nanomaterials as promising water oxidation catalysts and their further implementation into light-driven systems appeared very recently. To cite a few examples cobalt oxide materials, such as cobalt nanoclusters in the mesopores of silica associated with $[Ru(bipy)_3]^{2+}$ photosensitizer [196], Co_3O_4 nanoparticles loaded onto Ni foam supports have been reported as anode material for water electrolysis [197] as well as a hybrid material consisting of Co_3O_4 nanocrystals grown on reduced graphene oxide as a high-efficient bi-functional catalyst for the oxygen reduction reaction and oxygen evolution reaction [198]. In the latter, the unusual catalytic activity arises from synergetic chemical coupling effects between Co_3O_4 and graphene. As another example, one may cite the work of F. Odobel *et al.* who obtained an absorbed-photon to current conversion efficiency of 45% (the highest value reported for a sandwich dye-sensitized solar cell based on a nanostructured metal oxide) using a peryleneimide sensitizer and a covalently linked peryleneimide-naphthalenediimide dyad chemisorbed on nanoporous nickel oxide electrodes [199]. Concerning photocatalytic systems containing metallic nanoparticles, gold and platinum NPs have appeared as attractive platforms for the photocatalytic production of hydrogen gas. Photocatalysis of H_2 has been mainly achieved by alcohol re-forming on the surface of Au and Pt NPs supported on semiconductor materials such as titania. But the major drawback to this type of H_2 photocatalysis is that the photons must have an energy greater than the band gap of the semiconductor material in order to produce a charge-separated state that is able to sustain the reaction: $2H^+ + 2e^- \rightarrow H_2$. Thus, some groups tried to find attractive alternatives to the titania-supported particles for the photocatalytic production of H_2, by developing photosynthetic complexes based on the covalent attachment of a known active photosystem, such as ferredoxin oxidoreductase to Au and PtNPs, as shown by H. Golbeck *et al.* [200]. The challenge is to transfer the electron from the photosystem to the NP surface within 100-ms time frame. In the same way, a highly efficient photocatalytic system for hydrogen evolution has been reported by S. Fukuzumi *et al.* based on a dihydronicotinamide coenzyme (NADH) as a sacrificial agent in an aqueous solution. Their photocatalytic system has been built by using water-soluble platinum nanoclusters functionalized with methyl viologen-alkanethiol and a simple electron-donor dyad (9-mesityl-10- methylacridinium ion) [201]. In summary, despite significant efforts, developing oxygen electrode catalysts with high activity at low cost remains a great challenge in the field of sunlight-driven water splitting reaction for the production of hydrogen. It is interesting to note that immobilization of molecular catalysts on nanostructured electrode surfaces or the linkage of various active photosystems to metallic nanoparticles might result in increased stability and/or increased activity. Finally, various nanomaterials presently know a renewed interest in this field of research.

To conclude this section, nanocatalysts appear of high interest in photocatalytic processes, as shown by the numerous works using them in the degradation of pollutants or the renewed attention they presently receive for water splitting reactions. As for other catalytic processes, the precise control of their characteristics (size, shape, surface area, composition, etc.) is a key point.

1.4.5
Enantioselective Catalysis

Enantioselective catalysis is an increasingly important method of providing enantiomeric compounds for the pharmaceutical and agrochemical industries. To date, heterogeneous catalysts have failed to match the industrial impact achieved by homogeneous systems. One successful approach to the creation of heterogeneous enantioselective catalysts has involved the modification of conventional metal particle catalysts by the adsorption of chiral molecules, as reviewed recently by C.J. Baddeley *et al.* [202].

The best known metal-based heterogeneous enantioselective catalytic reactions are the hydrogenation of activated C=C bonds [203] (mainly using palladium catalysts modified with cinchona alkaloids), α-ketoesters, [204, 205] (mainly using Pt catalysts modified with cinchona alkaloids and related modifiers) and β-ketoesters [206] (mainly using Raney Ni systems modified with tartaric acid). The reagents (e.g., methyl pyruvate or methylacetoacetate) are close to planar and it is widely believed that: (i) the surface catalyzed reaction involves the dissociation of H_2 into H(ads); (ii) the adsorbed atomic hydrogen attack the pro-chiral reagent from beneath the molecular adsorbate; and (iii) the direction of attack results in the formation of the chiral center with the C—OH bond being oriented away from the surface. The chirality of the ultimate product is therefore thought to be determined by the adsorption geometry of the pro-chiral reagent, thanks to the adsorption of a chiral molecule. The cinchonidine is the most known chiral inductor employed to achieve methylpyruvate hydrogenation. Due to its large size, the cinchonidine could act as an enzyme by encapsulating the reactant. Numerous studies on Pt/cinchonidine systems, in particular from A. Baiker *et al.*, report that the inductor interacts with the reactant molecules through H-bonding thus stabilizing a prochiral adsorption geometry of the reactant to the detriment of the other configuration [204]. Nevertheless, there is still a debate concerning the detailed structure of such an adsorption complex. It appears that theoretical calculations [density functional theory (DFT)] could be very helpful to rationalize the role of any supramolecular docking structure between a substrate and a chiral inductor modified-nanoparticle in inducing chirality, as recently published by A. Baiker *et al.* [207]. Recently, a highly active and enantioselective heterogeneous asymmetric catalyst was fabricated by confining Pt nanoparticles that are modified with cinchonidine within the channels of carbon nanotubes [208]. A high turnover frequency (TOF) and enantioselectivity are achieved when using this catalyst for the asymmetric hydrogenation of α-ketoesters.

Tartaric acid or aminoacids (alanine and glutamic acid) are other modifiers, which are mainly used with Ni catalytic systems, but their mode of action is less understood. Intensive research on the metal-based heterogeneous enantioselective catalytic hydrogenation reactions continues to get more efficient, more stable or more 'green' nanocatalysts [209–211].

The C—C coupling reactions in their enantioselective version are another type of reaction in which nanocatalysts are evaluated, mainly using Pd nanoclusters stabilized with ligands that are common in homogeneous catalysis. To cite a few examples, interesting results were obtained with chiral xylofuranoside diphosphites-modified Pd NPs in allylic alkylation of *rac*-3-acetoxy-1,3-diphenyl-1-propene with dimethyl malonate [212]. These particles displayed a high selectivity, the reaction mainly proceeding with one enantiomer of the substrate, hence demonstrating a very high degree of kinetic resolution. For comparison, the same reaction catalyzed by a corresponding molecular species accommodating the same diphosphite ligand and in conditions of dilution where the rates of both catalysts matched proceeded at similar rates with both enantiomers. When changing the ligands and/or the substrates, these systems were found to be dependent both upon the ligands and the substrates in a much more dramatic way than corresponding molecular systems [213]. As another example, H. Fujihara *et al.* reported small PdNPs (1.2–1.7 nm) stabilized with various chiral mono- and bisphosphine (binap, segphos and diop derivatives) ligands as active nanocatalysts in Suzuki–Miyaura cross-coupling reactions at room temperature. Depending on the ligand, the results in terms of yield and enantioselectivity varied with *ee* up to 58% [214, 215]. More recently, they described FePd magnetic nanoparticles modified with chiral binap ligand as a recoverable catalyst for the same asymmetric coupling reaction [216]. This work was the first example of an asymmetric catalytic reaction using chiral magnetic FePdNPs. The interest comes from the easy separation of the catalyst from the reaction mixture by simple application of a magnetic field while keeping the catalytic properties.

To summarize, interesting results in enantioselective catalysis have been obtained by using chirally modified metal surfaces. However, and in contrast to the chiral homogeneous catalysts, the main problem faced in exploiting such systems is the multitude of possible active sites that make them more complicated and most of the time less successful catalysts. The use of chiral modifiers able to interact with the metal surface and/or the prochiral substrate through hydrogen bonding interactions appeared as an attractive way which allowed in some extent to control chirality and further to optimize enantioselectivity. Numerous researchers all over the world tend to find the right catalytic systems to increase the enantioselectivity in a given reaction, by taking inspiration from homogeneous systems that varies the metallic nanoparticles and the chiral inductor, but most of the time, the results obtained are not satisfactory as they are not general. Clearly, the development of convenient nanocatalysts for enantioselective catalysis remains a challenge and a better understanding of the chirality in nanocatalysis could help to succeed in this task. To cite a recent example, R. Jin *et al.* demonstrated that the use of NMR spectroscopy is very useful to probe the chirality of well-defined gold nanoclusters [217]. They could differentiate chiral nanoclusters from nonchiral ones by characterizing carefully the stabilizing ligands. These encouraging results show that a precise characterization of the nanocatalysts at a molecular level could give precious information on their potential in enantioselective catalysis.

1.5
Conclusions and Perspectives

The field of catalysis science is often criticized as being *ad hoc* and empirical. In that respect, the impact of nanoscience on catalysis, which has been already discussed by scientists from the heterogeneous catalysis community [218, 219], is obvious. Creative use of the modern methods of nanomaterial synthesis coupled with a deep understanding of fundamental molecular surface chemistry and advances in characterization methods has yielded significant progress in some of the most important and challenging problems in the field. This knowledge principally contributes to the development of model catalysts, which can however operate under mild conditions, but also to the design of more robust supported new catalysts.

Thus, since the pioneering work in the field already mentioned in the introduction of this chapter, impressive progress has been realized. We will give below some representative examples, which have been published in top international academic journals. As far as the characterization techniques are concerned, it is now possible: (i) to perform *in situ* studies of chemistry and structure of catalysts in reactive environments [220] [3]; (ii) to get atom-resolved TEM imaging of dynamic shape changes in NPs [221, 222] [4, 5], and to visualize by HRTEM *cis-/trans*-isomerization of molecules [223]; (iii) to use the metallic tip of a scanning tunneling microscope as a hydrogenation catalyst [224, 225], or to use the same tip to create a chemical bond between CO molecules and an iron atom to form Fe(CO) and then $Fe(CO)_2$ [226]; and (iv) to reveal the existence of catalytic regular oscillatory patterns at the nanoscale, by combining field ion microscopy with video techniques and chemical probing [227]. The use of physical techniques such as atomic layer deposition has also made significant progress, allowing the preparation of stable subnanometer platinum clusters deposited in alumina membranes [228].

The huge potential of nanochemistry has also encouraged the development of new or improved synthetic pathways to produce high performance catalytic nanomaterials. Significant progress has been made in controlling monometallic nanoparticle size and shape [229–231], particularly for gold nanocatalysis [232–234]. New strategies to improve nanoparticle stability have been developed [235–237]. The substitution of noble metals by inexpensive nano counterparts has shown promising results, particularly in the field of electrocatalysis [238, 239]. The composition of the nanocatalyst should also be taken into account, and many examples have shown that bimetallic nanocatalysis can bring significant advantages, provided a good tuning of the structure is achieved [240–243]. For these particles one should keep in mind that restructuring and metal segregation may occur under catalytic conditions [244]. New opportunities for improved performance arise when the multifunctionality inherent in catalytic processes, including molecular transport of reactants and products, is rethought in the light of architectures designed and fabricated from the appropriate nanoscale building blocks (nanoparticle/nanosupport), including the use of 'nothing' (void space) and deliberate disorder as design components [245]. The precise design of the nanosupport and the controlled localization of the active phase allows catalyst with specific properties to develop [246, 247].

In parallel, the extraordinary progress in DFT calculations for surface processes has created the possibility of computer-based catalyst design [248, 249]. Computational chemistry has been developed to the point that it can be used to greatly assist experimentalists to screen synthesis routes, to measure properties of materials that are difficult to perform under normal conditions, or to understand various mechanisms. However, nanocatalyst studies are relatively scarce [250]. The reason lies in the difficulty of accurately describing such complex systems, which exhibit a nanoscale metallic core with partially mobile organic molecules on its surface, not so easy to characterize experimentally. More investigation in this area is clearly needed. These ongoing progresses should rapidly enabling catalyst researchers in academia and industry to achieve the goal of catalysis by design.

For the next generation of catalysts, it should clearly have a reduced environmental impact. This will require entirely new catalysts: catalysts for new processes, more active and more selective catalysts and preferably catalysts that are made from earth-abundant elements. This represents a formidable challenge and it will demand an ability to design new catalytic materials well beyond our present capabilities. The ultimate goal is to have enough knowledge of the factors determining catalytic activity to be able to tailor catalysts atom-by-atom.

Beside the obvious scientific impact of nanoscience and nanotechnology on catalysis, we would like to stress the fact that nanoscience and nanotechnology have also been a tremendous lever for two communities, the coordination chemists of homogeneous catalysis and the solid-state chemists of heterogeneous catalysis, to meet and collaborate efficiently on common objectives. This is perhaps the more important contribution that nanoscience and nanotechnology have brought today to catalysis.

Finally, and to be complete, we should also stress that this new discipline combines well-known as well as new areas and concepts of both heterogeneous and homogeneous catalysis, such as:

- description of the chemical bond from both molecular and solid-state point of view
- surface science molecular analysis
- surface coverage, surface restructuring, nonequilibrium self-organizational behavior
- coordination and organometallic chemistry
- nanochemistry, supramolecular chemistry, self-organization
- nanomaterials (eco)toxicity.

just to cite some of them. Nowadays, few scientists have a general background in all of these areas. It is thus important to open new academic programs in order to form the catalysis community of tomorrow.

References

1 Cole-Hamilton, D. (2003) Homogeneous catalysis – new approaches to catalyst separation, recovery, and recycling. *Science*, **299** (5613), 1702–1706.

2 Dunworth, W.P. and Nord, F.F. (1954) Noble metal-synthetic polymer catalysts and studies on the mechanism of their action, in *Advances in Catalysis and Related Subjects* (ed. W.G. Frankenburg), Academic Press Inc., New York, pp. 125–143.

3 Freund, P.L. and Spiro, M. (1985) Colloidal catalysis: the effect of sol size and concentration. *J. Phys. Chem.*, **89**, 1074–1077.

4 Thomas, J.M. (1988) Colloidal metals: past, present and future. *Pure & Appl. Chem.*, **60** (10), 1517–1528.

5 Thomas, J.M. (2010) The advantages of exploring the interface between heterogeneous and homogeneous catalysis. *Chem. Cat. Chem.*, **2**, 127–132.

6 Astruc, D. (2008) *Nanoparticles and Catalysis* (ed. D. Astruc), Wiley-VCH Verlag GmbH, Weinheim, Germany, p. 1.

7 Klabunde, K.J. (2001) Introduction to nanotechnology, in *Nanoscale Materials in Chemistry* (ed. K.J. Klabunde), John Wiley & Sons, Inc., Hoboken, USA, pp. 1–13.

8 Boudart, M., Aldag, A., Benson, J.E., Doughart, N., and Harkins, C.G. (1966) On specific activity of platinum catalysts. *J. Catal.*, **6** (1), 92–99.

9 Che, M. and Bennett, C.O. (1989) The influence of particle size on the catalytic properties of supported metals. *Adv. Catal.*, **36**, 55–172.

10 Van Santen, R.A. (2009) Complementary structure sensitive and insensitive catalytic relationships. *Acc. Chem. Res.*, **42** (1), 57–66.

11 Tao, A.R., Habas, S., and Yang, P.D. (2008) Shape control of colloidal metal nanocrystals. *Small*, **4** (3), 310–325.

12 Feldheim, D.L. (2007) The new face of catalysis. *Science*, **316** (5825), 699–700.

13 Cuenya, B.R. (2010) Synthesis and catalytic properties of metal nanoparticles: size, shape, support, composition, and oxidation state effects. *Thin Solid Films*, **518**, 3127–3150.

14 Campbell, C.T., Parker, S.C., and Starr, D.E. (2002) The effect of size-dependent nanoparticle energetics on catalyst sintering. *Science*, **298** (5594), 811–814.

15 Sun, C.Q. (2007) Size dependence of nanostructures: Impact of bond order deficiency. *Prog. Solid State Chem.*, **35** (1), 1–159.

16 Pan, L., Gu, M., Ouyang, G., and Sun, C.Q. (2010) Behind the quantum and size effects: broken-bond-induced local strain and skin-depth densified quantum trapping of charge and energy. *Key Eng. Mater.*, **444**, 17–45.

17 Schmid, G. (2001) Metals, in *Nanoscale Materials in Chemistry* (ed. K.J. Klabunde), John Wiley & Sons, Inc., Hoboken, USA, pp. 15–59.

18 Schmid, G. (2004) *Nanoparticles, from Theory to Applications* (ed. G. Schmid), Wiley-VCH Verlag GmbH, Weinheim, Germany, pp. 1–3.

19 de Jongh, L.J. (1994) *Physics and Chemistry of Metal Cluster Compounds* (ed. L.J. de Jongh), Kluwer Academic Publishers, Dordrecht, pp. 1–39; Schmid, G. (1994) *Clusters and Colloids, from Theory to Applications* (ed. G. Schmid), Wiley-VCH Verlag GmbH, Weinheim, Germany.

20 Rodriguez, J.A. and Fernandez-Garcia, M. (eds) (2007) *Synthesis, Properties and Applications of Oxides Nanomaterials*, John Wiley & Sons, Inc., Hoboken, USA.

21 Fernández-García, M., Martínez-Arias, A., Hanson, J.C., and Rodriguez, J.A. (2004) Nanostructured oxides in

chemistry: characterization and properties. *Chem. Rev.*, **104**, 4063–4104.

22 Tosheva, L. and Valtchev, V.P. (2005) Nanozeolites: synthesis, crystallization mechanism, and applications. *Chem. Mater.*, **17** (10), 2494–2513.

23 Nampi, P.P., Warrier, K.G.K., Padmanabhan, M., and Wunderlich, W. (2009) High surface area sol-gel alumina-titania nanocatalyst. *J. Sol.-Gel. Sci. & Techn.*, **52** (1), 88–96.

24 Nampi, P.P., Moothetty, P., Wunderlich, W., Berry, F.J., Mortimer, M., Creamer, N.J., and Warrier, K.G. (2010) High-surface-area alumina-silica nanocatalysts prepared by a hybrid sol-gel route using a boehmite precursor. *J. Am. Ceram. Soc.*, **93** (12), 4047–4052.

25 Yang, Q., Choi, H., Al-Abed, S.R., and Dionysiou, D.D. (2009) Iron-cobalt mixed oxide nanocatalysts: Heterogeneous peroxymonosulfate activation, cobalt leaching, and ferromagnetic properties for environmental applications. *Appl. Catal. B: Environ.*, **88** (3–4), 462–469.

26 Shiju, N.R. and Guliants, V.V. (2009) Recent developments in catalysis using nanostructured materials. *Appl. Catal. A: Gen.*, **356**, 1–17.

27 Ayyub, P., Palkar, V.R., Chattopadhyay, S., and Multani, M. (1995) Effect of crystal size reduction on lattice symmetry and cooperative properties. *Phys. Rev. B.*, **51** (9), 6135–6138.

28 Samsonov, V.M., Sdobnyakov, N.Yu., and Bazulev, A.N. (2003) On thermodynamic stability conditions for nanosized particles. *Surf. Sci.*, **532–535**, 526–530.

29 Millot, N., Aymes, D., Bernard, F., Niepce, J.C., Traverse, A., Bouree, F., Cheng, B.L., and Perriat, P. (2003) Particle size dependency of ternary diagrams at the nanometer scale: evidence of TiO_2 clusters in Fe-based spinels. *J. Phys. Chem. B*, **107** (24), 5740–5750.

30 Schoiswohl, J., Kresse, G., Surnev, S., Sock, M., Ramsey, M.G., and Netzer, F.P. (2004) Planar vanadium oxide clusters: two-dimensional evaporation and diffusion on Rh(111). *Phys. Rev. Lett.*, **92**, 206103.

31 Zhang, H. and Bandfield, J.F. (1998) Thermodynamic analysis of phase stability of nanocrystalline titania. *J. Mater. Chem.*, **8** (9), 2073–2076.

32 Song, Z., Cai, T., Chang, Z., Liu, G., Rodriguez, J.A., and Hrbek, J. (2003) Molecular level study of the formation and the spread of MoO_3 on Au (111) by scanning tunneling microscopy and X-ray photoelectron spectroscopy. *J. Am. Chem. Soc.*, **125** (26), 8059–8066.

33 McHale, J.M., Auroux, A., Perrota, A.J., and Navrotsky, A. (1997) Surface energies and thermodynamic phase stability in nanocrystalline aluminas. *Science*, **277** (5327), 788–791.

34 Ayyub, P., Multani, M., Barma, M., Palkar, V.R., and Vijayaraghavan, R. (1988) Size-induced structural phase transitions and hyperfine properties of microcrystalline Fe_2O_3. *J. Phys. C: Solid State Phys.*, **21** (11), 2229–2237.

35 Garvie, R.C. and Goss, M.F. (1986) Intrinsic size dependence of the phase transformation temperature in zirconia microcrystals. *J. Mater. Sci.*, **21** (4), 1253–1257.

36 Hernandez-Alonso, M.D., Hungria, A.B., Coronado, J.M., Martmez-Arias, A., Conesa, J.C., Soria, J., and Fernandez-Garcia, M. (2004) Confinement effects in quasi-stoichiometric CeO_2 nanoparticles. *Phys. Chem. Chem. Phys.*, **6** (13), 3524–3529.

37 Balashov, E.M., Dalidchik, F.I., and Shub, B.R. (2008) Structure and electronic properties of imperfect oxides and nanooxides. *Russ. J. Phys. Chem. B, Focus on Physics*, **2**, 840–855.

38 Chiang, Y.-M., Lavik, E.B., Kosacki, I., Tuller, H.L., and Ying, J.H. (1997) Nonstoichiometry and electrical conductivity of nanocrystalline CeO_{2-x}. *J. Electroceram.*, **1** (1), 7–14.

39 Franke, M.E., Koplin, T.J., and Simon, U. (2006) Metal and metal oxide nanoparticles in chemiresistors: does the nanoscale matter? *Small*, **2** (1), 36–50.

40 Lutic, D., Sanati, M., and Lloyd Spetz, A. (2007) Gas Sensors, in *Synthesis, Properties and Applications of Oxides Nanomaterials* (eds J.A. Rodriguez and M. Fernandez-Garcia), John Wiley & Sons, Inc., Hoboken, USA, pp. 411–450.

41 Wang, C., Yin, L., Zhang, L., Xiang, D., and Gao, R. (2010) Metal oxide gas sensors: sensitivity and influencing factors. *Sensors*, **10**, 2088–2106.

42 Maekawa, H., Tanaka, R., Sato, T., Fujimaki, Y., and Yamamura, T. (2004) Size-dependent ionic conductivity observed for ordered mesoporous alumina-LiI composite. *Solid State Ionics*, **175** (1–4), 281–285.

43 Rodriguez, J.A. (2001) Orbital-band interactions and the reactivity of molecules on oxide surfaces: from explanations to predictions. *Theor. Chem. Acc.*, **107** (3), 117–129.

44 Bochenkov, V.E. and Sergeev, G.B. (2010) Adsorption, catalysis, and reactions on the surfaces of metal nano-oxides. *Catalysis in Industry*, **2** (1), 1–10.

45 Netzer, F.P. (2010) 'Small and beautiful' – The novel structures and phases of nano-oxides. *Surf. Sci.*, **604**, 485–489.

46 Colón-Ibáñez, G., Belver-Coldeira, C., and Fernández-García, M. (2007) Nanostructured Oxides in Photo-Catalysis, in *Synthesis, Properties and Applications of Oxides Nanomaterials* (eds J.A. Rodriguez and M. Fernandez-Garcia), John Wiley & Sons, Inc., Hoboken, USA, pp. 491–562.

47 Jeevanadam, P. and Klabunde, K.J. (2007) Adsorbents, in *Synthesis, Properties and Applications of Oxides Nanomaterials* (eds J.A. Rodriguez and M. Fernandez-Garcia), John Wiley & Sons, Inc., Hoboken, USA, pp. 381–410.

48 Yang, K., Lin, D., and Xing, B. (2009) Interactions of humic acid with nanosized inorganic oxides. *Langmuir*, **25** (6), 3571–3576.

49 Hu, J., Chen, L., and Richards, R. (2009) Properties, synthesis and applications of highly dispersed metal oxide catalysts, in *Metal Oxide Catalysis* (eds S.D. Jackson and J.S. Hargreaves), Wiley-VCH Verlag GmbH, Weinheim, Germany.

50 Huber, B., Koskinen, P., Häkkinen, H., and Moseler, M. (2006) Oxidation of magnesia-supported Pd-clusters leads to the ultimate limit of epitaxy with a catalytic function. *Nature Mater.*, **5** (1), 44–47.

51 Castillejos, E., Bacsa, R., Guerrero-Ruiz, A., Rodríguez-Ramos, I., Datas, L., and Serp, P. (2011) Catalytic activity of gold supported on ZnO tetrapods for the preferential oxidation of carbon monoxide under hydrogen rich conditions. *Nanoscale*, **3**, 929–932.

52 Bromley, S.T., Moreira, I.P.R., Neyman, K.M., and Illas, F. (2009) Approaching nanoscale oxides: models and theoretical methods. *Chem. Soc. Rev.*, **38**, 2657–2670.

53 Patil, K.C., Hegde, M.S., Rattan, T., and Aruna, S.T. (eds) (2008) *Chemistry of Nanocrystalline Oxide Materials, Combustion Synthesis, Properties and Applications*, World Scientific Publishing Co. Pte. Ltd.

54 Schaub, R., Wahlström, E., Rønnau, A., Lægsgaard, E., Stensgaard, I., and Besenbacher, F. (2003) Oxygen-mediated diffusion of oxygen vacancies on the TiO_2(110) surface. *Science*, **299** (5605), 377–379.

55 Xu, M., Fujita, D., Kajiwara, S., Minowa, T., Li, X., Takemura, T., Iwai, H., and Hanagata, N. (2010) Contribution of physicochemical characteristics of nano-oxides to cytotoxicity. *Biomaterials*, **31**, 8022–8031.

56 Fernández-García, M. and Rodriguez, J.A. (2008) Metal oxide nanoparticles, in *Nanomaterials: Inorganic and Bioinorganic Perspectives* (eds C.M. Lukehart and R.A. Scott), Encyclopedia in Inorganic Chemistry, vol. 5, John Wiley & Sons, Ltd, Chichester, UK, pp. 453–474.

57 Inagaki, M. and Feiyu, K. (2006) *Carbon Materials Science and Engineering: From Fundamental to Applications*, Tsinghua University Press.

58 Serp, P. and Figueiredo, J.L. (2009) *Carbon Materials for Catalysis*, John Wiley & Sons, Inc., Hoboken, USA.

59 Goldshleger, N.F. (2001) Fullerenes and fullerenes-based materials in catalysis. *Fullerene Sci. Technol.*, **9** (3), 255–280.

60 Serp, P. (2009) Carbon nanotubes and nanofibers in catalysis, in: *Carbon Materials for Catalysis* (eds P. Serp and J.L. Figueiredo), John Wiley & Sons, Inc., Hoboken, USA, pp. 309–372.

61 Machado, B. and Serp, P. (2012) Graphene materials in catalysis. *Catal. Sci. Technol.*, **2**, 54–75.

62 Dumitrică, T., Landis, C.M., and Yakobson, B.I. (2002) Curvature-induced polarization in carbon nanoshells. *Chem. Phys. Lett.*, **360** (1–2), 182–188.

63 Charlier, J.C. (2002) Defects in carbon nanotubes. *Acc. Chem. Res.*, **35**, 1063–1069.

64 Haddon, R.C. (1992) Electronic structure, conductivity and superconductivity of alkali metal doped (C60). *Acc. Chem. Res.*, **25** (3), 127–133.

65 Kostov, M.K., Cheng, H., Cooper, A.C., and Pez, G.P. (2002) Influence of carbon curvature on molecular adsorptions in carbon-based materials: a force field approach. *Phys. Rev. Lett.*, **89** (14), 146105.

66 Ruffieux, P., Gröning, O., Bielmann, M., and Gröning, P. (2004) Hydrogen chemisorption on sp^2-bonded carbon: Influence of the local curvature and local electronic effects. *Appl. Phys. A: Mater. Sci. & Process.*, **78** (7), 975–980.

67 Chen, W., Pan, X., Willinger, M.G., Su, D.S., and Bao, X. (2007) Tuning of redox properties of iron and iron oxides via encapsulation within carbon nanotubes. *J. Am. Chem. Soc.*, **129** (23), 7421–7426.

68 Barborini, E., Piseri, P., Milani, P., Benedek, G., Ducati, C., and Rob, J. (2002) Negatively curved spongy carbon. *Appl. Phys. Lett.*, **81** (18), 3359–3361.

69 Serp, P. and Castillejos, E. (2010) Catalysis in carbon nanotubes. *ChemCatChem*, **2** (1), 41–47.

70 Pan, X. and Bao, X. (2011) The effects of confinement inside carbon nanotubes on catalysis. *Acc. Chem. Res.*, **44** (8), 553–562.

71 Goettmann, F. and Sanchez, C. (2007) How does confinement affect the catalytic activity of mesoporous materials? *J. Mater. Chem.*, **17**, 24–30.

72 Santiso, E.E., Buongiorno Nardelli, M., and Gubbins, K.E. (2008) A remarkable shape-catalytic effect of confinement on the rotational isomerization of small hydrocarbons. *J. Chem. Phys.*, **128**, 034704.

73 Santiso, E.E., Kostov, M.K., George, A.M., Buongiorno Nardelli, M., and Gubbins, K.E. (2007) Confinement effects on chemical reactions—Toward an integrated rational catalyst design. *Appl. Surf. Sci.*, **253** (13), 5570–5579.

74 Coq, B., Planeix, J.M., and Brotons, V. (1998) Fullerene-based materials as new support media in heterogeneous catalysis by metals. *Appl. Catal. A: Gen.*, **173** (2), 175–183.

75 Zhang, J., Su, D.S., Blume, R., Schlögl, R., Wang, R., Yang, X., and Gajović, A. (2010) Surface chemistry and catalytic reactivity of a nanodiamond in the steam-free dehydrogenation of ethylbenzene. *Angew. Chem. Int. Ed. Engl.*, **49** (46), 8640–8644.

76 Vershinin, N.N., Efimov, O.N., Bakaev, V.A., Aleksenskii, A.E., Baidakova, M.V., Sitnikova, A.A., and Ya Vul', A. (2010) Detonation nanodiamonds as catalyst supports. *Fullerenes, Nanotubes and Carbon Nanostructures*, **19** (1–2), 63–68.

77 Zhu, S. and Xu, G. (2010) Single-walled carbon nanohorns and their applications. *Nanoscale*, **2** (12), 2538–2549.

78 Astruc, D. (2008) *Nanoparticles and Catalysis* (ed. D. Astruc), Wiley-VCH Verlag GmbH, Weinheim, Germany, p. 2.

79 Roucoux, A. (2005) Stabilized noble metal nanoparticles: an avoidable family of catalysts for arene derivative hydrogenation, surface and interfacial organometallic chemistry and catalysis, in *Surface and Interfacial Organometallic Chemistry and Catalysis*, Topics in Organometallic Chemistry, vol. 16 (eds C. Copéret and B. Chaudret), Springer, pp. 262–279.

80 Astruc, D., Lu, F., and Ruiz Aranzaes, J. (2005) Nanoparticles as recyclable catalysts: a frontier field between homogeneous and heterogeneous catalysis. *Angew. Chemie Int. Ed.*, **44**, 7852–7872.

81 Abbet, S. and Heiz, U. (2004) Nanocatalysis, in *The Chemistry of Nanomaterials* (eds C.N.R. Rao, A. Müller, and A.K. Cheetham), Wiley-VCH Verlag GmbH, Weinheim, Germany, pp. 551–588.

82 Roucoux, A. and Patin, H. (2002) Reduced transition metal colloids: a novel family of reusable catalysts? *Chem. Rev.*, **102**, 3757–3778.
83 Cushing, B.L., Kolesnichenko, V.L., and O'Connor, C.J. (2004) Recent advances in the liquid-phase syntheses of inorganic nanoparticles. *Chem. Rev.*, **104**, 3893–3946.
84 Park, J., Joo, J., Kwon, S.G., and Tyeon, T. (2007) Synthesis of monodisperse spherical nanocrystals. *Angew. Chemie Int. Ed*, **46** (25), 4630–4660.
85 Philippot, K. and Chaudret, B. (2007) Organometallic derived metals, colloids and nanoparticles, in *Comprehensive Organometallic Chemistry III* (Eds-in-Chief R.H. Crabtree and M.P. Mingos), Elsevier, pp. 71–99.
86 Kulkarni, G.U., Thomas, P.J., and Rao, C.N.R. (2004) Mesoscopic assembly and other properties of metal and semiconductor nanocrystals, in *The Chemistry of Nanomaterials* (eds C.N.R. Rao, A. Müller, and A.K. Cheetham), Wiley-VCH Verlag GmbH, Weinheim, Germany, pp. 58–93.
87 Niu, W. and Xu, G. (2011) Crystallographic control of noble metal nanocrystals. *Nano. Today*, **6** (5), 265–285.
88 Mondloch, J.E., Bayram, E., and Finke, R.G. (2012) A review of the kinetics and mechanisms of formation of supported-nanoparticle heterogeneous catalysts. *J. Mol. Catal. A: Chemical*, **355**, 1–38.
89 Burda, C., Chen, X., Narayanan, R., and El-Sayed, M.A. (2005) Chemistry and properties of nanocrystals of different shapes. *Chem. Rev.*, **105** (4), 1025–1102.
90 Li, Y.M. and Somorjai, G. (2010) Nanoscale advances in catalysis and energy applications. *Nano. Lett.*, **10** (7), 2289–2295.
91 Sau, T.K. and Rogach, A.L. (2010) Nonspherical noble metal nanoparticles: colloid-chemical synthesis and morphology control. *Adv. Mater.*, **22**, 1805–1825.
92 Uzio, D. and Berhault, G. (2010) Factors governing the catalytic reactivity of metallic nanoparticles. *Catal. Rev.: Sci. Eng.*, **52**, 106–131.
93 Tao, A.R., Habas, S., and Yang, P. (2008) Shape control of colloidal metal nanocrystals. *Small*, **4** (3), 310–325.
94 Semagina, N. and Kiwi-Minsker, L. (2009) Recent advances in the liquid-phase synthesis of metal nanostructures with controlled shape and size for catalysis. *Catal. Rev. Sci. Eng.*, **51**, 147–217.
95 Piccolo, L., Valcarcel, A., Bausach, M., Thomazeau, C., and Uzio, D. (2008) Tuning the shape of nanoparticles to control their catalytic properties: selective hydrogenation of 1,3-butadiene on Pd/Al_2O_3. *Phys. Chem. Chem. Phys.*, **10**, 5504–5506.
96 Ren, J. and Tilley, R.D. (2007) Preparation, self-assembly, and mechanistic study of highly monodispersed nanocubes. *J. Am. Chem. Soc.*, **129**, 3287–3291.
97 Watt, J., Young, N., Haigh, S., Kirkland, A., and Tilley, R.D. (2009) Synthesis and structural characterization of branched palladium nanostructures. *Adv. Mater.*, **21**, 2288–2293.
98 Axet, M.R., Philippot, K., Chaudret, B., Cabié, M., Giorgio, S., and Henry, C. (2011) TEM and HRTEM evidence for the role of ligands in the formation of shape-controlled platinum nanoparticles. *Small*, **7**, 235–241.
99 Zhang, Y.W., Grass, M.E., Kuhn, J.N., Tao, F., Habas, S.E., Huang, W.Y., Yang, P.D., and Somorjai, G.A. (2008) Highly selective synthesis of catalytically active monodisperse rhodium nanocubes. *J. Am. Chem. Soc.*, **130**, 5868–5869.
100 Tsung, C.K., Kuhn, J.N., Huang, W., Aliaga, C., Hung, L.I., Somorjai, G.A., and Yang, P.D. (2009) Sub-10 nm platinum nanocrystals with size and shape control: catalytic study for ethylene and pyrrole hydrogenation. *J. Am. Chem. Soc.*, **131**, 5816–5822.
101 Nakamula, I., Yamanoi, Y., Imaok, T., Yamamoto, K., and Nishihara, H. (2011) A uniform bimetallic rhodium/iron nanoparticle catalyst for the hydrogenation of olefins and nitroarenes. *Angew. Chemie Int. Ed.*, **50** (26), 5830–5833.

102 Dingsheng, W. and Yadong, L. (2011) Bimetallic nanocrystals: liquid-phase synthesis and catalytic applications. *Adv. Mater.*, **23**, 1044–1060.

103 Novio, F., Philippot, K., and Chaudret, B. (2010) Location and dynamics of CO coordination on Ru nanoparticles: A solid state NMR study. *Catal. Lett.*, **140**, 1–7.

104 Lara, P., Rivada-Wheelaghan, O., Conejero, S., Poteau, R., Philippot, K., and Chaudret, B. (2011) Ruthenium nanoparticles stabilized by *N*-heterocyclic carbenes: ligands location and influence on reactivity. *Angew. Chemie Int. Ed.*, **50**, 12080–12084.

105 Gonzalez-Galvez, D., Nolis, P., Philippot, K., Chaudret, B., and van Leeuwen, P.W.N.M. (2012) Phosphine-stabilized ruthenium nanoparticles: The effect of the nature of the ligand in catalysis. *ACS Catalysis*, **2** (3), 317–321.

106 Schmidt, E., Kleist, W., Krumeich, F., Mallat, T., and Baiker, A. (2010) Platinum nanoparticles: the crucial role of crystal face and colloid stabilizer in the diastereoselective hydrogenation of cinchonidine. *Chem. Eur. J*, **16** (7), 2181–2192.

107 Sonström, P. and Bäumer, M. (2011) Supported colloidal nanoparticles in heterogeneous gas phase catalysis: on the way to tailored catalysts. *Phys. Chem. Chem. Phys.*, **13**, 19270–19284.

108 Yan, N., Xiao, C., and Kou, Y. (2010) Transition metal nanoparticle catalysis in green solvents. *Coord. Chem. Rev.*, **254**, 1179–1218.

109 Geldbach, T.J. and Dyson, P.J. (eds) (2005) *Metal-Catalysed Reactions in Ionic Liquids*, Springer.

110 Hallett, J.P. and Welton, T. (2011) Room-temperature ionic liquids: solvents for synthesis and catalysis. 2. *Chem. Rev.*, **111** (5), 3508–3576.

111 Dupont, J. and Scholten, J.D. (2010) On the structural and surface properties of transition-metal nanoparticles in ionic liquids. *Chem. Soc. Rev.*, **39**, 1780–1804.

112 Moreno-Mañas, M. and Pleixats, R. (2004) Applications of fluorous compounds in materials chemistry: 12.2 Fluorous nanoparticles, in *Handbook of Fluorous Chemistry* (eds J.A. Gladysz, D.P. Curran, and I.T. Horváth), Wiley-VCH Verlag GmbH, Weinheim, Germany, pp. 491–507.

113 Chechik, V. and Crooks, R.M. (2000) Dendrimer-encapsulated Pd nanoparticles as fluorous-phase-soluble catalysts. *J. Am. Chem. Soc.*, **122** (6), 1243–1244.

114 Moreno-Mañas, M., Pleixats, R., and Villarroya, S. (2001) Preparation and characterization of silane-stabilized, highly uniform, nanobimetallic Pt–Pd particles in solid and liquid matrixes. *Organometallics*, **20**, 4524–4528.

115 Moreno-Mañas, M., Pleixats, R., and Tristany, M. (2005) Gold nanoparticles entrapped in heavily fluorinated compounds. *J. Fluorine Chem.*, **126**, 1435–1438.

116 Niembro, S., Vallribera, A., and Moreno-Mañas, M. (2008) Star-shaped heavily fluorinated aromatic sulfurs: stabilization of palladium nanoparticles active as catalysts in cross-coupling reactions. *New J. Chem.*, **32** (1), 94–98.

117 Tristany, M., Moreno-Mañas, M., Pleixats, R., Chaudret, B., Philippot, K., Guari, Y., Matsura, V., and Lecante, P. (2009) Self-Assembled platinum nanoparticles into heavily fluorinated templates: reactive gas effect on the morphology. *New J. Chem.*, **33**, 1529–1534.

118 Marre, S., Park, J., Rempel, J., Guan, J., Bawendi, M.G., and Jensen, K.F. (2008 Supercritical continuous-microflow synthesis of narrow size distribution quantum dots. *Adv. Mater.*, **20**, 4830–4834.

119 Cansell, F. and Aymonier, C. (2009) Design of functional nanostructured materials using supercritical fluids. *J. Supercrit. Fluids*, **47**, 508–516.

120 Keilitz, J., Nowag, S., Marty, J.-D., and Haag, R. (2011) Chirally modified platinum nanoparticles stabilized by dendritic core-multishell architectures for the asymmetric hydrogenation of ethyl pyruvate. *Adv. Synth. Catal.*, **352**, 1503–1511.

121 Badetti, E., Caminade, A.-M., Majoral, J.-P., Moreno-Mañas, M., and Sebastián,

R.M. (2008) Palladium(0) nanoparticles stabilized by phosphorus dendrimers containing coordinating 15-membered triolefinic macrocycles in periphery. *Langmuir*, **24**, 2090–2101.

122 Jansat, S., Pelzer, K., García-Antón, J., Raucoules, R., Philippot, K., Maisonnat, A., Chaudret, B., Guari, Y., Mehdi, A., Reyé, C., and Corriu, R.J.P. (2007) Synthesis of new Ru@SiO_2 composite nanomaterials and their application as catalytic filters for selective gas detection. *Adv. Funct. Mater.*, **17**, 3339–3347.

123 Kormann, H.-P., Schmid, G., Pelzer, K., Philippot, K., and Chaudret, B. (2004) Gas phase catalysis by metal nanoparticles in nanoporous alumina membranes. *Z. Anorg. Allg. Chem.*, **630**, 1913–1918.

124 Castillejos, E., Debouttière, P.-J., Roiban, L., Solhy, A., Martinez, V., Kihn, Y., Ersen, O., Philippot, K., Chaudret, B., and Serp, P. (2009) An efficient strategy to drive nanoparticles into carbon nanotubes and the remarkable effect of confinement on their catalytic performance. *Angew. Chemie Int. Ed.*, **48** (14), 2529–2533.

125 Rosario-Amorin, D., Wang, X., Gaboyard, M., Clérac, R., Nlate, S., and Heuzé, K. (2009) Functionalized core-shell superparamagnetic nanoparticles: magnetically recoverable and reusable catalysts for Suzuki C—C cross-coupling reactions. *Chem. Eur. J.*, **15**, 12636–12643.

126 Costa, N.J.S., Kiyohara, P.K., Monteiro, A.L., Coppel, Y., Philippot, K., and Rossi, L.M. (2010) A single-step procedure for the preparation of palladium nanoparticles and a phosphine-functionalized support as catalyst for Suzuki cross-coupling reactions. *J. Catal.*, **276** (2), 382–389.

127 Somorjai, G.A., Beaumont, S.K., and Alayoglu, S. (2011) Determination of molecular surface structure, composition, and dynamics under reaction conditions at high pressures and at the solid-liquid interface. *Angew. Chemie Int. Ed.*, **50** (43), 10116–10129.

128 Isaeva, V.I. and Kustov, L.M. (2010) The application of metal-organic frameworks in catalysis. *Petroleum Chemistry*, **50** (3), 167–180.

129 Chen, A., Zhang, W., Li, X., Tan, D., Han, X., and Bao, X. (2007) One-pot encapsulation of Pt nanoparticles into the mesochannels of SBA-15 and their catalytic dehydrogenation of methylcyclohexane. *Catal. Lett.*, **119** (1–2), 159–164.

130 Ugo, R., Psaro, R., Roberto, D., and Basset, J.M. (eds) (2009) *Modern Surface Organometallic Chemistry*, Wiley-VCH Verlag GmbH, Weinheim, Germany.

131 Ichikawa, M. (2000) 'Ship-in-Bottle' catalyst technology. *Platinum Metals Rev.*, **44** (1), 3–14.

132 Terashima, T., Mes, T., de Greef, T.F.A., Gillissen, M.A.J., Besenius, P., Palmans, A.R.A., and Meijer, E.W. (2011) Single-chain folding of polymers for catalytic systems in water. *J. Am. Chem. Soc.*, **133**, 4742–4745.

133 Grzelczak, M., Vermant, J., Furst, E.M., and Liz-Marzan, L.M. (2010) Directed self-assembly of nanoparticles. *ACS Nano*, **4** (7), 3591–3605.

134 Matsushita, T., Shiraishi, Y., Horiuchi, S., and Toshima, N. (2007) Synthesis and catalysis of polymer-protected Pd/Ag/Rh trimetallic nanoparticles with a core-shell structure. *Bull. Chem. Soc. Japan*, **80** (6), 1217–1225.

135 Yamada, Y.M.A. and Uozumi, Y. (2006) A solid-phase self-organized catalyst of nanopalladium with main-chain viologen polymers: α-alkylation of ketones with primary alcohols. *Org. Lett.*, **8** (7), 1375–1378.

136 Haryono, A. and Binder, W.H. (2006) Controlled arrangement of nanoparticle arrays in block-copolymer domains. *Small*, **2** (5), 600–611.

137 Libuda, J. and Freund, H.-J. (2005) Molecular beam experiments on model catalysts. *Surf. Sci. Reports*, **57** (7–8), 157–298.

138 Karakhanov, E.A., Maximov, A.L., Skorkin, V.A., Zolotukhina, A.V., Smerdov, A.S., and Tereshchenko, A.Yu. (2009) Nanocatalysts based on dendrimers. *Pure Appl. Chem.*, **81** (11), 2013–2023.

139 Astruc, D. and Chardac, F. (2001) Dendritic catalysts and dendrimers in catalysis. *Chem. Rev.*, **101**, 2991–3031.

140 Méry, D. and Astruc, D. (2006) Metallodendritic catalysts: recent progress. *Coord. Chem. Rev.*, **250** (15–16), 1965–1979.

141 Zhong, C.-J., Luo, J., Fang, B., Wanjala, B.N., Njoki, P.N., Loukrakpam, R., and Yin, J. (2010) Nanostructured catalysts in fuel cells. *Nanotechnology*, **21**, 062001.

142 Qiao, Y. and Li, C.M. (2011) Nanostructured catalysts in fuel cells. *J. Mater. Chem.*, **21**, 4027–4036.

143 Basri, S., Kamarudin, S.K., Daud, W.R.W., and Yaakub, Z. (2010) Nanocatalyst for direct methanol fuel cell (DMFC). *Int. J. Hydrogen Energy*, **35** (15), 7957–7970.

144 Liu, H., Song, C., Zhang, L., Zhang, J., Wang, H., and Wilkinson, D.P. (2006) A review of anode catalysis in the direct methanol fuel cell. *J. Power Sources*, **155** (2), 95–110.

145 Antolini, E. (2009) Carbon supports for low-temperature fuel cell catalysts. *Appl. Catal. B: Environ.*, **88**, 1–24.

146 Wang, J.X., Inada, H., Wu, L., Zhu, Y., Choi, Y., Liu, P., Zhou, W.-P., and Adzic, R.R. (2009) Oxygen reduction on well-defined core-shell nanocatalysts: particle size, facet, and Pt shell thickness effects. *J. Am. Chem. Soc.*, **131** (47), 17298–17302.

147 Perez-Alonso, F.J., McCarthy, D.N., Nierhoff, A., Hernandez-Fernandez, P., Strebel, C., Stephens, I.E.L., Nielsen, J.H., and Chorkendorff, I. (2012) The effect of the size on the oxygen electroreduction activity of mass-selected platinum nanoparticles. *Angew. Chem. Int. Ed.*, **51**, 4641–4643.

148 Zhang, X.Y., Lu, W., Da, J.Y., Wang, H.T., Zhao, D.Y., and Webley, P.A. (2009) Porous platinum nanowire arrays for direct ethanol fuel cell applications. *Chem. Commun.*, (2), 195–197.

149 Sun, S., Zhang, G., Geng, D., Chen, Y., Li, R., Cai, M., and Sun, X. (2011) A highly durable platinum nanocatalyst for proton exchange membrane fuel cells: multiarmed starlike nanowire single crystal. *Angew. Chem. Int. Ed.*, **50**, 422–426.

150 Lu, S.F., Pan, J., Huang, A.B., Zhuang, L., and Lu, J.T. (2008) Alkaline polymer electrolyte fuel cells completely free from noble metal catalysts. *Proc. Natl. Acad. Sci. USA*, **105** (52), 20611–20614.

151 Zhou, Q., Li, C.M., Li, J., and Lu, J.T. (2008) Electrocatalysis of template-electrosynthesized cobalt-porphyrin/polyaniline nanocomposite for oxygen reduction. *J. Phys. Chem. C*, **112** (47), 18578–18583.

152 Zhou, Q., Li, C.M., Li, J., Cui, X.Q., and Gervasio, D. (2007) Template-synthesized cobalt porphyrin/polypyrrole nanocomposite and its electrocatalysis for oxygen reduction in neutral medium. *J. Phys. Chem. C*, **111** (30), 11216–11222.

153 Maillard, F., Simonov, P.A., and Savinova, E.R. (2009) Carbon materials as supports for fuel cells electrocatalysts, in *Carbon Materials for Catalysis* (eds P. Serp and J.L. Figueiredo), John Wiley & Sons, Inc., Hoboken, USA, pp. 429–480.

154 Bai, H., Li, C., Shi, G. (2011) Functional composite materials based on chemically converted graphene. *Adv. Mater.*, **23**, 1089–1115.

155 Jafri, R.I., Arockiados, T., Rajalakshmi, N., and Ramaprabhu, S. (2010) Nanostructured Pt dispersed on graphene-multiwalled carbon nanotube hybrid nanomaterials as electrocatalyst for PEMFC. *J. Electrochem. Soc.*, **157**, B874–B879.

156 Kašpar, J. and Fornasiero, P. (2003) Nanostructured materials for advanced automotive de-pollution catalysts. *J. Solid State Chem.*, **171**, 19–29.

157 Trudeau, M.L. (2007) Nanostructured materials for gas reactive applications, in *Nanostructured Materials: Processing, Properties and Applications*, 2nd edn (ed. C.C. Koch), William Andrew Publishing, pp. 365–437.

158 Wang, R., Crozier, P.A., and Sharma, R. (2010) Nanoscale compositional and structural evolution in ceria zirconia during cyclic redox treatments. *J. Mater. Chem.*, **20**, 7497–7505.

159 Wang, R., Crozier, P.A., Sharma, R., and Adams, J.B. (2008) Measuring the redox activity of individual catalytic

nanoparticles in cerium-based oxides. *Nano Lett.*, **8** (3), 962–967.
160 Glatzel, P., Singh, J., Kvashnina, K.O., and van Bokhoven, J.A. (2010) *In situ* characterization of the 5d density of states of Pt nanoparticles upon adsorption of CO. *J. Am. Chem. Soc.*, **132**, 2555–2557.
161 Nagai, Y., Hara, N., Dohmae, K., Ikeda, Y., Takagi, N., Tanabe, T., Guilera, G., Pascarelli, S., Newton, M.A., Kuno, O., Jiang, H., Shinjoh, H., and Matsumoto, S. (2008) In situ redispersion of platinum autoexhaust catalysts: an on-line approach to increasing catalyst lifetimes? *Angew. Chem Int. Ed.*, **47** (48), 9303–9306.
162 Guilera, G., Gorges, B., Pascarelli, S., Vitoux, H., Newton, M.A., Prestipino, C., Nagai, Y., and Hara, N. (2009) Novel high-temperature reactors for in situ studies of three-way catalysts using turbo-XAS. *J. Synchrotron Radiation*, **16** (5), 628–634.
163 Zhang, J., Kumagai, H., Yamamura, K., Ohara, S., Takami, S., Morikawa, A., Shinjoh, H., Kaneko, K., Adschiri, T., and Suda, A. (2011) Extra-low-temperature oxygen storage capacity of CeO_2 nanocrystals with cubic facets. *Nano Lett.*, **11** (2), 361–364.
164 Liu, X., Zhou, K., Wang, L., Wang, B., and Li, Y. (2009) Oxygen vacancy clusters promoting reducibility and activity of ceria nanorods. *J. Am. Chem. Soc.*, **131** (9), 3140–3141.
165 Esch, F., Fabris, S., Zhou, L., Montini, T., Africh, C., Fornasiero, P., Comelli, G., and Rosei, R. (2005) Electron localization determines defect formation on ceria substrates. *Science*, **309** (5735), 752–755.
166 Kirchhoff, M., Specht, U., and Veser, G. (2005) Engineering high-temperature stable nanocomposite materials. *Nanotechnology*, **16** (7), S401–S408.
167 Cao, A. and Veser, G. (2010) Exceptional high-temperature stability through distillation-like self-stabilization in bimetallic nanoparticles. *Nature Mater.*, **9**, 75–81.
168 Kumar, P., Ketzel, M., Vardoulakis, S., Pirjola, L., and Britter, R. (2011) Dynamics and dispersion modeling of nanoparticles from road traffic in the urban atmospheric environment – A review. *J. Aerosol Sci.*, **42** (9), 580–603.
169 Wilkinson, K.E., Palmberg, L., Witasp, E., Kupczyk, M., Feliu, N., Gerde, P., Seisenbaeva, G.A., Fadeel, B., Dahlén, S.-E., and Kessler, V.G. (2011) Solution-engineered palladium nanoparticles: model for health effect studies of automotive particulate pollution. *ACS Nano*, **5** (7), 5312–5324.
170 Xia, T., Kovochich, M., Liong, M., Mädler, L., Gilbert, B., Shi, H., Yeh, J.I., Zink, J.I., and Nel, A.E. (2008) Comparison of the mechanism of toxicity of zinc oxide and cerium oxide nanoparticles based on dissolution and oxidative stress properties. *ACS Nano*, **2** (10), 2121–2134.
171 Alegret, S. and Merkoçi, A. (2003) Composite and biocomposite materials for electrochemical sensing, in *Integrated Analytical Systems* (ed. S. Alegret), Elsevier, Amsterdam, pp. 377–412; Alegret, S. and Merkoçi, A. (2007) Electrochemical sensor analysis, in *Comprehensive Analytical Chemistry* (eds S. Alegret, and A. Merkoçi,), Elsevier B.V.
172 Lutic, D., Sanati, M., and Lloyd Spetz, A. (2007) Gas sensors, in *Synthesis, Properties and Applications of Oxide Nanomaterials* (eds J.A. Rodríguez and M. Fernández-Garcia), John Wiley & Sons, Inc., Hoboken, USA, pp. 411–490.
173 Lantto, V. (1992) Semiconductors gas sensors based on SnO_2 thick films, in *Gas Sensors, Principle, Operation and Developments* (ed. G. Sberveglieri), Kluwer Academic, Dordrecht, Germany, pp. 117–168.
174 Batzill, M. and Diebold, U. (2005) The surface and materials science of tin oxide. *Prog. Surf. Sci.*, **79**, 47–154.
175 Barsan, N., Schweizer-Berberich, M., and Göpel, W. (1999) Fundamental and practical aspects in the design of nanoscaled SnO_2 gas sensors: a status report. *Fresenius J. Anal. Chem.*, **365**, 287–304.
176 Göpel, W. (1992) Future trends in the development of gas sensors, in *Gas Sensors, Principle, Operation and Developments* (ed. G. Sberveglieri), Kluwer Academic, Dordrecht, Germany, pp. 365–401.

177 Erades, L., Grandjean, D., Nayral, C., Soulantica, K., Chaudret, B., Menini, P., Parret, F., and Maisonnat, A. (2006) Organometallic approach for platinum and palladium doping of tin and tin oxide nanoparticles: structural characterisation and gas sensor investigations. *New J. Chem.*, **30**, 1026–1035.

178 Fau, P., Sauvan, M., Trautweiler, S., Nayral, C., Erades, L., Maisonnat, A., and Chaudret, B. (2001) Nanosized tin oxide sensitive layer on a silicon platform for domestic gas applications. *Sens. Actuators B*, **78** (1–3), 83–88.

179 Heilig, A., Barsan, N., Weimar, U., Schweizer-Berberich, M., Gardner, J.W., and Göpel, W. (1997) Gas identification by modulating temperatures of SnO_2-based thick film sensors. *Sens. Actuators B*, **43** (1–3), 45–51.

180 Parret, F., Menini, P., Martinez, A., Soulantica, K., Maisonnat, A., and Chaudret, B. (2006) Improvement of micromachined SnO_2 gas sensors selectivity by optimised dynamic temperature operating mode. *Sens. Actuators B*, **118** (1–2), 276–282.

181 Cabot, A., Arbiol, J., Cornet, A., Morante, J.R., Chen, F., and Liu, M. (2003) Mesoporous catalytic filters for semiconductor gas sensors. *Thin Solid Films*, **436**, 64–69.

182 Jansat, S., Pelzer, K., García-Antón, J., Raucoules, R., Philippot, K., Maisonnat, A., Chaudret, B., Guari, Y., Medhi, A., Reyé, C., and Corriu, R.J.P. (2007) Synthesis of new $RuO_2@SiO_2$ composite nanomaterials and their application as catalytic filters for selective detection of gas sensors. *Adv. Funct. Mater.*, **17**, 3339–3347.

183 Matsura, V., Guari, Y., Reyé, C., Corriu, R.J.P., Tristany, M., Jansat, S., Philippot, K., Maisonnat, A., and Chaudret, B. (2009) Novel $Ru@SiO_2$ and $RuO_2@SiO_2$ hybrid nanomaterials: from their facile synthesis to their application as catalytic filters for gas sensors. *Adv. Funct. Mater.*, **19** (23), 3781–3787.

184 Roy, S. and Basu, S. (2002) Improved zinc oxide film for gas sensor applications. *Bull. Mater. Sci* **25** (6), 513–515.

185 Gupta, S.K., Joshi, A., and Kaur, M. (2010) Development of gas sensors using ZnO nanostructures. *J. Chem. Sci.*, **122** (1), 57–62.

186 Singh, N., Gupta, R.K., and See Lee, P. (2011) Gold-nanoparticle-functionalized In_2O_3 nanowires as CO gas sensors with a significant enhancement in response. *ACS Appl. Mater. Interfaces*, **3** (7), 2246–2252.

187 Zhang, T., Mubeen, S., Myun, N.V., and Deshusses, M.A. (2008) Recent progress in carbon nanotube-based gas sensors. *Nanotechnology*, **19**, 332001.

188 Star, A., Joshi, V., Skarupo, S., Thomas, D., and Gabriel, J.C.P. (2006) Gas sensor array based on metal-decorated carbon nanotubes. *J. Phys. Chem. B*, **110** (42), 21014–21020.

189 Yamazoe, N. (2005) Toward innovations of gas sensor technology. *Sensors and Actuators B*, **108** (1–2), 2–14.

190 Colón-Ibáñez, G., Belver-Coldeira, C., and Fernández-García, M. (2007) Nanostructured oxides in photo-catalysis, in *Synthesis, Properties and applications of Oxide Nanomaterials* (eds J.A. Rodríguez and M. Fernández-Garcia), John Wiley & Sons, Inc., Hoboken, USA, pp. 491–562.

191 Kaur, R. and Pal, B. (2012) Size and shape dependent attachments of Au nanostructures to TiO_2 for optimum reactivity of Au-TiO_2 photocatalysis. *J. Mol. Catal. A: Chemical*, **355**, 39–43 and references therein.

192 Tryba, B., Morawski, A.W., and Inagaki, M. (2003) Application of TiO_2-mounted activated carbon to the removal of phenol from water. *Apply. Catal. B: Environ.*, **41**, 427–433.

193 Janus, M., Kusiak, E., and Morawski, A.W. (2009) Carbon modified TiO_2 photocatalyst with enhanced adsorptivity for dyes from water. *Catal. Lett.*, **131**, 506–511.

194 Wang, W., Serp, P., Kalck, P., and Faria, J.L. (2004) Photocatalytic degradation of phenol on MWCNT and titania composite catalysts prepared by a modified sol-gel method. *Apply. Catal. B: Environ.*, **56**, 301–308.

195 Artero, V., Chavarot-Kerlidou, M., and Fontecave, M. (2011) Splitting water

with cobalt. *Angew. Chemie Int. Ed.*, **50**, 7238–7266.

196 Jiao, F. and Frei, H. (2009) Nanostructured cobalt oxide clusters in mesoporous silica as efficient oxygen-evolving catalysts. *Angew. Chemie Int. Ed.*, **48**, 1841–1844.

197 Esswein, A.J., McMurdo, M.J., Ross, P.N., Bell, A.T., and Tilley, T.D. (2009) Size-dependent activity of Co_3O_4 nanoparticle anodes for alkaline water electrolysis. *J. Phys. Chem. C*, **113**, 15068–15072.

198 Liang, Y., Li, Y., Zhou, J., Wang, J., Regier, T., and Dai, H. (2011) Co_3O_4 nanocrystals on graphene as a synergistic catalyst for oxygen reduction reaction. *Nature Mater.*, **10**, 780–786.

199 Morandeira, A., Fortage, J., Edvinsson, T., Le Pleux, L., Blart, E., Boschloo, G., Hagfeldt, A., Hammarstro, L., and Odobel, F. (2008) Improved photon-to-current conversion efficiency with a nanoporous p-type NiO electrode by the use of a sensitizer-acceptor Dyad. *J. Phys. Chem. C*, **112**, 1721–1728.

200 Grimme, R.A., Lubner, C.E., Bryant, D.A., and Golbeck, J.H. (2008) Photosystem I/molecular wire/metal nanoparticle bioconjugates for the photocatalytic production of H_2. *J. Am. Chem. Soc.*, **130**, 6308–6309, and references therein.

201 Kotani, H., Ohkubo, K., Takai, Y., and Fukuzumi, S. (2006) Viologen-modified platinum clusters acting as an efficient catalyst in photocatalytic hydrogen evolution. *J. Phys. Chem. B*, **110**, 24047–24053.

202 Baddeley, C.J., Jones, T.E., Trant, A.G., and Wilson, K. (2011) Fundamental investigations of enantioselective heterogeneous catalysis. *Top. Catal.*, **54**, 1348–1356.

203 Blaser, H.U., Malan, C., Pugin, B., Spindler, F., Steiner, H., and Studer, M. (2003) Selective hydrogenation for fine chemicals: recent trends and new developments. *Adv. Synth. Catal.*, **345** (45), 103–151.

204 Studer, M., Blaser, H.U., and Exner, C. (2003) Enantioselective hydrogenation using heterogeneous modified catalysts: an update. *Adv. Synth. Catal.*, **345** (45), 45–65.

205 Baiker, A. (2005) Reflections on chiral metal surfaces and their potential for catalysis. *Catal Today*, **100**, 159.

206 Izumi, Y. (1983) Modified Raney nickel catalyst: Heterogeneous enantiodifferentiating (asymmetric) catalyst. *Adv. Catal.*, **32**, 215–271.

207 Schmidt, E., Bucher, C., Santarossa, G., Mallat, T., Gilmour, R., and Baiker, A. (2012) Fundamental insights into the enantioselectivity of hydrogenations on cinchona-modified platinum and palladium. *J. Catal.*, **289**, 238–248.

208 Chen, Z., Guan, Z., Li, M., Yang, Q., and Li, C. (2011) Enhancement of the performance of a platinum nanocatalyst confined within carbon nanotubes for asymmetric hydrogenation. *Angew. Chemie Int. Ed.*, **50** (21), 4913–4917.

209 Osawa, T., Kizawa, T., Lee, I.-Y.S., Kitamura, T., Inoue, Y., and Borovkov, V. (2011) Durability enhancement of chirally modified metallic nickel catalysts for enantioselective hydrogenation. *Catalysis Communications*, **15**, 15–17.

210 Hong, J., Lee, I., and Zaera, F. (2011) Cinchona alkaloids tethered on porous silica as enantioselctive heterogeneous catalysts. *Top. Catal.*, **54**, 1340–1347.

211 Guyonnet Bilé, E., Cortelazzo-Polisini, E., Denicourt-Nowicki, A., Sassine, R., Launay, F., and Roucoux, A. (2012) Chiral ammonium-capped Rhodium(0) nanocatalysts: synthesis, characterization, and adavances in asymmetric hydrogenation in neat water. *ChemSusChem.*, **5**, 91–101.

212 Jansat, S., Gómez, M., Philippot, K., Muller, G., Guiu, E., Claver, C., Castillón, S., and Chaudret, B. (2004) A case for enantioselective allylic alkylation catalysed by palladium nanoparticles. *J. Am. Chem. Soc.*, **126**, 1592–1593.

213 Favier, I., Gómez, M., Muller, G., Axet, R., Castillón, S., Claver, C., Jansat, S., Chaudret, B., and Philippot, K. (2007) Pd catalytic species containing chiral diphosphites: towards a discrimination between molecular and colloidal catalysts. *Adv. Synth. Catal.*, **349**, 2459–2469.

214 Sawai, K., Tatumi, R., Nakahodo, T., and Fujihara, H. (2008) Asymmetric Suzuki-Miyaura coupling reactions catalyzed by chiral palladium nanoparticles at room temperature. *Angew. Chemie Int. Ed.*, **47**, 6917–6919.

215 Tatumi, R., Akita, T., and Fujihara, H. (2006) Synthesis of small palladium nanoparticles stabilized by bisphosphine BINAP beraing an alkyl cahin and their palladium nanoparticle-catalyzed carbon-carbon coupling reactions under room temperature. *Chem. Commun.*, (31) 3349–3351.

216 Mori, K., Kondo, Y.T., and Fujihara, H. (2009) Synthesis and characterization of FePd magnetic nanoparticles modified with chiral BINAP ligand as a recoverable catalyst vehicle for the asymmetric coupling reaction. *Phys. Chem. Chem. Phys.*, **11**, 8949–8954.

217 Qian, H., Zhu, M., Gayathri, C., Gil, R.R., and Jin, R. (2011) Chirality in gold nanoclusters probed by NMR spectroscopy. *ACS Nano.*, **5** (11), 8935–8942.

218 Bell, A.T. (2003) The impact of nanoscience on heterogeneous catalysis. *Science*, **299** (5613), 1688–1691.

219 Schlögl, R. and Abd Hamid, S.B. (2004) Nanocatalysis: mature science revisited or something really new? *Angew. Chem. Int. Ed.*, **43**, 1628–1637.

220 Tao, F. and Salmeron, M. (2011) In situ studies of chemistry and structure of materials in reactive environments. *Science*, **331** (6014), 171–174.

221 Hansen, P.L., Wagner, J.B., Helveg, S., Rostrup-Nielsen, J.R., Clausen, B.S., and Topsøe, H. (2002) Atom-resolved imaging of dynamic shape changes in supported copper nanocrystals. *Science*, **295** (5562), 2053–2055.

222 Tao, F., Dag, S., Wang, L.-W., Liu, Z., Butcher, D.R., Bluhm, H., Salmeron, M., and Somorjai, G.A. (2010) Break-up of stepped platinum catalyst surfaces by high CO coverage. *Science*, **327** (5967), 850–853.

223 Liu, Z., Yanagi, K., Suenaga, K., Kataura, H., and Iijima, S. (2007) Imaging the dynamic behaviour of individual retinal chromophores confined inside carbon nanotubes. *Nature Nanotech.*, **2**, 422–425.

224 McIntyre, B.J., Salmeron, M., and Somorjai, G.A. (1994) Nanocatalysis by the tip of a scanning tunneling microscope operating inside a reactor cell. *Science*, **265** (5177), 1415–1418.

225 Müller, W.T., Klein, D.L., Lee, T., Clarke, J., McEuen, P.L., and Schultz, P.G. (1995) A strategy for the chemical synthesis of nanostructures. *Science*, **268** (5208), 272–273.

226 Lee, H.J. and Ho, W. (1999) Single-bond formation and characterization with a scanning tunneling microscope. *Science*, **286** (5445), 1719–1722.

227 McEwen, J.-S., Gaspard, P., Visart de Bocarmé, T., and Kruse, N. (2009) Nanometric chemical clocks. *Proc. Natl. Acad. Sci. USA*, **106** (9), 3006–3010.

228 Vajda, S., Pellin, M.J., Greeley, J.P., Marshall, C.L., Curtiss, L.A., Ballentine, G.A., Elam, J.W., Catillon-Mucherie, S., Redfern, P.C., Mehmood, F., and Zapol, P. (2009) Subnanometre platinum clusters as highly active and selective catalysts for the oxidative dehydrogenation of propane. *Nature Mater.*, **8**, 213–216.

229 Ahmadi, T.S., Wang, Z.L., Green, T.C., Henglein, A., and El-Sayed, M.A. (2002) Shape-controlled synthesis of colloidal platinum nanoparticles. *Science*, **272** (5270), 1924–1925.

230 Lee, I., Morales, R., Albiter, M.A., and Zaera, F. (2008) Synthesis of heterogeneous catalysts with well shaped platinum particles to control reaction selectivity. *Proc. Natl. Acad. Sci. USA*, **105** (40), 15241–15246.

231 Feldheim, D.L. (2007) The new face of catalysis. *Science*, **316** (5825), 699–700.

232 Herzing, A.A., Kiely, C.J., Carley, A.F., Landon, P., and Hutchings, G.J. (2008) Identification of active gold nanoclusters on iron oxide supports for CO oxidation. *Science*, **321** (5894), 1331–1335.

233 Corma, A. and Serna, P. (2006) Chemoselective hydrogenation of nitro compounds with supported gold catalysts. *Science*, **313** (5785), 332–334.

234 Chen, M.S. and Goodman, D.W. (2004) The structure of catalytically active gold on titania. *Science*, **306** (5694), 252–255.

235 Joo, S.H., Park, J.Y., Tsung, C.-K., Yamada, Y., Yang, P., and Somorjai, G.A. (2009) Thermally stable Pt/mesoporous silica core-shell nanocatalysts for high-temperature reactions. *Nature Mater.*, **8**, 126–131.

236 Farmer, J.A. and Campbell, C.T. (2010) Ceria maintains smaller metal catalyst particles by strong metal-support bonding. *Science*, **329** (5994), 933–936.

237 Boyen, H.-G., Kästle, G., Weigl, F., Koslowski, B., Dietrich, C., Ziemann, P., Spatz, J.P., Riethmüller, S., Hartmann, C., Möller, M., Schmid, G., Garnier, M. G., and Oelhafen, P. (2002) Oxidation-resistant gold-55 clusters. *Science*, **297** (5586), 1533–1536.

238 Lefèvre, M., Proietti, E., Jaouen, F., and Dodelet, J.-P. (2009) Iron-based catalysts with improved oxygen reduction activity in polymer electrolyte fuel cells. *Science*, **324** (5923), 71–74.

239 Wu, G., More, K.L., Johnston, C.M., and Zelenay, P. (2011) High-performance electrocatalysts for oxygen reduction derived from polyaniline, iron, and cobalt. *Science*, **332** (6028), 443–447.

240 Liu, H., Jiang, T., Han, B., Liang, S., and Zhou, Y. (2009) Selective phenol hydrogenation to cyclohexanone over a dual supported Pd-lewis acid catalyst. *Science*, **326** (5957), 1250–1252.

241 Lim, B., Jiang, M., Camargo, P.H.C., Cho, E.C., Tao, J., Lu, X., Zhu, Y., and Xia, Y. (2009) Pd-Pt bimetallic nanodendrites with high activity for oxygen reduction. *Science*, **324** (5932), 1302–1305.

242 Kesavan, L., Tiruvalam, R., Ab Rahim, M.H., bin Saiman, M.I., Enache, D.I., Jenkins, R.L., Dimitratos, N., Lopez-Sanchez, J.A., Taylor, S.H., Knight, D.W., Kiely, C.J., and Hutchings, G.J. (2011) Solvent-free oxidation of primary carbon-hydrogen bonds in toluene using Au-Pd alloy nanoparticles. *Science*, **331** (6014), 195–199.

243 Tedsree, K., Li, T., Jones, S., Chan, C.W. A., Yu, K.M.K., Bagot, P.A.J., Marquis, E.A., Smith, G.D.W., and Tsang, S.C.E. (2011) Hydrogen production from formic acid decomposition at room temperature using a Ag-Pd core-shell nanocatalyst. *Nature Nanotech.*, **6**, 302–307.

244 Tao, F., Grass, M.E., Zhang, Y., Butcher, D.R., Renzas, J.R., Liu, Z., Chung, J.Y., Mun, B.S., Salmeron, M., and Somorjai, G.A. (2008) Reaction-driven restructuring of Rh-Pd and Pt-Pd core-shell nanoparticles. *Science*, **322** (5903), 932–934.

245 Rolison, D.R. (2003) Catalytic nanoarchitectures – the importance of nothing and the unimportance of periodicity. *Science*, **299** (5613), 1698–1701.

246 Crossley, S., Faria, J., Shen, M., and Resasco, D.E. (2010) Solid nanoparticles that catalyze biofuel upgrade reactions at the water/oil interface. *Science*, **327** (5961), 68–72.

247 Pan, X., Fan, Z., Chen, W., Ding, Y., Luo, H., and Bao, X. (2007) Enhanced ethanol production inside carbon-nanotube reactors containing catalytic particles. *Nature Mater.*, **6**, 507–511.

248 Nørskov, J.K., Abild-Pedersena, F., Studta, F., and Bligaard, T. (2011) Density functional theory in surface chemistry and catalysis. *Proc. Natl. Acad. Sci. USA*, **108** (3), 937–943.

249 Nørskov, J.K., Bligaard, T., Rossmeisl, J., and Christensen, C.H. (2009) Towards the computational design of solid catalysts. *Nature Chem.*, **1**, 37–46.

250 Tao, F., Dag, S., Wang, L.-W., Liu, Z., Butcher, D.R., Bluhm, H., Salmeron, M., and Somorjai, G.A. (2010) Break-up of stepped platinum catalyst surfaces by high CO coverage. *Science*, **327** (5967), 850–853.

2
Metallic Nanoparticles in Neat Water for Catalytic Applications

Audrey Denicourt-Nowicki and Alain Roucoux

2.1
Introduction

In the drive towards the development of sustainable and economically viable transition metal processes, much effort has recently been devoted to the use of benign reaction media (e.g., water, ionic liquids, supercritical CO_2) [1, 2], as well as efficient and reusable catalytic materials [3, 4]. In such a context, nanometer-sized metal species, namely metallic nanoparticles (NPs), have appeared as an unavoidable family of catalysts at the border between mononuclear metal complexes and heterogeneous bulk metals [5, 6]. They have emerged as sustainable alternatives to conventional homogeneous catalysts [7, 8], potentially showing high activities and selectivities and possessing the ease of catalyst recovery of heterogeneous catalysts. In fact, nanoclusters are catalysts of tremendous interest because of their high surface area-to-volume ratios and consequently, a high number of potential active sites, which could greatly improve their catalytic activity [9–12]. For that purpose, the controlled and reproducible synthesis of well-defined and stable metallic NPs is essential to tune their catalytic activities, selectivities and recycling ability [13, 14]. For their preparation, various methods have been reported according to the 'organic' or 'aqueous' nature of the catalytic media, the nature of the precursor and also of the protective agent (surfactants, cyclodextrins (CDs), polymers or ligands) [15–17]. Very recently, Kou *et al.* [18] have reviewed the pertinent combination between nanocatalysis and green solvents such as water, ionic liquids, fluorous or supercritical solvents. Among these, water represents an interesting alternative for organic reactions for economic reasons [19–21]. Moreover, another important criterion for catalytic applications is that water could enable the catalyst recovery/recycling through a biphasic approach thanks to its low miscibility with most organic compounds [22, 23]. Furthermore, owing to its highly polar character, original reactivities and selectivities in catalytic reactions could be expected in water.

In this chapter, we will focus exclusively on the development of nanoparticle-based catalysts in colloidal suspensions for applications in neat water. Supported nanocatalysts will not be considered. Based on pertinent but non-exhaustive examples of water-soluble nanoparticles, we propose to rationalize the key aspects

Nanomaterials in Catalysis, First Edition. Edited by Philippe Serp and Karine Philippot.

in terms of advantages and constraints for the future development of nanoparticles in neat water and their interest in biphasic catalysis.

2.2
Synthesis of Nanoparticles in Water: The State of The Art

'*The best solvent is no solvent and if a solvent is needed the water is preferred. Water is nontoxic, nonflammable, abundantly available and inexpensive. Moreover, owing to its highly polar character one can expect novel reactivities for organometallic catalysis in water.*' With these few words, R.A. Sheldon summed up a critical review [24] concerning the fundamental advantages of the development of catalysis in water. Well-known in the homogeneous domain, the concept of liquid–liquid biphasic catalysis has largely been applied in various reactions through the development of water-soluble homogeneous catalysts containing mostly hydrophilic phosphorus derivatives [23]. In a water–organic biphasic system, the catalyst is dissolved in the aqueous phase while the reactants and final products remain in the organic phase (Figure 2.1). Finally, the catalyst can be recovered and recycled by simple phase separation.

The concept of pure biphasic liquid–liquid system is also used when a liquid substrate constitutes the only organic phase (no solvent is added). In this case, the final extraction can be carried out by addition of an organic solvent to separate the products from the catalytic aqueous phase before the recycling process.

Noble metal nanoparticles have been considered as the most promising catalytic systems due to their unique properties at the frontier between those of homogeneous and heterogeneous systems. During the last three decades, their development for various catalytic applications has known a great interest in the scientific community to achieve new activities and selectivities. This interest is reflected by a large number of books [11, 15, 25–28], reviews [8, 14, 16, 17, 29, 30] and publications devoted to their synthesis, characterization, physico-chemical properties and applications. In the context of green chemistry, the choice of the reaction medium has more recently received an increasing attention to limit the use of volatile organic solvents. For that purpose, the development of active nanocatalysts in unusual reaction media has become a priority. Among the greener solvents which

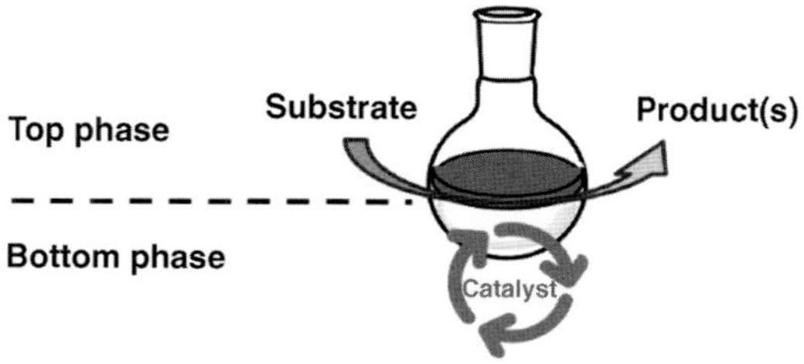

Figure 2.1 Concept of biphasic liquid–liquid catalysis.

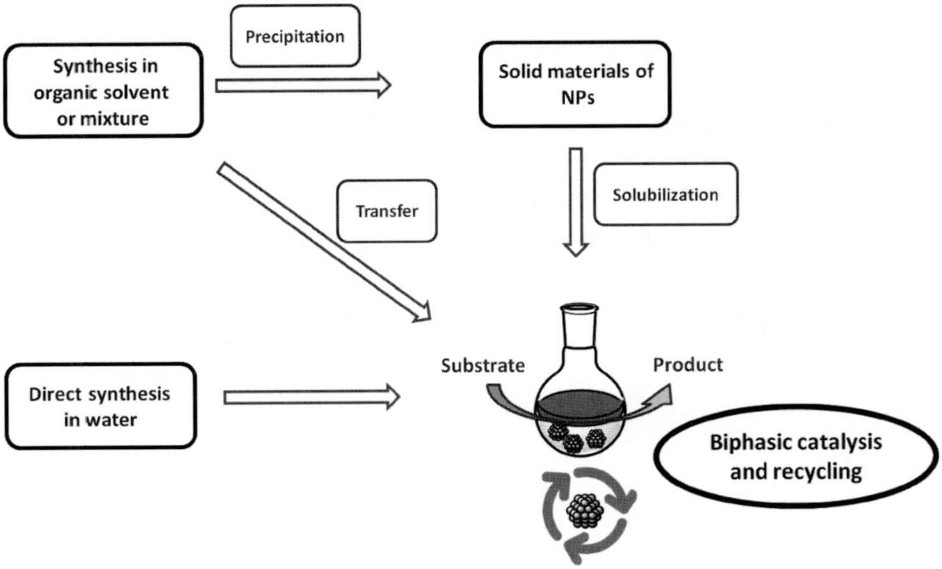

Figure 2.2 Strategies to obtain water-soluble NPs.

also provide biphasic approaches such as fluorous solvents, ionic liquids or even supercritical fluids, water has appeared as a pertinent reaction media. The use of colloidal metallic NPs finely dispersed in water has become an effective approach for biphasic catalysis with a targeted objective concerning the recycling as a sustainable technology. This smart approach has been less described in the literature, up to its recent attention for environmental reasons, owing to difficulties of reproducing the synthesis of NPs and keeping them stable during the catalysis. Consequently the preparation of the NPs and their use as catalysts in neat water have rarely been reported in reviews. The crucial parameter, which can be a dramatic hindrance, remains the disposal of easily modified water-soluble protective agents to prepare and to stabilize aqueous suspensions of NPs. In most cases, highly water-soluble compounds such as ionic species, polymers and derivatives or amphiphilic compounds were used to prevent aggregation and to facilitate recycling.

Several strategies have been developed to obtain metallic nanospecies in aqueous phase such as direct synthesis [11], phase transfer [31–35] and redispersion of powders of nanomaterials [5] (Figure 2.2).

Obviously, the direct synthesis of NPs in water is particularly appropriate given the cheap and easy access of water-soluble precursors such as commercial metal salts. Developed by many groups experienced in organosoluble homogeneous catalysis, the synthesis of NPs in organic solvents has been largely described for catalytic investigations such as hydrogenation, C—C coupling and oxidation reactions, and more recently this know-how has begun to be extrapolated to water. According to the pertinent choice of the added capping agent, one-pot nanoparticle transfer in water could be achieved. An alternative approach consists in the use of a ligand

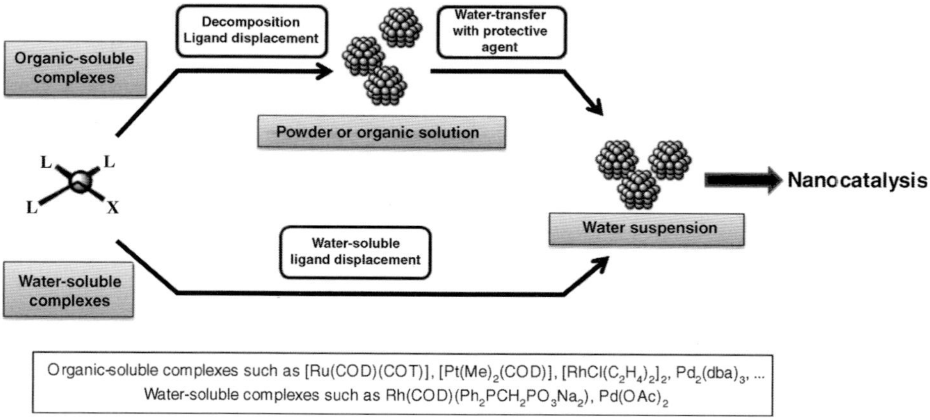

Figure 2.3 Organometallic approaches for the preparation of water-soluble NPs.

which displays solubility in both organic media and water. In this case, the NPs were prepared from organometallic complexes in organic conditions, then precipitated and transferred into water.

Based on an organometallic approach, two routes have been reported to prepare water-soluble NPs according to the use of organic- or water-soluble complexes (Figure 2.3). First, starting from water-soluble precursors such as [Rh(COD)($Ph_2PCH_2CH_2PO_3Na_2$)] [36] or $Pd(OAc)_2$ [37] the ligand displacement provides an aqueous suspension of nanoclusters, which is stabilized by the same ligand or by addition of a more water-soluble compound such as ammonium salts, thus enabling various stabilizing effects (steric or coulombic repulsion) to be combined. Secondly, some examples also report the ligand displacement from organic-soluble complexes such as [Ru(COD)(COT)], [$PtMe_2$(COD)] or [$RhCl(C_2H_4)]_2$. Recently, Philippot and co-workers have described a strategy based on the use of both organic- and water-soluble phosphine-based ligand such as 1,3,5-triaza-7-phosphaadamantane (PTA) [38, 39] or alkyl sulfonated diphosphines to get aqueous colloidal solutions of NPs [40]. The organometallic complexes were decomposed in tetrahydrofuran (THF) in the presence of the ligand which favors, after removal of the solvent, the transfer of the NPs powder into water. This strategy displays several specificities such as the control of the particle dispersity, size, shape and the nature of the chemical species present at their surface. Moreover, the amount of added ligand is a key-point as it allows the control of the particles' growth and therefore of their size. This approach could be particularly pertinent in the search of original selectivities in nanocatalysis. However, it requires a preliminary step with the delicate synthesis of the organometallic precursors in anaerobic conditions, which could be a limitation for future industrial applications. Catalytic investigations of these systems in water are reported in the following part of this chapter.

Among strategies to prepare NPs in water, the reduction of transition metal salts in solution is the most widely used method to generate aqueous suspensions of colloids (Figure 2.4). A wide range of reducing agents are available such as

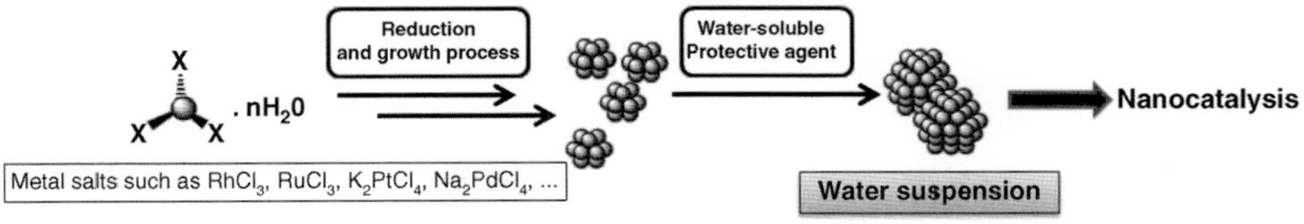

Figure 2.4 Salt approach for the synthesis of water-soluble NPs.

hydrogen, hydrides or salts such as sodium borohydride or sodium citrate. Based on the fact that the water-solubility of the metal core is controlled by the protective agents and not due to the solvent, the advantage of this approach is undoubtedly the great diversity and modularity of highly water-soluble stabilizers providing (i) electrostatic stabilization; (ii) steric protection based on the use of polymers, dendrimers or CDs; and (iii) electrosteric stabilization generated by surfactants. Without contradiction, the oldest stabilizer used to protect colloids in water is the trisodium citric salt [41].

This easy strategy provides some degree of control over the particle size, affording efficient and reproducible nanocatalysts in terms of activity and selectivity in catalytic applications. Nevertheless, salts and by-products could remain in contact with the particle surface, thus passivating them, leading to the production of surface oxides or hydroxides and potentially modifying their reactivity in catalysis. In most cases, these nanometer-sized metal particles efficiently act as nanocatalysts if no aggregation occurs. To avoid this phenomenon and also to facilitate the recycling, pertinent protective agents should be chosen according to the catalytic applications and the reaction conditions. In this chapter, pertinent but non-exhaustive examples are described according to the choice of the protective agents.

To conclude, and considering environmental and economical aspects, the synthesis of metallic NPs in water remains of high importance. Indeed, as the separation of expensive transition metal catalysts from substrates and products remains a keypoint for industrial applications, it is necessary to develop simple methods for the preparation of appropriate metal nanocatalysts. The synthesis of colloidal suspensions appears to be more difficult from an organometallic approach than from a metal salt, but recent results are promising. However, the use of more sophisticated protective agents for the direct synthesis of NPs in water offers new opportunities. At this stage, whatever the strategy is, the use of NPs in neat water displays a great potential of reactivities associated with a good reproducibility in a wide range of catalytic applications.

2.3
Water-Soluble Protective Agents and their use in Nanocatalysis

One of the most challenging issues in the synthesis of metallic nanoparticles remains the formation of well-defined and size-controllable particles [42], as their catalytic activity and selectivity are drastically influenced by their morphological characteristics [43]. However, the stability of the particles during the catalytic reaction is also of great importance. Thus, among the various parameters to control the particle morphology, the use of a capping agent, judiciously chosen according to the reaction medium, is crucial to obtain well-controlled NPs and to avoid their aggregation during the catalytic process. Hereafter, we describe the various water-soluble protective agents, namely surfactants, CDs, polymers or ligands used to stabilize NPs for investigation in catalytic processes in neat water.

2.3.1
Electrosteric Stabilization by Surfactants

Surfactants are well-suited as protective agents of metallic nanospecies in neat water as they bind strongly enough to the particle surface through electrosteric interactions, thus avoiding particles aggregation.

The group of Roucoux has largely described the stabilization of Rh, Ru and Ir NPs by easily-synthesized ionic surfactants for catalytic hydrogenation reactions in biphasic liquid–liquid medium [44–46]. Various *N*-alkyl-*N*-(2-hydroxyethyl) ammonium salts (Figure 2.5) possessing mono- or poly-hydroxylated polar heads and bearing a lipophilic alkyl chain containing $n = 12$–18 carbon atoms could be prepared with various counter-ions X such as halogens (Br^-, Cl^-, I^-, F^-), mesylate ($CH_3SO_3^-$), tetrafluoroborate (BF_4^-), hydrogen carbonate (HCO_3^-) and triflate ($CF_3SO_3^-$) [47–50]. They were obtained by quaternization of *N,N*-dimethylethanolamine with the appropriate functionalized halogenoalkanes in the Hydroxy Ethyl Ammonium (HEA) series [51] or of hexadecylamine with chloroethanol for THEA16Cl [49] or by anionic metathesis [48, 50].

All the ammonium compounds with a lipophilic chain of more than 12 carbon atoms display a surfactant behavior and self-aggregate into micelles above the critical micellar concentration (cmc). The cmc values decrease while increasing the hydrocarbon chain length and modifying the counter-ion in the following order: $F^- > Cl^- > Br^- > CH_3SO_3^- > BF_4^- \approx HCO_3^- > CF_3SO_3^- > I^-$ in the HEA16 series. The characteristics of some surfactants are gathered in Table 2.1.

HEA12X y = 6
HEA14X y = 8
HEA16X y = 10
HEA18X y = 12

with X = F, Cl, Br, I, CH_3SO_3, BF_4, HCO_3, CF_3SO_3

THEA16Cl

Figure 2.5 Ammonium surfactants used as protective agents of metallic nanospecies.

Table 2.1 Physico-chemical properties of ammonium derivatives and their predicted aggregate geometry.

Surfactant	CMC[a)] (mmol L^{-1})	γ_{CMC}[b)] (mN m^{-1})	A[c)] ($Å^2$)	CPP[d)]	Geometry
$HEA16CF_3SO_3$	0.25	32.0	49.6	0.42	rods
$HEA16BF_4$	0.62[e)]	30.3	37.3	0.55	rods
$HEA16HCO_3$	0.6	40.3	46.6	0.45	rods
HEA16F	2.12	43.2	87.1	0.24	spheres
HEA16Cl	1.23	40.5	33.4	0.63	bilayers
THEA16Cl	0.73	35	53	0.39	rods

a) cmc = critical micellar concentration measured at 303 K.
b) γ_{CMC} = surface tension at cmc.
c) A = area of polar head per molecule.
d) CPP = critical packing parameter, spherical micelles for CPP < 1/3, cylindrical or rod-like micelles for 1/3 < CPP < 1/2 and bilayers with CPP > 1/2.
e) cmc measured at 303 K.

The colloidal rhodium(0) suspensions were prepared in water by chemical reduction of rhodium trichloride with sodium borohydride in a dilute aqueous solution of ammonium surfactant (Figure 2.6). In the HEA series, the authors showed that only surfactants bearing a sufficiently lipophilic chain with 16 or 18 carbons give rise to stable monodispersed colloidal suspensions. Moreover, a molecular surfactant/metal ratio of two, proved to be a good compromise to maintain efficiently the NPs within the aqueous phase, to avoid aggregation and to achieve a good catalytic activity. Thermogravimetric analyses showed that the surfactants self-organize in a bilayer structure at the particle's surface (Figure 2.6), as previously observed by Chen [52] and El-Sayed [53] with ammonium salts-protected Cu and Ni NPs or Au nanorods. The authors suggest that the inner layer of surfactants interacts with the

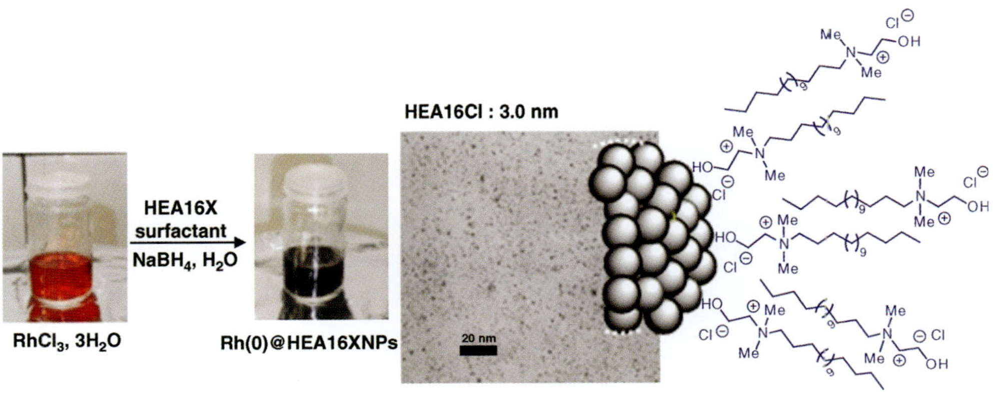

Figure 2.6 Synthesis of Rh(0)@HEA16X NPs and a representation of the bilayer structure on the NPs surface.

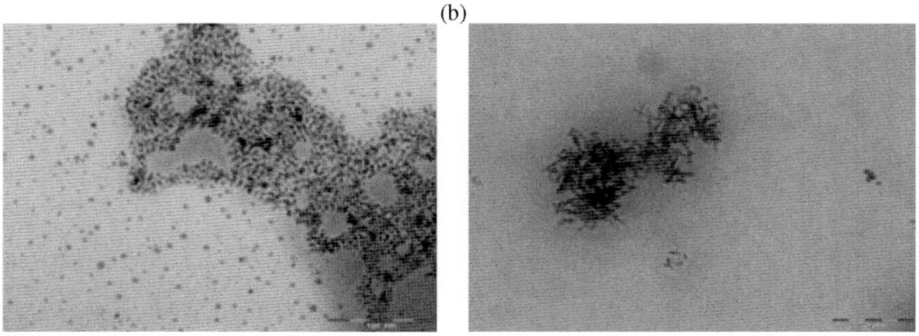

Figure 2.7 TEM micrographs of (a) Rh(0)@HEA16F worm-like and isolated NPs and (b) Rh(0)@HEA16BF_4 NPs worm-like structures in water. (Adapted from ref. [50])

NPs surface via the charged groups and is connected to the outer layer through hydrophobic interactions, while the polar groups of the outer layer are directed towards the aqueous solution.

Transmission electron microscopy (TEM) studies (Figure 2.7) demonstrated that the counter-ion has an influence on the particle sizes and morphologies: classical spherical geometries for those prepared with THEA16Cl or HEA16X with $X^- =$ Br^-, Cl^- and worm-like structures for HEA16X with $X^- = BF_4^-$ and $CF_3SO_3^-$. Moreover, the HEA16F surfactant leads to two types of morphology with spherical and elongated NPs, probably due to a weaker stabilization of the spherical particles inducing a sintering of the nanospecies.

Finally, from dynamic light scattering (DLS) measurements (hydrodynamic diameter and potential zeta values), the colloids present, for each surfactant, an apparently positive charge in solution ranging around 40–50 mV, which is in agreement with cationic protective agents. The measured values show the important role of the electrostatic repulsion (coulombic interaction) for the stabilization of nanoclusters.

The catalytic systems obtained, based on metallic NPs stabilized by ammonium surfactants were active in the hydrogenation of benzene and arene derivatives under atmospheric pressure, at room temperature in biphasic liquid/liquid conditions. Indeed the Rh(0)@HEA16X system acting as micellar nanoreactors was found to be pertinent for this reaction with high turnover frequencies (TOF). The results concerning the benzene hydrogenation depending on the protective agent of Rh(0) NPs are gathered in Table 2.2.

The catalytic activity was mostly influenced by a fine combination of the counter-ion and the polar head which could modify the physico-chemical properties of the colloidal suspension, such as the particle size and morphology and the potential zeta value.

It is worth underlining that no hydrogenation occurred with HEA16I as surfactant owing to the simultaneous redox formation of iodine as a poison for nanocatalysts. The selective hydrogenation of disubstituted arenes such as xylene isomers leads in all cases to the major formation of the *cis*-product, thus favoring the fast

Table 2.2 Benzene hydrogenation using various ammonium-protected Rh(0) NPs[a)].

Entry	Surfactant	Time (h)	TOF[b)] (h^{-1})
1	HEA16Cl	3.6	83
2	HEA16Br	5.3	57
3	HEA16F	6	50
4	$HEA16CH_3SO_3$	3.7	81
5	$HEA16BF_4$	3.7	81
6	$HEA16HCO_3$	6	50
7	$HEA16CF_3SO_3$	3	100
8	THEA16Cl	1	300

a) Reaction conditions: Rh (3.8×10^{-5} mol), Ammonium surfactant (7.6×10^{-5} mol), substrate (3.8×10^{-3} mol), H_2O (10 mL), 1 bar H_2, room temperature.
b) TOF = Turnover frequency defined as mol of H_2 per mol of Rh per h.

hydrogenation of the substrate and disfavoring its desorption during the catalytic process. In most cases, an increase in hydrogen pressure gave rise to an activation of the catalytic suspension and the authors showed that HEA16Cl-protected iridium(0) and ruthenium(0) nanoparticles are efficient in arene hydrogenation under higher pressures [45, 46]. Furthermore, in all cases the nanocatalysts could easily be separated from the reaction mixture by simple extraction of the product with the appropriate solvent (alkanes or ether). The durability of the catalytic system was checked by recycling the catalyst in several successive hydrogenation reactions. Similar TOFs were achieved in agreement with the absence of metal leaching and the efficiency of the ammonium surfactant to maintain the nanospecies within the aqueous phase. To conclude, ammonium salts possessing various counter-ions when used as capping agents of Rh(0) NPs provide new micellar nanoreactors for arene hydrogenation under mild conditions in biphasic liquid–liquid media.

Parallel to the use of HEA salts in the hydrogenation reactions, other catalytic applications using HEA16Cl-protected nanocatalysts were developed. After their preparation by the same methodology based on the chemical reduction of metallic salts by sodium borohydride in the presence of HEA16Cl [54], Rh(0) and Pd(0) NPs were studied in the tandem dehalogenation/hydrogenation reaction of various halogenated arenes. First, the chlorinated derivatives are more reactive than the brominated ones and no reaction was observed with iodobenzene owing to an inhibition of the catalytic system. Second, the kinetics and the selectivity of the reaction were greatly influenced by the nature of the metal and the hydrogen pressure (Table 2.3).

Thus, the aqueous palladium(0) suspension was able to cleave the C—Cl bond efficiently and was reusable in two successive runs, whereas the rhodium(0) nanospecies were active for the tandem dehalogenation–hydrogenation reaction. More recently, the colloidal rhodium(0) suspension was investigated in the hydrodechlorination of halogenoanisoles as model substrates of endocrine disrupting compounds, leading to the formation of relatively nontoxic saturated product and

Table 2.3 Effect of the metal nature on the hydrodechlorination of chlorobenzene derivatives[a]. (Adapted from ref. [54]).

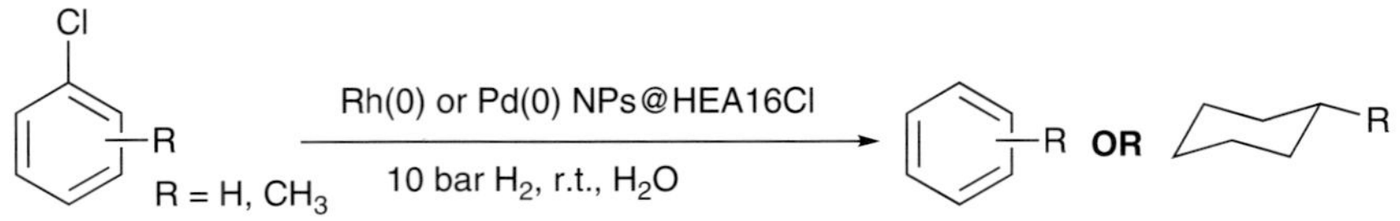

Metal	Substrate	Product	Conv. (%)[b]	Reaction Time (h)
Pd	Chlorobenzene	Benzene	100	1.5
Rh	Chlorobenzene	Cyclohexane	100	1.7
Pd	4-Chlorotoluene	Toluene	>90	6
Rh	4-Chlorotoluene	4-Methylcyclohexane	100	7.3

a) Reaction conditions: palladium or rhodium (3.4×10^{-5} mol), substrate (3.4×10^{-3} mol), Pd/HEA16Cl = 10, Rh/HEA16Cl = 2, substrate/metal = 100, 10 bar H_2, r.t., 10 mL water.
b) Determined by GC analysis.

added value cyclohexanone in liquid phase [55]. However, in some cases, the aggregation of the NPs was observed and could be easily circumvented by their deposit on inorganic supports [55, 56].

Finally, the activity of the catalytic Pd(0)@HEA16Cl system was also studied in the carbon–carbon coupling reaction, more particularly in the Suzuki reaction in neat water [57]. Using a HEA16Cl/Pd molar ratio of 25 to efficiently maintain the catalytic species within the aqueous phase, mild conditions (333 K, 1 equivalent of base) were applied for the coupling of bromobenzene with sodium tetraphenylborate.

Based on the interesting results obtained in reduction reactions, the enantioselective hydrogenation of ethylpyruvate was carried out in biphasic media using modified cinchonidine platinium(0) NPs produced from H_2PtCl_6 reduction in the presence of the HEA16Cl surfactant [58]. The aqueous suspensions showed efficient activity for the asymmetric hydrogenation of ethylpyruvate under hydrogen pressure (40 bar H_2), leading to the formation of (*R*)-ethyl lactate with up to 55% *e.e* (Figure 2.8). Interestingly, the aqueous phase containing the Pt(0) nanocatalyst could be used for several runs with no loss in activity and enantioselectivity. TEM

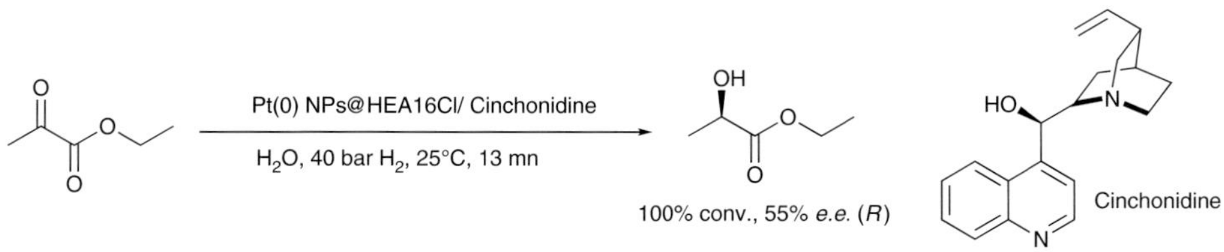

Figure 2.8 Asymmetric hydrogenation of ethylpyruvate in neat water using cinchonidine modified Pt(0) NPs.

investigations on the recovered catalyst revealed that the catalytic system contains nanoparticles with an average diameter of about 3.25 nm, with 73% of the NPs being between 2.22 and 3.97 nm and presenting a dendritic structure, without consequences on the stereodifferentiation.

Following their approach based on the electrosteric stabilization of nanospecies in neat water, the same authors developed a new family of water-soluble and optically active *N*-methylephedrium salts (*N*MeEph12X) that could be used as efficient protective agents of NPs in neat water and as chiral inducer in asymmetric reactions [59]. These surfactants bearing a dodecyl alkyl chain can possess various counter-ions ($X^- = Br^-$, HCO_3^-, (*S*)-(-)-lactate$^-$). The nanocatalysts obtained with mean sizes of about 2.5 nm were evaluated in the asymmetric hydrogenation of ethylpyruvate in neat water. The influence of the nature of the counter-ion of the surfactant, the hydrogen pressure and the use of an external chiral inducer (cinchonidine) was investigated. The results are gathered in Table 2.4.

The Rh(0)@(-)-*N*MeEph12X catalytic system obtained provides efficient results in terms of activity for the hydrogenation of ethylpyruvate with a promising asymmetric induction up to 13% *e.e.* using optically active *N*MeEph12X salts ($X^- = Br^-$, (*S*)-lactate$^-$). Moreover, the enantiomeric excess was slightly improved up to 18% by the use of an external chiral inducer such as (−)-cinchonidine.

The group of Le Bras described the reactivity of palladium(0) NPs, $Pd_{OAc,N}$, easily obtained from $Pd(OAc)_2$ and an usual ammonium salt, *n*-Bu_4NBr, for the regio- and chemoselective hydrogenolysis of benzylic epoxides [37]. The authors reported that among the various reaction media used (organic solvants or ionic liquids) for the hydrogenolysis reaction of styrene oxide, water was found to be the best

Table 2.4 Asymmetric hydrogenation of ethylpyruvate with Rh(0)@(-)-*N*MeEph12X[a]. (Adapted from ref. [59]).

Surfactant	P (bar H_2)	Additive	T (h)	Conv.[b] (%)	*e.e.*[b] (%)
(-)-*N*MeEph12Br	1	—	5	3	0
(-)-*N*MeEph12Br	30	—	1.5	100	2 (*R*)
(-)-*N*MeEph12Br	40	—	1	100	12 (*R*)
(-)-*N*MeEph12Br	60	—	1	100	2 (*R*)
(-)-*N*MeEph12HCO_3	40	—	1.5	100	4 (*R*)
(-)-*N*MeEph12(*S*)-lactate	40	—	1	100	13 (*R*)
(+)-*N*MeEph12Br	40	—	1	100	7 (*S*)
(-)-*N*MeEph12Br	40	(-)-cinchonidine	1	100	15 (*S*)
(-)-*N*MeEph12(*S*)-lactate	40	(-)-cinchonidine	1	100	18 (*S*)

a) Reaction conditions: catalyst (3.8×10^{-5} mol), surfactant (7.6×10^{-5} mol), substrate (3.8×10^{-3} mol), water (10 mL), temperature (293 K), stirred at 1500 mn^{-1}.
b) Substrate conversion and selectivity determined by gas chromatography analysis using a chiral column: Chiralsil-Dex CB 30m, $T_{oven} = 363$ K.

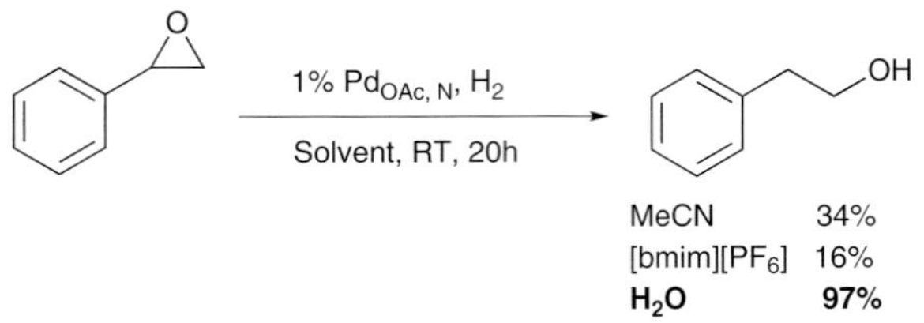

Figure 2.9 Hydrogenolysis of styrene oxide with Pd(0) NPs – Influence of the solvent.

medium allowing an excellent yield in alcohol (Figure 2.9). They observed that the dispersion of NPs in a water–oil system led to the formation of 'black aqueous droplets' in which palladium is probably adsorbed on the surface. The system could be easily recycled over five runs when performed with 0.5 eq. of n-Bu_4NBr as additive. The same catalyst was also used for the chemoselective hydrogenation of olefins in water [60].

Surfactant-protected Pd(0) NPs were also used in carbon–carbon coupling reactions. For example, Zhou *et al.* [61] reported an aqueous Heck reaction carried out under ultrasonic irradiation and catalyzed by *in situ* formed Pd(0) NPs. In fact, the authors observed the formation of *in situ* palladium(0) nanospecies by mixing $PdCl_2$, methylacrylate, tetra-n-butylammonium bromide (TBAB) in water followed by sonication. They suggested that the methylacrylate, which was required for the particles formation, probably acts as a reducing agent. The obtained Pd(0) nanospecies with sizes of about 6–16 nm were efficient catalysts in the Heck reaction of various aryl iodides and olefins (Table 2.5), affording the desired products with good yields.

More recently, the group of Özkar reported the use of sodium laurate (dodecanoate) for the stabilization of rhodium(0) nanoclusters [62]. These nanoclusters were prepared by reduction of the metallic salt by dimethylamine-borane in the presence of sodium laurate at room temperature. TEM and Fourier-transform infrared (FTIR) analyses showed the formation of highly dispersed rhodium(0) NPs with a mean size of 5.2 ± 2.7 nm and the presence of laurate anion most probably at the particle surface. The water-soluble laurate-protected Rh(0) nanoclusters were found to be highly active catalyst in the hydrolysis of ammonia-borane even at low catalyst concentration and low temperature (298 K), providing unprecedented catalytic lifetime and exceptional catalytic activity total turnover number (TTO) of 80 000 mol H_2/mol Rh and TOF of 200 mol H_2/mol Rh min), compared to laurate-capped Ru(0) nanoclusters (TTO value of 5900 mol H_2/mol Ru and TOF of 75 mol /mol Ru min) [63].

In summary, surfactants have been widely investigated as protective agents for the preparation of metallic nanospecies for biphasic catalysis in water. In many cases, they provide a sufficient hydrophilic character to maintain the catalytic species within the aqueous phase. These surfactants give rise to micelles that constitute interesting nanoreactors for the synthesis of size-controlled NPs due to the confinement of the particles inside the micelle cores. The obtained stable colloidal solutions can be easily used as nanocatalysts.

Table 2.5 Heck reaction promoted by ultrasonic aqueous media[a]. (Adapted from ref. [61]).

ArX	Olefin	Product	Time (h)	Yield(%)[b]
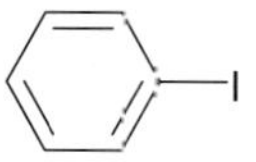	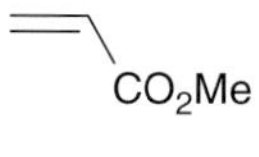	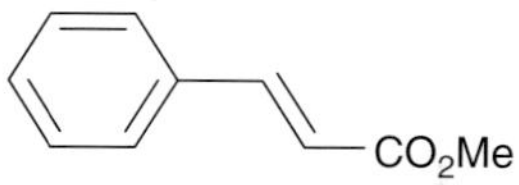	4.5	86
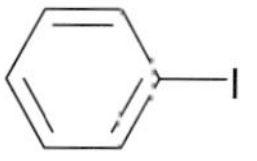	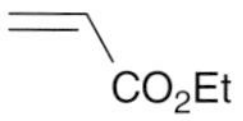	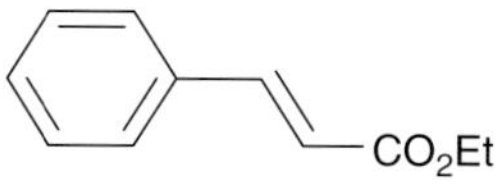	5	83
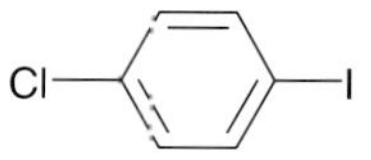	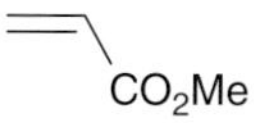	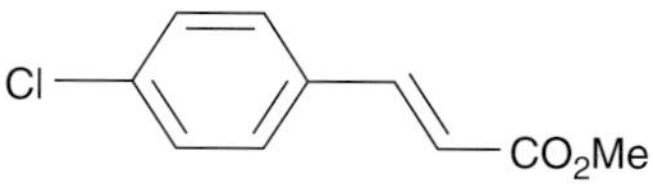	4.5	93
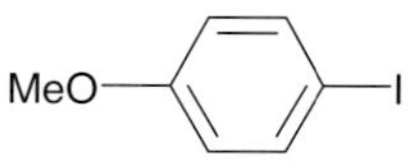	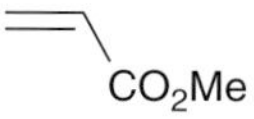	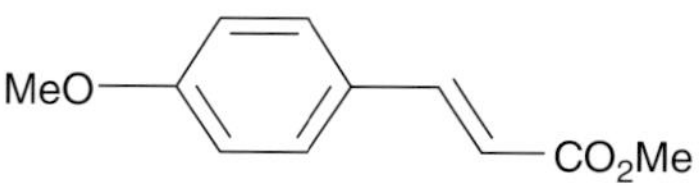	6	76

a) Reaction conditions: ArX (1 mmol), olefin (2 mmol), TBAB (1 mmol), Na_2CO_3 (3 mmol), $PdCl_2$ (0.02 mmol), water (3 mL). The mixtures were sonicated at ambient temperature (298 K) in running water bath for the corresponding time.
b) Isolated yield.

2.3.2
Steric Stabilization by Cyclodextrins

Cyclodextrins have proved to be promising candidates as protective agents for nanospecies [64], generating steric stabilization via hydrophobic interactions [65], and also as potential transfer shuttle to improve biphasic processes. Moreover, these cyclic oligosaccharides could be easily modulated (size of the cavity, functionalization on the primary and secondary rings) in order to finely tune their physicochemical properties (solubility, surface activity. . .) [66]. The CD-capped nanocatalysts have also proved to be original systems as the inclusion of the substrate in a constrained environment could modify the reactivity and the selectivity of these reactions. In this part, we describe their use as protective agents for metallic NPs in various catalytic applications such as hydrogenation and cross-coupling reactions. These reactions are described hereafter and classified according to the nature of the CDs.

2.3.2.1 Hydrogenation Reactions

To our knowledge, native CDs have only very recently been used as protective agents of Au(0) NPs for the reduction of *p*-nitrophenol in neat water [67]. The direct reduction of chloroauric acid by $NaBH_4$ in the presence of α-CD in an alkaline aqueous solution enables a controlled synthesis of the nanocatalysts with adjustable size in the range of 10–50 nm and narrow size distributions. The reaction rate of the reduction of nitro groups was inversely proportional to the particle diameter according to the surface area available for catalysis. More recently, H. Yamashita *et al.* described the first example of FePt NPs displaying water-compatibility, magnetic properties and catalytic activity under aqueous conditions in the presence of γ-CDs [68]. The formation of an inclusion complex between surface-bound organic compounds and γ-CDs drastically improved the dispersity of these FePt NPs having a Fe-rich core and a Pt-rich shell and stabilized by oleic acid and oleylamine in water (Figure 2.10) by. The surface coverage of γ-CDs on the FePt nanoparticles, with an average diameter of 2.5 nm, was roughly estimated to be about 60% by CHN elemental analysis.

These superparamagnetic FePt NPs showed an enhanced catalytic activity in water rather than in organic media in the hydrogenation of allyl alcohol under 1 bar H_2 (Figure 2.11) due to a host-guest complexation of the organic substrate with γ-CDs on FePt NPs. Moreover, the system was easily recycled, applying an external magnet and also active in the reduction of *p*-nitrophenol into *p*-aminophenol.

The stabilization of metallic NPs by chemically modified CDs, such as thiolated or methylated CDs (Figure 2.12), has also been investigated.

First, thio-functionalized CDs were synthesized and efficiently used as protective agents of Au(0), Pt(0) and Pd(0) NPs without passivating their surface [69]. The metallic nanospecies were prepared by reduction of the metallic sodium salts with BH_4^- in the presence of per-6-thio-β-CD (HS-β-CD) in a mixture of DMSO-H_2O (1/4, v/v) [70]. The FTIR spectrum revealed that the CD was chemisorbed on the NPs surface as the weak S–H stretching IR band, which was clearly observed in the spectrum of free CD, disappeared in the NP spectrum. Moreover, nuclear magnetic resonance (^1H NMR) spectroscopy also provides experimental evidence of the presence of HS-β-CD, as the proton resonances from the CD receptors appear as broad

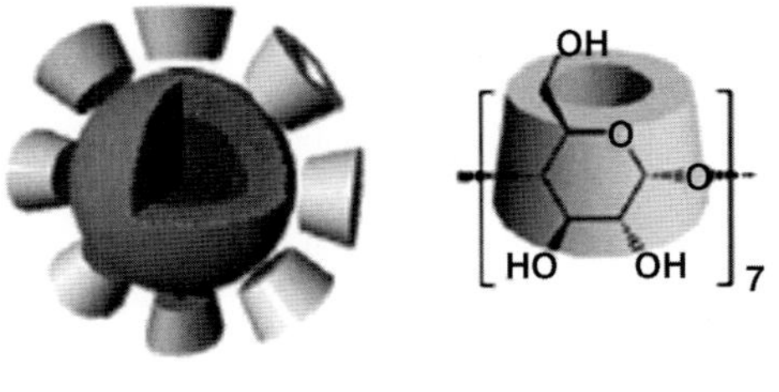

Figure 2.10 Schematic illustration of FePt NP and γ-CD. The core is Fe rich (dark gray) and the shell is Pt rich (light gray). (Adapted from ref. [68]).

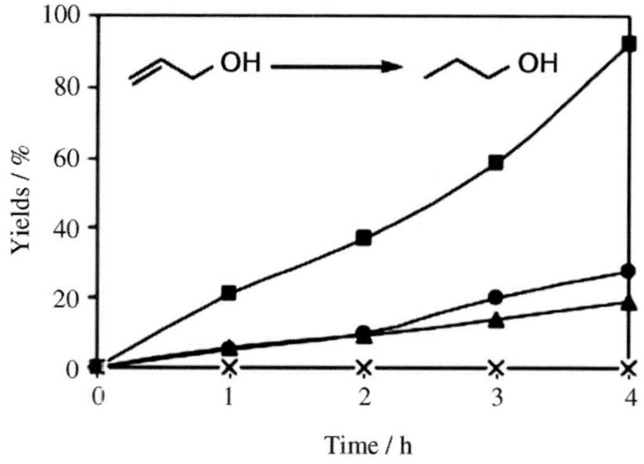

Figure 2.11 Hydrogenation of allylic alcohol using γ-CDs-capped $Fe_{core}Pt_{shell}$ NPs in water (■), $Fe_{core}Pt_{shell}$ NPs protected by oleic acid/oleylamine in water (•), $Fe_{core}Pt_{shell}$ NPs capped by oleic acid/oleylamine in THF (▲) and γ-CDs-stabilized $Fe_{shell}Pt_{core}$ NPs in water (×). (Reproduced with permission from [68]. Copyright © (2009) Royal Society of Chemistry.)

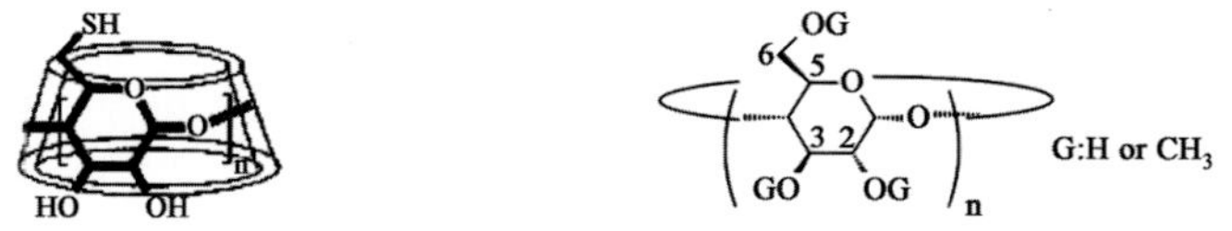

Perthiolated CDs **Methylated CDs**

n	Abbreviation	Abbreviation	n	Carbon bearing the OCH_3group	Average number of OH group substituted per glucopyranose unit
6	SH-α-CD	RAME-α-CD	6	2, 3, 6	1.8
7	SH-β-CD	RAME-β-CD(1.8)	7	2, 3, 6	1.8
		RAME-β-CD(0.7)	7	2	0.7
8	SH-γ-CD	RAME-γ-CD	8	2, 3, 6	1.8

Figure 2.12 CDs used as protective agents of metallic nanospecies (Schemes of CDs).

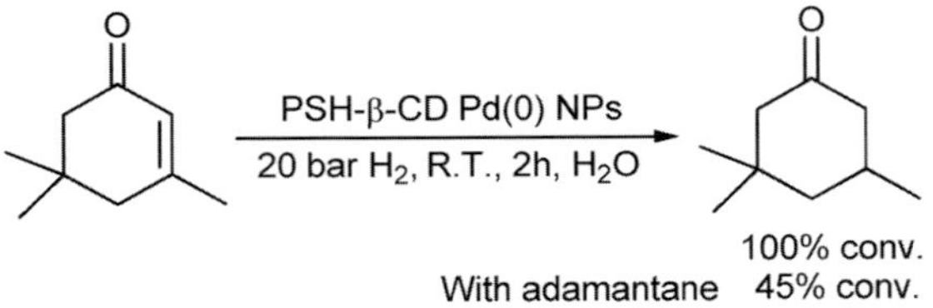

Figure 2.13 Catalytic C=C hydrogenation of isophorone with PSH-β-CD Pd(0) NPs in aqueous media in the presence or not of adamantane. (Adapted from [74])

peaks, a result of their relative proximity to the metal particles, as already reported [71, 72]. The Pt(0) or Pd(0) NPs obtained, respectively with an average diameter of 14.1 ± 2.2 and 15.6 ± 1.3 nm, behaved as active and recoverable catalysts for the hydrogenation of carbon–carbon [73] or carbon–nitrogen [74] double bonds in aqueous media. Moreover, the authors demonstrated that the catalytic activity of the NPs could be modulated by host-guest interactions between the surface immobilized CDs and properly chosen hosts. In the same way, Liu and *et al.* used perthiolated β-CD immobilized on Pd(0) NPs as efficient phase-transfer catalysts in aqueous biphasic reactions of various hydrophobic compounds such as selective C=C double bond hydrogenation of unsaturated carbonyl compounds and C=N double bond hydrogenation of imines [74]. In these reactions, a synergistic effect between CDs (a phase transfer catalyst) and Pd(0) NPs (hydrogenation catalyst) on the overall catalytic process was demonstrated, using adamantane as a competitive substrate in the hydrogenation of isophorone (Figure 2.13). Indeed, isophorone was completely hydrogenated within 2h with PSH-β-CD-Pd(0) NPs in water, whereas the TOF significantly decreased as adamantane competed with the substrate to the available CD binding sites on Pd(0) NPs.

The low affinity of CD-capped NPs for hydrophobic substrates could also be solved by using randomly methylated CDs (RAME-CDs), which improve the affinity of CD-protected NPs for hydrophobic substrates and help the mass transfer between the organic and the catalyst-containing phase [75]. Various methylated CDs, which are different in terms of the size cavity and the substitution degree, were efficiently used to stabilize aqueous suspensions of metallic Ru(0) NPs with an average particle size around 2.5 nm (Figure 2.14) by chemical reduction of ruthenium chloride. In that case, the RAME-CDs are weakly adsorbed on the metal surface and their mass transfer ability remains high.

The nanocatalysts obtained in this way proved to be very active in the hydrogenation of various olefins, and particularly α- or β-pinene, leading to the major formation of industrially interesting *cis*-pinane with high diastereoisomeric excesses [76]. The observed selectivities were higher than those with ammonium-protected Ru(0) catalysts, thus showing the beneficial effect of the CDs cavity on the diastereoselectivity. The nanoheterogeneous systems capped with various RAME-CDs also showed efficient activities in the hydrogenation of aromatic compounds under biphasic conditions at room temperature and atmospheric hydrogen pressure. Moreover, interesting chemoselectivities were observed in the hydrogenation of various

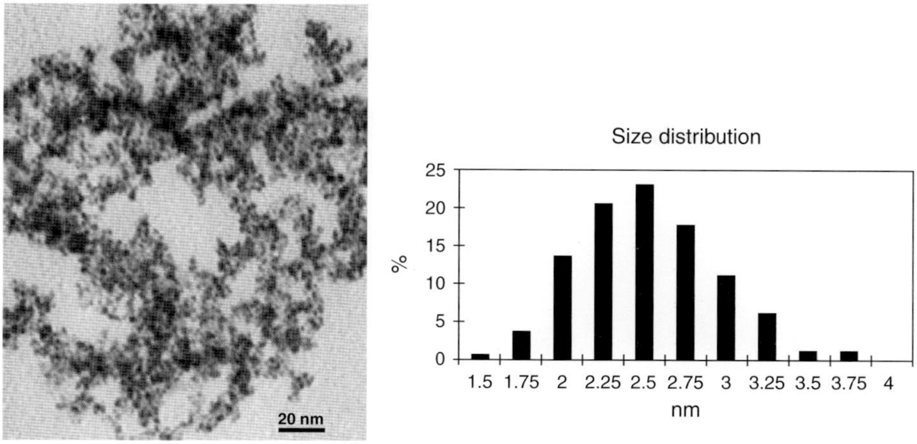

Figure 2.14 TEM micrograph and size distribution of RAME-β-CD-stabilized Ru(0) NPs. (Reproduced with permission from [76]. Copyright © (2007) Royal Society of Chemistry.)

substituted arene derivatives since the hydrogenation could be easily controlled by the relevant choice of cavity and substitution degree (SD) of the CD (Table 2.6).

The authors proposed a mechanism based on the dual role of the CDs as protective agent and supramolecular shuttle (Figure 2.15). First, the CD acts as a steric stabilizer through hydrophobic-hydrophobic interactions between Ru(0) NPs and CDs [77] or interactions between Ru(0) NPs and hydroxyl groups of native CDs as previously reported for gold NPs [78]. Secondly, the CDs play a role of mass transfer agent between the organic phase and the surface of the water-soluble NPs by forming inclusion complexes with the substrate.

Since several ammonium salts are known to strongly interact with β-CD derivatives [79], the catalytic hydrogenation of aromatic compounds was also carried out

Table 2.6 Hydrogenation of monosubstituted arene compounds with Me-CD Ru(0) NPs[a]. (Adapted from ref. [75]).

Substrate	Cyclodextrin (SD)	Product (Conv. %)[b]	t (h)	TOF (h^{-1})[c]
Styrene	RAME-α-CD	Ethylbenzene (100)	10	10
Styrene	RAME-β-CD (0.7)	Ethylcyclohexane (100)	11	9
Styrene	RAME-β-CD (1.8)	Ethylbenzene (100)	11	9
Styrene	RAME-γ-CD	Ethylcyclohexane (100)	24	4
Propylbenzene	RAME-β-CD (0.7)	Propylcyclohexane (100)	11	9
Propylbenzene	RAME-β-CD (1.8)	Propylbenzene	—	—
Ethylbenzoate	RAME-β-CD (0.7)	Ethylcyclohexylate (65)	24	3
Ethylbenzoate	RAME-β-CD (1.8)	Ethylbenzoate	—	—

a) Hydrogenation conditions: catalyst (1.4×10^{-5} mol), Me-CD (1.4×10^{-4} mol), substrate (1.4×10^{-3} mol), 1 bar H_2, r.t., stirred at 1500 min^{-1}, 10 mL water.
b) Determined by GC analysis.
c) TOF defined as number of mol of substrate per mol of ruthenium per hour.

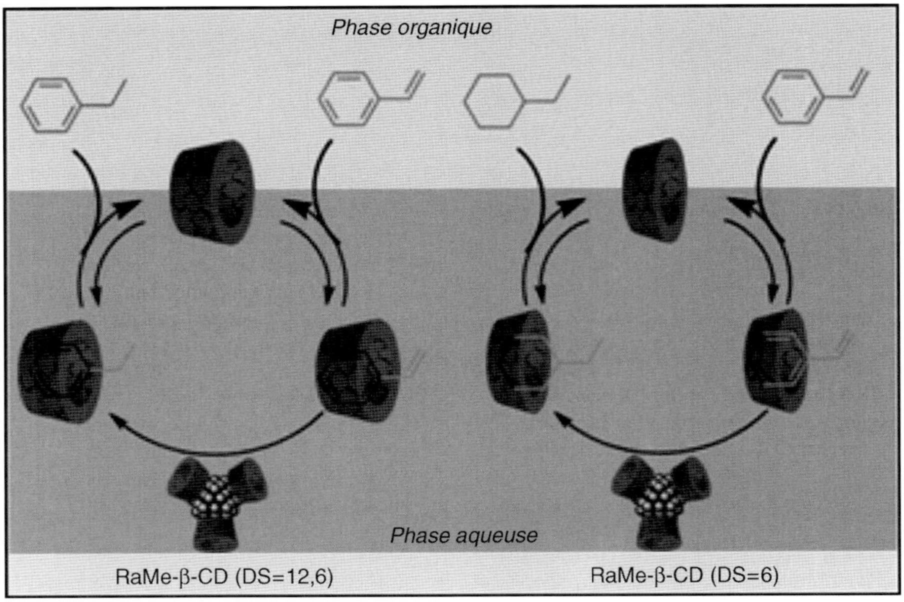

Figure 2.15 RAME-CD: Protective agents and mass transfer additives in NPs-catalyzed processes.

in neat water with Ru(0) NPs protected by 1:1 inclusion complexes formed between the chloride salt of *N,N*-dimethyl-*N*-hexadecyl-*N*-(2-hydroxyethyl)-ammonium (HEA16Cl) and RAME-β-CD [80]. The formation of inclusion complexes between HEA16Cl and RAME-β-CD in aqueous media was checked by NMR spectroscopic studies, showing strong dipolar contacts between the alkyl chain protons of the ammonium salt and the protons located inside the CD cavity. These monodispersed particles with an average size of 4 nm, organized in dendritic structures as observed by TEM, proved to be more active than the NPs stabilized by classical surfactants in arenes hydrogenation (Table 2.7).

Table 2.7 Catalytic hydrogenation of various benzene derivatives with Ru(0) NPs[a]. (Adapted from ref. [80]).

Protective agent	Substrate	t (h)	Product (%)[b]	TOF (h^{-1})[c]
HEA16Cl	Anisole	24	Methoxycyclohexane (27)	3.4
RAME-β-CD/HEA16Cl	Anisole	24	Methoxycyclohexane (82)	10.2
HEA16Cl	Toluene	24	Methylcyclohexane (18)	2.2
RAME-β-CD/HEA16Cl	Toluene	24	Methylcyclohexane (81)	10.1

a) Reaction conditions: Ru(0) (3.8×10^{-5} mol), HEA16Cl (7.6×10^{-5} mol) or HEA16Cl (3.8×10^{-5} mol) + RAME-β-CD (3.8×10^{-5} mol), substrate (mol/mol) = 100, 1 bar H_2, 293 K, 10 mL water, stirred at 1500 rpm.
b) Determined by GC analysis.
c) Based on hydrogen consumption.

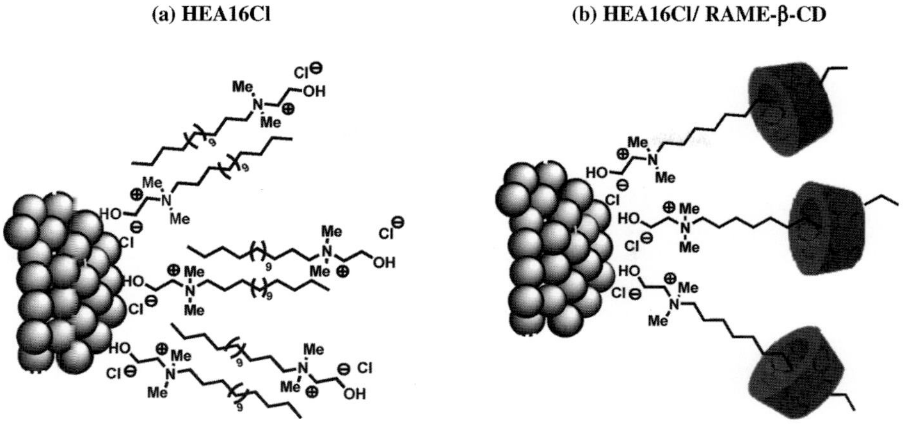

Figure 2.16 Dynamic organizations of the capping agents around the NPs.

The dynamic organization of the protective agents around the NPs could explain the differences observed in the catalytic activity. In the case of the ammonium salts, as already described by Chen [81] and El-Sayed [53], the surfactant self-organizes around the particles forming a double layer: the internal layer is linked to the particle surface via the charged groups and is connected to the external layer via hydrophobic interactions (Figure 2.16a). In opposition, for the RAME-β-CD/HEA16Cl inclusion complex, the proposed schematic model shows that the polar head of the surfactant is directed towards the metallic surface and that the methylated CD interacts with the nonpolar alkyl chain of the surfactant (Figure 2.16b).

The stabilization by the RAME-β-CD/HEA16Cl inclusion complex could greatly improve the mass transfer between the metallic surface and the aqueous phase and also the diffusion of the substrate towards the metallic surface.

2.3.2.2 Carbon–Carbon Coupling Reactions

Cyclodextrin-capped palladium(0) NPs have proved to be efficient catalysts for carbon–carbon coupling reactions such as the Heck, Suzuki or Sonogashira reactions in neat water. The main approaches are described below.

First, various native CDs were used as efficient protective agents of Pd(0) clusters [82] prepared by alcoholic reduction of $PdCl_2$. No significant difference in the clusters morphologies was noticed according to the nature of the CD. These CD@Pd(0) nanoclusters were found to be active toward the Suzuki–Miyaura cross-coupling reaction between *o*-, *m*-, *p*-iodophenols and phenylboronic acid at room temperature in neat water with moderate to excellent yields (Table 2.8). It appeared that the catalytic activities depend on the CD molecules present on the particles surface.

Perthiolated CDs have been reported as efficient protective agents of Pd(0) NPs for the Sonagashira reaction in water at room temperature [83]. The Pd(0) nanospecies with an average size of 3.2 ± 0.8 nm (Figure 2.17) are highly soluble in

Table 2.8 Cross-coupling reaction of iodophenol and phenylboronic acid in H_2O catalyzed by Pd(0)@CD clusters. (Adapted from ref. [82]).

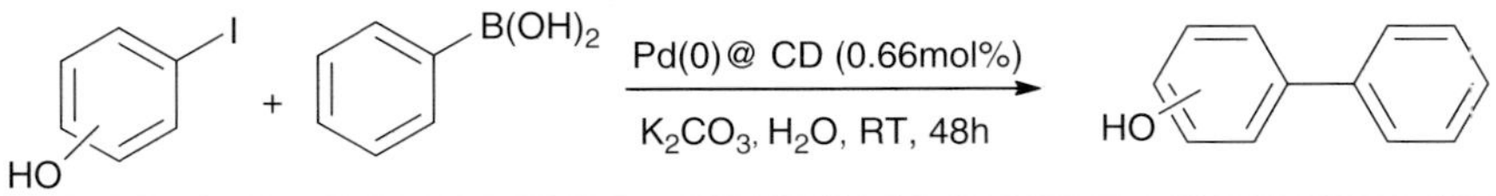

CD	Yield (%)		
	ortho	meta	para
α-CD	83	87	85
β-CD	75	90	77
γ-CD	64	89	84

water and the modification of the particle surface by the capping agents was checked by NMR studies. As determined by elemental analysis and TEM measurement, the β-CDs cover 50% of the total surface of each particle.

Interestingly, these nanospecies could be easily isolated and dissolved again in water without losing their stability. They efficiently catalyzed the Sonogashira reaction of various aryl iodides and terminal alkynes in water at room temperature even in the absence of copper co-catalyst and phosphine ligands to afford the products with excellent isolated yields (Table 2.9). The good yields obtained with hydrophobic compounds is attributed to the presence of β-CDs on Pd(0) NPs, which act not only as a protective agent to avoid aggregation but also as a supramolecular shuttle for small nonpolar organic substrates in aqueous media [84]. In fact, the inclusion of hydrophobic substrates in the CD cavity was demonstrated using adamantane (Entry 4) as an inert competing substrate in the Sonogashira reaction, leading to a great decrease in catalytic activity.

More recently, Malta *et al.* described a new approach for the synthesis of Pd (0) NPs in water from a renewable source, 2-hydroxypropyl-α-cyclodextrin (α-HPCD), which acts here both as a reductive and capping agent [85]. The nanospecies obtained in this way were well-characterized by DLS, TEM and X-ray photoelectron spectroscopy (XPS), which showed the formation of spherical

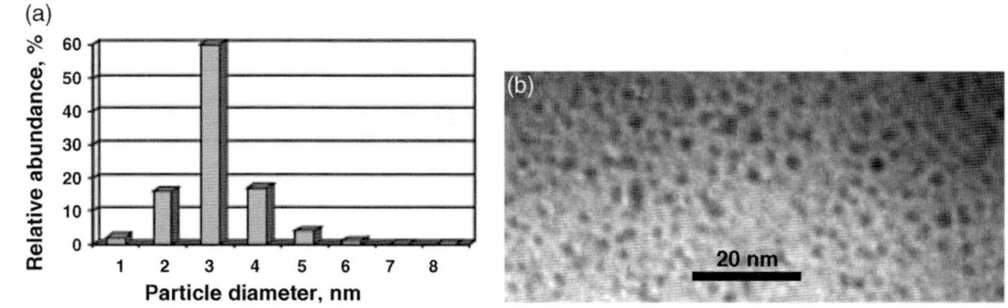

Figure 2.17 Size distribution and TEM micrograph of β-capped Pd NPs. (Adapted from ref. [83]).

Table 2.9 Sonogashira reaction catalyzed by β-CD-capped Pd NPs in water[a]. (Adapted from ref. [83]).

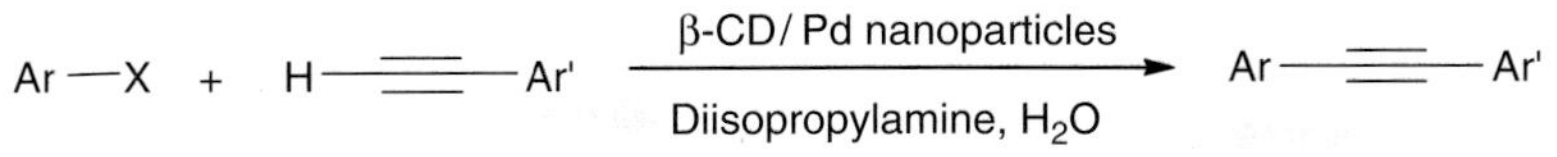

Ar'	Ar	X	Product	Isolated Yield (%)
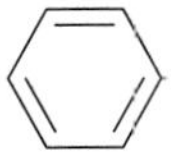	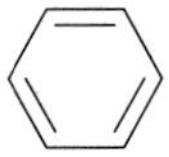	I	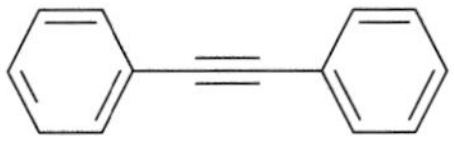	89
		I	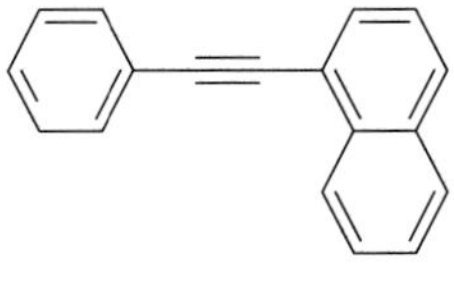	92
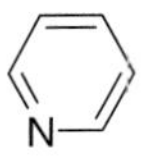	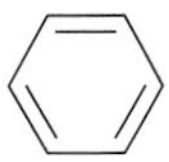	I	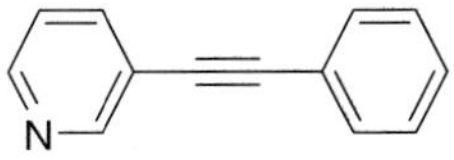	88
	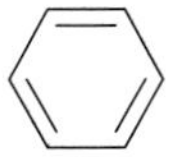	I		18[b]

a) Conditions: aryl halide (2 mmol), terminal alkyne (2.5 mmol), 10 mol% β-CD/Pd, 20 h, r.t.
b) In the presence of adamantane.

particles in the size range 2–7 nm. FTIR and ^1H NMR spectroscopy did not reveal the presence of covalent bonds between CDs and Pd(0) NPs, suggesting that the CD is only physically adsorbed on the NP surface through hydrophobic interactions. These α-HPCD capped Pd(0) NPs were evaluated in Suzuki, Heck and Sonogashira reactions in neat water, providing good yields and selectivities under low Pd loadings (0.5–0.01 mol%). The results obtained in the Heck reactions between various halogenoarenes and olefins in aqueous conditions are reported in Table 2.10.

The Sonogashira reactions between aryl halides and terminal acetylenes gave moderate to good yields under mild conditions in neat water without any copper co-catalyst (Table 2.11).

The Suzuki–Miyaura cross-coupling of aryl- or heteroarylboronic acids and halogenoarenes also proceeded with low Pd loadings of 0.01 mol%, led to good yields in the respective biaryl products (Table 2.12).

Remarkably, the nanocatalysts showed significant stability since the aqueous phase remained active for four subsequent runs. Thus, effective hydrophobic interactions [86] between NPs and the α-HPCD primary face provide the colloidal stability. This supramolecular arrangement could explain the high performances of this

Table 2.10 Aqueous Heck coupling using CD-capped Pd(0) NPs[a]. (Adapted from ref. [85]).

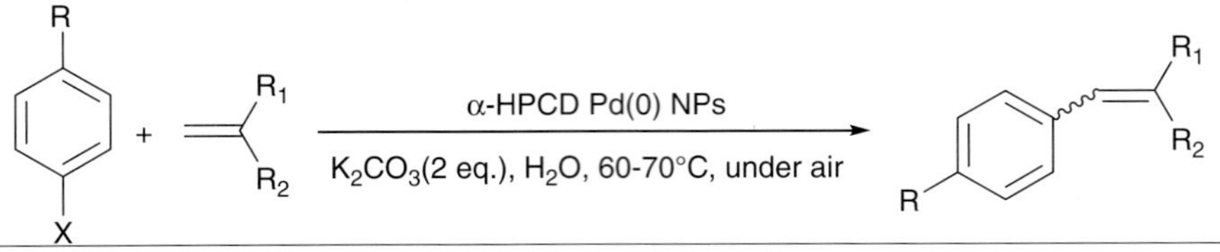

Ar-X	R_1	R_2	Pd load (mol%)	Isolated Yield (%)[b]	TON
C_6H_5I	Ph	H	0.05	100 (100/0)	2000
C_6H_5Br	Ph	H	0.05	81(100/0)	1620[c]
C_6H_5Cl	Ph	H	0.05	21 (100/0)	420[c]
C_6H_5Br	Ph	Me	0.05	78(91/9)	1560[c]
C_6H_5Br	CO_2Bu	Me	0.05	71(88/12)	1420[c]
4-$CO_2MeC_6H_4Br$	CO_2Me	H	0.05	63(100/0)	1260

a) Reactions were performed on a 1 mmol-scale in neat water (5 mL) under air for 24 h.
b) Determined by GC-MS through the corrected normalization of the peak areas (E/Z ratio is shown in parenthesis).
c) At 353 K.

nanocatalyst, as some degree of recognition between the capping CD hosts and the substrates would be favored.

In summary, CDs proved to be efficient as protective agents for nanospecies as well as mass transfer agents in aqueous-organic biphasic systems. Several relevant results were obtained in various catalytic reactions including reduction or cross-coupling reactions according to the pertinent choice of the CD. However, a synergistic effect between CDs (phase transfer promoters) and NPs (catalysts) on the overall catalytic process could be achieved. Finally, the field of metallic NPs in aqueous catalysis is in permanent development and undoubtedly the CDs will find a key place in this race thanks to their modularity. Moreover, modified CDs are

Table 2.11 Aqueous Sonogashira coupling using CD-capped Pd(0) NPs[a]. (Adapted from ref. [85]).

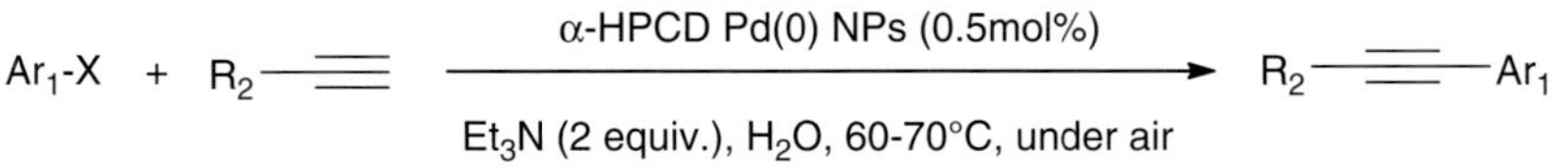

Ar_1-X	R_2	Yield (%)[b]	TON
C_6H_5I	Ph	93	186
C_6H_5Br	Ph	78	156
C_6H_5I	C_6H_{13}	85	170
$O_2NC_6H_4I$	Ph	82	176
$MeOC_6H_4I$	Ph	88	176

a) Reactions were performed on a 1 mmol-scale in neat water (5 mL) under air for 24 h.
b) Determined by GC-MS through the corrected normalization of the peak areas.

Table 2.12 Aqueous Suzuki–Miyaura coupling using CD-capped Pd(0) NPs[a]. (Adapted from ref. [85]).

$$\text{Ar}_1\text{-X} + \text{Ar—B(OH)}_2 \xrightarrow[\text{K}_2\text{CO}_3\text{ (2 eq.), H}_2\text{O, 60°C, under air}]{\alpha\text{-HPCD Pd(0) NPs (0.01mol\%)}} \text{Ar}_2\text{—Ar}_1$$

Ar_1-X	**Ar_2**	**Yield (%)[b]**	**TON**
C_6H_5I	Ph	100	10 000
C_6H_5I	2-thiophenyl	89	8900
C_6H_5Br	Ph	82	8200
C_6H_5Br	2-furyl	79	7900

a) Reactions were performed on a 1 mmol-scale in neat water (5 mL) under air for 24 h.
b) Determined by GC-MS through the corrected normalization of the peak areas.

particularly promising in enantioselective applications which still remain a challenge in nanocatalysis.

2.3.3
Steric Stabilization by Polymers and Derivatives

Organic polymers and unusual derivatives such as oligomers or dendrimers are often used for the stabilization of metallic NPs [87, 88]. The adsorption of these molecules at the particle surface provides a protective layer and in the interparticular space, the sterical environment of the adsorbed macromolecules reduces their mobility. The result is an osmotic repulsion to restore the equilibrium by diluting the macromolecules and thus separating the particles. Hirai and Toshima have published many papers on the synthesis of transition metallic NPs through alcoholic reduction of metal salts in the presence of usual polymers like polyvinyl alcohol (PVA), poly(methyl vinyl ether) (PMVE) or poly(*N*-vinyl-2-pyrrolidone) (PVP) in methanol (or ethanol)/water mixtures or in pure alcohols. Consequently, in most cases, the catalytic reactions were performed in alcoholic mixtures. In the context of 'green chemistry,' the use of water as solvent is a subject of considerable interest and several strategies based on water-soluble polymers were explored to obtain soluble NPs for catalysis in pure water. Typical examples of various catalytic applications such as hydrogenation, C—C coupling and oxidation are reported thereafter.

Catalytic investigations of poly(*N*-vinyl-2-pyrrolidone) (PVP) embedded rhodium NPs in aqueous hydrogenation were performed by Choukroun *et al.* [89] Water-soluble [Rh-PVP] NPs with irregular sizes in the range of 2–3 nm were obtained by reaction of $[RhCl(C_2H_4)_2]_2$ with two equivalents of Cp_2V as reducing agent and in the presence of K30-PVP and explored in the hydrogenation of benzene in water. The complete benzene conversion with a substrate/catalyst ratio of 2000 was observed under 7 bar H_2 and at 303 K after 8 h, giving rise to a TOF of 675 h^{-1}

(related to the consumed H_2). No influence of the amount of water on the catalytic activity was observed.

More recently, PVP was also used to prepare size-controlled gold NPs under microwave in aqueous solutions. The obtained colloids with a size in the range of 6.7–22 nm were controlled by the molecular weight of PVP molecules and the PVP to Au(III) molar ratio. The PVP-Au NPs were stable for more than three months in ambient conditions and exhibited excellent catalytic activity for the reduction of nitroarenes by $NaBH_4$ in water [90]. Gold NPs (<2 nm) protected by poly(*N*-vinyl-2-pyrrolidone) but prepared by reduction of $AuCl_4^-$ with $NaBH_4$ and PVP were also described by Tsukuda [91, 92] as catalysts toward homocoupling of phenylboronic acid (Figure 2.18) and benzylic alcohols oxidation in water under aerobic conditions.

In the presence of PVP, Yan *et al.* prepared water-soluble ruthenium(0) nanoclusters with a diameter of 2.0 nm by reduction of ruthenium trichloride hydrate under 2.0 MPa H_2 for the hydrogenation of carbon monoxide in water to produce hydrocarbons. The Fischer–Tropsch synthesis in pure liquid water with a TOF of $12.9\,h^{-1}$ was realized over the nanocatalyst at 423 K under 2.0 MPa H_2 and 1.0 MPa CO [93].

Catalytically active Pt(0) NPs stabilized by an ammonium polymer were also reported in water by Lahiri [94]. The use of poly(diallyldimethylammonium chloride) as stabilizer was investigated in several hydrogenation reactions including >C=C<, >C=O and nitro groups in $ArNO_2$ and in various media such as water, methanol and a water/toluene (1/1) mixture. Two platinum precursors such as the carbonyl cluster $[Pt_{30}(CO)_{60}]^{2-}$ and $[PtCl_6]^{2-}$ were compared in the reduction of several substrates (Figure 2.19).

TEM observations showed that the NPs derived from the $[Pt_{30}(CO)_{60}]^{2-}$ cluster are much smaller than those obtained from $[PtCl_6]^{2-}$. Interestingly, the reactivity of the various tested substrates was found to be higher in water than in methanol or in a water/toluene mixture. These catalysts were also investigated in the oxidation of alcohols in water by the same team [95]. The performances in the oxidation of benzyl alcohol and 1-phenylethanol were compared with those of classical heterogeneous platinum catalysts. Pt(0) NPs derived from the cluster were found to be considerably more active than other reported water-soluble Pt(0) nanocatalysts or heterogeneous catalysts.

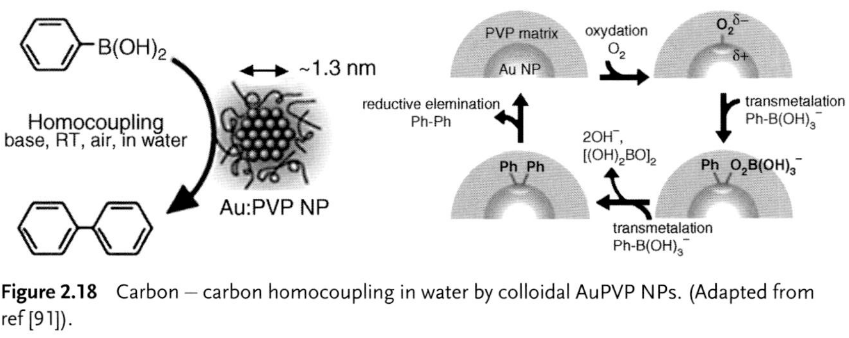

Figure 2.18 Carbon – carbon homocoupling in water by colloidal AuPVP NPs. (Adapted from ref [91]).

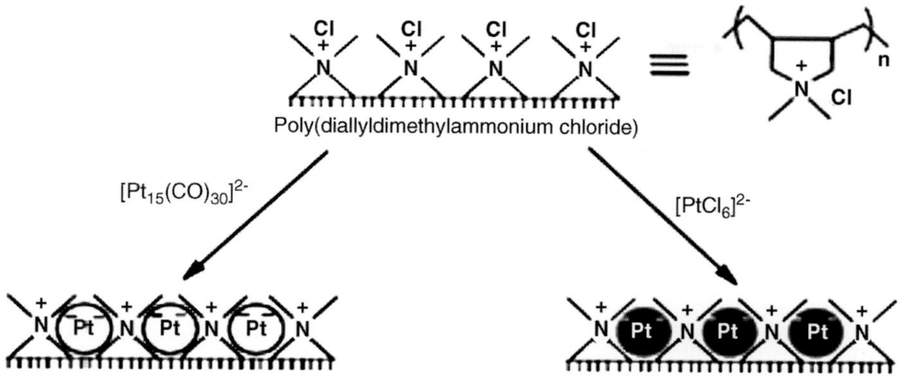

Figure 2.19 Pt NPs stabilized by the poly(diallyldimethylammonium chloride) for catalytic applications in water. (Adapted from ref. [94]).

The use of alkylated polyethylenimine (PEI) as protective agent of palladium(0) NPs was proposed by Neumann. The efficient dispersion of the so-obtained Pd(0) NPs in water allowed the aqueous biphasic hydrogenation of hydrophobic alkenes (Figure 2.20). Various primary olefins (1-hexene, 1-heptene, 1-octene, 1-decene, styrene) and cyclic alkenes (cyclohexene, cyclooctene, cyclododecene) gave very high yields up to 99% of the corresponding alkanes in mild conditions (5 h, 353 K) [96].

Another interesting approach was described by Zhang and coworkers with the synthesis of polymeric microreactors for the hydrogenation of olefins under organic/aqueous biphasic conditions (Figure 2.21) [97].

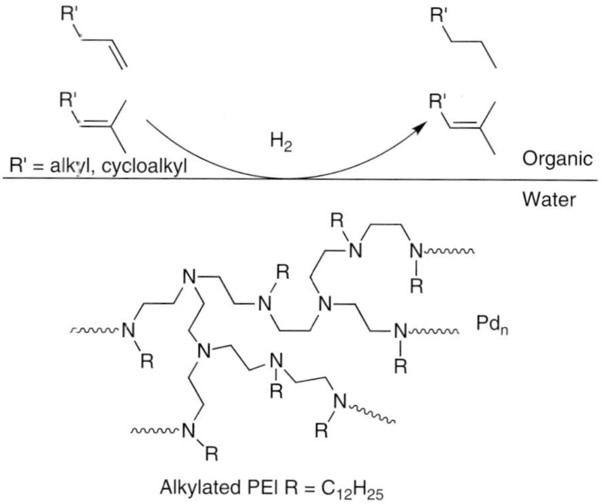

Figure 2.20 Pd(0) NPs protected by alkylated polyethyleneimine as aqueous biphasic catalysts. (Reproduced with permission from [96]. Copyright © (2006) American Chemical Society.)

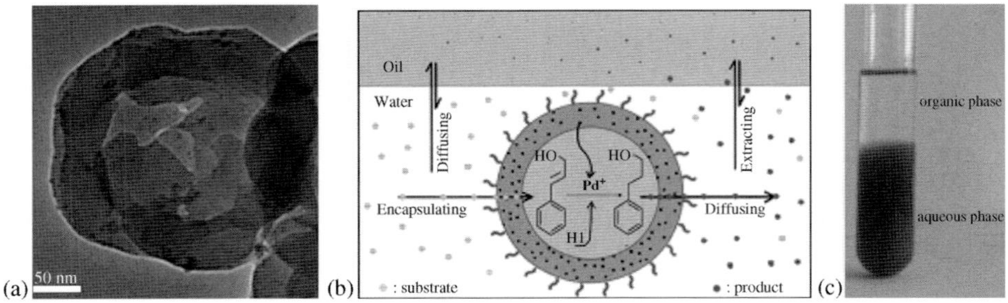

Figure 2.21 (a) TEM image of the microreactors PS-*co*-PAEMA-*co*-PAM (b) Hydrogenat on in water within the microreactor under organic/aqueous biphasic conditions (c) Biphasic catalytic system. (Adapted from ref. [97]).

The microreactors are composed of three parts: (1) the outer corona of polyacrylamide (PAM), which is hydrophilic and keeps the microreactors suspended in aqueous phase; (2) the cross-linked and hydrophobic wall of poly[styrene-co-2-(acetoacetoxy)-ethylmethacrylate] (PS-co-PAEMA), which maintains the body of the hollow microsphere; and (3) 3.9 nm Pd(0) NPs, which are immobilized in the wall of the hollow microspheres (Figure 2.22). The hydrogenation of typical aliphatic, aromatic and alicyclic olefins within the microreactors under organic/aqueous biphasic conditions at 308 K was investigated, showing excellent catalytic activities (Table 2.13).

Adapted microreactors poly[styrene-*co*-2-(acetoacetoxy)ethyl methacrylate-*co*-methyl acrylic acid] (PS-*co*-PAEMA-*co*-PMAA) which contained a pH-responsive shell of PMAA segment were also described by Zhang [98]. This colloid supported palladium catalyst was synthesized by loading 3-nm Pd NPs into the pH-responsive colloid of core–shell microspheres and proved to be efficient for the Suzuki and Heck reactions performed in water (Figure 2.22). Pertinently, the palladium catalyst dispersed in aqueous medium could be simply separated and recovered by adjusting the pH of the aqueous medium.

The Suzuki–Miyaura cross-coupling reaction of aryl halides with arylboronic acids in water was also reported by Guo with poly(2-aminothiophenol)-stabilized gold NPs [99]. The stable nanocatalysts with sizes around 1.0 nm and a narrow size distribution were synthesized by a chemical redox reaction between $HAuCl_4$ and the PATP polymer. The cross-coupling of phenylboronic acid and aryl chlorides, more economical and less reactive than bromides or

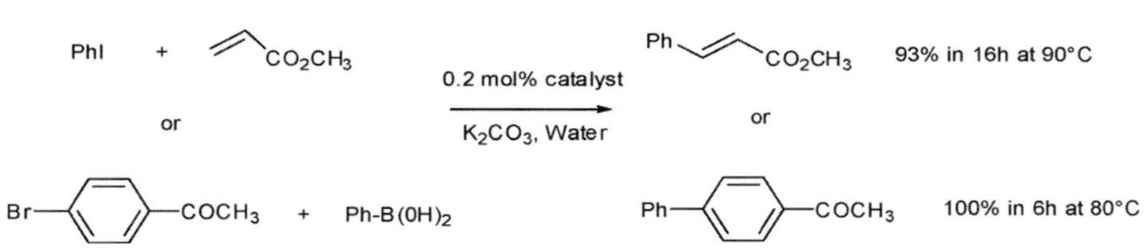

Figure 2.22 Heck and Suzuki coupling reactions catalyzed by pH-responsive colloid supported palladium catalysts.

Table 2.13 Olefin hydrogenation within the microreactor under organic/aqueous biphasic conditions[a)].

Substrate	Conditions Time [h]	Yield [%]	TOF [h^{-1}]
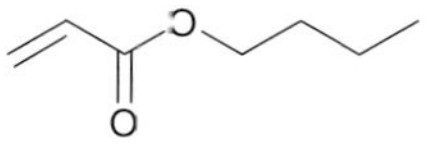	1	100	2150
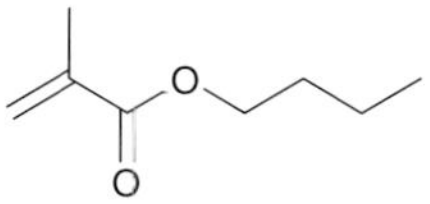	3	100	93
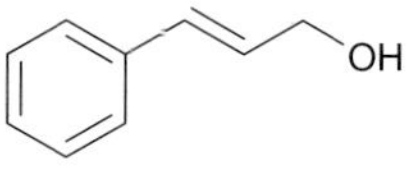	14	99.5	219
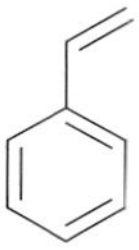	2	100	1449
	2	100	1136

a) Reaction conditions: aqueous dispersion of the microreactor (20 mL) containing the Pd catalyst (1.67×10^{-5} mol); substrate (3.33×10^{-2} mol); 308 K; 1 atm H_2. S/C = 2000

iodides derivatives, was studied at 353 K in water (Figure 2.23). The size of the NPs and the amount of PATP stabilizer were found crucial for determining the catalytic activity of gold NPs. Pleixats and coworkers recently investigated the Suzuki cross-coupling with a new PEG-tagged material as an efficient stabilizer for water-soluble palladium NPs. The nanomaterial, obtained by an organometallic approach, proved to be active in the coupling of a range of aryl and heteroaryl substrates and the recyclability of the nanocatalyst for aryl iodides and bromides was also established [100].

Due to their embedding effect, dendrimers, like other polymers, are monodispersed macromolecules and constitute an attractive stabilization mode for the

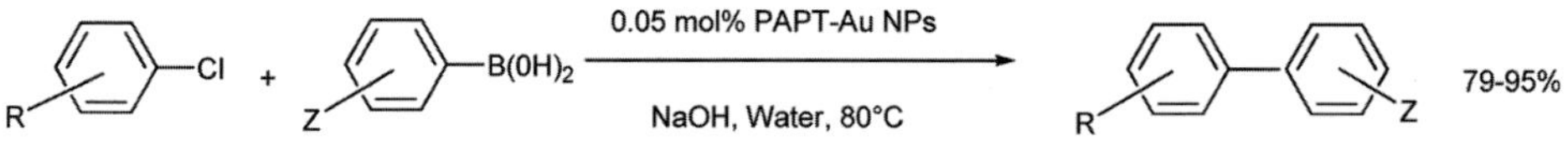

R= H, p-CO_2H, p-CHO, p-OCH_3, p-OCH_3

Z = H, m-NO_2, m-OCH_3, p-OCH_3

Figure 2.23 Suzuki–Miyaura coupling of aryl halides and arylboronic acids with gold NPs.

synthesis of metallic NPs [88, 101–103]. Dendrimers can act as template and protective agent thanks to functionalized cavities, which can entrap and stabilize metallic nanospecies. The formation of monometallic NPs inside dendrimers could be performed by two methods: the direct reduction of dendrimer-encapsulated metal ions or the displacement of less-noble metal clusters with more noble elements [104]. Metallic NPs, which are introduced inside or at the periphery of the dendrimer, were originally used as nanoreactors by the groups of Crooks, Tomalia and Esumi. The team of Crooks is involved in the synthesis and the use of poly(amidoamine) (PAMAM) and more particularly hydroxyl-terminated derivatives for the preparation of spherical dendrimer-encapsulated Pd and Pt NPs with a size around 1.4 nm [105]. The hydrogenation activities of the obtained colloids were evaluated in water for a simple olefin (allyl alcohol) and an electron-deficient alkene (*N*-isopropyl acrylamide). Dendrimer-stabilized Pd NPs showed higher catalytic activities than Pt for olefin hydrogenation in aqueous solution with TOF up to 372 and 218 h^{-1} for *N*-isopropyl acrylamide and allyl alcohol respectively.

Recently, Astruc and Bouquillon [106] described the use of a water-soluble glycodendrimer containing nine terminal modified xylose branches as protective agent of Pd and Pt NPs for the catalysis (Figure 2.24).

The TEM investigations showed that the NPs have a diameter of 4.7 ± 0.4 nm for Pd and of 14.0 ± 3.0 nm for Pt. Preliminary catalytic hydrogenation studies in water using these NPs, without any co-solvent, were performed with isophorone and (*R*)-(+)-pulegone as prochiral substrates under atmospheric hydrogen pressure and at room temperature. For a molar ratio of 100, only 86 and 95% of conversion determined by GLC were obtained in 26 and 50 h respectively. Moreover, no stereoselectivity was observed even in the presence of the chiral sugar. The same group also reported water-soluble 1,2,3-triazolyl sulfonate dendrimers-stabilized Pd NPs for olefin hydrogenation in water and Suzuki coupling reaction in EtOH/H_2O mixture [107]. Several generations of 1,2,3-triazolylsulphonate dendrimers (G0, G1 and

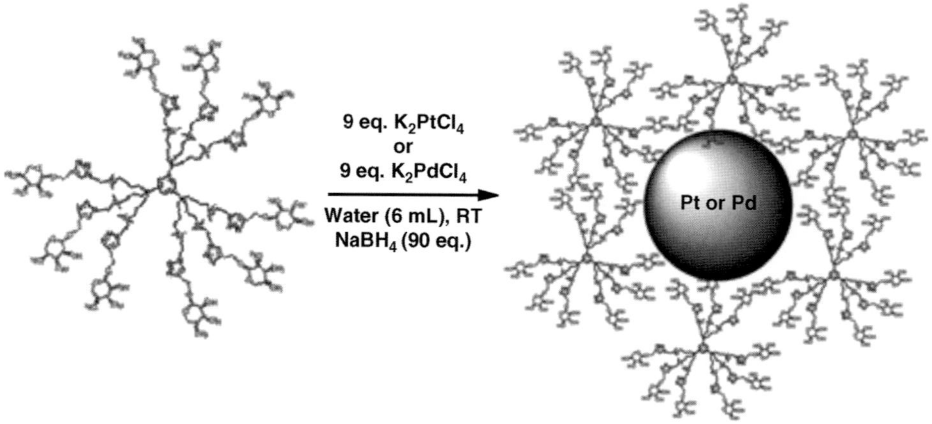

Figure 2.24 Formation and stabilization of Pd and Pt NPs in water. (Adapted from ref. [106])

G2) were used to stabilize Pd NPs with sizes of 2.3, 2.8 and 3.0 nm respectively. Dendrimer-protected Pd NPs with a size around 2.3 nm (G0 generation) were successfully used in the hydrogenation of allylic alcohol under ambient conditions (298 K, 1 atm H_2) in water providing a TOF of 8088 mol H_2/mol Pd/h. The most efficient catalyst is the G0 generation according to the size of the Pd NPs.

In both cases, usual polymers or unusual derivatives as dendrimers provide an efficient stabilization of various metallic NPs with small sizes, avoiding aggregates through steric effects. Whatever the water-soluble protective agents, the obtained nanocatalysts proved to be efficient active species in many catalytic applications. Undoubtedly, the development of original water-soluble polymers as protective agent for nanospecies will be an active research area in the future and the easy synthesis of these compounds will remain a priority.

2.3.4
Steric Stabilization by Ligands

The use of ligands derived from homogeneous catalysis as protective agents for metallic NPs stabilization is more and more investigated given their potentialities to tune the particles surface state through their chemical influence. Based on traditional organosoluble catalysts, many works described in organic solvents generally concern usual organic ligands as protective agents of NPs such as thiols, amines, carboxylic acids or phosphines. In some cases, the particle transfer from an organic medium to water was demonstrated through the use of a phase transfer agent [108–110]. With the aim of developing nanocatalysis in neat water without post-adding a phase transfer agent, the essential difficulties rely on finding water-soluble ligands giving rise to sufficiently stable but also active species. At this date, only a few articles have described the direct preparation of ligand-capped NPs for catalytic applications in water.

Among the water-soluble ligands recently used, stabilizers containing P atoms such as phosphines are the most studied. For example, W. Kläui [36] observed that the introduction of a phosphonate substituent (PO_3^{2-}) significantly increases the solubility of the phosphane ligand $Na_2[Ph_2PCH_2CH_2PO_3]$. The corresponding water-soluble rhodium(I) complex could be reduced by hydrogen providing rhodium nanoclusters with a narrow size distribution in the range of 2–6 nm, depending on the reaction conditions (Figure 2.25). In this approach, isolated and redispersible rhodium clusters in water without coagulation thanks to the electrostatic repulsion of the covering double negative charge of the phosphonate ligands were evaluated in the hydrogenation of hex-1-ene and benzene at 293 K under 20 bar H_2. The results show that small clusters exhibited the best catalytic activity. The TOF of the hydrogenation of 1-hexene catalyzed by 2 nm rhodium clusters is about 5700 h^{-1}. Under the same conditions, the benzene hydrogenation was complete giving a TOF of 80 h^{-1}. Finally, nitrobenzene was also reduced into aniline with a TOF of 600 h^{-1} at 333 K and 80 bar H_2, using 3 nm-sized clusters.

The catalytic formation of C—C bonds catalyzed by NPs in water was also investigated by several groups [100, 111, 112]. Recently, a new phosphine with

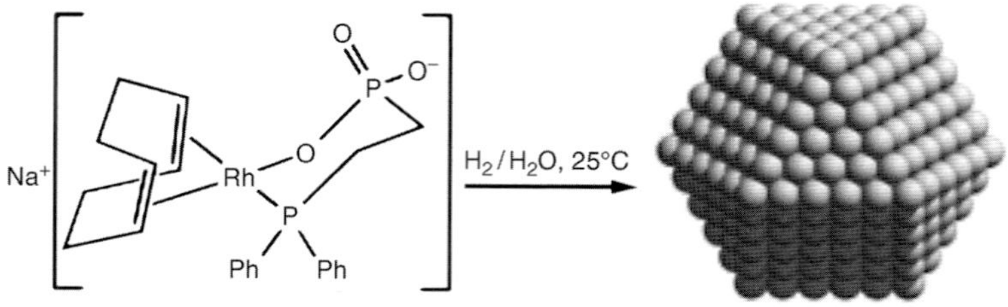

Figure 2.25 Synthesis of phosphinophosphonic acid-stabilized rhodium clusters. (Adapted from ref. [36])

zwitterionic liquid based on imidazolium sulfonate linked ethylene glycol moiety as a protective ligand for palladium and gold NPs was reported by Fujihara (Figure 2.26) [113]. The water soluble phosphine-stabilized palladium catalyst proved to be an efficient catalyst for the Suzuki cross-coupling reaction in an aqueous medium.

In the same manner, Wang *et al.* reported an original organometallic approach based on a Pd(II) complex [114]. The ionic water–soluble nitrogen ligand 1-[*N*,*N*′,*N*″-trimethyl-(4-butyl) ammonium]-2-(2-pyridyl) imidazole chloride (Figure 2.27), which acts as a chelate, proved to be efficient for the stabilization of palladium NPs in pure water.

The palladium NPs with a size of 3.0 ± 0.6 nm were synergistically stabilized through the coordination of the nitrogen-containing group at the NPs surface and the electrostatic stabilization which was provided by the ammonium polar group and directed towards the aqueous solution. The authors described high catalytic activity for the Suzuki–Miyaura reaction of aryl chlorides and aryl bromides at 393 K in neat water. Nevertheless, mercury drop test and poison experiments of the catalytic solution showed Pd(0) particles as the true catalytically active species and NPs formation after the catalytic reaction as a final step.

An organometallic approach for the synthesis of ruthenium and platinum NPs and their transfer into water using an aminophosphine derivative as protective agent was recently reported by Philippot [39]. 1,3,5-triaza-7-phosphaadamantane (PTA), a well-known ligand for its solubility both in organic solvents and water, has been largely used for the formation of water-soluble homogeneous complexes. Taking profit of this advantage, ruthenium and platinum NPs were obtained by decomposition of a THF solution of the organometallic precursor [Ru(COD)(COT)] or [Pt(Me)$_2$(COD)] respectively in the presence of PTA under dihydrogen (Figure 2.28).

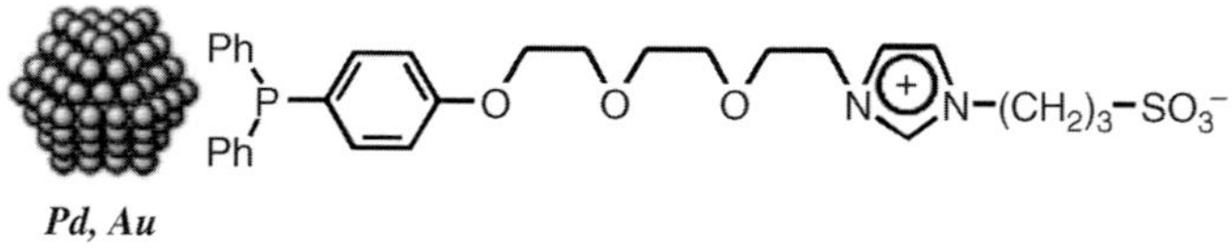

Figure 2.26 Pd and Au NPs stabilized by a water soluble functionalized phosphine. (Adapted from ref. [113])

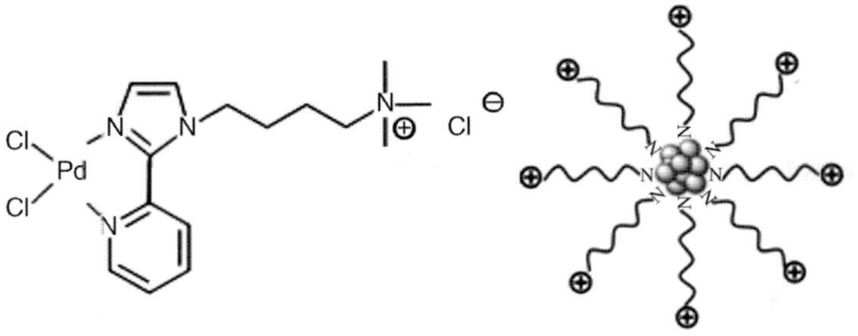

Figure 2.27 (a) Water–soluble nitrogen ligand and its palladium chelating complex as precursor of NPs. (b) Representation of Pd NPs stabilized by electrostatic and coordination effect. (Adapted from ref. [114])

After pentane addition and filtration, NPs were easily dissolved in water without any change in dispersion and in mean diameter (1.4 ± 0.2 nm and 1.1 ± 0.2 nm, Figure 2.29a and b respectively). The optimized conditions such as $T = 343$ K and [PTA]/[Ru] = 0.8 led to reproducible and stable aqueous colloidal solutions for several weeks. The coordination of PTA on the particle surface, which represents the driving force for the NPs phase transfer and stabilization in water, was evidenced by NMR ^{1}H and ^{13}C studies.

Finally, the water-soluble PTA-stabilized Ru and Pt NPs were evaluated as catalysts in pure biphasic liquid–liquid (neat water/substrate) conditions. The aqueous Ru@PTA and Pt@PTA suspensions were investigated in the hydrogenation of model olefins and arenes to probe their surface state as well as their stabilities under catalytic conditions (Table 2.14). Philippot and co-workers demonstrate that these nanocatalysts are active despite the change of environment they underwent after their dissolution in water. Indeed, these catalysts appeared efficient in mild conditions with conversions comparable to those of metallic NPs described in the literature [38].

More recently, in collaboration with others groups, Philippot described the use of water-soluble alkyl sulfonated diphosphines as ligands of ruthenium NPs [40]. According to a similar procedure to the one followed with PTA, the so-obtained NPs were redispersed into water providing an aqueous suspension of Ru colloids. Various diphosphines-TS [115] were investigated based on the number of carbon atoms between the two phosphorous atoms (Figure 2.30).

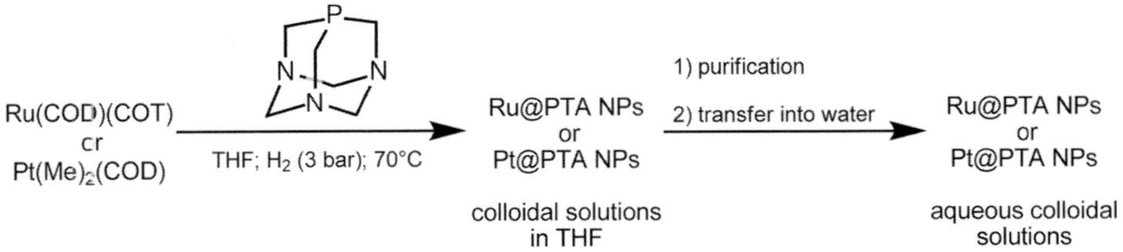

Figure 2.28 Synthesis of ruthenium and platinum NPs stabilized by 1,3,5-triaza-7-phosphaadamantane (PTA). (Adapted from ref. [39])

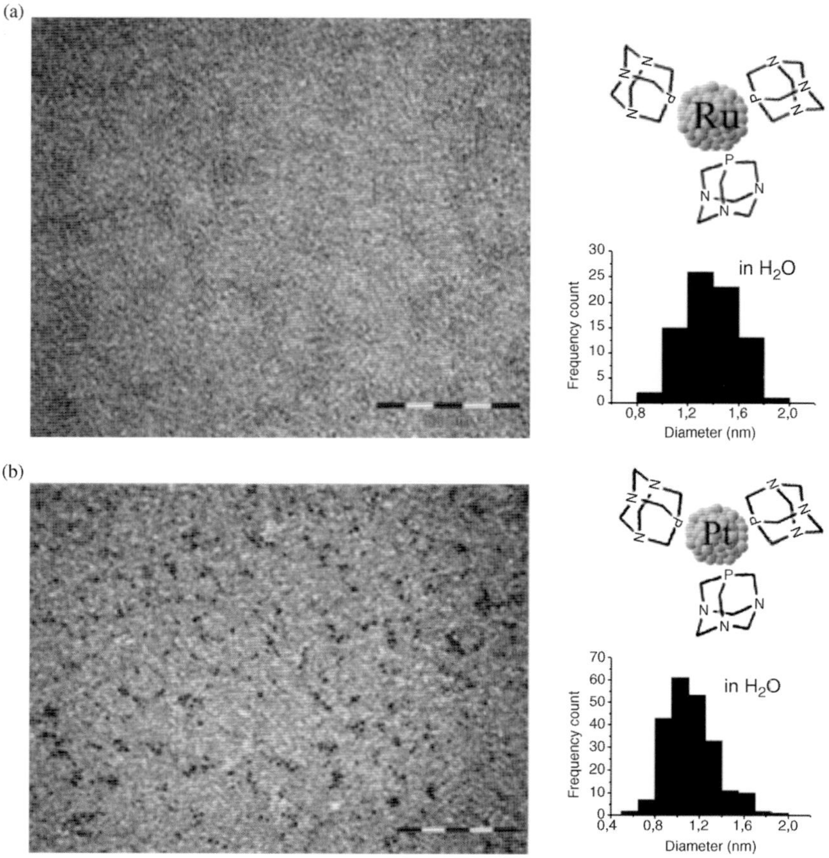

Figure 2.29 TEM micrographs in H_2O of PTA-stabilized Ru (a) and Pt (b) NPs. Scale bar 50 nm. (Adapted from ref. [38])

The hydrogenation of the organometallic precursor [Ru(COD)(COT)] in THF with different diphosphines-TS/metal ratio was investigated. TEM analysis reveals the presence of well-isolated spherical particles with a mean size in the range 1.2–1.5 nm. No change in dispersion and mean diameters was observed after the transfer in water. The DLS studies enable the ligand shell thickness to be estimated to about 0.15 nm.

These aqueous suspensions of ruthenium were investigated in the hydrogenation of unsaturated substrates in biphasic liquid–liquid conditions. The performances of diphosphine-stabilized Ru(0) NPs were first evaluated in the hydrogenation of styrene (Table 2.15). In all cases, the complete conversion in ethylcyclohexane was observed after 40 h and under atmospheric hydrogen pressure. The kinetic influence of the alkyl chain length was significantly observed after 1 and 20 h. The reactivity decreases in the following order dppb-TS > dppp-TS > dppe-TS and could be explained by a higher flexibility providing a suitable dynamic coordination mode at the particle surface in

Table 2.14 Olefins and arenes hydrogenation with Ru and Pt NPs in water[a]. (Adapted from ref. [38]).

Catalyst	Substrate	P (bar H_2)	Time (h)	Conversion (%)
Pt@PTA[b]	Octene	1	2	100
Pt@PTA[c]	Toluene	10	16	100
Pt@PTA[c]	*m*-Methylanisole	10	16	100
Ru@PTA[d]	Dodecene	1	5	100
Ru@PTA[d]	Octene	10	1	100
Ru@PTA[e]	Toluene	10	16	100
Ru@PTA[e]	*m*-Methylanisole	10	16	60

a) Reaction conditions: [substrate]/[metal] = 100, T = 293 K stirred at 1500 mn^{-1}.
b) Pt (0.0119 mmol); substrate (1.19 mmol).
c) Pt (0.0122 mmol); substrate (1.22 mmol).
d) Ru (0.0159 mmol); substrate (1.59 mmol).
e) Reaction conditions: Ru (0.020 mmol); substrate (2 mmol).

solution. The significant decrease of the necessary reaction time up to 2 h could be observed under 10 bar of H_2. Moreover, the efficient recyclability of the aqueous phase was checked with the Ru@dppb-Ts system justifying the stability of the nanocatalysts.

The best Ru@dppb-Ts system was also investigated in the hydrogenation of the lipophilic tetradecene. Due to its very low water-solubility, the authors observed only 7% of tetradecane after 1 h and a longer reaction time of 40 h was required for total conversion. In the same manner, the hydrogenation of acetophenone provides 100% conversion in 20 h with selectivities into phenylethan-1-ol (84%) and cyclohexylethan-1-ol (15%). The totally hydrogenated product (cyclohexylethan-1-ol) could only be quantitatively obtained under 10 bar of H_2 after 2 h.

Finally, these results justify the judicious interest in preparing water-soluble NPs with an organometallic approach. Taking advantage of the ligand solubility, the

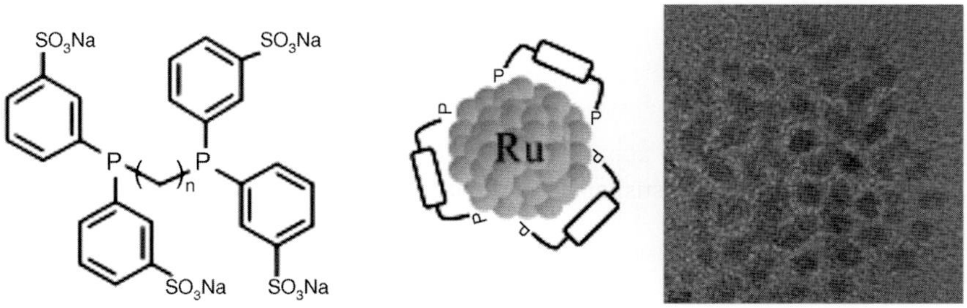

n = 2 : 1,4-bis [(di-m-sulfonatophenyl)phosphino]ethane or dppe-TS
n = 3 : 1,4-bis [(di-m-sulfonatophenyl)phosphino]propane or dppp-TS
n = 4 : 1,4-bis [(di-m-sulfonatophenyl)phosphino]butane or dppb-TS

Figure 2.30 Ruthenium NPs stabilized by dppe-TS, dppp-TS and dppb-TS and HRTEM image of Ru@dppb-Ts (mean size = 1.5 nm).

Table 2.15 Hydrogenation of styrene with dppb-TS, dppp-TS and dppe-TS-stabilized Ru nanoparticles in water[a)]. (Adapted from ref. [40]).

Alkyl sulfonated diphosphine	Time(h)	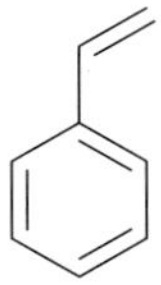	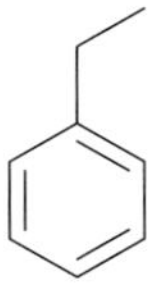	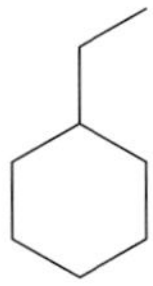
dppb-TS	1	25	75	0
dppb-TS	20	0	45	55
dppb-TS	40	0	3	97
dppp-TS	1	59	40	1
dppp-TS	20	0	47	53
dppp-TS	40	0	2	98
dppe-TS	1	76	24	1
dppe-TS	20	0	41	59
dppe-TS	40	0	1	99

a) Reaction conditions: ruthenium (3.9×10^{-5} mol), [Styrene]/[Ru] = 100, [diphosphine-TS]/[Ru] = 0.1, 1 bar H_2, 293 K, 10 ml water.

isolated NPs as a powder were easily dispersed in water. Nevertheless two constraints have to be mentioned: the preparation of the adapted organometallic precursors and the disposal of a ligand which is both organic- and water-soluble.

2.4 Conclusion and Perspectives

In the drive towards sustainable and environment-friendly chemistry, catalytic processes are undoubtedly more suitable and greener than stoichiometric applications considering the 12 green chemistry principles, and the development of catalytic systems based on nanospecies particularly contributes to this breakthrough. Moreover, in the current green context, the choice of the solvent is also of paramount importance. Consequently, among the cleaner solvents or fluids, water appears to be the most promising as a nontoxic, nonflammable, available and inexpensive reaction medium. Besides its properties, such as a highly polar character that provides unexpected reactivities and selectivities, the use of water as a benign solvent in liquid–liquid biphasic media remains an industrially attractive methodology which has already found broad applications. The traditional aqueous/organic biphasic systems have been intensively investigated since the beginning of 1980s but insufficient attention has been focused on pure biphasic water/liquid reactants (water containing the catalyst) which could be more targeted in the future.

For economic and environmental reasons, the use of commercial, cheap and less toxic metal precursors will be preferred to develop NPs active species in one pot. As a consequence, the use of metal salts appears better adapted than that of organometallic complexes which requires first the synthesis of the metal precursors, classically in organic media and sometimes time-consuming. Nevertheless, a greener approach for nanocatalysis should not be analyzed only step by step (precursor/synthesis/catalysis) but as a global view of the process. Today, the direct synthesis of well-stabilized water-soluble NPs starting from metal salts for catalytic applications in neat water could appear as the best target.

The catalytic activity of metallic NPs in terms of activity, selectivity and lifetime is drastically dependent on several morphological specificities such as the dispersity, stability, size, shape, organization of the particles and the nature of the chemical species present at their surface. Very relevant results were obtained due to the doggedness of the researchers motivated by the competition and industrial stakes. The design of new nanocatalysts affording higher performances in terms of activities, selectivities and recyclabilities, respectively expressed in TOF and TTO, remains a major concern. Moreover, important reactions such as C—H activation or asymmetric catalysis, still remain a challenge in the field of nanocatalysis and are relatively unexplored in neat water.

To conclude, the development of novel strategies for the synthesis of pertinent well-defined and -controlled nanomaterials for catalysis has attracted the attention of numerous groups all over the world and recent efforts to optimize nanoparticles chemistry are very promising. The bridge between homogeneous and heterogeneous catalysis, an image which is often used to describe the NPs approach, could be the most developed catalysis field since the last century.

References

1 Liu, S. and Xiao, J. (2007) Toward green catalytic synthesis–transition metal-catalyzed reactions in non-conventional media. *J. Mol. Cat. A.: Chemical*, **270**, 1–43.

2 Sheldon, R.A. (2008) E factors, green chemistry and catalysis: An odyssey. *Chem. Commun.*, 3352–3365.

3 Beletskaya, I.P. and Kustov, L.M. (2010) Catalysis as an important tool of green chemistry. *Russ. Chem. Rev.*, **79**, 441–461.

4 Uozumi, Y. (2010) Green chemistry – A new paradigm of organic synthesis. *Synlett*, 1988–1989.

5 Schmid, G. (ed.) (2004) *Nanoparticles*, Wiley-VCH Verlag GmbH, Weinheim, Germany.

6 Mori, K. and Yamashita, H. (2010) Progress in design and architecture of metal nanoparticles for catalytic applications. *Phys. Chem. Chem. Phys.*, **12**, 14420–14432.

7 Somorjai, G.A., Frei, H., and Park, J.Y. (2009) Advancing the frontiers in nanocatalysis, biointerfaces, and renewable energy conversion by innovations of surface techniques. *J. Am. Chem. Soc.*, **131**, 16589–16605.

8 Polshettiwar, V. and Varma, R.S. (2010) Green chemistry by nano-catalysis. *Green Chem.*, **112**, 743–754.

9 Wilcoxon, J.P. and Abrams, B.L. (2006) Synthesis, structure and properties of metal nanoclusters. *Chem. Soc. Rev.*, **35**, 1162–1194.

10 Narayanan, R. and El-Sayed, M. (2008) Some aspects of colloidal nanoparticle stability, catalytic activity, and recycling potential. *Top. Catal.*, **47**, 15–21.

11 Astruc, D. (ed.) (2008) *Nanoparticles and Catalysis*, Wiley-VCH Verlag GmbH, Weinheim, Germany.
12 Roucoux, A., Nowicki, A., and Philippot, K. (2008) Rhodium and ruthenium nanoparticles in catalysis, in *Nanoparticles and Catalysis* (ed. D. Astruc), Wiley-VCH Verlag GmbH, Weinheim, Germany, pp. 349–388.
13 El-Sayed, M.A. (2001) Some interesting properties of metals confined in time and nanometer space of different shapes. *Acc. Chem. Res.*, **34**, 257–264.
14 Astruc, D., Lu, F., and Aranzaes, J.R. (2005) Nanoparticles as recyclable catalysts: The frontier between homogeneous and heterogeneous catalysis. *Angew. Chem. Int. Ed.*, **44**, 7852–7872.
15 Chaudret, B. and Philippot, K. (2007) Organometallic derived metals, colloids, and nanoparticles, in *Comprehensive Organometallic Chemistry III* (eds H.C. Robert and D.M.P. Mingos), Elsevier, Oxford, pp. 71–99.
16 Pachón, L.D. and Rothenberg, G. (2008) Transition-metal nanoparticles: synthesis, stability and the leaching issue. *Appl. Organomet. Chem.*, **22**, 288–299.
17 Roucoux, A., Schulz, J., and Patin, H. (2002) Reduced transition metal colloids: A novel family of reusable catalysts? *Chem. Rev.*, **102**, 3757–3778.
18 Yan, N., Xiao, C.X., and Kou, Y. (2010) Transition metal nanoparticle catalysis in green solvents. *Coordin. Chem. Rev.*, **254**, 1179–1218.
19 Lindstrom, U.M. (2002) Stereoselective organic reactions in water. *Chem. Rev.*, **102**, 2751–2772.
20 Wei, G.-T., Yang, Z., Lee, C.-Y., Yang, H.-Y., and Wang, C.R.C. (2004) Aqueous-organic phase transfer of gold nanoparticles and gold nanorods using an ionic liquid. *J. Am. Chem. Soc.*, **126**, 5036–5037.
21 Li, C.-J. and Chen, L. (2006) Organic chemistry in water. *Chem. Soc. Rev.*, **35**, 68–82.
22 Li, C.J. and Chan, T.H. (eds) (1997) *Comprehensive Organic Reaction in Aqueous Media*, John Wiley & Sons, Inc., New York, USA.
23 Cornils, B. and Hermann, W.A. (eds) (2005) *Aqueous-Phase Organometallic Catalysis: Concepts and Applications*, 2nd edn, Wiley-VCH Verlag GmbH, Weinheim, Germany.
24 Sheldon, R.A. (2005) Green solvents for sustainable organic synthesis: state of the art. *Green Chem.*, **7**, 267–278.
25 Feldheim, D.L. and Foss, C.A. (eds) (2002) *Metal nanoparticles: Synthesis, Characterization, and Applications*, Marcel Dekker, New York.
26 Roucoux, A. (2005) Stabilized noble metal nanoparticles: An unavoidable family of catalysts for arene derivatives hydrogenation, in *Topics in Organometallic Chemistry: Surface and Interfacial Organometallic Chemistry and Catalysis* (eds C. Copéret and B. Chaudret), Springer GmbH, Heidelberg, pp. 261–279.
27 Corain, B., Schmid, G., and Toshima, N. (eds) (2007) *Metal Nanoclusters in Catalysis and Materials Science: The Issue of Size Control*, Elsevier.
28 Roucoux, A. and Philippot, K. (2007) Homogeneous hydrogenation: Colloids – hydrogenation with noble metal nanoparticles, in *The Handbook of Homogeneous Hydrogenation* (eds J.G.d. Vries and C.J. Elsevier), Wiley-VCH Verlag GmbH, Weinheim, Germany, pp. 217–256.
29 Gual, A., Godard, C., Castillon, S., and Claver, C. (2010) Soluble transition-metal nanoparticles-catalysed hydrogenation of arenes. *Dalton Trans.*, **39**, 11499–11512.
30 Zahmakran, M. and Ozkar, S. (2011) Metal nanoparticles in liquid phase catalysis; from recent advances to future goals. *Nanoscale*, **3**, 3462–3481.
31 Schmid, G., Klein, N., Korste, L., Kreibig, U., and Schönauer, D. (1988) Large transition metal clusters—VI. Ligand exchange reactions on $Au_{55}(PPh_3)_{12}Cl_6$—the formation of a water soluble Au55 cluster. *Polyhedron*, **7**, 605–608.
32 Gittins, D.I. and Caruso, F. (2001) Spontaneous phase transfer of nanoparticulate metals from organic to

aqueous media. *Angew. Chem. Int. Ed.*, **40**, 3001–3004.
33 Gittins, D.I. and Caruso, F. (2002) Biological and physical applications of water-based metal nanoparticles synthesised in organic solution. *ChemPhysChem*, **3**, 110–113.
34 Li, K., Wang, Y., Jiang, J., and Jin, Z. (2010) Thermoregulated phase-transfer rhodium nanoparticle catalyst for hydrogenation in an aqueous/organic biphasic system. *Catal. Commun.*, **11**, 542–546.
35 Simard, J., Briggs, C., Boal, A.K., and Rotello, V.M. (2000) Formation and pH-controlled assembly of amphiphilic gold nanoparticles. *Chem. Commun.*, 1943–1944.
36 Glöckler, J., Klützke, S., Meyer-Zaika, W., Reller, A., García-García, F.J., Strehblow, H.-H., Keller, P., Rentschler, E., and Kläui, W. (2007) With phosphinophosphonic acids to nanostructured, water-Soluble, and catalytically active rhodium clusters. *Angew. Chem. Int. Ed.*, **46**, 1164–1167.
37 Thiery, E., Le Bras, J., and Muzart, J. (2007) Palladium nanoparticles-catalyzed regio- and chemoselective hydrogenolysis of benzylic epoxides in water. *Green Chem.*, **9**, 326–327.
38 Debouttière, P.J., Coppel, Y., Denicourt-Nowicki, A., Roucoux, A., Chaudret, B., and Philippot, K. (2012) PTA-stabilized ruthenium and platinum nanoparticles: characterization and investigation in aqueous biphasic hydrogenation catalysis. *Eur. J. Inorg. Chem.*, **8**, 1229–1236.
39 Debouttiere, P.J., Martinez, V., Philippot, K., and Chaudret, B. (2009) An organometallic approach for the synthesis of water-soluble ruthenium and platinum nanoparticles. *Dalton Trans.*, 10172–10174.
40 Guerrero, M., Roucoux, A., Denicourt-Nowicki, A., Bricout, H., Monflier, E., Collière, V., Fajerweg, K., and Philippot, K. (2012) Alkyl sulfonated diphosphines-stabilized ruthenium nanoparticles as efficient nanocatalysts in hydrogenation reactions in biphasic media. *Catal. Today*, **183**, 34–41.
41 Turkevich, J., Stevenson, P.C., and Hillier, J. (1951) A study of the nucleation and growth processes in the synthesis of colloidal gold. *Discuss. Faraday Soc.*, **11**, 55–75.
42 Lee, H., Habas, S.E., Kweskin, S., Butcher, D., Somorjai, G.A., and Yang, P. (2006) Morphological control of catalytically active platinum nanocrystals. *Angew. Chem. Int. Ed.*, **45**, 7824–7828.
43 Somorjai, G.A. and Park, J.Y. (2008) Molecular factors of catalytic selectivity. *Angew. Chem. Int. Ed.*, **47**, 9212–9228.
44 Schulz, J., Roucoux, A., and Patin, H. (1999) Unprecedented efficient hydrogenation of arenes in biphasic liquid–liquid catalysis by re-usable aqueous colloidal suspensions of rhodium. *Chem. Commun.*, 535–536.
45 Nowicki, A., Le Boulaire, V., and Roucoux, A. (2007) Nanoheterogeneous catalytic hydrogenation of arenes: Evaluation of the surfactant-stabilized aqueous Ruthenium(0) colloidal suspension. *Adv. Synth. Catal.*, **349**, 2326–2330.
46 Mevellec, V., Roucoux, A., Ramirez, E., Philippot, K., and Chaudret, B. (2004) Surfactant-stabilized aqueous iridium(0) colloidal suspension: An efficient reusable catalyst for hydrogenation of arenes in biphasic media. *Adv. Synth. Catal.*, **346**, 72–76.
47 Schulz, J., Levigne, S., Roucoux, A., and Patin, H. (2002) Aqueous rhodium colloidal suspension in reduction of arene derivatives in biphasic system: a significant physico-chemical role of surfactant concentration on catalytic activity. *Adv. Synth. Catal.*, **344**, 266–269.
48 Roucoux, A., Schulz, J., and Patin, H. (2003) Arene hydrogenation with a stabilised aqueous rhodium(0) suspension: A major effect of the surfactant counter-anion. *Adv. Synth. Catal.*, **345**, 222–229.
49 Hubert, C., Denicourt-Nowicki, A., Guegan, J.P., and Roucoux, A. (2009) Polyhydroxylated ammonium chloride salt: a new efficient surfactant for nanoparticles stabilisation in aqueous media. Characterization and

application in catalysis. *Dalton Trans.*, 7356–7358.
50 Guyonnet Bilé, E., Sassine, R., Denicourt-Nowicki, A., Launay, F., and Roucoux, A. (2011) New ammonium surfactant-stabilized rhodium(0) colloidal suspensions: Influence of novel counter-anions on physico-chemical and catalytic properties. *Dalton Trans.*, **40**, 6524–6531.
51 Schulz, J., Roucoux, A., and Patin, H. (2000) Stabilized rhodium(0) nanoparticles: a reusable hydrogenation catalyst for arene derivatives in a biphasic water-liquid system. *Chem. Eur. J.*, **6**, 618–624.
52 Chen, S. and Kimura, K. (1999) Synthesis and characterization of carboxylate-modified gold nanoparticle powders dispersible in water. *Langmuir*, **15**, 1075–1082.
53 Nikoobakht, B. and El-Sayed, M.A. (2001) Evidence for bilayer assembly of cationic surfactants on the surface of gold nanorods. *Langmuir*, **17**, 6368–6374.
54 Léger, B., Nowicki, A., Roucoux, A., and Rolland, J.-P. (2007) Competitive hydrogenation/dehalogenation of halogenoarenes with surfactant-stabilized aqueous suspensions of rhodium and palladium colloids: A major effect of the metal nature. *J. Mol. Cat. A.: Chemical*, **266**, 221–225.
55 Hubert, C., Guyonnet Bilé, E., Denicourt-Nowicki, A., and Roucoux, A. (2011) Tandem dehalogenation-hydrogenation reaction of halogenoarenes as model substrates of endocrine disruptors in water: Rhodium nanoparticles in suspension vs. on silica support. *Appl. Catal. A-Gen.*, **394**, 215–219.
56 Mévellec, V., Nowicki, A., Roucoux, A., Dujardin, C., Granger, P., Payen, E., and Philippot, K. (2006) A simple and reproducible method for the synthesis of silica-supported rhodium nanoparticles and their investigation in the hydrogenation of aromatic compounds. *New J. Chem.*, **30**, 1214–1219.
57 Denicourt-Nowicki, A., Romagné, M.-L., and Roucoux, A. (2008) *N*-(2-hydroxyethyl)ammonium derivatives as protective agents for Pd(0) nanocolloids and catalytic investigation in Suzuki reactions in aqueous media. *Catal. Commun.*, **10**, 68–70.
58 Mévellec, V., Mattioda, C., Schulz, J., Rolland, J.-P., and Roucoux, A. (2004) Enantioselective hydrogenation of ethyl pyruvate in biphasic liquid–liquid media by reusable surfactant-stabilized aqueous suspensions of platinum nanoparticles. *J. Catal.*, **225**, 1–6.
59 Guyonnet Bilé, E., Denicourt-Nowicki, A., Sassine, R., Beaunier, P., Launay, F., and Roucoux, A. (2010) *N*-methylephedrium salts as chiral surfactants for asymmetric hydrogenation in neat water with rhodium(0) nanocatalysts. *ChemSusChem*, **3**, 1276–1279.
60 Callis, N.M., Thiery, E., Le Bras, J., and Muzart, J. (2007) Palladium nanoparticles-catalyzed chemoselective hydrogenations, a recyclable system in water. *Tetrahedron Lett.*, **48**, 8128–8131.
61 Zhang, Z., Zha, Z., Gan, C., Pan, C., Zhou, Y., Wang, Z., and Zhou, M.-M. (2006) Catalysis and regioselectivity of the aqueous Heck reaction by Pd(0) nanoparticles under ultrasonic irradiation. *J. Org. Chem.*, **71**, 4339–4342.
62 Durap, F., Zahmakıran, M., and Özkar, S. (2009) Water soluble laurate-stabilized rhodium(0) nanoclusters catalyst with unprecedented catalytic lifetime in the hydrolytic dehydrogenation of ammonia-borane. *Appl. Catal. A-Gen.*, **369**, 53–59.
63 Durap, F., Zahmakıran, M., and Özkar, S. (2009) Water soluble laurate-stabilized ruthenium(0) nanoclusters catalyst for hydrogen generation from the hydrolysis of ammonia-borane: High activity and long lifetime. *Int. J. Hydrogen Energy*, **34**, 7223–7230.
64 Denicourt-Nowicki, A. and Roucoux, A. (2010) Noble metal nanoparticles stabilized by cyclodextrins: a pertinent partnership for catalytic applications. *Curr. Org. Chem.*, **14**, 1266–1283.
65 Rekharsky, M.V. and Inoue, Y. (1998) Complexation thermodynamics of cyclodextrins. *Chem. Rev.*, **98**, 1875–1917.

66 Khan, A.R., Forgo, P., Stine, K.J., and D'Souza, V.T. (1998) Methods for selective modifications of cyclodextrins. *Chem. Rev.*, **98**, 1977–1996.

67 Huang, T., Meng, F., and Qi, L.M. (2009) Facile synthesis and one-dimensional assembly of cyclodextrin-capped gold nanoparticles and their applications in catalysis and surface-enhanced Raman scattering. *J. Phys. Chem. C*, **113**, 13636–13642.

68 Mori, K., Yoshioka, N., Kondo, Y., Takeuchi, T., and Yamashita, H. (2009) Catalytically active, magnetically separable, and water-soluble FePt nanoparticles modified with cyclodextrin for aqueous hydrogenation reactions. *Green Chem.*, **11**, 1337–1342.

69 Liu, J., Alvarez, J., and Kaifer, A.E. (2000) Metal nanoparticles with a knack for molecular recognition. *Adv. Mater.*, **12**, 1381–1383.

70 Alvarez, J., Liu, J., Roman, E., and Kaifer, A.E. (2000) Water-soluble platinum and palladium nanoparticles modified with thiolated beta-cyclodextrin. *Chem. Commun.*, 1151–1152.

71 Liu, J., Ong, W., Román, E., Lynn, M.J., and Kaifer, A.E. (2000) Cyclodextrin-modified gold nanospheres. *Langmuir*, **16**, 3000–3002.

72 Terrill, R.H., Postlethwaite, T.A., Chen, C.-h., Poon, C.-D., Terzis, A., Chen, A., Hutchison, J.E., Clark, M.R., and Wignall, G. (1995) Monolayers in three dimensions: NMR, SAXS, thermal, and electron hopping studies of alkanethiol stabilized gold clusters. *J. Am. Chem. Soc.*, **117**, 12537–12548.

73 Liu, J., Alvarez, J., Ong, W., Roman, E., and Kaifer, A.E. (2001) Tuning the catalytic activity of cyclodextrin-modified palladium nanoparticles through host-guest binding interactions. *Langmuir*, **17**, 6762–6764.

74 Mhadgut, S.C., Palaniappan, K., Thimmaiah, M., Hackney, S.A., Torok, B., and Liu, J. (2005) A metal nanoparticle-based supramolecular approach for aqueous biphasic reactions. *Chem. Commun.*, 3207–3209.

75 Nowicki, A., Zhang, Y., Leger, B., Rolland, J.P., Bricout, H., Monflier, E., and Roucoux, A. (2006) Supramolecular shuttle and protective agent: a multiple role of methylated cyclodextrins in the chemoselective hydrogenation of benzene derivatives with ruthenium nanoparticles. *Chem. Commun.*, 296–298.

76 Denicourt-Nowicki, A., Ponchel, A., Monflier, E., and Roucoux, A. (2007) Methylated cyclodextrins: an efficient protective agent in water for zerovalent ruthenium nanoparticles and a supramolecular shuttle in alkene and arene hydrogenation reactions. *Dalton Trans.*, 5714–5719.

77 Liu, Y.L., Male, K.B., Bouvrette, P., and Luong, J.H.T. (2003) Control of the size and distribution of gold nanoparticles by unmodified cyclodextrins. *Chem. Mater.*, **15**, 4172–4180.

78 Huang, Y.J., Li, D., and Li, J.H. (2004) beta-cyclodextrin controlled assembling nanostructures from gold nanoparticles to gold nanowires. *Chem. Phys. Lett.*, **389**, 14–18.

79 Tutaj, B., Kasprzyk, A., and Czapkiewicz, J. (2003) The spectral displacement technique for determining the binding constants of β-cyclodextrin - alkyltrimethylammonium inclusion complexes. *J. Inclus. Phenom. Mol.*, **47**, 133–136.

80 Hubert, C., Denicourt-Nowicki, A., Roucoux, A., Landy, D., Leger, B., Crowyn, G., and Monflier, E. (2009) Catalytically active nanoparticles stabilized by host-guest inclusion complexes in water. *Chem. Commun.*, 1228–1230.

81 Wu, S.H. and Chen, D.H. (2004) Synthesis and stabilization of Ni nanoparticles in a pure aqueous CTAB solution. *Chem. Lett. (Jpn.)*, **33**, 406–407.

82 Sakurai, H., Hirano, T., Negishi, Y., Tsunakawa, H., and Tsukuda, T. (2002) Palladium clusters stabilized by cyclodextrins catalyze Suzuki-Miyaura coupling reactions un water. *Trans. Mater. Res. Soc. Jpn.*, **27**, 185–188.

83 Xue, C.H., Palaniappan, K., Arumugam, G., Hackney, S.A., Liu, J., and Liu, H.Y. (2007) Sonogashira reactions catalyzed by water-soluble, beta-cyclodextrin-

capped palladium nanoparticles. *Catal. Lett.*, **116**, 94–100.

84 Breslow, R. and Dong, S.D. (1998) Biomimetic reactions catalyzed by cyclodextrins and their derivatives. *Chem. Rev.*, **98**, 1997–2011.

85 Senra, J.D., Malta, L.F.B., Costa, M.E.H. M.d., Michel, R.C., Aguiar, L.C.S., Simas, A.B.C., and Antunes, O.A.C. (2009) Hydroxypropyl-alpha-cyclodextrin-capped palladium nanoparticles: active scaffolds for efficient carbon-carbon bond forming cross-couplings in water. *Adv. Synth. Catal.*, **351**, 2411–2422.

86 Ng, C.H.B., Yang, J., and Fan, W.Y. (2008) Synthesis and self-assembly of one-dimensional sub-10nm Ag nanoparticles with cyclodextrin. *J. Phys. Chem. C*, **112**, 4141–4145.

87 Bronstein, L.M., Matveeva, V.G., and Sulman, E.M. (2008) Nanoparticulate catalysts based on nanostructured polymers, in *Nanoparticles and Catalysis* (ed. D. Astruc), Wiley-VCH Verlag GmbH, Weinheim, Germany, pp. 93–128.

88 Chandler, B.D. and Gilbertson, J.D. (2008) PAMAM dendrimer templated nanoparticle catalysts, in *Nanoparticles and Catalysis* (ed. D. Astruc), Wiley-VCH Verlag GmbH, Weinheim, Germany, pp. 129–160.

89 Pellegatta, J-.L., Blandy, C., Collière, V., Choukroun, R., Chaudret, B., Cheng, P., and Philippot, K. (2002) Catalytic investigation of rhodium nanoparticles in hydrogenation of benzene and phenylacetylene. *J. Mol. Cat. A.: Chemical*, **178**, 55–61.

90 Kundu, S., Wang, K., and Liang, H. (2009) Size-selective synthesis and catalytic application of polyelectrolyte encapsulated gold nanoparticles using microwave irradiation. *J. Phys. Chem. C*, **113**, 5157–5163.

91 Tsunoyama, H., Sakurai, H., Ichikuni, N., Negishi, Y., and Tsukuda, T. (2004) Colloidal gold nanoparticles as catalyst for carbon-carbon bond formation: Application to aerobic homocoupling of phenylboronic acid in water. *Langmuir*, **20**, 11293–11296.

92 Tsunoyama, H., Sakurai, H., Negishi, Y., and Tsukuda, T. (2005) Size-specific catalytic activity of polymer-stabilized gold nanoclusters for aerobic alcohol oxidation in water. *J. Am. Chem. Soc.*, **127**, 9374–9375.

93 Xiao, C.-X., Cai, Z.-P., Wang, T., Kou, Y., and Yan, N. (2008) Aqueous-phase Fischer–Tropsch synthesis with a ruthenium nanocluster catalyst. *Angew. Chem. Int. Ed.*, **47**, 746–749.

94 Maity, P., Basu, S., Bhaduri, S., and Lahiri, G.K. (2007) Superior performance of a nanostructured platinum catalyst in water: hydrogenations of alkenes, aldehydes and nitroaromatics. *Adv. Synth. Catal.*, **349**, 1955–1962.

95 Maity, P., Gopinath, C.S., Bhaduri, S., and Lahiri, G.K. (2009) Applications of a high performance platinum nanocatalyst for the oxidation of alcohols in water. *Green Chem.*, **11**, 554–561.

96 Vasylyev, M.V., Maayan, G., Hovav, Y., Haimov, A., and Neumann, R. (2006) Palladium nanoparticles stabilized by alkylated polyethyleneimine as aqueous biphasic catalysts for the chemoselective stereocontrolled hydrogenation of alkenes. *Org. Lett.*, **8**, 5445–5448.

97 Lan, Y., Zhang, M., Zhang, W., and Yang, L. (2009) Enhanced Pd-catalyzed hydrogenation of olefins within polymeric microreactors under organic/aqueous biphasic conditions. *Chem. Eur. J.*, **15**, 3670–3673.

98 Zheng, P. and Zhang, W. (2007) Synthesis of efficient and reusable palladium catalyst supported on pH-responsive colloid and its application to Suzuki and Heck reactions in water. *J. Catal.*, **250**, 324–330.

99 Han, J., Liu, Y., and Guo, R. (2009) Facile synthesis of highly stable gold nanoparticles and their unexpected excellent catalytic activity for Suzuki-Miyaura cross-coupling reaction in water. *J. Am. Chem. Soc.*, **131**, 2060–2061.

100 Mejías, N., Pleixats, R., Shafir, A., Medio-Simón, M., and Asensio, G. (2010) Water-soluble palladium nanoparticles: click synthesis and applications as a recyclable catalyst in Suzuki cross-

couplings in aqueous media. *Eur. J. Org. Chem.*, **26**, 5090–5099.

101 Fréchet, J.M.J. and Tomalia, D.A. (eds) (2001) *Dendrimers and Other Dendritic Polymers*, John Wiley & Sons, Ltd, Chichester, UK.

102 Astruc, D. (2004) La catalyse métallodendritique: une contribution efficace à la chimie verte. *C. R. Chimie*, **8**, 1101–1107.

103 Astruc, D. (2010) Palladium catalysis using dendrimers: molecular catalysts versus nanoparticles. *Tetrahedron: Asymmetr.*, **21**, 1041–1054.

104 Crooks, R.M., Chechnik, V., Lemon, B.I., Sun, L., Yeung, L.K., and Zhao, M. (2002) Synthesis, characterization, and applications of dendrimer-encapsulated metal and semiconductor nanoparticles, in *Metal Nanoparticles: Synthesis, Characterization and Applications* (eds D.L. Feldheim and C.A. Foss Jr.), Marcel Dekker, New-York, p. 262.

105 Zhao, M. and Crooks, R.M. (1999) Homogeneous hydrogenation catalysis with monodisperse, dendrimer-encapsulated Pd and Pt nanoparticles. *Angew. Chem. Int. Ed.*, **38**, 364–366.

106 Gatard, S., Liang, L., Salmon, L., Ruiz, J., Astruc, D., and Bouquillon, S. (2011) Water-soluble glycodendrimers: synthesis and stabilization of catalytically active Pd and Pt nanoparticles. *Tetrahedron Lett.*, **52**, 1842–1846.

107 Ornelas, C., Ruiz, J., Salmon, L., and Astruc, D. (2008) Sulphonated 'click' dendrimer-stabilized palladium nanoparticles as highly efficient catalysts for olefin hydrogenation and Suzuki coupling reactions under ambient conditions in aqueous media. *Adv. Synth. Catal.*, **350**, 837–845.

108 Eklund, S.E. and Cliffel, D.E. (2004) Synthesis and catalytic properties of soluble platinum nanoparticles protected by a thiol monolayer. *Langmuir*, **20**, 6012–6018.

109 Kostelansky, C.N., Pietron, J.J., Chen, M.-S., Dressick, W.J., Swider-Lyons, K.E., Ramaker, D.E., Stroud, R.M., Klug, C.A., Zelakiewicz, B.S., and Schull, T.L. (2006) Triarylphosphine-stabilized platinum nanoparticles in three-dimensional nanostructured films as active electrocatalysts. *J. Phys. Chem. B*, **110**, 21487–21496.

110 Giuffrida, S., Ventimiglia, G., Callari, F.L., and Sortino, S. (2006) Straightforward light-driven synthesis of ultrasmall, water-soluble monolayer-protected platinum nanoparticles. *Eur. J. Inorg. Chem.*, **2006**, 4022–4025.

111 Balanta, A., Godard, C., and Claver, C. (2011) Pd nanoparticles for C—C coupling reactions. *Chem. Soc. Rev.*, **40**, 4973–4985.

112 Fihri, A., Bouhrara, M., Nekoueishahraki, B., Basset, J.-M., and Polshettiwar, V. (2011) Nanocatalysts for Suzuki cross-coupling reactions. *Chem. Soc. Rev.*, **40**, 5181–5203.

113 Akiyama, T., Ibata, C., and Fujihara, H. (2010) Water-soluble palladium and gold nanoparticles functionalized by a new phosphine with zwitterionic liquid based on imidazolium sulfonate linked ethylene glycol moiety. *Heterocycles*, **80**, 925–931.

114 Zhou, C., Wang, J., Li, L., Wang, R., and Hong, M. (2011) A palladium chelating complex of ionic water-soluble nitrogen-containing ligand: the efficient precatalyst for Suzuki-Miyaura reaction in water. *Green Chem.*, **13**, 2100–2106.

115 Lecomte, L. and Sinou, D. (1990) Phosphines Sulfonées Asymétriques. V. Cinétique De Sulfonation des phosphines. *Phosphorus Sulfur*, **53**, 239–251.

3
Catalysis by Dendrimer-Stabilized and Dendrimer-Encapsulated Late-Transition-Metal Nanoparticles

Didier Astruc, Abdou Diallo, and Catia Ornelas

3.1
Introduction

Dendrimers are well-defined macromolecules with specific topologies [1–8]; therefore the use of nanoparticle (NP) stabilization by dendrimers, either by NP encapsulation [9, 10] or peripheral stabilization [11] is of great interest. Crooks, who pioneered the area, remarkably and extensively developed it with the commercial polyamidoamine (PAMAM) dendrimers serving as generation-dependent nanofilters for encapsulated palladium nanoparticles (PdNPs) [12–14].

Nanoparticle stabilization by dendrimers can be compared with previous NP stabilization by polymers for catalytic applications. Polymers containing heteroatoms have long been shown to stabilize catalytically active PdNPs. Catalysis by such polymer-stabilized NPs was efficiently exploited in the 1990s, in particular by the groups of Bönnemann [14], Reetz [15] and Toshima [16], and included carefully designed polymer-protected core-shell heterobimetallic nanocatalysts [14–17]. The advantage of dendrimers over polymers is their perfectly defined and specific topology including the choice of the generation number and location of intradendritic precursor metal cation complexes of metal atoms that will eventually form the metallic NP. With both classes of polymers and dendrimers, it is also possible to synthesize protected heterobimetallic particles of various types represented in Figure 3.1.

With dendrimers, the stabilized or encapsulated NPs can be used for several types of catalysis. In order to apply them to homogeneous catalysis, the choice of the nature of the peripheral groups will provide solubility in the adequate solvent. Heterogeneous catalysis involves immobilizing the dendrimer with the NP onto a solid support or generating the NP by reduction of the metal cation subsequent to fixation on the support. Immobilization on electrodes allows electrocatalytic reactions (in particular O_2 reduction) to be carried out, whereas high-temperature heterogeneous catalysis requires removal of the dendrimer frame.

Nanomaterials in Catalysis, First Edition. Edited by Philippe Serp and Karine Philippot.

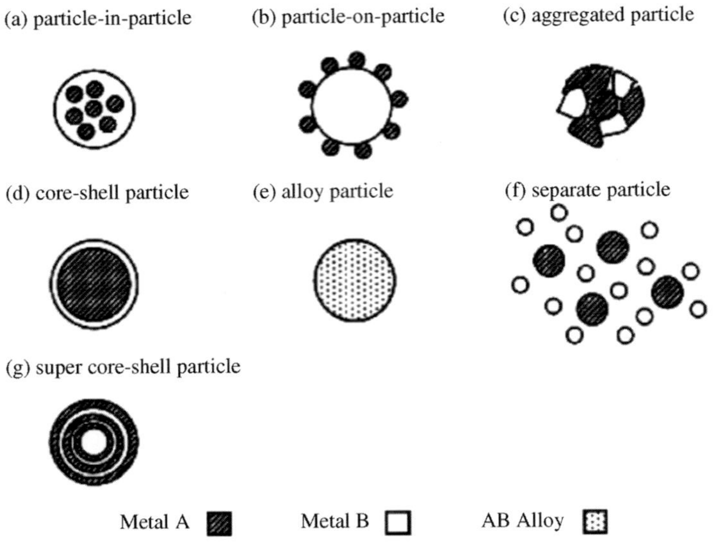

Figure 3.1 Various types of heterobimetallic transition-metal nanoparticles. (Reproduced with permission from [16c]. Copyright © (2003) American Chemical Society.)

3.2 Synthesis

The synthesis of dendrimer-encapsulated metallic NPs for instance consists in two steps (Figure 3.2). The first involves the complexation of ligand sites in the dendrimers by transition metal cations, and in the second step these cationic metal species are reduced by $NaBH_4$ to metal atoms that further agglomerate into a metallic NP inside the dendrimer. These NPs are called dendrimer-encapsulated nanoparticles, (DENs). When the dendrimers are too small, formation of dendrimer-stabilized metallic nanoparticles (DSNs) is possible, because the NPs are then protected at their periphery by several dendrimers. Crooks' group and others have used Tomalia's PAMAM dendrimers [2] in the synthesis of mono- and bimetallic DENs in the 1–2 nm size range with Cu, Pd, Pt, Au, PdCu, Au@Ag,

Figure 3.2 Principle of the generation of DENs from dendrimers, metal salts (here with the example of $K_2PdCl_4^-$ and BH_4^-. (Reproduced with permission from [18a]. Copyright © (2003) American Chemical Society.)

Pd@Au and PtPd [11–13, 18]. Oxidizable monometallic DENs with Ni [19], Sn [20] and Fe [21], and DENs with NiSn alloys [22] have also been synthesized in this way.

Evidence for encapsulation of the NP in a single dendrimer in a DEN is provided by transmission electron microscopy (TEM) and small-angle X-ray scattering. TEM studies show the high degree of DEN monodispersity and the relationship between the particle size and dendrimer-to-metal ion ratio used is their synthesis, which confirms the DEN topology. Further characterization methods include electron-electron force microscopy and atomic force microscopy (EFM and AFM respectively), X-ray methods [extended X-ray absorption fine structure (EXAFS), energy-dispersive X-ray spectroscopy (EDS), X-ray photoelectron spectroscopy (XPS)] and other spectroscopic [nuclear magnetic resonance (NMR), electron spin resonance (ESR), ultra-violet–visible (UV–vis), attenuated total reflection-infrared (ATR-IR)] and electrochemical methods, especially cyclic voltammetry. PAMAM DENs are prepared using aqueous PAMAM solutions at pH 2–5 at which all the peripheral amino groups are selectively protonated in order to avoid dendrimer crosslinking via terminal amino ligand groups and promote complexation by metal cations of the nonprotonated interior tertiary amine groups only (Figure 3.3). Alternatively,

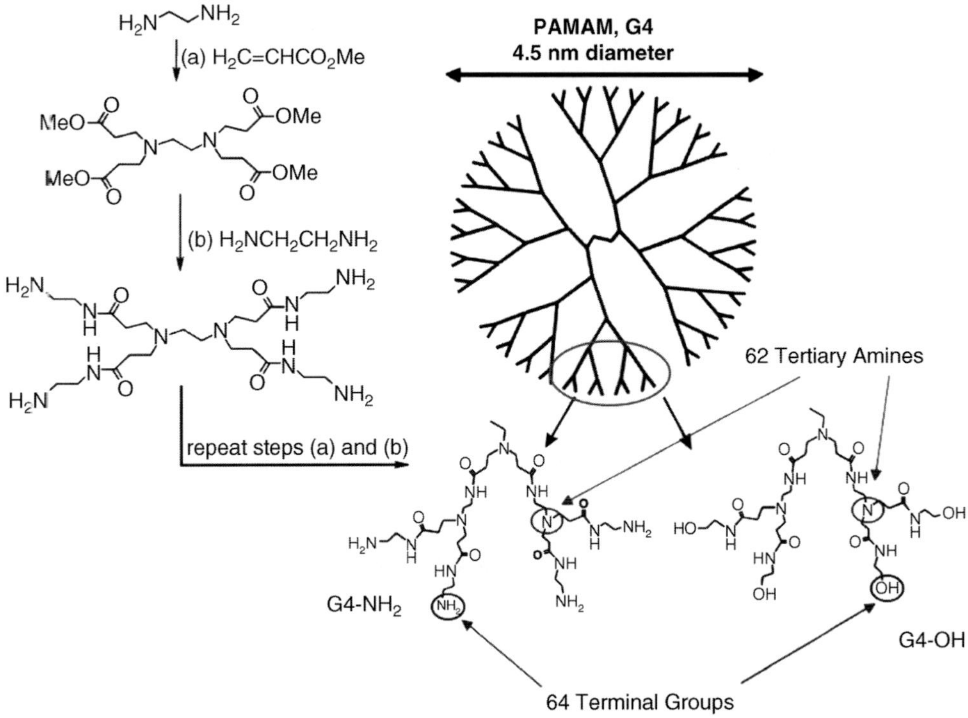

Figure 3.3 Synthesis of PAMAM dendrimers terminated by either amino or hydroxy groups for further syntheses of PAMAM DENs by complexation of interior amino groups at pH 2–5 and further BH_4^- reduction. (Reproduced with permission from [13]. Copyright © (2005) American Chemical Society.)

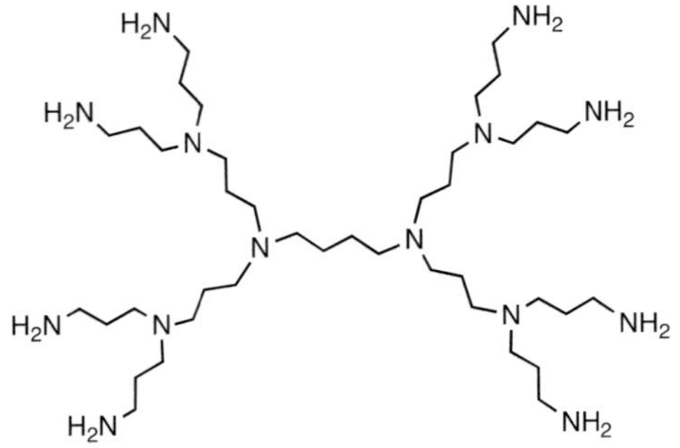

Figure 3.4 G2 polypropylene imine (PPI) dendrimers. The eight primary amine termini are preferentially protonated.

the PAMAM can be functionalized with terminal hydroxy groups, but the synthesis of amino-terminated DENs is useful for further branching of the DENs via these groups to other materials, including for biomedical applications. Although the size of the DENs is dependent on the metal-to-dendrimer ratio, it is independent of the dendrimer generation if there are enough dendritic ligands in the interior. With very large PAMAM dendrimers such as G10, several AuNPs were observed in a single dendrimer [23].

Another family of dendrimers that has often been used for the synthesis of DENs is the polypropyleneimine (PPI) dendrimers (Figure 3.4) [24]. Such dendrimers modified at their periphery with hexanoyl or palmitoyl groups encapsulate Cu^{2+} from $CHCl_3$/MeOH (4: 1, v/v) solutions, and the amount of Cu^{2+} is four times greater than in the corresponding PAMAM dendrimers and cannot be explained by a fixed metal-ion complexation stoichiometry. Pd^{2+} can also form PPI DENs in this way [25].

Yamamoto's group has reported the family of rigid phenylazomethine dendrimers and stepwise complexation (from core to peripheral layers) of the interior ligands with $SnCl_2$ and $FeCl_3$ (Figure 3.5) [26–32].

Among the synthetic routes to DENs, the redox displacement reaction also called galvanic displacement, driven by the difference of electrochemical potentials between the two metals involved, is especially elegant and useful. For instance, Cu DENs react with the cations of a more noble metal such as Ag^+ to form Ag DENs while the ions Cu^{2+} resulting from this reaction may remain trapped within the dendrimer:

$$Cu + 2\,Ag^+ \rightarrow Cu^{2+} + 2\,Ag \tag{3.1}$$

Crooks' group has reported this method for the preparation of Au, Ag, Pd, Pt [18, 33] and Au@Pt PAMAM DENs [34], and likewise Chandler's group has reported the galvanic synthesis of $Pt_{16}Au_{16}$ DSNs [35]. Crooks' group has also used underpotential deposition, that is, reduction of cations at potentials more positive than their standard potential, a method providing high surface selectivity. Core@shell NPs with a single-atom layer shell or eventually submonolayer shell have been synthesized. For instance, this method afforded deposition of a single layer of Cu onto Pt DENs [34].

The heterobimetallic NPs are of special interest for catalysis and can be synthesized by galvanic displacement, co-complexation or sequential loading as shown in Figure 3.6.

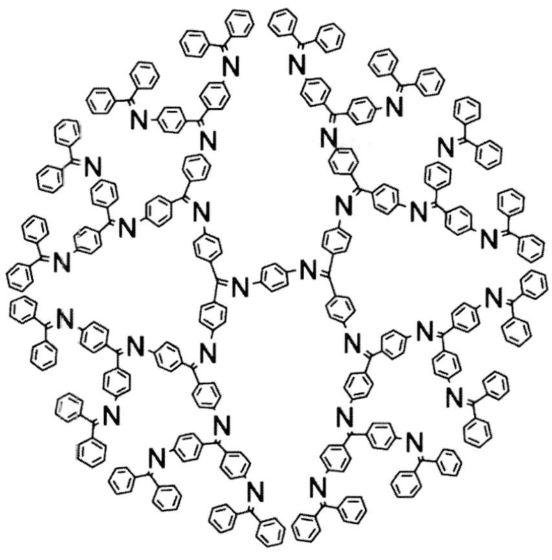

Figure 3.5 Yamamoto's G4 phenylazomethine dendrimer.

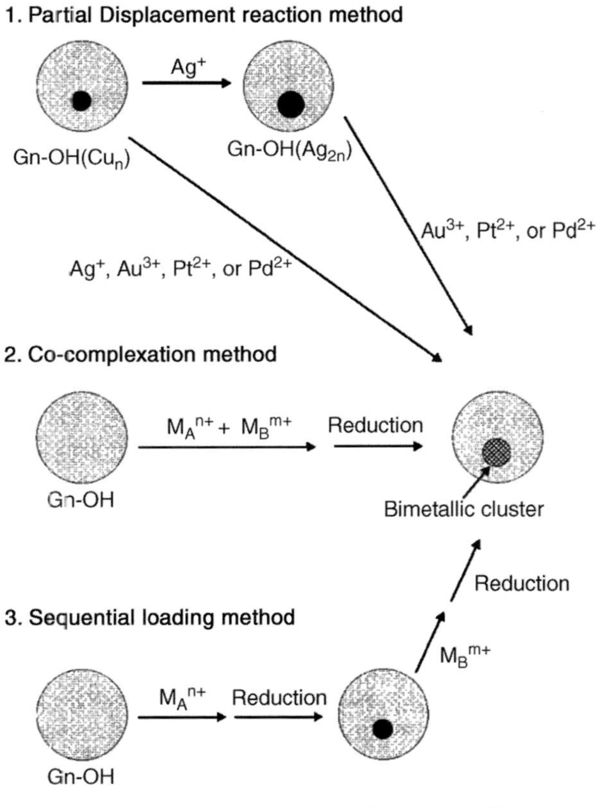

Figure 3.6 Three methods of synthesis of heterobimetallic NPs. (Reproduced with permission from [[illegible]1]. Copyright © (2001) American Chemical Society.)

Besides DENs and DSNs, other dendrimer–NP assemblies have been synthesized by coating NPs with dendrons [36–38]. The dendrons can bear various functionalities at their periphery for applications to catalysis (see below), molecular recognition [36–38] and materials science [39–41] including optics [42] and nanomedicine [43]. Thus this area is very broad and beyond the scope of this chapter, especially because the NP surface covered with dendrons is believed in principle not to be suitable for ready access of substrates on the NP surface for further catalytic transformation.

3.3
Homogeneous Catalysis with DENs Generated from PAMAM and PPI Dendrimers

3.3.1
Olefin and Nitroarene Hydrogenation

Catalysis with dendrimers emerged in the early 1990s when molecular catalysts were branched at various sites (core, branching points, tethers, termini) of dendrimers. This very popular area has been reviewed [44–50] and we will only concentrate here on NP catalysis. Indeed NPs are among the most efficient and selective catalysts [51–54]. They are usually synthesized by reduction of a transition metal salt, and the generated metal (0) atoms agglomerate, but this agglomeration can be stopped at a certain point in the presence of various stabilizers such as ligands, polymers, surfactants, ionic liquids or solid supports such as oxides, and so on. A problem, however, resides in the stabilization of the NP surface without blocking access of substrates to this surface, thus a fine balance between stabilization and surface access must be obtained. Localization of a NP inside a dendrimer brings about an elegant solution to this problem. Another important aspect is the control of the size and shape of the NPs that is best achieved upon encapsulation inside dendrimers. Catalysis with DENs was pioneered by Crooks in 1999 with Gn-OH PAMAM Pd and Pt DENs ($n = 4$–8) for the hydrogenation of allylic alcohol and *N*-isopropylacrylamide in water (Figure 3.7) [55]. The G4-OH Pd_{40}NP is much more efficient than the G6 and G8 DENs, because the latter serve as nanofilters [55, 56] to some extent inhibiting the penetration of the *N*-isopropylacrylamide substrates inside the dendrimer, in which the catalytically active NP is located. On the other hand, linear alkenes penetrate more easily, resulting in a much smaller decrease in activity [55, 56]. When G4-NH_2 PAMAM dendrimers were functionalized with various epoxide termini of increasing sizes, the hydrogenation catalysis results showed that the DENs functionalized with bulkier epoxides were less efficient catalysts than those having less bulky epoxides [13]. Molecular rulers containing a cyclodextrin stopper and allyl groups spanned by alkyl chains having different sizes were used to estimate the length between the DEN surface and the dendrimer surface, and for G4-OH Pd_{40} DENs such a length was estimated to 0.7 ± 0.2 nm [12, 13].

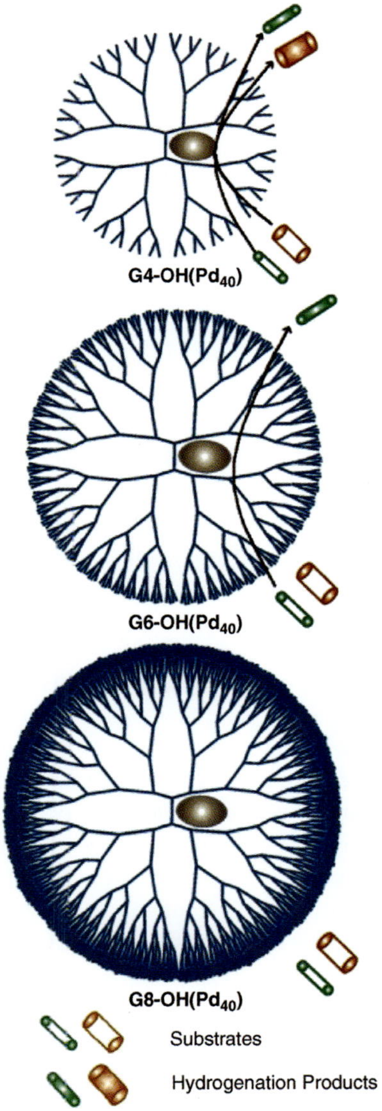

Figure 3.7 PAMAM G4, G6 and G8 Pd DENs for size-selective catalysis. (Reproduced with permission from [55b]. Copyright © (2001) American Chemical Society.)

Addition of dodecanoic acid to G4-NH_3^+ formed at pH 2 afforded electrostatic attachment leading to transfer of the DEN into the organic phase and efficient catalytic hydrogenation by the DENs in toluene [57]. Likewise, electrostatic attachment of a perfluoroether group provided DEN transfer into a fluorous solvent. Stirring or heating generated an emulsion and subsequent efficient catalysis proceeded at the interface between the organic and fluorous phases. Settling of the phases after the reaction leaves the organic product in the organic solution and the DEN catalyst in the fluorous phase, which allows the catalyst to be recycled [58].

The Yamamoto and Nishihara groups prepared PAMAM G4-OHRh DENs to catalyze olefin and nitroarene hydrogenation with a metal-ion/dendrimer ratio of 60 [59a] Yang's group also examined this reaction with bimetallic DENs (see Section 3.3.3) [59b]. G1-3 dihydroxybenzylalcohol-based dendrimers stabilized 14–35-nm sized Ag DSNs that catalyzed choloronitro-benzene hydrogenation at 20 bar H_2 and 413 K [59c].

3.3.2
PdNP-Catalyzed Carbon–Carbon Cross Coupling

Catalysis by Pd DENs of carbon–carbon cross-coupling reactions has been carried out with PAMAM and PPI dendrimers by several groups [11–13, 60–66]. PPI Pd DENs containing perfluoroether groups catalyzed the Heck reaction between iodobenzene and *n*-butyl acrylate with 100% selectivity at 363 K, which is superior to yields and selectivities obtained with other PdNPs [11, 60]. The fluorinated ponytail functionalized DENs also allow carrying out PdNP-catalyzed Heck coupling between aryl halides and methacrylate in supercritical CO_2; moreover the highly unfavored 2-phenylacrylic methyl ester isomer is exclusively obtained at 5000 psi and 348 K, whereas *trans*-cinnimaldehyde is obtained with 97% selectivity otherwise [60]. Although G4-OH DENs were found to be more stable than G2-OH and G3-OH DENs, the lower-generations DENs were also more active catalysts [62]. This finding disclosed a crucial problem in catalysis by DENs, that is, the correct balance between catalytic efficiency and stability requires careful search for a given dendrimer series, and it is tedious to maintain both advantages of optimized efficiency and stability. In some instances involving the Stille coupling, DEN size increase may be an indication of PdNP leaching as the DEN was shown to increase during catalytic reaction [67]. This point was debated using PAMAM Pd_{147} G6-OH DENs by carefully controlled monitoring of the stability under H_2 for prolonged periods of time, and leaching was only observed in the presence of N_2 or air, but about half of the PdNPs oxidized in this way were retained inside the dendrimer [68].

3.3.3
Heterobimetallic Catalysts

Heterobimetallic DENs are either alloys DENs (noted M^1M^2 DENs) or core@shell DENs. Turnover frequency (TOF) for the hydrogenation of allylic alcohol with Pd-rich heterobimetallic PdPt DENs is found to be significantly higher than that of physical mixtures of the single-metal analogs having the same percentage of the two metals [69]. This was attributed to positive synergistic effects (Figure 3.8) [70].

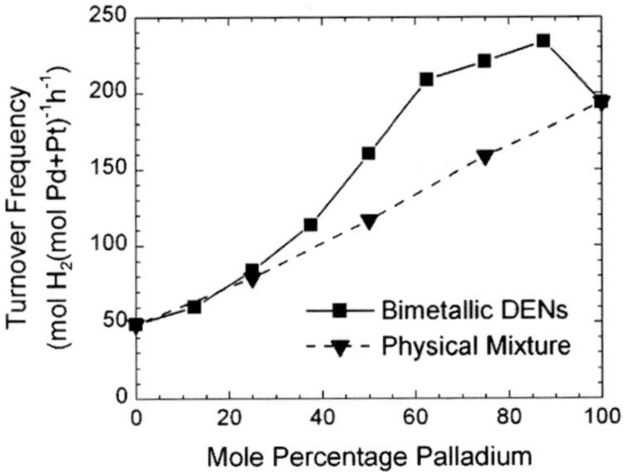

Figure 3.8 Turnover frequency (TOF) vs. mole v% of $PdCl_4^{2-}$ used to synthesize DENs. Allyl alcohol: 0.2 M, [Pd + Pt] = 0.095 mM; substrate/metal ratio: 2100/1; T = 292 K. (Reproduced with permission from [69a]. Copyright © (2003) American Chemical Society.)

Chung and Rhee also independently reported in 2003 synergistic effects for the hydrogenation of cyclohexene by PdPt DENs [71] and the partial hydrogenation of 1,3-cyclooctadiene by PdRh DENs (Figure 3.9) [72].

Au@PtNPs stabilized by Fréchet-type polyarylester dendrons showed higher catalytic activity in hydrogenation of nitrotoluenes to anilines compared with

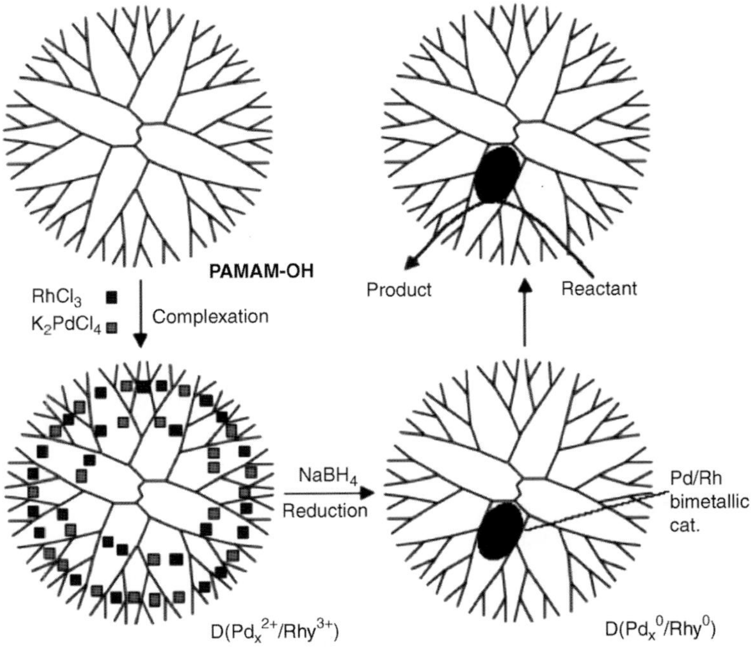

Figure 3.9 Strategy used by Chung and Rhee for the preparation and catalytic use of PdRh DEN alloys with PAMAM-OH dendrimers for the partial hydrogenation of 1,3-cyclooctadiene by PdRh DENs. (Reproduced with permission from [72]. Copyright © (2003) Elsevier.)

monometallic PtNPs or a mixture of Pt and AuNPs, which was attributed to the decreased electronic density on the Pt shell arising from the influence of the Au core [59b]. Similar effects were invoked for the better activity of heterobimetallic DEN catalysts compared with monometallic DENs hydrogenation of *p*-nitrophenol [69], selective hydrogenation of 1,3-cyclooctadiene [69b], hydrodechlorination of 1,2-dichloroethane [69c] and CO oxidation [69d–g].

3.4
Highly Efficient 'click'-Dendrimer-Encapsulated and Stabilized Pd Nanoparticle Pre-Catalysts

The commercial PAMAM and PIP dendrimers have brought about useful concepts in Pd catalysis owing to their easy access, which often resulted in improved selectivity in hydrogenation of unsaturated substrates. These dendrimers had not been specially designed for catalysis, however. In 1993, we reported the synthesis of arene-cored dendrimers by $CpFe^{+}$-induced *nona*-allylation of mesitylene [73]. Generation growth with $1 \rightarrow 3$ connectivity [74] was carried out in a divergent way using selective hydrosilylation with $HSiMe_2CH_2Cl$ followed by nucleophilic substitution by a phenolate triallyl dendron producing dendrimers terminated by 3^{n+2} allyl branches ($n = 0$–7) [75–77]. Variations of the divergent growth included cross metathesis with an acrylate-functionalized dendron at the focal point [78] or 'click' chemistry between a propargyl-modified dendron at the focal point and dendritic cores terminated by azido groups [79]. The 1,2,3-triazolyl dendrimers obtained by 'click' chemistry were ideal ligands for Pd^{II} [80] and Au^{III} [81]. When these click dendrimers were terminated by 1,2,3-triazolyl ligands connected to ferrocenyl (Fc) groups, the Fc groups were used as redox sensors for the recognition and titration of Pd^{II} ions. This electrochemical recognition was straightforward using cyclic voltammetry, because all the peripheral redox Fc groups appeared equivalent in a single wave that is both chemically and electrochemically reversible [80]. Various transition-metal cations and oxo-anions (including ATP^{2-}) could be sensed with selectivity, as monitored by the shift of the redox potential of the Fc wave. This titration technique was simple and useful to determine the number of Pd^{II} cations encapsulated in the 'click' dendrimers of various generations (Figure 3.10, [82, 83]).

There are various known modes of coordination of Pd^{II} with triazole ligands, and the monohapto mode was confirmed by X-ray crystal structure determination [84]. Reduction of the G_1 (27 Fc) and G_2 (81 Fc) dendritic-Pd^{II} complexes using $NaBH_4$ or methanol provided PdNPs for which the sizes, determined by TEM, corresponded to the theoretical number of Pd atoms according to the one-to-one stoichiometry determined by electrochemical titration of the Pd^{II} precursors. This result was indicative of intra-dendritic PdNP formation and encapsulation. On the other hand, the PdNPs formed from the G_0 dendrimer (9 Fc) were large. This small dendrimer cannot encapsulate PdNPs, but stabilization still occurs by locating dendrimer around the PdNPs (Figure 3.11). The fact that these G_0-'click'-dendrimer-stabilized PdNPs are large confirms that the PdNP size is independent of

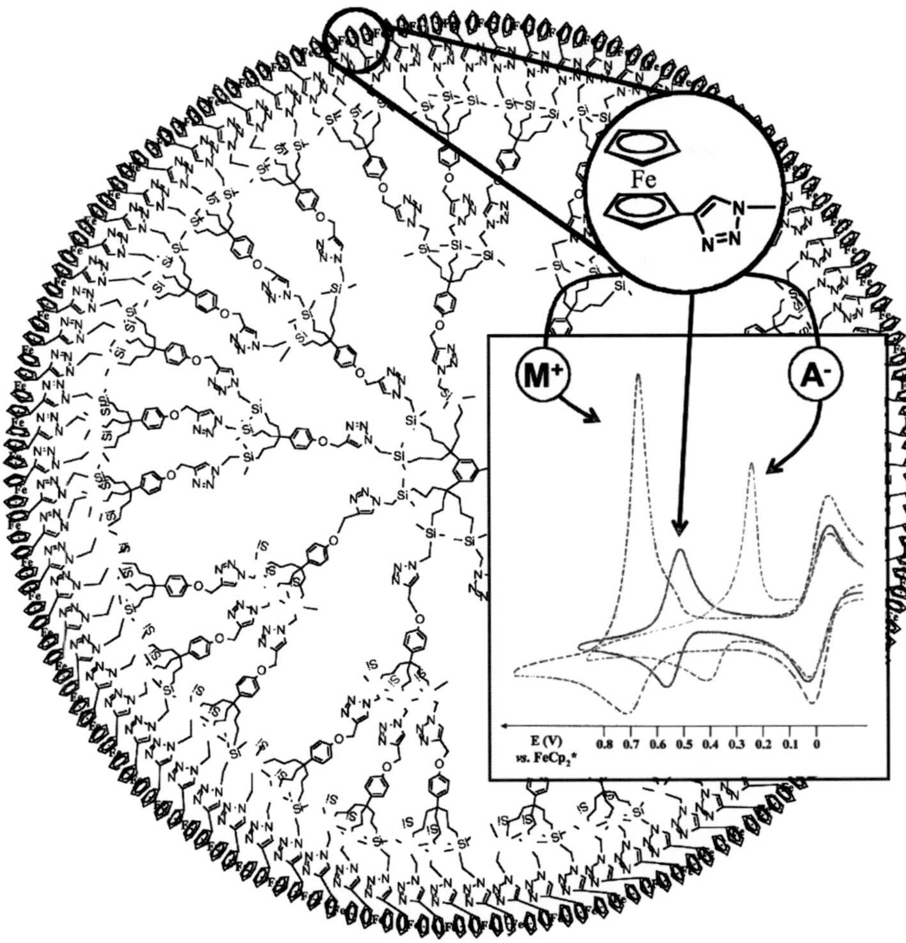

Figure 3.10 Cyclic voltammetry: recognition of both oxo-anions and transition-metal cationic acetonitrile complexes by a second-generation 'click' ferrocenyl dendrimer.

the dendrimer size when the latter is too small (Figure 3.12). Thus the smallest PdNPs are those formed from the G_1 dendrimer containing 27 Fc groups, 36 triazolyl rings and encapsulating PdNPs that contain 36 Pd atoms.

Selective hydrogenation of dienes to monoenes was readily achieved under ambient conditions for small dienes [85], but large steroidal dienes remained unreacted, in accord with their lack of ability to reach the PdNP surface. The rates (TOFs) and turnover numbers (TON)s of hydrogenation were all the larger as the PdNPs were smaller, as expected from previous results with polymer-stabilized PdNPs [16] according to a mechanism that involves mechanistic steps of the hydrogenation on the PdNP surface.

Whereas hydrogenation catalysis proceeds at the PdNP surface, as shown above, and therefore depends on the PdNP size, the catalysis of Suzuki–Miyaura C—C

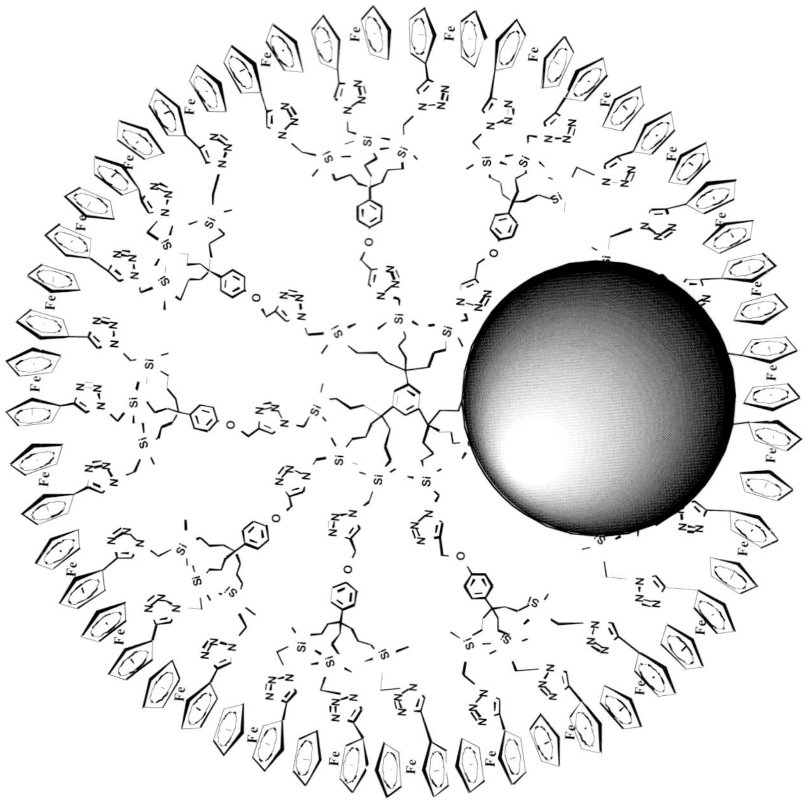

Figure 3.11 Synthesis of 'click'-ferrocenyl dendrimer-encapsulated PdNPs.

coupling [86] between PhI and $PhB(OH)_2$ was carried out at room temperature and does not depend on the PdNP size and whether its stabilization is intra- or interdendritic. This shows that the dendrimer is not involved in the rate-limiting step of the reaction. The dendrimer-stabilized PdNPs work identically whatever their size, and the TONs increase upon decreasing the amount of catalyst from 1% down to 1 ppm or upon dilution of the reaction solution. Thus, the efficiency of the catalyst is remarkable at homeopathic amount (54% yield at 298 K with 1 ppm equivalent of Pd atom, that is, TON = 540 000) and a quantitative yield is not even reached (75% yield) with 1% equivalent Pd atom [87]. The 'homeopathic' catalysis was already observed for the Heck reaction at 423 K and was rationalized [88] by de Vries according to a leaching mechanism involving detachment of Pd atoms from the PdNP subsequent to oxidative addition of the organic halide PhI on the PdNP surface [89–92]. This mechanism is established for high-temperature reactions due to decomposition of the Pd catalyst to naked PdNPs, but it is less expected for a room-temperature reaction. The ease of the room-temperature reaction must be due, however, to the lack of ligation onto the dendrimer-stabilized PdNPs, that therefore can easily undergo oxidative addition of PhI at their surface, which provokes the

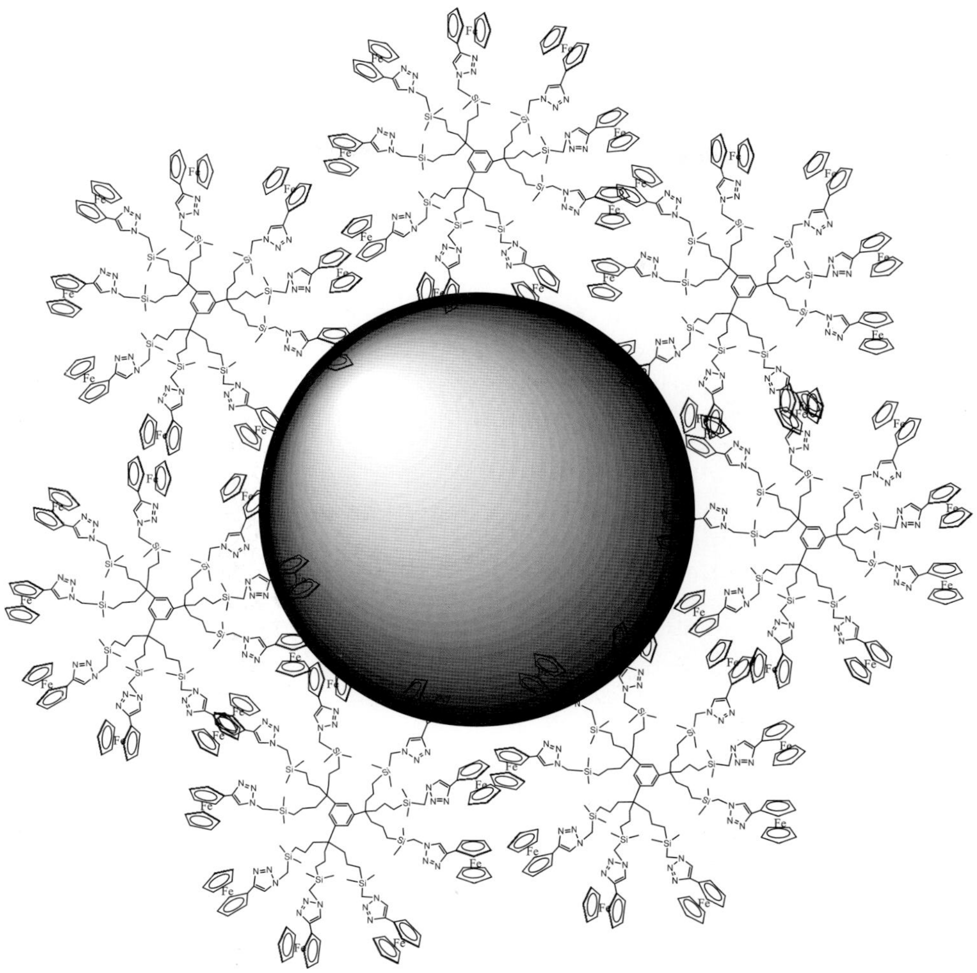

Figure 3.12 Pd nanoparticle surrounded and stabilized by several small G_0 nonaferrocenyl dendrimers.

leaching of Pd atoms. These isolated Pd atoms are apparently extraordinarily reactive in solution, because they do not bear ligands other than the very weakly coordinating solvent molecules. The limit in their efficiency is reached when these atoms or small clusters are trapped by their mother NP, if the solution is moderately concentrated. This trapping mechanism that inhibits catalysis is all the less efficient as the catalyst is more diluted in the solution. Therefore it is not efficient under extremely diluted solutions, whereas it strongly inhibits catalysis at relatively high concentrations. It is likely that this concept can be extended to other PdNP-catalyzed C–C bond formation reactions (Figure 3.13).

Analogous 'click'-dendrimer-stabilized PdNPs with other termini including sulfonate providing solubility in water were also active in aqueous medium for

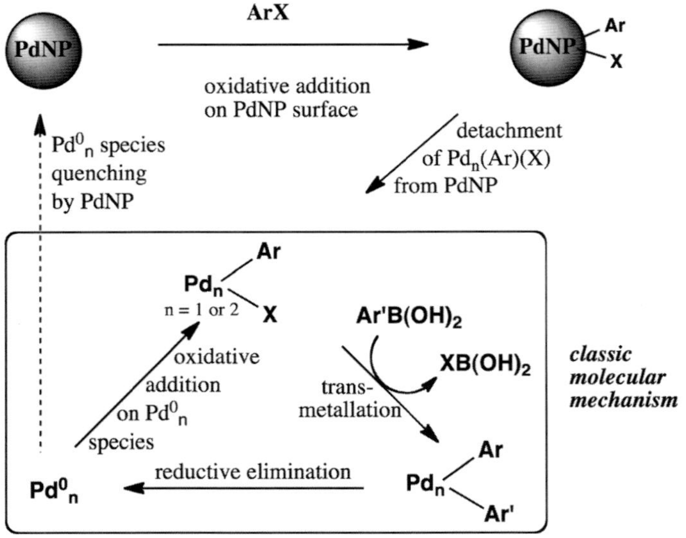

Figure 3.13 Leaching mechanism in the 'homeopathic' catalysis of Suzuki–Miyaura C—C coupling at ambient temperature between PhI and $PhB(OH)_2$ by 'click' ferrocenyl dendrimer-stabilized PdNPs. (Reproduced with permission from [87]. Copyright © 2007 Wiley-VCH Verlag GmbH & Co.)

hydogenation and Suzuki–Miyaura coupling reaction with high TOF and TON values [93], as were also related 'click'-polymer-stabilized PdNPs [94].

It has been shown by Crooks' group that it is possible to extract DENs from PAMAM dendrimers using thiols as potential stabilizing ligands for NPs [95]. Likewise, the G_1-dendrimer-encapsulated PdNPs can be extracted by hexanethiol to yield PdNP-cored hexanethiol stars that also catalyze the Suzuki–Miyaura reaction,

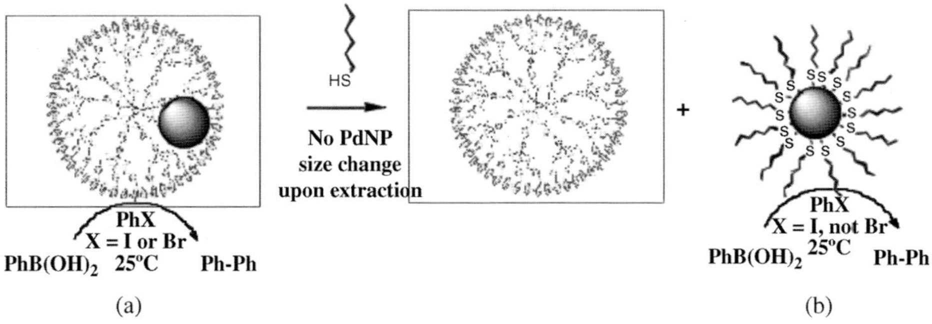

Figure 3.14 Extraction of 'click'-dendrimer-encapsulated PdNPs from the dendrimer with hexanethiol leading to hexanethiolate-PdNPs. (a) Suzuki cat: no PDNP size-dependence; homeopathic down to 1 pm Pd but yield never quantitative due to quenching active Pd atoms by PdNP; Pd black slowly forms. (b) Suzuki cat: no PDNP size-dependence; less active than dendrimer–PdNP; yield quantitative; air and water stable, recyclable; Pd black slowly forms, but only if PhX reacts (X = I, not Br).

under ambient conditions, between phenylboronic acid and iodobenzene, but not bromobenzene contrary to the G_1-dendrimer-encapsulated PdNPs. PdNP-cored decanethiolate species were formerly found to be air and water stable and good catalysts for the latter Suzuki–Miyaura reaction (Figure 3.14) [87]. Thus, the thiolate ligands are not a poison for this catalysis, but the PdNPs are not as free in the presence of the alkylthiolate ligands as in the dendrimer-stabilized PdNPs that are extremely active catalysts [96].

3.5 Heterogeneous Catalysis

Another promising area is that of dendrimer-encapsulated nanoparticles in heterogeneous catalysis pioneered by Chandler [97, 98]. Although the PAMAM dendrimer frame does not block access to the DEN in aqueous solution, it collapses around the DEN upon drying, and subsequent gas-phase reactions proceed very slowly or not at all [99], with few exceptions that involve extraction with silica and further treatment [100]. Thus the procedure consists in immobilizing the dendrimer template onto a solid support, followed by removing the dendrimer. The problem is that removing the dendrimer thermally leads to increase of both the NP size and NP size distribution. PdNPs supported on mica or highly-oriented pyrolytic graphite were calcined at 903 K forming large aggregates resulting from considerable PdNP size increase [101]. Chandler's group was able to deposit DENs onto commercial silica supports followed by calcination at 573 K under O_2/He, which allowed the original NP size around 2 nm to be retained. The technique was also extended to other catalytically useful metals (mostly Pt and Au) and to bimetallic nanoparticles including PdAuNPs [35, 97, 98, 102, 103]. Calcination of the dendrimer has also been carried out before reduction of the metal cations [104–107]. For instance, impregnation of high surface area carbon black with G4-OH Pt^{2+} Ru^{3+} complex solution, drying and activation by H_2 at 673 K removed all the PAMAM dendrimer and reduces Pt^{2+} and most of the Ru^{3+} to zero valence yielding a PtRu alloy that catalyzed methanol electro-oxidation with activity 60% greater than commercial 20%wt PtRu catalyst [107].

Adsorption properties of Pt DENs for CO and NO were examined and it was shown that untreated catalysts themselves were capable of CO uptake, but only slowly, in a diffusion-controlled process, and into a weak adsorption state. Either O_2 or H_2 pretreatment is necessary for stronger adsorption [108]. CO oxidation catalyzed by Pt/TiO_2 catalysts prepared from PAMAM G4-OH DENs was found to be independent of the activation protocol used, whereas toluene hydrogenation on this catalyst was sensitive to the different activation pretreatments, the most active catalysts resulting from short, low-temperature (423 K) H_2 treatment [109].

A valuable technique consists in the incorporation of the DEN into sol–gel matrixes which, for instance, minimized AuNP and PdNP growth upon calcination of PAMAM DEN dendrimer [110]. Another approach consisted in synthesizing the dendrons on the walls of a mesoporous SBA-15 or MCF-17 silica either by noncovalent (supramolecular) interactions or by covalent attachment [100] or formation of dendrons or dendrimers on the walls of the silica pores and subsequent NP formation [111a,b]. Following amine functionalization of SBA-15 with APTES in toluene, four consecutive sequences of a Michael addition reaction and amidation of ester moieties were carried out to synthesize the dendronized SBA-15 followed by treatment with aqueous K_2PtCl_6, then reduction with excess $NaBH_4$ to produce 1.2 to 2.6-nm sized Pt DENs. Catalysis by these Pd DENs of *p*-nitrophenol reduction to aminophenol was achieved using $NaBH_4$ at 298 K. The catalyst was reused five times preserving good catalytic performances [112a]. Likewise, silica NPs with an average size of 12 nm were aminopropylated by spraying an ethanol solution of 3-aminotriethoxysilane, and subsequent synthesis of hyperbranched G3-PAMAM was achieved by grafting methyl acrylate then ethylene diamine alternatively. Pd DSNs of 10–20 nm size were prepared from this dendronized silica NPs in the presence of the ionic liquid butylmethylimadazolium hexafluorophosphate in tetrahydrofuran (THF), and were efficient for the Suzuki–Miyaura coupling in 50% aqueous ethanol between aryl bromides and triflates and arylboronic acid under mild conditions and could be recycled five times after centrifugation [112b]. PAMAM Pd DEN dendrimers were shown to be efficient and recyclable catalysts for the Suzuki–Miyaura coupling between *p*-bromoacetophenone and phenylboronic acid upon polymerization-induced phase separation leading to a material with 9.9 nm pores; in this process, polymerization was achieved upon copolymerization with ethylene glycol dimethylacrylate [112c]. Immobilization of surface-grafted Au G2 PPI DSNs onto crosslinked poly(4-vinylpyridine) beads yielded a catalyst with high NP polydispersity that was used for the hydrogenation of 4-nitrophenol [112d].

Extraction of Au DENs from G-OH or G-alkyl PAMAM dendrimers by thiols followed by deposition onto commercial titania and CO oxidation catalysis led to results that compared favorably with standard Au test catalysts [113].

3.6
Electrocatalysis

Crooks' group initially showed that monolayers of DENs terminated with OH groups could be immobilized on Au surfaces and used for catalysis of O_2 electroreduction despite weak stability on the electrode [114]. Later, covalent linking between amine-terminated Pd and Pt DENs was achieved by reaction with acid-terminated mercaptoundecanoic acid attached to Au surfaces, yielding durable films. The insulating SAM prevented electrocatalysis, however [115], but the problem was solved by grafting G4-OH PAMAM Pt_{40} DENs onto glassy carbon electrodes at positive potential upon cycling three times from 0 to 1.0 V vs. Ag/AgCl, 3 M NaCl (Figure 3.15) [116].

The effect of Pt DEN size [117] and elemental composition (PdPt DENs) [118] on the kinetics of electrocatalytic O_2 reduction was also investigated, showing the strong size influence and the positive effect of the presence of Pd in PdPt

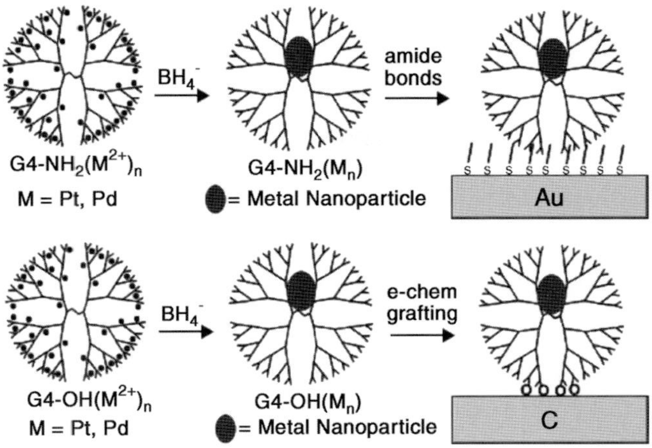

Figure 3.15 Immobilization of Pt onto surfaces for electrocatalysis (a) top: covalent grafting of G4-NH_2 onto the carboxylic acid termini or thiol chains on Au surface; (b) bottom: grafting of G-OH on glassy carbon electrode by cycling at positive potentials. (Reproduced with permission from [13]. Copyright © (2005) American Chemical Society.)

DENs. This reaction was also electrocatalyzed on a Pt DEN-modified nitrogen-doped carbon nanotube electrode [119]. Carbon nanofiber electrodes with immobilized PAMAM Pt DENs exhibited electrocatalytic activity in O_2 reduction. Such catalysts might potentially be usable in fuel cells because good catalytic activity can be achieved at low Pt loading [120]. Electrocatalytic hydrazine oxidation was obtained with DEN modified glassy carbon surface [121]. Alternatively, this reaction was examined with Pt DEN-modified carbon nanotube electrodes [122].

3.7 Conclusion and Outlook

Dendrimer encapsulation of late-transition-metal NPs (in DENs) has appeared as a method of choice to synthesize very small-sized nanoparticles that are the most active in catalysis. The PAMAM dendrimer frame has been shown by Crooks' group and others to serve as an excellent nanoreactor. In these PAMAM DENs, the dendrimer topology affords the property of a nanofilter, providing substrate selectivity and peripheral groups adapted to specific solvents, such as aqueous and fluorous media. The concept has also been somewhat extended by several research groups to the PPI dendrimer family, and by Yamamoto's group to phenylazomethine dendrimers that also showed great promise in this respect.

We have designed 'click' dendrimers with $1 \rightarrow 3$ connectivity containing several generations of triazole groups that can serve as ligands for late-transition-metal cations. Reduction of these 'click' dendrimer-Pd^{II} complexes has led to the formation of precise PdNPs that were shown to be very efficient catalysts for olefin hydrogenation and Suzuki–Miyaura cross-coupling reactions in organic solvent or 50% aqueous ethanol depending on the nature of the solubilizing dendrimer termini. An 'homeopathic' leaching mechanism could be proposed for the Suzuki–Miyaura reaction, based on the lack of dependence of dendrimer generation and type of stabilization (DEN vs. DSN), modest yields at relatively high catalyst concentration and remarkably large increase of TOF upon dilution down to 1 ppm Pd (TOF = 530 000). Thus, the dendrimer-mediated catalysis strategies have provided a deep insight into the mechanism and properties of PdNPs.

Applications to heterogeneous catalysis have been developed in two directions. First, high-temperature heterogeneous catalysts using Pd, Pt, Au and alloy or core@shell heterobimetallic NPs, especially with the seminal work of the Chandler group, demonstrated successful results in spite of the problem of removal of the thermally unstable PAMAM dendrimer template.

Another type of application of DENs to heterogeneous catalysis developed by Crooks' group that is attracting increasing attention, is the electrocatalytic redox reaction of small molecules such as O_2 upon immobilization of the DENs onto electrodes, preferably by covalent derivatization using the functional groups of the dendrimer termini.

In conclusion, dendrimers have served as excellent templates for key applications to various areas of homogeneous and heterogeneous catalysis, providing selective, tailored and very efficient catalysts. This young area is thus promising for future applications involving green nanocatalysis, fuel cells and well-defined materials with new efficient and selective catalysts.

Acknowledgments

Financial support from the Fundação para a Ciência e a Tecnologia (FCT), Portugal (Ph.D. grant to C.O.), Université Bordeaux 1, the Centre National de la Recherche Scientifique (CNRS), and the Agence Nationale pour la Recherche (ANR) are gratefully acknowledged.

References

1 Newkome, G.R., Moorefield, C.N., and Vögtle, F. (2001) Dendrimers and Dendrons, in *Concepts, Syntheses, Applications*, Wiley-VCH Verlag GmbH, Weinheim, Germany.

2 Tomalia, D.A. and Fréchet, J.M.J. (eds) (2001) *Dendrimers and Other Dendritic Polymers*, John Wiley & Sons, Ltd, Chichester, UK.

3 Astruc, D. (ed.) (2003) *Dendrimers and Nanosciences*, C. R. Chimie, Elsevier, Paris, vol **6** (8–10), pp. 709–1208.

4 Vögtle, F., Richardt, G., Werner, N., and Rackstraw, A. (eds) (2009) *Dendrimer Chemistry: Concepts, Syntheses, Properties, Applications*, Wiley-VCH Verlag GmbH, Weinheim, Germany.

5 (a) Newkome, G.R. and Shreiner, C. (2010) Dendrimers derived from 1 → 3 branching motifs. *Chem. Rev.*, **110**, 6338–6442; (b) Bronstein, L.M., and Shifrina, Z.B. (2011) Dendrimers as encapsulating, stabilizing, or directing agents for inorganic nanoparticles. *Chem. Rev.*, **111**, 5301–5344.

6 Astruc, D., Boisselier, E., and Ornelas, C. (2010) Dendrimers designed for functions: from physical, photophysical, and supramolecular properties to applications in sensing, catalysis, molecular electronics, photonics, and nanomedicine. *Chem. Rev.*, **110**, 1857–1959.

7 Caminade, A.-M., Turin, C.-O., Laurent, R., Ouali, A., and Delavaux-Nicot, B. (2011) *Dendrimers, Towards Catalytic, Material and Biomedical Uses*, John Wiley & Sons, Ltd, Chichester, UK.

8 Campagna, S., Ceroni, P., and Puntoriero, F. (2011) *Designing Dendrimers*, John Wiley & Sons, Inc., Hoboken, USA.

9 Zhao, M., Sun, L., and Crooks, R.M. (1998) Preparation of Cu nanoclusters within dendrimer templates. *J. Am. Chem. Soc.*, **120**, 4877–4878.

10 Balogh, L. and Tomalia, D.A. (1998) Synthesis of zerovalent copper nanoclusters. *J. Am. Chem. Soc.*, **120**, 7355–7356.

11 Crooks, R.M., Zhao, M., Sun, L., Chechik, V., and Yeung, L.K. (2001) Dendrimer-encapsulated metal nanoparticles: synthesis, characterization, and applications to catalysis. *Acc. Chem. Res.*, **34**, 181–190.

12 Niu, Y. and Crooks, R.M. (2003) Dendrimer-encapsulated metal nanoparticles and their applications to catalysis. *C. R. Chimie*, **6**, 1049–1059.

13 (a) Scott, R.W.J. Wilson, O.M., and Crooks, R.M. (2005) Synthesis, characterization, and applications of dendrimer-encapsulated nanoparticles. *J. Phys. Chem. B*, **109**, 692–704; (b) Myers, V.S., Weier, M.G., Carino, E.V., Yancey, D.F., Pande, S., and Crooks, R.M. (2011) Dendrimer-encapsulated nanoparticles: New synthetic and characterization methods and catalytic applications. *Chem. Sci.*, **2**, 1632–1646.

14 (a) Bönnemann, H., Brijoux, W., Brinkmann, R., Dinjus, E., Fretzen, T., Joussen, B., and Korall, J. (1992) Highly dispersed metal clusters and colloids for the preparation of active liquid-phase hydrogenation catalysts. *J. Mol. Catal.*, **74**, 323–333; (b) Bönnemann, H. and Brijoux, W. (1996) *Active Metals: Preparation, Characterization, Applications* (ed. A. Fürstner), Wiley-VCH Verlag GmbH, Weinheim, Germany, pp. 339–379.

15 (a) Reetz, M.T. and Helbig, W. (1994) Size-selective synthesis of nanostructured transition metal clusters. *J. Am. Chem. Soc.*, **116**, 7401–7402; (b) Reetz, M.T. and Quaiser, S.A. (1995) A new method for the preparation of nanostructured metal clusters. *Angew. Chem. Int. Ed. Engl.*, **34**, 2240–2241; (c) Reetz, M.T., Helbig, W., and Quaiser, S.A. (1996) *Active Metals: Preparation, Characterization, Applications* (ed. A. Fürstner), Wiley-VCH Verlag GmbH, Weinheim, Germany, pp. 279–297; (d) Reetz, M.T. and Westermann, E. (2000) Phosphane-free palladium-catalyzed coupling reactions: the decisive role of Pd nanoparticles. *Angew. Chem., Int. Ed.*, **39**, 165–168.

16 (a) Toshima, N. and Yonezawa, T. (1998) Bimetallic nanoparticles-novel materials for chemical and physical applications. *New J. Chem.*, **22**, 1179–1201; (b) Toshima, N. (1996) *Fine Particles Sciences and Technology – From Micro- to New Particles* (ed. E. Pellizzetti), Kluwer, Dordrecht, pp. 371–383; (c) He, J-.H., Ichinose, I., Kunitake, T., Nakao, A., Shiraishi, Y., and Toshima, N. (2003) Facile fabrication of Ag-Pd bimetallic nanoparticles in ultrathin TiO_2-gel films: nanoparticle morphology and catalytic activity. *J. Am. Chem. Soc.*, **125**, 11034–11040.

17 Astruc, D. (ed.) (2007) *Nanoparticle and Catalysis*, Wiley-VCH Verlag GmbH, Weinheim, Germany.

18 (a) Scott, R.W.J., Ye, H.C., and Crooks, R.M. (2003) Synthesis, characterization, and stability of dendrimer-encapsulated palladium nanoparticles. *Chem. Mater.*, **15**, 3873–3878; (b) Pande, S., Weir, M.G., Zaccheo, B.A., and Crooks, R.M. (2011) Synthesis, characterization, and electrocatalysis using Pt and Pd dendrimer-encapsulated nanoparticles

prepared by galvanic exchange. *New J. Chem.*, **35**, 2054–2060.

19 Knecht, M.R., Garcia, J.C., and Crooks, R.M. (2006) Synthesis, characterization, and magnetic properties of dendrimer-encapsulated nickel nanoparticles containing <150 atoms. *Chem. Mater.*, **18**, 5039–5044.

20 Gates, A.T., Nettleton, E.G., Myers, V.S., and Crooks, R.M. (2010) Synthesis and characterization of NiSn dendrimer-encapsulated nanoparticles. *Langmuir*, **26**, 12994–12999.

21 Knecht, M.R. and Crooks, R.M. (2007) Magnetic properties of dendrimer-encapsulated iron nanoparticles containing 55 and 147 atoms. *New J. Chem.*, **31**, 1349–1353.

22 Alexeev, O.S., Siani, A., Lafaye, G., Williams, C.T., Ploehn, H.J., and Amidis, M.D. (2006) EXAFS Characterization of dendrimer-Pt nanocomposites used for the preparation of Pt/γ-Al_2O_3 catalysts. *J. Phys. Chem. B*, **110**, 24903–24914.

23 Gröhn, F., Bauer, B., Akpalu, Y.A., Jackson, C.L., and Amis, E.J. (2000) Dendrimer templates for the formation of gold nanoclusters. *Macromolecules*, **33**, 6042–6050.

24 Bosman, A.W., Janssen, H.M., and Meijer, E.W. (1999) About dendrimers: Structure, physical properties, and applications. *Chem. Rev.*, **99**, 1665–1688.

25 Niu, Y. and Crooks, R.M. (2003) Preparation of dendrimer-encapsulated metal nanoparticles using organic solvents. *Chem. Mater.*, **15**, 3463–3467.

26 Higushi, M., Shiki, S., Ariga, S., and Yamamoto, K. (2001) First synthesis of phenylazomethine dendrimer ligands and structural studies. *J. Am. Chem. Soc.*, **123**, 4414–4420.

27 Yamamoto, K., Higushi, M., Shiki, S., Tsuruta, M., and Chiba, H. (2002) Stepwise radial complexation of imine groups in phenylazomethine dendrimers. *Nature*, **415**, 509–511.

28 Satoh, N., Cho, J.S., Higushi, M., and Yamamoto, K. (2003) Novel triarylamine dendrimers as a hole-transport material with controlled metal-assembling function. *J. Am. Chem. Soc.*, **125**, 8104–8105.

29 Higushi, M., Shiki, S., and Yamamoto, K. (2000) Novel phenylazomethine dendrimers: Synthesis and structural properties. *Org. Lett.*, **2**, 3079–3082.

30 Cho, J.-S., Takanashi, K., Higushi, M., and Yamamoto, M. (2005) Phenylazomethine dendrimer complexes as novel hole-transporting materials of organic light-emitting diodes. *Synth. Metals*, **150**, 79–82.

31 Nakashjima, T., Satoh, N., Albrecht, K., and Yamamoto, K. (2008) Interface modification on TiO_2 electrode using dendrimers in dye-sensitized solar cells. *Chem. Mater.*, **20**, 2538–2543.

32 Satoh, N., Nakashima, T., Kamikura, K., and Yamamoto, K. (2008) Quantum size effect in TiO_2 nanoparticles prepared by finely controlled metal assembly on dendrimer templates. *Nature Nanotechnol.*, **3**, 106–111.

33 Zhao, M. and Crooks, R.M. (1999) Intradendrimer exchange of metal nanoparticles. *Chem. Mater.*, **11**, 3379–3385.

34 (a) Yancey, D.F. Carino, E.V., and Crooks, R.M. (2010) Electrochemical synthesis and electrocatalytic properties of Au@Pt dendrimer-encapsulated nanoparticles. *J. Am. Chem. Soc.*, **132**, 10988–10989; (b) Carino, E.V. and Crooks, R.M. (2011) Characterization of Pt@Cu core@shell dendrimer-encapsulated nanoparticles synthesized by Cu underpotential deposition. *Langmuir*, **27**, 4227–4235.

35 Lang, H., Maldonado, S., Stevenson, K.J., and Chandler, B.D. (2004) Synthesis and characterization of dendrimer templated supported bimetallic Pt-Au nanoparticles. *J. Am. Chem. Soc.*, **126**, 12949–12956.

36 Labande, A. and Astruc, D. (2000) Colloids as redox sensors: recognition of $H_2PO_4^-$ and HSO_4^- by amidoferrocenylalkylthiol-gold nanoparticles. *Chem. Commun.*, 1007–1008.

37 Daniel, M.-C., Ruiz, J., Nlate, S., Blais, J.-C., and Astruc, D. (2003) Nanoscopic assemblies between supramolecular redox active metallodendrons and gold nanoparticles: Syntheses, charaterization and selective recognition of $H_2PO_4^-$.

HSO_4^- and adenosine-5-triphosphate (ATP^{2-}) anions. *J. Am. Chem. Soc.*, **125**, 2617–2628.

38 Daniel, M.-C. and Astruc, D. (2004) Gold nanoparticles: Assembly, supramolecular chemistry, quantum-size related properties, and applications towards biology, catalysis and nanotechnology. *Chem. Rev.*, **104**, 293–346.

39 Donnio, B., Garcia-Vazquez, P., Gallani, J.L., Guillon, D., and Terrazzi, E. (2007) Dendronized ferromagnetic gold nanoparticles self-organized in a thermotropic cubic phase. *Adv. Mater.*, **19**, 3534–3539.

40 Buathong, S., Ung, D., Daou, T.J., Ulhaq-Bouillet, C., Pourroy, G., Guillon, D., Ivanova, L., Bernhardt, I., Begin-Colin, S., and Donnio, B. (2009) Thermal, magnetic and luminescent properties of dendronized ferrite nanoparticles. *J. Phys. Chem. C.*, **113**, 12201–12212.

41 Guo, W.H., Li, J.J., Wang, A., and Peng, X.G. (2003) Luminescent CdSe/CdS core/shell nanocrystals in dendron boxes:superior chemical, photochemical and thermal stability. *J. Am. Chem. Soc.*, **125**, 3901–3909.

42 Locklin, J., Patton, D., Deng, S., Baba, A., Milan, M., and Advincula, R.C. (2004) Conjugated oligothiophene-dendron-capped CdSe nanoparticles: Synthesis and energy transfer. *Chem. Mater.*, **16**, 5187–5193.

43 Martin, A.L., Bernas, L.M., Rutt, B.K., Foster, P.J., and Gillies, E.R. (2008) Enhanced cell uptake of superparamagnetic iron oxide nanoparticles functionalized with dendritic guanidines. *Bioconjugate Chem.*, **19**, 2375–2384.

44 Oosterom, G.E., Reek, J.N.H., Kamer, P.C.J., and van Leeuwen, P.W.N.M. (2001) Transition metal catalysis using functionalized dendrimers. *Angew. Chem., Int. Ed.*, **40**, 1828–1849.

45 Astruc, D. and Chardac, F. (2001) Dendritic catalysts and dendrimers in catalysis. *Chem. Rev.*, **101**, 2991–3031.

46 Dijkstra, H.P., van Klink, G.P.M., and van Koten, G. (2002) The use of ultra- and nanofiltration techniques in homogeneous catalyst recycling. *Acc. Chem. Res.*, **35**, 798–810.

47 Gade, L. (ed.) (2006) *Dendrimer Catalysis*, Springer, Heidelberg.

48 Helms, B. and Fréchet, J.M.J. (2006) The dendrimer effect in homogeneous catalysis. *Adv. Synth. Catal.*, **348**, 1125–1148.

49 (a) de Jesus, E. and Flores, J.C. (2008) Dendrimers: solutions for catalyst separation and recycling-a review. *Ind. Eng. Chem. Res.*, **47**, 7968–7981; (b) Andres, R., de Jesus, E., and Flores, J.C. (2007) Catalysts based on palladium dendrimers. *New J. Chem.*, **31**, 1161–1181; (c) Peng, X., Pan, Q., and Rempel, G.L. (2008) Bimetallic dendrimer-encapsulated nanoparticles as catalysts: A review of the research advances. *Chem. Soc. Rev.*, **37**, 1619–1628.

50 Astruc, D., Boisselier, E., and Ornelas, C. (2010) Dendrimers designed for functions: from physical, photophysical and supramolecular properties to applications in sensing, catalysis, molecular electronics, photonics and nanomedicine. *Chem. Rev.*, **110**, 1857–1959.

51 (a) Lewis, L.N. (1993) *Chem. Rev.*, **93**, 2693–2730; (b) Bradley, J.S. (1994) *Clusters and Colloids*, Wiley-VCH Verlag GmbH, Weinheim, Germany.

52 Bönnemann, H. and Richards, R. (2001) Nanoscopic metal particles – Synthetic methods and potential applications. *Eur. J. Inorg. Chem.*, **10**, 2455–2480.

53 (a) Roucoux, A. Schulz, J., and Patin, H. (2002) Reduced transition metal colloids: A novel family of reusable catalysts? *Chem. Rev.*, **102**, 3757–3778; (b) Bronstein, L.M. (2004) *Encyclopedia of Nanoscience and Nanotechnology* (ed. H.S. Nalwa), vol. **7**, American Scientific Publishers, Los Angeles, (c) Zaera, F. (2009) Regio-, stereo-, and enantioselectivity in hydrocarbon conversion on metal surfaces. *Acc. Chem. Res.*, **42**, 1152–1160; (d) Chung, Y.-M. and Rhee, H.-K. (2004) Industrial exploitation of photocatalysis: Progress, perspectives and prospects. *Catal. Surv. Asia*, **8**, 211–223; (e) Bronstein, L.M., Sidorov, S. N., and Valetsky, P.M. (2004)

Nanostructured polymeric systems as nanoreactors for nanoparticle formation. *Russ. Chem. Rev.*, **73**, 501–515; (f) Chung, Y.-M. and Rhee, H.-K. (2004) Internal/external use of dendrimer in catalysis. *Korean J. Chem. Eng.*, **21**, 81–97.

54 (a) Astruc, D., Lu, F., and Ruiz, J. (2005) Nanoparticle catalysis. *Angew. Chem., Int. Ed.*, **44**, 7852–7872; (b) Astruc, D. (2007) Palladium nanoparticles as efficient green homogeneous and heterogeneous carbon-carbon coupling pre-catalysts: a unifying view. *Inorg. Chem.*, **46**, 1884–1894.

55 (a) Zhao, M. and Crooks, R.M. (1999) Homogeneous hydrogenation catalysis with monodisperse, dendrimer-encapsulated Pd and Pt nanoparticles. *Angew. Chem., Int. Ed.*, **38**, 364–366; (b) Niu, Y.H., Yeung, L.K., and Crooks, R.M. (2001) Size-selective hydrogenation of olefins by dendrimer- encapsulated palladium nanoparticles. *J. Am. Chem. Soc.*, **123**, 6840–6846.

56 Zhao, M., Tokuhisa, H., and Crooks, R. M. (1997) Molecule-sized gates based on surface-confined dendrimers. *Angew. Chem., Int. Ed.*, **36**, 2596–2598.

57 Chechik, V., Zhao, M., and Crooks, R.M. (1999) Self-assembled inverted micelles prepared from a dendrimer template: phase transfer of encapsulated guests. *J. Am. Chem. Soc.*, **121**, 4910–4911.

58 Chechik, V. and Crooks, R.M. (2000) Dendrimer-encapsulated Pd nanoparticles as fluorous phase-soluble catalysts. *J. Am. Chem. Soc.*, **122**, 1243–1244.

59 (a) Nakamula, I., Yamanoi, Y., Yonezawa, T., Imaoka, T., Yamamoto, K., and Nishihara, H. (2008) Nanocage catalysts-rhodium nanoclusters encapsulated with dendrimers as accessible and stable catalysts for olefin and nitroarene hydrogenations. *Chem. Commun.*, 5716–5718; (b) Zhang, W., Li, L., Du, Y., Wang, X., and Yang, P. (2009) Gold/platinum bimetallic core/shell nanoparticles stabilized by a Fréchet-type dendrimer: preparation and catalytic hydrogenations of phenylaldehydes and nitrobenzenes. *Catal. Lett.*, **127**, 429–436; (c) Sutton, A., Franc, G. and Kakkar, A. (2009) Silver metal nanoparticles: Facile dendrimer-assisted size-controlled synthesis and selective catalytic reduction of chloronitrobenzenes. *J. Polym. Sci. A: Polym. Chem.*, **47**, 4482–4493.

60 Yeung, L.K. and Crooks, R.M. (2001) Heck heterocoupling within a dendritic nanoreactor. *Nano Lett.*, **1**, 14–17.

61 Yeung, L.K., Lee, C.T., Johnston, K.P., and Crooks, R.M. (2001) Catalysis in supercritical CO_2 using dendrimer-encapsulated palladium nanoparticles. *Chem. Commun.*, 2290–2291.

62 Li, Y. and El-Sayed, M.A. (2001) The effect of stabilizers on the catalytic activity and stability of Pd colloidal nanoparticles in the Suzuki reactions in aqueous solution. *J. Phys. Chem. B*, **105**, 8938–8943.

63 Rahim, E.H., Kamounah, F.S., Fredericksen, J., and Christensen, J.B. (2001) Heck reactions catalyzed by PAMAM-dendrimer encapsulated Pd(0) nanoparticles. *Nano Lett.*, **1**, 499–501.

64 Pittelkow, M., Moth-Poulsen, K., Boas, U., and Christensen, J.B. (2003) Poly (Amidoamine)-dendrimer-stabilized Pd (0) nanoparticles as a catalyst for the Suzuki reaction. *Langmuir*, **19**, 7682–7684.

65 Ooe, M., Murata, M., Mizugaki, T., Ebitani, K., and Kaneda, K. (2004) Supramolecular catalysts by encapsulating palladium complexes within dendrimers. *J. Am. Chem. Soc.*, **126**, 1604–1605.

66 Lemo, J., Heuzé, K., and Astruc, D. (2006) Inorganic Mo 6X 14 clusters and their functionalization. *Inorg. Chim. Acta*, **359**, 4909–4911.

67 Bernechea, M., de Jesús, E., and López-Mardomingo, C. (2009) Dendrimer-encapsulated Pd nanoparticles versus palladium acetate as catalytic precursors in the Stille reaction in water. *Inorg. Chem.*, **48**, 4491–4496.

68 Carino, E.V., Knecht, M.R., and Crooks, R.M. (2009) Quantitative analysis of the stability of Pd dendrimer-encapsulated nanoparticles. *Langmuir*, **25**, 10279–10284.

69 (a) Scott, R.W., Datye, A.K., and Crooks, R.M. (2003) Bimetallic palladium-platinum dendrimer-encapsulated catalysts. *J. Am. Chem. Soc.*, **125**, 3708–3709; (b) Scott, R.W.J., Wilson, O.M., Oh, S.K., Kenik, E.A., and Crooks, R.M. (2004) Bimetallic palladium-gold dendrimer-encapsulated catalysts. *J. Am. Chem. Soc.*, **126**, 15583; (c) Xie, H., Howe, J.Y., Schwartz, V., Monnier, J.R., Williams, C.T., and Ploehn, H.J. (2008) Hydrodechlorination of 1,2-dichloroethane catalyzed by dendrimer-derived Pt-Cu/SiO_2 catalysts. *J. Catal.*, **259**, 111; (d) Auten, B.J., Lang, H., and Chandler, B.D. (2008) Dendrimer templates for heterogeneous catalysts: bimetallic Pt-Au nanoparticles on oxide supports. *Appl. Catal. B: Environment*, **81**, 225–235; (e) Weir, M.G., Myers, V.S., Frenkel, A.I., and Crooks, R.M. (2010) *In situ* X-ray absorption analysis of ~1.8 nm dendrimer-encapsulated Pt nanoparticles during electrochemical CO oxidation. *ChemPhysChem*, **11**, 2942–2950; (f) Chandler, B.D., Long, C.G., Gilbertson, J. D., Fursell, C.J., Vijayaraghavan, G., and Stevenson, K.J. (2010) Enhanced oxygen activation over supported bimetallic Au-Ni catalysts. *J. Phys. Chem.*, **114**, 11498–11508; (g) Bernechea, M., Garcia-Rodriguez, S., Terreros, P., de Jesus, E., Fierro, F.L.G., and Rojas, S. (2011) Synthesis of core-shell PtRu DENs. Relevance as electrocatalysts for CO oxidation. *J. Phys. Chem. C*, **115**, 1287–1294.

70 Toshima, N., Yonezawa, T., and Kushihashi, K. (1993) Polymer-protected palladium–platinum bimetallic clusters: preparation, catalytic properties and structural considerations. *J. Chem. Soc., Faraday Trans.*, **89**, 2537–2543.

71 Chung, Y.M. and Rhee, H.K. (2003) Pt-Pd bimetallic nanoparticles encapsulated in dendrimer nanoreactor. *Catal. Lett.*, **85**, 159–164.

72 Chung, Y.M. and Rhee, H.K. (2003) Partial hydrogenation of 1,3-cyclooctadiene using dendrimer-encapsulated Pd–Rh bimetallic nanoparticles. *J. Mol. Catal.*, **206**, 291–298.

73 Moulines, F., Djakovitch, L., Boese, R., Gloaguen, B., Thiel, W., Fillaut, J.-L., Delville, M.-H., and Astruc, D. (1993) Organometallic molecular trees as multi-electron and proton reservoirs: $CpFe^+$ induced nona-allylation of mesitylene and phase-transfer catalyzed synthesis of a redox active nona-iron complex. *Angew. Chem., Int. Ed.*, **32**, 1075–1077.

74 Newkome, G.R., Yao, Z., Baker, G.R., and Gupta, V.K. (1985) Micelles. Part 1. Cascade molecules: a new approach to micelles. A [27]-arborol. *J. Org. Chem.*, **50**, 2003–2004.

75 Sartor, V., Djakovitch, L., Fillaut, J.-L., Moulines, F., Neveu, F., Marvaud, V., Guittard, J., Blais, J.-C., and Astruc, D. (1999) Organoiron routes to a new dendron for fast dendritic syntheses using divergent and convergent methods. *J. Am. Chem. Soc.*, **121**, 2929–2930.

76 Ruiz, J., Lafuente, G., Marcen, S., Ornelas, C., Lazare, S., Cloutet, E., Blais, J.-C., and Astruc, D. (2003) Construction of giant dendrimers using a tripodal buiding block. *J. Am. Chem. Soc.*, **125**, 7250–7257.

77 Astruc, D. and Ruiz, J. (2010) Organoiron-mediated dendrimer syntheses with 1 à 3 connectivity and applications. Tetrahedron Report No. 903, *Tetrahedron*, **66**, 1769–1785.

78 Ornelas, C., Méry, D., Cloutet, E., Ruiz, J., and Astruc, D. (2008) Cross olefin metathesis for the selective functionalization, ferrocenylation, and solubilization in water of olefin-terminated dendrimers, polymers and gold nanoparticles and for a divergent dendrimer construction. *J. Am. Chem. Soc.*, **130**, 1495–1506.

79 Ornelas, C., Ruiz, J., Cloutet, E., Alves, S., and Astruc, D. (2007) Click assembly of 1,2,3-triazole-linked dendrimers including ferrocenyl dendrimers that sense Both oxo-anions and metal cations. *Angew. Chem., Int. Ed.*, **46**, 872–877.

80 Ornelas, C., Salmon, L., Ruiz, J., and Astruc, D. (2008) Dendrimers: synthesis, redox sensing of $Pd(OAc)_2$, and remarkable catalytic hydrogenation activity of precise Pd nanoparticles

stabilized by 1,2,3-triazole-containing dendrimers. *Chem. -Eur. J.*, **14**, 50–64.

81 Boisselier, E., Diallo, A.K., Salmon, L., Ornelas, C., Ruiz, J., and Astruc, D. (2010) Encapsulation and stabilization of gold nanoparticles with 'click' polyethyleneglycol dendrimers. *J. Am. Chem. Soc.*, **132**, 2729–2742.

82 Astruc, D., Ornelas, C., and Ruiz, J. (2008) Ferrocenyl-terminated dendrimers: design for applications in molecular electronics, molecular recognition and catalysis. *J. Inorg. Organomet. Polym. Mater.*, **18**, 4–17.

83 Astruc, D., Ornelas, C., and Ruiz, J. (2008) Metallocenyl dendrimers and their applications in molecular electronics, sensing and catalysis. *Acc. Chem. Res.*, **41**, 841–856.

84 Badèche, S., Daran, J.-C., Ruiz, J., and Astruc, D. (2008) Synthesis and coordination chemistry of ferrocenyl-1,2,3-triazolyl ligands. *Inorg. Chem.*, **47**, 4903–4908.

85 Ornelas, C., Salmon, L., Ruiz, J., and Astruc, D. (2007) Catalytically efficient palladium nanoparticles stabilized by 'click' ferrocenyl dendrimers. *Chem. Commun.*, 4946–4948.

86 Suzuki, A. (2002) *Modern Arene Chemistry* (ed. D. Astruc), Wiley-VCH Verlag GmbH, Weinheim, Germany, pp. 53–106.

87 Diallo, A.K., Ornelas, C., Salmon, L., Ruiz, J., and Astruc, D. (2007) Homeopathic catalytic activity and atom-leaching mechanism in the Miyaura-Suzuki reactions under ambient conditions using precise 'click' dendrimer-stabilized Pd nanoparticles. *Angew. Chem., Int. Ed. Engl.*, **46**, 8644–8648.

88 Beletskaya, I.P. (2000) The Heck reaction as a sharpening stone of palladium catalysis. *Chem. Rev.*, **100**, 3009–3066.

89 de Vries, G.J. (2006) A unifying mechanism for all high-temperature Heck reactions. The role of palladium colloids and anionic species. *Dalton. Trans.*, 421–429.

90 de Vries, J.G. and de Vries, A.H.M. (2003) The power of high-throughput experimentation in homogeneous catalysis research for fine chemicals. *Eur. J. Org. Chem.*, 799–811.

91 de Vries, A.H.M., Parlevliet, F.J., Schmeder-van de Vondervoort, L., Mommers, J.H.M., Henderickx, H.J.W., Walet, M.A.N., and de Vries, J. (2002) A practical recycle of a ligand-free palladium catalyst for Heck reactions. *Adv. Synth. Catal.*, **344**, 996–1002.

92 Phan, N.T.S., van der Sluis, M., and Jones, C.J. (2006) On the nature of the active species in palladium catalyzed Mizoroki-Heck and Suzuki–Miyaura couplings – homogeneous or heterogeneous catalysis, a critical review. *Adv. Syn. Catal.*, **348**, 609–679.

93 Ornelas, C., Ruiz, J., Salmon, L., and Astruc, D. (2008) Sulphonated 'click' dendrimer-stabilized palladium nanoparticles as highly efficient catalysts for olefin hydrogenation and Suzuki coupling reactions under ambient conditions in aqueous media. *Adv. Syn. Catal.*, **350**, 837–845.

94 Ornelas, C., Diallo, A.K., Ruiz, J., and Astruc, D. (2009) 'Click' polymer-supported palladium nanoparticles as highly efficient catalysts for olefin hydrogenation and Suzuki coupling reactions under ambient conditions. *Adv. Syn. Catal.*, **351**, 2147–2154.

95 Garcia-Martinez, J.C. and Crooks, R.M. (2004) Extraction of Au nanoparticles having narrow size distributions from within dendrimer templates. *J. Am. Chem. Soc.*, **126**, 16170–16178.

96 Lu, F., Ruiz, J., and Astruc, D. (2004) Palladium-dodecanethiolate nanoparticles as stable and recyclable catalysts for the Suzuki–Miyaura reaction of aryl halides under ambient conditions. *Tetrahedron Lett.*, 9443–9445.

97 Chandler, B.D. and Gilbertson, J.D. (2006) Dendrimer-encapsulated bimetallic nanoparticles – Synthesis, characterization, and applications to homogeneous and heterogeneous catalysis. *Top. Organomet.*, **20**, 97–120.

98 Chandler, B.D. and Gilbertson, J.D. (2007) *Nanoparticles and Catalysis*, (ed D. Astruc), Wiley-VCH Verlag GmbH, Weinheim, Germany, pp. 129–160.

99 Albiter F M.A., Crooks, R.M., and Zaera, F. (2010) Adsorption of carbon monoxide on dendrimer-encapsulated platinum nanoparticles: liquid versus gas phase. *J. Phys. Chem. Lett.*, **1**, 38–40.

100 (a) Huang, W., Kuhn, J.N., Tsung, C.-K., Zhang, Y., Habas, S.E., Yang, P., and Somorjai, G.A. (2008) Dendrimer templated synthesis of one nanometer Rh and Pt particles supported on mesoporous silica: Catalytic activity for ethylene and pyrrole hydrogenation. *Nano Lett.*, **8**, 2027–2034; (b) Witham, C. A., Huang, W., Tsung, C.-K., Kuhn, J.N., Somorjai, G.A., and Toste, F.D. (2010) Converting homogeneous to heterogeneous in electrophilic catalysis using monodisperse metal nanoparticles. *Nature Chem.*, **2**, 36–41; (c) Kuhn, J.N., Huang, W., Tsung, C.-K., Zhang, Y., and Somorjai, G.A. (2008) Structure sensitivity of carbon-nitrogen ring opening: Impact of platinum particle size from below 1 to 5nm upon pyrrole hydrogenation product selectivity over monodisperse platinum nanoparticles loaded onto mesoporous silica. *J. Amer. Chem. Soc.*, **130**, 14026–14027.

101 Sun, L. and Crooks, R.M. (2002) Dendrimer-mediated immobilization of catalytic nanoparticles on flat, solid supports. *Langmuir*, **18**, 8231–8236.

102 Long, C.G., Gilbertson, J.D., Vijayaraghavan, G., Stevenson, K.J., Pursell, C.J., and Chandler, B.D. (2008) Kinetic evaluation of highly active supported gold catalysts prepared from monolayer protected clusters: an experimental Michaelis-Menten approach for determining the oxygen binding constant during CO oxidation catalysis. *J. Am. Chem. Soc.*, **130**, 10103–10115.

103 Lafaye, G., Siani, A., Marécot, P., Amiridis, M.D., and Williams, C.T. (2006) Particle size control in dendrimer-derived supported ruthenium catalysts. *J. Phys. Chem. B.*, **110**, 7725–7731.

104 Ruckenstein, E. and Yin, W. (2000) SiO_2-poly(amidoamine) dendrimer inorganic/organic hybrids. *J. Polym. Sci. A* **38**, 1443–1449.

105 Velarde-Ortiz, R. and Larsen, G. (2002) A poly(propylene imine) (DAB-Am-64) dendrimer as Cu2+ chelator for the synthesis of copper oxide clusters embedded in sol-gel derived matrixes. *Chem. Mater.*, **14**, 858–866.

106 Choi, H.C., Kim, W.W., and Dai, H. (2002) Delivery of catalytic metal species onto surfaces with dendrimer carriers for the synthesis of carbon nanotubes with narrow diameter distribution. *J. Phys. Chem. B*, **106**, 12361–12365.

107 Gu, Y., Wu, G., Hu, X., Chen, D.A., Hansen, T., Loye, H.-C., and Ploehn, H.J. (2009) PAMAM-stabilized Pt-Ru nanoparticles for methanol electro-oxidation. *J. Power Sources*, **195**, 425–434.

108 Albiter, M.A. and Zaera, F. (2010) Adsorption properties of supported platinum catalysts prepared using dendrimers. *Langmuir*, **26**, 16204–16210.

109 Crump, C.J., Gilbertson, J.D., and Chandler, B.D. (2008) CO oxidation and toluene hydrogenation by Pt/TiO_2 catalysts prepared from dendrimer encapsulated nanoparticle precursors. *Topics Catal.*, **49**, 233–240.

110 Boyen, H.-G., Kästle, G., Weigl, F., Koslowski, B., Dietrich, C., Zieman, P., Spazt, J.P., Riethmüller, S., Hartmann, C., Möller, M., Schmid, G., Garrnier, M. G., and Oelhafen, P. (2002) Oxidation-resistant gold-55 clusters. *Science*, **297**, 1533–1536.

111 (a) Jiang, Y. and Gao, Q. (2006) Heterogeneous hydrogenation catalysis over recyclable Pd(0) nanoparticle catalysts stabilized by PAMAM-SBA-15 organic-inorganic hybrid composites. *J. Am. Chem. Soc.*, **128**, 716–717; (b) Zheng, Z., Li, H., Liu, T., and Cao, R. (2010) Monodisperse noble metal nanoparticle stabilized in SBA-15: Synthesis, characterization and application in microwave-assisted Suzuki–Miyaura coupling reaction. *J. Catal.*, **270**, 268–274; (c) Krishnan, G.R. and Sreekumar, K. (2010) Synthesis and characterization of polystyrene supported catalytically active poly(amidoamine) dendrimer-palladium nanoparticle conjugates. *Soft Mater.*, **8**, 114–129.

112 (a) Li, H., Lü, J., Zheng, Z., and Cao, R. (2011) An efficient and reusable silica/dendrimer supported platinum catalyst for electron transfer reactions. *J. Colloid Interface Sci.*, **353**, 149–155; (b) Hagiwara, H., Sasaki, H., Tsubokawa, N., Hoshi, T., Tsuda, T., and Kuwabata, S. (2010) Immobilization of Pd on nanosilica dendrimer as SILC: highly active and sustainable cluster catalyst for Suzuki–Miyaura reaction. *Synlett*, **13**, 1990–1996; (c) Ogasawara, S. and Kato, S. (2010) Palladium nanoparticles captured in microporous polymers: a tailor-made catalyst for heterogeneous carbon cross-coupling reactions. *J. Am. Chem. Soc.*, **132**, 4608–4613; (d) Murugan, E. and Rangasamy, R. (2010) Synthesis, characterization, and heterogeneous catalysis of polymer-supported poly(propyleneimine) dendrimer stabilized gold nanoparticle catalyst. *J. Polym. Sci. Polym. Chem.*, **48**, 2525.

113 (a) Korkosz, R.J., Gibertson, J.D., Prasifka, K.S., and Chandler, B.D. (2007) Dendrimer templates for supported Au catalysts. *Catal. Today*, **122**, 370–377; (b) Scott, R.W.J., Sivadinarayana, C., Wilson, O.M., Yan, Z., Goodman, D.W., and Crooks, R.M. (2005) Titania-supported PdAu bimetallic catalysts prepared from dendrimer-encapsulated nanoparticle precursors. *J. Am. Chem. Soc.*, **127**, 1380–1381.

114 Zhao, M. and Crooks, R.M. (1999) Dendrimer-encapsulated Pt nanoparticles: Synthesis, characterization, and applications to catalysis. *Adv. Mater.*, **11**, 217–220.

115 Oh, S.-K., Kim, Y.-G., He, H., and Crooks, R.M. (2003) Synthesis, characterization, and surface immobilization of metal nanoparticles encapsulated within bifunctionalized dendrimers. *Langmuir*, **19**, 10420–10425.

116 Ye, H. and Crooks, R.M. (2005) Electrocatalytic O_2 reduction at glassy carbon electrodes modified with dendrimer-encapsulated Pt nanoparticles. *J. Am. Chem. Soc.*, **127**, 4930–4934.

117 Ye, H., Crooks, J.A., and Crooks, R.M. (2007) The effect of particle size on the kinetics of the electrocatalytic oxygen reduction reaction catalyzed by Pt dendrimer-encapsulated nanoparticles. *Langmuir*, **23**, 11901–11906.

118 Ye, H. and Crooks, R.M. (2007) Effect of elemental composition of PtPd bimetallic nanoparticles containing an average of 180 atoms on the kinetics of the electrochemical oxygen reduction reaction. *J. Am. Chem. Soc.*, **129**, 3627–3633.

119 Vijayaraghavan, G. and Stevensen, K.J. (2007) Synergistic assembly of dendrimer templated platinum catalysts on nitrogen-doped carbon nanotube electrodes for oxygen reduction. *Langmuir*, **23**, 5279–5282.

120 Ledesma-Garcia, J., Escalante Garcia, K. J., Rodriguez, I.L., Chapman, F.J., and Godinez, T.W. (2008) Immobilization of dendrimer-encapsulated platinum nanoparticles on pretreated carbon-fiber surfaces and their application for oxygen reduction. *J. Appl. Electrochem.*, **38**, 515–522.

121 Kim, T.H., Choi, H., Go, B.R., and Kim, J. (2010) Modification of a glassy carbon surface with amine-terminated dendrimers and its application to electrocatalytic hydrazine oxidation. *Electrochem. Commun.*, **12**, 788–791.

122 Chen, Y., Xu, Q., Gao, H., and Zhu, N. (2009) Dendrimer-encapsulated Pd nanoparticles anchored on carbon nanotubes for electro-catalytic hydrazine oxidation. *Electrochem. Commun.*, **11**, 1329–1332.

4
Nanostructured Metal Particles for Catalysts and Energy-Related Materials

Helmut Bönnemann, Guram Khelashvili, Josef Hormes, Timna-Joshua Kühn, and Wolf-Jürgen Richter

4.1
General Survey

After three decades of intense global research one can legitimately ask the question, what has nanoscience contributed to catalysis? When comparing the micrographs of conventional catalysts obtained via classic wet impregnation with those of so-called 'designed nanocatalysts,' industrial catalyst makers tend to state that they have been 'nano' for some time. Indeed, they sometimes find the procedures of 'nanochemical catalyst synthesis' highly sophisticated and 'unnecessarily complicated.' Nevertheless, recent research has given us a greater understanding of the properties of nanoparticles, allowing us to tailor them better for specific catalytic applications, in particular with regard to selectivity and specificity. In this article, we report on the latest advances in the field and describe some of the newest developments.

According to a recent definition given by Schlögl [1], a 'nanocatalyst' may be described as a 'hierarchical system of basic structural units (active sites) with defined assembling strategies for multiple length scales.' Quintessentially, the contribution of nanochemistry to catalysis is the ability to fabricate tailored nanoparticles of dimensionally hierarchical supramolecular structure that are metastable under reaction conditions because a significant fraction of the total number of atoms is coordinatively unsaturated. In contrast to traditional catalysis research – which has tended to be based on trial and error strategies – the key advantage of 'nanocatalysis' is that the application of theoretical tools, combined with high throughput screening, feed-back from physical analysis evidence, and performance test data, allow for the first time, the rational design of a catalyst and the precise control of the kinetics of chemical transformations. The catalytic activity and selectivity depend on the size, shape, and composition of the active nanoparticles. A factor of large practical importance is longevity of the catalyst, which is brought about by stabilizing the active nanocenters via capping agents, their deposition on supports, encapsulation of the particles by sol–gel materials, and special entrapment methods leading to so-called 'yolk shell' catalyst systems. In recent years the

Nanomaterials in Catalysis, First Edition. Edited by Philippe Serp and Karine Philippot.

use of pre-prepared colloidal metallic nanoparticles of controlled size, shape, and composition has opened the door to the synthesis of better defined catalytic materials based on nanochemical methodologies [2]. Most nanocatalysts emerging from intense basic research over recent years are still model systems, which do not have immediate practical applications. However, it is realistic to expect that exploitation of this basic knowledge should enable the industrial fabrication of high-performance nanostructured catalysts within the next few years. A number of good candidates are discussed below in Section 4.2.

The basic features of the 'nanoconcept' are:

- size control ('monodispersity')
- shape effects
- tailored composition
- 'embedding' of active nanoparticles
- support effects.

The state-of-the-art wet chemical syntheses of highly dispersed metallic nanoparticles, i.e., metal colloids, have been extensively covered in a number of recent review articles and book chapters, which we shall not duplicate here. In essence, there are three main methods of preparation based on colloid chemistry techniques in solution, *viz* chemical reduction of metal salt precursors, electrochemical synthesis, and controlled decomposition of organometallic compounds and metal complexes. The resulting metallic nanoparticles are stabilized by polymers, for example poly(vinylpyrrolidone) (PVP), surfactants or ligands.

The issue of size control is discussed in great detail in Ref. [3]. In this context, nanoparticles are considered 'monodisperse' if the nanoparticle size deviates by less than 15% from the average. Monodispersity is necessary in order to be able to study size-selectivity relations. Also, particles having equal sizes are less susceptible to Ostwald ripening (i.e. the larger particles grow at the expense of the smaller ones). Even subnanometer nanoparticles, for example monodisperse Pt_{13} clusters, may be prepared in high yields in a one-pot reaction via wet chemistry [4]. The electronic structure of subnanometer sized nanoparticles can control catalytic activity [5, 6]. Subnanometer size effects on the direct propylene epoxidation reaction have recently been studied using silver by Schlögl *et al.* [7]. It was found that unpromoted, size-selected Ag_3 clusters and ~3.5-nanometer Ag nanoparticles on alumina supports can catalyze this reaction with only a negligible amount of carbon dioxide formation and with high activity at low temperatures. Oxidized silver trimers are more active and selective in the epoxidation reaction, owing to their open-shell electronic structures. These results suggest that new architectures based on ultra-small silver may provide highly efficient catalysts for propylene epoxidation. Subnanometer sized Pt clusters were also shown to outperform the best olefin epoxidation catalysts, in both activity (up to two orders of magnitude higher turnover frequencies) and selectivity [8].

The dependence of particle size on Pt–Pt bond length has recently been demonstrated [9]. The different aspects of the control of particle shape to yield, for example cubes or tetrahedral, are comprehensively addressed in a seminal review by El-Sayed [10], and the crucial influence of shape on nanocatalysis has been shown [11]. In systematic study the authors demonstrated that while monodisperse Pt tetrahedra of 5.0 ± 0.1 nm size are active catalysts in the Suzuki reaction, near spherical Pt particles of similar size (5.2 ± 0.1 nm) size are totally inactive. This qualitatively confirms the hypothesis that the edge and corner atoms are the active sites in many catalytic centers [12]. The influence of the shape of the metal particles has also been shown to be crucial in electron transfer catalysis. The electron transfer reaction between hexacyanoferrate (III) ions and thiosulfate ions has been studied by comparing the results of well-shaped (nearly spherical, tetrahedral and cubic) Pt nanoparticles. The tetrahedral nanoparticles that exposed mainly (111) facets and also a high fraction of surface atoms on edges and corners were found to be the most catalytically active [13]. Nanometallic particles having tailored bi- or even trimetallic constituents can be obtained via organometallic nanochemistry [14], as well as by the stepwise alcohol reduction of metal salt precursors, which was developed by Toshima [15, 16]. A classical application of bimetallic particles in nanocatalysis consists in Pt/Ru fuel cell catalysts [17]. This issue will be discussed in Section 4.2.1. The general influence of the composition of metallic nanoparticles on their catalytic performance is discussed in detail (see Section 4.2.3 of ref. [2]).

The embedding' of active nanoparticles, i.e., encasing the active catalyst inside a protecting environment such as sol–gel matrices, helps to avoid unwanted sintering, especially in nanocatalyst applications at high temperatures. In heterogeneous catalysis a sharp drop in activity is often observed as a result of sintering of both the metal and the corresponding support, since the nanosized components are nonequilibrium systems, and tend to rearrange to more stable (less active) arrangements. A way out of this problem is to isolate individual nanoparticles using protective shells to give core-shell type structures. Schüth *et al.* [2] have termed core@shell for these systems to indicate the structure of such materials. As reported by Somorjai *et al.*, an outer shell of 17 nm-thick mesoporous silica prevents sintering of 14 nm Pt particles in air, even at temperatures as high as 1023 K [19]. The three-step synthesis of Pt@mSiO_2 is summarized in Figure 4.1a. (1) shows the synthesis of Pt nanoparticles with tetradecyltrimethylammonium bromide (TTAB) as the capping agent; (2) the silica polymerization around the Pt cores to form the as-synthesized Pt@SiO_2 mesostructures; and (3) subsequent calcination to remove the TTAB molecules and yield Pt@mSiO_2. Figure 4.1b illustrates how the Pt cores were encased within the silica shells at temperatures up to 1023 K in air. Using ethylene hydrogenation as the test reaction, the authors have shown that neither the access of reacting substrates nor the exit of product molecules is hampered by the mesoporous silica shells. High thermal stability, as well as uniform mesoporous shell structures have also been observed for a number of similar systems, such as Au@SiO_2 [20], Ag@SiO_2 [21], Ni@SiO_2 [22], Au@SnO_2 [23], and Pd@CeO_2 [24]. Recently, high-temperature stable Fe/FeN_x@porousSiO_2 core shell catalysts for efficient ammonia decomposition have been obtained [25]. In summary, these results indicate that nanoparticle@oxide core shell systems are very promising candidates for future large-scale applications in industrial catalysis.

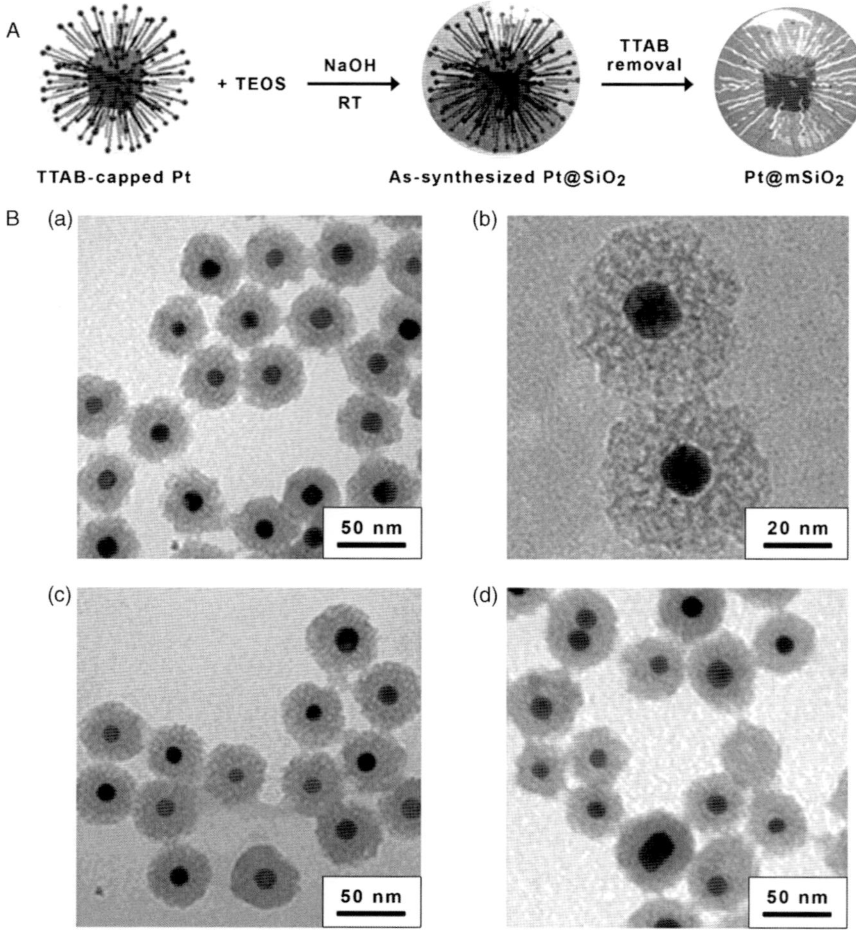

Figure 4.1 (A) Scheme showing the synthesis of Pt@mSiO_2 nanoparticles, (B) Thermal stability of Pt@mSiO_2 nanoparticles. (a)–(d), TEM images of Pt@mSiO_2 nanoparticles after calcination at 623 K (a, b), 823 K (c) and 1023 K (d). (Reproduced with permission from [19]. Copyright © (2009) Am Nature Publishing Group (NPG).)

In 'yolk shell catalysts' (represented by a comma and a blank from the core symbol, i.e., 'core, @shell') the core particle is not densely surrounded by the protective shell, but rather loosely encapsulated in a hollow sphere. These systems have attracted significant attention in recent research activities since the thin, porous, and hollow shells provide improved accessibility of the reactants to the insulated active nanoparticle. One of the first examples of a yolk shell material of catalytic relevance was the synthesis of Pt, @CoO, published in 2004 by Alivisatos *et al.* [26]. A more efficient and high-temperature stable Au (15–17 nm), @ZrO_2 yolk shell

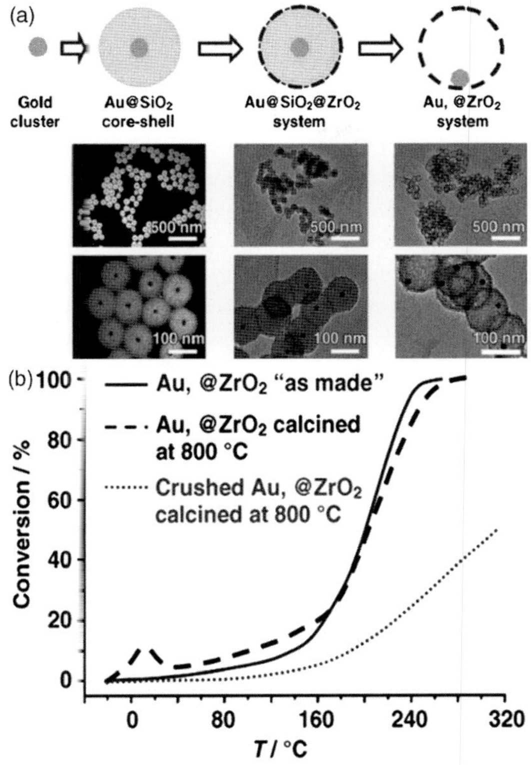

Figure 4.2 (a) General scheme of the synthetic process and TEM images (left: dark field; middle and right: bright field) of the products obtained after each step, (b) Catalytic performance of the different samples. The activities of the catalysts for the oxidation of CO were measured in a plug-flow reactor. (Adapted with permission from [2]. Copyright © (2011) Royal Society of Chemistry.)

nanocatalyst was prepared by Schüth *et al.* using the four step synthesis shown in Figure 4.2a [27]. The activities of the catalyst were tested for the oxidation of CO using a plug-flow reactor. As illustrated in Figure 4.2b, the thermal stability of the Au, @ZrO_2 system was found to be excellent and no growth of the gold particles was observed, even after treatment up to 1073 K. In contrast, crushing the zirconia shell caused mobility of the gold particles and sintering of the gold, resulting in a sharp decrease in activity. Subsequent research has confirmed that 'nanoreactors' with a core of ca. 6 nm gold nanoparticles and a mesoporous zirconia or titania shell are highly efficient and long-term stable catalysts [28, 29].

In the meantime, a whole series of yolk shell type nanocatalysts with a plethora of different yolks and shells has been reported, illustrating the rapid development of this nanocatalyst research field. Examples are Ni, @SiO_2, [30], Au, @Fe_2O_3 [31], Pt, @hmc (hmc = hollow mesoporous carbon) [32], Rh, @hmc [33], and Au,

@hmc [34]. In general, nanometal catalysts encapsulated in mesoporous shells were found to combine remarkable catalytic activity with excellent anti-aggregation properties. In order to enhance the specific catalyst loading, multiple core-shell nanocomposite catalysts have recently been developed. In this case several concentric layers are present. The nomenclature of this type of structure is indicated by separating the different layers by @ from each other, for example core1@core2@-shell, where core1 is the one closest to the nanoparticle. One example of this type of rather sophisticated nanocatalyst system is described in [35].

Although the industrial application of such special structures is still hampered by the complicated synthetic procedures, which are difficult to scale up, the practical value of solid catalysts using highly dispersed metallic nanoparticles supported on carbon or oxide carriers has already been demonstrated [36]. Worthy of note here are magnetically recoverable nanocatalysts [37, 38]. Selected examples of systems that are of significant relevance for energy-related catalysis will be discussed in Section 4.2.1. Another field of practical use is that of cross-coupling reactions, where metallic colloids are deployed in quasi-homogeneous phase (see Section 4.2.2). In Section 4.3 the impact of nanosized materials on energy-related processes is discussed, while Section 4.4 gives an overview of recent progress in the physical characterization of nanomaterials achieved using evidence from X-ray absorption spectroscopy (XAS).

4.2 Nanostructured Clusters and Colloids as Catalyst Precursors

4.2.1 Selected Applications in Energy-Related Processes

For a broader view of energy-related catalysis the reader should consult Volume 6 of the *Handbook of Heterogeneous Catalysis* [39], which describes all aspects of the subject in great detail. In order to demonstrate the impact of recent achievements in nanoscience in the field of catalysis, we restrict our focus here on three specific examples, namely the Fischer–Tropsch process, fuel cell catalysts, and the partial oxidation of methane.

4.2.1.1 Size-Selective Fischer–Tropsch Nanocatalysts

In 1923 Franz Fischer and Hans Tropsch at the Kaiser Wilhelm Institute for Coal Research in Mülheim, Germany (presently the Max-Planck-Institut für Kohlenforschung) first published the catalytic conversion of mixtures of CO and H_2 to hydrocarbons [40]. The gaseous mixture of CO and H_2, now known as synthesis gas or syngas, was prepared by treating coal with oxygen and steam at high temperatures. The catalytic CO hydrogenation – thereafter generally referred to as the 'Fischer–Tropsch (FT) process' – yields a whole range of valuable products, for example high-quality (i.e. sulfur- and nitrogen-free) diesel fuel, gasoline or alternatively alkenes and linear alkanes, depending on whether iron, cobalt, or ruthenium is used in the form of solid catalyst employed. By 1938 owing to their then abundant supply of coal, there were nine FT plants in operation at different places in Germany making fuels synthetically using the first generation of cobalt-based catalysts. Following enormous petroleum reserves being discovered in the Middle East in the 1950s and the subsequent wave of cheap crude oil, all coal-based FT plants in Germany were shut down. Since 70% of the capital costs stem from syngas production, they were uneconomic at the time. Interestingly, in South Africa, where suitable coal resources are available near the surface and hence cheaply available, the technology survived. This was particularly fueled by the oil embargo imposed on the country during apartheid times. The oil price crises of the 1970s led to the development of a second generation of Fischer–Tropsch catalysts. Also methane-based Fischer–Tropsch technology – commonly referred to as the gas-to-liquids (GTL) process – brought a major renaissance in the technology. New FT plants based on efficiently produced syngas from natural gas were built, for example in Malaysia. Both the coal- and gas-to-liquids processes are now well established and relied upon by industry. Currently, close to 50% of South Africa's transportation fuel is produced by this method and experts expect that in the near future commercial GTL chemistry could account for some 10% of the global diesel fuel supply [41]. From an ecological point of view, syngas production based on biomasses-to-liquid (BTL) processes could become an attractive option [42].

In the meantime, a third generation of FT catalysts has been developed by de Jong [43], Martinez [44], and van Steen [45]. 'Nano for Energy' is the relevant catchword here. Back in the 1990s in a systematic study of the influence of particle size on FT catalyst performance Iglesia found that reducing the cobalt catalyst's particle size from approximately 200 nm to 10 nm led to an increase in FT catalytic activity [46–48]. The true breakthrough came when de Jong *et al.*, using an inert carbon nanofiber as a neutral support material for the dispersion of cobalt nanoparticles of different sizes, found a very sharp maximum in activity at 6 nm (cf. Figure 4.3) [43]. When comparing the catalyst activity systematically in terms of the intrinsic activity referred to unit metal surface area, i.e., turnover frequency (TOF), de Jong's data show that at a syngas pressure of 1 bar, the TOF does not vary with particle size as the size is reduced from 25 nm to 6 nm. When the particle diameter is decreased further, a sharp drop in TOF occurs and, in addition, a decrease in product selectivity is observed. When the TOF is measured at a syngas pressure of 35 bars, the 'falloff point' occurs at a particle size of 8 nm. Larger particles make inefficient use of cobalt, and smaller particles are less useful catalysts.

The reason why TOF falls off so sharply for particles smaller than 6 or 8 nm still needs to be established. One possibility is that oxidation may degrade the catalysts. This would affect the smaller nanoparticles preferentially because they are more readily oxidized than the larger ones. This explanation is supported by thermodynamic calculations carried out by Eric van Steen, who showed that under common FT reaction conditions, cobalt particles smaller than 4.4 nm in diameter are likely to be oxidized [49]. Side reactions that form cobalt carbide surface species may also trigger catalyst deactivation.

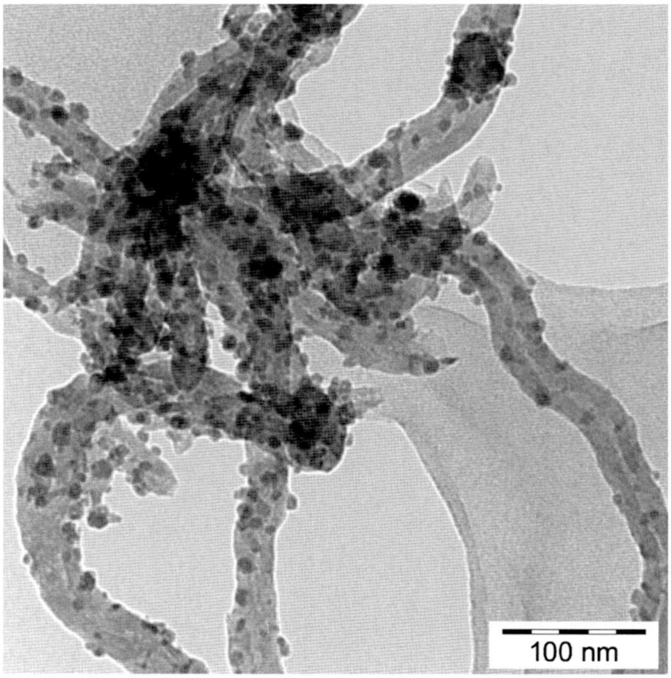

Figure 4.3 TEM image of HDP9 showing particles with sizes of ca 14 nm decorating the carbonfibers. (Reproduced with permission from [43]. Copyright © (2006) American Chemical Society.)

Commercial catalysts, in contrast, usually have a particle size close to 20 nm. It has been estimated that decreasing the particle size of today's industrial catalysts could improve the efficiency of cobalt usage by a factor of two to four. This significant finding illustrates the high impact that particle size has on the catalytic TOF. Taking a closer look at de Jong's study [43], however, reveals that size is not the only factor that controls FT catalyst activity. Another important aspect is the shape of the active nanoparticles. A key to understanding the catalytic mechanism was provided by an X-ray investigation showing that cobalt's coordination number changes as the particles are exposed to syngas. This suggests that the catalyst surfaces undergo reconstruction and become facetted owing to the influence of the reagents. Reconstruction of crystal surfaces generally leads to the formation of surface steps and such steps have been identified by several researchers as those specific surface sites that promote chemical reactions. Consequently, de Jong and coworkers propose that particles with 6 or 8 nm diameters have the optimum size because they are the smallest particles that have domains large enough to stabilize the monoatomic steps created by CO-induced surface reconstruction. In this context, van Santen and coworkers have found that monoatomic steps on ruthenium, which has structural and catalytic properties similar to those of cobalt, significantly lower the activation barrier to CO dissociation, a crucial step in Fischer–Tropsch catalysis [50] However, size and shape of the active nanoparticles are not the only factors to trigger the efficiency of FT catalysts. Metal–support interactions are another important factor. It has been demonstrated by a combination of XAS techniques and catalytic activity studies that under typical Fischer–Tropsch reaction conditions cobalt particles with diameters less than 10 nm interact strongly with common catalyst supports. In the presence of steam the cobalt(0) particles are easily oxidized to cobalt oxide species, such as CoO. These oxides, in turn, can react irreversibly with an alumina support to form cobalt–aluminate species which in turn, swallow up the catalyst particles [51]. Subsequent to the de Jong's seminal findings, international research activities have developed the field further. Martinez has studied the cobalt particle size effect via *in situ* spectroscopic characterization of reversed micelle-synthesized nanocobalt catalysts and has proposed a facile catalyst synthesis that is performed in the liquid phase [44, 52]. Van Steen has reported the preparation of supported nanosized cobalt oxide and fcc cobalt crystallites [53]. Very recently this group has also developed a novel synthesis route for Co, @SiO_2 egg-yolk FT catalysts. Accordingly, egg-yolk type FT catalysts were prepared by leaching the shell of a homogeneously impregnated, reduced catalyst with nitric acid followed by neutralization [54]. FT nanocatalysts in which metallic Co, Ru, Fe, Pd and Rh particles are disposed in pure ionic liquids have been recently disclosed in the patent literature [55]. Thus, in the area of Fischer–Tropsch synthesis, basic research has produced a large body of evidence that confirms that nanochemical methodologies can help catalyst manufacturers to significantly improve their products; recent advances achieved in Fischer–Tropsch syntheses are reported in Ref [56].

4.2.1.2 Nanocatalysts for Fuel Cell Devices

Fuel cells convert chemical energy into electrical energy directly at the interface between the electron conducting electrodes and the electrolyte. A general survey of all types of fuel cells is given in [57]. The specific impact of nanotechnology on the development of fuel cells has recently been reviewed by Heinzel and König [58]. Nanoscience has made significant contributions to the advancement of fuel cell devices in two areas. One is membrane materials and the other is the development of nanosized electrocatalysts, which consist of precious metal particles in the range of 1–5 nm supported on carbon. In the field of membranes the introduction of nanoscale hydrophilic inorganic materials, such as silica, in the polymer membranes led to the deployment of composite membranes that helped to improve the water management in the cell. This is a crucial factor that determines the ion conductivity and the overall efficiency of the fuel cell. Recent developments have been extensively reviewed in [59–61]. In order to improve the durability of fuel cell catalysts and prevent their degradation, multiwalled carbon nanotubes have been proposed as an alternative support material to carbon black [62]. Here we restrict our attention to the low-temperature proton exchange membrane fuel cell (PEMFC) and the direct methanol fuel cell (DMFC), in both of which the application of nanocatalysts is of critical importance.

In low-temperature PEMFCs the working principle is anodic oxidation of hydrogen and cathodic reduction of oxygen at ionomer-bonded, carbon-supported Pt metal catalysts. Since humidified perfluorosulfonic acid is mostly used as the electrolyte, the working temperature is limited to a maximum of ca. 363 K.

The main electrochemical reactions are:

$$\text{Anode}: \ H_2 \rightarrow 2\,H^+ + 2\,e^-$$

$$\text{Cathode}: \ \tfrac{1}{2}\,O_2 + 2\,e^- \rightarrow O^{2-}$$

$$\text{Overall}: \ H_2 + \tfrac{1}{2}\,O_2 \rightarrow H_2O$$

In theory, this reaction should yield a cell voltage of 1.23 V. However in practice, cell voltages of 1 V (at zero current) and 0.5 V during operation are reached. For technical applications therefore a number of single cells must be connected to give a 'cell stack.' Because of their ability to rapidly function at ambient temperature, low-temperature PEMFCs are the first choice for automotive and other mobile applications. The ideal feed for PEMFCs is pure hydrogen compressed at 70 MPa. In practice, chemical hydrogen carriers, such as methane, propane/butane, gasoline, or methanol, are used which have to be catalytically converted at $\approx$1073 K to H_2-rich reformates. Since the raw reformate gas still contains 8–10% CO, this would immediately poison the surface of the Pt catalyst. In two subsequent water gas shift reactors working at $\approx$723 K and $\approx$523 K the CO concentration is reduced to below 1%, which is still too high for PEMFC applications, where H_2 with a CO content of $<$10 ppm is required. An additional, highly selective CO oxidation catalyst is therefore required, which – for thermodynamic reasons – should even work at ambient temperatures. In his pioneering work, Haruta found as early as 1989 that when nanosized Au is deposited as hemispherical nanoparticles (2–10 nm) on selected metal oxides such as α-Fe_2O_3, Co_3O_4, TiO_2 or ZrO_2, the system exhibits surprisingly high catalytic activity for CO oxidation, even at 200 K [63]. In a detailed study on the oxidation mechanism, Haruta found that at temperatures below 200 K, the TiO_2 surfaces and the perimeter interfaces around Au particles are covered with carbonate species formed by surface reactions of CO. Specifically, the reaction of CO with O_2 occurs only on the step, edge and corner sites of the Au nanoparticles, with apparent activation energies of almost zero kJ/mol. From this Haruta concluded that when the Au particles are small enough, the catalytic activity can be detected at any temperature [64]. Owing to ongoing applications for PEMFCs running on cleaned reformate gas, hydrogen low temperature CO oxidation over supported metallic nanoparticles has become one of the most extensively studied systems in heterogeneous nanocatalysis in recent years [65–69]. In these studies, highly dispersed gold catalysts, with particle sizes of 2–5 nm, were supported on reducible metal oxides, in particular on TiO_2, Fe_2O_3, CeO_2, Co_3O_4. Recently, Schüth and coworkers have introduced small colloidal gold nanoparticles deposited on $MgFe_2O_4$ nanocrystals [70] or $Mg(OH)_2$ and MgO as the CO oxidation catalysts, which are efficient at very low temperatures [71]. It has long been believed that gold nanoparticles of 2–5 nm are the active species for

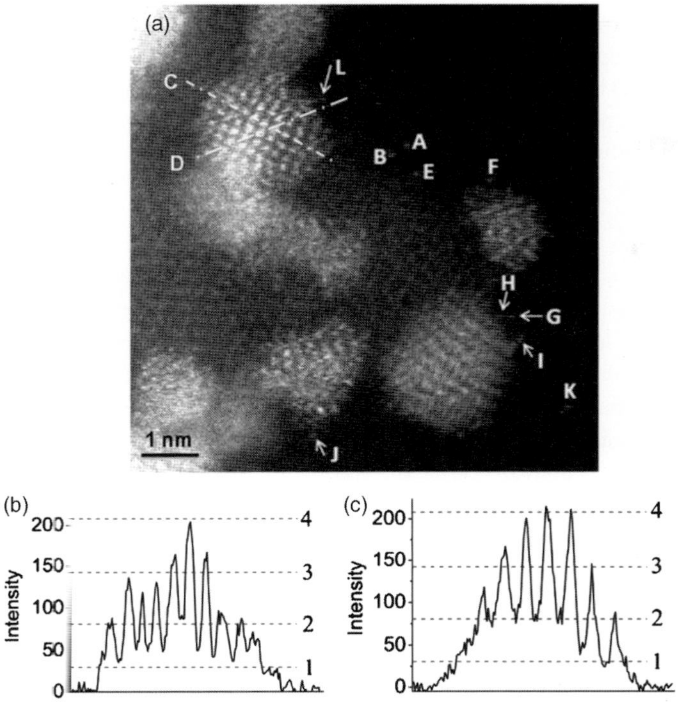

Figure 4.4 (a) HAADF-STEM image of the Au/FeOx catalyst. Different single atoms (A, B, E–L) are marked. (b) and (c) The intensity profiles along C and D marked by dashed lines in Figure 4.4(a). (Reproduced with permission from [73]. Copyright © (2010) Wiley-VCH Verlag GmbH.)

CO oxidation. However, Chen and Goodman reported that gold bilayers are more active than the monolayers [72]. In a very detailed high angle annular dark field scanning transmission electron microscopy (HAADF-STEM) study using highly dispersed Au particles supported on iron oxide as their model, F. Schüth's group has recently addressed the question of how small the gold nanoparticles must be to give an efficient CO oxidation catalyst [73]. One representative gold cluster with a diameter of 18 Å is shown in Figure 4.4a. Single gold atoms marked as A, B, E–L in the figure are also visible. The intensity of the gold atom (L) that is closest to the cluster was chosen as the standard. Its intensity is also near to the average value. Figure 4.4b and c show the intensity profiles of lines C (along [100]) and D (along [110]) after subtracting their respective backgrounds. Based on these results, the authors concluded that the brightest atom columns contain (4 ± 1) atoms along the beam direction. These results rule out the possibility that the gold cluster has a bilayer structure. In summary, these HAADF analyses revealed that the optimum size of the active gold nanoclusters should be between 1 nm and $\approx$0.5 nm and that bilayer structures are not a prerequisite.

The influence of special support materials has also been studied. Using a ceria-modified meso-/macroporous binary metal oxide carrier (CeO_2/TiO_2–ZrO_2) a new catalytic Au system has been developed for the water gas shift reaction [74]. In practice, AUROLITE^{TM} (3 wt% Au on TiO_2) is frequently used as a reference catalyst for comparative activity studies in room temperature CO oxidation. The system allows a reduction in the CO content in contaminated H_2 from 2000 ppm to <1 ppm [75]. Low-temperature PEMFCs (operating at ≈353 K) are dependent on noble metal [platinum group metals (PGM)] catalysts. Much research work has been devoted to the development of non-noble metal systems in order to reduce the price of these cells [76–84]. Since anodic oxidation of pure (or highly purified) hydrogen, containing less than 5 ppm CO, shows only low overpotentials at the electrode and the anode process is much faster than the oxygen reduction at the cathode, anodic oxidation does not usually limit the fuel cell performance. The best catalysts are 2–5 nm platinum particles deposited on carbon supports (loading: 0.45 mg cm^{-2}). Small residues of CO in the feed can be oxidized at the Pt surface [85] and it has been shown that shape-selectivity is an issue here [86]. The challenge in the design of nanochemical catalysts for low-temperature PEMFCs is to maintain the catalyst activity when CO contaminated reformate H_2 is used as the fuel. In 2002 Ralph and Hogarth demonstrated that nanosized (2–3 nm) PtRu alloys can tolerate up to 10 ppm CO contamination in the feed while still performing the H_2 oxidation at high current densities and low anode potentials (<50 mV) [87]. This led to the manufacture of 'CO tolerant' Pt/Ru electrodes that can operate under realistic conditions, i.e., at 353 K using a feed containing up to 40 ppm CO. It is now generally accepted that CO adsorbed on Pt atoms is continuously oxidized to CO_2 by the so-called bifunctional mechanism on neighboring Ru–OH species [88, 89]:

$$\text{Pt-(CO)}_{\text{ads}} + \text{Ru-(OH)}_{\text{ads}} \rightarrow \text{Pt/Ru} + CO_2 + H^+ + e^-$$

The influence of atomic scale intermixing in Pt/Ru alloying on the synergistic effect between Pt and Ru for the CO tolerance of bimetallic nanocatalysts has been studied in detail by a combination of physical analytical studies [90]. In addition, the function of PtRu-based electrocatalysts has been studied using *in situ*X-ray absorption near edge structure (XANES) spectroscopy at operating temperature [91, 92]. Further progress in CO tolerance has been achieved by using surface-engineered Pt/WC anode nanocatalysts [93] or carbon-supported PtRu/WO_x ternaries [94]. All in all, it can be stated that with the advent of CO-tolerant nanocatalyst systems the problems of the anodic hydrogen oxidation have been mainly solved.

The oxygen reduction reaction (ORR) at the cathode, in contrast, is still the object of intense research. With pure Pt catalysts the ORR kinetics are very slow, resulting in large kinetic losses, even at low current densities. Early on, it was noted that Pt alloys exhibit approximately threefold higher activities than pure Pt catalysts supported on carbon [95]. Over the decades this observation has prompted a flood of papers in which nanosized bi- and even trimetallic catalyst particles are described as showing remarkably high ORR activities. It should be noted here that a detailed comparison of the different catalyst systems is hampered in some cases by improper activity assignments of Pt/C benchmark systems. Currently, the best PEMFC cathode catalysts are Pt_xCo_{1-x} alloys with $x \approx 0.5$–0.75. Examining the ORR specific activities of a carbon-supported 20 wt% Pt/C catalyst and a 20 wt% Pt_3Co/C alloy by the rotating disk electrode (RDE) method, Markovic *et al.* found that under identical conditions the nanosized Pt_3Co alloy exhibits a twofold higher efficiency than pure Pt [96]. A series of nanosized electrocatalysts having well-defined particle structures and widely variable catalyst surface compositions has become accessible using organometallic particle synthesis methods combined with size and shape control by surfactants or aluminumalkyls [97, 98]. A special class of effective ORR electrocatalysts consists in Se-modified carbon-supported Ru nanoparticles. High-resolution electron microscopy shows the formation of core-shell particles, comprising a hexagonally packed Ru core and a Ru selenide shell with lamellar morphology. It was shown that modification of Ru nanoparticles with Se considerably enhances their electrocatalytic activity in the oxygen reduction reaction, which is explained by the role that Se plays in inhibiting surface oxidation [99–102]. Since Pt is a very expensive noble metal, the so-called 'Pt skin' concept has been developed, whereby platinum monolayers are deposited on non-noble metal substrates [103]. In fact, a platinum monolayer deposited on a Pd(111) surface exhibits a twofold higher specific activity than pure Pt(111). Using the colloidal precursor approach, Pt skin on Cu core nanocatalysts on a carbon support showing excellent ORR activities have been obtained [104]. Core-shell type $PtCo_3$/C ORR catalysts have also been obtained via colloidal nanometallic precursors and subsequent heat treatment between 623 K and 1273 K under reductive conditions. The catalysts were activated by cyclic voltammetry, which caused partial oxidative leaching of cobalt. Subsequent RDE measurements showed a maximum mass activity for samples tempered at 1073 K. These gave a 2.4 times higher activity than commercial $PtCo_x$/C. Transmission electron microscopy (TEM) images revealed that the particle size of the $PtCo_3$/C catalysts remains almost constant until a temperature of 873 K is reached. X-ray diffraction (XRD) measurements further showed that fct- and fcc-PtCo phases are present in samples with the highest catalytic activities [105]. A review of over more than 100 articles related to heat-treated Fe- and Co-N/C catalysts is available in [106]. Recently, Luo *et al.* reviewed the structural features of PtCo- and PtNi-based ORR catalysts [107].

As early as 2007 Strasser *et al.* established the structure–activity relationships of Pt–Co alloy electrocatalysts in gas diffusion electrode layers [108]. Later, they developed a very fruitful synthetic approach for binary PtM_3 (M = Cu, Co, Ni) and ternary $PtNi_3M$ (M = Cu, Co,Fe, Cr) electrocatalysts which they coined the 'dealloyed core-shell methodology' [109]. This technique was recently reviewed [110].

The active form of the catalyst is obtained by voltammetric dealloying of non-noble metal-rich Pt alloy precursors. In the dealloying process, the less noble precursor component, for example Cu, is selectively removed from the surface of the precursor alloy particles, giving a Pt-enriched particle shell. Single fuel cell tests of the resulting nanocatalysts at the cathode of PEMFCs showed up to six times higher ORR reactivities than those of conventional pure Pt fuel cell catalysts. In addition, the stability of dealloyed nanoparticle catalysts was found to be superior to that of pure Pt particles. Based on X-ray structural and compositional studies a core-shell particle structure was assumed to be the active form of the catalyst, consisting of a Pt-enriched particle shell surrounding a Pt alloy core. Currently, this catalyst family is one of the most active PEMFC ORR systems reported in the literature [111–113]. When it comes to the long-term stability of PEMFCs, the nature of the carbon support is a significant factor [114]. Specifically, Savinova has demonstrated that novel carbon supports derived from the Sibunit^{TM} family possess a number of attractive properties for fuel cell applications. Catalyst degradation as well as compositional and morphological changes at the platinum cathode while the PEMFC system is in operation has been studied by Gasteiger [115]. Despite intense research efforts in the past few years, the low durability of PEMFCs is still the Achilles heel of the envisaged future large-scale application of hydrogen-driven fuel cells in the transportation sector, where the electric power sources have to compete with highly optimized combustion engines running on average for 3000 h.

In contrast to the PEMFC applications discussed above, direct methanol fuel cells (DMFCs) with liquid methanol/air feed have been developed as alternative power supplies for small devices, such as cell phones or laptops that run at less than 50 W. The key advantages of DMFCs are the use of a liquid fuel and the relative simplicity of the cell. At the anode, methanol is completely oxidized giving CO_2:

$$CH_3OH + H_2O \rightarrow CO_2 + 6\,H^+ + 6\,e^-$$

The electrochemical reaction at the cathode is:

$$3/2\,O_2 + 6\,H^+ + 6\,e^- \rightarrow 3\,H_2O$$

Overall:

$$3/2\,O_2 + CH_3OH \rightarrow CO_2 + 2\,H_2O$$

In practice, DMFCs yield open circuit voltages of $\approx$0.7 V and 0.4–0.3 V during operation. While the key components used for DMFCs are similar to those for PEMFCs, additional problems arise from the fact that the polymer membranes in contact with methanol show significant crossover (i.e., transfer) of methanol and water from the anode side to the cathode. This leads to appreciable losses in efficiency due to the simultaneous occurrence of methanol oxidation and oxygen reduction at the air electrode. Nanotechnology is able to make a contribution to the development of improved membranes by providing better chemical and mechanical stability and reducing methanol crossover. One option that has been investigated is the insertion of a porous sheet of poly(tetrafluoroethylene) (PTFE) in order to support the membrane material mechanically [116, 117] Nanomaterials have also been used to synthesize novel inorganic/organic composite materials [118]. An advanced example is the high-temperature stable polybenzimidazole membrane material, PBI, which is usually doped with phosphoric acid [119, 120]. Despite the progress made in the last few years, the field is still fully developing. The current aims are to enhance the *in situ* generation of hydrogen from methanol and to increase the basic fuel cell reaction kinetics by incorporating nanoparticles into the membranes. When inorganic nanoparticles with a high affinity to water, such as silica, are introduced into the electrolyte polymers, the water retention of the membrane is improved. Nanocomposite membranes consisting of SiO_2/polyethylenoxide have been prepared to serve this purpose [121]. More recently, hybrid membranes, in which the nanoscale platelet structure of the inorganic particles is fully conserved, have been produced [122]. The inorganic additives help to increase the rigidity of the membrane and reduce the methanol transfer. Layered titanium- and zirconium-phosphate structures, for example, have been investigated with regard to proton conductivity [123].

Much work has been devoted to making improvements in the catalysts in order to achieve optimally engineered DMFC devices in which the performance is limited only by kinetic and ohmic losses. Unfortunately, since the degree of catalyst use in DMFCs is considerably lower than in PEMFCs high noble metal loadings in the $mg\,cm^{-2}$ range are required. At such high loadings the carrier material can even be omitted since the noble metal coating on the membrane usually has a sufficient thickness and lateral electronic conductivity. State-of-the-art anode catalysts use 60–85 wt% loadings of Pt_1Ru_1 on carbon or Pt_1Ru_1 black, while Pt black or 85 wt% Pt on C is used at the cathode [124, 125]. Colloidal nanoparticulate catalyst precursors with controlled sizes and tailored compositions have been introduced in DMFC devices to improve the performance and – at the same time – to reduce the amount of expensive noble metal required. The enhanced activity of PtRu DMFC anode electrodes compared with pure Pt particles is again explained by the *bifunctional mechanism* discussed above. The main factors that govern the anode catalyst activity are perfect alloy formation, surface structure, and a good dispersion of the nanosized particles. Because high noble metal loadings on carbon carriers are essential for good performance during methanol oxidation, it is important to avoid agglomeration of the deposited nanoparticles, for example by separating the particles from each other. This can be achieved by the so-called '*colloidal precursor method*' whereby colloidal protective shells, such as solvents, surfactants or polymers prevent particle agglomeration (for details see: [14, 126]. For example, the co-reduction of $PtCl_2$ and $RuCl_3$ dissolved in THF by $LiBH_4$ leads to a stable colloidal solution of bimetallic PtRu alloy nanoparticles with a well-defined particle size of 1.7 nm. This colloidal precursor is used for the preparation of Pt_1Ru_1 DMFC anode catalysts by subsequent addition of the carbon carrier and removal of the solvent. Such catalysts with a loading of $3\,mg\,cm^{-2}$ of noble metal at the DMFC anode showed a relatively high activity ($160\,mW\,cm^{-2}$ at 343 K, dry oxygen and 2-molar methanol solution) [127]. The *colloidal precursor method* has been applied in a comparative study to elucidate the influence of different metal salts and reducing agents as the starting materials on the activity of the resulting electrocatalysts. The catalysts were first characterized *ex situ* by energy dispersive X-ray (EDX) analysis and XRD and *in situ* CO stripping and cyclic voltammetry. Then the methanol oxidation activities of three 30 wt% PtRu/C catalysts of the same nominal composition (Pt:Ru, 1 : 1 a/o) were measured under steady-state conditions at 22 and 333 K.

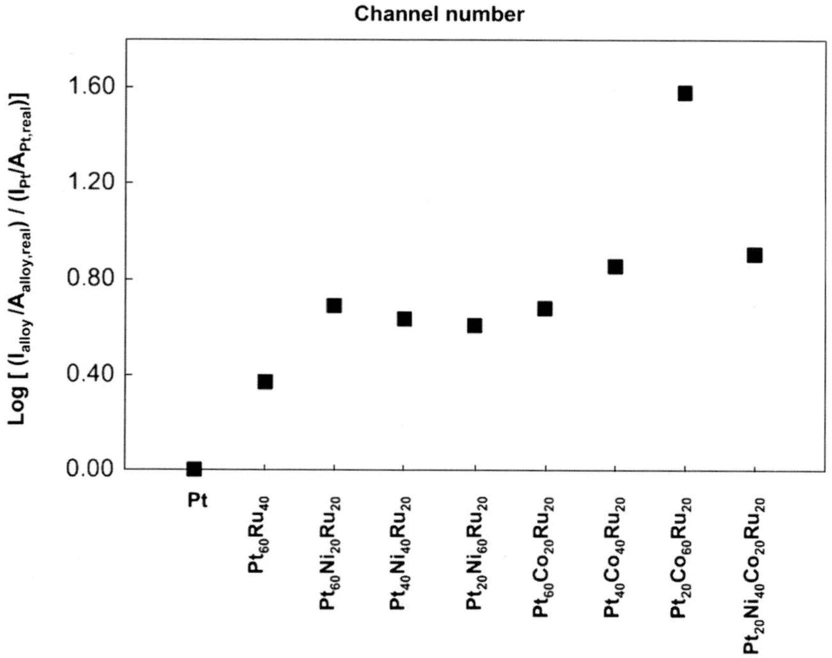

Figure 4.5 Specific activities (determined in 1 M CH_3OH at 298 K and 450 mV vs. RHE) of ternary and quaternary Pt alloys referenced to pure Pt. (Reproduced with permission from [131]. Copyright © (2003) American Chemical Society.)

The experimental rate data can be described by a kinetic model, which includes methanol adsorption on the Pt sites, formation of CO-containing adsorbed species, OH_{ads} formation on the Ru sites, and a heterogeneous surface reaction between CO-adsorbate and OH_{ads}. The differences in catalytic activity were explained using the kinetic model parameters identified from the experimental data [128]. The fact that PtRu catalysts provide significantly higher methanol electro-oxidation activity than pure Pt has prompted a lot of work to further improve the mass activity (defined as $mA\,mg\,Pt^{-1}$ at a given overpotential) by examining other bi- or even plurimetallic Pt–alloy systems with the help of combinatorial techniques [129, 130]. The study presented in Figure 4.5 [131] is an excellent example of the benefits of this research. According to the figure, by far the highest activity compared with pure Pt(0) is exhibited by $Pt_{20}Co_{60}Ru_{20}$ alloy particles, which exhibit a ≈30-fold higher activity than pure Pt particles. It is, however, unclear whether a carbon-supported high surface area $Pt_{20}Co_{60}Ru_{20}$ alloy catalyst is really superior to an optimized PtRu bulk alloy catalyst measured under similar conditions [132]. As far as the cathode side of DMFCs is concerned, the main area requiring improvement is in the field of voltage loss at the Pt electrode due to methanol crossover. Ru(0) is a methanol-tolerant catalyst but its ORR activity is far below that of platinum [133]. In 2005 Salgado and Antolini [134, 135] established that binary PtCo and PtNi alloys on carbon are very effective methanol-tolerant oxygen-reduction electrocatalysts. In 2007 a study on the activity, selectivity, and methanol tolerance of novel carbon-supported Pt and Pt_3M (M = Ni, Co) cathode catalysts indicated that the use of nanocatalysts with very small particles sizes and a narrow size distribution (1.7 ± 0.4 nm) is a possible way of achieving reasonably good ORR activities with an improved methanol tolerance at the DMFC cathode [136]. Nonetheless, the development of cathode catalysts with improved ORR performance and methanol tolerance still remains one of the major challenges in low-temperature fuel cell electrocatalysis.

4.2.1.3 Partial Methane Oxidation with NO

The partial oxidation of methane (POM) by NO to syngas (CO and H_2) is particularly interesting from an environmental point of view since nitrogen oxides are extremely harmful pollutants. As the majority of catalysts are self-poisoned by the oxygen resulting from the direct decomposition of NO, hydrocarbons provide an effective route to remove the poisoning oxygen from the surface of the catalyst. In the case of methane, CO and H_2 are formed, which is also economically favorable. While Green *et al.* found Ru/Al_2O_3 to be an active and selective POM catalyst [137]. Balint *et al.* [138] were the first to realize that NO can be used as the oxidant in POM. By preparing size-defined colloidal Ru nanoparticles (both 4.8 and 5.8 nm) and subsequent deposition on alumina, Balint *et al.* were able to show that at temperatures up to 723 K methane conversion is strongly size sensitive (Figure 4.6). Over a 12% Ru/Al_2O_3 catalyst having a larger Ru particle size (average: 5.8 nm) methane is selectively converted to CO_2 at 723 K.

In contrast, methane was oxidized at 723 K to both CO and CO_2 over a 6% Ru/Al_2O_3 catalyst with smaller Ru nanoparticles (average: 4.8 nm). This catalyst

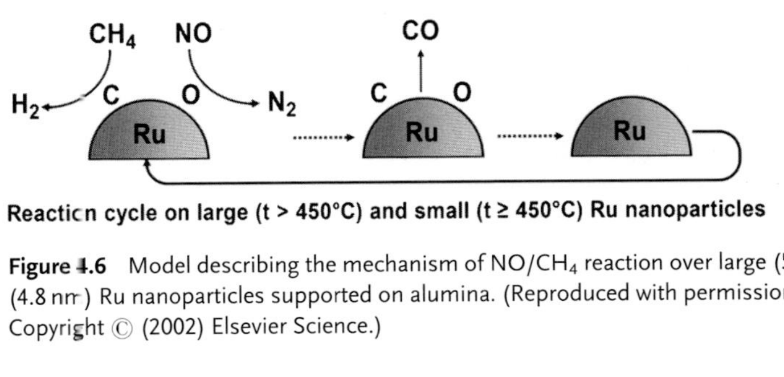

Figure 4.6 Model describing the mechanism of NO/CH_4 reaction over large (5.8 nm) and small (4.8 nm) Ru nanoparticles supported on alumina. (Reproduced with permission from [138]. Copyright © (2002) Elsevier Science.)

was rapidly deactivated by either O_2 or NO, whereas the 12% Ru/Al_2O_3 catalyst with larger Ru particles (5.8 nm) proved to be more resistant to oxygen. Above 723 K, however, the size sensitivity disappears completely. Thus at 873 K, both types of Ru/Al_2O_3 catalyst gave the best reaction selectivity to CO (≈80%) and H_2 (≈78%), and a good long-term stability was also observed.

4.2.2 Nanocatalysts for Specific Organic Reactions

In 2008 BASF launched a lead-free cis-selective semi-hydrogenation nano-catalyst that converts alkynes to cis-alkenes (NanoSelect LF 200™). Monodispersed Pd crystallites (with sizes of around 7 nm) uniformly distributed on a support of activated carbon or titanium silicate as the support provide similar activity and selectivity performance as found with Lindlar Catalysts, which usually contain 5 wt % palladium and lead, which is toxic and ecologically undesirable (see Figure 4.7) [139, 140].

Clearly, the employment of nanometer-sized metal particles greatly increases the metal surface area which in turn boosts the catalytic activity. Schlögl *et al.* have recently found that intermetallic Pd/Ga core-shell nanostructures where palladium is exclusively surrounded by gallium atoms are highly selective semihydrogenation catalysts for acetylene which outperform established systems [142]. Special nano-sized Au systems have also proved to be very selective oxidation catalysts. In

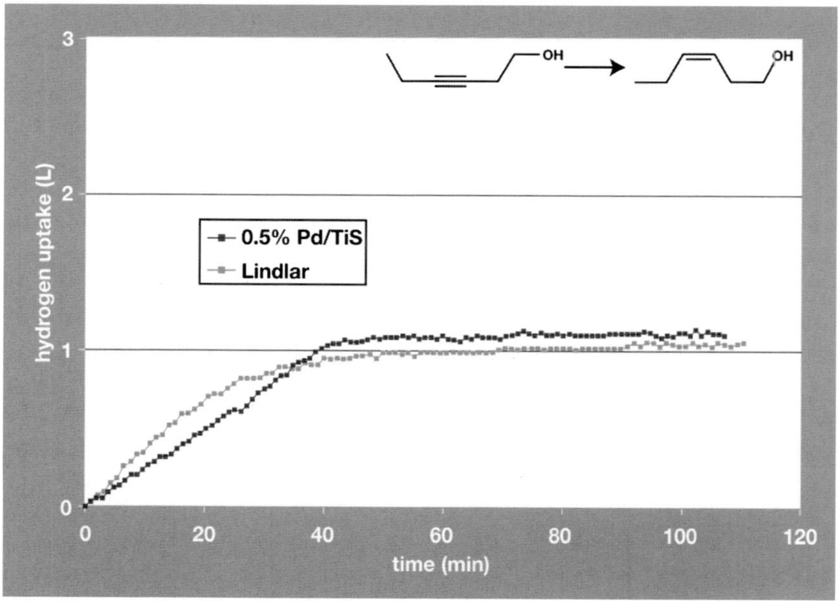

Figure 4.7 Comparison of catalytic activity of NanoSelect™ LF 200™ versus a conventional Lindlar catalyst. (Reproduced with permission from [141]. Copyright © 2012 BASF.)

particular, Inoue and Yamamoto performed the conversion of organosilanes into silanols (which are valuable building blocks for silicon-based polymers) with H_2O in high yield using nanoporous gold as the oxidation catalyst [143]. A catalyst class of its own – sometimes referred to as 'quasi homogeneous systems' – consists in unsupported nanostructured metal colloids dispersed in the form of organosols in organic solvents or dissolved in water as hydrosols. In order to prevent aggregation, the colloidal nanoparticles are usually stabilized using capping agents. If the capping agent is, however, too effective, the catalytic activity is significantly reduced because the active surface sites are blocked off. El-Sayed *et al.* have established a reasonable correlation between the catalytic activity of nanoparticles with different shapes and the fraction of atoms located on corners or edges [13]. Among the vast number of chemical reactions where transition metal nanocatalysts in colloidal solution have been found to be effective (see Ref. [2]), the so-called 'cross-coupling reactions,' namely the Suzuki and the Heck reactions, are the most important ones. The Suzuki reaction is the reaction of an aryl or vinyl boronic acid with an aryl or vinyl halide to afford a C—C bond, such as in biaryls. The Heck reaction is also a C—C bond formation reaction; however, in this case it is reaction of an unsaturated halide or triflate with an alkene. One example is the arylation of alkenes with aryl halides. Whether the Heck and Suzuki reactions truly occur on nanoparticles or atoms removed from the nanoparticles under the reaction conditions is still under debate. For the Suzuki coupling palladium nanoparticles [144–150] have been extensively studied. Other metals such as Ru [150, 151], Cu, [151] and bimetallic nanoparticles [151] were also found to be effective. A marked size effect was observed in the Suzuki reaction between phenylboronic acid and iodobenzene when using Pd nanoparticles of different diameters (3.0, 3.9, 5.2, and 6.6 nm) stabilized by PVP [152]. By comparing PVP, poly(amidomine) (PANAM) and dendrimers as the capping agents it has been found that the nature of the stabilizer also plays an important role during the Suzuki reaction [153]. The Heck cross-coupling reaction in 'quasi homogeneous' phase has predominantly been studied using colloidal palladium nanoparticles as the catalyst [154–156, 158–161]. Using PVP-stabilized Pd nanoparticles of mean sizes from 1.7 to 3.7 nm as model catalysts, it has been shown that the Heck coupling of *p*-bromobenzaldehyde with butyl acrylate is structure sensitive. A quantitative relationship was found between the initial reaction rates and the number of low coordinated ('defect') Pd surface atoms calculated using statistics for fcc cubo-octahedra (see Figure 4.8) [154].

In addition to PVP, other phosphine-free capping agents have been successfully deployed in Heck cross-coupling reactions as alternative colloidal shells to protect the Pd nanocatalysts. Examples here are PAMAM–dendrimer encapsulated palladium nanoparticles [159] and colloidal Pd stabilized with 1,5-bis[4,4b-bis(perfluorooctyl)phenyl]-1,4-pentadien-3-one, which were found to be efficient and recoverable catalysts in the absence of phosphine ligands [155]. In addition, G-3 dendrons with their highly unpassivated palladium surface (i.e., free from oxide or nitride protection layer) [157] and Pd nanoparticles encapsulated in dendrimers [161] were active in Heck reactions. In addition, Pd nanoparticles [158] and trimetallic gold–silver–palladium nanoparticles [161] have been successfully tested.

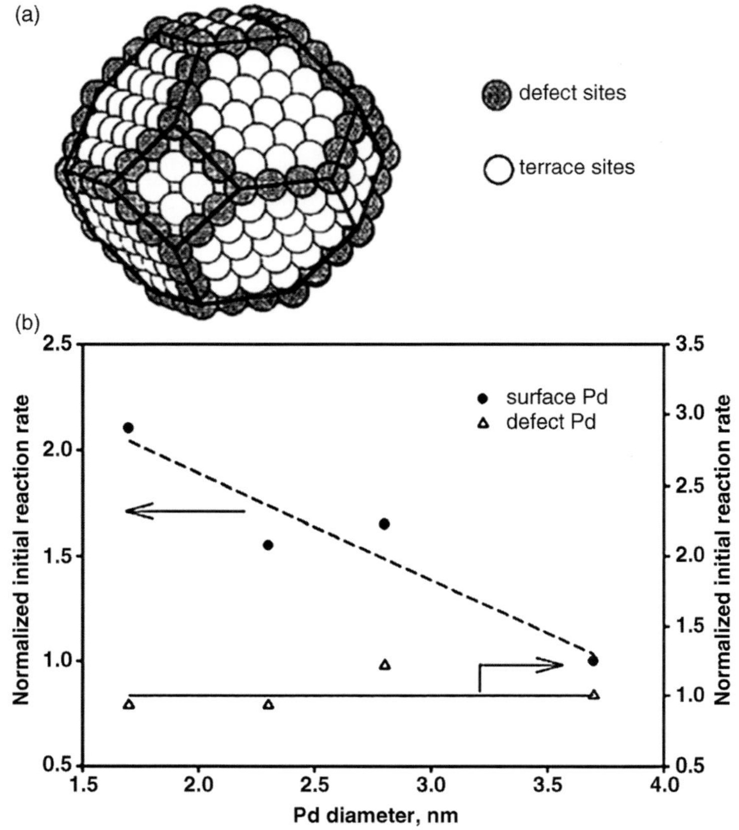

Figure 4.8 Structure sensitivity of a Heck reaction: (a) low-coordinated defect Pd surface atoms as active sites; (b) initial reaction rates normalized to that of the catalyst with the largest crystallite size and calculated per total surface Pd atoms and per defect atoms. (Reproduced with permission from [154]. Copyright © (1999) American Chemical Society.)

4.3
Nanostructured Materials in Energy-Related Processes

4.3.1
Nanomaterials for High-Performance Solar Cells

The *first generation* of photovoltaic devices for converting light into electricity (efficiency: 4.5%) were based on the monocrystalline silicon p–n junctions [162]. The *second generation*, introduced in 1976, relied on amorphous silicon thin film coatings, which brought both a significant cost reduction and, for the first time, the possibility of large-scale industrial production, at, however, the expense of low solar

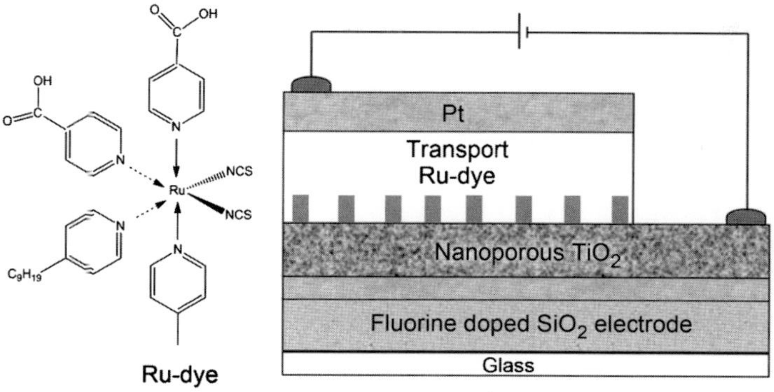

Figure 4.9 Standard set-up of a rutheniumbipyridyl-based dye-sensitized solar cell (DSSC).

energy conversion efficiencies ($\approx$1.1%). In the past decade a more advanced *third generation* of photovoltaics has been developed using nanomaterials, which combine low production costs with relatively high performance (8–10% under standard solar light conditions). The two main classes of *third-generation* photovoltaics are dye-sensitized solar cells (DSSC), first described by Grätzel and now generally referred to as the *Grätzel Cell* [163–165] and organic/polymer cells, where the key energy conversion function is based on quantum dot (QD) materials.

Figure 4.9 illustrates a standard DSSC device. All constituent elements of the cell are sandwiched between two transparent conducting oxide (TCO) electrodes. The electron generating element (dye) and the charge carrier (TiO_2) in DSSCs are separated, so that the absorption and the electron transfer occur in different parts of the cell. Standard DSSCs use a thin layer ($< 10\,\mu m$) of n-type semiconductor titanium dioxide particles (-20 nm) treated with a dye (usually a ruthenium–bipyridyl-type complex which is designed to attach itself to the TiO_2). Following excitation by the solar photons, the dye generates electrons and injects them into the TiO_2 phase. The oxidized dye is subsequently reduced by I^- ions in the electrolyte ($3\,I^- \rightarrow I_3^- + 2e^-$). The TiO_2/dye layer is attached to the front electrode, which is a TCO glass coated with fluorine doped tin oxide (FTO). The counter electrode (CE) is also TCO glass covered with a small amount of platinum (5–10 $\mu g\,cm^{-2}$) which catalyzes the cathodic reduction of triiode to iodide ($I_3^- + 2e^- \rightarrow I^-$). In QD-based cells the photons are harvested by cadmium chalkogenide particles, for example 7 nm CdSe, which are anchored to TiO_2 by organic surfactants (Figure 4.10) [166, 167]. The enormous impact of all kinds of nanostructured materials on the development of practical photovoltaic devices has been thoroughly reviewed by Tsakalakos [168, 169].

Despite intense research efforts in recent years devoted to the further development of DSSCs, there are still two major challenges to overcome. Both presently prevent upscaling and industrial production. The first is the cost and the second is the long-term durability of the cells. TCO substrates have been identified as the major contribution to the cost, making up about 60% of the total cost of the complete cell. The longevity of the cell is limited by the volatility of electrolyte liquids,

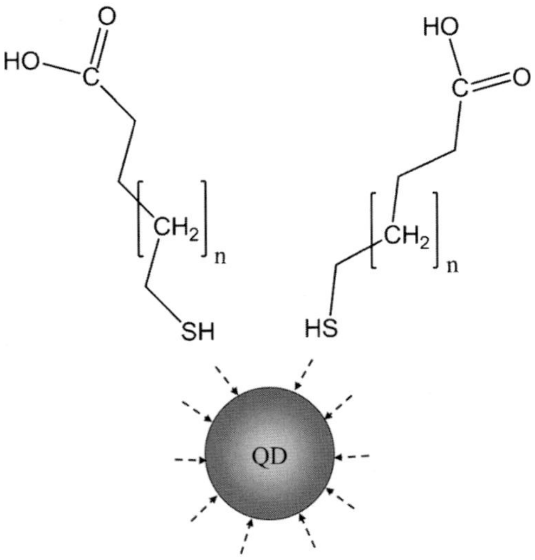

Figure 4.10 Quantum dots (QD) surrounded with organic surfactants anchoring titania.

since under real outdoor illumination, the temperature of solar cells can easily reach 323–343 K and containment of the electrolyte poses a problem.

In an integrated approach these problems were addressed by 12 partners from universities and research institutes in a feasibility study of the so-called monolithic DSSC set-up, which is entirely based on screen printing. [170]. In contrast to conventional DSSCs, only one TCO glass is required in the monolithic set-up, since the two electrodes are situated on the same surface (Figure 4.11). The cathode electrode (CE) is a carbon layer covered by Pt nanoparticles as the catalyst, which is screen printed and insulated from the front electrode by a thin ZrO_2 layer. Both electrodes are thus attached to the front TCO glass. For the use in semi-transparent DSC modules, an optimized polyol process for nanosized Pt has been developed, which allows the preparation of screen printed, transparent catalytic platinum layers showing an extremely low charge transfer resistance (0.25 Ohm cm^2) [171, 172].

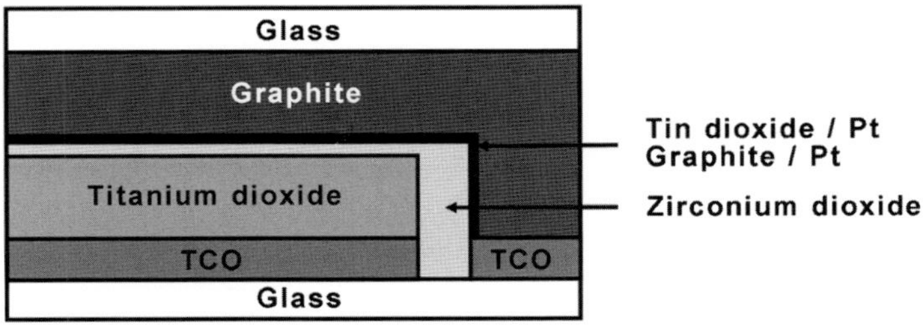

Figure 4.11 Cross section scheme of the 'monolithic set-up' device. (Reproduced with permission from [169]. Copyright © (2007) Elsevier Science.)

In order to improve the cell durability, various types of nonvolatile imidazolium-based binary ionic liquid electrolytes have been optimized with respect to diffusion-limited currents and charge transfer resistances. In addition, quasi-solid-state electrolytes have also been successfully tested by using inorganic (SiO_2) physical gelators. Sealing of the modules was achieved in a fusing step by the soldering of glass frit layers. With this printable monolithic set-up DSSC modification, a solar efficiency of 6.0% has been achieved on a 2.0 cm^2 active cell area. In the future, precise control of the surface modification and shape of the nanoparticles may lead to well-ordered nanostructures that improve the charge transport and increase absorbance properties in hybrid dye solar cell systems. Very recently, Yang *et al.* using 70 nm gold particles at the CE have reported an enhanced cell efficiency of 6.24% [173]. The large-scale application of QD-dependent solar cells, however, is currently hampered by the strong toxicity of II–VI semiconducting particles, namely the cadmium chalcogenides.

4.3.2 Nanocomposites for Batteries

Since the rechargeable lithium-ion (Li-ion) battery was introduced to the markets in 1991 by Sony [174, 175] billions of batteries have been marketed for portable electronic equipment. The working principle is illustrated in Figure 4.12. Lithium is intercalated in the graphite anode in the form of Li_xC_6. During discharge electrons are produced by Li, and Li^+ ions migrate to a layered cobalt or manganese oxide cathode giving for example $Li_{1-x}CoO_2$. The energy stored in an average 3.8 V voltage lithium cell is ca. 180 Wh kg^{-1} which is five times higher than a conventional lead–acid battery. This is sufficient to power small portable devices, but hopeless in

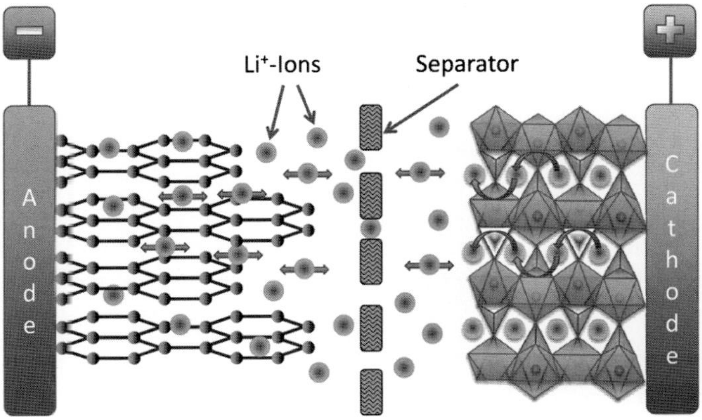

Figure 4.12 Working principle of a rechargeable Li^+-ion-Battery: Elemental Li atoms (encircled) intercalated between graphite layers at the anode supply electrons giving Li^+ ions which migrate through the membrane to the layered cobalt- or manganese-oxide cathode. (Reproduced with permission from [187]. Copyright © (2011) the authors.)

larger systems and is in no way able to compete with the energy density of gasoline (13 000 Wh kg^{-1}) currently used to drive combustion engines in vehicles [176]. Consequently, the actual research goal is the development of low-cost, safe, highly rechargeable Li batteries which are able to store ≈200 Wh kg^{-1} at >4 V voltage for use in the area of battery-powered electric vehicles [177]. There is also a need to reduce the carbon footprint of Li-ion batteries from the current 70 kg CO_2 per kWh [178].

The energy density of a battery can be optimized by enhancing the chemical potential difference between anode and cathode, minimizing the mass or the volume of the reactants during the electron supply, and through improving the electrolyte chemistry. The arrival of nanomaterials provided notable progress in improving the design of Li-ion batteries, which have recently been reviewed [179–186].

The search for Li-ion batteries with high power density has led to design and improvement of new intercalation electrodes at which graphite is replaced by silicon and silicon–carbon nanocomposites as the intercalation host for Li. In theory, alloying lithium with silicon at the anode to the limiting composition $Li_{44}Si$ would result in an enormous capacity of 4200 m Ah g^{-1}, which can be compared with 350 m Ah g^{-1} obtained with conventional graphite anodes [188]. In practice, however, lithiation of crystalline bulk silicon causes a volume expansion of 311% which leads to the complete amorphization of the solid-state Si host [189]. After only a few charge and discharge cycles, these dramatic volume changes degrade the electrode and lead to complete breakdown of the battery. Various approaches to producing silicon nanocomposites as the host for Li have been used to overcome this large volume change and have resulted in better capacity retention and cycle life for Si anodes. Recent research results targeted at obtaining long-term stable high-capacity Li anodes have comprehensively been reviewed by Appleby *et al.* [190]. During the past few years Schulz, Wiggers *et al.* have developed a straightforward gas-phase synthesis of nanoscale silicon, which provides an economic route to silicon–graphite nanocomposites using the decomposition of SiH_4 in a so-called hot-wall-reactor (HWR) at 1273 K and 300 mbar (Figure 4.13) [187, 191, 192]. High resolution transmission electron microscopy (HRTEM) investigations show that the individual Si particles (12 nm) are polycrystalline with a high crystallinity throughout the particles. To investigate the suitability for battery applications, silicon from the HWR process has been used to prepare silicon–graphite nanocomposite electrodes. The electrodes were composed of 60 wt% graphite, 20 wt% silicon, 12 wt% of the commercial carbon black additive 'Super P™' as the conducting agent, and 8 wt% carboxymethylcellulose (CMC) as the binder. The resulting dispersion was coated on a copper foil current collector.

Figure 4.14 illustrates the performance of the Si/C nanocomposite electrodes tested in a cycling study where the capacity was limited to 80% of the theoretical maximum.

The nancomposite electrode shows full durability during 100 charge–discharge cycles without significant degradation, at efficiency of about 95%. This result is claimed to be among the best available in current literature for silicon-containing

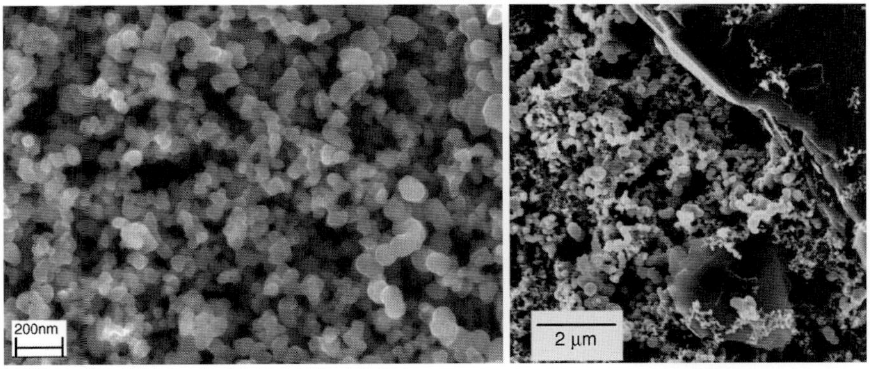

Figure 4.13 SEM image of an as-prepared silicon nanoparticles (left) and of the anode consisting of micron-sized graphite flakes, 'Super P™,' binder, and silicon nanoparticles. (Reproducedwith permission from [191]. Copyright © (2011) the Hosokawa Power Technology Foundation.)

nanocomposite electrodes. In a different approach Cui *et al.* have grown Si nanowires directly on the current collector and have found no pulverization or breakdown into smaller particles during the charge–discharge cycling process (Figure 4.15) [193]. Since there is facile strain relaxation in the nanowires, their diameter and length can increase during the cycling without causing breaking or amorphization. Furthermore, this nanowire anode design connects each silicon wire directly with the current collector, enabling an efficient 1D electron transport along the length of every nanowire.

In order to further increase the potential of $Li_{4.4}Si$ in broad commercial battery applications, Zhang *et al.* at the Lawrence Berkeley National Laboratory have designed novel graphene/Si multilayer structure anodes to be used in advanced half and full Li-ion cells [194]. Between 1990 and 2009, 21451 scientific papers and

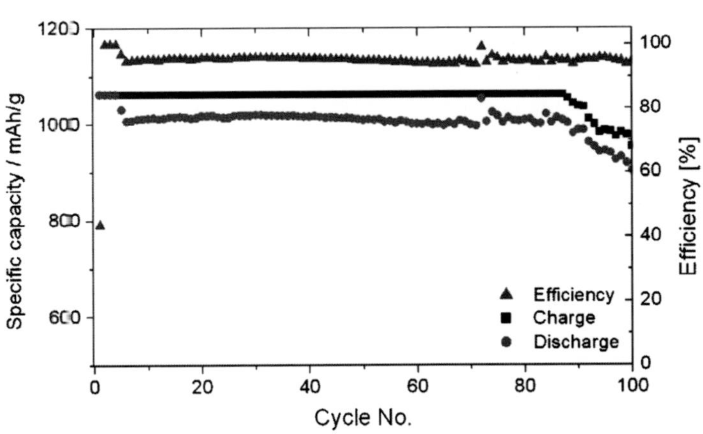

Figure 4.14 Cycling study of the graphite/silicon composite electrode. (Reproducedwith permission from [191]. Copyright © (2011) the Hosokawa Power Technology Foundation.)

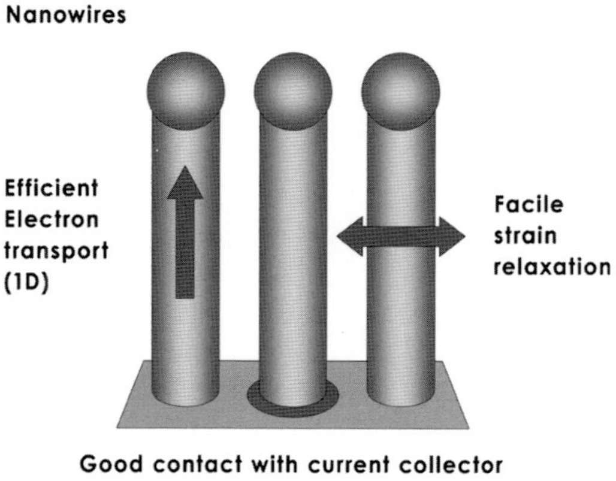

Figure 4.15 Silicon nanowires directly grown on the current collector. (Reproduced with permission from [193]. Copyright © (2008.) Nature Publishing Group (NPG).)

28940 patent documents have been published in the field of battery research [195]. Despite of these enormous efforts including recent breakthroughs, automotive market experts do not expect to see battery-powered e-cars in everyday traffic before the year 2020.

4.3.3
Applications for Energy and Hydrogen Storage

Hydrocarbons are currently the major source of energy for storage and transportation. Electricity is regarded as the energy carrier of the future, particularly when alternative energy sources, such as biomass derived feedstocks, solar energy, and wind power systems, will become of increasing importance. Electricity has the disadvantage that it cannot be easily stored. Since the production and consumption of electricity have to be balanced, an efficient buffer is needed for the intermediate storage of electricity. Because of hydrogen's physical and chemical properties, the conversion of electricity into hydrogen as a high-energy-content storage medium (heating value: 120 MJ kg^{-1}) is regarded as the best option. There are four major options to store hydrogen, depending on whether stationary or mobile applications are required:

1) gaseous hydrogen compressed in cylinders at 35–70 MPa at ambient temperature
2) liquid hydrogen at 0.5–1 MPa at 20–30 K
3) solid-state absorbers (hydrides or high-surface structures)
4) hydrogenation of CO_2 into methane (to be stored in the public gas supply systems)

The overall field of industrial hydrogen production has been covered in recent reviews by Roosen [196] and Tiwari [197] while Schüth *et al.* have summarized the role of hydrogen as a promising energy carrier in future energy systems [198]. The specific impact of nanotechnology on hydrogen production and storage is discussed in the sections below.

4.3.3.1 Nano for Hydrogen Production

Today 95% of hydrogen is produced from natural gases via catalytic steam reforming of methane [199] while methanol and steam reforming of fossil fuels is still expected to play a key role in a future economy [200]. Nanostructured catalyst systems can help to improve the existing processes considerably. For example, Pd@CeO_2/Al_2O_3 catalysts providing H_2 in high yields have been prepared using core-shell Pd@CeO_2 nanostructures as the starting material [201]. Other nanocatalysts have been successfully tried in the steam reforming of fossil fuels. Nanostructured zirconia powders, for example, have been used to catalyze the steam reforming of bioethanol [202] and a ZrO_2-modified Ni–nano-CaO system, prepared using the so-called nanocasting route, has been deployed in sorption-enhanced reforming for hydrogen production [203] while indium oxide nanocatalysts have already been commercialized [204]. The use of biomass energy as a renewable energy source has received particular attention [205]. For tar removal in biomass steam gasification, nano-Ni-La-Fe/γ-Al_2O_3, a supported trimetallic catalyst has been developed to enhance the quality of the gas produced [206]. With the help of this catalyst the efficiency of tar removal during gasification at 1073 K reached 99%, while the H_2 content increased markedly to over 10 vol%. A similarly structured catalyst, nano-NiO/γ-Al_2O_3, gave very good hydrogen yields in rice husk biomass air–steam gasification [207]. Clearly, the extremely small particle size, contributed greatly both to the gas quality and the yield [208]. Nanomaterials have also been used to improve the direct splitting of water to give hydrogen and oxygen:

$$H_2O \rightarrow H_2 + 1/2O_2$$

Here the separation of hydrogen and oxygen at >2500 K is the most severe problem. Porous ceramic zirconia membranes were developed using the sol–gel method toachieve this [209]. A process based on the SnO_2/SnO water-splitting cycle has been studied [210]. SnO nanoparticles resulting from the solar reduction of SnO_2 are hydrolyzed efficiently in the temperature range 773–873 K to give H_2 in yields over 90%.

Nanotechnology has also been applied to the direct electrolysis of water. Nanofabricated Ni–ceramic structures produced by a well-controlled electrodeposition process in a nanoporous alumina template were used as a cathode catalyst for the stable and effective hydrogen evolution in alkaline water [211]. Highly dispersed nanosized Ni particles were shown to be an alternative H_2 electrode material for high temperature electrolysis of water and steam. Their main function is to reduce the contact resistance between the samaria-doped ceria cathode and the electrolyte [212]. Nanosized metal oxide semiconductors, for example ZnO, TiO_2, $CuBO_2$, and $CuAlO_2$ [213–215] have been studied to improve the photocatalytic hydrogen production [216–219].

Photoelectrochemical and photocatalytic water decomposition are both very suitable candidates for the application of advanced high-surface nanomaterials. Nanocrystalline semiconductor electrodes are a very good example of this [220]. Accordingly, photoelectrodes made from nanocrystalline semiconductor materials such as colloidal TiO_2, proved to be very efficient because their electron injection time is several thousand times faster than the charge recombination process [221]. Nanoporous iron(III) oxide (hematite) has also been successfully used for photoelectrolysis and photocatalysis processes. The pore diameters of the nanoporous structure in this case ranged from 50 to 250 nm with a pore depth around 500 nm. The crystallized Fe_2O_3 nanoporous film, having a 2.2 eV band gap, exhibited a net photocurrent density of 0.51 mA cm^{-2} in 0.5M H_2O_2 + 1M NaOH at 0.6 V versus Ag/AgCl under simulated AM 1.5 sunlight [222]. The advantage of TiO_2 nanotubes has also been tested in photoelectrolysis applications and photocatalysis [223–226].

To conclude this section, it is expected that future hydrogen production will have to become independent of fossil sources which, in essence, means that direct water splitting will have to be developed further. Specifically, the current state-of-the-art methods of water splitting (photoelectrolysis and photocatalytic water decomposition) leave much room for significant improvements and nanotechnology will almost certainly play an important role here.

4.3.3.2 Nano for Hydrogen Storage

It has been stated [198] that the stepwise introduction of a future hydrogen economy will first take place in the field of stationary applications while transportation applications, for example fuel cell electric vehicles running on hydrogen, will have to come later because currently there is no fully satisfactory solution available for onboard hydrogen storage systems. For the time being, the physical storage systems mentioned above, namely compressed gaseous hydrogen in cylinders (35–70 MPa) [227–229] are the most developed. For large-scale energy storage, however, a so-called 'E-power to Gas' concept has been proposed as one of the future solutions [230, 231]. Since highly developed countries like Germany already have large storage capacities available for natural gas, electricity coming from renewable sources such as wind could first be converted into hydrogen (via electrolysis with 70% efficiency) and then via the hydrogenation of CO_2 to methane:

$$6\,H_2 + CO_2 \rightarrow CH_4 + 2\,H_2O$$

Both in the electrolysis step (see Section 4.3.3.1) and in the subsequent catalytic hydrogenation of CO_2, where effective nano-nickel catalysts have recently been developed [232], nanochemistry can significantly contribute to enhancing the currently rather poor overall efficiency of the proposed concept.

For all kinds of zero-carbon-emission electric vehicles, the onboard energy storage system is of critical importance [233]. In automotive applications there are currently two non-hydrocarbon options under debate. As is illustrated in Figure 4.16, there are large energy density differences between hydrocarbons, compressed hydrogen, and Li-ion batteries. As a consequence, pure battery-electric systems are

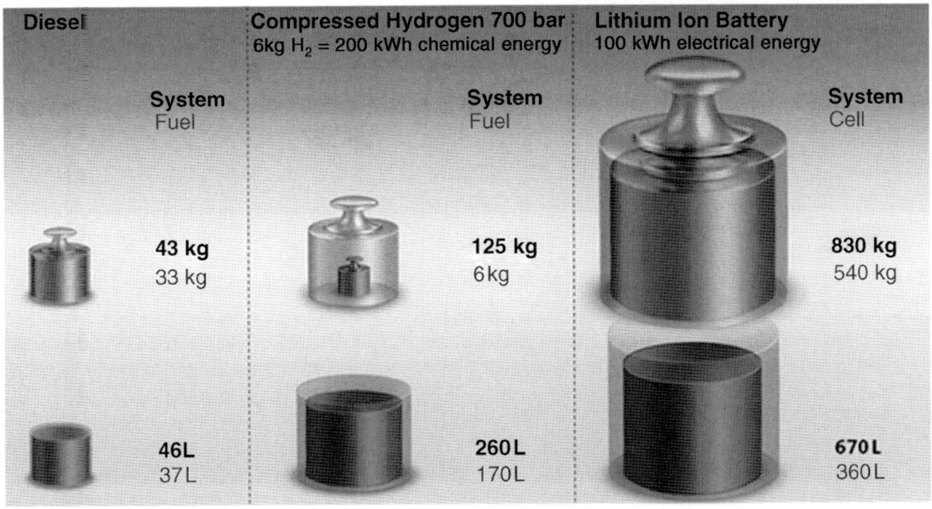

Figure 4.16 The weight and volume of various fuels and tank systems required for a 500 km range vehicle. (Reproduced with permission from [198]. Copyright © (2009) Wiley-VCH Verlag GmbH.)

only really suitable for small urban vehicles with ranges up to 150 km while for longer distances fuel cell cars running on hydrogen is an attractive alternative. To reach a mobility range of ca. 500 km, however, 4–7 kg of hydrogen must be stored on board [234].

Hydrogen must be stored in cylindrical vessels in compressed gaseous or liquefied form. Three carbon fiber tanks containing 4.3 kg of hydrogen allow a range of 320 km, and the storage system can be refilled within three minutes. Since there is limited space in existing vehicles for this type of hydrogen storage device and a wish to avoid the necessity of transporting high pressure hydrogen as the energy carrier intense research activity has been devoted to the development of new materials for the reversible low-pressure absorption of hydrogen. A complete overview on solid-state hydrogen storage systems that are currently under consideration as alternatives to compressed or liquefied hydrogen can be found in recently published books edited by H.S. Nalwa [235] and M. Hirscher [236] Basically, all future energy storage systems for transport applications face the constraints of limited available space in the vehicles, the specific requirements of a stop-and-go propulsion system, the demands of costs, overall capacity, and refueling time. In the best scenario, one could imagine a concept whereby the car is constructed around a nanostructured porous hydrogen storage material. To date two strategies are under active investigation to reach this goal. One option is chemical storage, for example in the form of hydrides; the alternative is physical storage using adsorbents. Reversible metal hydride systems make interesting candidates for onboard hydrogen storage and have been intensely studied over the last 20 years [198, 236]. MgH_2 offers a hydrogen storage capacity of 7.7 wt% H_2, good reversibility during the cycling process, and low material costs. However, MgH_2 is a high-temperature metal hydride (equilibrium pressure of 0.1 MPa at 573 K) which runs at temperatures far above the operation temperature of PEMFCs, which lies around 363 K. In order to improve the kinetics of the desorption and absorption of hydrogen and the cycle stability, MgH_2-based V_2O_5 or Nb_2O_5 nanocomposites produced via ball-milling exhibiting extremely short hydrogen absorption and desorption times (<2 min) have been investigated [237, 238]. Downsizing the material from bulk to nanodimensions results in considerable improvements, because the thermodynamics are strongly dependent on the size of magnesium clusters. Particles with sizes of only of 0.9 nm, comprising only 20 magnesium atoms, are far less stable than the bulk, which results in reduction of the desorption temperature to approximately 473 K [239]. Another approach is so-called melt infiltration of Mg into nanoporous carbon to afford magnesium nanoparticles of less than 2 nm [240]. All in all, MgH_2 is not an attractive hydrogen storage material, but since it has the very high energy density of 9 MJ per kg of magnesium, it may still be a candidate material for the reversible storage of high amounts of thermal energy over its working temperature range [241]. Another good candidate for reversible hydrogen storage are the complex hydrides of the type $A_xMe_yH_z$, where cation A is a metal of the first or second row of the periodic table (or a transition metal) and Me represents either boron or aluminum; here the hydrogen is covalently bonded [242]. Complex hydrides, such as Mg_2FeH_6 or $Al(BH_4)_3$, offer very high volumetric hydrogen densities (up to 150 kg m^{-3}) compared with liquid hydrogen (70 kg m^{-3}) [243]. However, because high decomposition temperatures are required to release H_2, these systems were not regarded as serious contenders for reversible hydrogen carriers until Bogdanović and Schwickardi [244] discovered in 1997 that doping $NaAlH_4$ with titanium compounds, such as $Ti(Obutyl)_4$ or $TiCl_3$, affords a novel reversible solid hydrogen storage material with short rehydrogenation times (of the order of minutes at 10 MPa and 373 K) in which a zerovalent Ti_{13} nanosystem is the active catalyst [244]. In practice, ball-milling highly disperses the catalyst precursors such as $TiCl_3$ within the hydride, leading to zerovalent titanium, the presence of which is necessary for achieving good kinetics in the solid-state hydrogen charge and discharge reactions [245–247]. Recently, the reactive ball-milling synthesis was further improved by starting from aluminum metal, NaH, and $TiCl_3$, which are cheap industrial chemicals, under a hydrogen pressure of 8 MPa [248]. Since these seminal findings, research interest in complex metal hydrides as reversible hydrogen storage materials has increased dramatically. Nanochemical strategies aimed at meeting hydrogen storage goals have been summarized in recent reviews by de Jongh [249] and Jensen [250].

A new class of porous nanomaterials for the adsorptive storage of hydrogen is the so-called metal–organic frameworks (MOFs) which comprise metal ion coordination centers that are connected by organic molecules, such as polyvalent carboxylic acids or nitrogen donor linkers and others. Yaghi has coined the principle of MOF synthesis as 'reticular design' [251]. These frameworks exhibit extremely low densities (0.2 g cm^{-3}) and exceedingly high surface areas which in combination allow an overall uptake of 4.7 wt% of hydrogen at 50 MPa pressure and 77 K [252]. The

Figure 4.17 Scanning electron microscopy (SEM) image of MOF-5 cubic crystals. The inset shows an enlarged view. (Reproduced with permission from [253]. Copyright © (2006) Wiley-VCH Verlag GmbH.)

formula unit for MOF-5, which is currently the best studied and most promising MOF structure for hydrogen storage, is $Zn_4O(BDC)_3$, where BDC is benzene dicarboxylate. Figure 4.17 shows the scanning electron micrograph of a typical single and polycrystalline cubic MOF-5 crystal (size ≈50–200 μm). This system is permanently porous (that is to say it does not collapse) and is consequently always available for the uptake of gases such as hydrogen or methane.

Since several members of the MOF family have the potential to achieve the gravimetric goal of 6 wt% hydrogen uptake, these structures have attracted a lot of research activity in the past few years which has resulted in a flood of papers. Recently, the field has been comprehensively reviewed in a book chapter by DeSoto and Dreyer [254]. The major drawback of MOF systems in reversible hydrogen storage application stem from the fact that the attractive values quoted above for the hydrogen uptake abilities can only be achieved at cryogenic temperatures (77 K) and at elevated pressures (≈50 MPa). Another limitation is the high heat of adsorption (5–7 kJ mol^{-1} H_2) which occurs during the charge–discharge cycle.

Another area where nanostructures are expected to make a substantial contribution to the problem of hydrogen storage is in the field of carbon–metal nanocomposite materials. Here, the reversible H_2 uptake is brought about by the so-called hydrogen spillover effect, which is observed for certain carbon–metal composite structures. The highest H_2 uptake reported so far is 4 wt% at 298 K and 100 MPa by Yang [255] for carbonized Pt/carbon catalysts mixed with highly microporous metal–organic frameworks. A thorough understanding of the system has, however, yet to be reached. The value of hydrogen spillover systems for practical storage devices seems to be questionable at the present time. For details see [256].

As far as the current state-of-the art is concerned, liquefied hydrogen at 20.2 K is the best option for large volume storage despite the inevitable evaporation losses and the relatively high energy required for the condensation process. Specialists agree that hydrogen pressurized at 70 MPa is currently the best solution for transport applications, for example electric cars driven by fuel cells [233]. That being said, solid-state nanomaterials still have the potential to outperform cryostorage or high-pressure technologies. As far as solid hydrogen storage materials based on hydrides are concerned, it appears that higher storage capacities can be achieved, whereas advances in adsorptive storage on high-surface-area materials has already come close to the theoretical limits. In this context, the substantial progress of theoretical investigations in recent years is worth mention [257–259]. These may provide the stimulus for future experimental breakthroughs leading to the development of completely novel nanostructured storage systems for hydrogen.

4.4
Characterization of Nanostructured Metallic Catalyst Precursors and their Interaction with Coatings and Supports Using X-ray Absorption Spectroscopy

The experimental results discussed in the previous paragraphs give empirical evidence as to how to optimize size, shape and composition of metallic nanocatalysts. There are also clear indications that the type of support (for example acidic vs. basic properties) influences the performance of a catalyst. However, all these data do not answer the basic question: which is the optimum electronic and geometric configuration that makes a specific metallic nanoparticle a good catalyst and what are suitable ways to tailor and optimize these properties? There is also experimental evidence that the catalytic interaction itself changes the properties of a catalyst. Thus, it is desirable to find ways to control these changes that influence the performance in a negative way. This task requires a detailed knowledge of the alterations of the electronic/geometric structure during operation and of the parameters that induce these changes, i.e., *in situ* and time-resolved investigations during the catalytic process are desirable whenever possible.

What is absolutely necessary for a better understanding of the properties of nanostructured materials as catalysts, are experimental data (and also theoretical calculations) that answer the basic questions that have to be addressed to achieve the above described basic understanding, as shown in the following. incomplete list:

- What is the geometric structure of the nanoparticle: size of the particles, bond lengths, crystal structure, and chemical composition in particular for bi- or multimetallic systems? What is the structure of a bimetallic system: statistical alloy, core shell structure, spatially separated systems (segregation)?

- What is the electronic structure of the nanoparticle: valency, band structure, size dependence of these properties? How does a second component (or a 'promotor' influence these properties?
- Is there any interaction of the metal core with the protective cover (capping molecules) and/or the support? How does that interaction modify the geometric and electronic properties of the nanoparticle?

In principle, this series of questions should begin by asking for the details of the chemical reaction that takes place during the wet chemical synthesis of metallic nanoparticles; this is of course a crucial question as this knowledge is necessary to tailor size/structure of nanoparticles. It is quite obvious from the few detailed investigations that are available until now that the 'old model' as suggested, for example by Turkevich *et al.* [260] is not realistic; it consists mainly of three steps: nucleation, growth, and agglomeration. Obviously each synthesis of nanoparticles is different regarding the steps involved – and perhaps even regarding the resulting nanoparticles. Here is a (again not complete) list of questions with respect to the details of the synthesis:

- What are the chemical steps (intermediate products) in the reaction from the molecular precursors to the metallic nanoparticles? Which ways are available to control the size and the (crystallographic) structure of the particles?
- What is the role of the capping molecules used as 'protecting layers' (for example surfactants, polymers or ligands) during synthesis? How do these capping molecules influence the geometric and electronic (and other) properties?
- How is it possible to functionalize the surface of metallic nanoparticles for a specific application in a defined way and how does that process influence the properties of the core?

Numerous experimental techniques are available to obtain some of the answers for the questions summarized above, as – at least in principle – most of the techniques used to characterize bulk materials, such as nuclear magnetic resonance (NMR) spectroscopy, UV–visible spectroscopy, Mössbauer spectroscopy, and ultraviolet photoelectron spectroscopy (UPS) or X-ray excited photoelectron spectroscopy (XPS) can also be applied to nanoparticles. However, because of the small size of the particles, most of the 'classical' methods requiring long range order i.e., crystallites with at least several complete unit cells, for example, XRD, only provide limited information because of peak broadening. The most valuable techniques for determining the 'macroscopic' particle properties, i.e., size and shape, are small angle X-ray scattering (SAXS) [261] and HRTEM [262, 263] and at least one of these techniques is applied in most cases to characterize nanoparticles after synthesis. Over the last few years special arrangements have been developed for both techniques that also allow the *in situ* investigation of synthetic reactions [264, 265]. In this chapter, only results obtained using XAS as the analytical technique will be discussed. Compared with the techniques mentioned before, XAS is very suitable for the investigation of nanoparticles, with two striking advantages: it does not require any long-range order (crystallinity) of the material and XAS spectra provide information about the electronic and the geometric properties of the particles at the same time.

In view of the huge number of publications dealing with metallic nanoparticles and XAS, the goal is not to give a comprehensive overview of the existing literature, but to highlight examples from about the last five years that demonstrate the enormous potential of the technique. From here, the interested reader can easily refer to the old literature. With a few exceptions, only examples from areas of application discussed in Section 4.2.1 (fuel cells, FT reaction, etc.) will be presented. In Section 4.4.1 the theoretical and experimental background of XAS will be described without digging too deep into theoretical discussions; the next sections will guide the reader from the easiest examples, i.e., the 'standard' XAS of monometallic particles (Section 4.4.2) via bimetallic systems (Section 4.4.3) and the investigation of the interaction of nanoparticles with their coatings and/or support (Section 4.4.4), to more advanced techniques such as resonant elastic and inelastic X-ray scattering and their applications to site and/or valency specific investigations (Section 4.4.5).

This review will lead the reader to realize that there are still a lot of open questions and many unresolved problems. Furthermore, there are even contradictory results for identical or very similar systems investigated by different groups. Thus, it is hardly possible at this time to come to conclusions of general validity. One of the reasons for this situation is an obvious lack of systematic investigations (size control, influence of capping molecules etc.) on the one side and the lack of closer collaboration between the chemists, who prepare the samples, and physicists, who analyze the samples, on the other. We hope that the reader will not be too disappointed reading these facts but might be inspired to start his own research: metallic nanoparticles are still an extremely important and fascinating area of research!

4.4.1
X-ray Absorption Spectroscopy (XANES and EXAFS) as an Analytical Tool for Nanostructures

X-ray absorption spectroscopy (XAS) is a technique that uses tunable monochromatic X-rays to probe the physical and chemical structure of matter at an atomic scale. X-rays are photons (light) with energies between ~500 eV and ~100 keV. In this energy range, the X-ray photons are mainly absorbed by the photoelectric effect. In this process, the X-ray is absorbed by a tightly bound electron of a so called 'inner shell' (or core shell) of an atom. Each time the X-ray energy equals the binding energy of such a core electron, a sharp absorption edge appears that interrupts the otherwise continuous decrease of the absorption coefficient $\mu(E)$ with increasing photon energy. As the binding energies of the inner shell electrons are characteristic for each element (Moseley's law), the absorption process provides element selectivity for different atomic species. These edges, arising from the ejection of strongly bound core electrons, are labeled K, L, M, etc., which corresponds to the principal quantum number $n = 1, 2, 3 \ldots$ of the main electron shell.

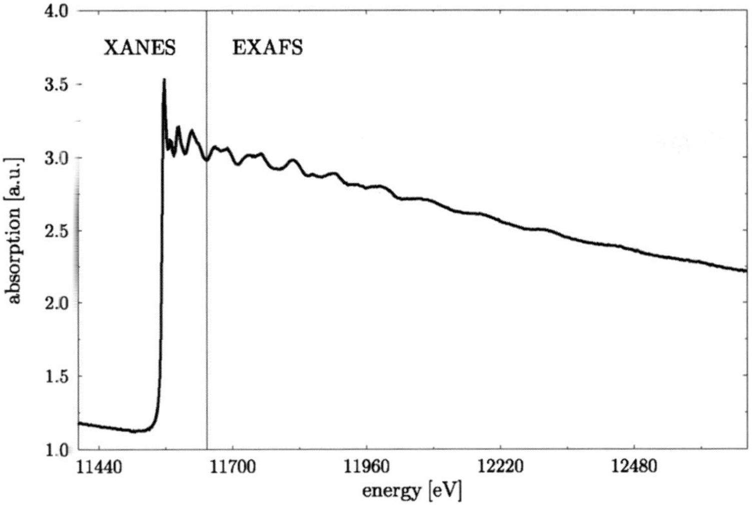

Figure 4.18 The L-III-edge-XAFS spectrum of a Pt foil showing the XANES and EXAFS regions.

The process following such an X-ray absorption event is the 'relaxation' of the excited atom by filling its core hole with a weaker bound electron, which is accompanied by either the emission of a photon or an Auger electron, carrying away the energy difference of the two atomic states involved in the process. The detailed study of the fluorescent X-ray radiation (or alternatively of the Auger electrons) that reflects the electronic structure of an atom species is called X-ray emission spectroscopy (XES) (or Auger electron spectroscopy (AES)). One can also make use of these emitted photons or electrons to gain XAS spectra, since the emission events are proportional to the absorption events.

Figure 4.18 shows a 'typical' X-ray absorption fine structure (XAFS) spectrum of a Pt foil measured at the L-III-edge. In general, the XAFS spectrum is divided into two parts as indicated in Figure 4.18: (1) the X-ray absorption near edge structure (XANES) that starts ~50 eV before the 'edge' (i.e., the jump of the absorption coefficient when the energy of the incoming photon is high enough to excite the corresponding inner shell electron into an empty orbital or into the continuum) and reaches up to ~50–100 eV above the edge, and (2) the extended X-ray absorption fine structure (EXAFS) that is visible up to ~1 keV above the edge.

XANES contains (mainly) information on the electronic structure of the local vicinity of the absorbing atom species. Hence, the energy and shape of the absorption edge depend sensitively on the chemical environment of the absorbing atom. By comparison, for example, of XANES spectra of an 'unknown compound' with those of appropriate reference compounds the chemical valence state can be determined and conclusions can be drawn about the type of chemical neighborhood, the local symmetry, and a possible charge transfer between binding partners. In a simplified model, resonances in the XANES region can often be assigned to transitions into specific molecular orbitals or into specific (valence) bands of a sample.

The extended X-ray absorption fine structure (EXAFS) arises from inner shell electrons that are excited into the continuum, i.e., they are ejected from the atoms. These photoelectrons are traveling through the material with a kinetic energy E_e equal to the excess energy $E_X - E_0$, where E_X is the energy of the X-rays and E_0 the (positive) binding energy of the inner shell electron. EXAFS is now best understood when moving from a description of the absorption process based on X-ray energies to one based on the kinetic energy or the corresponding wavenumber of the ejected electron:

$$k_e = \sqrt{\frac{2m}{\hbar^2}(E_x - E_0)},$$

where $E_e = (h/2\pi)^2 \, k_e^2$.

According to de Broglie, a wavelength $\lambda_e = 2\pi/k_e$ can be assigned to the photoelectron. Thus, the ejected photoelectron can be viewed as an outgoing spherical wave, which is scattered at the neighboring atoms – very similar, for example, to water waves – partially resulting in a second backscattered electron wave. The backscattered and the outgoing electron waves interfere at the position of the absorbing atom – again similar to water waves. This interference that depends on the wavelength of the outgoing electron wave and the distance between the excited atom and the backscattering atom modulates the absorption coefficient $\mu(E)$, giving rise to an increase when the interference is constructive or a decrease when it is destructive. This oscillating fine structure is superposed onto the smooth energy dependent absorption coefficient of an isolated atom as can be seen in Figure 4.18.

With a few (reasonable) assumptions and simplifications, for example, assuming that there is not too much static disorder in the sample (which is, however, not always true in very small particles!) the resulting interference pattern $\chi(k)$ can be described analytically by:

$$\chi(\mathrm{k}) = \sum_{i=1}^{n} S_{0,i}^2(k)\left(\frac{N_i}{kR_i^2}\right) f_i(k) e^{-2k^2\sigma_i^2} e^{\frac{-2R_i}{\lambda_i}} \sin(2kR_i + 2\delta_l(k) + \Phi_i(k))$$

The sum in this equation is over coordination shells i with N_i identical backscattering atoms.

S_0^2 is less than 1 and describes the reduction of the EXAFS – amplitude due to multi-electron effects which are not included in the simple picture described above;

$f_i(k)$ is the backscattering amplitude describing the potential of an atom to 'reflect' the electron wave. As the run of f as a function of k depends on the type of atom it is in principle possible to determine the type of backscattering atom from the measured $\chi(k)$ function;

R_i is the radial distance from the excited atom to the neighboring atoms in coordination shell i;

λ_i is the mean free path describing the reduction in coherence of the electron wave due to inelastic scattering processes and the lifetime of the excited final state.
$e^{-2k^2\sigma_i^2}$ corresponds to the Debye–Waller factor known from XRD. However, for EXAFS the mean square relative displacement between the central atom and the backscattering atoms has to be considered whereas in the diffraction formalism the mean square displacement of each type of atom has to be taken into account;
Φ_i is the phase – function of the respective backscattering atoms, and δ_l is the phase shift of the central atom.

The analysis of the phase of the interference pattern provides very accurate information on radial distances R_i between the absorbing atom and the neighbor atoms in the first coordination shells. From the amplitude of the interference structure information on the type and the number of atoms N_i in the respective coordination shell can be derived. This latter information is much less accurate than the information about the radial distances. Details of the EXAFS – theory and the process of data analysis are discussed extensively in the literature [266–268]. As the energy dependence of the absorption cross section is measured, the tunability of synchrotron radiation is a crucial precondition for XAFS spectroscopy. When compared with traditional diffraction techniques, XAFS spectroscopy has a few decisive advantages for the investigation of nanoparticles:

XAFS is nondestructive and element specific
no long range order is required
XAFS is extremely sensitive (ppm-range or even better)
there are no special requirements for the samples; samples can be solid, liquid, or in the gas phase
quantitative analysis of spectra is possible
no vacuum is required
time-resolved *in situ* experiments are possible.

The theory of XAFS spectroscopy, the various modes for measuring XAFS spectra, the data reduction processes, and numerous applications from a broad range of research areas are described in several text books and review articles, and in the proceedings of the corresponding conferences [269–274].

The analysis of EXAFS spectra does not directly provide information about the size, shape or crystallographic structure of a nanoparticle. However, the coordination numbers that are derived from an EXAFS analysis are independent from a specific model and can in principle be used for the characterization of nanoparticles regarding their structure and geometry. It is well documented that at least the size of a particle can be estimated from the analysis of the first coordination shell (assuming that the particle is spherical); however, it is possible to obtain much more detailed information from the analysis of higher coordination shells. The coordination numbers obtained from the EXAFS analysis for the first neighboring shells are compared with the corresponding numbers derived from polyhedral cluster models. The model structure that best matches the EXAFS data, yields the

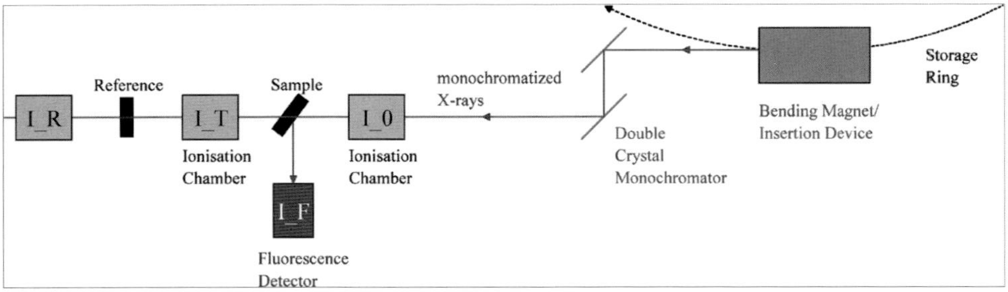

Figure 4.19 Scheme of the experimental set-up used for measuring XAFS spectra in transmission or fluorescence mode.

three dimensional structure and the shape of the nanoparticle. For the examination of bimetallic systems, an analysis of the EXAFS spectra obtained at a suitable edge for both metals is desirable as this provides the information about the size, composition and distribution of each species. The potential of EXAFS as a tool for determining the structure of nanoparticles was recognized very early, for example by Lytle and coworkers [275, 276]. However, it was not used very often as the quality of the data (and/or the analysis) was not sufficient to analyze the higher coordination shells. The various aspects that are crucial for obtaining reliable structural information from EXAFS spectra, for example the structural and compositional homogeneity of the sample and the effects of surface disorder, are discussed in recent publications from Yevick and Frenkel [277], Beale and Weckhuysen [278], and Frenkel *et al.* [279].

A typical experimental set-up for measuring X-ray absorption spectra in the transmission and/or fluorescence mode, as it is used with sometimes minor modifications in all synchrotron radiation laboratories, is shown in Figure 4.19. The white light emitted from a storage ring – either from a bending magnet or an insertion device – enters a double crystal X-ray monochromator employing two perfect single crystals. Changing the Bragg angle in the monochromator varies the photon energy. The primary intensity I_0 falling on the sample and the intensity I_T transmitted through the sample are measured by ionization chambers. The preamplified currents of the ionization chambers are digitized and stored in a microcomputer system. By dividing the two signals and taking the logarithm of the result, a value directly proportional to the desired absorption coefficient is obtained. To eliminate the influences of beam instabilities on the energy calibration of the monochromator, very often a third ionization chamber is used and a reference sample (in many cases a metal foil) is put between the second and the third chamber. The transmission spectrum of this reference sample is now recorded simultaneously with the spectrum of the actual sample. Thus, it is possible to determine for example the energy position of the absorption edges relative to each other to better than 0.1 eV.

Instead of measuring the absorption of a sample, also the relaxation products, i.e., Auger electrons or fluorescence photons (I_F) can be detected by a fluorescence detector as is indicated in Figure 4.19. Fluorescence XAFS spectra are recorded, for example, for diluted samples where the concentration of the atoms of interest is below 1 at%.

One of the major advantages of X-rays for spectroscopic applications is their penetration strength that allows *in situ* experiments to monitor chemical reactions (for example, catalytic reactions) when suitable '*in situ* cells' are available. In combination with the high intensity of synchrotron radiation sources, also time-resolved *in situ* experiments can be carried out. A 'typical' *in situ* cell used for the investigation for Fischer–Tropsch catalysts is described by Girardon *et al.* [280]. A more advanced design again for the investigation of catalyst in a Fischer–Tropsch reaction is described in a recent publication by Rochet *et al.* [281]. This group also uses a special monochromator design (a so-called QEXAFS monochromator [282]) that allows time-resolved measurements down to the second and even the subsecond time domain. Special *in situ* cells have also been developed for the XAS investigation of fuel cells. One typical example is described in a paper by Principi *et al.* [283]. Overviews of *in situ* studies of heterogeneous catalysts using X-ray diffraction and X-ray absorption techniques were very recently published by Frenkel *et al.* [284] and by Grunwaldt and Baiker [285]. An introduction into the use of *in situ* X-ray absorption spectroscopy for fuel cell research is given in an article by Croze *et al.* [286].

Over the last few years several optical components have been developed/optimized for focusing X-rays, for example, Fresnel zone plates, X-ray lenses, capillary optics, and Kirkpatrick–Baez mirror systems. Some of these systems provide a spatial resolution of a few micrometers (and sometimes even nanometers) and allow spatially resolved XAFS experiments. Using a so-called scanning transmission X-ray microscope (STXM) with a spatial resolution of about 15 nm, de Smit *et al.* observed *in situ* the reduction behavior of a single Fe-based catalyst particle [287]. In two review articles, Grunwaldt *et al.* give an in depth overview of the challenges and opportunities measuring spatially resolved XAFS spectra for catalytic systems [288, 289].

4.4.2
The Electronic and Geometric Properties of Monometallic Systems

There are hundreds of publications in the literature describing the characterization of metallic nanoparticles for catalytic and energy-related applications by using XAS. Thus, this section cannot to cover this vast literature, but will rather focus on a few examples that emphasize the potential – and the limitations – of the technique. As a 'typical' example we choose Co nanoparticles for our discussion. Figure 4.20 shows typical Co-K-XANES spectra of metallic Co in comparison with various Co oxides. In Figure 4.20a the spectra of metallic Co (valency 0) is compared with CoO (valency +2). Significant differences are observed in the 'edge region' (metallic Co has here a strong shoulder), at the 'white line,' which is very intense for CoO and hardly visible for Co and in the region of the so called 'shape resonances' at higher energies that reflect differences in the geometric structure. All differences in the edge and white line region can be interpreted by changes in the density of states of the 3d and 4s orbitals of Co. In Figure 4.20b, the Co-K-XANES spectra of the stable

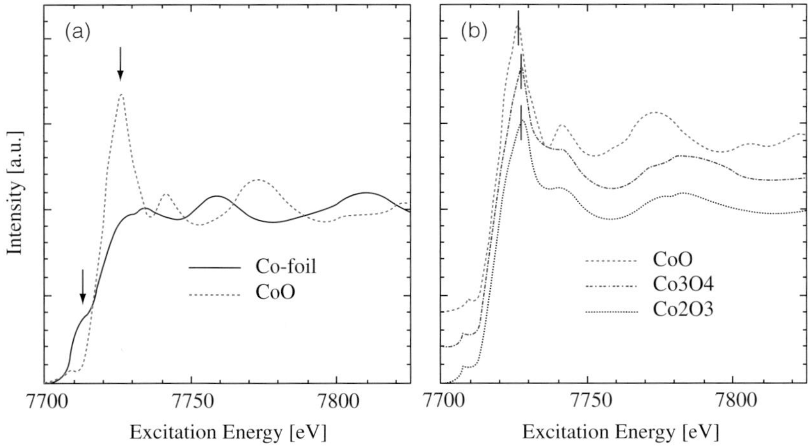

Figure 4.20 Co-K-XANES spectra of metallic Co and various Co-oxides (from [290].)

oxides are compared. As a function of the increase of the formal valency (+2, +2/ + 3 and +3) these spectra show a successive shift of the white line to higher energies. All these spectra demonstrate quite clearly the chemical sensitivity of XANES spectra.

As was discussed in Section 4.2.1 Co nanoparticles are still the best commercial catalysts for the Fischer–Tropsch reaction. Moreover, Co nanoparticles belong to the class of 'magnetic nanoparticles' which can be manipulated using magnetic fields. In previous years these particles have been the focus of intense research because they show tremendous potential for a broad range of applications such as biomedicine, magnetic resonance imaging, magnetic particle imaging, data storage, environmental remediation, and optical filters. Co nanoparticles are also very interesting objects for XAS as they can be observed in three crystallographic phase-s: hcp, fcc, and β-manganese (the so-called 'ε-phase') with, for example, very different electronic and magnetic properties. In a review Murray *et al.* [291] discussed the different magnetic properties of these three phases in great detail. For bulk Co at high temperature an additional phase exists, namely bcc, that has not yet been observed for nanoparticles. Though the geometric arrangement of the various phases is quite similar (at least for the first coordination shell), XANES is sensitive enough to clearly distinguish between them. This is shown in Figure 4.21, where the theoretical spectra for the four stable phases are shown calculated by using the FEFF 8 code [292].

Cheng *et al.* are one of several groups that used X-ray absorption spectroscopy (EXAFS and XANES measurements) to characterize the size dependence of the electronic and geometric properties of Co nanoparticles [293]. They synthesized three samples containing 3-, 5-, and 12-nm Co nanoparticles via thermodecomposition and used XANES and EXAFS spectra recorded at the Co-K-edge to probe the structure of these particles. As usual, different combinations of surfactants were used in order to obtain the different sizes. Figure 4.22 shows the XANES spectra of

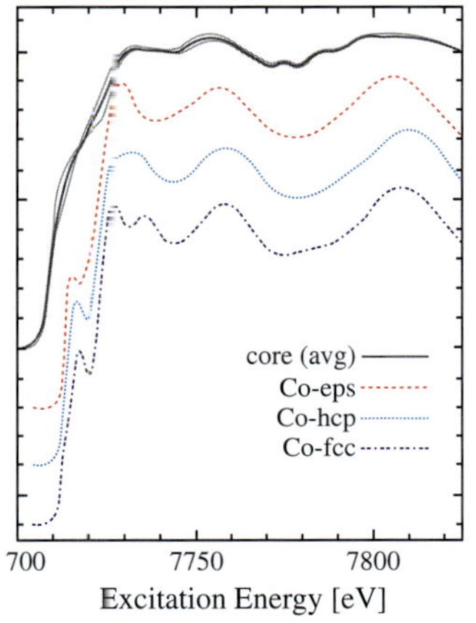

Figure 4.21 Calculated (by using FEFF8) Co-K-XANES spectra of the different crystallographic forms of metallic Co-nanoparticles (from [290].)

the three samples synthesized under anaerobic conditions in comparison with the spectra of CoO and a Co foil measured as references. The spectra of the three nanoparticles are significantly different with the 12 nm spectrum having the strongest similarity to the bulk spectrum; the spectra of the two other samples show some similarity with the spectrum of CoO. The observed systematic changes could be a pure 'size effect,' i.e., the XANES spectra reflect changes in the electronic and/or

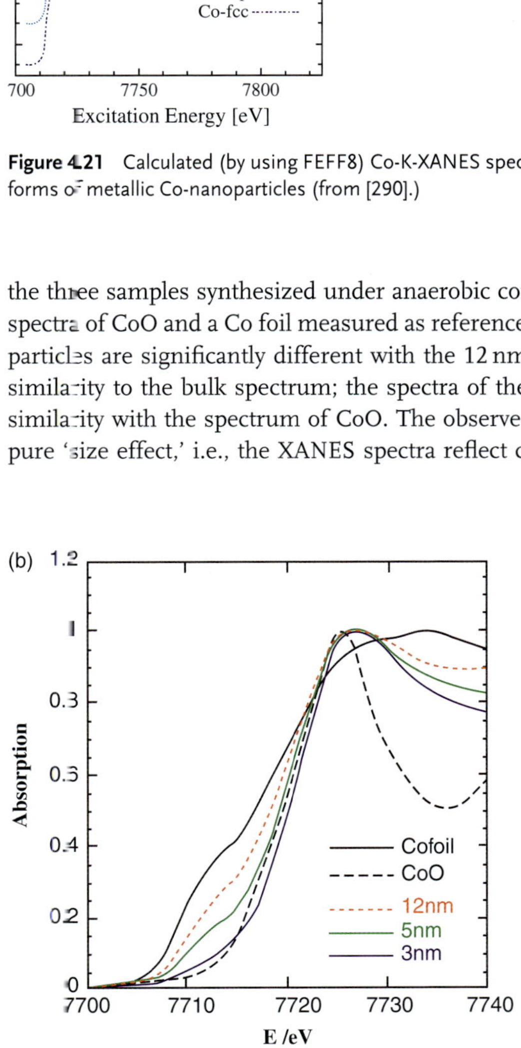

Figure 4.22 Co-K-XANES spectra of Co-nanoparticles of different sizes in comparison with the spectra of reference compounds (Co metal and CoO) (Reproduced with permission from [293]. Copyright © (2004) Elsevier Science.)

geometric structure as a function of the size; the authors themselves assume that the particles – though prepared under anaerobic conditions – are no longer in the metallic state but could be oxidized. However, comparing the spectra in Figure 4.22 with the calculated spectra in Figure 4.21 some of the observed differences could also be explained by assuming different crystallographic phases for the different particles, for example fcc instead of hcp – a fact that was not seriously considered in the publication.

The authors also observed significant differences in the EXAFS spectra of the three samples of particles and also in comparison with Co, CoO, and $CoCO_3$. What the authors didn't consider in their interpretation of the spectra is the changing ratio between bulk and surface atoms and the possible interaction of the surface atoms with the surfactant(s) that were used for the synthesis. For the 3 nm particles ~35% of all atoms are 'surface atoms' and for 5 nm particles that are still more than 20% (for 12 nm actually less than 10%) and at least part of these surface atoms could interact with the surfactants changing their valency from 0 to +2. These effects, i.e., the interaction of surface atoms with the coating shell, were confirmed, for example, in a study by Margeat *et al.* [294]. This group synthesized Co nanoparticle of ~3 nm diameter from two different precursors but using the same capping agent. They observed not only different magnetic properties of these particles but also significant differences in their geometric structure, which they related to the difference in the chemical environment at the surface of the nanoparticles. An observation in the same direction was reported by Kraum and Baerns [295]. These authors observed a dependence of the activity of a Co-based catalyst for Fischer–Tropsch synthesis on the type of cobalt precursor: an metal organic precursor, such as cobalt(III)acetyl acetonate resulted in a significant increase of carbon monoxide (CO) conversion compared with a catalyst made from cobalt nitrate. No rational explanation was given for these observations.

The sensitivity of the electronic and geometric structure of Co nanoparticles as a function of the surfactant used was also investigated by Bönnemann *et al.* [296]. They observed significant differences in the Co-K-XANES spectra of particles dispersed in vacuum oil and KorantinSH mineral oil. These differences show up not just in the near edge structure that is sensitive to changes in the electronic structure but also in the region of the shape resonances at higher energies that are sensitive to changes in the coordination geometry. Comparing these results with FEFF8 calculations comparable to those shown in Figure 4.21, the authors concluded that the sample dispersed in vacuum oil is present in the fcc phase whereas the particles dispersed in KorantinSH are a mixture of particles with fcc and bcc structures.

Using transmission electron microscopy Shukla *et al.* investigated the effects of various surfactants (mainly amines) and mixtures of other surfactants on the shape of cobalt nanoparticles that were synthesized by thermal decomposition of a cobalt containing precursor [297]. By using various surfactants, they could synthesize spherical, triangular, rod-like, and hexagonal particles. They observed not only a dependence on the type of surfactant but also a dependence on the temperature at which the precursor was added to the reaction mixture. No rational explanation of their observation could be given.

The examples for the XAS analysis of Co nanoparticles discussed here show not only the sensitivity of this technique and the challenges of providing a reliable interpretation of the corresponding data, but also emphasize the obviously strong interaction of Co nanoparticles with for example surfactants and/or support material. This topic will be discussed in more detail in Section 4.4.4.

As has been mentioned repeatedly, Co nanoparticles are the most efficient catalysts available for industrial Fischer–Tropsch synthesis. Accordingly, there are numerous papers using XAS to characterize Co nanoparticles used as catalysts in this specific reaction and to understand their interaction with various promoters and support materials. A short overview of this research with a focus on the relation between structure of the Co catalysts and their catalytic performance is given by A.Y. Khodakov [298]. Khodakov comes to the conclusion that the catalytic performance of Co in the FT reaction depends not only on size and crystallographic phase of the Co nanoparticles but also on the support material, the texture of that support, and, last but not least, also on the various promoters applied. He also emphasizes the transient behavior of the Co particles during the reaction and he concludes that structural modifications could be due to a broad range of experimental parameters from impurities in the feed, presence of water, carbon dioxide, heavier hydrocarbons and other organic compounds in the reaction products, to temperature control and the sintering of cobalt. All these effects can at least in principle be investigated by XAFS spectroscopy and some of those investigations will be discussed below.

A detailed study of the effect of water on Co/Al_2O_3 catalysts and the correlated reoxidation phenomena was carried out by Jacobs *et al.* [51] using XANES and EXAFS measurements. The authors investigated two catalysts: a 15% Co/Al_2O_3 with a cluster size of about 6 nm and a 25% Co/Al_2O_3 with a cluster size of about 12 nm. For both catalysts a strong impact is observed of H_2O on the CO conversion and on product selectivity. The detailed analysis especially of the EXAFS spectra of the catalysts shows that there are significant differences in the reaction of the two catalysts when exposed to up to 30% of H_2O. For the 15%Co/Al_2O_3 catalyst the addition of 25% H_2O caused a decrease in the Co–Co coordination number for the four coordination shells that were analyzed in this study. This is clear evidence for a surface reoxidation effect that was not reversible when the water was removed. The analysis of the XANES spectra indicates the formation of a cobalt–aluminate species, i.e., a reaction of the nanoparticle with the support. Also for the 25% Co/Al_2O_3 catalyst the addition of 25% H_2O results in a drop of the Co–Co coordination. However here, stopping the water supply resulted not only in a recovery of the coordination numbers but an increase exceeding the original numbers that were observed. This is an indication for a sintering process. Here analysis of the XANES spectra indicates the formation of CoO as a surface oxidation effect.

A more detailed investigation of the correlation between the size of cobalt nanoparticles and the water oxidation process was carried out by Saib *et al.* [299]. For this study well defined Co/SiO_2 model catalysts were synthesized with an average size of the Co nanoparticles of 4, 13 and 28 nm. In a first step the reductive behavior of these catalysts was studied using *in situ* XANES spectroscopy at the Co-K-edge.

After calcination the two larger particles contained mainly Co in the form of Co_3O_4 and for the smallest particle Co occurred as a mixture between CoO and Co_2SiO_4 as could be concluded from the corresponding XANES spectra. The larger particles were reduced in two steps: Co_3O_4 to CoO and then CoO to Co metal. The small particles were reduced in one step from CoO to Co metal; a reduction of Co_2SiO_4 was not observed at temperatures up to 773 K. These results were also quantified by using the usual linear combination fitting technique for XANES spectra. In a second step, XANES *in situ* experiments were used to study the oxidation behavior. Also here, the corresponding Co-K-XANES spectra revealed significant differences. For the largest particles (28 nm) only very little oxidation (5%) was observed; the medium sized particles (13 nm) were found to undergo significant oxidation already at low temperatures; the smallest particles (4 nm) showed no oxidation at all for temperatures up to 673 K. As a possible explanation the encapsulation of the 4 nm particles with silica is discussed; this encapsulation was actually confirmed using HRTEM when the reduction temperature of the crystallites was increased to about 973 K. Unfortunately, these results cannot be transferred directly to the above discussed system Co/Al_2O_3 so that there are still a lot of open questions when it comes to the deactivation processes caused by water – just to mention one of the many parameters that influence the FT reaction.

Co FT catalysts very often contain small amounts of metal promoter (typically noble metals, 0.05–0.1% weight) and/or oxide promoters (zirconia, lantania, cerium oxide, with 1–10% weight). Many studies demonstrate that the addition of even very small amounts of noble metals (specifically Ru, Rh, Pt, and Pd) have a strong impact on the performance of the Co catalysts. An overview of these mainly empirical observations is given in a review by Diehl and Khodakov [300]. Chu *et al.* [301] have carried out a very detailed comparison of monometallic and Pt-promoted cobalt alumina catalysts by using a broad range of techniques – among others also XAS. In this study, the Co nanoparticles were promoted with 0.1 wt% of Pt. After calcination, the as synthesized nanoparticles had a diameter of about 9 nm independent of the promotion and for both types of particles with and without promotor Co-K-XANES spectra were identical and identical to the one of Co_3O_4. Temperature programmed reduction (TPR) experiment showed differences in the calcination temperature of the catalysts but also a significant enhancement of both cobalt reduction steps ($Co_3O_4 \rightarrow CoO$ and $CoO \rightarrow Co$) by the Pt promotion. Co reducibility was relatively low in the monometallic, nonpromoted system but significantly enhanced in the 0.1 wt% Pt-promoted system. This is obvious from the Co-K-XANES spectra shown in Figure 4.23.

The spectrum of the Co/Al_2O_3 system reduced at 886 K is still more or less identical to the one of CoO (compare spectra 1 and 2 in Figure 4.23) whereas the spectrum of the promoted system $CoPt/Al_2O_3$ is very similar to the one of metallic cobalt (spectra 3 and 4 in Figure 4.23). From the spectra is it not really possible to decide in which crystallographic form the metallic particles occur. It was also reported that the promotion significantly increased the FT reaction rate and the selectivity. Unfortunately these findings also remain entirely empirical as no plausible explanations are available why such a low concentration of a promoter implies

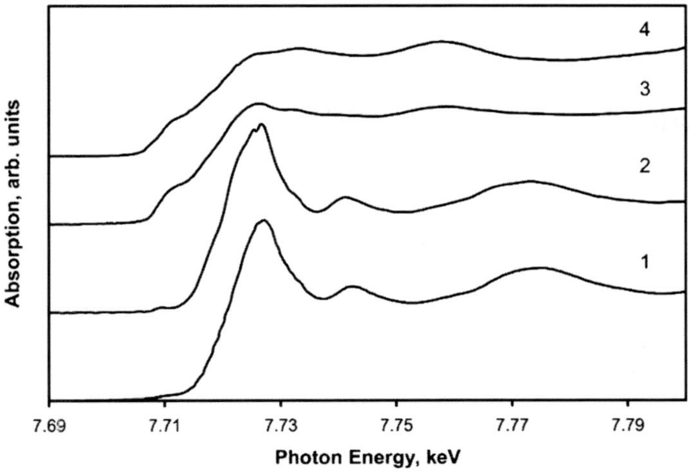

Figure 4.23 Co-K-XANES spectra of (1) CoO, (2) reduced Co/Al_2O_3 - 613, (3) reduced $CoPt/Al_2O_3$ – 613 and (4) metallic cobalt. (Reproduced with permission from [301]. Copyright © (2007) Elsevier Science.)

such a dramatic change the chemical structure and the performance of the Co catalyst. In a recent follow-up study the Khodakov group used *in situ* time-resolved XRD/XANES/EXAFS measurements for the identification of the active species of the Pt-0.1 wt%-promoted Co catalyst [302, 303]. For the reduced catalyst (H_2 at 623 K) metallic Co was once again observed. The XANES spectra taken at the Pt-L-III edge indicated also metallic Pt though the shape of the edge was different from that of a Pt foil. The EXAFS data recorded at the Pt-L-III edge could be analyzed assuming a bimetallic structure, i.e., Pt atoms surrounded by 12 Co atoms. Based mainly on *in situ* XRD data, the authors concluded, that the Co-fcc phase is dominant in the reduced catalyst and that the catalytically more active Co-hcp phase develops over time by consecutive CO and H_2 treatment. As the nonpromoted system was not investigated in these studies, the mechanism in which the Pt promotion modifies the structure and the performance of the catalyst remains still an open question.

There are several more publications investigating the effect of promoters on the structure and the catalytic performance of Co in FT catalysis using X-ray absorption techniques. Cronauer *et al.*, for example, investigated iridium as a promoter [304] and Hong *et al.* looked at zirconia [305]; however these studies again give no indication about the effectiveness of the promoters. Jacobs *et al.* carried out a more systematic study of the promotion of a 15%Co/Al_2O_3 catalyst by Cu, Ag and Au [306]. This group could observe significant impact of the various promoters on the catalytic performance; however, the EXAFS experiment carried out with these systems showed only Co–Co and Me–Me (Me = Cu, Ag, Au) coordinations suggesting that the metals tend to form different phases. Thus, here the mechanisms with which the promoters influence catalytic properties also remain unclear.

4.4.3
The Geometric and Electronic Structure of Bimetallic Systems

Adding a second metal to a metallic nanoparticle gives new opportunities to modify/control the chemical and physical properties of the particles. Thus, bimetallic particles are used in a broad range of applications especially in catalysis. A comprehensive review article by Guczi *et al.* [307], for example, just summarizes the results obtained for bimetallic cobalt based catalysts with a focus on the correlation between structure and reactivity/selectivity as determined by the second metal. The basic assumption of adding a second metal to a metallic catalyst is that the properties are modified by alloying. However, this process depends not only on the inherent properties of the individual metals such as bulk crystal structure and electronegativity but also on the details of the synthetic process and, in the case of catalysts, finally also on the reaction conditions that can change the structure of the particles completely. Thus, depending on all these parameters a lot of structures are observed starting from 'real alloys' with a statistical distribution of atoms, and core-shell systems, up to particles with strong segregation of the components and finally 'bi-clusters' in which the metals are completely separated [278]. This rather complex situation prevents the successful applications of simple thermodynamic guidelines for tailoring bimetallic catalysts as suggested, for example, by Fernandez *et al.* [308]. However, there are some recent examples of using density functional theory (DFT) calculations for modeling the structure of bimetallic systems and thus optimizing the catalytic performance. A demonstration of this approach is the comparison of core-shell Ru–Pt nanoparticles with Ru/Pt alloys and mixtures of monometallic Ru and Pt nanoparticles as catalyst for the oxidation of CO [309] and a systematic study M/Pt core shell particles (M = Ru, Rh, Ir, Pd, and Au) as catalysts for the same reaction by the same group [310].

XAS is a very powerful tool to investigate such bimetallic systems: XANES provides (mainly) the information about changes of the electronic structure for both components while EXAFS spectra provide the geometric information for determining the structure of the particles. Methods of using EXAFS and XANES for a quantitative determination of the degree of alloying and/or the atomic distribution are, for example, described in a paper by Hwang *et al.* [311] and also the review by Beale and Weckhuysen [278]. To demonstrate the sensitivity of XAS to structural details, the latter authors have simulated EXAFS spectra for bimetallic Pt–Pd structures based on data from Pt and Pd foils. The Fourier transform for the first coordination shell of the various simulated structures are shown in Figure 4.24.

In the Pt-L-III FT spectra (a) two main contributions can be clearly distinguished at 2.73 Å and at 2.46 Å respectively in contrast to the Pd-K-edge-FT spectra with one strong peak at 2.69 Å and just a shoulder to the low R side. The intensity of high R-peaks is connected to the degree of monometallic bonds (Pt–Pt and Pd–Pd respectively) whereas the low R peak in the Pt-FT spectrum reflects the degree of bimetallic bonds and thus also the degree of alloy formation. Another review by Antoniak [312] summarizes recent developments in the field of EXAFS

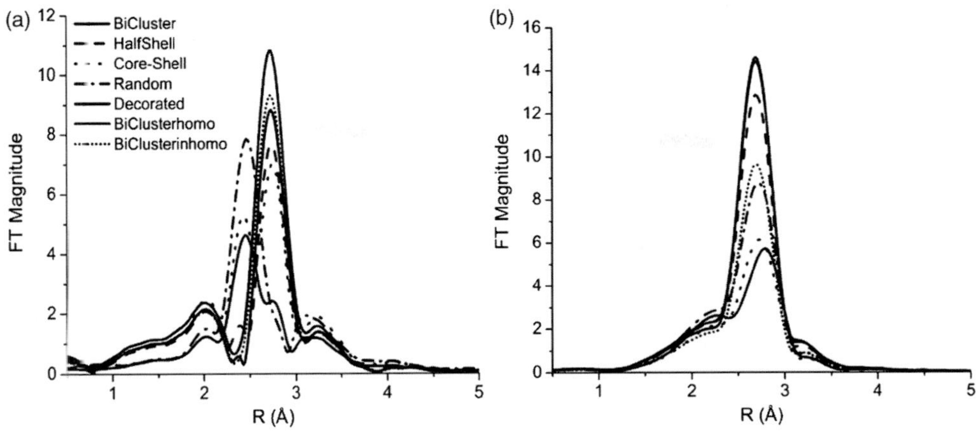

Figure 4.24 Influence of the bimetallic structure on the EXAFS FT data at the Pt-L-III edge (a) and the Pd-K-edge (b). (Reproduced with permission from [278]. Copyright © (2010) Royal Society of Chemistry.)

measurements and the analysis methods for the structural investigation of bimetallic nanoparticles. In particular the standard analysis that is based on Fourier transforms is compared with a recently developed new technique based on wavelet transforms that shows some advantages for bimetallic systems.

The behavior of the Pt-FT in bimetallic systems as predicted by Beale and Weckhuysen [278] was actually observed in an EXAFS/XANES study [313] of the Pt–Ru core-shell/alloy nanoparticles mentioned before [309]. However, the behavior of the Ru-K-edge-FT was different from that predicted for Pd. By combining EXAFS, XRD, and pair distribution function analysis the authors could develop a rather comprehensive picture of the geometric structure of their particles. The 4 nm Pt–Ru core-shell particles, for example, had a highly distorted *hcp* Ru core and a 1–2 monolayers thick crystalline Pt shell with a fcc structure slightly distorted as compared with bulk Pt. The XANES spectra that were also recorded at the Pt-L-III and the Ru-K-edge showed only minor differences for the two samples but an increased white line intensity as compared with the bulk metals. The authors tried to explain this observation by some degree of oxidation. No rational explanation is provided for the different catalytic performance of the samples.

As already discussed in Section 4.2.1.2, for PEMFCs, carbon-supported platinum nanoparticles are the 'standard' catalysts. From the bimetallic systems, Pt–Co is one of the systems that show the highest increase in catalytic activity as compared with pure Pt. Kozinkin *et al.* investigated with X-ray diffraction, X-ray emission, and X-ray absorption spectroscopy such carbon-supported catalysts of the type Pt_xCo/C with different compositions ($x = 1$–3) and an average size between 2 and 4 nm [314]. From the FT of the EXAFS spectra of the Pt-L-III edge, that are similar to the FT for metallic Pt, the authors conclude that the structure in the nanoparticles is very similar to the one in bulk Pt metal, i.e., fcc. The FT of the EXAFS spectra measured for the three particles at the Co-K-edge are different and different as compared with metallic cobalt. From a detailed analysis they conclude that Co atoms interact with a light atom (such as O or C) in the first coordination shell, with other Co atoms in the second coordination shell and with platinum in the third shell. These observations are confirmed by $CoK_{\beta5}$ X-ray emission spectra. The authors conclude that the particles are bimetallic because of the observed Pt–Co bond and that Co atoms interact with the carbon support. Loukrakpam *et al.* also investigated Pt–M nanoparticles (M = Co and Ni) by XRD, TEM, and XAS regarding their electronic/geometric structure and their electrocatalytic properties [107]. For Pt–Co they investigated two sizes $Pt_{73}Co_{27}$ with 1.8 nm and $Pt_{70}Co_{30}$ with 6.7 nm and for Pt–Ni they investigated $Pt_{78}Ni_{22}$ with 1.5 nm and $Pt_{71}Ni_{29}$ with 5.0 nm and all the corresponding particles on a carbon support. For Pt–Co/C they observe a decreasing catalytic activity as a function of increasing particle size and for Pt–Ni/C there is a slight increase as a function of increasing size! For both systems the white line in the Pt-L-III XANES spectra of the nanocatalysts (which is due to the electron transition from $2p_{3/2}$ to $5d_{5/2}$) has a slightly higher intensity than a metal foil. These changes indicate a decrease of the filling of the 5d orbital that could be caused either by some oxidation or by hybridization between Pt and the second metal.

Recently Schulenburg *et al.* [315] investigated a $PtCo_3$/C catalysts using XRD, TEM and nano-EDX. The authors found a strong dependence of the catalytic activity from heat treatment under reductive conditions between 623 and 1273 K with the highest activity at ~1073 K. From 873 K on, they observed an increase in particle size and a broadening of the particle size distribution. During the heat treatment, they observe also the formation of specific phases such as fcc-Pt, Pt_3Co, fct-PtCo, and fcc-PtCo with fct- and fcc-PtCo phases being present in the samples with the highest catalytic activity. The same nanocatalyst together with a commercial system (Pt_3Co/C from TKK) was also investigated by Kühn using XAS [290]. Figure 4.25 shows the Co-K-XANES spectra and Figure 4.26 the corresponding Pt-L-III XANES spectra of these catalysts.

In Figure 4.25a it can be seen that significant changes occur upon tempering the catalyst to 1073 K. From Figure 4.25b it is clear that Co_3Pt/C-asprep is almost identical to CoC-asprep, indicating that the observed strong white line is due to bonds with carbon. Unwanted oxidation, can be ruled out in this case because the nanoparticles were handled strictly air-protected all the time and because significant changes in the XANES spectra were observed when the particles were deliberately exposed to air for a few days. From the spectra in Figure 4.25b a formal oxidation state similar to Co_3O_4 can be assigned for Co_3Pt/c-asprep, i.e., a valency between +2 and +3. When oxidized (Co_3Pt/C-350-ox), Co has an even higher valency, at least +3 as compared with Co_2O_3. Co_3Pt/C-800 is obviously in a metallic state as its 'edge shoulder' at about 7713 eV resembles that of a Co foil. However, the catalyst has a different crystallographic structure, recognizable in the differences of the white line and the shape resonances at higher energies.

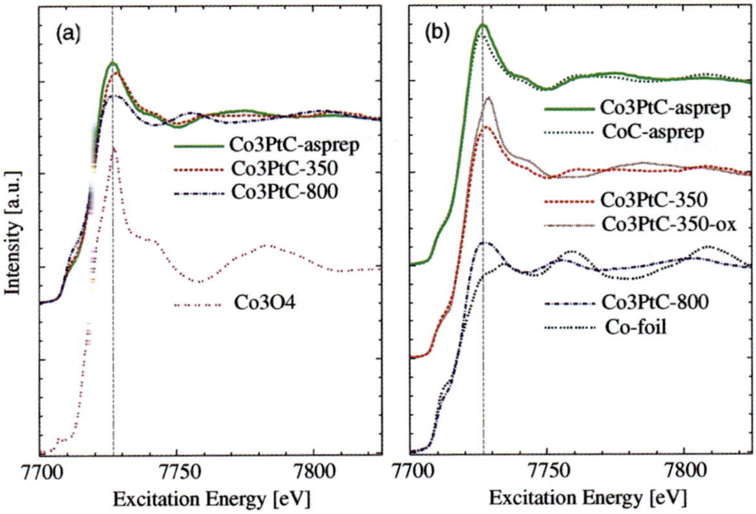

Figure 4.25 Co-K-XANES spectra of Co_3Pt/C nanoparticle catalyst as prepared and tempered to 623 K and 1073 K in comparison with spectra from reference compounds (from [290].)

From the Pt-L-III XANES in Figure 4.26a it is again obvious that electronic and structural changes take place in the Co_3Pt/C-800 system whereas hardly any changes are visible between the as-prepared and the 623 K system. With respect to the ORR most active nanoparticles – Co_3Pt/C-800 – exhibit the lowest intensity white line, which is also true comparing these particles to a commercial Pt_3Co/C catalyst (from TKK), as shown in the lower part of Figure 4.26a. This lowering of the white line intensity is accompanied by a slight shift to lower energies, indicated by an arrow in Figure 4.26a. It can be attributed to an increased occupancy of the 5d electron levels, as the result of a lowered 5d band center, which is an ORR activity increasing effect as was shown by Xu *et al.* based on DFT calculations [316]. The lowering of the 5d band center is related to a 'weakening' of the metal–oxygen bond strength, which is induced both by a shortened Pt–Pt bond distance and by the 'ligand effect' [317, 318]. The former is a result of strained surface layers due to the different lattice constant of the second metal (Co in this case), and the latter comprises the interactions of the ligands (as part of a coating or substrate) with the metal(s), leading to changes in the chemical properties of the surface [317]. The weakened metal–oxygen bond strength allows for a faster ORR, as the adsorbed O_2 will be less tightly bound, but nonetheless dissociates. An additional benefit of these effects is the reduced sensitivity to CO poisoning, as CO will also be bound less tightly to the surface. The observed shortening of the Pt–Pt bond (to 2.697 Å) for the Co_3Pt/C-800 system and also the Pt–C interaction could be verified by the simultaneous analysis of the Co and Pt EXAFS spectra. Kühn assumes that the differences in the findings regarding improved catalytic activity between XRD measurements [315] and XANES/EXAFS experiments originate in the different 'sensitivities' of the two methods, a fact, which is very often ignored in the literature: XAS

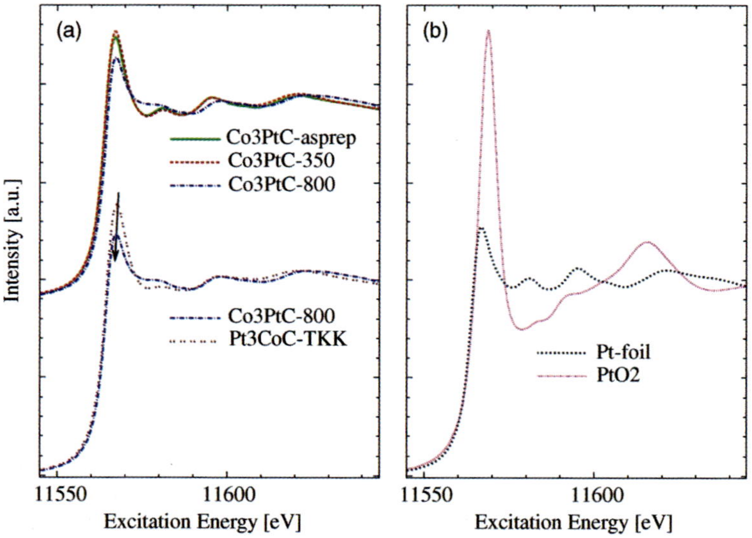

Figure 4.26 Pt-L-III XANES spectra of the Co_3Pt/C nanoparticle catalyst as prepared and tempered to 623 K and 1073 K and Pt-L-III XANES spectra of reference compounds (from [290].)

probes all particles regardless of their crystallinity (or noncrystallinity) whereas XRD probes only the crystalline part of a sample.

As was mentioned before, many catalysts are not 'stable' under working conditions, but they undergo structural changes. In a mini review Zafeiratos *et al.* discuss the most important phenomena: phase separation and surface segregation in reactive environments [319]. Lai *et al.* have investigated the dealloying mechanism using XAS and STEM [320]. From the analysis of their EXAFS spectra they conclude that initially the Pt–Co nanoparticles have a Pt-enriched core and a Co enriched shell. The chemical dealloying process dissolves mainly Co—Co bonds – a process leading to smaller particles with a rough Pt shell. When the dealloying process continues cobalt under the skin layer is dissolved leading finally to a 'collapse' of the particle. Uemura *et al.* carried out an *in situ* time-resolved XAS study of the structural changes of Pt_3Sn and PtSn nanoparticles under working conditions [321].

For several Pt–Cu fuel cell catalysts, Strasser *et al.* investigated the improved electrochemical performance of these dealloyed catalysts and they found that dealloying results in a core-shell structure with a Pt-rich shell [113]. In accord with the above discussed observations from Kühn [290] the authors come to the conclusion that the lattice strain in this shell results in a shift of the electronic band structure of Pt and a weakening chemisorption of oxygenated species. Both observations: the formation of a Pt-enriched shell in an acidic environment and the improved catalytic performance of a Pt_3Co nanocatalyst were confirmed by a recent investigation by Carlton *et al.* using subnanometer aberration-corrected scanning transmission electron microscopy [322].

There is striking evidence that bimetallic systems have in many cases significantly better catalytic properties that the monometallic catalyst. There are also strong indications that changes in the electronic structure of the catalytic active component are responsible for these improvements. However, it is still not clear which geometric structure (alloy, core-shell system, etc.) has the best properties and how these favorable properties can be systematically enhanced.

4.4.4
The Specific Interaction of Metallic Nanoparticles with Coatings and Supports

In this section two additional aspects of using metallic nanoparticles as catalysts are discussed: the 'strong metal–support interaction' (SMSI) and modification of the properties of nanoparticles by the interaction with their coating.

The metal–support interaction in catalysis by supported metal clusters and nanoparticles is a phenomenon that has been well known for more than 25 years [323]. Also the advantage of XAS for the investigation of this interaction was recognized very early and the first reviews were published in the early 1990s [324, 325]. In general, the changes observed in the catalytic activity are ascribed to changes in the electronic properties of the nanoparticles induced by the specific interaction. Several models have been developed to describe these effects in more detail; most of them based on the assumption of electron transfer between support and catalysts, but also alternative approaches such as a decrease in the ionization potential of the catalyst due to SMSI are discussed in the literature [326].

Shishido *et al.* investigated the reduction behavior of Pt dispersed as Pt^{4+} on TiO_2 and Al_2O_3 surfaces in a reductive atmosphere using time-resolved *in situ* XANES/EXAFS measurements at the Pt-L-III edge [327]. By using an energy dispersive X-ray monochromator the experiments had a time-resolution better than 1 s per spectrum. On TiO_2, Pt^{4+} was effectively reduced at room temperature whereas on Al_2O_3 even at 473 K the reduction was very slow. The authors try to explain their findings not directly by a metal–support interaction but by assuming different adsorption mechanisms of H_2 and thus also a different supply of H atoms for the two support materials.

Gonzalez *et al.* investigated a Pt catalyst supported on TiO_2, CeO_2 and Ce-modified TiO_2 for the water gas shift (WGS) reaction [328]. For the Ce-TiO_2 catalysts the Ce/Ti atomic ratio was 0.032. The authors observe significant differences in the WGS activity and the thermal stability of the Pt catalysts depending on the support: Pt/CeO_2 is the least active and Pt/Ce-TiO_2 the most active and at the same time the most stable. Figure 4.27 shows the Pt-L-III XANES spectra for the three systems in reduced form and under WGS conditions. All spectra were measured *in situ*. Hardly any changes are observed for the TiO_2 supported system between the freshly reduced catalyst and the one under working conditions; however there are significant (and different) changes for both Ce-containing supports. For the reduced Pt–CeO_2 system the XANES spectrum is very similar to the one for the Pt foil and the spectrum for the working catalyst shows a reduced intensity for the

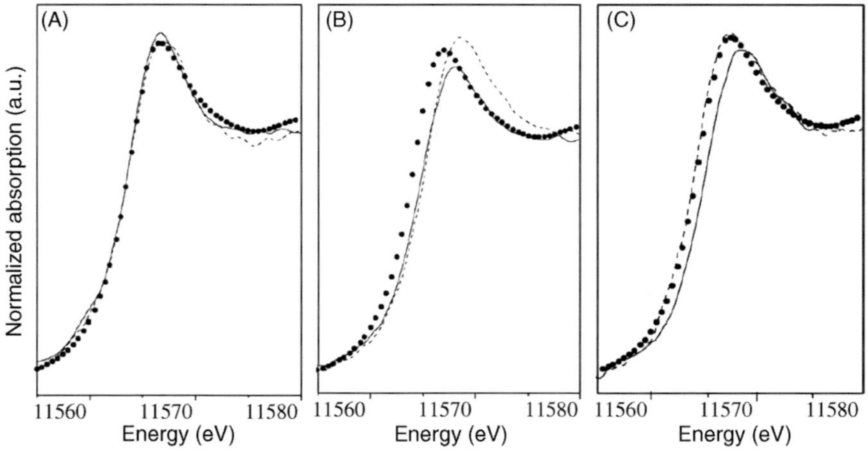

Figure 4.27 Pt L-III XANES spectra of: (A) Pt/TiO_2, (B) $Pt/Ce\text{-}TiO_2$, (C) Pt/CeO_2, • = Pt – foil, - - - = reduced catalyst under CO at 473 K, ________ = catalysts under WGS conditions at 523 K. (Reproduced with permission from [328]. Copyright © (2010) Elsevier Science.)

white line and a shift to higher energies; these findings could be explained by an increased electron density of Pt caused by an electronic transfer from the support to the Pt centers. For the $Pt\text{-}Ce\text{-}TiO_2$ already the reduced system shows a shift of the Pt-L-III XANES spectrum to higher energies as compared with metallic Pt; for the working catalyst a reduction of the white line intensity is observed similar to the one for the Pt/CeO_2 system.

Including Pt-L-III EXAFS and Ce-L-III XANES results into their interpretation, the authors conclude that Ce in TiO_2 is highly dispersed and strongly interacting with the Ti atoms. A modification in the reducibility of the support in the Pt/Ce-TiO_2 system is proposed to be the cause for the better activity and stability of this sample.

Kim *et al.* investigated in great detail the effect of silica characteristics on the catalytic activity of Pt catalysts supported on silica [329]. They found that the catalytic activity of Pt in the oxidation of CO was more or less independent from the properties of the silica but that this activity improved by the incorporation of titania into the silica and the treatment with hydrogen peroxide. However, the chemical state of the silica influenced the coordination state of the titania incorporated into them. Without a detailed explanation, the authors conclude that the high dispersion of Pt, the presence of a stable titania layer, and the high acidity of H_2O_2-Pt/Ti-silica are responsible for a high catalytic activity.

Ho *et al.* investigated the influence of a nanostructured $Ti_{0.7}Mo_{0.3}O_2$ support on the ORR of Pt [330]. They report a significant increase in ORR current density as compared with commercial Pt/C and PtCo/C catalysts. From their XAS results they conclude that this enhancement results from an electron transfer from the support to Pt leading to a shift in the d-band center of the surface Pt atoms.

It is generally accepted that reduced metal surface sites are the active sites for FT synthesis when using Co-based catalysts. This finding leads to the conclusion that high surface area oxides would be the best support as they would support a high dispersion of small Co nanoparticles. However, these conclusions are obviously not completely valid. In a comprehensive study, Jacobs *et al.* investigated the influence of support type (here SiO_2 and Al_2O_3), cobalt loading and noble metal promoter on the reduction behavior of cobalt oxide particles using *in situ* temperature-programmed EXAFS/XANES experiments [331]. They found that the more weakly interacting support, SiO_2, yielded larger cobalt particles with a smaller cobalt active metallic surface area. The stronger interacting support Al_2O_3 resulted in a lower extend of reduction of cobalt oxides but lead to much smaller Co particles and thus to a higher metallic surface area. The addition of noble metal promoters significantly decreased the temperature required for reducing the cobalt oxides supported on Al_2O_3. In a follow-up study the same group confirmed these findings by investigating the detailed influence of these supports on the reaction kinetics of the FT reaction [332].

There are a few studies investigating also the effect of 'unconventional' supports on the FT catalytic activity of Co catalysts. Gnanamani *et al.* studied a molecular sieve (KIT-6) as a support. Based on the Co-K-XANES spectra they conclude that the molecular sieve improves the reducibility of the cobalt oxides and based on their EXAFS results they suggest that the Co crystals are slightly smaller than for the Co/SiO_2 system [333]. Zhang *et al.* investigated multiwall nanotubes as support for the Co catalyst [334]. It was observed that carbon nanotubes allowed preparation of highly dispersed cobalt catalysts with high reducibility and high catalytic activity. The diameter of the nanotubes had no significant influence on the catalytic activity.

In spite of all these efforts, the nature of the metal–support interaction still remains unclear as is demonstrated by inconsistent results that have also been published in the recent literature (cf. conclusion in [332]). The analyses of XANES and EXAFS spectra very often give indications for the origin of the observed metal–support interaction that (most likely) leads to electronic changes in the corresponding catalytic active metal. Most studies conclude that the metal valence band is shifted by the SMSI [326] as was also observed for bimetallic systems [290]. However, also here systematic studies that vary just one parameter at a time (the support material) are still missing.

For the creation of metallic nanoparticles two general approaches have been established: the 'top-down' and the 'bottom-up' pathway. In the top-down approach nanoparticles are formed by the structural decomposition of a solid body, i.e., going from large entities to nanosized ones. Ball milling and lithography are two of the techniques used for this approach. In bottom-up approaches nanoparticles are created by nucleation of atoms or molecules. The fundamental process of all wet chemical syntheses is the creation of smallest fragments from a precursor material in a solution, which then cluster to form finally nanoparticles. To control the overall chemical process and thereby the nanoparticles'

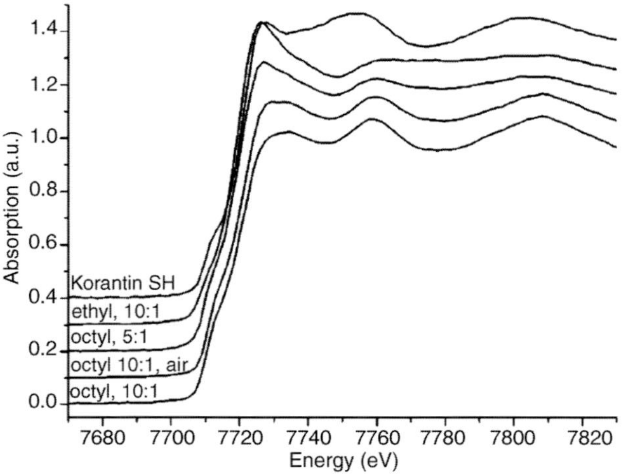

Figure 4.28 Co-K-XANES spectra of Co nanoparticles obtained *via* slow thermolysis of dicobaltocto-carbonyl coated with AlR_3 (R = ethyl, octyl) in various concentration ratios with Korantin SH. (Reproduced with permission from [335]. Copyright © (2005) American Institute of Physics.)

size and shape, a so-called stabilizing agent (or reactant/surfactant) is used, which has the following important functions (compare [335]): (i) preventing agglomeration so that stable particles can be formed; (ii) controlling the kinetics of nucleation, thereby determining the size of the particles; (iii) selectively being attached to different crystal surfaces during the growth to control the particle shape; and (iv) determining the particle's polarity.

An important point is that the surfactants can contain functional groups which interact weakly or strongly with the nanoparticles, thereby preserving the electronic properties of the 'naked' nanoparticles, or altering them by charge redistribution (on the nanoparticles surface) so that new (wanted) properties can arise, like self-assembling (compare also [336]). For a long time, the crucial influence of these capping molecules on the properties of – especially very small – nanoparticles was ignored. In a short review from 2005, Hormes *et al.* highlighted the influence of these coatings on the electronic, magnetic and geometric properties of cobalt nanoparticles [335]. They could show that not only the Co-K-XANES spectra (reflecting the electronic properties of the particles) were different for different coatings but that, for example, for aluminum organically stabilized Co particles the Co:AlR_3 ratio and the length of the alkyl chain had a significant influence on the XANES spectra as can be seen in the corresponding spectra in Figure 4.28.

Because of their importance for biomedical applications and as catalysts Au nanoparticles and their interaction with different capping systems were investigated by several authors. Thus, we will depart from the general structure of this subsection and will include Au nanoparticles in our discussion. Lopez-Cartes investigated the properties of very small Au nanoparticles coated with tetralkylammonium salt, an alkanethiol, and a thiol-derivatized neoglycoconjugate using Au-L-III XANES and EXAFS measurements [337]. They found significant changes in the Au d-hole density as a function of the capping system: for very small nanoparticles capped with thiol-derivatized molecules they observe an increase in d-hole density due to a Au–S charge transfer. By analyzing the corresponding EXAFS data, they observed that thiol functionalization introduced a distortion of the fcc structure caused by the particle–coating interaction. Some of these observations contradict the results reported by Zhang and Sham [336]. These authors investigated alkanethiolate-capped Au nanoparticles with nominal sizes of 1.6, 2.4, and 4.0 nm. They report that the d-charge at the Au atom site depletes and that the lattice contracts relative to bulk Au as the size of the nanoparticles decreases.

In a recent study, Marchetti *et al.* investigated the influence of alkylamines with different chain length on the size and the electronic/geometric properties of Au nanoparticles [338]. They found that the size of the Au nanoparticles decreased with increasing alkylamine chain length; however, in contrast to the results reported by Hormes *et al.* [335] for Co nanoparticles they do not observe significant differences in the Au-L-III spectra of the Au particles capped with hexylamine, dodecylamine, and octadecylamine. This group also used XAFS spectroscopy to investigate the various steps of the synthesis with respect to the coordination geometry and the valency of Au.

Liu and Zhang have investigated the influence of various capping molecules on the electronic and geometric structure of bimetallic AuPd nanoparticles [339]. Based on the analysis of the Au-L-III EXAFS spectra they conclude that the geometric structure of the bimetallic nanoparticle can be tailored by the type of capping molecule from random alloy to core-shell structures. From the Au-L-III XANES spectra, they conclude that together with the geometric structure, the d-electron density of the Au atoms can also be tailored.

There are a lot of strong indications that the types of capping molecules modify/determine the properties of metallic nanoparticles; however, there are also strong indications that at least in some cases (mainly for core-shell particles) even the thickness of the coating influences the properties. This was already emphasized in the above-mentioned article by Hormes *et al.* [335] when the different blocking temperatures of Co nanoparticles as a function of the thickness of the Au coating were discussed. In a recent paper, Chen *et al.* investigated the effects of the Pt shell thickness on the atomic structure of Ru–Pt core shell nanoparticles [340]. They varied the Pt coating between 1.5 and 3.6 atomic layers and found that nanoparticles with 1.5 atomic layers showed the optimum performance as catalyst. This system showed also the highest density of d-band vacancies. The authors give two reasons for the observed electronic modifications: lattice strain and the heteroatom electron orbital interaction (i.e., a ligand effect). The importance of lattice strain for the activity of fuel cell catalysts was also emphasized by Strasser and coworkers in their study of Pt–Cu alloy nanoparticles [113]. Their Cu–Pt core shell systems had Pt layer thicknesses between 0.6 and 1.0 nm so that the strain effect should be solely responsible for the observed exceptional reactivity of their particles that the authors correlate with a shift of the oxygen 2p and Pt 5d antibonding state from above to below the Fermi level that is weakening the adsorbate bonds.

4.4.5
Resonant Elastic and Inelastic X-ray Scattering: Site and/or Valency Specific Spectroscopy

In the previous sections a broad range of recent examples of applications of XAFS spectroscopy for the investigation of various synthetic parameters (capping molecules, support materials, promoters etc.) have been discussed, that determine the electronic/geometric and thus also the catalytic properties of metallic nanoparticles. 'Standard' XAFS spectroscopy is without any doubt a very powerful tool for the investigations of these properties. However, nanoparticles are no easy target for spectroscopic investigations as there are at least three different 'types' of atoms with respect to the first coordination shell. There are atoms 'inside' the particles having the normal bulk number of neighbors and there are surface atoms that have a reduced number of neighbors both exhibiting a valency of 0. Furthermore, there are surface atoms that interact directly with the coating (capping molecules) – may it be via a chemical bond or just via physi- or chemisorptions – that have a valency $\neq 0$ (what makes nanoparticles to mixed-valency compounds). In real world systems there is in most cases a fourth site: the surface atoms that interact with the support. By characterizing those sites of different valency separately, it should be possible to gain additional evidence about the interaction between the coating/the support and the 'interior' that could lead to the new and desired properties of nanoparticles. However, although one of the special properties of XAS is its sensitivity to the immediate environment of the absorbing atom, a standard XAS experiment just gives a spectrum that averages over all these different sites. In some cases it is possible to extract the desired information from the spectra by fitting the measured spectra with those of well characterized reference compounds but very often this information is not very reliable as there are too many free fitting parameters. To obtain the information wanted for the different atomic species, valency-selective XAS can be used, which is possible by using X-ray emission spectroscopy (XES) sometimes also called highly-resolved fluorescence-detected XAS (HRFD-XAS). In this type of experiment, it is not the usual Kα fluorescence that is detected, but fluorescence from higher orbitals (for example K$\beta_{1,3}$ fluorescence from the M-shell in the case of Co) that are chemically more sensitive. The origin of the 'selectivity' of HRFD-XAS spectra is found in the chemical sensitivity of emission lines, which is connected, in the case of transition metals like Co, to the 3d valence band. This valence band overlaps with the initial state of the emission process, i.e., 3p in the case of K$\beta_{1,3}$ emission, resulting in a 3p–3d exchange interaction. The K$\alpha_{1,2}$ emission, which has 2p as initial state that sparsely overlaps with 3d, is much less sensitive. Emission starting from the 3d level itself, or even from higher valence states (4p or ligand orbitals), are very sensitive albeit considerably less intense. If the excitation energy is scanned through a 'resonance,' i.e., a 1s or 2p core state, the process described here is called 'resonant inelastic X-ray scattering (RIXS).' While

X-ray absorption probes the unoccupied electron states, X-ray emission is complementary as it reflects the occupied ones. In the case of resonant excitation both processes are strongly coupled.

This kind of valency-selective XAS experiment was performed for the first time by Peng *et al.* [341] and Grush *et al.* [342] on manganese mixed-valency complexes. In both experiments, (partial) valency and site-selectivity, respectively, was achieved, though the bad signal-to-noise ratio was a serious obstacle. A few years later, this type of experiment was resumed with significant technical improvements, for example, by Glatzel *et al.* [343]. By using some numerical procedures these authors could extract from the measured spectra site specific EXAFS spectra for Fe in 'Prussian Blue.' With this improved experimental set-up the investigations of the Mn oxidation states in photosystem II have also been repeated with a significant better signal-to-noise ratio [344]. The processing of the (partially site-selective) HRFD-XAS spectra by numerical procedures is necessary to get pure site-selective spectra as the respective emission lines always partially overlap.

The basic experimental set-up for measuring HRFD-XAS spectra is similar to that shown in Figure 4.19, but instead of a semiconductor detector as fluorescence detector (that has an energy resolution $\Delta E > 100$ eV) a secondary crystal spectrometer with several crystals is used that typically has an energy resolution of about 1 eV. Such a high resolution set-up can of course also be used advantageously for 'standard' XAS experiments when the Kα (for 3d elements) or Lα (for 4d elements) fluorescence radiation is detected as the lifetime broadening of the 'initial state' for the fluorescence transition is much narrower than for the absorption transition. By applying this detection technique Safonova *et al.*, for example, could identify the CO adsorption site in a supported Pt catalyst [344]. Figure 4.29 shows the direct comparison of spectra detected using low resolution (total fluorescence detection) and HERFD. The spectral features in the HERFD XANES spectra are significantly sharper compared with standard spectra and they show a clear doublet structure for Pt/Al_2O_3 in a 1% CO/He atmosphere. By theoretical calculations, this feature was shown to be characteristic for CO adsorbed in the atop position on the platinum surface.

In a follow-up investigation Singh *et al.* studied the structure of a working Pt/Al_2O_3 catalyst combining *in situ*, time-resolved, and HERFD XAS with kinetic measurements using mass spectrometry [346]. In contrast to the general claim that metallic platinum is the active surface species for CO even in an oxygen-rich environment, the authors conclude from their HERFD XAS and QEXAFS data that oxidized platinum plays an active role in generating high catalytic activity. Based on their spectra, the authors can distinguish a low and a high activity regime: Pt with adsorbed CO was observed for the low activity regime and oxydic platinum in the high activity regime.

As a typical example of a HRFD experiment, Figure 4.30 shows the 1s3p RIXS plane of CoO, the high resolution Co K$\beta_{1,3}$ RXES, and the HRFD-Co-K-XANES spectrum of CoO that have been extracted from the RIXS plane. Figure 4.31 shows the K$\beta_{1,3}$ emission spectra of Co metal and CoO (as powder). The K$\beta_{1,3}$ peak of

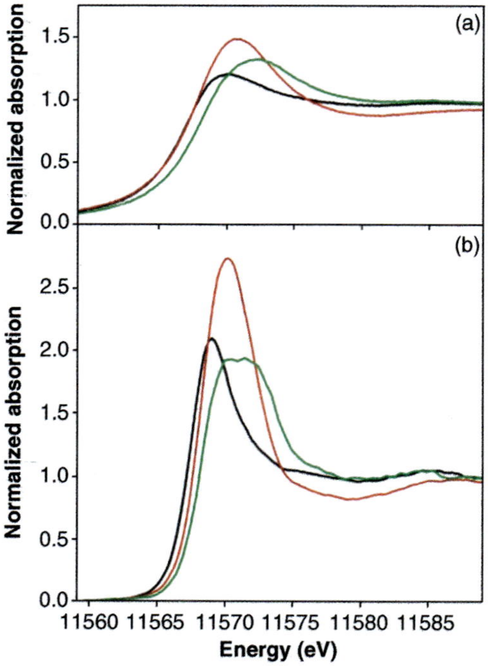

Figure 4.29 Pt-L-III XANES spectra of 5 wt% Pt/Al_2O_3 after reduction in the presence of different atmospheres: He (black), $He(O_2)$ (red), and 1% CO/He (green) measured using total fluorescence detection (a) and HERFD XAS spectroscopy (b). (Reproduced with permission from [345]. Copyright © (2006) American Chemical Society.)

CoO is shifted about 1 eV to higher energies with respect to Co as can be seen in Figure 4.31. For mixed valence compounds hence, one can extract HRFD-XAS spectra, for example at the peak positions of the different valence compounds that are present to gain partly (because of the overlap of the peaks) valency-selective spectra.

Kühn *et al.* have used RIXS for the investigation of smoothly oxidized cobalt nanoparticles [347]. They developed a numerical procedure that allows extracting site-selective XANES spectra from the high resolution Kβ fluorescence spectra. For the Co particles under investigation two sites could be characterized: the metallic core with a mixture of Co-ε and Co-hcp as crystal structure and the CoO shell with valency +2 and cubic as well as hexagonal phases and a mixture of O and most probably C as ligands. In principle, this technique allows also the investigation of the interaction between nanoparticles and their coatings.

Kβ emission spectroscopy and also valence-to-core emission spectra have been used by Beckwith *et al.* to investigate a series of mononuclear Mn compounds [348]. The Kβ main lines were found to be mainly sensitive to the oxidation state of Mn, whereas the valence to core region showed more sensitivity to the chemical

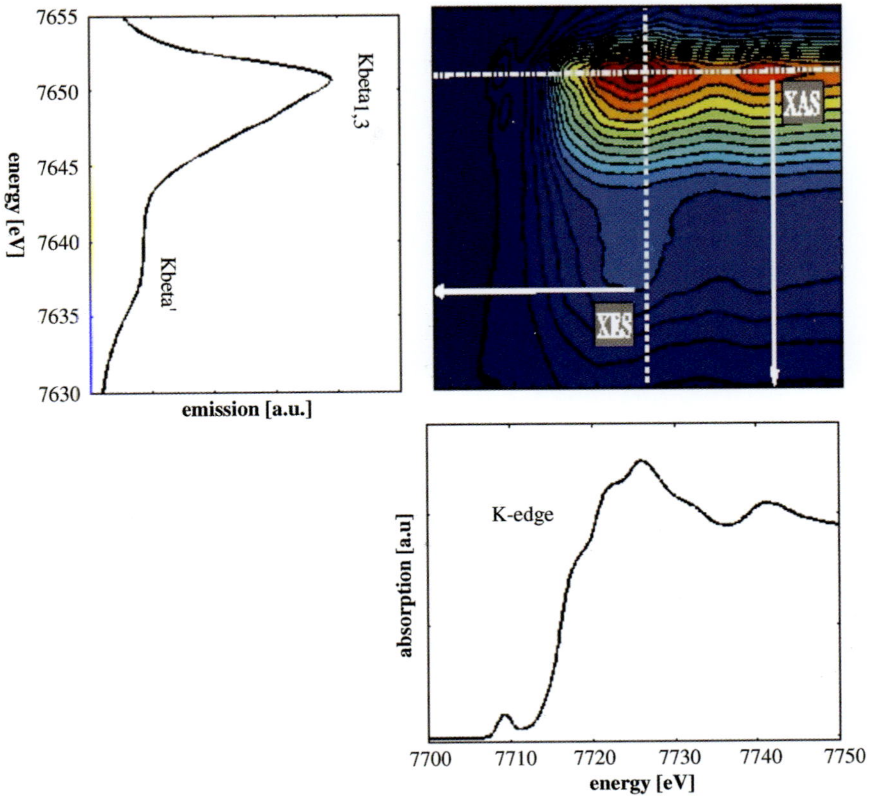

Figure 4.30 1s3p RIXS plane of CoO (top right); high resolution Co Kβ RXES spectrum extracted at excitation energy of 7726.0 eV (top left) (vertical white dashed line in RIXS plane), and Co K-edge HRFD-XANES spectrum extracted at emission energy of 7750.8 eV (bottom right) (horizontal white dashed line in RIXS plane) (from [290] with permission.)

environment of the Mn centers, i.e., to the identity of the ligands as well as metal–ligand bond lengths. A DFT approach has been used to simulate the valence-to-core spectra and to assess the various contributions to the energies and the intensities of spectral features. These results also clearly indicate the opportunities of studying the interaction between surface atoms of nanoparticles and capping molecule in a more direct way.

An interesting survey of XAS techniques and the opportunities given by RIXS in homogeneous catalysis was published by Bauer and Gastl [349]. As a special example they discuss in detail the iron-catalyzed Michael addition reaction. These authors show that it is possible to identify the catalytically active compound formed in the first step of this reaction by using *in situ* RIXS spectroscopy. These results highlight once again the tremendous potential of RIXS to determine the exact d-electron structure of a catalyst and also charge transfer complexes.

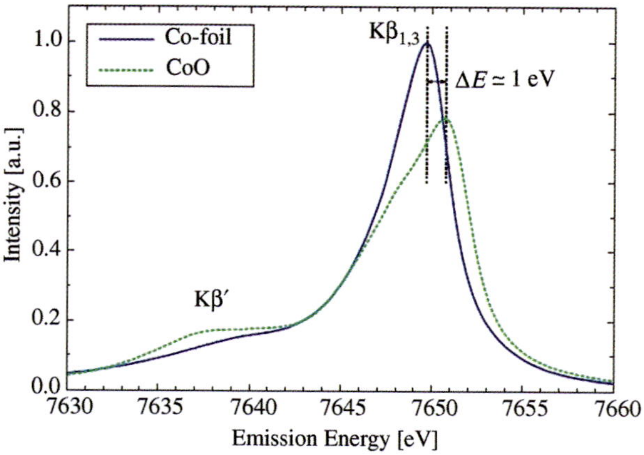

Figure 4.31 Normalized $K\beta_{1,3}$ emission spectrum of Co-foil and CoO (as powder). ΔE denotes the chemical shift due to the different valencies of the two samples (from [290] with permission.)

Just recently, RIXS was used for the *in situ* characterization of the adsorption of CO on Pt nanoparticles already mentioned above, i.e., to elucidate the mechanism of CO poisoning of Pt nanocatalyst [350]. By comparing RIXS data of supported Pt nanoparticles in metallic form and with CO adsorbed with suitable model calculations, the authors can assign the stop site of the Pt surface as the most likely adsorption site for CO; they can also show that the CO adsorption changes the electronic structure of Pt and that the center of mass of the 5d band is shifted to lower energies. This shift to lower energies decreases also the tendency of the nanoparticles to form bonds with other reactants. Thus, the adsorption of CO poisons the surface of the catalyst by both modifying the catalytic activity and blocking suitable surface sites.

A quite comprehensive overview of the XES and RIXS experiments that have been carried out for catalytic systems is given in a tutorial review by Singh *et al.* [351] and also in a review by Lee [352]. Both reviews also give an overview of recent time-resolved, *in situ* XAS experiments.

The *in situ* RIXS experiments discussed above are coming very close to the final goal of all catalysis research: the characterization of the electronic and geometric structure of the catalytically active atoms and changes of these properties occurring in the course of the catalytic reaction!

Acknowledgments

The authors warmly thank Dr. Richard Goddard, Max-Planck-Institut für Kohlenforschung, for critical perusal of the manuscript and careful linguistic revision. We also thank Mr. Rainer Brinkmann, Dipl.-Ing., Max-Planck-Institut für Kohlenforschung, for adapting and, in several cases, redrafting the Figures for this work. Special thanks are due to STREM CHEMICALS, Inc. for financial support.

References

1 Schlögl, R. and Abd Hamid, S.B. (2004) Mature science revisited or something really new? *Angew. Chem. Int. Ed.*, **43** (13), 1628–1637.

2 Jia, C.-J. and Schüth, F. (2011) Colloidal metal nanoparticles as a component of designed catalyst. *Phys. Chem. Chem. Phys.*, **13** (7), 2457–2487.

3 Corain, B., Schmid, G., and Toshima, N. (eds) (2008) *Metal Nanoclusters in Catalysis and Materials Science*, Elsevier, Amsterdam.

4 Wen, F. and Bönnemann, H. (2005) Preparation of Pt^{13} clusters in the presence of trialkylaluminium. *Appl. Organomet. Chem.*, **19** (7), 827–829.

5 Dietsche, R., Lim, D.C., Bubek, M., Lopez-Salido1, I., Ganteför, G., and Kim, Y.D. (2008) Comparison of electronic structures of mass-selected Ag clusters and thermally grown Ag islands on sputter-damaged graphite surfaces. *Appl. Phys.*, **90** (3), 395–398.

6 Kaden, W.E., Wu, T., Kunkel, W.A., and Anderson, S.L. (2009) Electronic structure controls reactivity of size-selected Pd clusters adsorbed on TiO_2 surfaces. *Science*, **326** (5954), 826–829.

7 Lei, Y., Mehmood, F., Lee, S., Greeley, J., Lee, B., Seifert, S., Winans, R.E., Elam, J. W., Meyer, R.J., Redfern, P.C., Teschner, D., Schlögl, R., Pellin, M.J., Curtiss, L.A., and Vajda, S. (2010) Increased silver activity for direct propylene epoxidation via subnanometer size effects. *Science*, **328** (5975), 224–228.

8 Vajda, S., Pellin, M.J., Elam, J.W., Marshall, C.L., Wians, R.A., and Meiwes-Broer, K.-H. (2010) Subnanometer and nanometer catalysts, method for preparing size-selected catalyst. US patent 2011/0045969 A1, filed Feb. 26, 2010 and issed Feb. 24, 2011.

9 Lei, Y., Jelic, J., Nitsche, L.C., Meyer, R., and Miller, J. (2011) Effect of particle size and adsorbates on the L_3, L_2 and L_1 X-ray absorption near edge structure of supported Pt nanoparticles. *Top. Catal.*, **54** (5–7), 334–348.

10 Burda, C., Chen, X.B., Narayanan, R., and El-Sayed, M. (2005) The chemistry and properties of nanocrystals of different shapes. *Chem. Rev.*, **105** (4), 1025–1102.

11 Narayan, R. and El-Sayed, M. (2005) Catalysis with transition metal nanoparticles in colloidal solution: Nanoparticle shape dependence and stability. *J. Phys. Chem. B*, **109** (26), 12663–127676.

12 Narayanan, R. and El-Sayed, M.A. (2005) Effect of colloidal nanocatalysis on the metallic nanoparticle shape. *Langmuir*, **21** (5), 2027–2033.

13 Narayanan, R. and El-Sayed, M.A. (2004) Shape-dependent catalytic activity of platinum nanoparticles in colloidal solution. *Nano Lett.*, **4** (7), 1343–1348.

14 Bönnemann, H., Nagabushana, K.S., and Richards, R.M. (2008) *Nanoparticles and Catalysis* (ed. D. Astruc), Wiley-VCH Verlag GmbH, Weinheim, Germany, pp. 49–91.

15 Toshima, N., Yan, H., and Shiraishi, Y. (2008) Metal nanoclusters, in *Catalysis and Materials* (eds B. Corain, G. Schmid, and N. Toshima), Science, Elsevier, Amsterdam, pp. 49–75.

16 Toshima, N. (2011) *Inorganic Nanoparticles* (eds C. Altavilla and E. Ciliberto), CRC Press, Boca Raton, USA, pp. 475–509.

17 Bönnemann, H. and Richards, R. (2003) *Catalysis and Electrocatalysis at Nanoparticle Surfaces* (eds A. Wieckowski, E. Savinova, and C. Vayenas), Marcel Dekker, New York, pp. 343–377.

18 Schüth, F. (2011) Controlled nanostructures for applications in catalysis. *Phys. Chem. Chem. Phys.*, **13** (7), 2447–2448.

19 Joo, S.H., Park, J.Y., Tsung, C.K., Yamada, Y., Yang, P.D., and Somorjai, G.A. (2009) Thermally stable Pt/mesoporous silica core–shell nanocatalysts for high-temperature reactions. *Nature Mater.*, **8** (2), 126–131.

20 Liu, S.H. and Han, M.Y. (2005) Synthesis, functionalization, and bioconjugation of monodisperse, silica-coated gold nanoparticles: robust

bioprobes. *Adv. Func. Mater.*, **15** (6), 961–967.

21 Graf, C., Vossen, D.L.J., Imhor, A., and van Blaaderen, A. (2003) A general method to coat colloidal particles with silica. *Langmuir*, **19** (17), 6693–6700.

22 Takenaka, S., Umebayashi, H., Tanabe, E., Matsune, H., and Kishida, M. (2007) Specific performance of silica-coated Ni catalysts for the partial oxidation of methane to synthesis gas. *J. Catal.*, **245** (2), 392–400.

23 Yu, K., Wu, Z., Zhao, Q., Li, B., and Xie, Y. (2008) High-temperature-stable Au@SnO_2 core/shell supported catalyst for CO oxidation. *J. Phys. Chem. C*, **112** (7), 2244–2247.

24 Cargnello, M., Montini, T., Polizzi, S., Wieder, N.L., Gorte, R.J., Graziania, M., and Fornasiero, P. (2010) Novel embedded Pd@CeO(2) catalysts: a way to active and stable catalysts. *Dalton Trans.*, **39** (8), 2122–2127.

25 Feyen, M., Weidenthaler, C., Güttel, R., Schlichte, K., Holle, U., Lu, A.H., and Schüth, F. (2011) High temperature stable iron based core-shell catalysts for ammonia decomposition. *Chem. Eur. J.*, **17** (2), 598–605.

26 Yin, Y.D., Rioux, R.M., Erdonmez, C.K., Hughes, S., Somorjai, G.A., and Alivisatos, A.P. (2004) Formation of hollow nanocrystals through the nanoscale kirkendall effect. *Science*, **304** (5671), 711–714.

27 Arnal, P.M., Comotti, M., and Schüth, F. (2006) High-temperature-stable catalysts by hollow sphere encapsulation. *Angew. Chem. Int. Ed.*, **45** (48), 8224–8227.

28 Güttel, R., Paul, M., and Schüth, F. (2010) Ex-post size control of high-temperature-stable yolk-shell Au@ZrO_2 catalysts. *Chem. Commun.*, **46** (6), 895–897.

29 Huang, X.Q., Guo, C.Y., Zuo, J.Q., Zheng, N.F., and Stucky, G.D. (2009) An assembly route to inorganic catalytic nanoreactors containing Sub-10-nm gold nanoparticles with anti-aggregation properties. *Small*, **5** (3), 361–365.

30 Park, J.C., Bang, J.U., Lee, J., Ko, C.H., and Song, H. (2010) Ni@SiO_2 yolk-shell nanoreactor catalysts: High temperature stability and recyclability. *J. Mater. Chem.*, **20** (7), 1239–1246.

31 Shevchenko, E.V., Bodnarchuk, M.I., Kovalenko, M.V., Talapin, D.V., Smith, R.K., Aloni, S., Heiss, W., and Alivisatos, A.P. (2008) Gold/Iron oxide core/hollow-shell nanoparticles. *Adv. Mater.*, **20** (22), 4323–4329.

32 Ikeda, S., Ishino, S., Harada, T., Okamoto, N., Sakata, T., Mori, H., Kuwabata, S., Torimoto, T., and Matsumura, M. (2006) Ligand-free platinum nanoparticles encapsulated in a hollow porous carbon shell as a highly active heterogeneous hydrogenation catalyst. *Angew. Chem. Int. Ed.*, **45** (42), 7063–7066.

33 Harada, T., Ikeda, S., Ng, Y.H., Sakata, T., Mori, H., Torimoto, T., and Matsumura, M. (2008) Rhodium nanoparticle encapsulated in a porous carbon shell as an active heterogeneous catalyst for aromatic hydrogenation. *Adv. Func. Mater.*, **18** (15), 2190–2196.

34 Galeano, C., Güttel, R., Paul, M., Arnal, P., Lu, A-H., and Schüth, F. (2012) Yolk-shell gold nanoparticles as model materials for support-effect studies in heterogeneous catalysis: Au,@C and Au, @ZrO2 for CO oxidation as an example. *Chem. Eur. J.*, **17**, 8434–8439

35 Ge, J., Zhang, Q., Zhang, T., and Yin, Y. (2008) Core–satellite nanocomposite catalysts protected by a porous silica shell: Controllable reactivity, high stability, and magnetic recyclability. *Angew. Chem. Int. Ed.*, **47** (46), 8924–8928.

36 Freund, A. and Panster, P. (1995) Shell catalyst, process for its preparation and its use, (to Degussa AG). European Patent 715 889 A2, filed No. 8, 1995 and Jun. 06. 1995.

37 Polshettiwar, V., Luque, R., Fihri, A., Zhu, H., Bouhrara, M., and Basset, J.-M. (2011) Magnetically recoverable nanocatalysts. *Chem. Rev.*, **111** (5), 3036–3075.

38 Park, J. and Kwon, M. (2008) Magnetic nanocatalyst composition and method for the preparation thereof. WO 2008/ 156320 A1, filed Jun. 19, 2008 and issued Dec. 24 2008.

39 Ertl, G., Knözinger, H., Schüth, F., and Weitkamp, J. (eds) (2008) *Handbook of Heterogeneous Catalysis*, 2nd edn, Wiley-VCH Verlag GmbH, Weinheim, Germany.

40 Fischer, F. and Tropsch, H. (1925) Verfahren zur Gewinnung mehrgliedriger Paraffinkohlenwasserstoffe aus Kohlenoxyden und Wasserstoff auf katalytischem Wege. German patent DE 484337, filed Jul. 22, 1925, issued Oct. 3, 1929.

41 Jacoby, M. (2006) Making fuels synthetically. *Chem. Eng. News*, **84** (23), 57–61.

42 Tijmensen, M.J.A. (2002) Exploration of the possibilities for production of Fischer–Tropsch liquids and power via biomass gasification. *Biomass Bioenerg.*, **23** (2), 129–152.

43 Bezemer, G.L., Bitter, J.H., Kuipers, H.P. C.E., Oosterbeek, H., Holewijn, J.E., Xu, X., Kapteijn, F., van Dillen, A.J., and de Jong, K.P. (2006) Cobalt particle size effects in the Fischer–Tropsch reaction studied with carbon nanofiber supported catalysts. *J. Am. Chem. Soc.*, **128** (12), 3956–3964.

44 Martinez, A. and Prieto, G. (2007) Breaking the dispersion-reducibility dependence in oxide-supported cobalt nanoparticles. *J. Catal.*, **245** (2), 470–476.

45 Bankhuizen, D., Mabaso, I., Welker, E.C., Claeys, M., van Steen, E., and Fletcher, J.C.Q. (2006) Experimental approaches to the preparation of supported metal nanoparticles. *Pure Appl. Chem.*, **78** (9), 1759–1769.

46 Iglesia, E., Soled, S.L., and Fiato, R.A. (1992) Fischer–Tropsch synthesis on cobalt and ruthenium. Metal dispersion and support effects on reaction rate and selectivity. *J. Catal.*, **137** (1), 212–224.

47 Iglesia, E., Soled, S.L., Fiato, R.A., and Via, G.H. (1994) Dispersion, support, and bimetallic effects in Fischer–Tropsch synthesis on cobalt catalysts. *Stud. Surf. Sci. Catal.*, **81**, 433–442.

48 Iglesia, E. (1997) Design, synthesis, and use of cobalt-based Fischer–Tropsch synthesis catalysts. *Appl. Catal. A*, **161** (1–2), 59–78.

49 van Steen, E., Claeys, M., Dry, M.E., van de Loosdrecht, J., Viljoen, E.L., and Visagie, J.L. (2005) Stability of nanocrystals: thermodynamic analysis of oxidation and re-reduction of cobalt in water/hydrogen mixtures. *J. Phys. Chem. B*, **109** (8), 3575–3577.

50 Ciobica, M. and van Santen, R.A. (2003) Carbon monoxide dissociation on planar and stepped Ru(0001) surfaces. *J. Phys. Chem. B*, **107** (16), 3808–3812.

51 Jacobs, G., Patterson, P.M., Das, T.K., Luo, M., and Davis, B.H. (2004) Fischer–Tropsch synthesis: effect of water on Co/Al_2O_3 catalysts and XAFS characterization of reoxidation phenomena. *Appl. Catal. A*, **270** (1–2), 65–76.

52 Prieto, G., Martinez, A., Concepción, P., and Moreno-Trost, R. (2009) Cobalt particle size effects in Fischer–Tropsch synthesis: Structural and *in situ* spectroscopic characterisation on reverse micelle-synthesised Co/ITQ-2 model catalysts. *J. Catal.*, **266** (1), 129–144.

53 Fischer, N., van Steen, E., and Claeys, M. (2011) Preparation of supported nano-sized cobalt oxide and fcc cobalt crystallites. *Catal. Today*, **171** (1), 174–179.

54 Zhuang, Y.Q., Claeys, M., and van Steen, E. (in preparation)

55 Dupont, J., Silva, D., Pavan, F.A., Machado, G., Teixeira, S.R., Soares, H.S., Belizario Dos Santos, A.C., and Aguiar, E.F.S. (2008) Metallic catalyst and method for the production of metallic catalyst. European Patent EP2082800 (A1), filed Jan. 24, 2008 and issued Aug. 6, 2009.

56 Galvis, H.M.T., Bitter, J.H., Chaitanya, B. K., Ruitenbeek, M., Dugulan, A.J., and de Jong, K.P. (2012) Supported iron nanoparticles as catalysts for sustainable production of lower olefins, *Science*, **335** (6070), 835–838.

57 Gasteiger, H.A. and Garche, J. (2008) *Handbook of Heterogeneous Catalysis*, 2nd edn (eds G. Ertl, H. Knözinger, F. Schüth, and J. Weitkamp), Wiley-VCH Verlag GmbH, Weinheim, Germany, pp. 3081–3121.

58 Heinzel, A. and König, U. (2009) *Nanostructured Materals Electrochemical Energy Production and Storage* (ed. E.R. Leite), Springer Science & Business Media LLC, pp. 163–167.

59 Jana, R.N. and Bhunia, H. (2008) Thermal stability and proton conductivity of silane based nanostructured composite membranes Original Research Article. *Solid State Ionics*, **178** (37–38), 1872–1878.

60 Peron, J., Shi, Z., and Holdcroft, S. (2011) Hydrocarbon proton conducting polymers for fuel cell catalyst layers. *Energy & Environmental Science*, **4** (5), 1575–1591.

61 Roloefs, K.S., Hirth, T., and Schiestel, T. (2010) Sulfonated poly(ether ether ketone)-based silica nanocomposite membranes for direct ethanol fuel cells. *J. Membr. Sci.*, **346** (1), 215–226.

62 Hasché, F., Oezaslan, M., and Strasser, P. (2010) Activity, stability and degradation of multi walled carbon nanotube (MWCNT) supported Pt fuel cell electrocatalysts. *Phys. Chem. Chem. Phys.*, **12** (46), 15251–15258.

63 Haruta, M., Yamada, N., Kobayashi, T., and Iijima, S. (1989) Gold catalysts prepared by coprecipitation for low-temperature oxidation of hydrogen and of carbon monoxide. *J. Catal.*, **115** (2), 301–302.

64 Haruta, M. (2002) Catalysis of gold nanoparticles deposited on metal oxides. *Cattech.*, **6** (3), 102–115.

65 Haruta, M. (2004) Gold as a novel catalyst in the 21st century: Preparation, working mechanism and applications. *Gold Bulletin*, **37** (1–2), 27–36.

66 Hutchings, G.J. (2007) New directions in gold catalysis. *Gold Bulletin*, **37** (1–2), 3–11.

67 Hashmi, A.S.K. and Hutchings, G.J. (2006) Gold catalysis. *Angew. Chem. Int. Ed.*, **45** (47), 7896–7936.

68 Min, B.K. and Friend, C.M. (2007) Heterogeneous gold-based catalysis for green chemistry: low-temperature CO oxidation and propene oxidation. *Chem. Rev.*, **107** (6), 2709–2724.

69 Perkas, N., Teo, J., Shen, S., Wang, Z., Highfield, J., Zhong, Z., and Gedanken, A. (2011) Supported Ru catalysts prepared by two sonication-assisted methods for preferential oxidation of CO in H_2. *Phys. Chem. Chem. Phys.*, **13** (4), 1590–15698.

70 Jiang, C.-J., Liu, Y., Schwickardi, M., Weidenthaler, C., Spliethoff, B., Schmidt, W., and Schüth, F. (2010) Small gold particles supported on $MgFe_2O_4$ nanocrystals as novel catalyst for CO oxidation. *Appl. Catal. A*, **386** (1–2), 94–100.

71 Jiang, C.-J., Liu, Y., Bongard, H., and Schüth, F. (2010) Very low temperature CO oxidation over colloidally deposited gold nanoparticles on $Mg(OH)_2$ and MgO. *J. Am. Chem. Soc.*, **132** (5), 1520–1522.

72 Chen, M.S. and Goodman, D.W. (2004) The structure of catalytically active gold on titania. *Science*, **306** (5694), 252–255.

73 Liu, Y., Jia, C.-J., Yamasaki, J., Terasaki, O., and Schüth, F. (2010) Highly active iron oxide supported gold catalysts for CO oxidation: how small must the gold nanoparticles be? *Angew. Chem.*, **122** (33), 5907–5911.

74 Idakien, V. and Tenchen, K. (2009) Gold nanoparticles supported on ceria-modified mesoporous–macroporous binary metal oxides as highly active catalysts for low-temperature water–gas shift reaction. *J. Mater. Sci.*, **44** (24), 6637–6643.

75 McPherson, J.S. and Thompson, D.T. (2009) Capabilities of AUROlite^{TM} catalyst. *The Strem Chemiker*, **24** (1), 21–24.

76 Wu, G., More, K.L., Johnston, C.M., and Zelenay, P. (2011) High-performance electrocatalysts for oxygen reduction derived from polyaniline, iron, and cobalt. *Science*, **332** (6055), 443–447.

77 Zhiani, M., Gasteiger, H.A., Piana, M., and Catanorchi, S. (2011) Comparative study between platinum supported on carbon and non-noble metal cathode catalyst in alkaline direct ethanol fuel cell (ADEFC). *Int. J. Hydrogen Energy*, **36** (8), 5110–5116.

78 Sanabria-Chinchilla, J., Aszawa, K., Sakamoto, T., Yamada, J., Tanaka, H., and Strasser, P. (2011) Noble metal-free hydrazine fuel cell catalysts: EPOC effect in competing chemical and electrochemical reaction pathways. *J. Am. Chem. Soc.*, **133** (14), 5425–5431.

79 Wu, G., Nelson, M.A., Mack, N.H., Ma, S., Sekhar, P., Garzon, F.H., and Zelenay, P. (2010) Titanium dioxide-supported non-precious metal oxygen reduction electrocatalyst. *Chem. Commun.*, **46** (40), 7489–7491.

80 Chung, H.T., Johnston, C.M., Artyushkova, K., Ferrandon, M., Myers, D.J., and Zelenay, P. (2010) Cyanamide-derived non-precious metal catalyst for oxygen reduction. *Electrochem. Commun.*, **12** (12), 1792–1795.

81 Morozan, A., Jousselme, B., and Palacin, S. (2011) Low-platinum and platinum-free catalysts for the oxygen reduction reaction at fuel cell cathodes. *Energy Environ. Sci.*, **4** (4), 1238–1254.

82 Jaouen, F., Proietti, E., Lefèvre, M., Chenitz, R., Dodelet, J.-P., Wu, G., Chung, H.T., Johnston, J.M., and Zelenay, P. (2011) Recent advances in non-precious metal catalysis for oxygen-reduction reaction in polymer electrolyte fuel cells. *Energy Environ. Sci.*, **4** (1), 114–130.

83 Johnston, C.M., Piela, P., and Zelenay, P. (2009) *Handbook of Fuel Cells*, vol. **5** (eds W. Vielstich, H.A. Gasteiger, and H. Yokokawa), John Wiley & Sons, Ltd, Chichester, UK, pp. 48–70.

84 Biddinger, E.J., von Deak, D., and Ozkan, U.S. (2009) Nitrogen-containing carbon nanostructures as oxygen-reduction catalysts. *Top. Catal.*, **52** (11), 1566–1574.

85 Maillard, F., Savinova, E.R., and Stimmig, U. (2007) CO monolayer oxidation on Pt nanoparticles: Further insights into the particle size effects. *J. Electroanal. Chem.*, **599** (2), 221–232.

86 Kinge, S., Urgeghe, C., de Battisti, A., and Bönnemann, H. (2008) Dependence of CO oxidation on Pt nanoparticle shape: a shape-selective approach to the synthesis of PEMFC catalysts. *Appl. Organomet. Chem.*, **22** (1), 49–54.

87 Ralph, T.R. and Hogarth, M.P. (2002) Catalysis for low temperature fuel cells. *Platinum Metals Rev.*, **46** (3), 117–135.

88 Gasteiger, H.A., Markovic, N.M., Ross, P.N., and Cairns, E.J. (1995) H_2 and CO electrooxidation on well-characterized Pt, Ru, and Pt-Ru. 2. Rotating Disk Electrode Studies of CO/H_2 Mixtures at 62 °C. *J. Phys. Chem.*, **99** (45), 16757–16767.

89 Watanabe, M. and Motoo, S. (1975) Electrocatalysis by ad-atoms: Part II. Enhancement of the oxidation of methanol on platinum by ruthenium ad-atoms. *J. Electroanal. Chem.*, **60** (3), 267–273.

90 Kaiser, J., Colmenares, L., Jusys, Z., Mörtel, R., Bönnemann, H., Köhl, G., Modrow, H., and Behm, R.J. (2006) On the role of reactant transport and (Surface) alloy formation for the CO tolerance of carbon supported PtRu polymer electrolyte fuel cell catalysts. *Fuel Cells*, **6** (3–4), 190–202.

91 Modrow, H., Köhl, G., Hormes, J., Bönnemann, H., Endruschat, U., and Mörtel, R. (2005) *In-situ* x-ray absorption spectroscopy investigations during the conditioning procedure of PtRu colloid catalysts. *Phys. Scripta*, **T115**, 671–674.

92 Brendebach, B., Bucher, S., Hormes, J., Bönnemann, H., Nagabhushana, K.S., Brinkmann, R., and Modrow, H. (2005) *In-situ* X-ray absorption near edge structure (XANES) spectroscopy of PtRu colloid catalysts at operating temperature. *Phys. Scripta*, **T115**, 773–775.

93 Colmenares, L., Jusys, Z., King, S., Bönnemann, H., and Behm, R.J. (2006) Synthesis, characterization and electrocatalytic performance of W surface modified, carbon supported Pt anode catalysts for low-temperature fuel cell applications. *J. New Mater. Electrochem. Sys.*, **9** (2), 107–120.

94 He, C., Kunz, H.R., and Fenton, J.M. (2003) Electro-oxidation of hydrogen with carbon monoxide on Pt/Ru-based ternary catalysts. *J. Electrochem. Soc.*, **150** (8), A1017–A1024.

95 Jalan, V. and Taylor, E.J. (1983) Importance of interatomic spacing in

catalytic reduction of oxygen in phosphoric acid. *J. Electrochem. Soc.*, **130** (11), 2299–2302.

96 Paulus, U.A., Wokaun, A., Scherer, G.G., Schmidt, T.J., Stamenkovic, V., Radmilovic, V., Markovic, N.M., and Ross, P.N. (2002) Oxygen reduction on carbon-supported Pt–Ni and Pt–Co alloy catalysts. *J. Phys. Chem. B*, **106** (16), 4181–4191.

97 Bönnemann, H., Brinkmann, R., Britz, P., Endruschat, U., Mörtel, R., Paulus, U. A., Feldmayer, G.J., Schmidt, T.J., Gasteiger, H.A., and Behm, R.J. (2000) Nanoscopic Pt-bimetal colloids as precursors for PEM fuel cell catalyst. *J. New Mater. Electrochem. Sys.*, **3** (3), 199–206.

98 Bönnemann, H. and Khelashvili, G. (2010) Efficient fuel cell catalysts emerging from organometallic chemistry. *Appl. Organomet. Chem.*, **24** (4), 257–268.

99 Zaikovskii, V.I., Nagabhushana, K.S., Kriventsov, V.V., Loponov, K.N., Cherepanova, S.V., Kvon, R.I., Bönnemann, H., Kochubey, D.I., and Savinova, E.R. (2006) Synthesis and structural characterization of se-modified carbon-supported Ru nanoparticles for the oxygen reduction reaction. *J. Phys. Chem. B*, **110** (13), 6881–6890.

100 Bogolowski, N., Nagel, T., Lanova, B., Ernst, S., Baltruschat, H., Nagabushana, K.S., and Bönnemann, H. (2007) Activity of selenium modified ruthenium-electrodes and determination of the real surface area. *J. Appl. Electrochem.*, **37** (12), 1485–1494.

101 Kulesza, P.P.J., Miecznikowski, K., Baranowska, B., Skunik, M., Kolary-Zurowska, A., Lewera, A., Karnicka, K., Chojak, M., Rutkowska, I., Fiechter, S., Bogdanoff, P., Dorbandt, I., Zehl, G., Hiesgen, R., Dirk, E., Nagabhushana, K. S., and Bönnemann, H. (2007) Electroreduction of oxygen at tungsten oxide modified carbon-supported RuSex nanoparticles. *J. Appl. Electrochem.*, **37** (12), 1439–1446.

102 Bogolowski, N., Nagel, T., Lanova, B., Ernst, S., Baltruschat, H., Nagabhushana, K.S., and Boennemann, H. (2007) Activity of selenium modified ruthenium-electrodes and determination of the real surface area. *J. Appl. Electrochem.*, **37** (12), 1485–1494.

103 Zhang, J., Mo, Y., Vukmirovic, M.B., Klie, R., Sasaki, K., and Adzic, R.R. (2004) Platinum monolayer electrocatalysts for O_2 reduction: Pt monolayer on Pd(111) and on carbon-supported Pd nanoparticles. *J. Phys. Chem. B*, **108** (30), 10955–10964.

104 Nagabushana, K.S., Weidenthaler, C., Hocevar, S., Strmcbnik, D., Garberscek, M., Antozzi, A.L., and Martelli, G.N. (2006) Preparation, characterization and properties of Pt-Cu Co-reduced and Pt-on-Cu skin type bimetallic carbon-supported (Vulcan XC72) electrocatalysts. *J. New Mater. Electrochem. Sys.*, **9** (2), 73–81.

105 Schulenburg, H., Müller, E., Khelashvili, G., Roser, I.T., Bönnemann, H., Wokaun, A., and Scherer, G.G. (2009) Heat-treated $PtCo_3$ nanoparticles as oxygen reduction catalysts. *J. Phys. Chem. C*, **113** (10), 4069–4077.

106 Bezerra, C.W.B., Zhang, L., Lee, K., Liu, H., Marques, A.L.B., Marques, E.P., Wand, H., and Zhang, J. (2008) A review of Fe–N/C and Co–N/C catalysts for the oxygen reduction reaction. *Electrochim Acta*, **53** (15), 4937–4951.

107 Loukrakpam, R., Luo, J., He, T., Chen, Y., Xu, Z., Njoki, P.N., Wanjala, B.N., Fang, B., Mott, D., Yin, Y., Klar, J., Powell, B., and Zhong, C.-J. (2011) Nanoengineered PtCo and PtNi catalysts for oxygen reduction reaction: An assessment of the structural and electrocatalytic properties. *J. Phys. Chem. C*, **115** (5), 1682–1694.

108 Koh, S., Leisch, J., Toney, M.F., and Strasser, P. (2007) Structure-activity-stability relationships of Pt–Co alloy electrocatalysts in gas-diffusion electrode layers. *J. Phys. Chem. C*, **111** (9), 3744–3752.

109 Strasser, P. (2009) *Handbook of Fuel Cells: Advances in Electrocatalysis, Materials, Diagnostics and Durability*, vol. **5 & 6** (eds W. Vielstich, H.A. Gasteiger, and H. Yokokawa), John Wiley & Sons Ltd., Chichester, West Sussex, UK, pp. 30–47.

110 Mani, P., Srivastava, R., and Strasser, P. (2011) Dealloyed binary PtM_3 (M=Cu, Co, Ni) and ternary $PtNi_3M$ (M=Cu, Co, Fe, Cr) electrocatalysts for the oxygen reduction reaction: Performance in polymer electrolyte membrane fuel cells. *J. Power Sources*, **196** (2), 666–673.

111 Yang, R., Strasser, P., and Toney, M.F. (2011) Dealloying of Cu_3Pt (111) studied by surface X-ray scattering. *J. Phys. Chem. C*, **115** (18), 9074–9080.

112 Oezaslan, M., Strasser, P., and Heggen, M. (2012) Size-dependent morphology of dealloyed bimetallic catalysts: linking the nano to the macro scale. *J. Am. Chem. Soc.*, **14** (1), 514–524.

113 Strasser, P., Koh, S., Anniyev, T., Greeley, J., More, K., Yu, C., Liu, Z., Kaya, S., Nordlund, D., Ogasawara, H., Toney, M. F., and Nilsson, A. (2010) Lattice-strain control of the activity in dealloyed core–shell fuel cell catalysts. *Nat. Chem.*, **2** (6), 454–460.

114 Kaiser, J., Simonov, P.A., Zaikovskii, V.I., Hartnig, C., Jorissen, L., and Savinova, E. R. (2007) Influence of carbon support on the performance of platinum based oxygen reduction catalysts in a polymer electrolyte fuel cell. *J. Appl. Electrochem.*, **37** (12), 1429–1437.

115 Chen, S., Gasteiger, H.A., Hayakawa, K., Tada, T., and Shao-Horn, Y. (2010) Platinum-alloy cathode catalyst degradation in proton exchange membrane fuel cells: nanometer-scale compositional and morphological changes. *J. Electrochem. Soc.*, **157** (1), A82–A97.

116 Springer, T.E., Zawodzinski, T.A., Wilson, M.S., and Gottesfeld, S. (1996) Characterization of polymer electrolyte fuel cells using AC impedance spectroscopy. *J. Electrochem. Soc.*, **143** (2), 587–599.

117 Bahar, B., Hobson, A.R., Kolde, J.A., and Zuckerbrod, D. (1995) Ultra-thin integral composite membrane. US patent 5,547,551, filed Mar. 15, 1995 and issued Aug. 16, 1996.

118 Nakao, N. and Yoshitake, M. (2003) *Handbook of Fuel Cells, Fundamentals, Technology and Applications*, vol. **3**, Part 1 (eds W. Vielstich, A. Lamm, and H. Gasteiger), John Wiley & Sons, Ltd, Chichester, UK, pp. 412–419.

119 Xiao, L., Zhang, H., Scanion, E., Ramanathan, L.S., Choe, E.-W., Rogers, D., Apple, T., and Benicewicz, B.C. (2005) High-temperature polybenzimidazole fuel cell membranes via a Sol–Gel process. *Chem. Mater.*, **17** (21), 5328–5333.

120 Schmidt, T.J., Simbeck, K., and Scherer, G.G. (2005) Influence of cross-linking on performance of radiation-grafted and sulfonated FEP 25 membranes in H_2-O_2 PEFC. *J. Electrochem. Soc.*, **152** (1), A93–A97.

121 Honma, I., Takeda, Y., and Bae, J.M. (1999) Protonic conducting properties of sol-gel derived organic/inorganic nanocomposite membranes doped with acidic functional molecules. *Solid State Ionics*, **120** (1–4), 255–264.

122 Sonnet, B., Jones, D.J., Rozière, J., Tichicaya, L., Alberli, G., Casciola, M., Masineli, L., Bauer, B., Peraio, A., and Ramunni, E. (2000) Hybrid organic–inorganic membranes for a medium temperature fuel cell. *J. New Mater. Electrochem. Sys.*, **3** (1), 87–92.

123 Bauer, F. and Willert-Porada, M. (2005) Characterisation of zirconium and titanium phosphates and direct methanol fuel cell (DMFC) performance of functionally graded Nafion(R) composite membranes prepared out of them. *J. Power Sources*, **145** (2), 101–107.

124 Hogarth, M.P. and Ralph, T.R. (2002) Catalysis for low temperature fuel cells. *Platinum Metals Rev.*, **46** (4), 146–164.

125 Thomas, S.C., Ren, X., Gottesfeld, S., and Zelenay, P. (2002) Direct methanol fuel cells: progress in cell performance and cathode research. *Electrochim. Acta*, **47** (22–23), 3741–3748.

126 Bönnemann, H. (2008) Chapter 2, in *Metal Nanoclusters in Catalysis and Materials Science* (eds B. Corain, G. Schmid, and N. Toshima), Elsevier, Amsterdam, pp. 21–48.

127 Lee, A., Park, K., Choi, J., Kwon, B., and Sung, Y. (2002) Nanoparticle synthesis and electrocatalytic activity of Pt alloys for direct methanol fuel cells.

J. Electrochem. Soc., **149** (10), A1299–A1304.

128 Vidakovic, T., Christov, M., Sundmacher, K., Nagabhushana, K.S., Fei, W., Kinge, S., and Bönnemann, H. (2007) PtRu colloidal catalysts: Characterisation and determination of kinetics for methanol oxidation. *Electrochim. Acta*, **52** (6), 2277–2284.

129 Mallouk, T.E. and Smotkin, E.S. (2003) *Handbook of Fuel Cells – Fundamentals, Technology and Applications*, vol. **2** (eds W. Vielstich, A. Lamm, and H.A. Gasteiger), John Wiley & Sons, Ltd, Chichester, UK, pp. 334–345.

130 Strasser, P., Gorer, S., Fan, Q., Chondroudis, K., Cendak, K., Giaquinta, D., and Devenney, M. (2004) *High Throughput Screening in Chemical Catalysis* (eds A. Hagemeyer, P. Strasser, and A.F. Jr. Volpe), Wiley-VCH Verlag GmbH, Weinheim, Germany, pp. 271–286.

131 Strasser, P., Fan, Q., Devenney, M., Weinberg, W.H., Liu, P., and Nùrskov, J. K. (2003) High throughput experimental and theoretical predictive screening of materials—A comparative study of search strategies for new fuel cell anode catalysts. *J. Phys. Chem. B*, **107** (40), 11013–11021.

132 Gasteiger, H.A., Markovic, N.M., Ross, P. N. Jr., and Cairns, E.J. (1993) Carbon monoxide electrooxidation on well-characterized platinum-ruthenium alloys. *J. Phys. Chem.*, **97** (46), 12020–10229.

133 Schmidt, T.J., Paulus, U.A., Gasteiger, H. A., Alonso-Vante, N., and Behm, R.J. (2000) Oxygen reduction on $Ru_{1.92}Mo_{0.08}SeO_4$, Ru/Carbon, and Pt/Carbon in pure and methanol-containing electrolytes. *J. Electrochem. Soc.*, **147** (7), 2620–2624.

134 Salgado, J.R.C., Antolini, E., and Gonzalez, E.R. (2005) Carbon supported Pt–Co alloys as methanol-resistant oxygen-reduction electrocatalysts for direct methanol fuel cells. *Appl. Catal. B*, **57** (4), 283–290.

135 Antolini, E., Salgado, J.R.C., dos Santos, A.M., and Gonzalez, E.R. (2005) Carbon-supported Pt-Ni alloys prepared by the borohydride method as electrocatalysts for DMFCs. *Electrochem. Solid State Lett.*, **8** (4), A226–A230.

136 Colmenares, L., Guerrini, E., Jusys, Z., Nagabhushana, K.S., Dinjus, E., Behrens, S., Habicht, W., Bönnemann, H., and Behm, R.J. (2007) Activity, selectivity, and methanol tolerance of novel carbon-supported Pt and Pt_3Me (Me=Ni, Co) cathode catalysts. *J. Appl. Electrochem.*, **37** (12), 1413–1427.

137 Tsang, S.C., Claridge, J.B., and Green, M. L.H. (1995) Recent advances in the conversion of methane to synthesis gas. *Catal. Today*, **23** (1), 3–15.

138 Balint, I., Miyazaki, A., and Aika, K. (2002) Methane reaction with NO over Alumina-Supported Ru nanoparticles. *J. Catal.*, **207** (1), 66–75.

139 Witte, P.T. (2010) Process for the preparation of an aqueous colloidal precious metal suspension. Patent Application WO 2009NL50039 20090127, filed Jun. 10, 2010 and issued Jul. 19, 2010.

140 Witte, P.T., de Groen, M., de Rooij, R.M., Bakermans, P., Donkervoort, J.G., Berben, P.H., and Geus, J.W. (2010) Highly active and selective precious metal catalysts by use of the reduction-deposition method. *Stud. Surf. Sci. Catal.*, **175**, 135–143.

141 BASF, *NanoSelect™ LF 200™ catalysts.* http://www.strem.com/uploads/resources/documents/basf_nanoselect.pdf (accessed 23 February 2012).

142 Armbruster, M., Konvir, K., Behrens, M., Teschner, D., Grin, Y., and Schlögl, R. (2010) Pd–Ga intermetallic compounds as highly selective semihydrogenation catalysts. *J. Am. Chem. Soc.*, **132** (42), 14745–14747.

143 Asao, N., Ishikawa, Y., Hatakeyama, N., Menggenbateer, Yamamoto, Y., Chen, M., Zhang, W., and Inoue, A. (2010) Nanostructured materials as catalysts: nanoporous-gold-catalyzed oxidation of organosilanes with water. *Angew. Chem.*, **122** (52), 10291–10293.

144 Li, Y. and El-Sayed, M.A. (2001) The effect of stabilizers on the catalytic activity and stability of pd colloidal nanoparticles in the Suzuki reactions in

aqueous solution. *J. Phys. Chem. B*, **105** (37), 8938–8943.

145 Narayanan, R. and El-Sayed, M.A. (2003) Effect of catalysis on the stability of metallic nanoparticles: Suzuki reaction catalyzed by PVP-palladium nanoparticles. *J. Am. Chem. Soc.*, **125** (27), 8340–8347.

146 Narayanan, R. and El-Sayed, M.A. (2004) Effect of colloidal catalysis on the nanoparticle size distribution: dendrimer-Pd vs PVP-Pd nanoparticles catalyzing the Suzuki coupling reaction. *J. Phys. Chem. B*, **108** (25), 8572–8580.

147 Moreno-Manas, M., Pleixats, R., and Villarroya, S. (2001) Fluorous phase soluble palladium nanoparticles as recoverable catalysts for Suzuki cross-coupling and Heck reactions. *Organometallics*, **20** (22), 4524–4528.

148 Kogan, V., Aizenshtat, Z., Popovitz-Biro, R., and Neumann, R. (2002) Carbon–Carbon and Carbon–Nitrogen coupling reactions catalyzed by palladium nanoparticles derived from a palladium substituted Keggin-type polyoxometalate. *Org. Lett.*, **4** (20), 3529–3532.

149 Gopidas, K.R., Whitesell, J.K., and Fox, M.A. (2003) Synthesis, characterization, and catalytic applications of a palladium-nanoparticle-cored dendrimer. *Nano Lett.*, **3** (12), 1757–1760.

150 Na, Y., Park, S., Han, S.B., Han, H., Ko, S., and Chang, S. (2004) Ruthenium-catalyzed Heck-type olefination and Suzuki coupling reactions: studies on the nature of catalytic species. *J. Am. Chem. Soc.*, **126** (1), 250–258.

151 Thathagar, M.B., Beckers, J., and Rothenberg, G. (2002) Copper-catalyzed Suzuki cross-coupling using mixed nanocluster catalysts. *J. Am. Chem. Soc.*, **124** (40), 11858–11859.

152 Li, Y., Boone, E., and El-Sayed, M.A. (2002) Size effects of PVP–Pd nanoparticles on the catalytic Suzuki reactions in aqueous solution. *Langmuir*, **18** (12), 4921–4925.

153 Narayanan, R. and El-Sayed, M.A. (2005) Catalysis with transition metal nanoparticles in colloidal solution: nanoparticle shape dependence and stability. *J. Phys. Chem. B*, **109** (26), 12663–12676.

154 Le Bars, J., Specht, U., Bradley, J.S., and Blackmond, D.G. (1999) A catalytic probe of the surface of colloidal palladium particles using Heck coupling reactions. *Langmuir*, **15** (22), 7621–7625.

155 Moreno-Manas, M., Pleixats, R., and Villarroya, S. (2001) Fluorous phase soluble palladium nanoparticles as recoverable catalysts for Suzuki cross-coupling and Heck reactions. *Organometallics*, **20** (22), 4524–4528.

156 Kogan, V., Aizenshtat, Z., Popovitz-Biro, R., and Neumann, R. (2002) Carbon–carbon and carbon–nitrogen coupling reactions catalyzed by palladium nanoparticles derived from a palladium substituted Keggin-type polyoxometalate. *Org. Lett.*, **4** (20), 3529–3532.

157 Kumar, V.K.R. and Gopidas, K.R. (2011) Palladium nanoparticle-cored G_1-dendrimer stabilized by carbon–Pd bonds: synthesis, characterization and use as chemoselective, room temperature hydrogenation catalyst. *Tetrahedron Lett.*, **52** (24), 3102–3105.

158 Tamami, B., Allahyari, H., Ghasemi, S., and Farjadian, F. (2011) Palladium nanoparticles supported on poly (*N*-vinylpyrrolidone)-grafted silica as new recyclable catalyst for Heck cross-coupling reactions. *J. Organomet. Chem.*, **696** (2), 594–599.

159 Rahim, E.H., Kamounah, F.S., Frederiksen, J., and Christensen, J.B. (2001) Heck reactions catalyzed by PAMAM-dendrimer encapsulated Pd(0) nanoparticles. *Nano Lett.*, **1** (9), 499–501.

160 Yeung, L.K. and Crooks, R.M. (2001) Heck heterocoupling within a dendritic nanoreactor. *Nano Lett.*, **1** (1), 14–17.

161 Tsai, S.H., Liu, Y.H., Wu, P.L., and Yeh, C.S. (2003) Preparation of Au–Ag–Pd trimetallic nanoparticles and their application as catalysts. *J. Mater. Chem.*, **13** (5), 978–980.

162 Chapin, D.M., Fuller, C.S., and Pearson, G.L. (1954) A New Silicon p – n Junction Photocell for converting solar radiation into electrical power. *J. Appl. Phys.*, **25** (5), 676–677.

163 O'Regan, B. and Grätzel, M. (1991) A low-cost, high-efficiency solar cell based on dye-sensitized colloidal TiO_2 films. *Nature*, **353** (6346), 737–740.

164 Bach, U., Lupo, D., Comte, P., Moser, J. E., Weissörtel, J.F., Salbeck, J., Spreitzer, H., and Grätzel, M. (1998) Solid-state dye-sensitized mesoporous TiO_2 solar cells with high photon-to-electron conversion efficiencies. *Nature*, **395** (6702), 583–585.

165 Grätzel, M. (2001) Photoelectrochemical cells. *Nature*, **414** (6861), 338–345.

166 Huynh, W.U., Dittmer, J.J., and Alivisatos, A.P. (2002) Hybrid nanorod-polymer solar cells. *Science*, **295** (5564), 2425–2427.

167 Huynh, W.U., Dittmer, J.J., and Liby, W.C. (2003) Controlling the morphology of nanocrystal–polymer composites for solar cells. *Adv. Func. Mater.*, **13** (1), 73–79.

168 Tsakalakos, L. (2008) Nanostructures for photovoltaics. *Mater. Sci. Eng. R*, **62** (6), 175–189.

169 Khelashvili, G., Behrens, S., Hinsch, A., Habicht, W., Schild, D., Eichhöfer, A., Sastrawan, R., Skupien, K., Dinjus, E., and Bönnemann, H. (2007) Preparation and characterization of low platinum loaded Pt:SnO_2 electrocatalytic films for screen printed dye solar cell counter electrode. *Thin Solid Films*, **515** (7–8), 4074–4079.

170 Hinsch, A., Behrens, S., Berginc, M., Bönnemann, H., Brandt, H., Drewitz, A., Einsele, F., Faßler, D., Gerhard, D., Gores, H., Haag, R., Herzig, T., Himmler, S., Khelashvili, G., Koch, D., Nazmutdinova, G., Opara-Krasovec, U., Putyra, P., Rau, U., Sastrawan, R., Schauer, T., Schreiner, C., Sensfuss, S., Siegers, C., Skupien, K., Wachter, P., Walter, J., Wasserscheid, P., Würfel, U., and Zistler, M. (2008) Material development for dye solar modules: results from an integrated approach. *Prog. Photovolt: Res. Appl.*, **16** (6), 489–501.

171 Khelashvili, G., Behrens, S., Weidenthaler, C., Vetter, C., Hinsch, A., Kern, R., Skupien, K., Dinjus, E., and Bönnemann, H. (2006) Catalytic platinum layers for dye solar cells: A comparative study Original Research Article. *Thin Solid Films*, **511–512**, 342–348.

172 Bönnemann, H., Khelashvili, G., Behrens, S., Hinsch, A., Skupien, K., and Dinjus, E. (2007) Role of the platinum nanoclusters in the iodide/triiodide redox system of dye solar cells. *J. Cluster Sci.*, **18** (1), 141–155.

173 Yang, J., You, J., Cen, C.-C., Hsu, W.-C., Tan, H.-R., Zhang, X.W., Hong, Z., and Yang, Y. (2011) Plasmonic polymer tandem solar cell. *ACS Nano*, **5** (8), 6210–6217.

174 Nagaura, T. and Tozawa, K. (1990) Lithium ion rechargeable battery. *Prog. Batteries Solar Cells*, **9**, 209–215.

175 Linden, D. and Reddy, T.B. (eds) (2002) *Handbook of Batteries*, 3rd edn, McGraw-Hill.

176 Stubenrauch, J. (2011) Speicher-Batterien für neue Stromnetze. *Chemie Plus*, **5**, 54–57.

177 Goodenough, J.B. and Kim, Y. (2011) Challenges for rechargeable batteries. *J. Power Sources*, **196** (16), 6688–6694.

178 Armani, A. and Tarascon, J.-M. (2008) Building better batteries. *Nature*, **451** (7179), 652–657.

179 Tarascon, J.-M. and Armand, M. (2001) Issues and challenges facing rechargeable lithium batteries. *Nature*, **414** (6861), 359–367.

180 Nazar, L.F., Goward, G., Leroux, F., Duncan, M., Huang, H., Kerr, T., and Gaubicher, J. (2001) Nanostructured materials for energy storage. *Int. J. Inorg. Mater.*, **3** (3), 191–200.

181 Tiarado, J.L. (2003) Inorganic materials for the negative electrode of lithium-ion batteries: state-of-the-art and future prospects. *Mater. Sci. Eng. R*, **40** (3), 103–136.

182 Fu, L.J., Liu, H., Li, C., Wu, Y.P., Rahm, E., Holze, R., and Wu, H.Q. (2005) Electrode materials for lithium secondary batteries prepared by sol-gel methods. *Prog. Mater. Sci.*, **50** (7), 881–928.

183 Shukla, A.K. and Prem Kumar, T. (2008) Materials for next-generation lithium batteries. *Curr. Sci.*, **94** (3), 314–331.

184 Goodenough, J.B. and Kim, Y. (2010) Challenges for rechargeable Li batteries. *Chem. Mater.*, **22** (3), 587–603.

185 Ellis, B.L., Lee, K.T., and Nazar, L.F. (2010) Positive electrode materials for Li-ion and Li-batteries. *Chem. Mater.*, **22** (3), 691–714.

186 Lee, K.T. and Cho, J. (2011) Roles of nanosize in lithium reactive nanomaterials for lithium ion batteries. *Nano Today*, **6** (1), 28–41.

187 Wiggers, H., Schulz, C., and Franke, M. (2011) Nanomaterialien für Hochleistungs-Batterien, *Unikate, Univ. Duisburg-Essen*, **39**, 30–41.

188 Chiang, Y.-M. (2010) Building a better battery. *Science*, **330** (6010), 1485–1486.

189 Limthongkul, P., Jang, Y.-I., Dudney, N. J., and Chiang, Y.-M. (2003) Electrochemically-driven solid-state amorphization in lithium-silicon alloys and implications for lithium storage. *Acta Mater.*, **51** (4), 1103–1113.

190 Kasavajjula, U., Wang, C.S., and Appleby, A.J. (2007) Nano- and bulk-silicon-based insertion anodes for lithium-ion secondary cells. *J. Power Sources*, **163** (2), 1003–1039.

191 Hülser, T., Schnurre, S.M., Wiggers, H., and Schulz, C. (2011) Gas-phase synthesis of nanoscale silicon as an economical route towards sustainable energy technology. *KONA: Powder and Particle Journal*, **29**, 191–207.

192 Hülser, T., Schnurre, S.-M., Wiggers, H., and Schulz, C. (2011) In preparation.

193 Chan, C.K., Peng, H., Liu, G., McIlwrath, K., Zhang, X.F., Huggins, R.A., and Cui, Y. (2008) High-performance lithium battery anodes using silicon nanowires. *Nature Nanotechnology*, **3** (1), 31–36.

194 Ji, L., Zheng, H., Ismach, A., Tan, Z., Xun, S., Lin, E., Battaglia, V., Srinvasan, V., and Zhang, Y. (2012) Graphene/Si multilayer structure anodes for advanced half and full lithium-ion cells. *Nano Energy*, **1** (1), 164–171.

195 von Delft, S. and Leker, J. (2012) Technische Revolution und wirtschaftliche Reaktion. *Nachrichten aus der Chemie*, **60** (2), 143–146.

196 Gehrke, H., Ruthardt, K., Mathiak, J., and Roosen, C. (2011) Hydrogen: A small molecule with large impact. *Oil Gas European Magazine*, **37**, 29–33.

197 Lang, Y., Arnepalli, R.R., and Tiwari, A. (2011) A review on hydrogen production: methods, materials and nanotechnology. *J. Nanosci. Nanotech.*, **11** (5), 3719–3739.

198 Felderhoff, M., Schüth, F., and Eberle, U. (2009) Chemical and physical solutions for hydrogen storage. *Angew. Chem. Int. Ed.*, **48** (36), 6608–6630.

199 Balat, H. and Kirtay, E. (2010) Metabolic pathway engineering for enhanced biohydrogen production. *Int. J. Hydrogen Energy*, **35** (17), 7404–7416.

200 Rostrup-Nielsen, J.R., Sehested, J., and Nùrskov, J.K. (2002) Hydrogen and synthesis gas by steam- and CO_2 reforming. *Adv. Catal.*, **47**, 65–139.

201 Cargnello, M., Wieder, N.L., Montini, T., Gorte, R.J., and Fornasiero, P. (2009) Synthesis of dispersible Pd@CeO_2 core-shell nanostructures by self-assembly. *J. Am. Chem. Soc.*, **132** (4), 1402–1409.

202 Machocki, A., Denis, A., Grzegorczyk, W., and Gac, W. (2010) Nano- and micro-powder of zirconia and ceria-supported cobalt catalysts for the steam reforming of bio-ethanol. *Appl. Surf. Sci.*, **256** (17), 5551–5558.

203 Schüth, F. (2003) Endo- and exotemplating to create high-surface-area inorganic materials. *Angew. Chem. Int. Ed.*, **42** (31), 3604–3622.

204 Umegaki, T., Kuratani, K., Yamada, Y., Ueda, A., Kuriyama, N., Kobayashi, T., and Xu, Q. (2008) Hydrogen production via steam reforming of ethyl alcohol over nano-structured indium oxide catalysts. *J. Power Sources*, **179** (2), 566–570.

205 Caputo, A.C., Palumbo, M., Pelagagge, P. M., and Scacchia, F. (2005) Economics of biomass energy utilization in combustion and gasification plants: Effects of logistic variables. *Biomass Bioenerg.*, **28** (1), 35–51.

206 Li, J. (2009) Development of a nano-Ni-La-Fe/Al_2O_3 catalyst to be used for syn-gas production and tar removal after biomass gasification. *BioResources*, **4** (4), 1520–1535.

207 Li, J., Liu, J., Liao, S., and Yan, R. (2010) Hydrogen-rich gas production by air–steam gasification of rice husk using

supported nano-NiO/γ-Al_2O_3 catalyst. *Int. J. Hydrogen Energy*, **35** (14), 7399–7404.

208 Li, J., Yin, Y., Zhang, X., Liu, J., and Yan, R. (2009) Hydrogen-rich gas production by steam gasification of palm oil wastes over supported tri-metallic catalyst. *Int. J. Hydrogen Energy*, **34** (22), 9108–9115.

209 Kogan, A. (2000) Direct solar thermal splitting of water and on-site separation of the products — IV. Development of porous ceramic membranes for a solar thermal water-splitting reactor. *Int. J. Hydrogen Energy*, **25** (11), 1043–1050.

210 Wegner, K., Ly, H.C., Weiss, R.J., Pratsinis, S.E., and Steinfeld, A. (2006) *In situ* formation and hydrolysis of Zn nanoparticles for H_2 production by the 2-step ZnO/Zn water-splitting thermochemical cycle. *Int. J. Hydrogen Energy.*, **31** (1), 55–61.

211 Yi, Y., Lee, J.K., Lee, H.J., Uhm, S., Nam, S.C., and Lee, J. (2009) A single-step approach to create nano-pottery structures for efficient water electrocatalysis. *Electrochem. Commun.*, **11** (11), 2121–2124.

212 Osada, N., Uchida, H., and Watanabe, M. (2006) Polarization behavior of SDC cathode with highly dispersed Ni catalysts for solid oxide electrolysis cells. *J. Electrochem. Soc.*, **153** (5), A816–A820.

213 Snure, M. and Tiwari, A. (2007) Synthesis, characterization, and green luminescence in ZnO nanocages. *J. Nanosci. Nanotech.*, **7** (1), 481–485.

214 Snure, M. and Tiwari, A. (2007) $CuBO_2$: A p-type transparent oxide. *Appl. Phys. Lett.*, **91** (9), 092123-0-092123-3.

215 Nada, A.A., Barakat, M.H., Hamed, H.A., Mohamed, N.R., and Veziroglu, T.N. (2005) Studies on the photocatalytic hydrogen production using suspended modified TO_2 photocatalysts. *Int. J. Hydrogen Energy*, **30** (7), 687–691.

216 Bak, T., Nowotny, J., Rekas, M., and Sorrell, C.C. (2002) Photo-electrochemical hydrogen generation from water using solar energy. Materials-related aspects. *Int. J. Hydrogen Energy*, **27** (10), 991–1022.

217 Koriche, N., Bouguelia, A., Aider, A., and Trari, M. (2005) Photocatalytic hydrogen evolution over delafossite $CuAlO_2$. *Int. J. Hydrogen Energy*, **30** (7), 693–699.

218 Kato, H. and Kudo, A. (2003) Photocatalytic water splitting into H_2 and O_2 over various tantalate photocatalysts. *Catal. Today*, **78** (1–4), 561–569.

219 Brahimi, R., Bessekhouad, Y., Bouguelia, A., and Trari, M. (2007) $CuAlO_2/TiO_2$ heterojunction applied to visible light H_2 production. *J. Photoch. Photobio. A*, **186** (2–3), 242–247.

220 Hagfeldt, A. and Graetzel, M. (1995) Light-induced redox reactions in nanocrystalline systems. *Chem. Rev.*, **95** (1), 49–68.

221 Moser, J.E. and Grätzel, M. (1993) Observation of temperature independent heterogeneous electron transfer reactions in the inverted Marcus region. *Chem. Phys.*, **176** (2–3), 493–500.

222 Prakasam, H.E., Varghese, O.K., Paulos, M., Mor, G.K., and Grimes, G.A. (2006) Synthesis and photoelectrochemical properties of nanoporous iron (III) oxide by potentiostatic anodization. *Nanotechnology*, **17** (17), 4285–4291.

223 Wu, J., Ju, Y.-D., and Jeng, C.-C. (2010) Promotion of generation of hydrogen by splitting of water by CNT over Pt/TiO_2 catalysts. *J. Nanosci. Nanotech.*, **10** (7), 4213–4220.

224 Ling, Y., Liao, J., Liu, X., Wu, X., and Bai, X. (2010) Enhanced hydrogen production on porous TiO_2 nanotube arrays. *J. Nanosci. Nanotech.*, **10** (11), 7020–7024.

225 Hong, W.S., Park, J.H., and Han, G.Y. (2009) Optimization of conditions for hydrogen production from anodized TiO_2 nanotube-based photoelectrochemical cells. *J. Nanosci. Nanotech.*, **9** (12), 7293–7297.

226 Wu, K.-R., Yeh, C.-W., Hung, C.-H., Cho, T.-P., and Liu, W.-J. (2009) Visible-light-responsive layered titanium Oxide/tin indium oxide catalysts for hydrogen production. *J. Nanosci. Nanotech.*, **9** (4), 3433–3440.

227 Yun, H.J., Lee, H., Joo, J.B., Kim, N.D., and Yi, J. (2011) Effect of TiO_2 nanoparticle shape on hydrogen evolution via water splitting. *J. Nanosci. Nanotech.*, **11** (2), 1688–1691.

228 Irani, R.S. (2002) Hydrogen storage: high-pressure gas containment. *MRS Bull.*, **27** (9), 680–684.

229 Wolf, J. (2002) Liquid-hydrogen technology for vehicles. *MRS Bull.*, **27** (9), 684–687.

230 Enertrag, Wind energy, https://www.enertrag.com (accessed 22 February 2012).

231 Fraunhofer, http://www.iwes.fraunhofer.de/en/highlights/_power-to-gas_renewablegasconnectsenergynetworks.html (24 July, 2012).

232 Vesselli, E., Schweicher, J., Bundhoo, A., Frennet, A., and Kruse, N. (2011) Catalytic CO_2 hydrogenation on nickel: novel insight by chemical transient kinetics. *J. Phys. Chem. C*, **115** (4), 1255–1260.

233 Eberle, U. (2011) Hydrogen for automotive applications and beyond. *Oil Gas European Magazine*, **37** (1), 34–39.

234 von Helmolt, R. and Eberle, U. (2007) Fuel cell vehicles: Status 2007. *J. Power Sources*, **165** (2), 833–843.

235 Nalwa, H.S. (ed.) (2009) *Nanomaterials for Energy Storage Applications*, American Scientific Publishers.

236 Hirscher, M. (ed.) (2010) *Handbook of Hydrogen Storage*, Wiley-VCH Verlag GmbH, Weinheim, Germany.

237 Huot, J. (2010) *Handbook of Hydrogen Storage* (ed. M. Hirscher), Wiley-VCH Verlag GmbH, Weinheim, Germany, pp. 81–116.

238 Khrussanova, M., Terzieva, M., Peshev, P., Konstanchuk, I., and Ivanov, E. (1989) Hydriding kinetics of mixtures containing some 3d-Transition metal oxides and magnesium. *Zeitschrift für Physikalische Chemie*, **164** (2), 1261–1266.

239 Dornheim, M., Doppiu, S., Barkhordarian, G., Boesenberg, U., Klassen, T., Gutfleisch, O., and Bormann, R. (2007) Hydrogen storage in magnesium-based hydrides and hydride composites. *Scripta Mater.*, **56** (10), 841–846.

240 Wagemans, R.W.P., van Lenthe, J.H., de Jongh, P.E., van Dillon, A.J., and de Jong, K.P. (2005) Hydrogen storage in magnesium clusters: Quantum chemical study. *J. Am. Chem. Soc.*, **127** (47), 16675–16680.

241 de Jongh, P.E., Wagemans, R.W.P., Eggenhuisen, T.M., Dauvillier, B.S., Radstake, P.B., Meeldijk, J.D., Geus, J. W., and de Jong, K.P. (2007) The preparation of carbon-supported magnesium nanoparticles using melt infiltration. *Chem. Mater.*, **19** (24), 6052–6057.

242 Bogdanović, B., Ritter, A., and Spliethoff, B. (1990) Active MgH_2Mg systems for reversible chemical energy storage. *Angew. Chem. Int. Ed.*, **29** (3), 223–234.

243 Weidenthaler, C. and Felderhoff, M. (2010) *Handbook of Hydrogen Storage* (ed. M. Hirscher), Wiley-VCH Verlag GmbH, Weinheim, Germany, pp. 117–157.

244 Bogdanović, B. and Schwickardi, M. (1997) Ti-doped alkali metal aluminium hydrides as potential novel reversible hydrogen storage materials. *J. Alloy. Compd.*, **253–254**, 1–9.

245 Bogdanović, B., Felderhoff, M., Kaskel, S., Pommerin, A., Schlichte, K., and Schüth, F. (2003) Improved hydrogen storage properties of Ti-Doped sodium alanate using titanium nanoparticles as doping agents. *Adv. Mater.*, **15** (12), 1012–1015.

246 Zaluski, L., Zaluska, A., and Strùm-Olsen, J.O. (1999) Hydrogenation properties of complex alkali metal hydrides fabricated by mechano-chemical synthesis. *J. Alloy. Compd.*, **290** (1–2), 71–78.

247 Huot, J., Liang, G., and Schultz, R. (2001) Mechanically alloyed metal hydride systems. *Appl. Phys. A*, **72** (2), 187–195.

248 Bellosta von Colbe, M., Felderhoff, M., Bogdanović, B., Schüth, F., and Weidenthaler, C. (2005) One-step direct synthesis of a Ti-doped sodium alanate hydrogen storage material. *Chem. Commun.*, **37**, 4732–4734.

249 de Jongh, P.E. and Adelhelm, Ph. (2010) Nanosizing and nanoconfinement: New strategies towards meeting hydrogen storage goals. *ChemSusChem (Chemistry & Sustainability, Energy & Materials)*, **3** (12), 1332–1348.

250 Jensen, C.M., Zidan, R., Mariels, N., Hee, A., and Hagen, C. (1999) Advanced

titanium doping of sodium aluminum hydride: Segue to a practical hydrogen storage material? *Int. J. Hydrogen Energy*, **24** (5), 461–465.

251 Yaghi, O.M., O'Keeffe, M., Ockwig, N.W., Chae, H.K., Eddaoudi, M., and Kim, J. (2003) Reticular synthesis and the design of new materials. *Nature*, **423**, 705–714.

252 Nielsen, T.K., Besenbacher, F., and Jensen, T.R. (2011) Nanoconfined hydrides for energy storage. *Nanoscale*, **3** (5), 2086–2098.

253 Panella, B., Hirscher, M., Pütter, H., and Müller, U. (2006) Hydrogen adsorption in metal–organic frameworks: Cu-MOFs and Zn-MOFs compared. *Adv. Func. Mater.*, **16** (4), 520–524.

254 DeSoto, S.M. and Dreyer, R. (2009) *Nanomaterials for Energy Storage Applications* (ed. H.S. Nalwa), American Scientific Publishers, pp. 180–213.

255 Li, Y.W. and Yang, R.T. (2006) Significantly enhanced hydrogen storage in metal-organic frameworks via spillover. *J. Am. Chem. Soc.*, **128** (3), 726–727.

256 Lueking, A.D. and Jain, P. (2009) *Nanomaterials for Energy Storage Applications* (ed. H.S. Nalwa), American Scientific Publishers, pp. 213–257.

257 Deng, W.Q., Xu, X., and Goddard, W.A. III (2004) New alkali doped pillared carbon materials designed to achieve practical reversible hydrogen storage for transportation. *Phys. Rev. Lett.*, **92** (16), 166103, (1–4).

258 Han, S.S. and Goddard, W.A. III (2007) Lithium-doped metal-organic frameworks for reversible H_2 storage at ambient temperature. *J. Am. Chem. Soc.*, **129** (27), 8422–8423.

259 Han, S.S., Deng, W.Q., and Goddard, W. A. III (2007) Improved designs of metal–organic frameworks for hydrogen storage. *Angew. Chem. Int. Ed.*, **46** (33), 6289–6292.

260 Turkevich, J., Stevenson, P.C., and Hillier, J. (1951) A study of the nucleation and growth in the synthesis of colloidal gold. *Discuss. Faraday Soc.*, **11**, 55–75.

261 Svergun, D.I., Shtykova, E.V., Volkov, V. V., and Feigin, L.A. (2011) Small-angle X-ray scattering, synchrotron radiation, and the structure of Bio- and nanosystems. *Crystallogr. Rep.*, **56**, 725–750.

262 Thüne, P.C., Westrate, C.J., Moodley, P., Saib, A.M., van de Loosdrecht, J., Miller, J.T., and Niemantsverdriet, J.W. (2011) Studying Fischer–Tropsch catalysts using transmission electron microscopy and model systems of nanoparticles on planar supports. *Catal. Sci. Technol.*, **1**, 689–697.

263 Cheng, G. and Hight Walker, A.R. (2010) Transmission electron microscopy characterization of colloidal copper nanoparticles and their chemical reactivity. *Anal. Bioanal Chem.*, **396**, 1057–1069.

264 de Jonge, N., Bigelow, W.C., and Veith, G.M. (2010) Atmospheric pressure scanning transmission electron microscopy. *Nano Lett.*, **10**, 1028–1031.

265 Polte, J., Erler, R., Thünemann, A.F., Sokolov, S., Ahner, T.T., Rademann, K., Emmerling, F., and Kraehnert, R. (2010) Nucleation and growth of gold nanoparticles studies via in situ small angle scattering at millisecond time resolution. *Nano*, **2**, 1076–1082.

266 Rehr, J.J. and Ankudinov, A.L. (2005) Progress in the theory and interpretation of XANES. *Coordination Chem. Rev.*, **249**, 131–140.

267 Rehr, J.J., Kas, J.J., Prange, M.P., Sorini, A.P., Takimoto, Y., and Vila, F. (2009) Ab initio theory and calculations of X-ray spectra. *Comptes rendus Physique*, **10**, 548–559.

268 Rehr, J.J., Kas, J.J., Vila, F.D., and Prange, M.P. (2010) Parameter-free calculations of X- ray spectra with FEFF9. *Phys. Chem. Chem. Phys.*, **12**, 5503–5513.

269 Bunker, G. (2010) *Introduction to XAFS, A Practical Guide to X-ray Absorption Fine Structure Spectroscopy*, Cambridge University Press.

270 Willmott, P. (2011) *An Introduction to Synchrotron Radiation: Techniques and Applications*, John Wiley & Sons, Ltd, Chichester, UK.

271 Modrow, H. (2004) Tuning nanoparticle properties – the X-ray absorption spectroscopy point of view. *Appl. Spec. Rev.*, **39**, 183–290.

272 Iwasawa, Y. (1996) *X-ray Absorption Fine Structure for Catalysis and Surfaces*, World Scientific.

273 Di Cicco, A. and Filipponi, A. (eds) (2009) The 14th International conference on X-ray absorption fine structure (XAFS14), 26–31 July 2009, Camerino, Italy, J. Phys. Conf. Ser. 190.

274 Russell, A.E. and Rose, A. (2004) X-ray absorption spectroscopy of low temperature fuel cell catalysts. *Chem. Rev.*, **104**, 4613–4635.

275 Lytle, F.W., Via, G.H., and Sinfelt, J. (1977) Determination of structure of Cu and Cu-Ru on carriers using X-ray absorption spectroscopy. *Erdöl Kohle Erdgas Petrochem*, **30**, 231.

276 Greegor, R.B. and Lytle, F.W. (1980) Morphology of supported metal clusters: Determination by EXAFS and chemisorption. *J. Catal.*, **63**, 476–486.

277 Yevick, A. and Frenkel, A.I. (2010) Effects of surface disorder on EXAFS modelling of metallic clusters. *Phys. Rev. B*, **81**, 115451.

278 Beale, A.M. and Weckhuysen, B.M. (2010) EXAFS as a tool to interrogate the size and shape of mono and bimetallic catalyst nanoparticles. *Phys. Chem. Chem. Phys.*, **12**, 5562–5574.

279 Frenkel, A.I., Yevick, A., Cooper, Ch., and Vasic, R. (2011) Modeling the structure and composition of nanoparticles by extended X-ray absorption fine-structure spectroscopy. *Annu. Rev. Anal. Chem.*, **4**, 23–39.

280 Girardon, J.-S., Khodakov, A.Y., Capron, M., Cristol, S., Dujardin, C., Dhainaut, F., Nikitenko, S., Meneau, F., Bras, W., and Payen, E. (2005) A new experimental cell for in situ and operando X-ray absorption measurements in heterogeneous catalysis. *J. Synchrotron Rad.*, **12**, 680–684.

281 Rochet, A., Moizan, V., Pichon, Ch., Diehl, F., Berliet, A., and Briois, V. (2011) *In situ* and *operando* structural characterisation of a Fischer–Tropsch supported cobalt catalyst. *Catal. Today*, **171**, 186–191.

282 Khalid, S., Ehrlich, S.N., Lenhard, A., and Clay, B. (2011) Hard X-ray QEXFS instrumentation with scan range 20 → 4000eV. *Nucl. Instr. Methods in Phys. Res. A*, **649**, 64–66.

283 Principi, E., Di Cicco, A., Witkowska, A., and Marassi, R. (2007) Performance of a fuel cell optimized for in situ X-ray absorption experiments. *J. Synchrotron Rad.*, **14**, 276–281.

284 Frenkel, A.I., Wang, Q., Marinkovic, N., Chen, J.G., Barrios, L., Sis, R., Lopez Camara, A., Estrella, A.M., Rodriguez, J. A., and Hanson, J.C. (2011) Combining X-ray absorption and X-ray diffraction techniques for in situ studies of chemical transformations in heterogeneous catalysis: advantages and limitations. *J. Phys. Chem. C*, **115**, 17884–17890.

285 Grunwaldt, J.-D. and Baiker, A. (2007) Time-resolved and operando XAS studies on heterogeneous catalysts – from the gas phase towards reactions in supercritical fluids. (eds B. Hedman and P. Pianetta), 3rd International Conference on X-ray absorption fine structure – XAFS13, AIP Conference Proceedings, 882, pp. 577–581.

286 Croze, V., Ettingshausen, F., Melke, J., Soehn, M., Stuermer, D., and Roth, Ch. (2010) The use of *in situ* X-ray absorption spectroscopy in applied fuel cell research. *J. Appl. Electrochem.*, **40**, 877–883.

287 de Smit, E., Swart, I., Creemer, J.F., Karunakaran, Ch., Bertwistle, D., Zandbergen, H.W., de Groot, F.M.F., and Weckhuysen, B.M. (2009) Nanoscale chemical imaging of the reduction behaviour of a single catalyst particle. *Angew. Chem. Int. Ed.*, **48**, 3632–3636.

288 Grunwaldt, J.-D., Kimmerle, B., Baiker, A., Boye, P., Schroer, Ch.G., Glatzel, P., Borca, C.N., and Beckmann, F. (2009) Catalyst at work: from integral to spatially resolved X-ray absorption spectroscopy. *Catal. Today*, **145**, 267–278.

289 Grunwaldt, J.-D. and Schroer, C.G. (2010) Hard and soft X-ray microscopy and tomography in catalysis: bridging the different time and length scales. *Chem. Soc. Rev.*, **39**, 4741–4753.

290 Kühn, J.-T. (2011) *Characterization of metallic nanoparticles by high-resolution X-ray absorption and X-ray emission spectroscopy*, PhD Thesis, Institute of Physics, Bonn University, 2011.

291 Murray, C.B., Sun, S.H., Doyle, H., and Betley, T. (2001) Monodisperse 3d transition-metal (Co, Ni, Fe) nanoparticles and their assembly into nanoparticle superlatices. *MRS Bull.*, **26**, 985–991.

292 Ankundinov, A.L., Ravel, B., Rehr, J.J., and Conradson, S.D. (1998) Real-space multiple- scattering calculation und interpretation of X-ray absorption near-edge structure. *Phys. Rev. B*, **58**, 7565–7576.

293 Cheng, G., Carter, J.D., and Guo, T. (2004) Investigation of Co nanoparticles with EXAFS and XANES. *Chem. Phys. Lett.*, **400**, 122–127.

294 Margeat, O., Amiens, C., Chaudret, B., Lecante, P., and Benfield, R.E. (2005) Chemical control of structural and magnetic properties of cobalt nanoparticles. *Chem. Mater.*, **17**, 107–111.

295 Kraum, M. and Baerns, M. (1999) Fischer–Tropsch synthesis: the influence of various cobalt compounds applied in the preparation of supported cobalt catalysts on their performance. *Appl. Catal. A*, **186**, 189–200.

296 Bönnemann, H., Brijoux, W., Brinkmann, R., Matoussevitch, N., Waldöfner, N., Palina, N., and Modrow, H. (2003) A size-selective synthesis of air stable colloidal magnetic cobalt nanoparticles. *Inorg. Chim. Acta*, **350**, 617–624.

297 Shukla, N., Svedberg, E.B., Ell, J., and Roy, A.J. (2006) Surfactant effects on the shapes of cobalt nanoparticles. *Mat. Lett.*, **60**, 1950–1955.

298 Khodakov, A.Y. (2009) Fischer–Tropsch synthesis: Relations between structure of cobalt catalysts and their catalytic performance. *Catal. Today*, **144**, 251–257.

299 Saib, A.M., Borgna, A., van de Loosdrecht, J., van Berge, P.J., Geus, J. W., and Niemantsverdriet, J.W. (2006) Preparation and characterisation of spherical Co/SiO_2 model catalysts with well-defined nano-sized cobalt crystallites and a comparison of their stability against oxidation with water. *J. Catal.*, **239**, 326–339.

300 Diehl, F. and Khodakov, A.Y. (2009) Promotion of COBALT Fischer–Tropsch catalysts with noble metals: a review. *Oil & Gas Science and Technology-Rev. IFP*, **64**, 11–24.

301 Chu, W., Chernavskii, P.A., Gengembre, L., Pankina, G.A., Fongarland, P., and Khodakov, A.Y. (2007) Cobalt species in promoted cobalt alumina-supported Fischer–Tropsch catalysts. *J. Catal.*, **252**, 215–230.

302 Sadeqzadeh, M., Karaca, H., Safonova, O.V., Fongarland, P., Chambrey, S., Roussel, P., Griboval-Constant, A., Lacroix, M., Curulla-Freee, D., Luck, F., and Khodakov, A.Y. (2011) Identification of the active species in the working alumina-supported cobalt catalysts under various conditions of Fischer–Tropsch synthesis. *Catal. Today*, **164**, 62–67.

303 Karaca, H., Safonova, O.V., Chambrey, S., Fongarland, P., Roussel, P., Griboval-Conatnt, A., Lacroix, M., and Khodakov, A.Y. (2011) Structure and catalytic performance of Pt-promoted alumina-supported cobalt catalysts under realistic conditions of Fischer–Tropsch synthesis. *J. Catal.*, **277**, 14–26.

304 Cronauer, D.C., Jacobs, G., Lingansio, L., Kropf, A.J., Elam, J.W., Christensen, S.T., Marshall, C.L., and Davis, B.H. (2011) CO Hydrogenation: Exploring iridium as a promoter for supported cobalt catalysts by TPR-EXAFS/XANES and reaction testing. *Catal. Lett.*, **141**, 968–976.

305 Hong, J., Chu, W., Chernavskii, P.A., and Khodakov, A.Y. (2010) Effects of zirconia promotion on the structure and performance of smaller and larger pore silica-supported cobalt catalysts for Fischer–Tropsch synthesis. *Appl. Catal. A*, **382**, 28–35.

306 Jacobs, G., Ribeiro, M.C., Ma, W., Ji, Y., Khalid, S., Sumodjo, P.T.A., and Davis, B.H. (2009) Group 11 (Cu, Ag, Au) promotion of $15\%Co/Al_2O_3$ Fischer–Tropsch synthesis catalysts. *Appl. Catal.*, **361**, 137–151.

307 Guczi, L., Boskovic, G., and Kiss, E. (2010) Bimetallic cobalt based catalysts. *Catal. Rev.: Science and Engineering*, **52**, 133–203.

308 Fernandez, J.L., Walsh, D.A., and Bard, A.J. (2005) Thermodynamics guidelines for the design of bimetallic catalysts for

oxygen electroreduction and rapid screening by scanning electrochemical microscopy. *J. Am. Chem. Soc.*, **127**, 357–365.

309 Alayoglu, S., Nilekar, A.U., Manrikakis, M., and Eichhorn, B. (2008) Ru-Pt core shell nanoparticles for preferential oxidation of carbon monoxide in hydrogen. *Nat. Mat.*, **7**, 222–338.

310 Nilekar, A.U., Alayoglu, S., Eichhorn, B., and Mavrikakis, M. (2010) Preferential CO oxidation in hydrogen: reactivity of core-shell nanoparticles. *J. Am. Chem. Soc.* **132**, 7418–7428.

311 Hwang, B.-J., Sarma, L.S., Chen, J.-M., Chen, C.-H., Shih, S.-C., Wang, G.-R., Liu, D.G., Lee, J.-F., and Tang, M.-T. (2005) Structural models and atomic distribution of bimetallic nanoparticles as investigated by X-ray absorption spectroscopy. *J. Am. Chem. Soc.*, **127**, 11140–11145.

312 Antoniak, C. (2011) Extended X-ray absorption fine structure of bimetallic nanoparticles. *Beilstein J. Nanotechnol.*, **2**, 237–251.

313 Alayoglu, S., Zavalij, P., Eichhorn, B., Wang, Q., Frenke, A.I., and Chupas, P. (2009) Structural and architectural evaluation of bimetallic nanoparticles: a case study of Pt-Ru core-shell and alloy nanoparticles. *Nano*, **10**, 3127–3137.

314 Konzinkin, A.V., Vlasenko, V.G., Kulikova, O.V., Shvachko, O.V., Kozinkin, Y.A., Vysochina, L.L., Guterman, V.E., and Zubavichus, Y.V. (2011) Electronic and atomic structure of platinum-cobalt nanocatalyst. *J. Struct. Chem.*, **52** (Suppl.), S76–S81.

315 Schulenburg, H., Müller, E., Khelashvili, G., Roser, T., Bönnemann, H., Wokaun, A., and Scherer, G.G. (2009) Heat-treated $PtCo_3$ nanoparticles as oxygen reduction catalysts. *J. Phys. Chem.*, **113**, 4069–4077.

316 Xu, Y., Ruban, A.V., and Mavrikakis, M. (2004) Adsorption and dissociation of O_2 on Pt-Co and Pt-Fe alloys. *J. Am. Chem. Soc.*, **126**, 4717–4725.

317 Kitchin, J.R., Nùrskov, J.K., Barteau, M. A., and Chen, J.G. (2004) Role of Strain and ligand effects in the modification of the electronic and chemical properties of bimetallic surfaces. *Phy. Rev. Lett.*, **93**, 156801– 1-156801-4.

318 Chen, S., Sheng, W., Yabuuchi, N., Ferreira, P.J., Allard, L.F., and Shao-Horn, Y. (2009) Origin of oxygen reduction reaction acitivity on Pt_3Co nanoparticles: atomically resolved chemical compositions and structures. *J. Phys. Chem. C*, **113**, 1109–1125.

319 Zafeiratos, S., Piccinin, S., and Teschner, D. (2012) Alloys in catalysis: phase separation and surface segregation phenomena in response to the reactive environment. *Catal. Sci. Technol.* doi: 10.1039/c2cy00487a

320 Lai, F.-J., Su, W.-N., Sarma, L.S., Liu, D.-G., Hsieh, C.-A., Lee, J.F., and Hwang, B.-J. (2010) Chemical dealloying mechanism of bimetallic Pt-Co nanoparticles and enhancement of catalytic activity toward oxygen reduction. *Chem. Eur. J.*, **16**, 4602–4611.

321 Uemura, Y., Inada, Y., Bando, K.K., Sasaki, T., Kamiuchi, N., Eguchi, K., Yagishita, A., Nomura, M., Tada, M., and Iwasawa, Y. (2011) *In situ* time-resolved XAFS study on the structural transformation and phase separation of Pt3Sn and PtSn alloy nanoparticles on carbon in the oxidation process. *Phys. Chem. Cem. Phys.*, **13**, 15833–15844.

322 Carlton, Ch.E., Chen, S., Ferreira, P.J., Allard, L.F., and Shao-Horn, Y. (2011) Sub- nanometer-resolution elemental mapping of Pt3Co nanoparticle catalyst degredation in proton-exchange membrane fuel cells. *J. Phys. Chem. Lett.*, **3**, 161–166.

323 Stevenson, S.A., Dumesic, J.A., Baker, R. T.K., and Ruckenstein, E. (eds) (1987) *Metal- Support Interaction in Catalysis, Sintering and Redispersion*, Van Nostrand Reinhold, New York.

324 Koningsberger, D.C. and Gates, B.C. (1992) Nature of the metal-support interface in supported metal catalysts: results from X-ray absorption spectroscopy. *Catal. Lett.*, **14**, 271–277.

325 Johnston, P. and Wells, P.B. (1995) EXAFS spectroscopy of supported metal catalysts. *Radiat. Phys. Chem.*, **45**, 393–412.

326 Mojet, B.L., Miller, J.T., Ramaker, D.E., and Koningsberger, D.C. (1999) A new model describing the metal-support interaction in nobel metal catalyst. *J. Catal.*, **186**, 373–386.

327 Shishido, T., Asakura, H., Amano, F., Sone, T., Yamazoe, S., Kato, K., Teramura, K., and Tanaka, T. (2009) In situ time-resolved energy dispersive XAFS study on reduction behaviour of Pt supported on TiO_2 and Al_2O_3. *Catal. Lett.*, **131**, 413–418.

328 Gonzales, I.D., Navarro, R.M., Wen, W., Marinkovic, N., Rodriguez, J.A., Rosa, F., and Fierro, J.L.G. (2010) A comparative study of the water gas shift reaction over platinum catalysts supported on CeO_2, TiO_2 and Ce-modified TiO_2. *Catal. Today*, **149**, 372–379.

329 Kim, M.-Y., Park, S.M., Park, J.-H., Shin, C.-H., Moon, W.-J., Sung, N.-E., and Seo, G. (2011) Platinum catalysts supported on silicas: effect of silica characteristics on their catalytic activity in carbon monoxide oxidation. *Reac. Kinet. Mech. Cat.*, **103**, 463–479.

330 Ho, V.T.T., Pan, Ch.-J., Rick, J., Su, W.-N., and Hwang, B.-J. (2011) Nanostructured $Ti_{0.7}Mo_{0.3}O_2$ support enhances electron transfer to Pt: high-performance catalyst for oxygen reduction reaction. *J. Am. Chem. Soc.*, **133**, 11716–11724.

331 Jacobs, G., Ji, Y., Davis, B.H., Cronauer, D., Kropf, A.J., and Marshall, C.L. (2007) Fischer–Tropsch synthesis: Temperature programmed EXAFS/XANES investigation of the influence of support type, cobalt loading, and noble metal promoter addition to the reduction behaviour of cobalt oxide particles. *Appl. Catal. A: Gen.*, **333**, 177–191.

332 Ma, W., Jacobs, G., Sparks, D.E., Gnanamani, M.K., Pendyala, V.R.R., Yen, C.H., Klettlinger, J.L.S., Tomsik, T.M., and Davis, B.H. (2011) Fischer–Tropsch synthesis: support and cobalt cluster size effects on kinetics over Co/SiO_2 catalysts. *Fuel*, **90**, 756–765.

333 Gnanamani, M.K., Jacobs, G., Graham, U.M., Ma, W., Pendyala, V.R.R., Ribeiro, M., and Davis, B.H. (2010) Studies on KIT-6 supported cobalt catalyst for Fischer–Tropsch synthesis. *Catal. Lett.*, **134**, 37–44.

334 Zhang, H., Lancelot, Ch., Chu, W., Hong, J., Khodakov, A.Y., Chernavskii, P. A., Zheng, J., and Tong, D. (2009) The nature of cobalt species in carbon nanotubes and their catalytic performance in Fischer–Tropsch reaction. *J. Mater. Chem.*, **19**, 9241–9249.

335 Hormes, J., Modrow, H., Bönnemann, H., and Kumar, C.S.S.R. (2005) The influence of various coatings on the electronic, magnetic and geometric properties of cobalt nanoparticles. *J. Appl. Phys.*, **97**, 10R1002.

336 Zhang, P. and Sham, T.K. (2003) X-ray studies of the structure and electronic behaviour of alkanethiolate-capped gold nanoparticles: the interplay of size and surface effects. *Phys. Rev. Lett.*, **90**, 245502.

337 Lopez-Cartes, C., Rojas, T.C., Litran, R., Martinez-Martinez, D., de la Fuente, J. M., Penades, S., and Fernandez, A. (2005) Gold nanoparticles with different capping systems: an electronic and structural XAS analysis. *J. Phys. Chem. B*, **109**, 8761–8766.

338 Marchetti, B., Joseph, Y., and Bertagnolli, H. (2011) Amine-capped gold nanoparticles: reaction steps during the synthesis and the influence of the ligand on the particle size. *J. Nanopart. Res.*, **13**, 3353–3362.

339 Liu, F. and Zhang, P. (2010) Tailoring the local structure and electronic property of AuPd nanoparticles by selecting capping molecules. *Appl. Phys. Lett.*, **96**, 043105.

340 Chen, T.-Y., Lin, T.-L., Luo, T.-J.M., Choi, Y., and Lee, J.-F. (2010) Effects of Pt shell thickness on the atomic structure of Ru-Pt core-shell nanoparticles for methanol electrooxidation applications. *Chem. Phys. Chem.*, **11**, 2383–2392.

341 Peng, G., de Groot, F.M.F., Haemaelaeinen, K., Moore, J.A., Wang, X., Grush, M.M., Hastings, J.B., Siddons, D.P., and Arnstrong, W.H. (1994) High-resolution manganese X- ray fluorescence spectroscopy, oxidation-state and spin-state sensitivity. *J. Am. Chem. Soc.*, **116**, 2914–2920.

342 Grush, M.M., Christou, G., Haemaelaeinen, K., and Cramer, S.P. (1995) Site selective XANES and EXAFS: a demonstration with manganese mixtures and mixed-valence complexes. *J. Am. Chem. Soc.*, **117**, 5895–5896.

343 Glatzel, P., Jacquament, L., Bergmann, U., de Groot, F.M.F., and Cramer, S.P. (2002) Site-selective EXAFS in mixed-valence compounds using high-resolution fluorescence detection: a study of iron in Prussian blue. *Inorg. Chem.*, **41**, 3121–3127.

344 Bergmann, U., Grush, M.M., Horne, C. R., DeMarois, P., Penner-Hanh, J.E., Yocum, C.F., Wright, D.W., Dube, C.E., Armstrong, W.H., Christou, G., Eppley, H.J., and Cramer, S.P. (1998) Characterization of the Mn oxidation states in photosystem II by Kβ X-ray fluorescence spectroscopy. *J. Phys. Chem. B*, **102**, 8350–8352.

345 Safonova, O.V., Tromp, M., van Bokhoven, J.A., de Groot, F.M.F., Evans, J., and Glatzel, P. (2006) Identification of CO adsorption sites in supported Pt catalysts using high-energy-resolution fluorescence detection X-ray spectroscopy. *J. Phys. Chem. B*, **110**, 16162–16164.

346 Singh, J., Alayon, E.M.C., Tromp, M., Safonova, O.V., Glatzel, P., Nachtegaal, M., Frahm, R., and van Bokhoven, J.A. (2008) Generating highly active partially oxidized platinum during oxidation of carbon monoxide over Pt/Al_2O_3: in situ, time-resolved, and high-energy-resolution X-ray absorption spectroscopy. *Angew. Chem. Int. Ed.*, **47**, 9260–9264.

347 Kühn, T.-J., Caliebe, W., Matoussevitch, N., Bönnemann, H., and Hormes, J. (2011) Site-selective X-ray absorption spectroscopy of cobalt nanoparticles. *Appl. Organometal. Chem.*, **25**, 577–584.

348 Beckwith, M.A., Roemelt, M., Collomb, M.-N., DuBoc, C., Weng, T.-C., Bergmann, U., Glatzel, P., Neese, F., and DeBeer, S. (2011) Manganese Kβ X-ray emission spectroscopy as a probe of metal-ligand interactions. *Inorg. Chem.*, **50**, 8397–8409.

349 Bauer, M. and Gastl, Ch. (2010) X-ray absorption in homogeneous catalysis research: the iron – catalysed Michael addition reaction by XAS, RIXS and multi-dimensional spectroscopy. *Phys. Chem. Chem. Phys.*, **12**, 5575–5584.

350 Glatzel, P., Singh, J., Kvashnina, K.O., and van Bokhoven, J.A. (2010) In situ characterization of the 5d density of states of Pt nanoparticles upon adsorption of CO. *J. Am. Chem. Soc.*, **132**, 2555–2557.

351 Singh, J., Lamberti, C., and van Bokhoven, J.A. (2010) Advanced X-ray absorption and emission spectroscopy: in situ catalytic studies. *Chem. Soc. Rev.*, **39**, 4754–4766.

352 Lee, A.F. (2012) Active site elucidation in heterogenous catalysis via in situ X-ray spectroscopies. *Aust. J. Chem.*, **65** (6), 615–623http://dx.doi.org/10.1071/CH11455.

5
Metallic Nanoparticles in Ionic Liquids – Applications in Catalysis

Isabelle Favier, David Madec, and Montserrat Gómez

5.1
Introduction

In this chapter, an overview of the main catalytic applications of metallic nanoparticles (MNPs) in ionic liquids (ILs) is presented, organized by the metal nature. In the first section, physico-chemical interactions between ILs and MNPs are examined based on different characterization techniques and theoretical models, followed by a selection of the most important contributions in catalysis.

The abbreviations of imidazolium-based ionic liquids (ImILs) containing *n*-alkyl substituents are indicated in Table 5.1; for those ionic liquids containing other kinds of substituents, the full name is given in the main text.

Imidazolium-based ionic liquids represent the most common molten salts used in catalysis, in particular for the stabilization of metallic nanoparticles [1]. ImILs exhibit polymeric hydrogen-bonded supramolecules or nanostructured materials with polar and nonpolar regions, even when mixed with other molecules [2]. Thus, 1-butyl-3-methylimidazolium hexafluorophosphate, $[C_1C_4Im][PF_6]$, has a melting point of 209 K and exhibits crystalline diffraction patterns at 253 K, consistent with a long range molecular ordering [3]. For some applications, ILs are used in combination with a surface such as in chromatography, supported liquid membranes [4] or supported ionic liquid phase (SILP) catalysis [5]. In this last case, an IL thin film is immobilized on a high surface area support, resulting in a solid material. However, if high amounts of ILs are combined with supports, such as silica [6] or carbon nanotubes [3, 7], gel confined ILs can be produced. The formation of interconnected support networks in ImILs should result from specific interactions, inducing different physico-chemical characteristics from the ones exhibited by neat ILs.

An influence of the IL-support interaction on the catalytic activity and IL thermostability has been observed in the studies aiming at the preparation of SILP catalysts on various supports [8]. A widespread characterization by thermogravimetric analysis (TGA), differential scanning calorimetry (DSC), X-ray diffraction (XRD) and solid nuclear magnetic resonance (NMR) analyses of thin films of $[C_1C_4Im]$ $[PF_6]$ immobilized on high surface area silica and γ-alumina amorphous supports,

Nanomaterials in Catalysis, First Edition. Edited by Philippe Serp and Karine Philippot.

Table 5.1 Abbreviations used in this chapter for the cations involved in ImILs containing *n*-alkyl substituents, R_1-R_3.

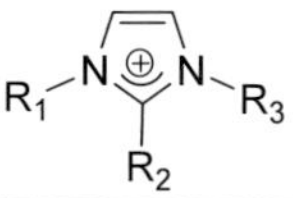

	R_1	R_2, R_3	Abbreviation
	C_mH_{2m+1}	H, C_nH_{2n+1}	$[C_mC_nIm]$
	C_mH_{2m+1}	C_nH_{2n+1}, C_pH_{2p+1}	$[C_mC_nC_pIm]$

evidenced a dramatic effect on the ionic liquid structure depending on the nature of the support. For the alumina composite $[C_1C_4Im][PF_6]@Al_2O_3$, a loss of the supramolecular arrangement of the ionic liquid occurs, while for the analogous silica composite $[C_1C_4Im][PF_6]@SiO_2$, ions interact strongly with the support, leading to a $[C_1C_4Im][PF_6]$ thermostability decrease [9]. In particular, the NMR analyses (^{1}H MAS, ^{13}C CPMAS, ^{19}F MAS, ^{29}Si CPMAS and ^{27}Al MAS NMR experiments) point to a loss of the structural order of the IL upon being supported on alumina, increasing the protons mobility of the imidazolium moiety, in agreement with previous works concerning the interaction of ammonium surfactants with disordered materials [10]; but for $[C_1C_4Im][PF_6]@SiO_2$, a strong interaction of the IL with silica was demonstrated for both imidazolium cation and hexafluorophosphate anion, by means of hydrogen bonds between silica, the anion and the cation analogously to the physisorption proposed for *N*-methylimidazolium chloride adsorbed on silica [11]. This different behavior between both supports leads to a decrease of the thermostability of the IL for $[C_1C_4Im][PF_6]@SiO_2$ composite as observed by TGA analyses.

The synthetic methodologies of MNPs in ILs applied in catalysis usually consist in the reduction of metal salts or preformed complexes (e.g., organometallic compounds) in the presence of reducing agents, such as molecular hydrogen, hydrides ($LiAlH_4$, $NaBH_4$), hydrazine, alcohols or thiols [1, 12]. Some ionic liquids act as reducing agents (e.g., hydroxylated imidazolium derivatives [13]), even leading to the oxidation of the imidazolium cation by the metal precursor, as observed in the formation of gold nanoparticles (AuNPs) starting with Au(III) species [14].

The nature of interaction between MNPs and ILs, as well as the influence of the ILs structural arrangements on the preparation of MNPs are next discussed. These aspects are crucial for the control of the size and shape of the as-prepared nanomaterials, and in consequence for their applications in catalysis in order to provide reproducible results. Structural characteristics of MNPs can be also tuned modifying the nature of the cation and/or anion of the IL.

5.2
Interactions between Ionic Liquids and Metallic Nanoparticles

Quaternary ammonium salts are the most widespread model which can explain the stabilization of metallic nanoparticles by salt effect. A well-established stabilization mode consists in the interstitial layer constituted by the anions between the metal core and the surfactant shell [15]. However, controversial opinion in the literature

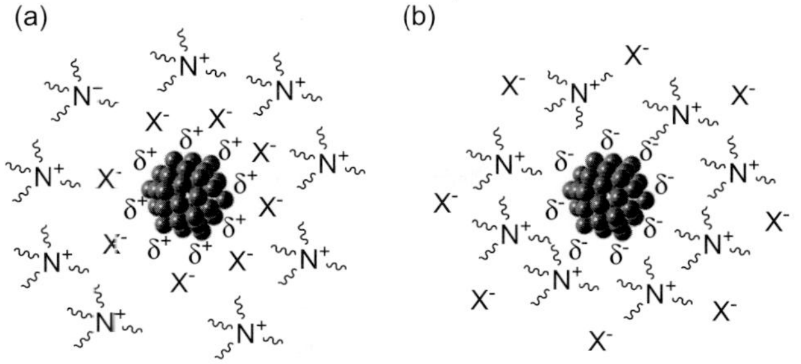

Figure 5.1 Representation of ionized MNPs at the surface: (a) positively and (b) negatively charged MNPs, surrounded by the corresponding shells of ionic species (black cluster denotes a metallic nanoparticle).

describes a direct interaction of the ammonium cation to the metal surface [16] (Figure 5.1).

Neat ImILs, in contrast to quaternary ammonium salts, exhibit organized structures, mainly due to hydrogen bonds [17], which induce structural directionality and polar and nonpolar regions [18]. The group of J. Dupont evidenced that ImILs look like polymeric supramolecules of general formula $\{[\text{Cation}_3\text{Anion}]^{2+} [\text{CationAnion}_3]^{2-}\}_n$. Besides these merely ionic interactions, π–π stacking arrangements between imidazolium rings are also present [2, 19].

All these interactions generate channel-like molecular assemblies (Figure 5.2). Depending on the anion and cation nature, the interaction energies can be very different as well as the organization of the IL [20]. Tsuzuki *et al.* have calculated the intermolecular energies following an *ab initio* methodology, showing that these arrangements are mostly due to the electrostatic attraction between ion pairs and the interaction of the anion in close contact with the C2-H of the imidazolium ring [21].

Figure 5.2 View of [DMBIm][PF_6] structure (DBMIm = 1,3-di-(*R*)-(+)-methylbenzyl imidazolium) along the crystallographic *y* axis. Hydrogen atoms are omitted for clarity [19a].

5.2.1
Stabilization Modes of Metallic Nanoparticles by Ionic Liquids

Metallic nanoparticles stabilized by ILs are generally considered as core-shell systems, mainly involving electrostatic and steric repulsion forces. Electrostatic stabilization is often based on Derjaguin, Landau, Verwey and Overbeek (DLVO) theory, which describes a balance between repulsive coulombic forces opposed to attractive van der Waals ones. The anions interact with the electrophilic surface of the nanoparticles, forming a layer at the surface of the MNP. The thickness of the counterions surrounding the MNP (Debye layers) affects the stability of the MNP. In general, the thinner the layers are, the greater the interparticle distance is [22]. But this anionic stabilization mode cannot explain all the reported observations. The most relevant works that elucidate the kind of interactions between MNPs and ILs are next briefly described.

5.2.1.1 DLVO Theory: Anionic Stabilization Mode

In the DLVO theory, the closest shell to the metallic surface is formed by anions, stabilizing the electrophilic MNP surface. Experimental analyses such as small angle X-ray scattering (SAXS), evidence the presence of the protective IL layer where the anions are closer to the metallic surface of the MNP. By X-ray photoelectron spectroscopy (XPS), slightly lower binding energies for platinum nanoparticles (PtNPs) stabilized by $[C_1C_2Im][BF_4]$ are found than those of bulk polycrystalline Pt, which points to an interaction between the metal and electron donor species, in this case BF_4^- [23]. For iridium nanoparticles (IrNPs), the interaction of the IL (with $[C_1C_4Im]$ as a cation) with the metal surface occurs through fluorine (for BF_4 and PF_6 anions) or oxygen (for CF_3SO_3 anion) atoms [24]. The XPS spectra for palladium nanoparticles (PdNPs) and PtNPs stabilized by $[C_1C_4Im][Y]$ ($Y = PF_6$, BF_4), indicate that the Pd3s binding energies correspond to two oxidation states, Pd(0) and Pd(II) due to Pd–Pd and Pd(II)–F bonds, respectively; for PtNPs, the Pt4f binding energies correspond to three oxidation states: Pt (0), Pt(II) and Pt(IV) related to Pt–Pt, Pt–F, Pt–O and Pt–O–F bonds [25].

Raman spectroscopy could demonstrate the direct interaction of the anion with the metallic surface of silver nanoparticles (AgNPs) when $[C_1C_4Im][BF_4]$ is used as IL; an intensity increase of the band corresponding to BF_4^- upon the addition of AgNPs, has been correlated with the interaction between the slightly positive polarized Ag surface and BF_4^-; this fact is in agreement with the corresponding XPS data [26].

Ultraviolet-visible (UV-vis) spectroscopy has been especially useful in the case of CdS nanoparticles. Thus, the absorption band of the NP shifts towards higher wavelength with the particle size increase, by the effect of the anion in the ionic liquids $[C_1C_4Im][Y]$ ($Y = MeSO_4$, BF_4, PF_6). In consequence, the stabilization of the nanoparticles seems to be dependent on the anion nature, leading to nanoparticles of different sizes [27]. Gold nanoparticles (AuNPs) have been prepared in the presence of 3,3′-disulfanylbis(hexane-1,6-diyl)-bis(1-methylimidazol-3-ium) dichloride in water. Anion exchange reactions (Cl substitution by Br, BF_4, PF_6 or I), monitored by UV-vis spectroscopy, led to the aggregation of MNPs, due to the increase of hydrophobic nature of the metallic surface [28]. For AgNPs stabilized by tetradecyltrimethylammonium bromide (TTAB) in water, ImILs ($[C_1C_4Im][PF_6]$, $[C_1C_4Im][BF_4]$) have been used for the phase transfer of MNPs towards the organic phase. In water, AgNPs are stabilized by bromine anions which are capped with a double layer of TTAB. A proposed mechanism for the phase transfer consists in the formation of a hydrophobic ion-pair between the cation of the stabilizer on the particle surface and the anion of the ionic liquid, which causes the change in the hydrophilic properties of the metallic surface favoring the phase transfer of MNPs into the organic phase [29].

A membrane coated by a solution of copper nanoparticles (CuNPs) in an IL was used to separate olefin/paraffin mixtures. The efficiency of the membrane depends on the aptitude of the anion of the IL to generate positively charged surface of the CuNPs which then acts as olefin carrier. The positive charge density measured for Cu surface by XPS gave the following trend: $[C_1C_4Im][BF_4] > [C_1C_4Im][PF_6] > [C_1C_2Im][BF_4]$, in agreement with the separation performance of the membrane [30].

PdNPs supported on an ordered alumina film were impregnated of $[C_1C_4Im][NTf_2]$. Infrared reflection absorption spectroscopy (IRAS) studies suggest that the anion interacts with the metallic surface by means of the SO_2 group of the anion [31].

All these analytical data tend to show that the surface of the MNP is slightly positively charged and in consequence, MNPs interact directly with the anion of ILs, even for those showing weak coordinating behavior. But other interactions, such as dispersion forces, hydrogen bonds, hydrophilicity, steric interactions or multiple charge counter-ions, are not taken into account by DLVO theory [22d,32].

5.2.1.2 **Steric Stabilization Mode**

Steric repulsion forces should be also considered for the stabilization of MNPs in ILs. Sometimes the addition of polymers to the MNP–IL system is necessary to improve the stabilization of MNPs. Silica nanoparticles (SiO_2NPs) grafted by poly (methylmethacrylate) are well-stabilized by $[C_1RIm]$ cations (R = alkyl), but not when non-functionalized silica particles are involved, showing that SiO_2NPs stabilization is due to steric effects rather than electrostatic factors [33].

For nickel nanoparticles (NiNPs), the increase of the alkyl side-chain of the imidazolium ring triggers a MNP size decrease; moreover, the distance between NiNPs decreases from $[C_1C_4Im][NTf_2]$ to $[C_1C_{14}Im][NTf_2]$ as evidenced by SAXS analysis [34].

5.2.1.3 **Cationic Stabilization Mode**

Dupont and co-workers have shown for AgNPs [35] and AuNPs [36] stabilized by ImILs, that the absorption bands of the cation in the Raman spectra show higher intensity and lower frequencies than those corresponding to neat ILs. In contrast, no changes could be observed for the signals relative to the anion. From these results, authors conclude that the cation $[C_1C_4Im]$ interacts with the MNP surface.

XPS analyses for AuNPs and PtNPs exhibit negative shifts for the binding energies of C1s and N1s, corresponding to the interaction between amino or carboxylic acid functions and/or imidazolium ring and PtNPs. The interaction of the negative surface of MNPs with the imidazolium fragment affects the photoemission of the atoms of the heterocycle. In addition, resonance Rayleigh scattering (RRS) analyses for AuNPs point to an aggregation decrease and in consequence a stability increase for MNPs stabilized by ILs in comparison to those stabilized by sodium citrate [37].

PdNPs synthesized in $[C_1C_4Im][PF_6]$ and $[C_1C_4Im][NTf_2]$ have been characterized by multinuclear diffusion ordered spectroscopy (DOSY) NMR technique. Even if the nanoparticles themselves cannot be detected through NMR, observation of the signals corresponding to solvent (in this case, methanol) and the ionic liquid, their diffusion coefficients (D) and their changes in the presence of nanoparticles, permit to draw significant assumptions about the interaction of PdNPs with the IL. Therefore, D_{anion}/D_{cation} for PdNPs dispersed in $[C_1C_4Im][PF_6]$ is higher than 1 (about 1.28), mainly due to a decrease of the D_{cation} value in relation to the neat IL, indicating that the cation interacts with the PdNP surface [38].

The fact that the imidazolium cation could form *N*-heterocyclic carbene by C–H activation of the acidic H in the C2 position reinforces the cation-metallic surface interaction [39]. The carbene formation by base deprotonation of the weak 2-H acid of the imidazolium was ruled out, because under D_2 atmosphere and in the presence of IrNPs, deuterium atoms are indifferently incorporated at the positions C2, C4, C5 and C8 of the imidazolium cation [40]. For IrNPs stabilized by $[C_1C_2Im][EtSO_4]$, XPS analysis shows that the binding energy of the C2 of the imidazolium ring shifts to lower value in relation to the neat IL, revealing an increase of its nucleophilic character. Due to the basic behavior of the anion, which interacts with the most acid proton of the imidazolium ring, the polarization of the C2–H bond leads to the formation of NHC–Ir bonds [41]. It was also shown that these carbene species could reduce $AgNO_3$ to form AgNPs through their nonbonding electron pair [42].

The fact that the side chain of imidazolium ring also affects the stabilization of MNPs, proves the proximity of the cation to the metallic surface. These chains could enhance the stability of MNPs due to additional interactions with the metallic surface. XPS measurements indicate lower binding energies for PtNPs in $[C_1RIm][BF_4]$ ($R = CH_2CH_2OH$) than in $[C_1C_4Im][BF_4]$, suggesting an additional electron transfer from the IL to the MNP [43]. For AuNPs and PtNPs, the interaction between the ImIL functionalized by an amino group has been proved by XPS analysis. The negative shifts of binding energies for C1s and N1s imply interactions between cation and MNP (M = Au, Pt) [37]. *In situ* deuterium labeling of ruthenium nanoparticles (RuNPs) stabilized by an ImIL proves that H/D exchange involves metallic surface, taking essentially place at the C2, C4 and C5 positions of the heterocycle; however, when unsaturated functional side groups are present in the imidazolium ring (such as butenyl or benzyl), they are reduced by the D_2 atmosphere, proving the proximity of the side chain to the metallic surface [44].

5.2.1.4 **Anionic and Cationic Stabilization Mode**

Both anion and cation of the IL can be found at the MNP surface. For TiO_2 nanocrystals stabilized by $[C_1C_4Im][Cl]$, N and Cl atoms were identified at the surface by XPS analysis [45]. For AuNPs and PtNPs stabilized by ImIL containing thiol functionalities located at the cation or anion, the MNP size decreases with the increase of the number of thiol functions. ^1H NMR studies reveal signal broadening of thiol fragments independently if they are located in the cationic or anionic part of the IL [46]. Rheological measurements were run in order to estimate the properties of ILs as dispersive medium for colloidal particles. The colloidal stability of hydrophilic SiO_2 nanoparticles with BF_4 anion-based ILs and $[C_1RIm][NTf_2]$ ($R = CH_2CH_2OH$) is due to repulsive solvation forces, because of the hydrogen bonds between the cation with OH end-group or BF_4 with the silanol surface. In consequence, agglomeration is observed for the methylated hydrophobic silica nanoparticles surface [47].

More recently, density functional theory (DFT) studies have been carried out to determine the interactions between $[C_1C_4Im][NTf_2]$ and RuNPs, showing that both anions and cations are close to the metallic surface. The orientation of the cation is perpendicular to the MNP surface with the methyl group closer to the metal and both butyl and CF_3 groups are far away from the metallic surface. The authors conclude that the stabilization of the RuNPs by ImILs is due to a template effect of the IL used as solvent more than by electrostatic or steric repulsions [48].

5.2.1.5 **Interactions of Ionic Liquids with Metal Oxide Nanoparticles (MONPs)**

For WO_3 nanoparticles, the existence of hydrogen bonds between the C2–H of the imidazolium ring and the oxygen of the metal oxide nanoparticles (MONPs) has been evidenced by Raman spectroscopy [49]. For α-Fe_2O_3 NPs stabilized in $[C_1C_4Im][Cl]$, FTIR spectra indicate the existence of strong hydrogen bonds between C2–H–O–(α-Fe_2O_3) [50]. For TiO_2 NPs, only the rutile form permits the adsorption of the IL, where the cations are perpendicularly arranged, simultaneously favoring both the π-stacking interaction between aromatic rings and hydrogen bonds between O(rutile)–H–C($C_1C_2Im^+$) [51] (Figure 5.3). In contrast, the

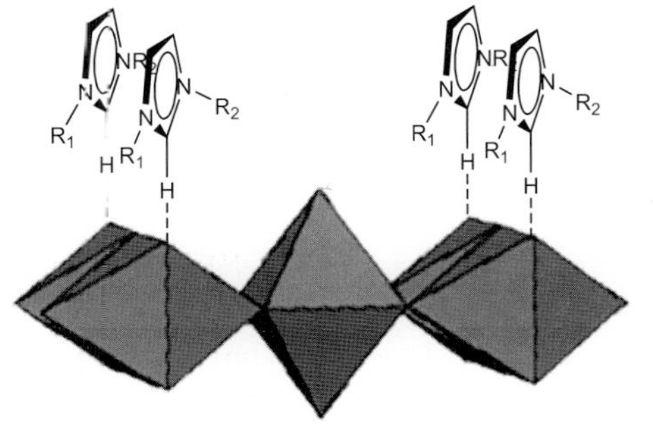

$R_1 = CH_2CH_3$; $R_2 = CH_3$

Figure 5.3 Model representing the adsorption of $[C_1C_2Im][Cl]$ at the rutile surface (gray octahedrons represent TiO_6 units of the rutile structure).

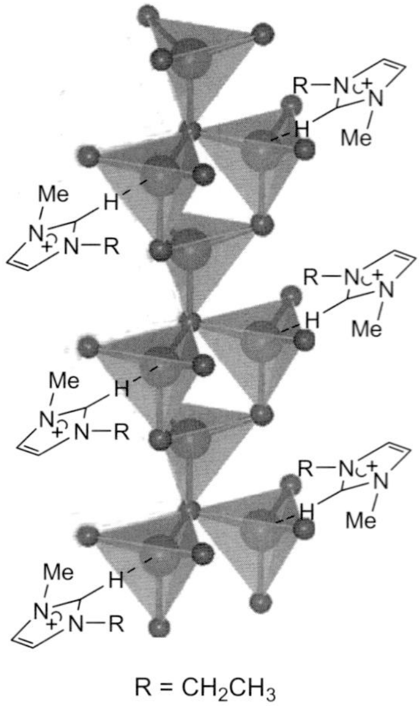

Figure 5.4 Schematic representation of 1D ZnO nanostructure in $[C_1C_2Im][BF_4]$ (gray tetrahedrons represent ZnO_4 units in the ZnO wurtzite structure).

anatase plane is toothed and the distance between O atoms of the TiO_6 octahedron is not in the space range to permit the π-stacking of the aromatic rings.

A mechanism has been proposed to explain the morphological changes of ZnO depending on the nature of the substituents in the ImIL. This suggestion is based on the fact that stronger hydrogen bonds exist between C2–H of the imidazolium ring and O–Zn, than between methyl substituted C2–Me imidazolium ring and ZnO nanoparticles (Figure 5.4).

Fourier-transform infrared (FTIR) analysis shows that the adsorption band related to the C2–H stretching, corresponding to the imidazolium cation ($[C_1C_2Im][BF_4]$), is broadened and weakened when ZnO nanorods are present. When strong attraction between ZnO and IL occurs, 1D growth (nanorods formation) is favored; however, when the attraction is weak, the nanoparticles formation are then formed, as could be observed for $[C_1C_1C_4Im][BF_4]$ [52].

5.2.2 Effect of Ionic Liquids on the Structures of Metallic Nanoparticles

As previously discussed, ImILs interact with the metal surface of the NP in different ways in order to stabilize them, avoiding their aggregation, mostly due to their supramolecular arrangement. In particular, a better morphological control in the synthesis of Au nanorods can be reached in IL than in aqueous solution of ammonium salts. The authors postulate that surface rearrangements in the nanorods trigger a lower overall free energy, which is more efficient in $[C_1C_2Im][EtSO_4]$ than in ammonium salts [53].

It is also well-established that ILs can also act as templates and influence the structural and surface properties of MNPs [1a,54]. Due to the low interfacial tension of ImILs, the nucleation rates of MNPs are higher in these ILs than in water, and in consequence small MNPs are generated, as observed in the synthesis of ZnCdS and CdS nanoparticles [55]. Morphological modifications of commercial ZnO nanoparticles have been also stated after their dispersion in $[C_1C_6Im][PF_6]$ at high temperature (553 K for 30 min), leading to ZnO nanorods and ZnF_2 needles [56]. The formation of Ag nanowires in $[C_1C_4Im][MeSO_4]$ monitored by transmission electron microscopy (TEM), shows firstly the formation of AgNPs and after the transition from nanoparticles to nanowires, indicating that the growth pathway of nanoparticles involves a self-organization of adjacent particles controlled by the IL [57].

When the synthesis of MNPs consists in the decomposition of nonpolar complexes, these metallic precursors are preferentially located in the nonpolar domains of IL (see above). The size of these nano domains (calculated by molecular dynamics simulation) depends on the interaction anion–cation in the IL, which can restrict the growth of the MNPs [18d]. A linear correlation has been found between the size of RuNPs (determined by TEM microscopy) and the length of the alkyl side chain of the imidazolium cation for *in situ* generated RuNPs [58]. In addition, the length of rutile nanorods increases with that of the alkyl chain length comparing butyl and C13 alkyl chain in C1-position of the imidazolium cation [51].

Depending on the temperature, the micro-domain size of the IL affects the size of the MNPs. Therefore, the size of RuNPs at 273 K is smaller than that observed at 298 K and 348 K, because the macromolecular organization of the IL is better maintained at low temperature [59].

The control of the morphology of MNPs by the template effect of the IL is important to modulate their physico-chemical properties. Thus, bismuth oxyiodine (BiOI) hollow microspheres are synthesized in $[C_1C_4Im][I]$, exhibiting higher photochemical activities than BiOI nanoplates prepared in aqueous solution. A mechanism for the synthesis of BiOI hollow structures is proposed, based on the formation of micelles by the IL cation. Time-dependent experiments monitored by SEM evidenced the transformation of flower-like structures formed at shorter times into BiOI hollow microspheres after 1 h at 413 K in $[C_1C_4Im][I]$ [60] (Figure 5.5).

Some contributions essentially attribute the control of the MNPs size in ILs to an anion effect. This work is based on the DLVO theory (see above) where the first inner shell is constituted by anions, consequently the anion would have the greatest

Figure 5.5 SEM images of the BiOI structures formed in $[C_1C_4Im][I]$ at different times at 413 K. (Adapted with permission from [60]. Copyright © (2011) American Chemical Society.)

influence on the MNPs stabilization, in particular on the MNPs size, because the anion molecular volume contributes to the size of the polar region in the IL [61]. For PtNPs in ImILs, smaller particles are obtained with PF_6 than with BF_4, because PF_6 is a stronger binding anion [25b]. The nature of the anion also influences the physico-chemical properties (viscosity, dielectric constant, melting point) of ILs. Therefore, small AuNPs (d_{mean}: 5–7 nm) are obtained in $[C_1C_2Im][MeSO_4]$ and $[C_1C_2Im][CF_3SO_3]$, but aggregates are formed when temperature increases at 453 K in $[C_1C_2Im][MeSO_4]$. However in $[C_1C_2Im][EtSO_4]$, AuNPs show a broad size distribution at 393 K (d_{mean}: 20 nm), but at 423 K the particles are mono-dispersed (d_{mean}: 15 nm). These observations prove that the temperature and anion nature influence the size and dispersion of the MNPs [62].

The effect of the cation on the MNPs size has also been the aim of some research work. The effect of the cation on the shape of ZnO nanostructures can be also influenced by the chain length in C1 position of the imidazolium ring (see above the influence of C2-H substitution in the formation of ZnO nanorods and nanoparticles). Therefore, the length of ZnO nanorods is shorter (200–600 nm) when $[C_1C_2Im][BF_4]$ is involved than that obtained in the case of $[C_1C_4Im][BF_4]$ (500–1500 nm), probably due to the steric hindrance of the alkyl chain [52].

It has also been proved that the anion and cation of the IL can influence the morphology of MNPs. Thus, for Y_2O_3 nanoparticles, if the length of the side chain of the imidazolium ring diminishes (from C_{17} to C_4), the thickness of nanosheets increases; changing $H_2PO_4^-$ by HSO_4^- as IL counter-anion, the structure adopts a coral-like arrangement [63]. IR data evidences that worm-like mesoporous silica NPs are formed in $[C_1C_4Im][BF_4]$, due both to the formation of hydrogen bond between the anion and the silica gel and to the π–π stack interaction between the imidazolium rings [64]. For iron nanoparticles (FeNPs), TEM analysis reveals that the increase of the alkyl side-chain length of the imidazolium ring, the anion nature and the concentration of the IL influence the morphology and size of FeNPs [65].

The concentration of the IL in aqueous solution also exhibits an important influence on the morphology of 2D metal nanostructures. For PdNPs stabilized by $[C_1C_2Im][BF_4]$, interconnected belt-like structures are obtained at low concentrations of IL (0.15 M); however at higher concentrations (0.4 M), uniform and well-dispersed holes are observed; in both cases, PdNPs are located at the air–water interface [23]. Under hydrothermal synthetic conditions, different hematite (α-Fe_2O_3) morphologies (nanoparticles, mesoporous hollow microspheres, microcubes and nanorods) can be observed. It was found that when the concentration of $[C_1C_4Im][Cl]$ increases the hematite switches from monodispersed MNPs to superstructures. The authors explain this phenomenon by the changes in viscosity and by the influence of the IL on the NPs dispersion [50]. It could be mentioned that for MNPs (M = Fe, Ru, Os, Fe_2O_3) prepared in $[C_1C_4Im][BF_4]$ starting from metal carbonyl complexes, their particle size is independent of the metal concentration (from 0.2 to 1 wt% of precursor in IL) but the thickness of the stabilizing shell around the MNP depends in fact on IL molecular ion volumes [66].

5.3
Catalytic Applications

5.3.1
Metallic Nanoparticles of Block p

Very few reports have been published concerning metallic nanoparticles in ILs that involve elements of block p. Besides bismuth oxyhalides, only some examples of tin alloys have been reported, such as Au/SnO_2 nanocomposites supported on multi-walled carbon nanotubes for biosensor purposes [67] and Pd/Sn solid-solution-coated electrodes as electrocatalysts for ethanol oxidation [68].

BiOX hollow microspheres have been recently synthesized in ImILs, which play a double role, both solvent and template for the synthesis of the inorganic material taking advantage of their organized structure [60, 69]. Therefore, the BiOBr microspheres formed at the mini-emulsion interface using a mixture of 2-methoxyethanol and [C_1C_{16}Im][Br], induce higher photocatalytic degradation of rhodamine B than that obtained using BiOBr nanoplates. The same trend was observed for the photoreduction of Cr(VI) species to give the corresponding Cr(III) ones, hypotoxic cation for the living organisms and environments [69]. Li and co-workers have correlated the photocatalytic activity on the degradation of methyl orange using BiOI hollow microspheres with their structure: high BET surface area, high surface-to-volume ratios, energy band structure and light absorbance [60] (see above, Section 5.2.2).

5.3.2
Metallic Nanoparticles of Block d and f

5.3.2.1 Early Transition Metals and Block f Metals

For lanthanides, actinides and the early transition metals of groups 3–7, only titanium oxide nanoparticles have been applied in catalysis. It is also important to note that CeO_2, TiO_2 and ZrO_2 nanomaterials become appropriate supports for metallic catalysts, where the ionic liquid acts as a template agent for the synthesis of the nanosupport and favors the dispersion of nanocatalysts on the oxide support, such as AgNP@TiO_2 [70], AuNP@CeO_2 [71], AuNP@TiO_2 [72] or PdNP@ZrO_2 [73].

The photocatalytic properties of titanium oxide are directly related to its structure; only the nanocrystalline systems are active, the amorphous material commonly being inefficient. The classic sol–gel synthetic methodology requires further treatments to obtain crystalline products. The independent pioneering works of Zhou [74] and Kimizuka [75] proved the preparation of anatase nanoparticles in ImILs ([C_1C_4Im][PF_6] and [C_1C_4Im][BF_4]) starting from $TiCl_4$ and $Ti(OBu)_4$ respectively. In the last three years, several works have appeared in the literature showing the photocatalytic potential of crystalline TiO_2 nanoparticles in the degradation of methyl orange and phenol derivatives. The catalysts prepared in ImILs show better catalytic performance than those prepared by classical methods, probably due to the control of the crystallinity degree of the nanomaterial and in consequence the control of their electronic structure [76], in agreement with the morphological and structural studies carried out by Zheng and co-workers concerning the assisted synthesis of titanium oxide nanoparticles in water/[C_1C_2Im][Br] mixtures [51] (see above, Section 5.2.1.5).

5.3.2.2 **Metallic Nanoparticles of Groups 8–9**

5.3.2.2.1 **Iron Nanoparticles** Janiak and co-workers reported the preparation of Fe, Ru and Os metallic nanoparticles by thermal or photolytic decomposition of the corresponding metal carbonyls [$Fe_2(CO)_9$], [$Ru_3(CO)_{12}$] and [$Os_3(CO)_{12}$] in [C_1C_4Im][BF_4], giving small and uniform sized MNPs (M = Ru, Os) of mean diameter about 1.5–2.5 nm. FeNPs lead to agglomerates as a result of their magnetic properties [66]. Iron oxide nanoparticles have been synthesized by thermal decomposition of [$Fe(CO)_5$] in an ionic liquid/DMF solvent system in the absence of any extra stabilizer. Several ImILs have been used, observing that [C_1C_8Im][BF_4] can regulate the shape and allow the preparation of anisotropic iron oxide nanoparticles, such as bar-like and wire-like nanoparticles [77]. However, the use of [C_1C_4Im][NTf_2] and oleic acid as oxidant, gives particles of mean diameter size 10.6 ± 1.6 nm, which precipitate in the reaction medium [78]. These results have been extended to the preparation of iron oxide nanorods and nanocubes in [C_1C_4Im][NTf_2] in the presence of various surfactants including oleic acid, oleylamine and 1,2-hexadecanediol. In particular, oleylamine and 1,2-hexadecanediol appear very important for the synthesis of Fe_2O_3 nanorods [79]. 1,2-Hexadecanediol has been used in the presence of a small amount of [C_1C_4Im][BF_4] and oleic acid in dibenzylether, generating mono-dispersed magnetite nanoparticles (d_{mean}: about 6 nm) under microwave heating [80]. Hydroxyapatite-encapsulated γ-Fe_2O_3 nanoparticles have been synthesized and functionalized with various basic ILs via silane chemistry in order to create surface basic sites. The basic magnetic nanoparticles generated in this way have been used as efficient heterogeneous catalysts for the aqueous Knoevenagel condensation reaction of aldehydes with malononitrile under mild conditions (Scheme 5.1). Both the basic magnetic nanoparticle carrier and supported basic IL seem to induce a positive cooperative effect, increasing the

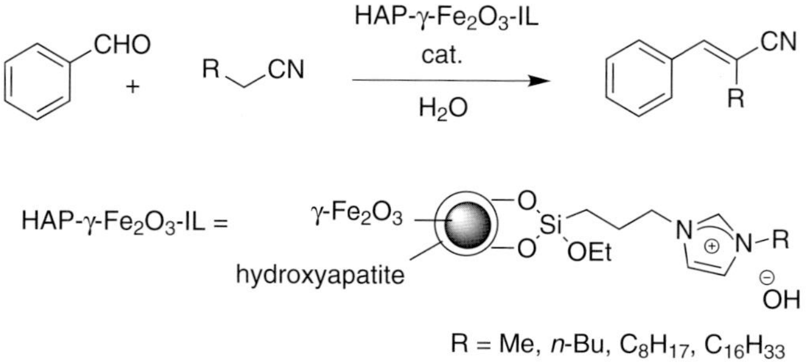

Scheme 5.1 Aqueous Knoevenagel condensation reaction catalyzed by hydroxyapatite-encapsulated γ-Fe_2O_3 nanoparticles using supported basic imidazolium derivatives.

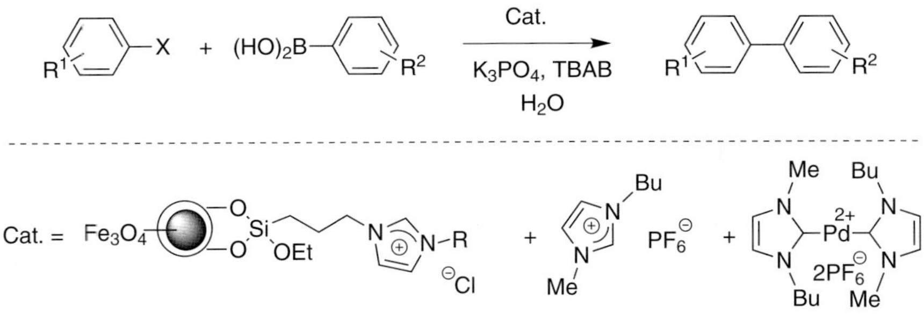

Scheme 5.2 Suzuki-Miyaura cross-coupling reactions catalyzed by Fe_3O_4 nanoparticles-ionic liquid matrix functionalized with *N*-heterocyclic carbene-palladium complex.

catalytic activity; this effect diminishes with the increase of the alkyl chain length of the imidazolium ring. Moreover, the magnetic catalyst can be easily isolated from the catalytic medium by applying an external magnetic field and reused at least four times without loss of catalytic activity [81].

$ZnFe_2O_4$ nanoparticles have been successfully used as photocatalyst for the degradation of phenol. These nanoparticles have been synthesized by microwave-hydrothermal ionic liquid treatment. The ionic liquid nature ($[C_1C_4Im][BF_4]$) and temperature have significant influence on the crystal phase [82].

AuNPs functionalized with iron(II) terpyridine complexes have been applied to develop integrated visible-light photocatalysts for H_2 generation from water, without using methyl viologen as electron relay. Nanoparticles are able to generate microsecond charge separated states in conventional organic solvent such as acetonitrile that even become much longer lived in $[C_1C_4Im][PF_6]$ [83].

An Fe_3O_4 nanoparticles–ionic liquid matrix functionalized with *N*-heterocyclic carbenes have been used as supports for the immobilization of palladium catalysts applied in Suzuki-Miyaura cross-coupling reactions, in water in the presence of tetrabutylammonium bromide (TBAB) as additive to increase the reactants solubility (Scheme 5.2). The catalyst can be recycled five times without significant loss of the catalytic activity after recovery by an external permanent magnet [84].

5.3.2.2.2 **Ruthenium Nanoparticles** Hydrogenation of olefins and arenes catalyzed by RuNPs has received increasing attention [85]. Dupont and co-workers reported in 2004 the preparation of RuNPs (d_{mean}: 2.0–2.5 nm) by hydrogen reduction of RuO_2 hydrate in $[C_1C_4Im][Y]$ ($Y = PF_6$, BF_4, OTf), showing a narrow size distribution. These nanoparticles have been applied in the hydrogenation of 1-hexene, giving high activities (up to a turnover of 320 000, based on the exposed Ru atoms at the metallic surface). RuNPs dispersed in $[C_1C_4Im][PF_6]$ could be reused up to ten times without any significant loss in the catalytic activity [86]. The same ionic liquids have been used in the preparation of RuNPs from decomposition of [Ru(cod)(cot)] (cod = 1,5-cyclooctadiene; cot = cycloocatriene) under hydrogen giving nanoparticles of mean diameter of 57 nm; these RuNPs have been applied in the partial hydrogenation of benzene to cyclohexene. The ternary diagram benzene/cyclohexene/$[C_1C_4Im][PF_6]$ indicates that a maximum of 1% cyclohexene concentration is attained at a 4% benzene concentration in the IL phase. This difference in solubility in the IL can be applied for the extraction of cyclohexene during benzene hydrogenation. A selectivity of about 39% in cyclohexene could be attained at very low benzene conversion. However, the cyclohexene yield and selectivity achieved are too low for industrial applications and are much lower than those obtained by Ru catalysts dispersed in water [87]. The formation of cyclohexene has been also detected at low conversion rates using RuNPs obtained by reduction of ruthenium dioxide under hydrogen pressure in $[C_1C_4Im][Y]$ ($Y = PF_6$, BF_4, OTf) [88].

Rossi and co-workers have reported the formation of stable RuO_2 nanoparticles (d_{mean}: 2–3 nm) with a narrow size distribution, by reaction of $RuCl_3$ with $NaBH_4$ in $[C_1C_4Im][PF_6]$. These small RuNPs show high catalytic activity either in solventless or liquid-liquid biphasic hydrogenation of various olefins and arenes under mild reaction conditions; they could be re-used up to 10 times in the hydrogenation of 1-hexene yielding a total turnover number of 175 000 [89]. The influence of substituents on mono-alkylbenzenes during hydrogenation reaction by RuNPs obtained by controlled decomposition of [Ru(cod)(cot)] dispersed in $[C_1C_4Im][PF_6]$ or by classical heterogeneous catalysts, has been rationalized by Dupont and co-workers. They postulate that the reaction constants for alkyl-substituent substrates can be expressed by steric factors and are independent of any other factors. Thus, bulky alkylbenzene substituents, for both transition MNPs and classical heterogeneous hydrogenation reactions, lower the overall hydrogenation rate, implying a more disturbed transition state compared to the initial state of the hydrogenation [90]. Concerning the effect of the coordination ability of the anion on the formation of RuNPs from $[Ru(2\text{-methylallyl})_2(cod)]$, smaller nanoparticles are obtained in ImILs containing the less coordinating NTf_2 anion than those containing BF_4 anion. These RuNPs could be used for liquid–liquid biphasic hydrogenation of arenes. Nevertheless, it was observed that the presence of two methyl groups on the aromatic ring restricts its hydrogenation, on the same level as one bulky substituent such as *tert*-butyl group, confirming that the liquid-liquid arene hydrogenation catalyzed by RuNPs in ImILs is highly sensitive to the steric hindrance of the aromatic substrate [91]. The relationship between size and catalytic performance of RuNPs generated by the decomposition of [Ru(cod)(cot)] under hydrogen pressure in $[C_1C_4Im][NTf_2]$, has been studied during the catalytic hydrogenation of 1,3-cyclohexadiene and cyclohexene. It has been observed that for the catalytic hydrogenation of both substrates, the catalyst activity increases with RuNPs size. On the contrary, the selectivity in the formation of cyclohexene versus cyclohexane for the hydrogenation of 1,3-cyclohexadiene, drops from 97 to 80% when RuNPs size increases from 1.1 to 2.9 nm. This behavior is in agreement with a mechanism involving a π-bond activation and a double coordination of the diene substrate, requiring neighboring surface atoms only found in facial positions on the larger NPs. These RuNPs could be reused with neither loss of activity nor significant agglomeration [92]. Few reports study the stabilizers influence during the preparation of RuNPs in ILs. Santini, Philippot, Chaudret and co-workers have described the synthesis of small-sized RuNPs (d_{mean}: about 1.2 nm) by

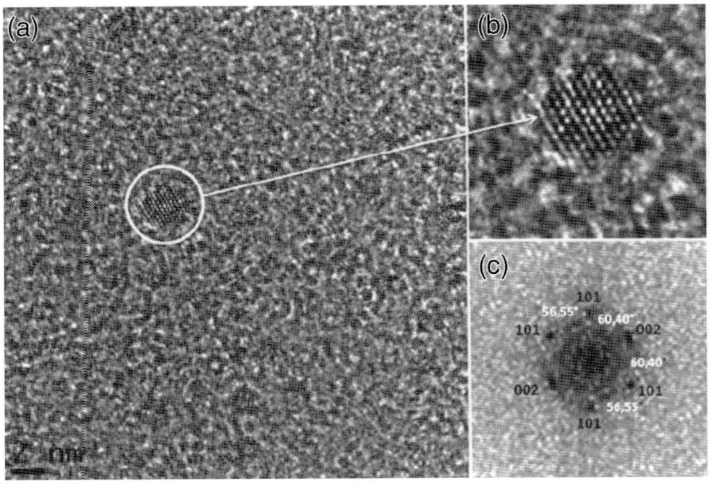

Figure 5.6 TEM analysis of RuNPs stabilized by octylamine in $[C_1C_4Im][NTf_2]$: (a) HRTEM image; (b) HRTEM image of one isolated RuNP with (c) the corresponding FFT image. (Reproduced with permission from [93]. Copyright © (2011) The Royal Society of Chemistry.)

decomposition of [Ru(cod)(cot)] under hydrogen pressure in $[C_1C_nIm][NTf_2]$ ($n = 2$, 4, 6, 8, 10) in the presence of amines as stabilizers, such as 1-octylamine and 1-hexadecylamine (Figure 5.6). ^{13}C and DOSY NMR experiments evidence that amines are coordinated to RuNPs surface. These RuNPs have been tested in toluene hydrogenation, giving moderate activity but without loss of reactivity after recycling (up to ten runs) [93].

Concerning the preparation of supported RuNPs in ILs, Janiak and co-workers have recently reported the synthesis of RuNPs supported on chemically modified graphene, from $[Ru_3(CO)_{12}]$ in $[C_1C_4Im][BF_4]$. Catalytic applications in cyclohexene and benzene hydrogenation give full conversion of the corresponding substrate to cyclohexane (with activities exhibiting up to $1570\,h^{-1}$ for cyclohexene hydrogenation) [94].

An example of asymmetric hydrogenation of aromatic ketones using preformed TPPTS-stabilized RuNPs (d_{mean}: 5 nm) in $[C_1C_nIm][TolSO_3]$ ($n = 2$, 4, 8, 12, 16) in the presence of the chiral diamine (1*R*,2*R*)-DPENDS, has been reported by Chen and co-workers (Scheme 5.3). They observed that the conversion and enantioselectivity decrease gradually when increasing the alkyl chain length. This fact could be probably due to the solubility decrease in IL of the water-soluble ligand, which agrees with the enantioselectivity increase when water is added. Moreover, the catalyst immobilized in the ionic liquid could be reused seven times without any significant enantioselectivity and activity loss [95].

Nitrile-functionalized IL $[C_1RIm][NTf_2]$ (R= 1-butyronitrile) led to small-sized RuNPs (d_{mean}: 2.2 ± 0.5 nm) from reduction of $[Ru(2\text{-methylallyl})_2(cod)]$ under hydrogen pressure. This catalytic system has been applied in the hydrogenation of nitrile containing aromatic compounds, giving selectively the corresponding imine derivatives. The nitrile groups in the IL side-chain coordinate to the metallic

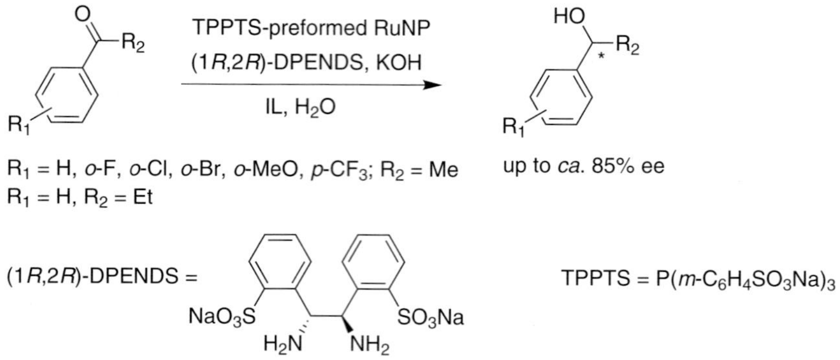

Scheme 5.3 Asymmetric hydrogenation of aromatic ketones catalyzed by preformed RuNPs in the presence of an optically pure inductor, (1*R*,2*R*)-DPENDS.

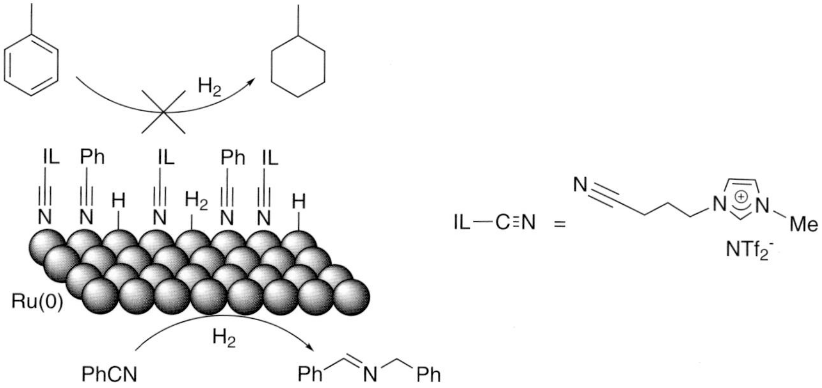

Scheme 5.4 Model depicting the selective hydrogenation of the nitrile function in aromatic derivatives promoted by RuNPs.

surface, hindering the approach of arenes and their further hydrogenation (Scheme 5.4) [96].

The solubility of biomass derived substrates in ILs allows ruthenium-catalyzed hydrogenation reactions, such as selective hydrogenation of furfuryl derivates [97], conversion of cellobiose to C_6-alcohols [98], conversion of cellulose into ethylene glycol [99] and hydrodeoxygenation of lignin-derived phenol following a tandem hydrogenation and dehydration process (Scheme 5.5) [100].

5.3.2.2.3 **Cobalt Nanoparticles** Very few reports have been published concerning the preparation of cobalt nanoparticles (CoNPs) in ILs and their applications in catalytic processes [101]. Janiak and co-workers reported the preparation of CoNPs in $[C_1C_4Im][BF_4]$ (d_{mean}: 14 ± 8 nm) and their evaluation in cyclohexene hydrogenation to cyclohexane, giving a poor catalytic activity [61c]. Dupont and

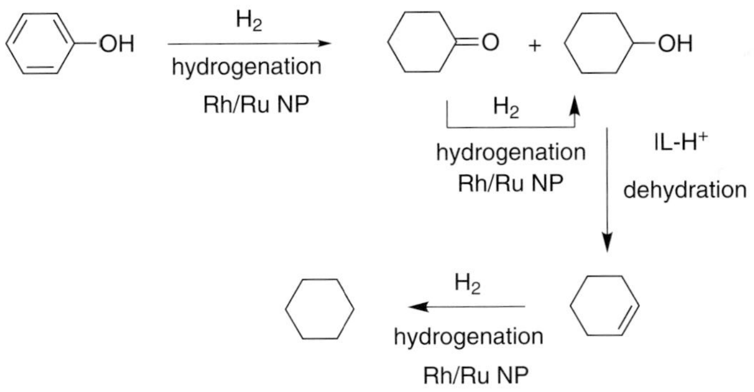

Scheme 5.5 Lignin-derived phenol in a tandem hydrogenation-dehydration process in $[C_1C_4Im][NTf_2]$.

co-workers have demonstrated that the thermal decomposition of $[Co_2(CO)_8]$ in $[C_1C_{10}Im][NTf_2]$ preferentially affords ligand-free cobalt nanocubes or spherical CoNPs basically depending on the type of the IL. The isolated CoNPs prepared in $[C_1C_{10}Im][NTf_2]$ are active catalysts in the Fischer–Tropsch synthesis, mainly affording hydrocarbons (8–26 carbon atoms with a mono-modal hydrocarbon distribution centered at C_{12}) in the liquid phase [102]. However, when CoNPs prepared in $[C_1C_4Im][NTf_2]$ were tested, a bimodal distribution centered at C_{12} and C_{21} was then obtained [103]. These different results permits to conclude that the different nanoparticle-size distribution (7.7 nm in $[C_1C_4Im][NTf_2]$ and 79 nm in $[C_1C_{10}Im][NTf_2]$) and also the structure of CoNPs (fcc arrangement in $[C_1C_4Im][NTf_2]$ and ε arrangement in $[C_1C_{10}Im][NTf_2]$) are directly related to the catalytic behavior.

5.3.2.2.4 **Rhodium and Iridium Nanoparticles** The major catalytic applications of rhodium nanoparticles (RhNPs) and IrNPs in ILs concern arenes and alkenes hydrogenation. Small RhNPs [104] and IrNPs [105] of 2–3 nm mean diameter have been reported by Dupont and co-workers in ImILs [90]. Mechanistic studies concerning the IrNPs formation in the presence of 1-decene prove an autocatalytic pathway [24, 106]. In contrast to IrNPs, preformed RhNPs exhibit significant agglomeration during the hydrogenation process, leading to a noteworthy loss of activity when the catalyst is recycled. A similar trend has been observed during the preparation of RhNPs by thermal decomposition of $[Rh_6(CO)_{16}]$ in $[C_1C_4Im][BF_4]$ or trifluoromethane sulfonate and in butyl-trimetylammonium trifluoromethane sulfonate [61c]. Evans' group has reported the recycling of RhNPs stabilized by poly (*N*-vinyl-2-pyrrolidone) (PVP) in $[C_1C_4Im][PF_6]$ in the hydrogenation of different substrates (1-alkenes, cyclohexene, benzene) [107]. The use of the ionic-liquid-like copolymer poly[(*N*-vinyl-2-pyrrolidone)-*co*-(1-vinyl-3-butylimidazolium chloride) as stabilizer of RhNPs (d_{mean}: 3 nm), leads to active catalysts for arenes hydrogenation (up to 20 000 of TTO for benzene) [108]; this approach has been improved combining hydroxyl-functionalized ILs and PVP, providing an efficient catalytic system for biphasic hydrogenation reactions [109]. Similar catalytic behavior was observed for

RhNPs (d_{mean}: 2.9 nm) stabilized by (*N*-vinyl-2-pyrrolidone)-*co*-(1-vinyl-3-butylimidazolium) chloride copolymer in $[C_1C_4Im][BF_4]$ for benzene hydrogenation [110].

Deposition of an IL phase containing RhNPs in $[C_1C_4Im][BF_4]$ (d_{mean}: 4 nm) on a silica network has led to efficient catalysts for hydrogenation of cyclohexene and 1-decene [111]. RhNPs in $[C_1C_4Im][NTf_2]$ dispersed with a sirup of cellulose acetate led to catalytically active membranes for cyclohexene hydrogenation [112].

The influence of additives, in particular poly-*N*-donor ligands on the stabilization of RhNPs (d_{mean}: 2–5 nm) in ILs, has been reported by Roucoux and co-workers. RhNPs prepared in the presence of ligands such as 2,2′-, 3,3′- 4,4′-bipyridines, 2,4,6-tri(2-pyridyl)-s-triazine (TPTZ) and tetra-2-pyridinylpyrazine (TPPZ) in various ILs, have been tested in the hydrogenation of arenes (Figure 5.7) [113]. For bipyridine derivatives, the best reactivity is obtained with 3,3′- and 4,4′-substituted ligands giving full conversion of styrene into ethylcyclohexane. The differences observed in terms of activity can be explained when a monodentate coordination mode is plausible, as for 4,4′-bipyridine: the steric hindrance decrease on the nanoparticle surface favors the substrate access, giving an efficient aromatic ring hydrogenation. The use of merely pyridine as a protective ligand is enough to avoid RhNP agglomeration, giving catalytically active RhNPs. An hybrid imidazolium-monofunctionalized bipyridine ligand has been conceived as protective agent for the stabilization of RhNPs in $[C_1C_4Im][PF_6]$, giving rise to an efficient and stable catalyst for arenes hydrogenation [114].

The same group has described the direct phase transfer of RhNPs from an aqueous solution to an ionic liquid phase based on surfactant counter anion exchange. This phase transfer was carried out by addition of $LiNTf_2$; the so-obtained RhNPs in IL (average size of 2.8 nm) are efficient in styrene hydrogenation, exclusively affording ethylbenzene [115].

Dupont and co-workers synthesized RhNPs in $[C_1C_4Im][BF_4]$ to be applied in hydroformylation of 1-alkenes under solvent-less conditions [116]; linear aldehydes were obtained using RhNPs (d_{mean}: 5 nm), giving up linear/branched ratio of 25 when Xantphos (9,9-dimethyl-4,5-bis(diphenylphosphino)xanthene) is used as

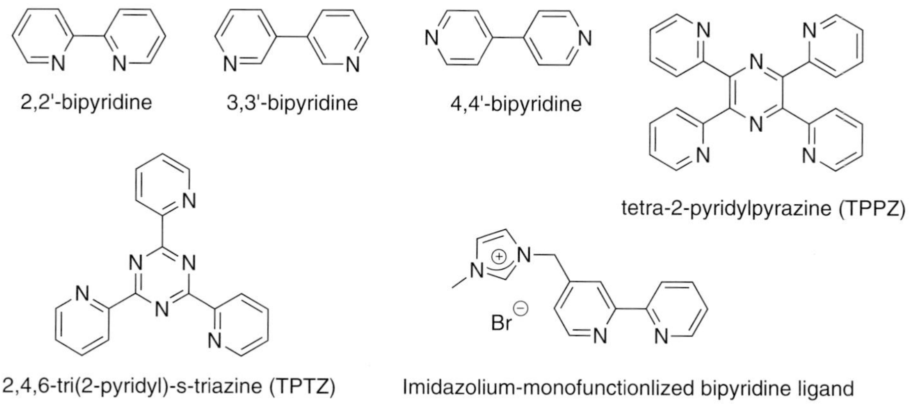

Figure 5.7 Poly-*N*-donor ligands used in the stabilization of RhNPs in ImILs.

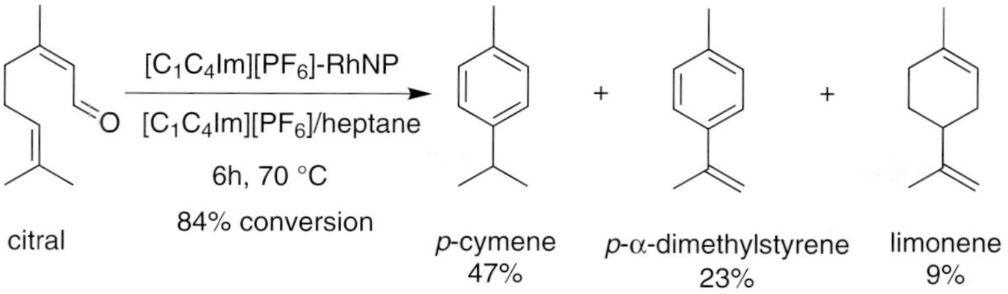

Scheme 5.6 IL-stabilized RhNPs applied in citral cyclodehydration.

ligand. Nanoparticles of less than 5 nm size afforded lower chemoselectivity than that observed with RhNPs displaying 5 nm size. In opposite, larger sized nanoparticles (about 15 nm) produced only small amounts of aldehydes similar to that observed with a classical Rh/C heterogeneous catalyst. This behavior permits to conclude that RhNPs are probably degraded under reaction conditions giving catalytically active mononuclear Rh-carbonyl species, in agreement with both the induction period observed in the reactions using RhNPs as catalytic precursors and different analyses [TEM, XRD, infrared (IR) and NMR].

$[C_1C_4Im][PF_6]$-stabilized RhNPs have been applied in the chemoselective citral cyclodehydration giving interesting starting compounds for the production of fine chemicals (Scheme 5.6) [117]. Under optimized conditions, mainly *p*-cymene and *p*-α-dimethylstyrene were obtained, together with limonene and other by-products.

IrNPs (d_{mean}: 2.3 ± 0.4 nm) generated by reduction of $[IrCl(cod)]_2$ in $[C_1C_4Im]$ $[PF_6]$ and then isolated by centrifugation, were tested in solvent-less or biphasic hydrogenation of ketones [118]. The reactions performed with IrNPs dispersed in $[C_1C_4Im][PF_6]$ took longer time than those performed under solvent-less conditions. The recovered ionic liquid catalytic dispersion could be reused at least 15 times without any significant activity loss. On the other hand, under solvent-less conditions, IrNPs started to lose their efficiency after the third recycle, pointing to that the IL provides a beneficial stabilization effect. Moreover, this catalytic system is selective towards the C=C double bond reduction of α,β-unsaturated cyclic ketones, such as cyclohexenone and isophorone, and the arene ring reduction over the carbonyl group of acetophenone.

IrNPs stabilized by trihexyltetradecylphosphonium methylsulfonate (d_{mean}: 3.5 nm) were obtained from the reduction of the air-stable hydrido-iridium carborane complex, $[Ir(H)(7,8\text{-}nido\text{-}C_2B_9H_{11})(PPh_3)_2]$, under hydrogen atmosphere [119]. The IrNPs prepared in this way were used as catalytic precursors for the borylation reaction starting with benzene derivatives and pinacolborane to form boric acids under microwave irradiation. Although this nanocatalyst is less active than Ir-based homogeneous catalysts, it could be recycled more than six times without significant activity loss.

In relation to the true catalytic nature when IrNPs are involved in hydrogenation processes, the work of Finke and co-workers concerning the hydrogenation of acetone catalyzed by IrNPs reveals that ImILs can poison the catalyst [120]. One of the main findings of this work proves that the addition of $[C_1C_4Im][PF_6]$ inhibits the formation of iridium nanoclusters from $[IrCl(cod)]_2$. In fact the added IL poisons the nuclei, which in turn lead to the formation of catalytically active nanoparticles.

5.3.2.3 **Metallic Nanoparticles of Group 10**

5.3.2.3.1 **Nickel Nanocatalysts: Hydrogenation Processes** The synthesis and full characterization of NiNPs in neat ImILs has been mostly carried out by the group of J. Dupont, through decomposition of $[Ni(cod)_2]$ under hydrogen atmosphere [34, 121]. This research has proved that $[C_1C_nIm][NTf_2]$ ($n = 4, 8, 10, 14, 16$) are templates for the preparation of NiNPs, showing a slight influence of the alkyl chain length of the IL in the diameter and size distribution of NiNPs obtained. SAXS analysis evidences an increase of range order with the increase of the number of carbon atoms of the alkyl side-chain, except for the hexadecyl derivative; the smaller nanoparticles (d_{mean}: 4.9 nm) are obtained in the more structural organized IL. EXAFS demonstrate the presence of a NiO phase, probably capping the Ni(0) core of the nanoparticles, and also the existence of a medium range order around the nanoparticles. NiNPs prepared in $[C_1C_4Im]$-based ILs efficiently catalyzed the hydrogenation of olefins. Thus, their catalytic activity for the cyclohexene hydrogenation is two orders of magnitude higher than that obtained using classical supported Ni catalyst [121a].

Hou and co-workers have recently prepared well-defined NiNPs in ethanol medium by reduction of nickel(II) acetate with hydrazine in the presence of amino-functionalized imidazolium ILs, $[C_1C_1RIm][Y]$ (R = 3-aminopropyl; Y = Br, OAc) [122]. Structural characterization of these materials showed that the NH_2-functionalized imidazolium derivative is adsorbed at the metallic surface, avoiding the metal oxidation as evidenced by XRD and XPS analyses. After isolation, these solid materials could be re-dispersed in water, giving high activities for the hydrogenation of several alkenes [122a] (cyclohexene, 2-propen-1-ol, styrene . . .) and allowing the catalyst recyclability (up to seven runs without significant activity loss). In particular, the selective hydrogenation of substituted nitro aromatic compounds preferentially led to the formation of the corresponding anilines [122b] (Scheme 5.7).

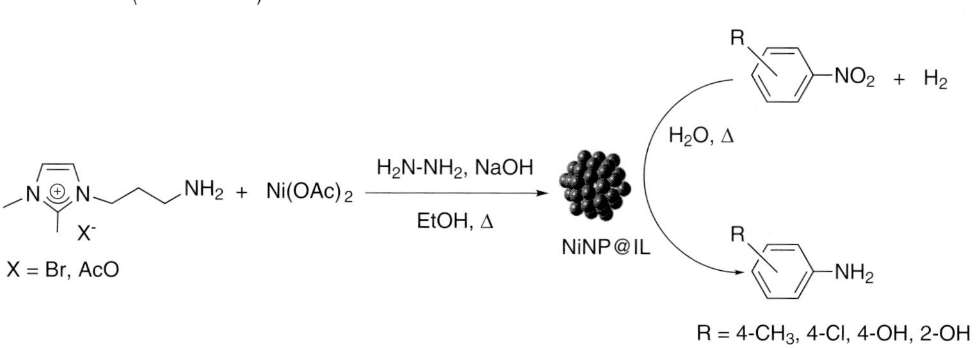

Scheme 5.7 Synthesis of NiNPs stabilized by 1-(3-aminopropyl)-2,3-dimethylimidazolium ILs and their catalytic applications in hydrogenation of nitro aromatic compounds.

5.3.2.3.2 **Palladium Nanoparticles** Palladium-catalyzed reactions have become an important tool in organic synthesis [123], owing to their high efficiency, selectivity and their ability to perform a range of interesting transformations [124]. In the last decade, the use of MNPs in catalysis has exploded and more than 40% of the reported works concerns PdNPs [125]. The immobilization of MNPs in ILs and on solid supports (inorganic oxides, polymers, magnetic nanoparticles, carbon nanotubes) turns into useful means to increase the stability of metal dispersions, in particular interesting for applications in catalyzed multi-step processes. PdNPs are one of the most used and efficient catalysts making C—C bonds and also other chemical transformations such as carbon-heteroatom bond formation, hydrogenation, carbonylation and oxidation processes. The most relevant contributions of PdNPs in ILs are next overviewed, mainly in C—C bond formation and hydrogenation processes.

5.3.2.3.2.1 **Carbon—Carbon Bond Formation Processes** Suzuki–Miyaura cross-couplings of aryl bromides and chlorides have been described by Caló and co-workers in the presence of PdNPs stabilized in quaternary ammonium salts under rather mild conditions [126]. PdNPs were formed by chemical reduction of $Pd(OAc)_2$ in the presence of the corresponding ammonium salts (d_{mean}: 3.3 nm). These nanocatalysts gave high yields (up to 90%), in particular using tetraheptylammonium bromide.

Gómez and co-workers have prepared ligand-free PdNPs in ImILs, giving star-like shaped interparticles organization (Scheme 5.8). These catalytic systems showed high activity in Suzuki-Miyaura couplings, allowing the recycling of the catalytic phase up to ten times without activity loss [127].

This group also described the use of different functionalized ligands derived from norborn-5-ene-2,3-dicarboxylic anhydride for the stabilization of Pd catalysts used in Suzuki-Miyaura reactions in $[C_1C_4Im][PF_6]$ [128] (Scheme 5.9). High conversions and yields were obtained when amine ligands containing hydroxyl groups were involved. TEM analyses after catalysis showed the formation of small particles (d_{mean}: 8 nm), in contrast to the agglomerates observed when particles were preformed, with the consequent decrease in the catalytic activity. Different tests,

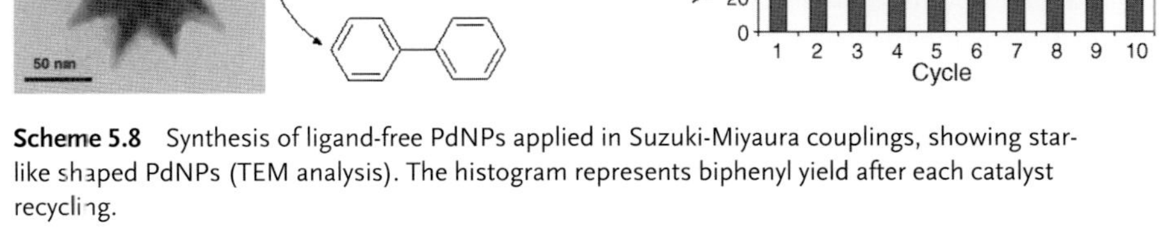

Scheme 5.8 Synthesis of ligand-free PdNPs applied in Suzuki-Miyaura couplings, showing star-like shaped PdNPs (TEM analysis). The histogram represents biphenyl yield after each catalyst recycling.

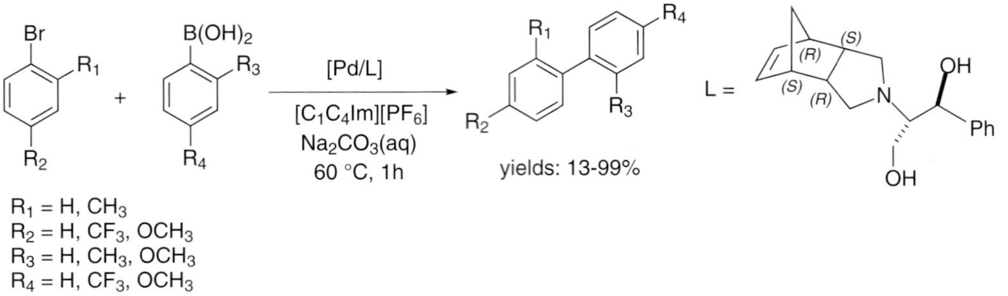

Scheme 5.9 PdNPs catalyzed Suzuki-Miyaura cross-couplings for the synthesis of biaryl derivatives.

including the correlation effect between solvent and ligand leading to molecular or colloidal catalysts, were carried out with the aim to identify the true nature of the catalyst, concluding that PdNPs formation is required to give a catalytic active system in Suzuki-Miyaura couplings, in agreement with the results obtained in organic solvents [129].

Since the pioneer work of Jeffery [130], Herrmann [131], Reetz [132] and Buchwald [133] reporting that the addition of ammonium salts leads to higher catalytic activities in the absence of ligand, the development of PdNP-catalyzed Heck–Mizoroki reaction has triggered much interest in the scientific community. Caló and co-workers have reported PdNP-catalyzed Heck–Mizoroki arylation of 1,1-disubstituted alkenes in ILs, studying different parameters affecting the regioselectivity of the coupling process [134]. They observed that the Heck-Mizoroki reaction of neutral or electron-rich aryl bromides with the 1,1-disubstituted olefins butyl methacrylate and α-methylstyrene catalyzed by PdNPs in a mixture of TBAB and acetate, as solvent and base respectively, led to a predominant formation of the terminal olefin, while the reaction of *p*-bromoacetophenone led to the internal one. The terminal olefins were efficiently converted into the more stable internal *E* isomers using tetrabutylammonium pivalate as catalyst. Moreover, this group reported the PdNP-catalyzed Heck-Mizoroki reaction of *trans*-cinnamates in TBAB as solvent [135]. PdNPs were formed from $Pd(OAc)_2$ dissolved in TBAB followed by the addition of tetrabutylammonium acetate.

Heck–Mizoroki reactions of aryl chlorides catalyzed by ligand-free palladium acetate in tetraalkylammonium ILs under aerobic and relatively mild conditions, have been reported by the same group [136]. This catalytic system appeared particularly very efficient in a one-pot sequential coupling of aryl dihalides to produce unsymmetrically disubstituted arenes using 1-bromo-4-chlorobenzene and two different olefins, selectively activating the C—Br and C—Cl bonds on the aromatic ring at two different temperatures (373 and 393 K) (Scheme 5.10).

The design of a task-specific ionic liquid (TSIL), based on an ethanolamine-functionalized quaternary ammonium salt, namely 4-di(hydroxyethyl) aminobutyl tributylammonium bromide, has been described by Wang *et al.* [137]. PdNPs, *in situ* formed from palladium acetate as evidenced by TEM analysis (d_{mean}: 4 nm), exhibited a high activity and recyclability for Heck–Mizoroki reaction (Scheme 5.11).

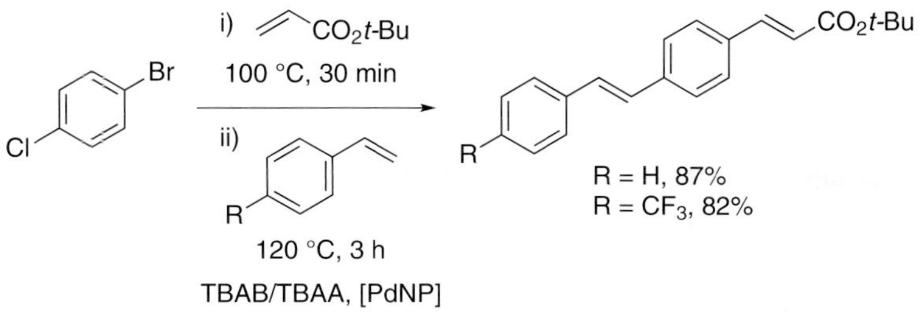

Scheme 5.10 PdNPs catalyzed a one-pot sequential Heck–Mizoroki coupling of aryl dihalides.

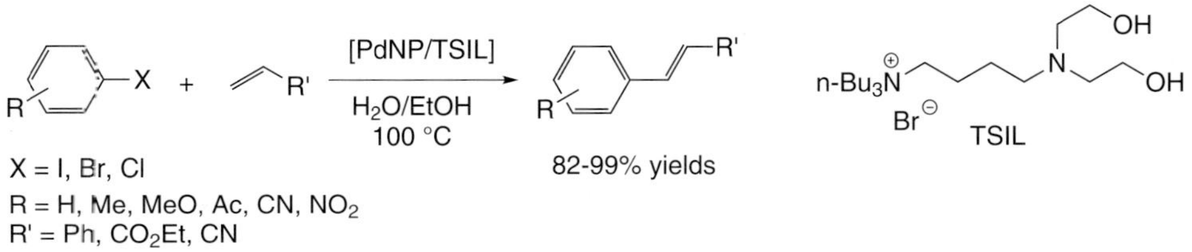

Scheme 5.11 Task-specific ionic liquid in Heck–Mizoroki reaction catalyzed by PdNPs.

Heck-Mizoroki olefinations of iodo-, bromo- and chloro-arenes gave the corresponding cross-coupling products in good to excellent yields. It is noteworthy that the catalytic phase could be recycled up to six times without significant activity loss.

PdNPs generated *in situ* in a microemulsion formed by water, $[C_1C_4Im][PF_6]$ and a surfactant (TX-100 = polyethylene glycol p-(1,1,3,3-tetramethylbutyl)-phenyl ether), which serves as reductant and stabilizer of the nanoparticles, are efficient catalysts in Heck-Mizoroki coupling reactions, allowing the catalyst recycling up to four times without significant activity loss [138].

Dyson and co-workers have developed nitrile-functionalized pyridinium [139a], imidazolium [139b] and pyrrolidinium [139c] ILs showing their benefit effect in C—C cross-coupling reactions in comparison with nonfunctionalized ILs, including desulfitative Heck–Mizoroki couplings of sulfonyl derivatives with alkenes [140]. The formation of PdNPs *in situ* during the catalytic process points to their role as catalyst reservoirs.

Hardacre and co-workers have investigated by XAFS the nature of palladium species generated in ILs during the Heck-Mizoroki coupling between iodobenzene and butylacrylate. These analyses indicate that nanoclusters of 0.8–1.6 nm diameter represent the main species formed during the catalytic process in different ILs (ImILs, ammonium derivatives and pyridinium salts) [141].

The immobilization of MNPs on functionalized polymers is one of the strategies followed to improve the catalyst reuse. Han and co-workers have prepared a divinylbenzene crosslinked polymer chemically modified by a functionalized ImIL, which has proved its efficiency for PdNP immobilization (Scheme 5.12).

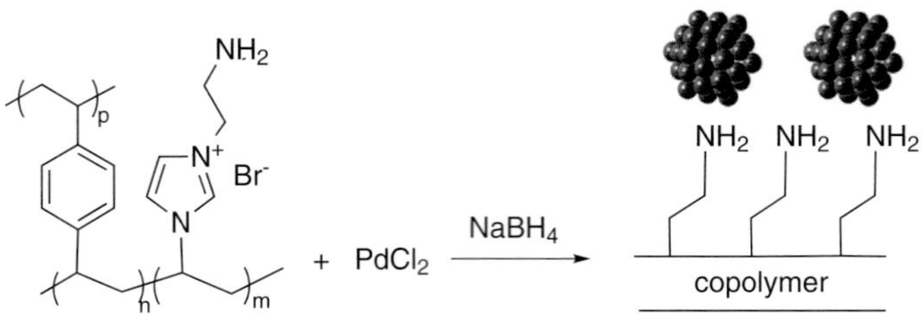

Scheme 5.12 Preparation of PdNPs immobilized in a divinylbenzene crosslinked polymer modified by an ImIL (black metallic clusters represent PdNPs).

This catalytic material is active for Heck-Mizoroki coupling reactions and recyclable for three consecutive runs [142, 143].

Sonogashira reactions catalyzed by PdNPs formed *in situ* at ambient conditions in acetone or $[C_4C_4Im][BF_4]$ as solvent under ultrasound irradiation have been reported [144]. TEM analysis showed the formation of polydispersed PdNPs varying in diameter from 3 to 8 nm using acetone and in the range of 10–20 nm using the IL. Reactions were significantly faster in acetone as compared to those in ILs, but reactions in ILs afforded the expected products in relatively higher yields and the IL catalytic phase could be recycled five times for the reaction of iodobenzene with ethynylbenzene with only a slight decrease in catalytic activity. No Glaser homo-coupling product was observed for all the reactions involving iodoarenes, however the homo-coupling product arising out of the terminal acetylene could be observed using less reactive bromoarenes (up to 7%). In addition, it could be demonstrated that ultrasound not only generated PdNPs but also enhanced their catalytic activity in the Sonogashira cross-coupling reaction (Scheme 5.13).

A room-temperature catalytic alternative to the Ullmann reaction has been recently reported via electro-reductive homo-coupling of haloarenes catalyzed by PdNPs [145]. This approach combines electrochemical reduction and PdNP catalysts in $[C_1C_8Im][BF_4]$. The use of the IL as solvent exhibits two main advantages: good conductor of electricity and stabilizer of PdNPs, avoiding the use of additional electrolytes and stabilizers. A variety of *p*-substituted aryl halides has been tested. Aryl iodides gave good yields with the exception of the electron-donating iodoaniline. Aryl bromides were also active, but surprisingly low conversion was observed for the electron-withdrawing bromobenzotrifluoride.

An efficient PdNP-catalyzed Ullmann-type reductive homo-coupling of aryl, vinyl and heteroaryl halides promoted by an aldehyde in tetraalkylammonium ILs under

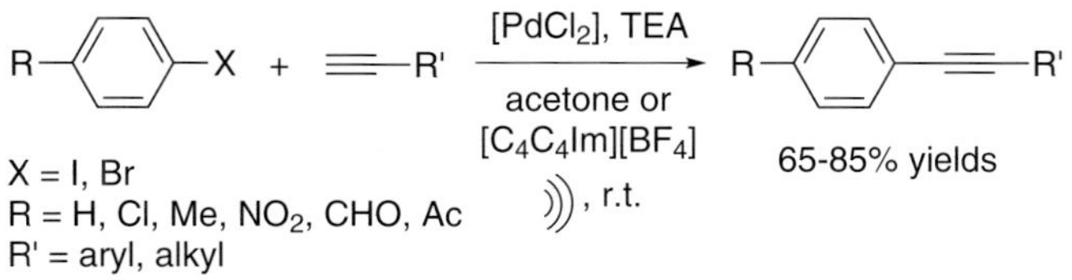

Scheme 5.13 Sonogashira C—C couplings catalyzed by PdNPs under ultrasound irradiation.

mild reaction conditions has been described by Caló *et al.* [146]. More recently, Xu and co-workers have developed a new catalyst for Ullmann homo-coupling of aryl chloride derivatives, based on PdNPs supported on graphene oxide which shows a good dispersion in [C_1RIm][HSO_4] (R = sulfobutyl). The biphasic IL/scCO_2 catalytic systems not only facilitate the products extraction and the catalyst recycling, but also play a kinetic role increasing the catalyst activity [147].

Ranu and co-workers have reported the synthesis of secondary and tertiary amines by allylic amination processes using palladium salts in the presence of ionic stabilizers such as TBAB; PdNPs were formed *in situ* during the catalytic reaction, pointing to be the catalytic active species [148].

5.3.2.3.2.2 **Hydrogenation Processes** The group of J. Dupont has demonstrated that PdNPs in ImILs can selectively hydrogenate alkenes and alkynes [25a]. Therefore, preformed PdNPs in [C_1C_4Im][BF_4] (d_{mean}: 4.9 nm) were applied in the hydrogenation of 1,3-butadiene, mainly giving 1-butene with a selectivity up to 97%; the formation of butane and *cis*-2-butene remained marginal. This selectivity points to a pronounced surface-like reactivity rather than single site catalytic behavior. The same group has recently proved that bigger PdNPs (d_{mean}: 7.3 nm), stabilized by [C_1RIm][NTf_2] (R = 1-butyronitrile), can induce different selectivity in the hydrogenation of alkynes depending on the catalytic conditions. Then, at low hydrogen pressure (1 bar) the catalyst exclusively gives the formation of the corresponding *Z*-alkene, while at higher pressure (4 bar), full hydrogenation is observed. The catalyst could be recycled at least four times without activity loss, proving its robustness under catalytic conditions, which can be favored by the interaction of the nitrile groups with the metallic surface (Scheme 5.14) [149].

Mono- and bi-metallic PdAuNPs, solubilized in [C_1C_4Im][PF_6] in the presence of PVP as stabilizer, have been studied in the hydrogenation of allyl alcohol, 1,3-cyclooctadiene, *trans*-cinnamaldehyde and 3-hexyn-1-ol [150]. In each case, the activity and selectivity could be tuned in relation to the composition of the bimetallic NPs, observing that particles containing high Pd proportion gave the highest activity. Monometallic PdNPs showed lower activity than bimetallic ones, while pure AuNPs were inactive.

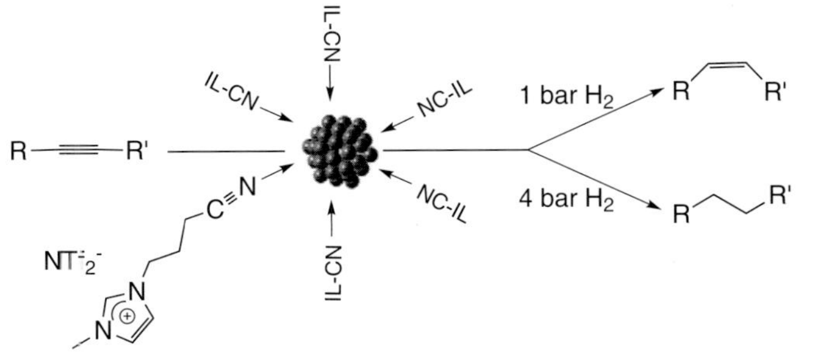

Scheme 5.14 PdNPs-catalyzed selective hydrogenation of alkynes (IL-CN = [C_1RIm][NTf_2] (R = 1-butyronitrile); black cluster corresponding to PdNPs).

The functionalized IL $[C_1C_1RIm][PF_6]$ (R = *N*,*N*-bis(2-pyridyl)-propylamido) efficiently stabilizes PdNPs (d_{mean}: 5–6 nm) [151]. This system is very active and selective towards the reduction of conjugated C=C bonds of cinnamaldehyde and citral, leading to 64% of hydrocinnamaldehyde and 93% of citronellal respectively. When 2,2′-dipyridylamine is used, the efficiency of the corresponding catalyst is lower than that obtained using PdNPs only stabilized by the IL, probably due to the strong coordination ability of this amine ligand. From this study, it appears that ILs are able to stabilize MNPs enhancing the activity and durability of the catalyst, without poisoning the metallic surface of nanoparticles.

PdNPs formed *in situ* from palladium salts in $[C_1C_4Im][NTf_2]$ and in the presence of silica aerogels have been prepared and applied in the hydrogenation of cinnamaldehyde [152]. Full substrate conversion is achieved under mild conditions, only giving hydrocinnamaldehyde. Upon catalyst recycling, the selectivity towards hydrocinnamaldehyde remained unchanged, even if the rate lowered (0.47 mM s^{-1} versus 0.28 mM s^{-1}).

PdNPs were deposited onto imidazolium bromide-functionalized ionic multi-walled carbon nanotube (MWCNTs) through hydrogen reduction of $Na_2[PdCl_4]$ in water [153]. Direct anion-exchange of the hydrophilic bromide anion with NTf_2^- and SbF_6^- salts led to the corresponding hydrophobic PdNP@IL-MWCNT, PdNP@IL(NTf_2)-MWCNT and PdNP@IL(SbF_6)-MWCNT. The combination of PdNPs supported onto ionic CNTs with ILs generates a new robust and recyclable ionic liquid-based system (up to 50 runs), active in the hydrogenation of *trans*-stilbene and related olefins (Scheme 5.15).

Small PdNPs (d_{mean}: about 1 nm) supported on functionalized IL silica have proved their efficiency in the hydrogenation of nitrobenzene derivatives to give the corresponding arylamines under solvent-free or neat water conditions [154].

The group of B. Feng has reported PdNPs formed in 2,2′-dipyridylamine-functionalized ImILs (d_{mean}: 5–6 nm) as efficient catalysts for olefins hydrogenation (cyclohexene, styrene, ethyl acrylate) [155]. They also described PdNPs stabilized by a poly(ethylene glycol)-functionalized dicationic ImIL. This amphiphilic IL forms micelles in water, both stabilizing the nanoparticles and promoting catalytic hydrogenation of styrene; the catalyst could be recycled at least eight times without activity loss [156].

5.3.2.3.2.3 **Dual Catalytic Behavior of Palladium Nanoparticles** The dual catalytic behavior of MNPs (as a reservoir of homogeneous species and surface-like reactivity, depending on the reaction conditions) becomes notably significant in an

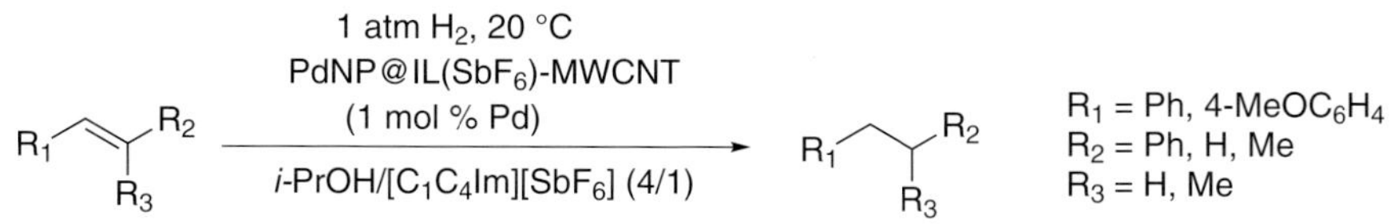

Scheme 5.15 Hydrogenation reaction of *trans*-stilbene derivatives using PdNP@IL(SbF_6)-MWCNTs as catalysts.

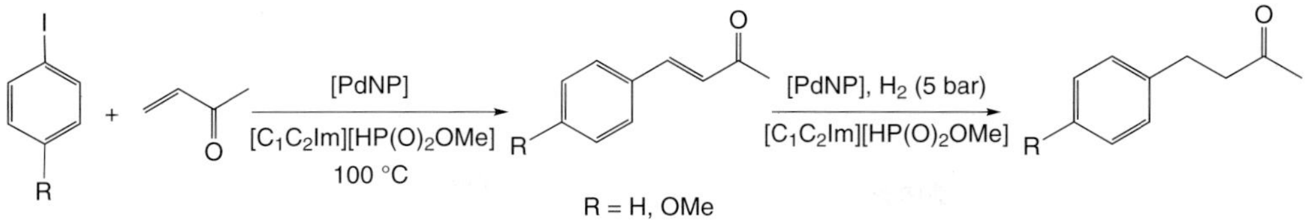

Scheme 5.16 Heck-Mizoroki/hydrogenation sequential process catalyzed by preformed PdNPs in $[C_1C_2Im][HP(O)_2OMe]$.

ionic liquid medium due to the enhanced stability of the nanoclusters. The group of M. Gómez has demonstrated the interest of such catalytic behavior by studying preformed ligand-free PdNPs in ImILs in a sequential process involving two benchmark reactions: Heck-Mizoroki coupling followed by hydrogenation reaction (Scheme 5.16) [157].

The use of $[C_1C_2Im][HP(O)_2OMe]$ as the ionic liquid enables PdNPs to regenerate after the C–C coupling reaction (the catalyst acts as a reservoir of molecular species) thanks to the low basicity of methyl hydrogenophosphonate anion, which prevents the formation of robust molecular Pd–carbene species by deprotonation of the imidazolium cation. The regenerated PdNPs exhibit a surface-like reactivity, allowing the hydrogenation of the olefinic bond.

The interest of researchers in using carbon nanotubes as catalysts supports has recently increased, mainly due to their mesoporous nature overcoming mass transfer limitations and to their well-defined and tunable structure allowing easily controlled modifications [158]. Gómez, Serp and co-workers have efficiently immobilized preformed PdNPs (prepared by the decomposition of $[Pd_2(dba)_3]$ under dihydrogen pressure and in the presence of 4-(3-phenylpropyl)pyridine as ligand) in ImILs, $[C_1C_4Im][PF_6]$ and $[C_1C_2Im][HP(O)_2(OMe)]$ on imidazolium-functionalized MWCNTs [159]. The PdNPs in $[C_1C_2Im][HP(O)_2(OMe)]$ exhibit a good dispersion, showing a mean diameter of 4.5 nm (Figure 5.8). Upon their immobilization on the functionalized MWCNTs [8], the catalytic materials were applied in the sequential Heck-Mizoroki/hydrogenation process described above (Scheme 5.17), being recycled up to four times without significant activity loss.

5.3.2.3.3 **Platinum Nanocatalysts: Hydrogenation and Oxidation Processes** Preformed PtNPs in ILs have found applications in both hydrogenation of unsaturated groups and oxidation reactions. The first PtNPs only stabilized by ILs were described by Dupont and co-workers, by decomposition of the organometallic precursor $[Pt_2(dba)_3]$ (where dba = dibenzylidenacetone) under hydrogen pressure using $[C_1C_4Im][PF_6]$ as solvent [160]. This methodology led to small (d_{mean}: 2–3 nm) and well-dispersed PtNPs showing a narrow size-distribution. These PtNPs were efficient catalysts for the hydrogenation of C=C bonds of olefins and aromatic compounds, such as benzene and toluene, giving the corresponding full hydrogenated compounds; the activities were higher than those exhibited by the classical PtO_2 Adams's catalyst. Structural characterization of those PtNPs stabilized by ImILs by SAXS analysis evidenced the formation of a semi-organized IL

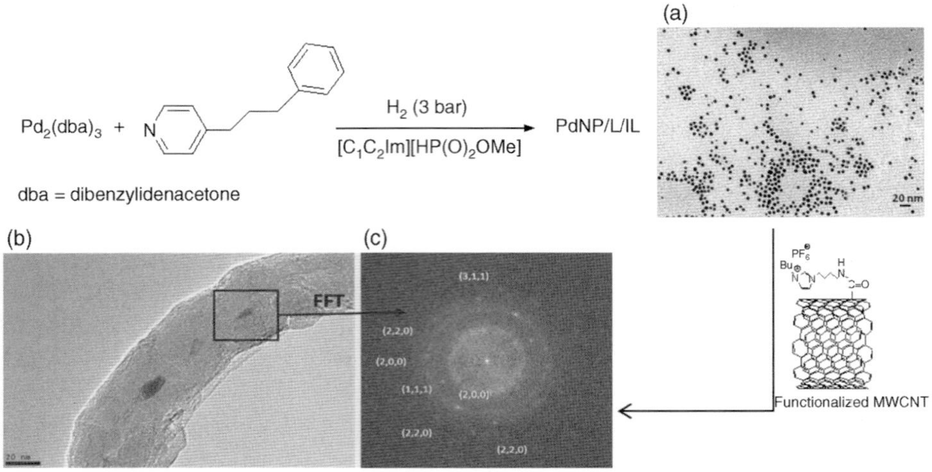

Figure 5.8 Synthesis of PdNPs in the presence of a pyridine derivative ligand in IL with the corresponding TEM image (a). TEM (b) and electron diffraction analysis (c) of PdNPs upon immobilization on functionalized MWCNTs.

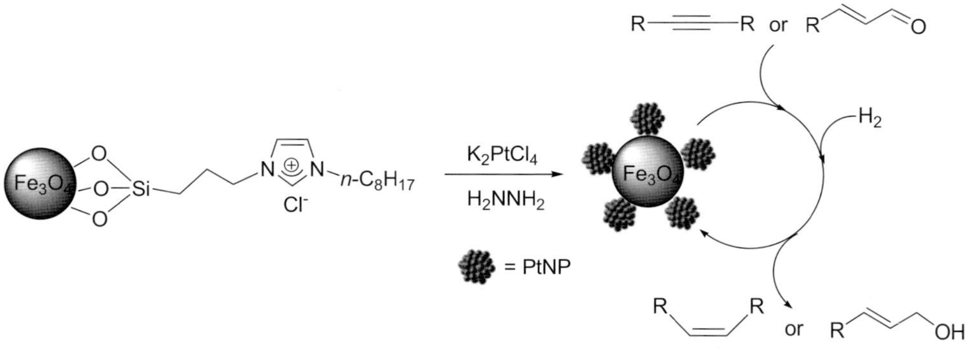

Scheme 5.17 Synthesis of PtNPs supported on ionic liquid functionalized Fe_3O_4 NPs and their applications in selective hydrogenation processes.

layer around the particles, showing the interactions between the IL and the metallic surface. In addition, it is important to underline the influence of the anion of $[C_1C_4Im]$-based ionic liquids: less coordinating anions (BF_4) gave more imperfect nanoparticles than those obtained when more coordinating anions were involved (PF_6, CF_3SO_3); these irregular shaped materials showed higher activities in the hydrogenation of cyclohexene, evidencing the importance of structural defects in catalysis [25b].

Alper and co-workers have functionalized magnetite nanoparticles with imidazolium groups in order to decorate them with PtNPs (d_{mean}: 2–2.5 nm); this catalytic material could be easily separated from the catalytic medium by applying an external magnetic field and then recycled (Scheme 5.17).

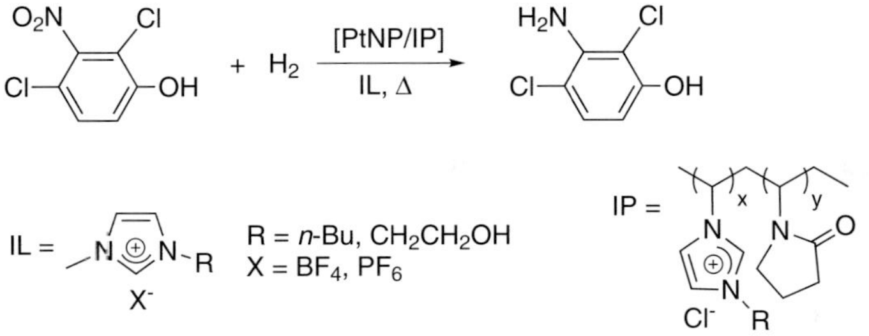

Scheme 5.18 PtNPs/ionic polymer-catalyzed selective hydrogenation of 2,4-dichloro-3-aminophenol in ImILs.

This catalyst is highly selective giving *cis*-alkenes by hydrogenation of the corresponding alkynes and allyl alcohols, hydrogenating α,β-unsaturated aldehydes [161].

Kou and co-workers have been interested in combining different effects in the stabilization of MNPs, in particular steric and electrostatic effects. Therefore, they tested the stabilization of isolated PtNPs synthesized in ethanol/water medium starting with $H_2[PtCl_6]$ as metal precursor and using PVP as stabilizer in $[C_1C_4Im]$ $[PF_6]$; this ionic liquid phase represents an efficient catalytic system for the hydrogenation of olefins, giving comparable activities as those obtained with PtNPs prepared in neat ILs [107]. Going on with this strategy, the use of ionic copolymers in $[C_1C_4Im][BF_4]$ as stabilizers for MNPs led to active catalytic systems [110]. PtNPs preformed in ethylene glycol (d_{mean}: 3.7 ± 0.2 nm) and dispersed in the ionic copolymer IP in ImILs gave efficient catalysts for the selective hydrogenation of aromatic chloronitro compounds into the corresponding aromatic chloroamines (Scheme 5.18) [43].

Preformed PtNPs synthesized by reduction of $H_2[PtCl_6]$ with $NaBH_4$ in water in the presence of amino-functionalized ionic liquid $[C_1RIm][Br]$ (R = 3-aminopropyl) and further isolated by centrifugation, were deposited on a glassy carbon electrode from an aqueous suspension. These nanoparticles proved their electrocatalytic activity towards the reduction of oxygen and oxidation of methanol, being good candidates to be used as efficient cathodes in fuel cells [162].

Recently, several groups have been interested in the deposition of PtNPs as well as bimetallic nanoparticles (PtRuNPs, AuPtNPs) on MWCNTs to be used as electrocatalysts for the oxidation of alcohols, because of the wide interest of Pt-based catalysts in fuel cells [163]. To avoid agglomeration and leaching of the metallic nanocatalysts, Chen and co-workers have deposited a film of an ionic polymer, formed by thermal initiation free radical polymerization of the monomer $[C_2RIm]$ $[Br]$ (R = vinyl) on MWCNTs; further addition of the appropriate metallic salt(s) in ethylene glycol leads to the formation of small and well-dispersed PtNPs or PtRuNPs in the ionic polymer [164]; these materials gave better catalytic performances in the electrooxidation of methanol than the corresponding nonsupported materials. Polymeric structures, such as polypyrrole nanotubes, have also found applications as support of PtNPs by grafting imidazolium groups at the amino functions present at the surface of the nanotubes [165].

5.3.2.4 Metallic Nanoparticles of Group 11

5.3.2.4.1 Copper Nanoparticles Copper nanoparticles (CuNPs) in ILs can be applied in the formation of carbon–carbon and carbon–heteroatom bonds in similar reactions to those described using palladium-based catalysts. Thus, Caló and coworkers reported that CuNPs, coming from the reaction of iodobenzene with copper bronze, can catalyze the reactions of aryl iodides and activated bromides with acrylates in TBAB as solvent and tetrabutylammonium acetate (TBAA) as base (Scheme 5.19) [166]. The system could be recycled six times showing an increase of both yields and reaction rates after the second cycle. It is suggested that the copper bronze undergoes, during the first two cycles, a slow dissolution as a consequence of the oxidative addition step, affording highly active CuNPs (average size of 4.2 nm), homogeneously dispersed in TBAB as evidenced by TEM analysis.

An effective Cu_2O nanoparticles/P(*o*-tol)$_3$ catalytic system for Stille cross-coupling reaction of aryl halides including aryl chlorides with organotins in TBAB, has been described; the catalyst could be recycled up to three times without activity loss [167].

Concerning the generation of carbon–heteroatom bonds, the foremost results have been obtained for the formation of organochalcogenides. CuO nanoparticles in $[C_1C_1C_4Im][BF_4]$ applied in carbon–sulfur cross-coupling reactions of thiols with aryl iodides, led to excellent yields; the catalyst could be recycled up to four times without activity loss [168] (Scheme 5.20). Analogous reactivity could be observed in cross-coupling reactions of aryl and alkyl bromides with diaryl diselenides in $[C_1C_4Im][BF_4]$ to generate various substituted arylselenides in good yields [169].

CuO nanoparticles in $[C_1C_4Im][PF_6]$ have been also used for the synthesis of selenoesters from acyl chlorides from diphenyl diselenide (Scheme 5.21) [170]. It is suggested that the reaction involves a heterogeneous process via surface CuO nanopowder catalyst.

CuNPs displaying 50–60 nm of mean diameter formed in a tetrazolium ionic liquid functionalized with nitrile and acid functions have been described as efficient

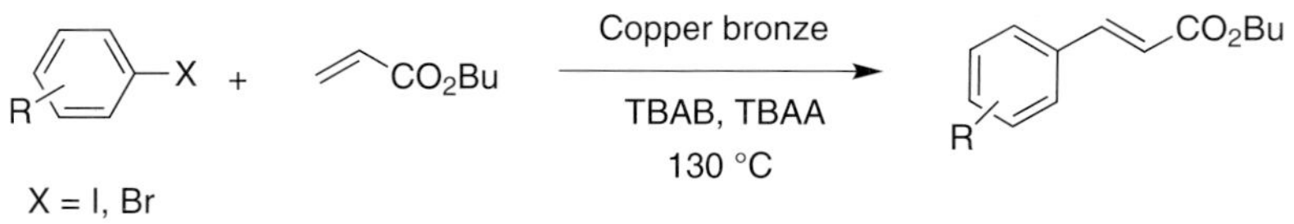

Scheme 5.19 Copper bronze catalyzed Heck reaction in TBAB.

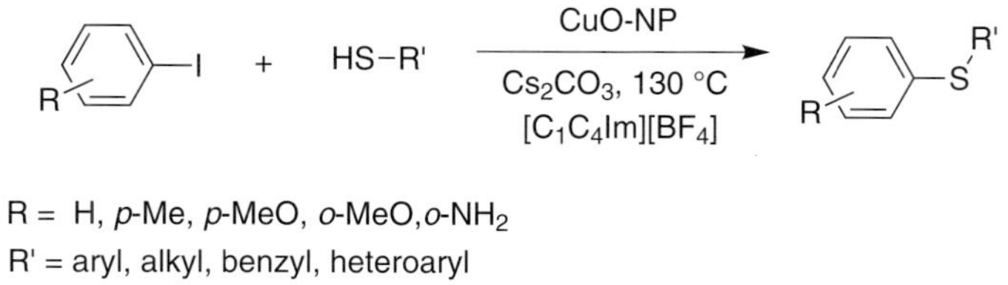

Scheme 5.20 Copper oxide nanoparticles in the synthesis of thioethers.

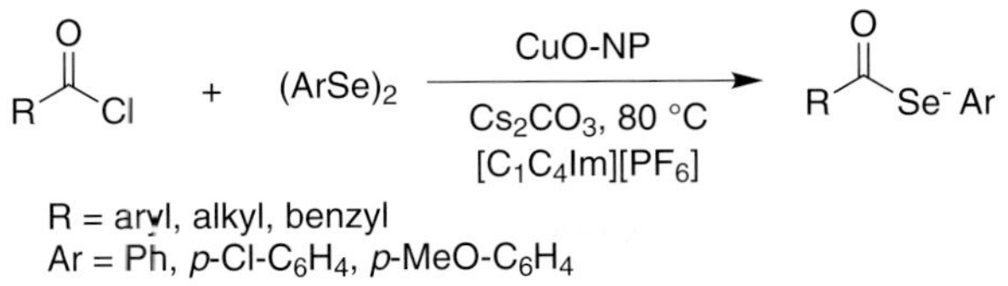

Scheme 5.21 Copper oxide nanoparticles applied in the synthesis of selenoesters.

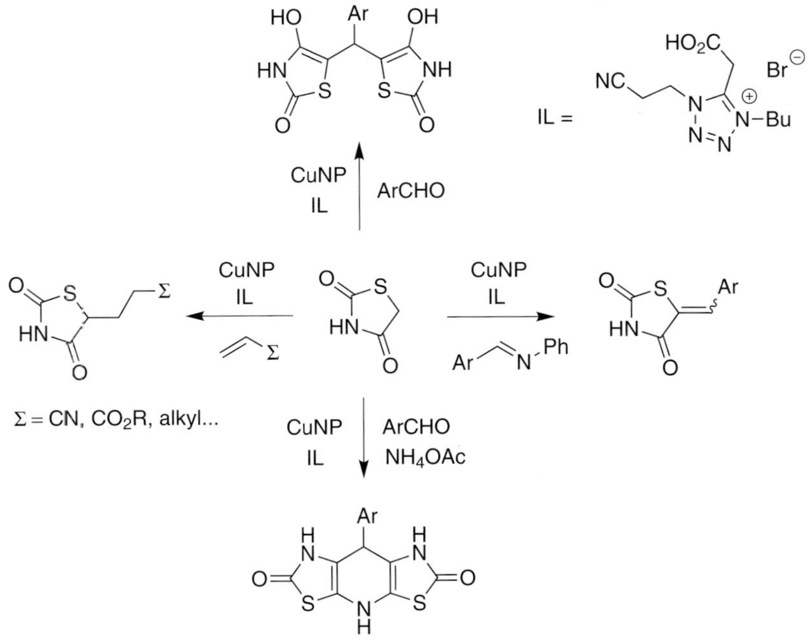

Scheme 5.22 CuNPs applied in catalytic reactions starting from thiazolidine-2,4-dione.

catalysts for condensation reactions (Scheme 5.22), such as the reaction of thiazolidine-2,4-dione with aldehydes [171], the three components reaction of thiazolidine-2,4-dione, aromatic aldehydes and ammonium acetate [172], the carba-Michael addition to acrylates [173] and the addition-elimination reaction between active methylene compounds and imines [174].

Spherical or cubical CuNPs prepared respectively in $[C_1C_4Im][BF_4]$ and $[C_1C_4Im][PF_6]$ by reduction of copper acetate, showed excellent catalytic activity in 1,3-dipolar cycloadditions (Huisgen reaction) between aryl and sugar based terminal alkynes and azides [175]. These CuNPs could be recycled up to five times.

5.3.2.4.2 **Gold Nanoparticles** Gold nanoparticles (AuNPs) in ILs have been extensively studied and can be considered as an easy to handle model for a better understanding of MNP–IL interactions (see above, Section 5.2.1) [13, 28, 36, 176].

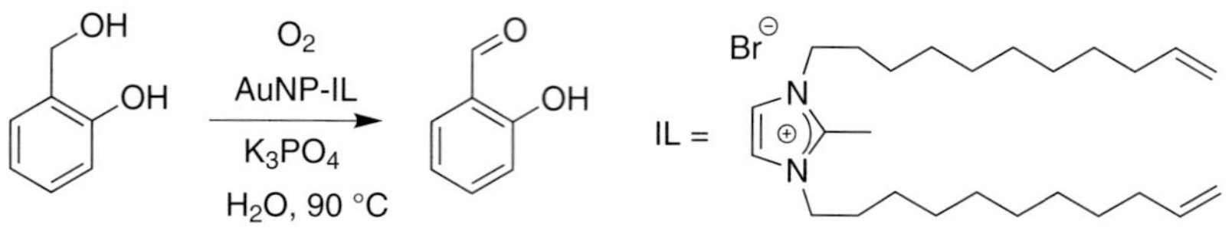

Scheme 5.23 Oxidation of 2-hydroxybenzyl alcohol to salicylaldehyde catalyzed by *N,N′*-bis(10-undecenyl)-2-methylimidazolium bromide.

Nevertheless, catalytic applications involving these numerous and various nanostructures are surprisingly sporadic. García's group reported the use of an original ImIL bearing two long-alkenyl chains with terminal C=C double bonds as a versatile monomer (*N,N′*-bis(10-undecenyl)-2-methylimidazolium bromide, Scheme 5.23) that photochemically polymerizes to give cell-like or spherical particles [177]. AuNPs (d_{mean}: 10 ± 5 nm) dispersed in this IL have been efficiently applied in the selective aerobic oxidation of 2-hydroxybenzyl alcohol to salicylaldehyde.

Small AuNPs (d_{mean}: 7 nm) have been prepared in $[C_1RIm][Br]$ (R = 3-aminopropyl), where the IL plays different roles, acting as solvent, MNPs stabilizer and reducing agent of $[AuCl_4]^-$ thanks to the terminal amine function of the ionic liquid. These AuNPs showed significant electrocatalytic activity towards O_2 reduction [178]. AuNPs formed in $[C_1C_4Im][BF_4]$ have been also described as efficient catalysts for the cyclopropanation of styrene with ethyldiazoacetate [179]. AuNPs stabilized by a task-specific oligomeric ionic liquid containing imidazolium and disulfide groups (Scheme 5.24) and miscible with $[C_1C_4Im][PF_6]$, have been applied in the epoxidation reaction of styrene [180]. Nevertheless gradual and fast deactivation of the catalyst was observed during the catalyst reuse due to agglomeration of AuNPs.

Bimetallic AuPdNPs in $[C_1C_4Im][PF_6]$ using PVP as stabilizer have been described as efficient catalysts in the hydrogenation reaction of 1,3-cyclooctadiene and 3-buten-1-ol, being more stable than the corresponding monometallic AuNPs [150, 181].

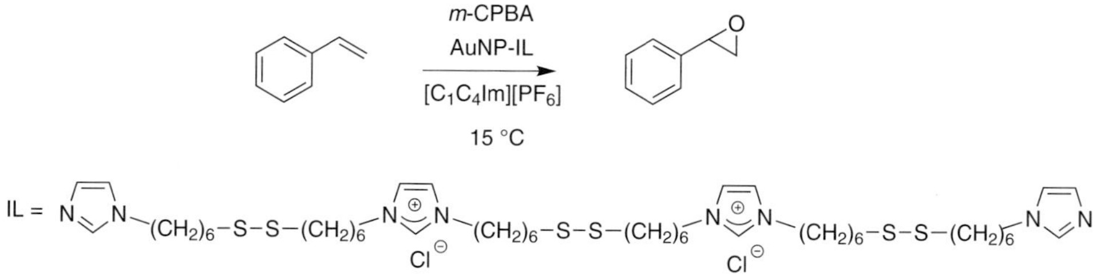

Scheme 5.24 Task specific imidazolium ionic liquid for epoxidation of styrene.

5.3.2.5 Metallic Nanoparticles of Group 12

5.3.2.5.1 Zinc and Cadmium Nanoparticles Zinc, cadmium and mercury are poorly represented in the preparation of MNPs in ILs for catalytic purposes. ZnO nanostructures in $[C_1C_nIm][BF_4]$ ($n=2$, 4) or $[C_1C_1C_4Im][BF_4]$ have been used in the degradation of rhodamine B by photocatalytic processes [52]. Spherical and flower-like ZnO nanostructures prepared in the presence of $[C_1C_nIm]$[ascorbate] ($n=2$, 6) have been described as efficient catalysts in the Friedel–Crafts acylation reaction of anthracene with benzoyl chloride [182]; the catalytic activity of flower-like ZnO nanostructures was slightly higher than the spherical counterpart. $[C_1C_4Im][BF_4]$ has been used for the synthesis of $ZnFe_2O_4$ nanoparticles applied in the photocatalytic process for the degradation of phenol [82]. ZnO nanopowder in ILs has been also used in the synthesis of diorganoselenides [183]. The photocatalytic activity of CdS nanoparticles in $[C_1C_2Im][EtSO_4]$ towards the photodegradation of methylene blue has been recently demonstrated by Habibi-Yangjeh and co-workers [55b,184].

5.4 Conclusions

Ionic liquids have become a convenient alternative to volatile organic solvents and a suitable medium to immobilize catalysts, due to their negligible vapor pressure, thermal stability and high polarity, exhibiting physico-chemical properties, easily tunable depending on the nature of the ions involved. Often, the ionic liquid phase containing the active catalytic species can be readily reused without significant loss of catalytic activity, as demonstrated by the pioneering work of Chauvin and co-workers in 1995 [185]. When metallic nanoparticles are concerned, the IL acts as an electrostatic and/or electro-steric stabilizer for the MNPs thus avoiding their agglomeration and in some cases, the leaching of atoms located at low coordination positions at the metallic surface. In order to decrease the amount of solvent working in homogeneous conditions, the methodology based on supported ionic liquid catalytic phase using various solids, has been successfully developed in the last years, frequently allowing a better immobilization of the catalyst than that observed in neat IL medium. With the aim to search a distinctive application of MNPs in relation to classical homogeneous and heterogeneous catalysts, MNPs can appear as the catalyst of choice for multistep syntheses which often require both molecular and surface-like catalytic behavior. Recent results in this area seem to be promising in order to progress in the applications of metallic nanocatalysts.

Acknowledgments

Authors thank Université Paul Sabatier and CNRS (Centre National de la Recherche Scientifique) for financial support.

References

1 (a) Dupont, J. and Scholten, J.D. (2010) On the structural and surface properties of transition-metal nanoparticles in ionic liquids. *Chem. Soc. Rev.*, **39**, 1780–1804; (b) Olivier-Bourbigou, H., Magna, L., and Morvan, D. (2010) Ionic liquids and catalysis: Recent progress from knowledge to applications. *Appl. Catal., A*, **373**, 1–56; (c) Antonietti, M., Kuang, D., Smarsly, B., and Zhou, Y. (2004) Ionic liquids for the convenient synthesis of functional nanoparticles and other inorganic nanostructures. *Angew. Chem. Int. Ed.*, **43**, 4988–4992.

2 Dupont, J. (2004) On the solid, liquid and solution structural organization of imidazolium ionic liquids. *J. Braz. Chem. Soc.*, **15**, 341–350.

3 Fukushima, T., Kosaka, A., Ishimura, Y., Yamamoto, T., Takigawa, T., Ishii, N., and Aida, T. (2003) Molecular ordering of organic molten salts triggered by single-walled carbon nanotubes. *Science*, **300**, 2072–2075.

4 De los Ríos, A.P., Hernández-Fernández, F.J., Tomás-Alonso, F., Palacios, J.M., Gómez, D., Rubio, M., and Víllora, G. (2007) A SEM-EDX study of highly stable supported liquid membranes based on ionic liquids. *J. Membr. Sci.*, **300**, 88–94.

5 Mehnert, C.P. (2005) Supported ionic liquid catalysis. *Chem. Eur. J.*, **11**, 50–56.

6 Ueno, K., Hata, K., Katakabe, T., Kondoh, M., and Watanabe, M. (2008) Nanocomposite ion gels based on silica nanoparticles and an ionic liquid: ionic transport, viscoelastic properties, and microstructure. *J. Phys. Chem. B*, **112**, 9013–9019.

7 Fukushima, T. and Aida, T. (2007) Ionic liquids of soft functional materials with carbon nanotubes. *Chem. Eur. J.*, **13**, 5048–5058.

8 Rodríguez-Pérez, L., Teuma, E., Falqui, A., Gómez, M., and Serp, P. (2008) Supported ionic liquid phase catalysis on functionalized carbon nanotubes. *Chem. Commun.*, 4201–4203.

9 Rodríguez-Pérez, L., Coppel, Y., Favier, I., Teuma, E., Serp, P., and Gómez, M. (2010) Imidazolium-based ionic liquids immobilized on solid supports: Effect on the structure and thermostability. *Dalton Trans.*, **39**, 7565–7568.

10 Wang, L.Q., Exarhos, G.J., and Liu, J. (1999) Nuclear magnetic resonance. Characterization of self-assembled nanostructured materials. *Adv. Mater.*, **11**, 1331–1341.

11 Lungwitz, R. and Spange, S. (2008) Structure and polarity of the phase boundary of *N*-methylimidazolium chloride/silica. *J. Phys. Chem. C*, **112**, 19443–19448.

12 Favier, I., Teuma, E., and Gómez, M. (2009) Palladium and ruthenium nanoparticles: reactivity and coordination at the metallic surface. *C. R. Chımie*, **12**, 533–545.

13 Kim, K.S., Choi, S., Cha, J.H., Yeon, S. H., and Lee, H. (2006) Facile one-pot synthesis of gold nanoparticles using alcohol ionic liquids. *J. Mater. Chem.*, **16**, 1315–1317.

14 Gao, Y., Voigt, A., Zhou, M., and Sundmacher, K. (2008) Synthesis of single-crystal fold nano- and microprisms using a solvent-reductant-template ionic liquid. *Eur. J. Inorg. Chem.*, 3769–3775.

15 (a) Modrow, H., Bucher, S., Hormes, J., Brinkmann, R., and Bönnemann, H. (2003) Model for chain length-dependent core-surfactant interaction in N(Alkyl)(4) Cl-stabilized colloidal metal particles obtained from X-ray absorption spectroscopy. *J. Phys. Chem. B*, **107**, 3684–3689; (b) Bucher, S., Hormes, J., Modrow, H., Brinkmann, R., Waldöfner, N., Bönnemann, H., Beuermann, L., Krischok, S., Maus-Friedrichs, W., and Kempter, V. (2002) Interaction between core and protection shell of N(butyl)(4) Cl^- and N(octyl)(4)Cl^- stabilized Pd colloids. *Surf. Sci.*, **497**, 321–332; (c) Ott, L.S. and Finke, R.G. (2006) Nanocluster formation and stabilization fundamental studies: Investigating 'solvent-only' stabilization en route to discovering stabilization by the traditionally weakly coordinating anion BF_4^- plus high

dielectric constant solvents. *Inorg. Chem.*, **45**, 8382–8393.

16 (a) Oskar, S. and Finke, R.G. (2002) Nanocluster formation and stabilization fundamental studies: Ranking commonly employed anionic stabilizers via the development, then application, of five comparative criteria. *J. Am. Chem. Soc.*, **124**, 5796–5810; (b) Roucoux, A., Schultz, J., and Patin, H. (2003) Arene hydrogenation with a stabilised aqueous rhodium(0) suspension: A major effect of the surfactant counter-anion. *Adv. Synth. Catal.*, **345**, 222–229; (c) Bönnemann, H., Brijoux, W., Brinkmann, R., Dinjus, E., Joußen, T., and Korall, B. (1991) Formation of colloidal transition-metals in organic phases and their application in catalysis. *Angew. Chem. Int. Ed. Engl.*, **30**, 1312–1314.

17 (a) Dong, K., Zhang, S., Wang, D., and Yao, X. (2006) Hydrogen bonds in imidazolium ionic liquids. *J. Phys. Chem. A*, **110**, 9775–9782; (b) Kölle, P. and Dronskowski, R. (2004) Hydrogen bonding in the crystal structures of the ionic liquid compounds butyldimethylimidazolium hydrogen sulfate, chloride, and chloroferrate(II, III). *Inorg. Chem.*, **43**, 2803–2809; (c) Elaiwi, A., Hitchcock, P.B., Seddon, K.R., Srinivasan, N., Tan, Y.-M., Welton, T., and Zora, J.A. (1995) Hydrogen bonding in imidazolium salts and its implications for ambient temperature halogenoaluminate(III) ionic liquid. *J. Chem. Soc, Dalton Trans.*, 3467–3472; (d) Holbrey, J.D., Reichert, W.M., and Rogers, R.D. (2004) Crystal structures of imidazolium bis (trifluoromethanesulfonyl)imide 'ionic liquid' salts: the first organic salt with a cis-TFSI anion conformation. *Dalton Trans*, 2267–2271; (e) Hardrace, C., Holbrey, J.D., McMath, S.E.J., Bowron, D.T., and Soper, A.K. (2003) Structure of molten 1,3-dimethylimidazolium chloride using neutron diffraction. *J. Chem. Phys*, **118**, 273–278.

18 (a) Gannon, T.J., Law, G., Watson, P.R., Carmichael, A.J., and Seddon, K.R. (1999) First observation of molecular composition and orientation at the surface of a room-temperature ionic liquid. *Langmuir*, **15**, 8429–8434; (b) Law, G., Watson, P.R., Carmichael, A.J., and Seddon, K.R. (2001) Molecular composition and orientation at the surface of room-temperature ionic liquids: Effect of molecular structure. *Phys. Chem. Chem. Phys.*, **3**, 2879–2885; (c) Chiappe, C. (2007) Nanostructural organization of ionic liquids: Theoretical and experimental evidences of the presence of well defined local structures in ionic liquids. *Monatsch. Chem.*, **138**, 1035–1043; (d) Canongia Lopes, J.N.A. and Pádua, A.A.H. (2006) Nanostructural organization in ionic liquids. *J. Phys. Chem. B*, **110**, 3330–3335.

19 (a) Consorti, C.S., Suarez, P.A.Z., De Souza, R.F., Burrow, R.A., Farrar, D.H., Lough, A.J., Loh, W., Da Silva, L.H.M., and Dupont, J. (2005) Identification of 1,3-dialkylimidazolium salt supramolecular aggregates in solution. *J. Phys. Chem. B*, **109**, 4341–4349; (b) Dupont, J., Suarez, P.A.Z., De Souza, R.F., Burrow, R.A., and Kintzinger, J.P. (2000) C-H-pi interactions in 1-n-butyl-3-methylimidazolium tetraphenylborate molten salt: Solid and solution structures. *Chem. Eur. J.*, **6**, 2377–2381; (c) Canongia Lopes, J.N.A., Costa Gomes, M.F., and Pádua, A.A.H. (2006) Nonpolar, polar, and associating solutes in ionic liquids. *J. Phys. Chem. B*, **110**, 16816–16818; (d) Triolo, A., Russina, O., Bleif, H.-J., and Di Cola, E. (2007) Nanoscale segregation in room temperature ionic liquids. *J. Phys. Chem. B*, **111**, 4641–4644; (e) Wang, Y. and Voth, G.A. (2005) Unique spatial heterogeneity in ionic liquids. *J. Am. Chem. Soc.*, **127**, 12192–12193; (f) Urrahata, S.M. and Ribeiro, M.C. (2004) Structure of ionic liquids of 1-alkyl-3-methylimidazolium cations: A systematic computer simulation study. *J Chem. Phys.*, **120**, 1855–1863.

20 Bini, R., Bortoloni, O., Chiappe, C., Pieraccini, D., and Siciliano, T. (2007) Development of cation/anion 'interaction' scales for ionic liquids through ESI-MS measurements. *J. Phys. Chem. B*, **111**, 598–604.

21 Tsuzuki, S., Tokuda, H., Hayamizu, K., and Watanabe, M. (2005) Magnitude and directionality of interaction in ion pairs of ionic liquids: Relationship with ionic conductivity. *J. Phys. Chem. B*, **109**, 16474–16481.

22 (a) Bradley, J.S. (1994) *Clusters and Colloids* (ed. G. Schmidt), VCH, Weinheim, pp. 469–473; (b) Bönnemann, H. and Richards, R.M. (2001) Nanoscopic metal particles – Synthetic methods and potential applications. *Eur. J. Inorg. Chem.*, (10), 2455–2480; (c) Roucoux, A., Schultz, J., and Patin, H. (2002) Reduced transition metal colloids: A novel family of reusable catalysts? *Chem. Rev.*, **102**, 3757–3778; (d) Ott, L.S. and Finke, R.G. (2007) Transition-metal nanocluster stabilization for catalysis: A critical review of ranking methods and putative stabilizers. *Coord. Chem. Rev.*, **251**, 1075–1100.

23 Chen, H. and Dong, S. (2007) Self-assembly of ionic liquids-stabilized Pt nanoparticles into two-dimensional patterned nanostructures at the air-water interface. *Langmuir*, **23**, 12503–12507.

24 Fonseca, G.S., Machado, G., Teixeira, S. R., Fecher, G.H., Morais, J., Alves, M.C. M., and Dupont, J. (2006) Synthesis and characterization of catalytic iridium nanoparticles in imidazolium ionic liquids. *J. Colloid Interface Sci.*, **301**, 193–204.

25 (a) Umpierre, A.P., Machado, G., Fecher, G.H., Morais, J., and Dupont, J. (2005) Selective hydrogenation of 1,3-butadiene to 1-butene by Pd(0) nanoparticles embedded in imidazolium ionic liquids. *Adv. Synth. Catal.*, **347**, 1404–1412; (b) Scheeren, C.W., Machado, G., Teixeira, S. R., Morais, J., Domingos, J.B., and Dupont, J. (2006) Synthesis and characterization of Pt(0) nanoparticles in imidazolium ionic liquids. *J. Phys. Chem. B*, **110**, 13011–13020.

26 Kang, S.W., Char, K., and Kang, Y.S. (2008) Novel application of partially positively charged silver nanoparticles for facilitated transport in olefin/paraffin separation membranes. *Chem. Mater.*, **20**, 1308–1311.

27 Biswas, K. and Rao, C.N.R. (2007) Use of ionic liquids in the synthesis of nanocrystals and nanorods of semiconducting metal chalcogenides. *Chem. Eur. J.*, **13**, 6123–6129.

28 Itoh, H., Naka, K., and Chujo, Y. (2004) Synthesis of gold nanoparticles modified with ionic liquid based on the imidazolium cation. *J. Am. Chem. Soc.*, **126**, 3026–3027.

29 Lee, C.-Y., Huang, C.-H., and Wei, G.-T. (2010) Behaviours of ionic liquids in the phase transfer of aqueous metal nanoparticles. *Colloids and Surfaces A: Physicochem. Eng. Aspects*, **367**, 24–30.

30 Han, K.I., Kang, S.W., Kim, J., and Kang, Y.S. (2011) Effect of ionic liquids on dissociation of copper flake into copper nanoparticles and its application to facilitated olefin transport membranes. *J. Membrane Sci.*, **374**, 43–48.

31 Sobota, M., Schmid, M., Happel, M., Amende, M., Maier, F., Steinrück, H.P., Paape, N., Wasserscheid, P., Maurin, M., Gottfried, J.M., and Libuda, J. (2010) Ionic liquid based model catalysis: interaction of [BMIM][Tf_2N] with Pd nanoparticles supported on an ordered alumina film. *Phys. Chem. Chem. Phys.*, **12**, 10610–10621.

32 (a) Boström, M. Williams, D.R.M., and Ninham, B.W. (2001) Specific ion effects: Why DLVO theory fails for biology and colloid systems. *Phys. Rev. Lett.*, **87**, 168103-1–168103-4; (b) Ninham, B.W. (1999) On progress in forces since the DLVO theory. *Adv. Colloid Interface Sci.*, **83**, 1–17.

33 Ueno, K., Inaba, A., Kondoh, M., and Watanabe, M. (2008) Colloidal stability of bare and polymer-grafted silica nanoparticles in ionic liquids. *Langmuir*, **24**, 5253–5259.

34 Migowski, P., Machado, G., Texeira, S.R., Alves, M.C.M., Morais, J., Traverse, A., and Dupont, J. (2007) Synthesis and characterization of nickel nanoparticles dispersed in imidazolium ionic liquids. *Phys. Chem. Chem. Phys.*, **9**, 4814–4821.

35 Rubim, J.C., Trindade, F.A., Gelesky, M. A., Aroca, R.F., and Dupont, J. (2008) Surface-enhanced vibrational spectroscopy of tetrafluoroborate 1-n-butyl-3-methylimidazolium (BMIBF(4)) ionic liquid on silver surfaces. *J. Phys. Chem. C*, **112**, 19670–19675.

36 Schrekker, H.S., Gelesky, M.A., Stracke, M.P., Schrekker, C.M.L., Machado, G., Teixeira, S.R., Rubim, J.C., and Dupont, J. (2007) Disclosure of the imidazolium cation coordination and stabilization mode in ionic liquid stabilized gold(0) nanoparticles. *J. Colloid Interface Sci.*, **316**, 189–195.

37 Zhang, H. and Cui, H. (2009) Synthesis and characterization of functionalized ionic liquid-stabilized metal (gold and platinum) nanoparticles and metal nanoparticle/carbon nanotube hybrids. *Langmuir*, **25**, 2604–2612.

38 Durand, J., Fernández, F., Barrière, C., Teuma, E., Gómez, K., González, G., and Gómez, M. (2008) DOSY technique applied to palladium nanoparticles in ionic liquids. *Magn. Reson. Chem.*, **46**, 739–743.

39 Arduengo, J. III, Dias, H.V.R., Harlow, R. L., and Kline, M. (1992) Electronic stabilization of nucleophilic carbenes. *J. Am. Chem. Soc.*, **114**, 5530–5534.

40 (a) Ott, L.S., Cline, M.L., Deetlefs, M., Seddon, K.R., and Finke, R.G. (2005) Nanoclusters in ionic liquids: Evidence for *N*-heterocyclic carbene formation from imidazolium-based ionic liquids detected by H-2 NMR. *J. Am. Chem. Soc.*, **127**, 5758–5759; (b) Scholten, J.D., Ebeling, G., and Dupont, J. (2007) On the involvement of NHC carbenes in catalytic reactions by iridium complexes, nanoparticles and bulk metal dispersed in imidazolium ionic liquids. *Dalton Trans.*, 5554–5560.

41 Bernardi, F., Scholten, J.D., Fecher, G.H., Dupont, J., and Morais, J. (2009) Probing the chemical interaction between iridium nanoparticles and ionic liquid by XPS analysis. *Chem. Phys. Lett*, **479**, 113–116.

42 Xu, H., Li, H., Xia, J., Yin, S., Luo, Z., Liu, L., and Xu, L. (2011) One-Pot Synthesis of visible-light-driven plasmonic photocatalyst Ag/AgCl in ionic lLiquid. *ACS Applied Mater. and Interfaces*, **3**, 22–29.

43 Yuan, X., Yan, N., Xiao, C., Li, C., Fei, Z., Cai, Z., Kou, Y., and Dyson, P.J. (2010) Highly selective hydrogenation of aromatic chloronitro compounds to aromatic chloroamines with ionic-liquid-like copolymer stabilized platinum nanocatalysts in ionic liquids. *Green Chem.*, **12**, 228–233.

44 Campbell, P.S., Santini, C.C., Bouchu, D., Fenet, B., Philippot, K., Chaudret, B., Pádua, A.A.H., and Chauvin, Y. (2010) A novel stabilisation model for ruthenium nanoparticles in imidazolium ionic liquids: in situ spectroscopic and labelling evidence. *Phys. Chem. Chem. Phys.*, **12**, 4217–4223.

45 Ding, K., Miao, Z., Hu, B., An, G., Sun, Z., Han, B., and Liu, Z. (2010) Study on the anatase to rutile phase transformation and controlled synthesis of rutile nanocrystals with the assistance of ionic liquid. *Langmuir*, **26**, 10294–10302.

46 Kim, K.-S., Demberelnyamba, D., and Lee, H. (2004) Size-selective synthesis of gold and platinum nanoparticles using novel thiol-functionalized ionic liquids. *Langmuir*, **20**, 556–560.

47 Ueno, K., Imaizumi, S., Hata, K., and Watanabe, M. (2009) Colloidal interaction in ionic liquids: effects of ionic structures and surface chemistry on rheology of silica colloidal dispersions. *Langmuir*, **25**, 825–831.

48 Pensado, A.S. and Pádua, A.A.H. (2011) Solvation and stabilization of metallic nanoparticles in ionic liquids. *Angew. Chem. Int. Ed.*, **50**, 8683–8687.

49 Chang, H.-C., Jiang, J.-C., Tsai, W.-C., Chen, G.-C., and Lin, S.H. (2006) The role of charge-enhanced C-H center dot center dot center dot O interactions in gel-like mixtures prepared from ionic liquids and tungsten(VI) oxide nanoparticles. *Chem. Phys. Lett.*, **427**, 310–316.

50 Lian, J., Duan, X., Ma, J., Peng, P., Kim, T., and Zheng, W. (2009) Hematite (alpha-Fe(2)O(3)) with various morphologies: ionic liquid-assisted synthesis, formation mechanism, and properties. *ACS Nano*, **3**, 3749–3761.

51 Zheng, W., Liu, X., Yan, Z., and Zhu, L. (2009) Ionic liquid-assisted synthesis of large-scale TiO_2 nanoparticles with

controllable phase by hydrolysis of $TiCl_4$. *ACS Nano*, **3**, 115–122.
52 Wang, L., Chang, L., Zhao, B., Yuan, Z., Shao, G., and Zheng, W. (2008) Systematic investigation on morphologies, forming mechanism, photocatalytic and photoluminescent properties of ZnO nanostructures constructed in ionic liquids. *Inorg. Chem.*, **47**, 1443–1452.
53 Ryu, H.J., Sanchez, L., Keul, H.A., Raj., A., and Bockstaller, M.R. (2008) Imidazolium-based ionic liquids as efficient shape-regulating solvents for the synthesis of gold nanorods. *Angew. Chem. Int. Ed.*, **47**, 7639–7643.
54 Bansal, V. and Bhargava, S.K. (2011) Ionic liquids as designer solvents for the synthesis of metal nanoparticles, in *Ionic Liquid, Theory and Applications* (ed. A. Kokorin), InTech, pp. 367–381.
55 (a) Esmaili, M. and Habibi-Yangjeh, A. (2010) Microwave-assisted preparation and characterization of Zn(1−x)Cd(x)S nanoparticles in the presence of an ionic liquid and their photocatalytic activities. *J. Alloys Compd.*, **496**, 650–655; (b) Barzegar, M., Habibi-Yangjeh, A., and Behboudnia, M. (2010) Ultrasonic-assisted preparation and characterization of CdS nanoparticles in the presence of a halide-free and low-cost ionic liquid and photocatalytic activity. *J. Phys. Chem. Solids*, **71**, 1393–1397.
56 Sanes, J., Carrión, F.-J., and Bermúdez, M.-D. (2009) ZnO-ionic liquid nanostructures. *Appl. Surf. Sci.*, **255**, 4859–4862.
57 Kim, T.Y., Kim, W.J., Hong, S.H., Kim, J. E., and Suh, K.S. (2009) Ionic-liquid-assisted formation of silver nanowires. *Angew. Chem. Int. Ed.*, **48**, 3806–3809.
58 Gutel, T., Santini, C.C., Philippot, K., Padua, A., Pelzer, K., Chaudret, B., Chauvin, Y., and Basset, J.-M. (2009) Organized 3D-alkyl imidazolium ionic liquids could be used to control the size of in situ generated ruthenium nanoparticles? *J. Mater. Chem.*, **19**, 3624–3631.
59 Gutel, T., Garcia-Antón, J., Peltzer, K., Philippot, K., Santini, C.C., Chauvin, Y., Chaudret, B., and Basset, J.-M. (2007) Influence of the self-organization of ionic liquids on the size of ruthenium nanoparticles: effect of the temperature and stirring. *J. Mater. Chem.*, **17**, 3290–3292.
60 Xia, J., Yin, S., Li, H., Xu, H., Yan, Y., and Zhang, Q. (2011) Self-assembly and enhanced photocatalytic properties of BiOI hollow microspheres via a reactable ionic liquid. *Langmuir*, **27**, 1200–1206.
61 (a) Redel, E. Thomann, R., and Janiak, C. (2008) First correlation of nanoparticle size-dependent formation with the ionic liquid anion molecular volume. *Inorg. Chem.*, **47**, 14–16; (b) Redel, E., Thomann, R., and Janiak, C. (2008) Use of ionic liquids (ILs) for the IL-anion size-dependent formation of Cr, Mo and W nanoparticles from metal carbonyl M(CO)(6) precursors. *Chem. Commun.*, 1789–1791; (c) Redel, E., Krämer, J., Thomann, R., and Janiak, C. (2009) Synthesis of Co, Rh and Ir nanoparticles from metal carbonyls in ionic liquids and their use as biphasic liquid-liquid hydrogenation nanocatalysts for cyclohexene. *J. Organomet. Chem.*, **694**, 1069–1075.
62 Khare, V., Li, Z., Mantion, A., Ayi, A.A., Sonkaria, S., Voelkl, A., Thuenemann, A. F., and Taubert, A. (2010) Strong anion effects on gold nanoparticle formation in ionic liquids. *J. Mater. Chem.*, **20**, 1332–1339.
63 Kowsari, E. and Faraghi, G. (2010) Synthesis by an ionic liquid-assisted method and optical properties of nanoflower Y_2O_3. *Mater. Res. Bull.*, **45**, 939–945.
64 Zhou, Y., Schattka, J.H., and Antonietti, M. (2004) Room-temperature ionic liquids as template to monolithic mesoporous silica with wormlike pores via a sol-gel nanocasting technique. *Nano Lett.*, **4**, 477–481.
65 Zhao, Y., Cui, G., Wang, J., and Fan, M. (2009) Effects of ionic liquids on the characteristics of synthesized Nano Fe(0) particles. *Inorg. Chem.*, **48**, 10435–10441.
66 Krämer, J., Redel, E., Thomann, R., and Janiak, C. (2008) Use of ionic liquids for the synthesis of iron, ruthenium, and osmium nanoparticles from their metal

carbonyl precursors. *Organometallics*, **27**, 1976–1978.

67 Li, F., Song, J., Li, F., Wang, X., Zhang, Q., and Han, D. (2009) Direct electrochemistry of glucose oxidase and biosensing for glucose based on carbon nanotubes@SiO_2-Au composite. *Biosens. Bioelectron.*, **25**, 883–888.

68 Jou, L.-H., Chang, J.-K., Whang, T.-J., and Sun, I.-W. (2010) Electrodeposition of palladium-tin alloys from 1-ethyl-3-methylimidazolium chloride-tetrafluoroborate ionic liquid for ethanol electro-oxidation. *J. Electrochem. Soc.*, **157**, D443–D449.

69 Cheng, H., Huang, B., Wang, Z., Qin, X., Zhang, X., and Dai, Y. (2011) One-pot miniemulsion-mediated route to BiOBr hollow microspheres with highly efficient photocatalytic activity. *Chem. Eur. J.*, **17**, 8039–8043.

70 Zhang, H., Li, X., and Chen, G. (2009) Ionic liquid-facilitated synthesis and catalytic activity of highly dispersed Ag nanoclusters supported on TiO_2. *J. Mater. Chem.*, **19**, 8223–8231.

71 Miao, S., Liu, Z., Miao, Z., Han, B., Ding, K., An, G., and Xie, Y. (2009) Ionic liquid-mediated synthesis of crystalline CeO2 mesoporous films and their application in aerobic oxidation of benzyl alcohol. *Microporous Mesoporous Mater.*, **117**, 386–390.

72 Avellaneda, R.S., Ivanova, S., Sanz, O., Romero-Sarria, F., Centeno, M.A., and Odriozola, J.A. (2009) Ionic liquid template TiO_2 nanoparticles as a support in gold environmental catalysis. *Appl. Catal. B: Environ.*, **93**, 140–148.

73 Cioffi, N., Faticanti, M., Ditaranto, N., De Rossi, S., Traversa, L., Monopoli, A., Nacci, A., Torsi, L., and Sabbatini, L. (2007) Analytical characterisation of Pd/ZrO_2 composite nanoparticles employed in heterogeneous catalysis. *Curr. Nanosci.*, **3**, 121–127.

74 Zhou, Y. and Antonietti, M. (2003) Synthesis of very small TiO_2 nanocrystals in a room-temperature ionic liquid and their self-assembly toward mesoporous spherical aggregates. *J. Am. Chem. Soc.*, **125**, 14960–14961.

75 Nakashima, T. and Kimizuka, N. (2003) Interfacial synthesis of hollow TiO_2 microspheres in ionic liquids. *J. Am. Chem. Soc.*, **125**, 6386–6387.

76 (a) Zhai, Y., Gao, Y., Liu, F., Zhang, Q., and Gao, G. (2007) Synthesis of nanostructured TiO_2 particles in room temperature ionic liquid and its photocatalytic performance. *Mater. Lett.*, **61**, 5056–5058; (b) Alammar, T., Birkner, A., Shekhah, O., and Mudring, A.-V. (2010) Sonochemical preparation of TiO_2 nanoparticles in the ionic liquid 1-(3-hydroxypropyl)-3-methylimidazolium-bis (trifluoromethylsulfonyl)amide. *Mater. Chem. Phys.*, **120**, 109–113; (c) Yarui, W., Weisheng, G., and Shu, L. (2011) Microwave assisted preparation and characterization of ionic liquid [Bmim] PF_6 and synthesis and characterization of titanium dioxide using [Bmim]PF_6 as medium. *Adv. Mater. Res.*, **183–185**, 1662–1666.

77 Lee, C.-M., Jeong, H.-J., Lim, S.T., Sohn, M.-H., and Kim, D.W. (2011) Synthesis of iron oxide nanoparticles with control over shape using imidazolium-based ionic liquids. *Appl. Mater. Interfaces*, **2**, 756–759.

78 Wang, Y., Matsimuk, S., Shen, R., and Yang, H. (2007) Synthesis of iron oxide nanoparticles using a freshly-made or recycled imidazolium-based ionic liquid. *Green Chem.*, **9**, 1051–1056.

79 Wang, Y. and Yang, H. (2009) Synthesis of iron oxide nanoparticles using a freshly-made or recycled imidazolium-based ionic liquid. *Chem. Eng. J.*, **147**, 71–78.

80 Hu, H., Yang, H., Huang, P., Cui, D., Peng, Y., Zhang, J., Lu, F., Lian, J., and Shi, D. (2010) Unique role of ionic liquid in microwave-assisted synthesis of monodisperse magnetite nanoparticles. *Chem. Commun.*, **46**, 3866–3868.

81 Zhang, Y., Zhao, Y., and Xia, C. (2009) Basic ionic liquids supported on hydroxyapatite-encapsulated γ-Fe_2O_3 nanocrystallites: An efficient magnetic and recyclable heterogeneous catalyst for aqueous Knoevenagel condensation. *J. Mol. Catal. A: Chemical*, **306**, 107–112.

82 Cao, S.-W., Zhu, Y.-J., Cheng, G.-F., and Huang, Y.-H. (2009) $ZnFe_2O_4$ nanoparticles: microwave-hydrothermal ionic liquid synthesis and photocatalytic property over phenol. *J. Hazard. Mater.*, **171**, 431–435.

83 Alvaro, M., Aprile, C., Ferrer, B., Sastre, F., and García, H. (2009) Photochemistry of gold nanoparticles functionalized with an iron(II) terpyridine complex. An integrated visible light photocatalyst for hydrogen generation. *Dalton Trans.*, 7437–7444.

84 Taher, A., Kim, J.-B., Jung, J.-Y., Ahn, W.-S., and Jin, M.-J. (2009) Highly active and magnetically recoverable Pd NHC catalyst immobilized on Fe_3O_4 nanoparticle - ionic liquid matrix for Suzuki reaction in water. *Synlett*, **15**, 2477–2482.

85 Geldbach, T.J. and Dyson, P.J. (2005) Searching for molecular arene hydrogenation catalysis in ionic liquid. *J. Organomet. Chem.*, **690**, 3552–3557.

86 Rossi, L.M., Machado, G., Fichtner, P.F. P., Teixeira, S.R., and Dupont, J. (2004) On the use of ruthenium dioxide in 1-n-butyl-3-methylimidazolium ionic liquids as catalyst precursor for hydrogenation reactions. *Catal. Lett.*, **92**, 149–155.

87 Silveira, E.T., Umpierre, A.P., Rossi, L. M., Machado, G., Morais, J., Soares, G.V., Baumvol, I.J.R., Teixeira, S.R., Fichtner, P.F.P., and Dupont, J. (2004) The partial hydrogenation of benzene to cyclohexene by nanoscale ruthenium catalysts in imidazolium ionic liquids. *Chem. Eur. J.*, **10**, 3734–3740.

88 Rossi, L.M. and Machado, G. (2009) Ruthenium nanoparticles prepared from ruthenium dioxide precursor: Highly active catalyst for hydrogenation of arenes under mild conditions. *J. Mol. Cat. A: Chemical*, **298**, 69–73.

89 Rossi, L.M., Dupont, J., Machado, G., Fichtner, P.F.P., Radtke, C., Baumvol, I.J. R., and Teixeira, S.R. (2004) Ruthenium dioxide nanoparticles in ionic liquids: synthesis, characterization and catalytic properties in hydrogenation of olefins and arenes. *J. Braz. Chem. Soc.*, **15**, 904–910.

90 Fonseca, G.S., Silveira, E.T., Gelesky, M. A., and Dupont, J. (2005) Competitive hydrogenation of alkyl substituted arenes by transition-metal nanoparticles: correlation with the alkyl-steric effect. *Adv. Synth. Catal.*, **347**, 847–853.

91 (a) Prechtl, M.H.G., Scariot, M., Scholten, J.D., Machado, G., Teixeira, S. R., and Dupont, J. (2008) Nanoscale Ru (0) particles: arene hydrogenation catalysts in imidazolium ionic liquids. *Inorg. Chem.*, **47**, 8995–9001; (b) Prechtl, M.H.G., Campbell, P.S., Scholten, J.D., Fraser, G.B., Machado, G., Santini, C.C., Dupont, J., and Chauvin, Y. (2010) Imidazolium ionic liquids as promoters and stabilising agents for the preparation of metal(0) nanoparticles by reduction and decomposition of organometallic complexes. *Nanoscale*, **2**, 2601–2606.

92 Campbell, P.S., Santini, C.C., Bayard, F., Chauvin, Y., Collière, V., Podgoršek, A., Costa Gomes, M.F., and Sá, J. (2010) Olefin hydrogenation by ruthenium nanoparticles in ionic liquid media: Does size matter? *J. Catal.*, **275**, 99–107.

93 Salas, G., Santini, C.C., Philippot, K., Collière, V., Chaudret, B., Fenet, B., and Fazzini, P.F. (2011) Influence of amines on the size control of in situ synthesized ruthenium nanoparticles in imidazolium ionic liquids. *Dalton Trans.*, **40**, 4660–4668.

94 Marquardt, D., Vollmer, C., Thomann, R., Steurer, P., Mülhaupt, R., Redel, E., and Janiak, C. (2011) The use of microwave irradiation for the easy synthesis of graphene-supported transition metal nanoparticles in ionic liquids. *Carbon*, **49**, 1326–1332.

95 Wang, J., Feng, J., Qin, R., Fu, H., Yuan, M., Chen, H., and Li, X. (2007) Asymmetric hydrogenation of aromatic ketones catalyzed by achiral monophosphine TPPTS-stabilized Ru in ionic liquids. *Tetrahedron: Asymmetr.*, **18**, 1643–1647.

96 Prechtl, M.H.G., Scholten, J.D., and Dupont, J. (2009) Tuning the selectivity of ruthenium nanoscale catalysts with functionalized ionic liquids: hydrogenation of nitriles. *J. Mol. Catal. A: Chemical*, **313**, 74–78.

97 Julis, J., Hölsher, M., and Leitner, W. (2010) Selective hydrogenation of biomass derived substrates using ionic liquid-stabilized ruthenium nanoparticles. *Green Chem.*, **12**, 1634–1639.

98 Yan, N., Zhao, C., Luo, C., Dyson, P.J., Liu, H., and Kou, Y. (2006) One-step conversion of cellobiose to C_6-alcohols using a ruthenium nanocluster catalyst. *J. Am. Chem. Soc.*, **128**, 8714–8715.

99 Zheng, M.-Y., Wang, A.-Q., Pang, J.-F., Wang, X.-D., and Zhang, T. (2010) Transition metal-tungsten bimetallic catalysts for the conversion of cellulose into ethylene glycol. *ChemSusChem.*, **3**, 63–66.

100 Yan, N., Yuan, Y., Dykeman, R., Kou, Y., and Dyson, P.J. (2010) Hydrodeoxygenation of lignin-derived phenols into alkanes by using nanoparticle catalysts combined with Brönsted acidic ionic liquids. *Angew. Chem. Int. Ed.*, **49**, 5546–5553.

101 (a) For preparation of cobalt nanoparticles, see the following selected references: Zou, D., Xu, C., Luo, H., Wang, L., and Ying, T. (2008) Synthesis of Co_3O_4 nanoparticles via an ionic liquid-assisted methodology at room temperature. *Mater. Lett.*, **62**, 1976–1978; (b) Wang, Y. and Yang, H. (2005) Synthesis of CoPt nanorods in ionic liquids. *J. Am. Chem. Soc.*, **127**, 5316–5317.

102 Scariot, M., Silva, D.O., Scholten, J.D., Machado, G., Teixeira, S.R., Novak, M.A., Ebeling, G., and Dupont, J. (2008) Cobalt nanocubes in ionic liquids: synthesis and properties. *Angew. Chem. Int. Ed.*, **47**, 9075–9078.

103 Silva, D.O., Scholten, J.D., Gelesky, M.A., Teixeira, S.R., Dos Santos, A.C.B., Souza-Aguiar, E.F., and Dupont, J. (2008) Catalytic gas-to-liquid processing using cobalt nanoparticles dispersed in imidazolium ionic liquids. *ChemSusChem.*, **1**, 291–294.

104 Fonseca, G.S., Umpierre, A.P., Fichtner, P.F.P., Teixeira, S.R., and Dupont, J. (2003) The use of imidazolium ionic liquids for the formation and stabilization of Ir_0 and Rh_0 nanoparticles: efficient catalysts for the hydrogenation of arenes. *Chem. Eur. J.*, **9**, 3263–3269.

105 Dupont, J., Fonseca, G.S., Umpierre, A. P., Fichtner, P.F.P., and Teixeira, S.R. (2002) Transition-metal nanoparticles in imidazolium ionic liquids: recyclable catalysts for biphasic hydrogenation reactions. *J. Am. Chem. Soc.*, **124**, 4228–4229.

106 Fonseca, G.S., Domingos, J.B., Nome, F., and Dupont, J. (2006) On the kinetics of iridium nanoparticles formation in ionic liquids and olefin hydrogenation. *J. Mol. Catal. A: Chemical*, **248**, 10–16.

107 Mu, X.-D., Evans, D.G., and Kou, Y. (2004) A general method for preparation of PVP-stabilized noble metal nanoparticles in room temperature ionic liquids. *Catal. Lett.*, **97**, 151–154.

108 Zhao, C., Wang, H.-Z., Yan, N., Xiao, C.-X., Mu, X.-D., Dyson, P.J., and Kou, Y. (2007) Ionic-liquid-like copolymer stabilized nanocatalysts in ionic liquids: II. Rhodium-catalyzed hydrogenation of arenes. *J. Catal.*, **250**, 33–40.

109 Yang, X.Y., Yan, N., Fei, Z., Crespo-Quesada, R.M., Laurenczy, G., Kiwi-Minsker, L., Kou, Y., Li, Y., and Dyson, P. J. (2008) Biphasic hydrogenation over PVP stabilized Rh nanoparticles in hydroxyl functionalized ionic liquids. *Inorg. Chem.*, **47**, 7444–7446.

110 Mu, X.-D., Meng, J.-Q., Li, Z.-L., and Kou, Y. (2005) Rhodium nanoparticles stabilized by ionic copolymers in ionic liquids: long lifetime nanocluster catalysts for benzene hydrogenation. *J. Am. Chem. Soc.*, **127**, 9694–9695.

111 Gelesky, M.A., Chiaro, S.S.X., Pavan, F. A., dos Santos, J.H.Z., and Dupont, J. (2007) Supported ionic liquid phase rhodium nanoparticle hydrogenation catalysts. *Dalton Trans.*, 5549–5553.

112 Gelesky, M.A., Scheeren, C.W., Foppa, L., Pavan, F.A., Dias, S.L.P., and Dupont, J. (2009) Metal nanoparticle/ionic liquid/cellulose: new catalytically active membrane materials for hydrogenation reactions. *Biomacromolecules*, **10**, 1888–1893.

113 (a) Léger, B., Denicourt-Nowicki, A., Roucoux, A., and Olivier-Bourbigou, H. (2008) Synthesis of bipyridine-stabilized

rhodium nanoparticles in non-aqueous ionic liquids: A new efficient approach for arene hydrogenation with nanocatalysts. *Adv. Synth. Catal.*, **350**, 153–159; (b) Léger, B., Denicourt-Nowicki, A., Olivier-Bourbigou, H., and Roucoux, A. (2008) Rhodium nanocatalysts stabilized by various bipyridine ligands in non-aqueous ionic liquids: influence of the bipyridine coordination modes in arene catalytic hydrogenation. *Inorg. Chem.*, **47**, 9090–9096; (c) Léger, B., Denicourt-Nowicki, A., Olivier-Bourbigou, H., and Roucoux, A. (2008) Rhodium colloidal suspensions stabilised by poly-N-donor ligands in non-aqueous ionic liquids: preliminary investigation into the catalytic hydrogenation of arenes. *ChemSusChem.*, **1**, 984–987.

114 Léger, B., Denicourt-Nowicki, A., Olivier-Bourbigou, H., and Roucoux, A. (2009) Imidazolium-functionalized bipyridine derivatives: a promising family of ligands for catalytical Rh(0) colloids. *Tetrahedron Lett.*, **50**, 6531–6533.

115 Mévellec, V., Léger, B., Mauduit, M., and Roucoux, A. (2005) Organic phase stabilization of rhodium nanoparticle catalyst by direct phase transfer from aqueous solution to room temperature ionic liquid based on surfactant counter anion exchange. *Chem. Commun.*, 2838–2839.

116 Bruss, A.J., Gelesky, M.A., Machado, G., and Dupont, J. (2006) Rh(0) nanoparticles as catalyst precursors for the solventless hydroformylation of olefins. *J. Mol. Cat. A: Chemical*, **252**, 212–218.

117 Quek, X.-Y., Guan, Y., van Santen, R.A., and Hensen, E.J.M. (2010) Ionic-liquid-stabilized rhodium nanoparticles for citral cyclodehydration. *ChemSusChem.*, **3**, 1264–1267.

118 Fonseca, G.S., Scholten, J.D., and Dupont, J. (2004) Iridium nanoparticles prepared in ionic liquids: an efficient catalytic system for the hydrogenation of ketones. *Synlett*, **9**, 1525–1528.

119 Yinghuai, Z., Chenyan, K., Peng, A.T., Emi, A., Monalisa, W., Louis, L.K.-J., Hosmane, N.S., and Maguire, J.A. (2008) Catalytic phenylborylation reaction by iridium(0) nanoparticles produced from hydridoiridium carborane. *Inorg. Chem.*, **47**, 5756–5761.

120 Ott, L.S., Campbell, S., Seddon, K.R., and Finke, R.G. (2007) Evidence that imidazolium-based ionic ligands can be metal(0)/nanocluster catalyst poisons in at least the test case of iridium(0)-catalyzed acetone hydrogenation. *Inorg. Chem.*, **46**, 10335–10344.

121 Migowski, P., Texeira, S.R., Machado, G., Alves, M.C.M., Geshev, J., and Dupont, J. (2007) Structural and magnetic characterization of Ni nanoparticles synthesized in ionic liquids. *J. Electron Spectrosc. Relat. Phenom.*, **156–158**, 195–199.

122 (a) Hu, Y., Yu, Y., Hou, Z., Yang, H., Feng, B., Li, H., Qiao, Y., Wang, X., Hua, L., Pan, Z., and Zhao, X. (2010) Ionic liquid immobilized nickel(0) nanoparticles as stable and highly efficient catalysts for selective hydrogenation in the aqueous phase. *Chem. Asian J.*, **5**, 1178–1184; (b) Hu, Y., Yu, Y., Zhao, X., Yang, H., Feng, B., Li, H., Qiao, Y., Hua, L., Pan, Z., and Hou, Z. (2010) Catalytic hydrogenation of aromatic nitro compounds by functionalized ionic liquids-stabilized nickel nanoparticles in aqueous phase: the influence of anions. *Sci. China Chem.*, **53**, 1541–1548.

123 Tsuji, J. (2004) *Palladium Reagents and Catalysts*, John Wiley & Sons, Ltd, Chichester, UK.

124 (a) Suzuki, A. (2005) Carbon-carbon bonding made easy. *Chem. Commun.*, 4759–4763; (b) Reetz, M.T. and de Vries, J.G. (2004) Ligand-free Heck reactions using low Pd-loading. *Chem. Commun.*, 1559–1563; (c) Zapf, A., Jackstell, R., Rataboul, F., Riermeier, T., Monsees, A., Fuhrmann, C., Shaikh, N., Dingerdisse, U., and Beller, M. (2004) Practical synthesis of new and highly efficient ligands for the Suzuki reaction of aryl chlorides. *Chem. Commun.*, 38–39.

125 Favier, I., Madec, D., Teuma, E., and Gomez, M. (2011) Palladium nanoparticles applied in organic synthesis as catalytic precursors. *Curr. Org. Chem.*, **15**, 3127–3174.

126 Caló, V., Nacci, A., Monopoli, A., and Montingelli, F. (2005) Pd nanoparticles as efficient catalysts for Suzuki and Stille coupling reactions of aryl halides in ionic liquids. *J. Org. Chem.*, **70**, 6040–6044.

127 Durand, J., Teuma, E., Malbosc, F., Kihn, Y., and Gómez, M. (2008) Palladium nanoparticles immobilized in ionic liquid: An outstanding catalyst for the Suzuki C—C coupling. *Catal. Commun.*, **9**, 273–275.

128 Fernández, F., Cordero, B., Durand, J., Muller, G., Malbosc, F., Kihn, Y., Teuma, E., and Gómez, M. (2007) Palladium catalyzed Suzuki C—C couplings in an ionic liquid: Nanoparticles responsible for the catalytic activity. *Dalton Trans.*, 5572–5581.

129 Sanhes, D., Raluy, E., Rétory, S., Saffon, N., Teuma, E., and Gómez, M. (2010) Unexcepected activation of carbon-bromide bond promoted by palladium nanoparticles in Suzuki C—C couplings. *Dalton Trans.*, **39**, 9719–9726.

130 Jeffery, T. (1994) Heck-type reactions in water. *Tetrahedron Lett.*, **35**, 3051–3054.

131 Herrmann, W.A., Alison, M., Fisher, J., Köcher, K., and Artus, G.R.J. (1995) Metal complexes of N-heterocyclic carbenes. A new structural principle for catalysts in homogeneous catalysis. *Angew. Chem. Int. Ed.*, **34**, 2371–2374.

132 (a) Reetz, M.T. Lohmer, G., and Schwickardi, R. (1998) New catalyst system for the Heck reaction of unreactive aryl halides Angew. *Chem. Int. Ed.*, **37**, 481–483; (b) Reetz, M.T. and Westermann, E. (2000) Phosphane-free palladium-catalyzed coupling reactions: The decisive role of Pd nanoparticles. *Angew. Chem. Int. Ed.*, **39**, 165–168.

133 Gürtler, C. and Buchwald, S.L. (1999) A phosphane-free catalyst system for the Heck arylation of disubstituted alkenes: application to the synthesis of trisubstituted olefins. *Chem. Eur. J.*, **5**, 3107–3112.

134 Caló, V., Nacci, A., Monopoli, A., Detomaso, A., and Iliade, P. (2003) Pd nanoparticle catalyzed Heck arylation of 1,1-disubstituted alkenes in ionic liquids. Study on factors affecting the regioselectivity of the coupling process. *Organometallics*, **22**, 4193–4197.

135 Caló, V., Nacci, A., and Monopoli, A. (2004) Regio- and stereo-selective carbon-carbon bond formation in ionic liquids. *J. Mol. Catal. A: Chemical*, **214**, 45–56.

136 Caló, V., Nacci, A., Monopoli, A., and Cotugno, P. (2009) Heck reactions with palladium nanoparticles in ionic liquids: Coupling of aryl chlorides with deactivated olefins. *Angew. Chem. Int. Ed.*, **48**, 6101–6103.

137 Wang, L., Li, H., and Li, P. (2009) Task-specific ionic liquid as base, ligand and reaction medium for the palladium-catalyzed Heck reaction. *Tetrahedron*, **65**, 364–368.

138 Zhang, G., Zhou, H., Hu, J., Liu, M., and Kuang, Y. (2009) Pd nanoparticles catalyzed ligand free- Heck reaction in ionic liquid microemulsion. *Green Chem.*, **11**, 1428–1432.

139 (a) Zhao, D., Fei, Z., Geldbach, T.J., Scopelliti, R., and Dyson, P.J. (2004) Nitrile-functionalized pyridinium ionic liquids: synthesis, characterization, and their application in carbon-carbon coupling reactions. *J. Am. Chem. Soc.*, **126**, 15876–15882; (b) Fei, Z., Zhao, D., Pieraccini, D., Ang, W.H., Geldbach, T.J., Scopelliti, R., Chiappe, C., and Dyson, P. J. (2007) Development of nitrile-functionalized ionic liquids for C—C coupling reactions: implication of carbene and nanoparticle catalysts. *Organometallics*, **26**, 1588–1598; (c) Cui, Y., Bionid, I., Chaubey, M., Yang, X., Fei, Z., Scopelliti, R., Hartinger, C.G., Li, Y., Chiappe, C., and Dyson, P.J. (2010) Nitrile-functionalized pyrrolidinium ionic liquids as solvents for cross-coupling reactions involving in situ generated nanoparticle catalyst reservoirs. *Phys. Chem. Chem. Phys.*, **12**, 1834–1841.

140 Dubbaka, S.R., Zhao, D., Fei, Z., Volla, C. R., Dyson, P.J., and Vogel, P. (2006) Novel syntheses of N-arylozazolidin-2-ones via tandem reactions of vinyl sulfonium salts. *Synlett*, 3155–3158.

141 Hamill, N.A., Hardrace, C., and McMath, S.E. (2002) *In situ* XAFS investigation of

palladium species present during the Heck reaction in room temperature ionic liquids. *Green Chem.*, **4**, 139–142.

142 Liu, G., Hou, M., Song, J., Jiang, T., Fan, H., Zhang, Z., and Han, B. (2010) Immobilization of Pd nanoparticles with functional ionic liquid grafted onto cross-linked polymer for solvent-free Heck reaction. *Green Chem.*, **12**, 65–69.

143 For other couplings using ionic liquid-functionalized supports, see also: Gruttadaurai, M., Liotta, L.F., Salvo, A.M. P., Giacalone, F., La Parola, V., Aprile, C., and Noto, R. (2011) Multi-layered, covalently supported ionic liquid phase (mlc-SILP) as highly cross-linked support for recyclable palladium catalysts for the Suzuki reaction in aqueous medium. *Adv. Synth. Catal.*, **353**, 2119–2130.

144 Gholap, A.R., Venkatesan, K., Pasricha, R., Daniel, T., Lahoti, R.J., and Srinivasan, K.V. (2005) Copper- and ligand-free Sonogashira reaction catalyzed by Pd(0) nanoparticles at ambient conditions under ultrasound irradiation. *J. Org. Chem.*, **70**, 4869–4872.

145 Durán Pachón, L., Elsevier, C.J., and Rothenberg, G. (2006) Electroreductive palladium-catalyzed Ullmann reactions in ionic liquids: scope and mechanism. *Adv. Synth. Catal.*, **348**, 1705–1710.

146 Caló, V., Nacci, A., Monopoli, A., and Cotugno, P. (2009) Palladium-nanoparticle-catalyzed Ullmann reactions in ionic liquids with aldehydes as the reductants. *Chem. Eur. J.*, **15**, 1272–1279.

147 (a) Cheng, J. Tang, L., and Xu, J. (2010) An economical, green pathway to biaryls: palladium nanoparticles catalyzed Ullmann reaction in ionic liquid/supercritical dioxide system. *Adv. Synth. Catal.*, **352**, 3275–3286; (b) Cheng, J., Zhang, G., Du, J., Tang, L., Xu, J., and Li, J. (2011) New role of graphene oxide as active hydrogen donor in the recyclable nanoparticles catalyzed Ullmann reaction in environmental friendly ionic liquid/supercritical carbon dioxide system. *J. Mater. Chem.*, **21**, 3485–3494.

148 Adak, L., Chattopadhyay, K., and Ranu, B. C. (2009) Palladium nanoparticle-catalyzed C—N bond formation. A highly regio- and seteroselective allylic amination by allyl acetates. *J. Org. Chem.*, **74**, 3982–3985.

149 Venkatesan, R., Prechtl, M.H., Scholten, J.D., Pezzi, R.P., Macahdo, G., and Dupont, J. (2011) Palladium nanoparticle catalysts in ionic liquids: synthesis, characterization and selective partial hydrogenation of alkynes to *Z*-alkenes. *J. Mater. Chem.*, **21**, 3030–3036.

150 Dash, P., Dehm, N.A., and Scott, R.W.J. (2008) Bimetallic PdAu nanoparticles as hydrogenation catalysts in imidazolium ionic liquids. *J. Mol. Catal. A: Chemical*, **286**, 114–119.

151 Hu, Y., Yang, H., Zhang, Y., Hou, Z., Wang, X., Qiao, Y., Li, H., Feng, B., and Huang, Q. (2009) The functionalized ionic liquid-stabilized palladium nanoparticles catalyzed selective hydrogenation in ionic liquid. *Catal. Commun.*, **10**, 1903–1907.

152 Anderson, K., Cortiñas Fernández, S., Hardacre, C., and Marr, P.C. (2004) Preparation of nanoparticulate metal catalysts in porous supports using an ionic liquid route, hydrogenation and C—C coupling. *Inorg. Chem. Commun.*, **7**, 73–76.

153 Chun, Y.S., Shin, J.Y., Song, C.E., and Lee, S.-G. (2008) Palladium nanoparticles supported onto ionic carbon nanotubes as robust recyclable catalysts in an ionic liquid. *Chem. Commun.*, 942–944.

154 Li, J., Shi, X.-Y., Bi, Y.-Y., Wei, J.-F., and Chen, Z.-G. (2011) Pd nanoparticles in ionic liquid brush: a highly active and reusable heterogeneous catalytic assembly for solvent-free or on-water hydrogenation of nitroarene under mild conditions. *ACS Catal.*, **1**, 657–664.

155 Hu, Y., Yu, Y., Hou, Z., Li, H., Zhao, X., and Feng, B. (2008) Biphasic hydrogenation of olefins by functionalized ionic liquid-stabilized palladium nanoparticles. *Adv. Synth. Catal.*, **350**, 2077–2085.

156 Zhu, W., Yang, H., Yu, Y., Hua, F., Li, H., Feng, B., and Hou, Z. (2011) Amphiphilic ionic liquid stabilizing palladium nanoparticles for highly

efficient catalytic hydrogenation. *Phys. Chem. Chem. Phys.*, **13**, 13492–13500.

157 Jansat, S., Durand, J., Favier, I., Malbosc, F., Pradel, C., Teuma, E., and Gómez, M. (2009) A single catalyst for sequential reactions: dual homogeneous and heterogeneous behaviour of palladium nanoparticles in solution. *ChemCatChem*, **1**, 244–246.

158 Serp, P., Corrias, M., and Kalck, P. (2003) Carbon nanotubes and nanofibers in catalysis. *Appl. Catal., A*, **253**, 337–358.

159 Rodriguez-Pérez, L., Pradel, C., Serp, P., Gómez, M., and Teuma, E. (2011) Supported ionic liquid phase containing palladium nanoparticles on functionalized multi-walled carbon nanotubes: catalytic materials for sequential Heck coupling/hydrogenation process. *ChemCatChem*, **3**, 749–754.

160 Scheeren, C.W., Machado, G., Dupont, J., Fichtner, P.F.P., and Teixeira, S.R. (2003) Nanoscale Pt(0) particles prepared in imidazolium room temperature ionic liquids: synthesis from and organometallic precursor, characterization, and catalytic properties in hydrogenation reactions. *Inorg. Chem.*, **42**, 4738–4742.

161 Abu-Reziq, R., Wang, D., Post, M., and Alper, H. (2007) Platinum nanoparticles supported on ionic liquid-modified magnetic nanoparticles: selective hydrogenation catalysts. *Adv. Synth. Catal.*, **349**, 2145–2150.

162 Li, F., Li, F., Song, J., Song, J., Han, D., and Niu, L. (2009) Green synthesis of highly stable platinum nanoparticles stabilized by amino-terminated ionic liquid and its electrocatalysts for dioxygen reduction and methanol oxidation. *Electrochem. Commun.*, **11**, 351–354.

163 (a) Mo, Z., Zhao, F., Xiao, F., and Zen, B. (2010) Preparation and characterization of AuPt alloy nanoparticle-multi-walled carbon nanotube-ionic liquid composite film for electrocatalytic oxidation of cysteine. *J. Solid State Electrochem.*, **14**, 1615–1620; (b) Yu, P., Qian, Q., Lin, Y., and Mao, L. (2010) *In situ* formation of three-dimensional uniform Pt/carbon nanotube nanocomposites from ionic liquid/carbon nanotube fuel matrix with enhanced electrocatalytic activity toward methanol oxidation. *J. Phys. Chem. C*, **114**, 3575–3579; (c) Chu, H., Shen, Y., Lin, L., Qin, X., Feng, G., Lin, Z., Wang, J., Liu, H., and Li, Y. (2010) Ionic-liquid-assisted preparation of carbon nanotube-supported uniform noble metal nanoparticles and their enhanced catalytic performance. *Adv. Funct. Mater.*, **20**, 3747–3752; (d) Guo, D.-J. (2010) Novel synthesis of PtRu/multi-walled carbon nanotube catalyst via a microwave-assisted imidazolium ionic liquid method for methanol oxidation. *J. Power Sources*, **195**, 7234–7237.

164 (a) Wu, B., Hu, D., Kuang, Y., Liu, B., Zhang, X., and Chen, J. (2009) Functionalization of carbon nanotubes by an ionic-liquid polymer: Dispersion of Pt and PtRu nanoparticles on carbon nanotubes and their electrocatalytic oxidation of methanol. *Angew. Chem. Int. Ed.*, **48**, 4751–4754; (b) Wu, B., Hu, D., Yu, Y., Kuang, Y., Zhang, X., and Chen, J. (2010) Stabilization of platinum nanoparticles dispersed on carbon nanotubes by ionic liquid polymer. *Chem. Commun.*, **46**, 7954–7956.

165 Qiu, L., Liu, B., Peng, Y., and Yan, F. (2011) Fabrication of ionic-liquid-functionalized polypyrrole nanotubes decorated with platinum nanoparticles and their electrocatalytic oxidation of methanol. *Chem. Commun.*, **47**, 2934–2936.

166 Caló, V., Nacci, A., Monopoli, A., Leva, E., and Cioffi, N. (2005) Copper bronze catalyzed Heck reaction in ionic liquids. *Org. Lett.*, **7**, 617–620.

167 Li, J.-H., Tang, B.-X., Tao, L.-M., Xie, Y.-X., Liang, Y., and Zhang, M.B. (2006) Reusable copper-catalyzed cross-coupling reactions of aryl halides with organotins in inexpensive ionic liquids. *J. Org. Chem.*, **71**, 7488–7490.

168 Schwab, R., Singh, D., Alberto, E.E., Piquini, P., Rodrigues, O.E.D., and Braga, A.L. (2011) C–S cross-coupling of thiols with aryl iodides under ligand-free conditions using nano copper oxide as a recyclable catalyst in ionic liquid. *Catal. Sci. Technol.*, **1**, 569–573.

169 Singh, D., Alberto, E.E., Rodrigues, O.E. D., and Braga, A.L. (2009) Eco-friendly cross-coupling of diaryl diselenides with aryl and alkyl bromides catalyzed by CuO nanopowder in ionic liquid. *Green Chem.*, **11**, 1521–1524.

170 Singh, D., Narayanaperumal, S., Gul, K., Godoi, M., Rodrigues, O.E.D., and Braga, A.L. (2010) Efficient synthesis of selenoesters from acyl chlorides mediated by CuO nanopowder in ionic liquid. *Green Chem.*, **12**, 957–960.

171 Singh, P., Katyal, A., Kalra, R., and Chandra, R. (2008) Copper nanoparticles in an ionic liquid: an efficient catalyst for the synthesis of bis-(4-hydroxy-2-oxothiazolyl)methanes. *Tetrahedron Lett.*, **49**, 727–730.

172 Singh, P., Katyal, A., Kalra, R., and Chandra, R. (2008) Copper nanoparticles in ionic liquid: An easy and efficient catalyst for the coupling of thiazolidine-2,4-dione, aromatic aldehyde and ammonium acetate. *Catal. Commun.*, **9**, 1618–1623.

173 Singh, P., Kumari, K., Katyal, A., Kalra, R., and Chandra, R. (2009) Copper nanoparticles in ionic liquid: an easy and efficient catalyst for selective carba-Michael addition reaction. *Catal. Lett.*, **127**, 119–125.

174 Singh, P., Kumari, K., Katyal, A., Kalra, R., and Chandra, R. (2009) Cu nanoparticles in ionic liquid: an easy and efficient catalyst for addition-elimination reaction between active methylene compounds and imines in an ionic liquid. *Catal. Lett.*, **130**, 648–654.

175 Raut, D., Wankhede, K., Vaidya, V., Bhilare, S., Darwatkar, N., Deorukhkar, A., Trivedi, G., and Salunkhe, M. (2009) Copper nanoparticles in ionic liquids: recyclable and efficient catalytic system for 1,3-dipolar cycloaddition reaction. *Catal. Commun.*, **10**, 1240–1243.

176 (a) Zhao, D., Fei, Z., Ang, W.H., and Dyson, P.J. (2006) A Strategy for the synthesis of transition-metal nanoparticles and their transfer between liquid phases. *Small*, **2**, 879–883; (b) Warren, S.C., Banholzer, M., Slaughter, L.S., Giannelis, E.P., DiSalvo, F.J., and Wiesner, U.B. (2006) Generalized route to metal nanoparticles with liquid behavior. *J. Am. Chem. Soc.*, **128**, 12074–12075; (c) de Vries, G.A., Brunnbauer, M., Hu, Y., Jackson, A.M., Long, B., Neltner, B.T., Uzun, O., Wunsch, B., and Stellacci, F. (2007) Divalent metal nanoparticles. *Science*, **315**, 358–361; (d) Dash, P. and Scott, R.W. (2009) 1-methylimidazole stabilization of gold nanoparticles in imidazolium ionic liquids. *Chem. Commun.*, 812–814; (e) Hatakeyama, Y., Okamoto, M., Torimoto, T., Kuwabata, S., and Nishikawa, K. (2009) Small-angle X-ray scattering study of Au nanoparticles dispersed in the ionic liquids 1-alkyl-3-methylimidazolium tetrafluoroborate. *J. Phys. Chem. C*, **113**, 3917–3922; (f) Redel, E., Walter, M., Thomann, R., Vollmer, C., Hussein, L., Scherer, H., Krüger, M., and Janiak, C. (2009) Synthesis, stabilization, functionalization and DFT calculations of gold nanoparticles in fluorous phases (PTFE and ionic liquids). *Chem. Eur. J.*, **15**, 10047–10049.

177 Buaki, M., Aprile, C., Dhakshinamoorthy, A., Alvaro, M., and García, H. (2009) Liposomes by polymerization of an imidazolium ionic liquid: use as microreactors for gold-catalyzed alcohol oxidation. *Chem. Eur. J.*, **15**, 13082–13089.

178 Wang, Z., Zhang, Q., Kuehner, D., and Niu, L. (2008) Green synthesis of 1–2nm gold nanoparticles stabilized by amine-terminated ionic liquid and their electrocatalytic activity in oxygen reduction. *Green Chem.*, **10**, 907–909.

179 Corma, A., Domínguez, I., Ródenas, T., and Sabater, M.J. (2008) Stabilization and recovery of gold catalysts in the cyclopropanation of alkenes within ionic liquids. *J. Catal.*, **259**, 26–35.

180 Luo, L., Yu, N., Tan, R., Jin, Y., Yin, D., and Yin, D. (2009) Gold nanoparticles stabilized by task-specific oligomeric ionic liquid for styrene epoxidation without using VOCs as solvent. *Catal. Lett.*, **130**, 489–495.

181 Dash, P., Miller, S., and Scott, R.W.J. (2010) Stabilizing nanoparticle catalysts

in imidazolium-based ionic liquids: A comparative study. *J. Mol. Catal. A: Chemical*, **329**, 86–95.

182 Raula, M., Rashid, H., Paira, T.K., Dinda, E., and Mandal, T.K. (2010) Ascorbate-assisted growth of hierarchical ZnO nanostructures: sphere, spindle, and flower and their catalytic properties. *Langmuir*, **26**, 8769–8782.

183 Naryanaperumal, S., Gul, K., Kawasoko, C.Y., Singh, D., Dornelles, L., Rodrigues, O.E.D., and Braga, A.L. (2010) Transition metal oxide nanopowder and ionic liquid: an efficient system for the synthesis of diorganyl selenides, selenocysteine and derivatives. *J. Braz. Chem. Soc.*, **21**, 2079–2087.

184 Taghvaei, V., Habibi-Yangjeh, A., and Behboudnia, M. (2010) Simple and low temperature preparation and characterization of CdS nanoparticles as a highly efficient photocatalyst in presence of a low-cost ionic liquid. *J. Iran. Chem. Soc.*, **7**, S175–S186.

185 Chauvin, Y., Mussmann, L., and Olivier, H. (1995) A novel class of versatile solvents for two-phase catalysis: hydrogenation, isomerization, and hydroformylation of alkenes catalyzed by rhodium complexes in liquid 1,3-dialkylimidazolium salts. *Angew. Chem. Int. Ed. Engl.*, **34**, 2698–2700.

6
Supported Ionic Liquid Thin Film Technology

Judith Scholz and Marco Haumann

6.1
Introduction

Heterogeneous catalysts still play the dominant role in large-scale industrial processes with more than 90% of all catalytic steps being catalyzed by solid materials. Although homogeneous and nanoparticle (NP) catalysts offer significant advantages with respect to activity and selectivity over heterogeneous catalysts, their elaborate recycling and separation from the product phase constitutes a major drawback of these, otherwise attractive, materials. The ultimate goal in catalysis engineering is therefore to combine the advantages of homogeneous or nanoparticle catalysis with the ease of separation that is achieved in classical heterogeneous catalysis [1–5].

In this context, the application of ionic liquids (ILs) as an alternative reaction media attracted growing interest in the last decade [6–10]. Their extremely low vapor pressure, combined with a tunable polarity and a high thermal stability, make them attractive solvents for biphasic ionic liquid–organic liquid catalysis. With the appropriate combination of anion and cation, solubility properties can be adjusted in such a way, that the molecular catalysts or the nanoparticle as well as the substrates readily dissolve in the ionic liquid, whereas the products form an extra phase, enabling easy catalyst recovery and product separation [11]. In addition, ionic liquids are remarkably suitable for the generation of metallic NPs that are stabilized against an otherwise strong tendency for particle aggregation and agglomeration (see Chapter 5) [12–14]. A protective layer around the NPs is formed by the IL, providing a steric and electronic protection against agglomeration [15–17]. Thus, the application of ILs in catalysis as solvent is of great interest for nanoparticle catalysis. If the IL is in contact with the surface of an inorganic or organic support material, highly efficient catalysts can be achieved that possess the macroscopic features of classical heterogeneous catalysts [18]. Based on the role of the ionic liquid, three different immobilization techniques can be distinguished, termed SILP, SCILL, and IL surface modification.

Nanomaterials in Catalysis, First Edition. Edited by Philippe Serp and Karine Philippot.

6.1.1
Supported Ionic Liquid Phase (SILP)

Conventional biphasic systems require large amounts of quite costly ILs, making potential industrial processes economically inefficient. Moreover, the high viscosity of ILs can induce mass transfer limitations in the case of a fast chemical reaction, causing only a minor part of the IL and the therein entrapped nanoparticles to take part in the catalytic reaction. The supported ionic liquid phase (SILP) concept represents a possibility to minimize these depicted negative effects of ILs in biphasic catalysis [19, 20]. In this technology, a thin film of IL, containing a molecular defined catalyst or metallic NPs, is immobilized on the surface of a highly porous, inert support material, schematically depicted in Figure 6.1a. The modification of the solid support material can be achieved by various methods such as physisorption or covalent anchoring of IL fragments [21]. The resulting material appears macroscopically as a solid, thus, it can be handled like a heterogeneous catalyst and be applied in continuously operated processes. During SILP catalysis, the feedstock molecules enter the residual pore space of the system, dissolve in the IL phase and react at the catalyst. The product molecules formed diffuse back, out of the void pore space and eventually leave the SILP particle. Using transition metal complexes, this concept has been successfully tested for several reactions including hydrogenation [22, 23], hydroformylation [19, 20, 24, 25], methanol carbonylation [26], Heck reactions [27], metathesis [28, 29] and hydroaminations [30].

It should be pointed out that SILP technology is ideally suited mainly for gas-phase reactions. In contrast, the application of SILP catalysts in liquid-phase reactions is often restricted by cross-solubility problems, mechanical removal of the IL film, or catalyst leaching from the film into the product mixture [31, 32]. The term nano-SILP is used in the context of NPs dissolved within the IL film with no contact between the NP and the surface. This can be achieved by reducing the metal precursor after the impregnation step. Examples are given in Section 6.2.1. If the NPs have been deposited prior to the coating step, these materials are labeled nano-SCILL.

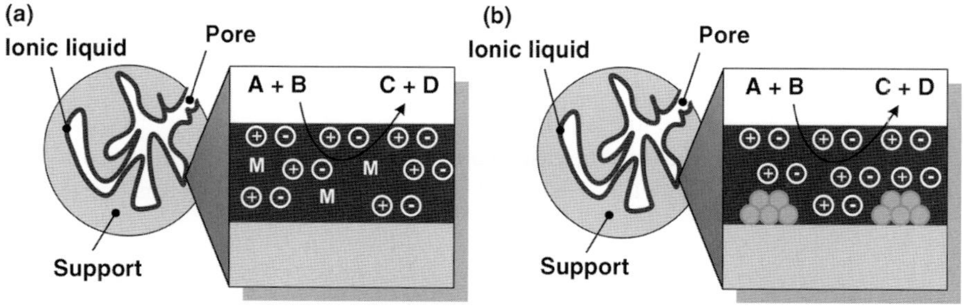

Figure 6.1 Schematic representation of (a) SILP with dissolved transition metal complexes (M) or nanoparticles (M) and (b) SCILL with surface bound nanoparticles.

6.1.2
Solid Catalysts with Ionic Liquid Layers (SCILL)

In strong analogy to the SILP technology, a concept called 'solid catalysts with ionic liquid layers' (SCILL) has been discussed in literature [33]. In this case, a solid heterogeneous catalyst is coated with a thin film of IL as shown in Figure 6.1b. In contrast to SILP catalysts, the support material is catalytically active itself and no homogeneous catalyst or dissolved nanoparticle is involved. It has been experimentally demonstrated that such systems may exhibit better selectivity and even higher activities than their uncoated analogs [34]. However, the origins of such selectivity and activity effects are yet unclear [35]. The IL may influence the catalytic performance in a twofold manner. On the one hand, it can directly interact with the active centers comparable to the behavior of a ligand. These so-called 'co-catalytic effects' have been extensively reported for catalytically active metallic NPs. Such interactions may even lead to decomposition of the IL under reaction conditions, with the co-adsorbed decomposition products further modifying the catalytic properties. On the other hand, the IL can modify the effective concentrations of the substrates and intermediates at the active sites, so that the solubility of liquids or gaseous reactants in the IL differs in a significant way from that in the liquid organic phase, causing a 'physical solvent effect' [36–38]. The SCILL technology has successfully been applied in various hydrogenation reactions, resulting in better selectivity and enhanced activities [33–35]. Examples of SCILL catalysis involving metallic NPs can be found in Section 6.2.2.

Similarly to the SILP technology, the application of the SCILL technique in the liquid-phase may also be affected by physical removal of the IL film due to convective liquid flow, potentially causing loss of activity and selectivity.

6.1.3
Ionic Liquid as Surface Modifier

Conventional solid supports for the immobilization of metallic NPs include for example metal oxides as well as carbons. However, especially in the absence of space confinement, particles in a narrow size range, which are necessary for sufficient catalytic activity, tend to aggregate and/or agglomerate due to their higher surface free energies [39]. Thus, the physico-chemical properties of the support material, such as the textural structures, compositions, and affinities for the respective metal play decisive roles for the overall stability of the supported metallic NP. In these materials, the amount of IL can be reduced further compared with SILP and SCILL systems, in the extreme case to only a monolayer or islands of IL coating the support [40]. The role of IL in these systems is to transfer a certain functionality of the IL to the support surface. In NP catalysis the main role of the IL is to anchor the nanoparticles on the support and to prevent agglomeration of the NPs during reaction.

Such thin films of ionic liquid can obviously have no significant influence on substrate solubility. The IL is usually anchored onto the support by chemisorption,

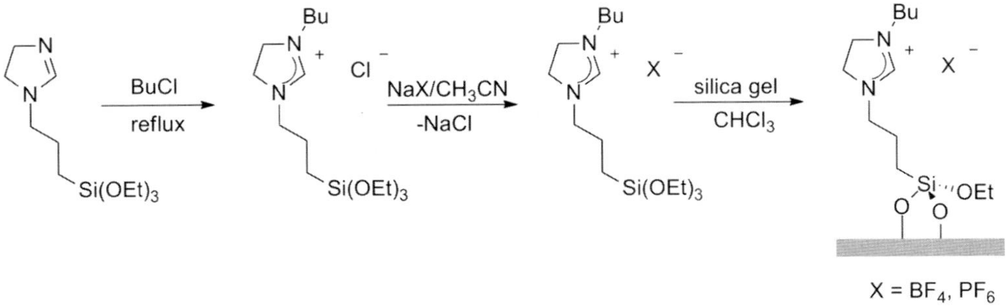

Scheme 6.1 Example of covalent bonding of ionic liquid like fragments on silica gel surface according to the original work by Mehnert [19].

involving a surface reaction between the ionic liquid's cation and the surface as depicted in Scheme 6.1 for the original work by Mehnert in 2003 [19]. Other procedures have been reported in the literature and will be highlighted in Section 6.2.3.

6.2 Nanoparticle Catalysis with Supported Ionic Liquids

6.2.1 Nanoparticles in SILP Systems (nano-SLIP)

The group of Virtanen studied the hydrogenation of citral and cinnamaldehyde (Scheme 6.2) using palladium nanoparticles (PdNPs) in SILP catalysts from a material as well as an engineering point of view [41]. The applied PdNPs were immobilized in different ILs on an activated carbon cloth (ACC) support with a high BET surface area of 1120 $m^2\,g^{-1}$. The ACC was impregnated in a mixture of IL and palladium acetate in dry acetone. The solvent was removed in an oven at 353 K,

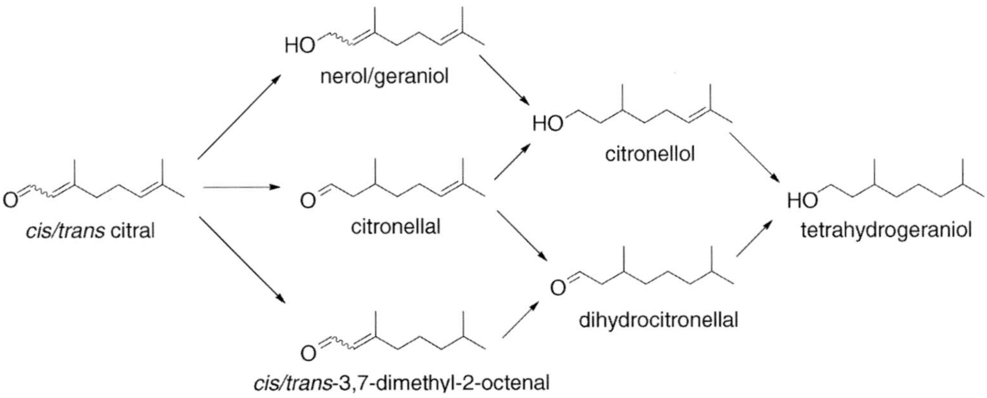

Scheme 6.2 Reaction network of citral hydrogenation.

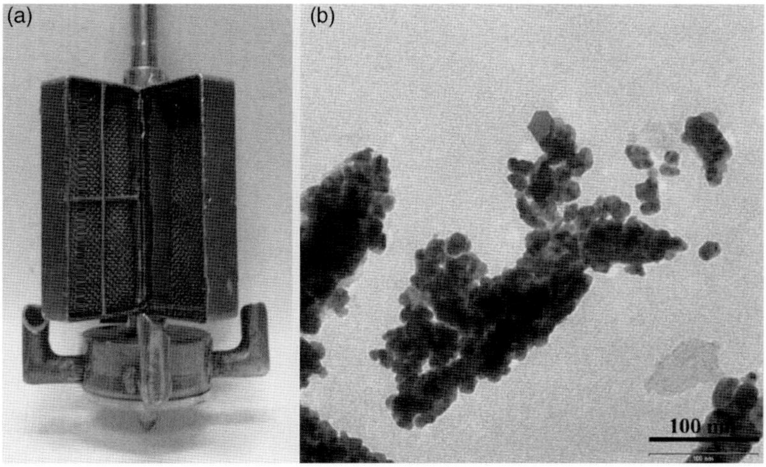

Figure 6.2 (a) Dedicated stirrer with gas-entrainment and ACC sample holder. (b) EFTEM picture of the supported ionic liquid catalyst containing palladium particles (black spots and areas) in $[NB_4MPy][BF_4]$ on ACC [41–45].

followed by reduction of the Pd at 393 K under H_2 flow. The cloth could be placed in a Carberry-type sample holder in order to minimize mass transport effects in the batch experiments.

X-ray photoelectron spectroscopy (XPS) results indicated that the oxidation state of the Pd varied, depending on the IL environment [42] Only in 1-butyl-3-methylimidazolium tetrafluoroborate $[BMIM][BF_4]$ were Pd^0 nanoparticles observed, while in 1-butyl-3-methylimidazolium hexafluorophosphate $[BMIM][PF_6]$, $[NB_4MPy][BF_4]$ and $[A336][PF_6]$ Pd^I species were detected, indicating formation of complexes between Pd and the IL. Energy-filtered transmission electron microscopy (EFTEM) experiments revealed the Pd particles to be in the range between 2 and 60 nm with predominant sizes around 10 nm as shown in Figure 6.2. Interestingly, the IL-coated catalysts exhibited higher initial reaction rates than the uncoated Pd on ACC, demonstrating that the thin film of IL does not constitute a large diffusion barrier but rather that the IL prevents agglomeration of the PdNPs under reaction conditions, ensuring a high dispersion of the catalyst on the surface. Main products were citronellal (CAL) and dihydrocitronellal with minor amounts of 3,7-dimethyl-octanal when using $[NB_4MPy][BF_4]$ as ionic liquid.

Deactivation was observed when using these SILP catalysts in consecutive batch reactions. Careful analysis of both the ionic liquid leaching by high-performance liquid chromatography (HPLC) and the loss of Pd metal by inductively coupled plasma mass spectrometry (ICP-MS) revealed that no significant amount of IL was removed from the ACC support after the second cycle. Pd leaching was below 50 ppb and could not explain the deactivation behavior [42]. It was assumed that the accumulation of hydrogenation products within the IL

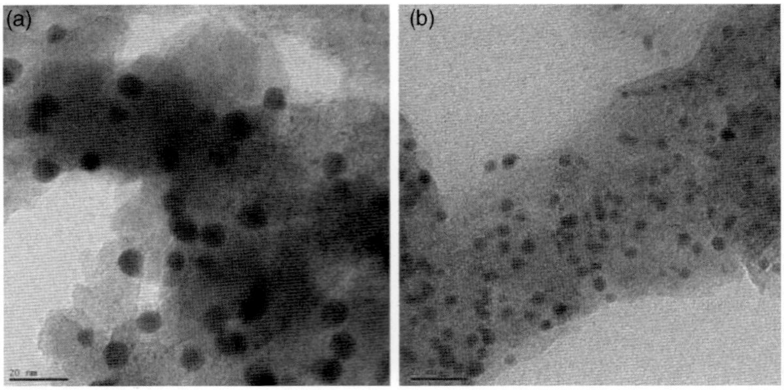

Figure 6.3 HRTEM micrographs: (a) 10 nm-sized PdNPs formed in [BMIM][PF_6]/CNF/SMF and (b) 5 nm-sized nanoparticles formed in [BMIMOH][NTf_2]/CNF/SMF [46].

film led to swelling of the film and clogging of transport pores in a similar way to that reported for gas-phase hydroformylation reactions [24e]. The authors modeled the reaction kinetics based on a Langmuir–Hinshelwood mechanism and included a deactivation term in their model to fit the data [43]. In a similar approach, they modeled the kinetics of cinnamaldehyde hydrogenation using the same SILP catalyst [44]. Addition of the Lewis acid $ZnCl_2$ to the IL film yielded a catalyst that produced significant amounts of menthols via ring closing reactions of citronellal [45].

Palladium nanoparticles of 5 and 10 nm were synthesized by Ruta *et al.* in the ILs [BMIM][PF_6] and [BMIMOH][NTf_2] (NTf_2 = bis(tri-fluoromethylsulfonyl)imide) supported on carbon nanofibers anchored to sintered metal fibers (CNF/SMF) [46]. The Pd(acac)$_2$ precursor in [BMIM][PF_6] was reduced by H_2 to yield particles of 10 nm average size, while heating of the same precursor in [BMIMOH][NTf_2] afforded 5 nm particles (Figure 6.3).

It was postulated that the OH functionality in the IL is responsible for a redox mechanism, leading to Pd^0. Both SILP catalysts were applied in the gas-phase hydrogenation of acetylene. Based on the coordination strength of the anion towards the nanoparticle, different rates were obtained with the less coordinating [BMIMOH][NTf_2] always giving the higher rates. More important, the selectivity for the desired ethene was high with 85% at 423 K for the SILP catalysts compared to 60% of the IL-free Pd on CNF/SMF under identical conditions. This selectivity increase was attributed to the lower ethene solubility at higher temperatures, making the consecutive reaction from ethene to ethane negligible. The catalysts even showed good activity and selectivity in the hydrogenation of acetylene in an ethene-rich feed that is common in ethene polymerization, making the SILP technology applicable for this industrially important reaction to reduce the amount of acetylene impurities.

In a follow-up study, a similar support was used by Kovtunov *et al.* to investigate the hydrogenation of propyne by parahydrogen-induced polarization (PHIP)

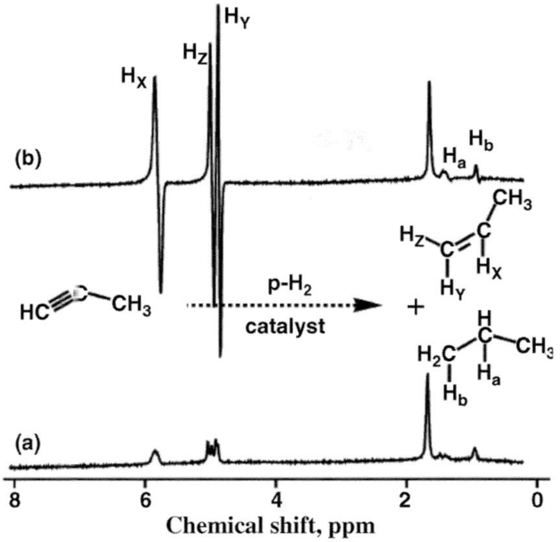

Figure 6.4 ^{1}H NMR spectra detected during the heterogeneous hydrogenation of propyne over a Pd^0/[BMIM][PF_6]/C catalyst with (a) normal hydrogen and (b) parahydrogen. The polarized protons of methyne and methylene groups of propylene are labeled as H_X, H_Y and H_Z and the polarized methylene and methyl protons of propane are labeled H_a, H_b. The spectra are shown on the same vertical scale [47].

experiments in the presence of Pd/[BMIM][PF_6]/CNF/MSF and Pd/[BMIMOH][NTf_2]/CNF/MSF [47]. At 403 K both SILP catalysts were active in the hydrogenation of propyne into propylene and propane. When normal H_2 was replaced with parahydrogen-enriched H_2 gas, the high-intensity polarized multiplets of propylene were observed for both catalysts as shown in Figure 6.4.

The presence of PHIP suggested that both hydrogen atoms of the H_2 molecule were added to the same substrate by the catalyst in a pairwise manner.

In an attempt to stabilize a higher number of PdNPs within the IL film, Gruttadauria *et al.* synthesized multilayered SILP materials as schematically shown in Scheme 6.3 [48]. Synthesis of the SILP material was straightforward, starting either from SiO_2 or SBA-15. The total polymerization of the bis-vinylimidazolium bromide resulted in the formation of a highly cross-linked layer of IL, as confirmed by ^{13}C- nuclear magnetic resonance (NMR). Palladium was introduced into this

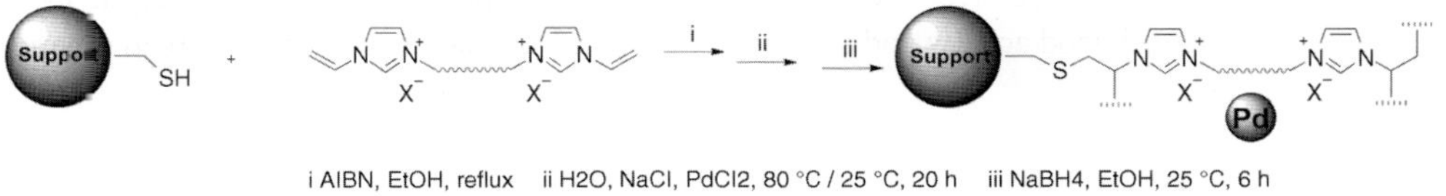

Scheme 6.3 Synthesis of multi-layered-SILP materials SiO_2-i and SBA-15-i and catalysts SiO_2-Pd-iii and SBA-15-Pd-iii.

Table 6.1 Recycling investigations for Suzuki coupling reaction of phenylboronic acid and 4-bromobenzaldehyde using multi-layerd ionic liquid Pd-SILP catalysts.

Entry	Cycle	Pd-SILP SiO_2/%[a)]	Pd-SILP SBA-15/%[a)]
1	1	88	88
2	2	88	85
3	3	86	83
4	4	85	77

a) Yield of biphenyl-4-carboxaldehyde. Reaction conditions: 1.34 g (11 mmol) phenylboronic acid, 1.85 g (10 mmol) 4-bromobenzaldehyde, 10 mg SILP catalyst, $EtOH/H_2O$, K_2CO_3, 298 K.

polymeric film of IL fragments by stirring the coated supports in an aqueous solution of $PdCl_2$. After impregnation, the solids were filtered off and no Pd was found in the filtrate. Reduction of the Pd species was achieved by using $NaBH_4$ in ethanol, yielding highly dispersed PdNPs with an average size of 2.5 nm, as detected by X-ray diffraction (XRD).

The as-prepared catalysts were tested in a variety of Suzuki coupling reactions. The catalysts exhibited high activity even under mild conditions of 0.1 mol% Pd and room temperature. At low conversions, no Pd leaching could be detected by blind tests, whereas at full conversion Pd leaching was found to be 2%. Recycling investigations were carried out using catalysts SiO_2-Pd-iii and SBA-15-Pd-iii in 0.1 mol% in the reaction between phenylboronic acid and 4-bromobenzaldehyde as summarized in Table 6.1.

The catalysts were recovered by centrifugation and could be reused for three consecutive runs. In the fourth run, a lower yield was observed for both catalysts. Post run XRD experiments indicated highly dispersed PdNPs of 3 nm (SBA-15) and 4 nm (SiO_2). While XRD patterns indicated a minor sintering of the PdNPs, both transmission electron microscopy (TEM) and XPS studies indicated that there is neither sintering of palladium particles nor leaching during reaction as the origin of the deactivation (Figure 6.5).

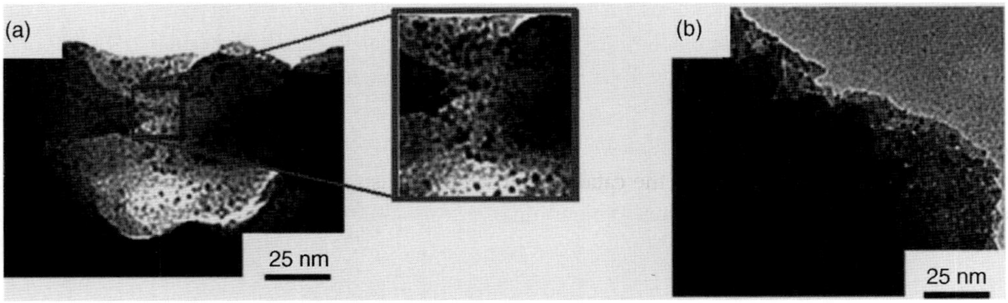

Figure 6.5 TEM images for (a) SBA-15-Pd-iii and (b) SiO_2-Pd-iii catalysts [48].

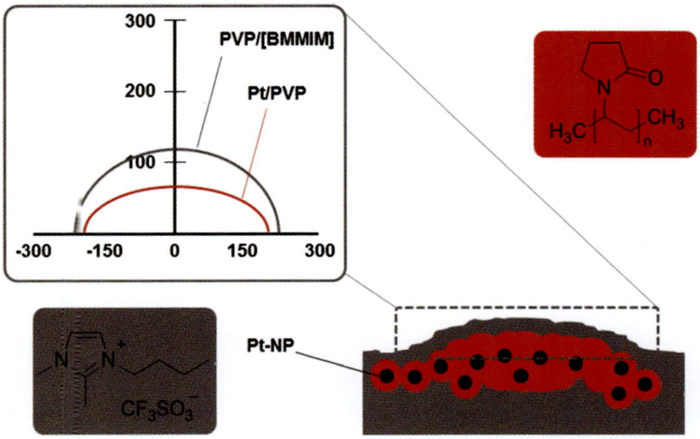

Figure 6.6 Graphical representation of the average surface structure of PtNPs embedded in PVP and ionic liquid [BMMIM][OTf] derived from NICISS data. Redrawn from [49].

Knapp *et al.* reported a similar approach to stabilize and immobilize platinum nanoparticles (PtNPs) in SILP systems with the help of a polymeric shell of polyvinylpyrrolidone (PVP) [49]. PtNPs were prepared by either the Busser or Teranishi method, yielding particles sizes around 2.2 nm. A silica support (100 μm) was added to a Pt/PVP/[BMMIM][OTf] solution. The suspension (containing 1 wt % Pt as NPs) was stirred at room temperature for 10 min and the volatile components were removed slowly by freeze-drying to give a free-flowing gray powder. The prepared SILP material was characterized completely be means of TEM, liquid and solid-state NMR, atomic force microscopy (AFM) and neutral impact collision ion scattering spectroscopy (NICISS). Based on the results from the NICISS the authors postulated an average thickness of the polymer layer around platinum to be 5 nm, corresponding to 5 monolayers of PVP. Additionally, they assumed that the IL surface is composed of local structures that are described by droplet-like ellipsoids as shown in Figure 6.6.

The PVP-encapsulated Pt aggregates float at the surface of the IL and are wetted by the IL. This mobility of the Pt particles markedly enhances the interaction with the IL in comparison to Pt particles supported on silica, exemplified by the activity of these materials in ethene hydrogenation at 303 K.

In order to immobilize rhodium nanoparticles (RhNPs), Gelesky and coworkers encapsulated RhNPs (4 nm) dispersed in [BMIM][BF_4] in a silica network via the sol-gel method under acid and basic conditions [50]. By that, the IL served not only as stabilizing media for the catalytic particles but also as template for the silica support during its preparation. Depending on the conditions (acidic or basic), the obtained xerogels differed in the encapsulated IL content. Acidic conditions afforded higher IL loading and xerogels with larger pore diameter, which in turn yielded in catalyst systems with higher catalyst activity in the hydrogenation of alkenes. Nonetheless, all SILP systems were more active than that constituted of isolated RhNPs or than the commercial Rh/C for the hydrogenation of alkenes. The catalytic material could be recovered and reused at least ten times without losing activity and showed no sign of agglomeration as observed from TEM analysis.

Recently, SILP technology was also implemented in enzymatic biosensors containing metallic NPs designed for quantitative analysis in pharmaceutical samples via square-wave voltammetry [51, 52]. In one system, gold nanoparticles (AuNPs) dispersed in [BMIM][PF_6] were immobilized in chitin chemically cross-linked with glyoxal and epichlorohydrin providing an excellent environment for enzyme immobilization. This modified biopolymer matrix was used as support for the immobilization of peroxidases obtained from pea, and employed to develop a biosensor for rosmarinic acid quantification [51]. In the other system, a matrix, comprising iridium nanoparticles and [BMIM][BF_4] supported in montmorillonate (MMT, see Section 6.2.3), was used to immobilize laccase and polyphenol oxidase and applied in a biosensor for the determination of rutin [52]. The as-prepared biosensors displayed high sensitivities, adequate selectivity, and good reproducibility. The efficient analytical performance of both sensors could be attributed to the effective immobilization of the enzymes in the modified chitin and MMT matrix, respectively, the significant contribution of the high conductivity of the ILs, the facilitation of electron transfer promoted by the NPs, and the inherent catalytic ability of the ILs and the metallic NPs. These modifiers provoked an increase in the current response compared to an unmodified electrode.

6.2.2
Nanoparticles in SCILL Systems (nano-SCILL)

The hydrogenation of citral (see Scheme 6.2) was used as model reaction for a detailed study on the influence of IL coating on supported PdNPs by the group of Claus. In one of their first papers, the authors compared the hydrogenation in a batch reactor using nonsupported, IL-coated catalysts with native Pd/C catalysts [53]. The IL was added either as additive or used as bulk solvent, when applying the native catalyst. The results from these experiments are summarized in Table 6.2.

When the native catalyst was used in *n*-hexane, almost equal amounts of citronellal and dihydrocitronellal were obtained. Using the hydrophilic IL 1-butyl-3-methylimidazolium dicyanamide [BMIM][DCA] as solvent improved the selectivity toward citronellal (CAL, see Scheme 6.2) to 97%. Interestingly, addition of only small amounts of [BMIM][DCA] to *n*-hexane and the native catalyst resulted in further improvement of the selectivity up to 99% CAL. The same result was observed when using the SCILL catalyst in *n*-hexane. Similar to the results reported by Jess [33], the IL seems to increase the selectivity by inhibiting the consecutive hydrogenation reactions. It should be noted that almost no diffusion limitation through the thin film of IL was observed, also in agreement with calculations from Jess.

Table 6.2 Hydrogenation of citral using different Pd catalysts and reaction conditions.

Entry	Reaction Type	Conversion X/%	S_{CAL}/%[a)]	S_{DHC}/%[b)]	S_{rest}/%[c)]
1	IL free	100	41	49	10
2	IL-coated[d)]	100	>99	<1	<1
3	IL as additive[e)]	42	>99	<1	<1
4	IL as solvent[f)]	100	97	1	2

Reaction conditions: 423 K, 200 mg catalyst, 10 barvH_2, $c_{citral,0} = 1.1\ mol\,l^{-1}$, *n*-hexane solvent, 6 h.
a) Selectivity toward citronellal.
b) Selectivity toward dihydrocitronellal.
c) Cumulative selectivity toward 3,7-dimethyl-2-octenal, isopulegol and others.
d) Catalyst with 50 w% Il loading.
e) 200 mg [BMIM][DCA] added to reaction mixture.
f) *n*-hexane replaced by [BMIM][DCA].

For the same reaction network, the authors applied a different catalytic support, namely Ru on Al_2O_3 [54]. Covering this catalyst with IL resulted in the preferred hydrogenation of the C=O bond rather than the C=C bond. The major products were therefore geraniol and nerol (see Scheme 6.2) as shown in Table 6.3 and Figure 6.7.

The native catalyst showed a lower selectivity for geraniol and nerol around 26% throughout the course of hydrogenation, while the SCILL type catalyst, regardless of type and amount of IL, yielded the intermediates in 43 to 46% selectivity. Again, interaction between the weakly basic IL and the metal center alters the chemical nature of the catalyst and leads to improved selectivity.

A detailed study on the interactions between IL and Pd/SiO_2 support included CO pulse chemisorption to measure the postulated interaction between the IL and the active center [34]. In the case of IL-coated catalysts, the CO chemisorption was lowered by a factor of three, indicating that the IL blocks active sites. Additionally, it

Table 6.3 Hydrogenation of citral catalyzed by Ru/Al_2O_3 with different ionic liquids as additives.

Entry	Ionic Liquid	Conversion X/%	S_{CAL}/%[a)]	S_{G+N}/%[b)]	S_{COL}/%[c)]	S_{rest}/%[d)]
1	IL free	>99	51	26	14	9
2	[BMIM][NTf_2]	94	37	43	13	7
3	[HMIM][NTf_2]	93	37	43	14	6
4	[BMIM][NTf_2][e)]	86	34	46	16	4

Reaction conditions: 373 K, 1 g catalyst, 70 bar H_2, $c_{citral,0} = 0.3\ mol\,l^{-1}$, *n*-hexane solvent, 6 h, 300 mg IL.
a) Selectivity toward citronellal.
b) Selectivity toward geraniol and nerol.
c) Selectivity toward citronellol.
d) Cumulative selectivity toward 3,7-dimethyl-2-octenal and others.
e) Catalyst with 50 w% IL loading; 500 mg IL.

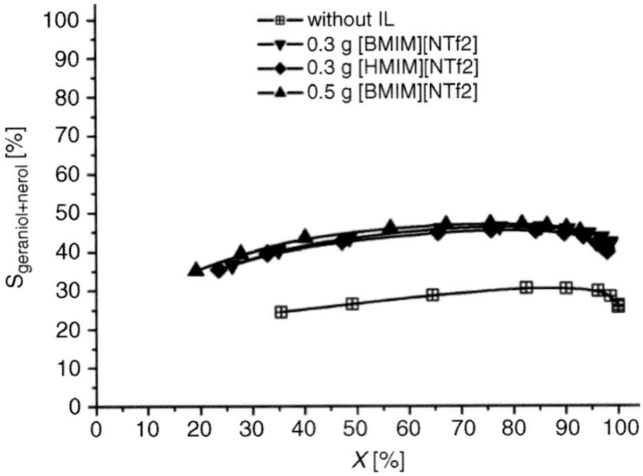

Figure 6.7 Selectivity-conversion plot of Ru/Al_2O_3 catalyzed citral hydrogenation with ionic liquids as additive, conditions: 1 g Ru/Al_2O_3, 373 K, 7 MPa H_2, 0.3 mol L^{-1} citral in *n*-hexane, 360 min (without IL), and 480 min (IL as additive) [54].

was concluded from Fourier-transform infrared (FTIR) spectroscopy that the dicyanamide [DCA] anion mainly interacts via the central nitrogen with the palladium as indicated in Table 6.4.

A variation of the IL revealed that fluorous based anions resulted in conversions around 70% with moderate selectivity of 60% toward the intermediate citronellal. In contrast, all [DCA]-based SCILL catalysts resulted in full citral conversion and selectivity between 81 and 99%, with [BMPL][DCA] showing the best results. The reactions were found not to be mass transfer limited in all cases.

In 2010, the group of Claus reported a detailed spectroscopic characterization of the Pd-SCILL materials using different ILs to study the influence of the IL modification of the palladium catalyst [35]. X-ray absorption fine structure spectrum (EXAFS), X-ray absorption near edge structure (XANES) and XPS studies revealed that the IL does not change the electronic state of the palladium, since constant Pd–Pd distances were measured. An increase, however, in the Pd–O distance in the presence of ILs was attributed to a better dispersion of the nanoparticles in the IL, leading to partial removal of the palladium from the silica surface. A lower coordination number indicated that the IL could not only stabilize the palladium, but

Table 6.4 Infrared data for [BMIM][DCA]-coated silica materials.

Entry	Ionic Liquid	$\nu_s + \nu_{as}$ (C—N)	ν_1 (C≡N)	ν_2 (C≡N)
1	[BMIM][DCA]/SiO_2	2235	2135	2195
2	Pd/[BMIM][DCA]/SiO_2	2260	2151	2210
3	$Pd^{II}[N(CN)_2]_2$	2305	2190	2236

prevented it from oxidation. In contrast to that, XPS studies indicated the appearance of a partially oxidized palladium, complexed by the IL anion. This ligand role of the anion is in agreement with the FTIR studies as shown in Table 6.4.

The hydrogen uptake of the Pd on silica decreased in the presence of the IL layer and was stronger influenced by the choice of anion than cation. The [DCA] anion gave not only lower hydrogen solubility compared with [NTf_2], but showed lower heats of hydrogen absorption as well, corresponding to a competing mechanism between the IL and the hydrogen for adsorption sites on the palladium. As postulated previously, this competition is believed to be the reason for increased selectivity toward the intermediate product.

Knapp *et al.* reported a similar spectroscopic study on IL-coated Pt on silica in 2009 [55]. The FTIR experiments suggested the interaction between the IL anion and the silica surface via hydrogen bonding. The cations were mainly found to interact with the platinum centers, as was confirmed by inelastic neutron scattering (INS) and solid-state MAS-NMR studies (Figure 6.8).

The Pt-SCILL catalyst was applied in ethene hydrogenation where a zero order in ethene and a first order in hydrogen were observed between 303 and 323 K. The results are in good agreement with literature mechanism and account for complete surface coverage of the Pt with ethene. This is noteworthy, since ethene and the IL cation compete for the binding sites and can help to improve SCILL catalysts in the future by tuning the IL–metal interaction in balance to substrate metal interactions.

The group of Lercher reported on the catalytic properties of alumina supported copper nanoparticles coated with a thin film of 1-butyl-2,3-dimethyl-imidazolium trifluoromethane sulfonate in the low-temperature water-gas shift (WGS)

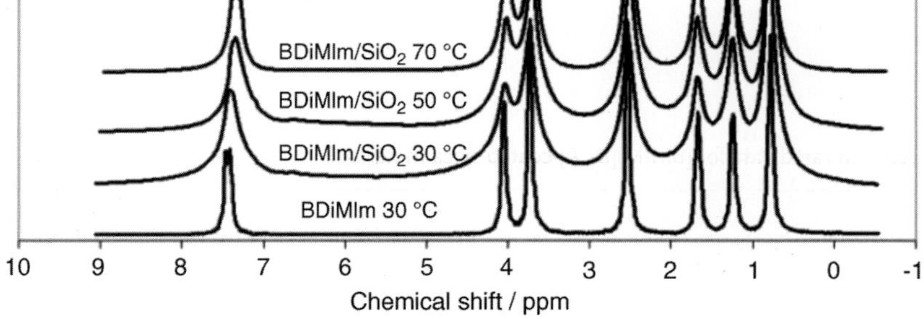

Figure 6.8 Solid state ^1H NMR spectra of neat [BMMIM][OTf] at 303 K, [BMMIM][OTf]/SiO_2 at different temperatures and Pt/[BMMIM][OTf]/SiO_2 at 303 K [55].

reaction [56]. By using an IL, the authors intended to improve the availability of water at the active sites and, consequently, to enhance the catalytic activity.

Compared with uncoated and commercial WGS systems, the SCILL catalysts exhibited a higher turn-over frequency in the water-gas shift reaction at low temperatures. In fact, the observed concentrations of the reactants on the IL-coated catalysts were in good agreement with the initial motivation of the modification of the supported NPs. Higher concentration of water in the proximity of the active sites due to the interaction of the IL with reaction carboxyl intermediates facilitated the decomposition of these species.

A detailed investigation on the interaction of the room temperature IL [BMIM][NTf_2] with the well-defined model catalysts $Pd/Al_2O_3/NiAl(110)$ under ultrahigh vacuum (UHV) conditions in the presence of CO has been reported in 2010 by the groups of Libuda and Steinrück [57]. From time-resolved infrared reflection absorption spectroscopy (IRAS) it was found that [BMIM][NTf_2] adsorbs molecularly on the PdNPs by chemical interactions of the [NTf_2] anion with the Pd particles via the SO_2 groups.

The ionic liquid competes with CO for adsorption sites of the Pd, especially replacing CO from particle edge and defect sites and from defect (particle corner) on-top sites as depicted in Figure 6.9. At higher temperatures, the IL decomposes on the Pd surface, with the decomposition products blocking or modifying most of the adsorption sites. It can be assumed that by this IL–Pd interaction the properties of the catalyst will be altered.

6.2.3
Nanoparticles in IL Surface Modified Systems

As the stabilization effects of ILs on NPs and the successful catalytic application of nanoparticles dispersed in ILs have been demonstrated in several studies (see previous sections), IL-functionalized materials appear to be suitable supports for NPs.

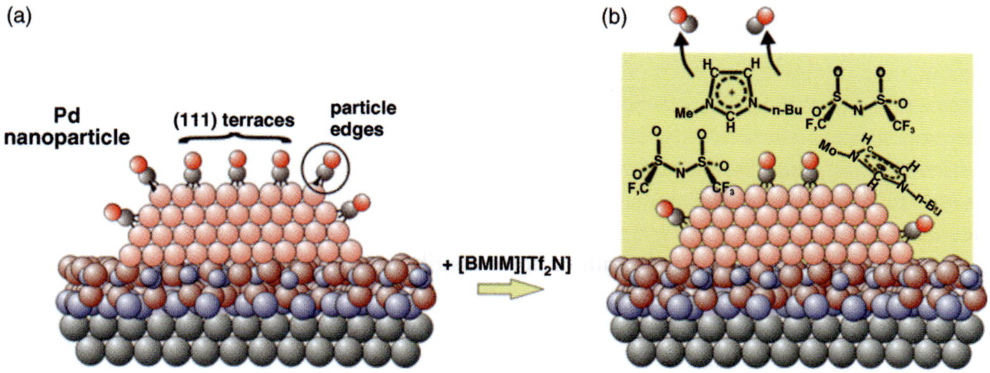

Figure 6.9 Schematic representation of a Pd nanoparticle on a model oxide surface in the presence of CO. Preferred adsorption sites of CO on pure Pd (a) and IL-coated Pd (a) are shown. Orientation of the IL on the surface determined from IRAS measurements [57].

By introducing IL-like functionalities in the support material, essential properties provided by the IL such as nucleophilicity, acidity or basicity are transferred to the material at the molecular and nanoscale level [58, 59]. Different IL-functionalized support materials have been reported and some will be discussed in detail in the following sections.

6.2.3.1 **Surface-Modified Ordered Meso-Porous Silica**

Ordered meso-porous silicas have attained increasing attention in recent years, not only due to their structural robustness and chemical inertness, but also by having controllable architecture, exceptionally high surface area, and uniform pore size [60]. However, having a highly negatively charged surface, untreated mesoporous silicas are not applicable for the deposition of, for example, AuNPs or PdNPs [61, 62]. By introducing functional groups onto the channel walls, the surface properties of periodic meso-porous organo-silicas (PMOs) can be modified in such a way that nanoparticles are stabilized and particle agglomeration is prevented. These positive effects were shown by Karimi *et al.* in Suzuki coupling reactions of a variety of haloarenes with arylboronic acids with supported PdNPs in water [63]. The immobilization of the PdNPs was achieved in the framework of alkylimidazolium IL-modified SBA-15 type PMO. Aside its high activity, the catalyst system could be reused four times without a considerable loss of catalytic activity. Hot filtration tests and selective catalyst poisoning as well as detailed TEM studies showed that although the IL-functionalized PMO nanostructure acted as reservoir for soluble Pd species, it also operated as a nanoscaffold to recapture the PdNPs into the meso-channels, thus preventing extensive agglomeration of the PdNPs.

Jin *et al.* synthesized a variety of novel periodic PMOs incorporated with bridging disulfide-imidazolium units (Scheme 6.4) which they successfully applied in the epoxidation of styrene and alkenes [64–66].

AuNPs with particle sizes down to 1.8 nm could be conveniently loaded in this support material via *in situ* reduction of chloroaurate ions by silanol groups on the pore surface of the PMO and successively captured by the disulfide groups. The as-supported AuNPs could be applied successfully in epoxidation reactions of styrene and alkenes and exceeded the catalytic performance of Au supported on meso-structured Al_2O_3 and Au-MgO under the same reaction conditions [67, 68]. The authors explain this superior catalytic activity with the presence of smaller AuNPs and their uniform distribution in the functionalized PMO in contrast to the non-functionalized supports since small AuNPs (~2 nm) are known to be the active sites with the highest activity [69]. In contrast to terminal mercapto-functionalized meso-porous silica, PMO materials with uniform distribution of organic moieties

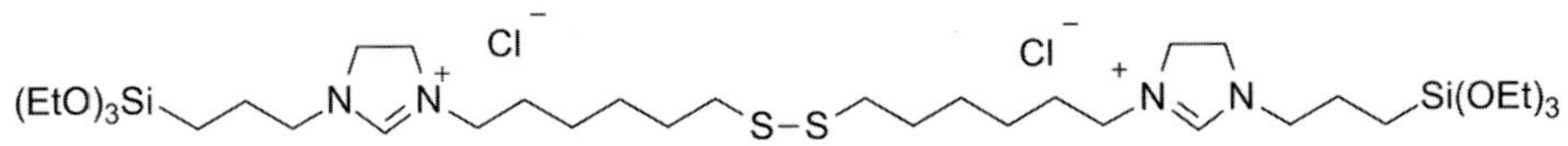

Scheme 6.4 Structure of a bridging organosilica precursor.

led to well-dispersed AuNPs and excellent catalytic performance in the epoxidation of styrene with H_2O_2 as oxidant (99.6% conversion, 93.3% selectivity to styrene oxide) [65]. Aside from the location and nature of the organic moieties, pore architecture of the PMO materials, as well as the method employed to implement the functionalities onto the support are strong influencing factors on the dispersion and particle size of AuNPs and subsequently on their catalytic activity.

Meso-porous silica with a 3D pore structure, MSU [70], appeared to be more susceptible for aggregation of AuNPs during synthesis and, hence, are less preferable as catalysts supported with 1D pore structures, namely MCM-41 [71] and SBA-15 [66, 72, 73]. These surface-modified POM materials with 1D pore structure synthesized via the post-synthesis route were prone to facilitate dispersion of AuNPs with smaller particle sizes than those prepared by the one-pot condensation method, but were also more vulnerable to metal aggregation and, by that, less stable in terms of their reusability [66]. Catalysts generated via the one-pot condensation route could be reused for up to nine times without any significant loss of both activity and selectivity.

6.2.3.2 **Surface Modified Nanocrystalline Metal Oxides**

Aerogel-prepared nanocrystalline magnesium oxide is a unique solid because of its highly ionic character, simple stoichiometry, and crystal structure. Its three-dimensional polyhedral structure and purely ionic nature leads to an inherently high surface per unit area, having high surface concentrations of edge/corner and various exposed crystal planes [74–78].

Kantam *et al.* exploited these distinctive properties and incorporated choline hydroxide, a basic IL, into this support material for the stabilization of ruthenium nanoparticles (RuNPs) [79, 80]. The resulting catalysts were tested in aerobic alcohol oxidation, as well as in transfer hydrogenation of carbonyl compounds.

The prepared catalyst showed high activity and selectivity in the oxidation reaction of a wide range of alcohols, with primary benzylic alcohols being the substrates converted with the fastest reaction rates, and it could be recovered and reused in up to seven cycles [80]. Likewise, the RuNPs supported on the choline hydroxide modified MgO generally exhibited a good performance in the transfer hydrogenation of various carbonyl compounds, such as acetophenone or 4-methylcinnamaldehyde, with KOH in 2-propanol [79].

Although a catalyst without the stabilizing IL also displayed excellent activity for both kinds of reactions, rapid deactivation during recycling was observed, probably caused by overoxidation of the active sites and considerable leaching of the active species. This, and the fact that a catalyst, which was calcined at 723 K after preparation, did not show any catalytic activity because of dissociation of the surface functionalized moiety of the solid matrix, clearly demonstrate that the IL is necessary for a stable performance of the RuNPs. TEM images of both, the fresh and used catalyst system with choline hydroxide, showed highly dispersed ruthenium particles (< 2 nm), indicating that the nanoparticles did not agglomerate and, instead, retained their properties.

Titanium oxide is an oxide most frequently used as a support in catalysis which is surely, to some extent, due to its good mechanical properties, low price and its association with the so-called strong-metal-support interaction effect when a noble metal is supported [81–83]. The production of TiO_2 via the sol–gel reaction usually employed results in highly amorphous titania which requires an additional calcination step to develop the anatase crystalline structure. However, the use of an IL as template solvent during TiO_2 production directly yields anatase or rutile structures without further calcination being necessary [84]. Avellaneda and co-workers reported on the preparation of IL-templated titania and its use in the gold-catalyzed oxidation of CO [85]. In contrast to numerous other reports, the applied IL, 1-(3-carboxipropyl)-3-methyl imidazolium chloride, was not removed from the resulting material, since the stabilizing influence of the IL on the AuNPs was to be investigated. The deposition of the AuNPs was achieved by two different methods: the standard Turkevich [86] method, involving a high temperature step, and the reduction of gold precursor by sodium borohydrate [87] in water as well as acetone at room temperature in order to prevent the risk of IL removal during the preparation. The presence of the IL seemed to stabilize the gold particle size in cold gold solutions independent of the solvent used. Whereas the Turkevich method of preparation resulted in large gold crystallites for which no effect of the IL could be observed and the catalytic activity in the CO oxidation was very low. The use of acetone as solvent during gold deposition resulted in a catalytically inactive catalyst due to the absence of a gold–titania interface and the production of defect-free gold crystallites, which do not offer active sites for CO oxidation [88]. The presence of the IL at the particle/support interface and the stabilized AuNPs caused an unusual catalytic behavior in the sample produced in aqueous solution. This behavior was characterized by strong dependence on the presence of CO in the mixture as well as changed kinetics and reaction rate of oxidation. The role of the IL seems to be, on the one hand, the electronic configuration stabilization of the AuNPs and, on the other hand, the *in situ* promotion of active sites.

6.2.3.3 IL-Functionalized Highly Cross-Linked Polymers as Support

Polymers are very attractive supports for metallic NPs due to some unusual advantages, such as synthetic flexibility and combination of hydrogen and ionic bonding [89–91]. To load nanoparticles onto the surface of a polymeric material, prefunctionalization is typically required. In addition, since common polymers have weak structure stability in organic solvents, high cross-linking of the polymeric support should be present. The group of Han developed a method to immobilize Pd/Au bimetallic NPs on highly cross-linked polymer microspheres as shown in Scheme 6.5 [92].

In this route, the cross-linker divinylbenzene and the IL, 1-vinyl-3-butylimidazolium chloride ([VBIM]Cl) were first copolymerized, resulting in cross-linked polymer microspheres. Palladium nanoparticles could be immobilized on the as-prepared support without any further modification via deposition of $PdCl_4$ and

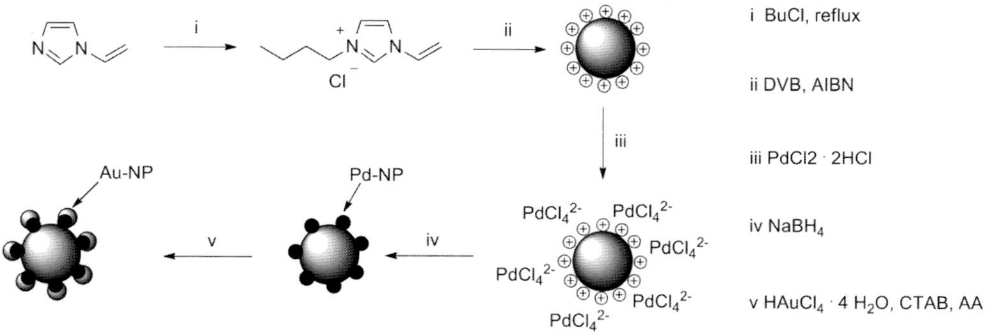

Scheme 6.5 Route to prepare the cross-linked polymer-supported bimetallic Pd/AuNPs.

subsequent reduction. The formed PdNPs were used as seeds for the production of AuNPs by a seeding growth method. The so-formed composites had excellent synergy for catalyzing hydrogenation of cyclohexene to cyclohexane with a four times shorter reaction time needed for full conversion than the monometallic supported system with palladium.

6.2.3.4 **Natural Clays with IL-Functionalization**

Natural clays are environmentally friendly materials with crystalline structures and relatively large surface areas, which can be used as catalyst supports. The group of Miao employed montmorillonate (MMT) [93], sepiolite [94] and attapulgite [95], all types of naturally occurring clay, as templates for IL-modified support material for the immobilization of ruthenium, palladium and rhodium nanoparticles.

MMT can be structurally defined as layers of negatively charged 2D silicate sheets, separated by interlayer cationic species with a high exchange ability for other cations [96]. The IL-functionalized MMT was obtained by exchanging the Na cations with the cations of the IL 1,1,3,3-tetramethylguanidinium trifluoroacetate ([TMG][TFA]). The as-prepared support had then been dispersed in an aqueous solution of $RuCl_3$ in order to adsorb Ru^{III} and form TMG-MMT/$RuCl_3$. After removal of the solvent RuNPs were generated by hydrogen treatment at a relative low temperature of 493 K. XPS measurements indicated a total conversion of Ru^{III} into Ru^0.

The obtained catalyst showed excellent activity and selectivity towards cyclohexane in the hydrogenation of benzene and outperformed traditional catalyst systems such as Ru/C and Ru/Al_2O_3. Additionally, it could be reused for five times without any significant loss of activity and selectivity.

Measurements with a transmission electron microscope equipped with an energy-dispersive X-ray spectrometer revealed two forms of nanoparticles existing in the catalytic system: some nanoparticles were present on the outer surfaces whereas others were embedded in the interlayers. Since nanoparticles on the MMT surface of a catalyst, which had been used five times in hydrogenation reactions, tend to aggregate, the authors believe that particles intercalated into the MMT interlayers represent the species with the highest activity.

Sepiolite is a microcrystalline hydrated magnesium silicate, which is structurally formed by an alternation of blocks, containing two silica sheets sandwiching a central sheet of magnesium oxide-hydroxide, and cavities. Silanol groups, present on the external surface of the silicate particles, are located at the edges of the channels and are directly accessible to reagents allowing the preparation of organic-inorganic materials containing different surface organic functions [97]. Again, [TMG][TFA] was used to introduce the IL-functionality into this clay and by that, convert the material into a highly suitable support for palladium nanoparticles [94]. XPS analysis and TEM examinations confirmed the existence of mainly Pd^0, with a small amount of its oxides, uniformly distributed on the support with a particle size of 5 nm. The as-prepared catalysts exhibited very high activities in hydrogenation reactions of alkenes (turnover frequencies of up to 10000 h^{-1}) strongly exceeding the catalytic performance of the commercially available catalyst Pd/C. Recovery and reuse of the catalyst did not cause activity loss.

Similar observations were made with RhNPs immobilized in 1,1,3,3-tetramethylguanidine lactate modified attapulgite material [95]. The activity of the composite for cyclohexene hydrogenation was investigated, which exhibited much higher efficiency compared to those of other catalysts, and the turnover frequencies reached 2700 h^{-1}.

6.2.3.5 **Carbon Nanotubes**

The group of Lee developed IL-coated carbon nanotubes for the immobilization and stabilization of PdNPs according to Scheme 6.6 [98]. The average size of the PdNPs formed was determined to be 10 nm using TEM, with 55.3 wt% Pd being deposited onto the IL-coated carbon nanofibers (CNF).

The counter ion of the ionic liquid fragment was varied by ion exchange techniques in order to change the solubility properties of the Pd–CNF materials in different solvent systems. The CNF material bearing the Br^- anion is easily dispersed in water, but the same system with NTf_2^- and SbF_6^- anions precipitated while all systems could be dispersed in alcohols. All catalysts showed high activity in *trans*-stilbene hydrogenation at 1 bar H_2. Since the catalyst system iii could be dispersed

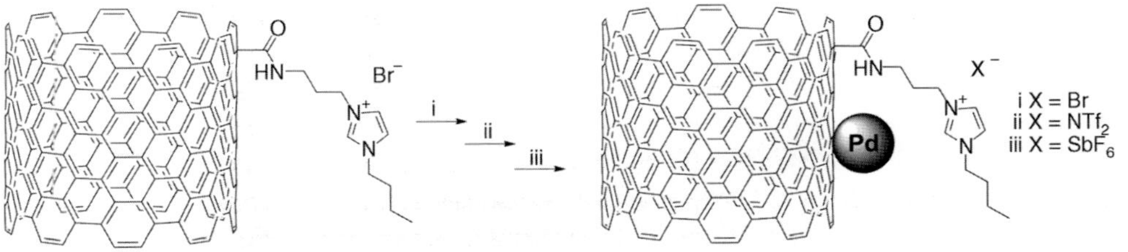

i H_2, 1 bar, Na_2PdCl_4, H_2O, 25 °C, 30 min ii $LiNTf_2$, H_2O, 25 °C, 12 h iii $NaSbF_6$, H_2O, 25 °C, 12 h

Scheme 6.6 Synthesis of palladium nanoparticles supported on imidazolium-functionalized multiwalled carbon nanotubes in water.

Table 6.5 Selected recycling studies in the hydrogenation of alkenes using PdNPs on ionic liquid functionalized CNF.

Entry	R_1	R_2	R_3	t_R/min[a]	Cycle[b]	Conversion/%
1	Ph	Ph	H	10	1–10	100
2	Ph	H	H	5	11–20	100
3	Ph	Me	H	10	21–20	100
4	Ph	Me	Me	10	31–40	100
5	4-MeOC6H4	H	H	10	41–50	100

Reaction conditions: 1.0 mmol substrate in ^{i}PrOH/[BMIM][SbF_6] (8 mL–2 mL), Pd/[BMIM][SbF_6]/CNF (1 mol% Pd), 1 bar H_2, 293 K.
a) Reaction time.
b) Catalyst recovered from previous run was used.

in the IL [BMIM][SbF_6], the recovery of the catalyst was achieved by simple phase decantation and the catalyst could be reused without loss in activity for 10 runs as shown in Table 6.5.

ICP-MS analysis of the ^{i}PrOH layer separated from each run indicated that no detectable Pd was leached out. Moreover, after 10 times of reuse, the recovered catalyst was still active, and therefore was subsequently reused for the hydrogenations of four additional substrates (entries 2–5 in Table 6.2) with total turnover number of 5000 with most of the PdNPs remaining on the surface after 50 recycling cycles.

The use of a polymerizable IL fragment *N*-allylimidazolium halide together with *N*-vinyl-*N*-butylimidazolium halide in the presence of AIBN resulted in a network of ILs coated on CNF surface (polyIL(X)-CNF) [99] (Figure 6.10). Within this network PdNPs have been immobilized and tested for the oxygen reduction reaction in $HClO_4$.

Variation of the anions of the imidazolium moiety influenced the oxygen reduction reaction kinetics significantly; attributed to interactions between the anions and the electrolytes affecting the local proton concentration around the catalyst surfaces.

6.2.3.6 **Miscellaneous Supports**

A xerogel has been used as support for IL coating, followed by PdNP formation [100]. The average particle diameter was found to be 70 nm. The catalyst was active in a number of Suzuki and Heck coupling reactions and could be reused at least five times.

IL-modified magnetic nanoparticles of Fe_2O_3 were used as support for the immobilization of PtNPs as depicted in Scheme 6.7 [101].

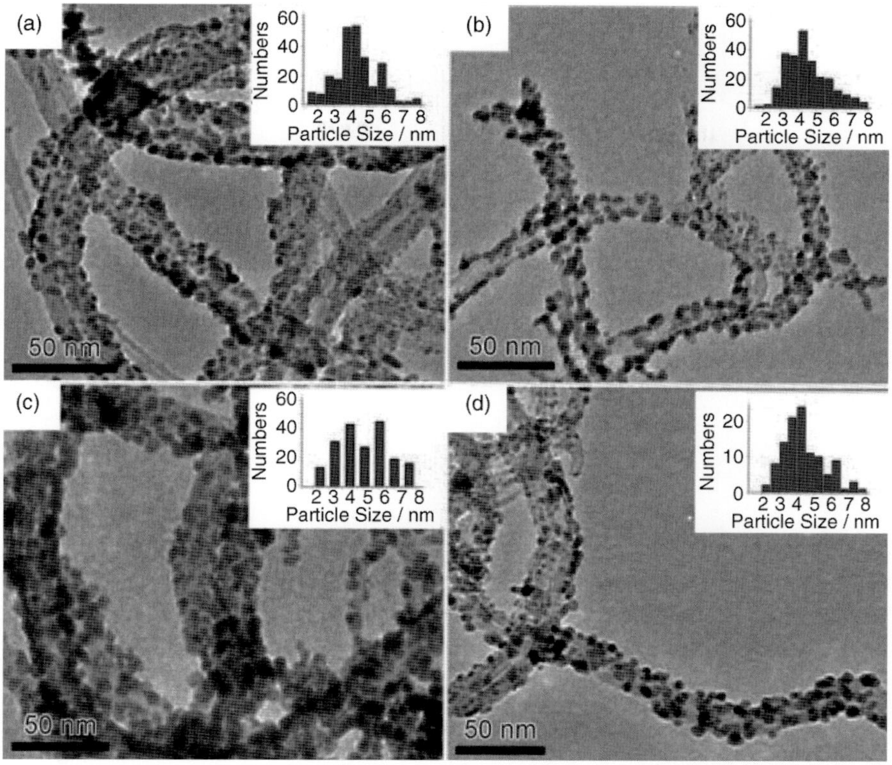

Figure 6.10 TEM images and size distribution histograms of different PdNPs supported on ionic liquid functionalized CNF [99].

Small size PtNPs of 2–2.5 nm were obtained, directly attached to the surface of the magnetic nanoparticle. These catalysts (Figure 6.11) were tested in the hydrogenation of α,β-unsaturated aldehydes and alkynes.

The catalyst was found to be selective for the hydrogenation of the alkynes to the corresponding alkenes, for example, diphenylacetylene could be converted selectively to stilbene, with high selectivity for the *cis* isomer. Only the aldehyde functionality was hydrogenated to the corresponding alcohol. Recycling of the catalyst was achieved by a magnet, followed by washing with methanol. After four consecutive runs, the loss in activity was less than 3%.

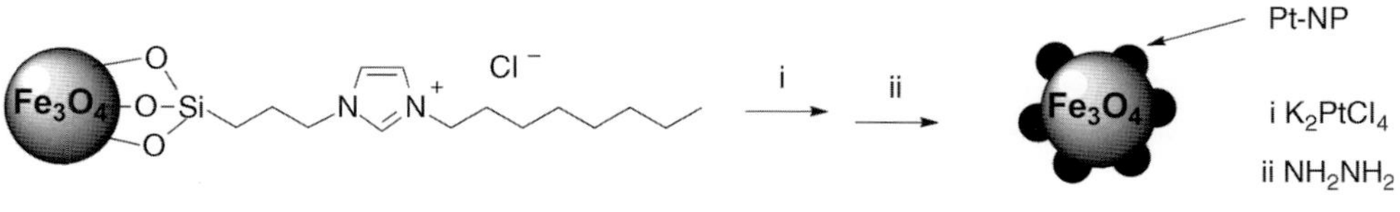

Scheme 6.7 Preparation of platinum nanoparticles supported on magnetite nanoparticles.

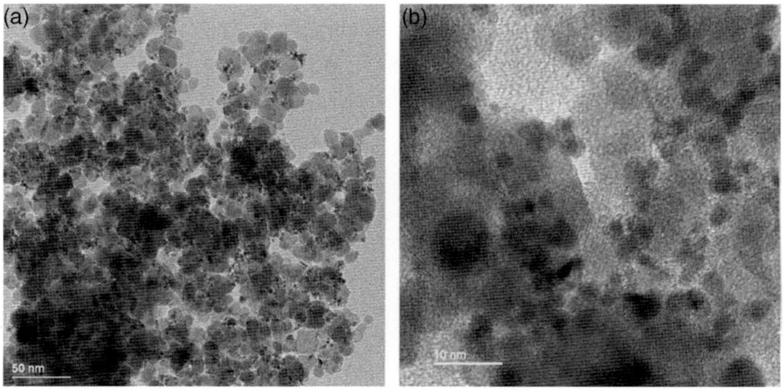

Figure 6.11 TEM images of PtNPs supported on ionic liquid-modified magnetic nanoparticles at different resolution: scale bar 50 nm (a), scale bar 10 nm (b) [101].

6.3
Benefits for Synthesis and Processes

With the help of ILs, either as catalyst phase (nano-SILP), liquid coating (nano-SCILL), or surface modifier, the attractive features of nanoparticle catalysis can be transferred from small-scale batch into larger-scale continuous processes. The fact that the nanoparticles are now immobilized on a rather small support volume, instead of a large liquid volume, enables simplified reactor concepts and improved recycling strategies [102]. The solid support material can be designed to match the

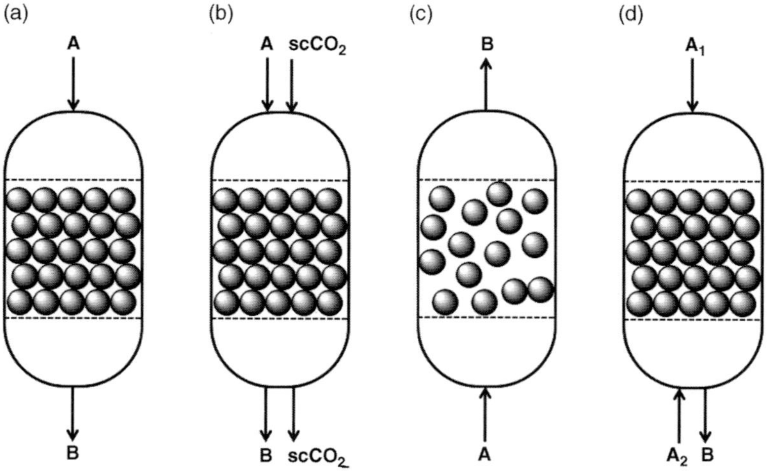

Figure 6.12 Possible reactor concepts for nano-SILP or nano-SCILL catalyzed reactions A → B. (a) Tubular-fixed bed reactor for gas- or liquid-phase reactions; (b) tubular fixed-bed reactor for gas-phase reactions with $scCO_2$ as mobile phase; (c) fluidized-bed reactor (gas-phase) or slurry reactor (liquid-phase); (d) trickle-bed reactor for liquid (A_1) gas (A_2) reactions.

requirements of industrial processes, especially good heat and mass transport as well as low pressure drop inside the reactor [103, 104]. The IL would allow efficient gas-phase processes, or, in case of high-boiling substrates, it could be combined with $scCO_2$ as mobile phase [25, 105]. Since ILs are insoluble in $scCO_2$, the leaching of ILs from the support is negligible and only has to be prevented if cross-solubility between the substrate or the products exist. Figure 6.12 gives an overview of possible reactor concepts for nanoparticle catalysis enhanced by supported ILs.

Beside these possible large-scale industrial applications, the use of a solid catalyst is beneficial on laboratory scale as well. The nano-SILP and nano-SCILL materials can be placed into a cartridge system so that the reaction can be carried out in flow mode rather than batch mode [106]. Liquid pumps or HPLC systems ensure exact dosing of substrates and the cartridge can be placed in a cooling or heating bath, if required [107]. Similar to the industrial scenario, the solid material can be recovered easily and reused several times.

6.4 Conclusion

The combination of nanoparticles and ionic liquids on supports is very attractive to improve catalytic processes. By the use of the appropriate IL, the metallic NP can be stabilized, agglomeration can be hindered or suppressed completely, and the local concentration of substrates/products at the catalyst surface can be adjusted, thereby enhancing the selectivity of the overall process. By this, an active, more stable and more selective catalyst system can be obtained, which is the ultimate goal in catalysis research. The field of nanoparticles immobilized in supported IL thin films is still in its infancy, but the promising results reported within the last five years are certainly encouraging. One should be aware that an IL coating does not constitute an advantage in all cases, but sometimes can lead to catalyst deactivation or mass transport limitations. Only in gas-phase applications is the leaching of the IL negligible, whereas in cases of liquid phase reactions with dispersed nano-SILP or nano-SCILL materials the cross-solubility of the IL in the solvent, the substrate or the product can result in significant and irreversible removal of the IL from the support. In cases where such liquid slurry type applications are desired, a covalent anchoring of the first IL monolayer on the support should be taken into consideration.

Besides this, nanoparticle catalysis in supported ionic liquids has tremendous potential both in small laboratory-scale and large industrial-scale applications.

References

1 Cole-Hamilton, D.J. (2003) Homogeneous catalysis–new approaches to catalyst separation, recovery, and recycling. *Science*, **299**, 1702–1706.

2 (a) Herrmann, W.A. and Kohlpaintner, C.W. (1993) Water-soluble ligands, metal complexes, and catalysts: Synergism of homogeneous and heterogeneous

catalysis. *Angew. Chem. Int. Ed.*, **32** (11), 1524–1544; (b) Tzschucke, C.C., Markert, C., Bannwarth, W., Roller, S., Hebel, A., and Haag, R. (2002) Modern separation techniques for the efficient workup in organic synthesis. *Angew. Chem. Int. Ed.*, **41** (21), 3964–4000.
3 Bailar, J.C. (1974) Heterogenizing homogeneous catalysts. *Catal. Rev.*, **10** (1), 17–36.
4 Cornils, B., Herrmann, W.A., Horvath, I.T., Leitner, W., Mecking, S., Olivier-Bourbigou, H., and Vogt, D. (eds) (2005) *Multiphase Homogeneous Catalysis*, Wiley-VCH Verlag GmbH, Weinheim, Germany.
5 Copéret, C., Chabanas, M., Saint-Arroman, R.P., and Basset, J.-M. (2003) Homogeneous and heterogeneous catalysis: bridging the gap through surface organometallic chemistry. *Angew. Chem. Int. Ed.*, **42** (2), 156–181.
6 Wasserscheid, P. and Keim, W. (2000) Ionic liquids—new 'Solutions' for transition metal catalysis. *Angew. Chem. Int. Ed.*, **39** (21), 3772–3789.
7 Hallett, J.P. and Welton, T. (2011) Room-temperature ionic liquids: solvents for synthesis and catalysis. 2. *Chem. Rev.*, **111** (5), 3508–3576.
8 Olivier-Bourbigou, H., Magna, L., and Morvan, D. (2010) Ionic liquids and catalysis: Recent progress from knowledge to applications. *Appl. Catal. A-Gen.*, **373** (1–2), 1–56.
9 Pârvulescu, V.I. and Hardacre, C. (2007) Catalysis in ionic liquids. *Chem. Rev.*, **107** (6), 2615–2665.
10 Wasserscheid, P. and Welton, T. (eds) (2007) *Ionic Liquids in Synthesis*, Wiley-VCH Verlag GmbH, Weinheim, Germany.
11 Haumann, M. and Riisager, A. (2008) Hydroformylation in room temperature ionic liquids (RTILs): Catalyst and process developments. *Chem. Rev.*, **108** (4), 1474–1497.
12 Astruc, D., Lu, F., and Aranzaes, J.R. (2005) Nanoparticles as recyclable catalysts: The frontier between homogeneous and heterogeneous catalysis. *Angew. Chem. Int. Ed.*, **44** (48), 7852–7872.
13 Migowski, P. and Dupont, J. (2007) Catalytic applications of metal nanoparticles in imidazolium ionic liquids. *Chem. Eur. J.*, **13** (1), 32–39.
14 Dupont, J., Fonseca, G.S., Umpierre, A. P., Fichtner, P.F.P., and Teixeira, S.R. (2002) Transition-metal nanoparticles in imidazolium ionic liquids: Recycable catalysts for biphasic hydrogenation reactions. *J. Am. Chem. Soc.*, **124** (16), 4228–4229.
15 Scheeren, C.W., Machado, G., Teixeira, S.R., Morais, J., Domingos, J.B., and Dupont, J. (2006) Synthesis and characterization of Pt(0) nanoparticles in imidazolium ionic liquids. *J. Phys. Chem. B*, **110** (26), 13011–13020.
16 Dupont, J. and Scholten, J.D. (2010) On the structural and surface properties of transition-metal nanoparticles in ionic liquids. *Chem. Soc. Rev.*, **39**, 1780–1804.
17 Fonseca, G.S., Machado, G., Teixeira, S.R., Fecher, G.H., Morais, J., Alves, M.C.M., and Dupont, J. (2006) Synthesis and characterization of catalytic iridium nanoparticles in imidazolium ionic liquids. *J. Colloid Interface Sci.*, **301**, 193–204.
18 Gu, Y. and Li, G. (2009) Ionic liquids-based catalysis with solids: State of the art. *Adv. Synth. Catal.*, **351** (6), 817–847.
19 Mehnert, C.P., Cook, R.A., Dispenziere, N.C., and Afeworki, M. (2002) Supported ionic liquid catalysis – a new concept for homogeneous hydroformylation catalysis. *J. Am. Chem. Soc.*, **124** (44), 12932–12933.
20 Riisager, A., Fehrmann, R., Flicker, S., van Hal, R., Haumann, M., and Wasserscheid, P. (2005) Very stable and highly regioselective supported ionic-liquid-phase (SILP) catalysis: Continuous-flow fixed-bed hydroformylation of propene. *Angew. Chem. Int. Ed.*, **44** (5), 815–819.
21 van Doorslaer, C., Wahlen, J., Mertens, P., Binnemans, K., and de Vos, D. (2010) Immobilization of molecular catalysts in supported ionic liquid phases. *Dalton Trans.*, **39**, 8377–8390.
22 Mehnert, C.P., Mozeleski, E.J., and Cook, R.A. (2002) Supported ionic liquid catalysis investigated for hydrogenation reactions. *Chem. Commun.*, 3010–3011.

23 Öchsner, E., Schneider, M.J., Meyer, C., Haumann, M., and Wasserscheid, P. (2011) Challenging the scope of continuous, gas-phase reactions with supported ionic liquid phase (SILP) catalysts—Asymmetric hydrogenation of methyl acetoacetate. *Appl. Catal. A-Gen.*, **399** (1–2), 35–41.

24 (a) Riisager, A., Fehrmann, R., Haumann, M., and Wasserscheid, P. (2006) Supported ionic liquid phase (SILP) catalysis: An innovative concept for homogeneous catalysis in continuous fixed-bed reactors. *Eur. J. Inorg. Chem.*, **2006** (4), 695–706; (b) Riisager, A., Wasserscheid, P., van Hal, R., and Fehrmann, R. (2003) Continuous fixed-bed gas-phase hydroformylation using supported ionic liquid-phase (SILP) Rh catalysts. *J. Catal.*, **219** (2), 452–455; (c) Riisager, A., Eriksen, K.M., Wasserscheid, P., and Fehrmann, R. (2003) Propene and 1-Octene hydroformylation with silica-supported, ionic liquid-phase (SILP) Rh-Phosphine catalysts in continuous fixed-bed mode. *Catal. Lett.*, **90** (3–4), 149–153; (d) Riisager, A., Fehrmann, R., Haumann, M., Gorle, B.S.K., and Wasserscheid, P. (2005) Stability and kinetic studies of supported ionic liquid phase catalysts for hydroformylation of propene. *Ind. Eng. Chem. Res.*, **44** (26), 9853–9859; (e) Haumann, M., Dentler, K., Joni, J., Riisager, A., and Wasserscheid, P. (2007) Continuous gas-phase hydroformylation of 1-butene using supported ionic liquid phase (SILP) catalysts. *Adv. Synth. Catal.*, **349** (3), 425–431; (f) Haumann, M., Jakuttis, M., Werner, S., and Wasserscheid, P. (2009) Supported ionic liquid phase (SILP) catalyzed hydroformylation of 1-butene in a gradient-free loop reactor. *J. Catal.*, **263** (2), 321–327; (g) Jakuttis, M., Schönweiz, A., Werner, S., Franke, R., Wiese, K.-D., Haumann, M., and Wasserscheid, P. (2011) Rhodium–phosphite SILP catalysis for the highly selective hydroformylation of mixed C4 feedstocks. *Angew. Chem. Int. Ed.*, **50** (19), 4492–4495; (h) Haumann, M., Jakuttis, M., Schönweiz, A., Franke, R., and Wasserscheid, P. (2011) Continuous gas-phase hydroformylation of a highly diluted technical C4 feed using supported ionic liquid phase catalysts. *ChemCatChem*, **3** (11), 1822–1827.

25 (a) Hintermair, U., Zhao, G., Santini, C.C., Muldoon, M.J., and Cole-Hamilton, D.J. (2007) Supported ionic liquid phase catalysis with supercritical flow. *Chem. Commun.*, 1462–1464; (b) Hintermair, U., Gong, Z., Serbanovic, A., Muldoon, M.J., and Cole-Hamilton, D.J. (2010) Continuous flow hydroformylation using supported ionic liquid phase catalysts with carbon dioxide as a carrier. *Dalton Trans.*, **39** (36), 8501–8510.

26 Riisager, A., Jorgensen, B., Wasserscheid, P., and Fehrmann, R. (2006) First application of supported ionic liquid phase (SILP) catalysis for continuous methanol carbonylation. *Chem. Commun.*, 994–996.

27 Hagiwara, H., Sugawara, Y., Isobe, K., Hoshi, T., and Suzuki, T. (2004) Immobilization of $Pd(OAc)_2$ in ionic liquid on silica: Application to sustainable Mizoroki–Heck reaction. *Org. Lett.*, **6** (14), 2325–2328.

28 (a) Hagiwara, H., Okunaka, N., Hoshi, T., and Suzuki, T. (2008) Immobilization of grubbs catalyst as supported ionic liquid catalyst (Ru-SILC). *Synlett*, **12**, 1813–1816; (b) Hagiwara, H., Nakamura, T., Okunaka, N., Hoshi, T., and Suzuki, T. (2010) Catalytic performance of ruthenium-supported ionic-liquid catalysts in sustainable synthesis of macrocyclic lactones. *Helv. Chim. Acta*, **93** (1), 175–182.

29 Scholz, J., Loekman, S., Szesni, N., Hieringer, W., Görling, A., Haumann, M., and Wasserscheid, P. (2011) Ethene-induced temporary inhibition of grubbs metathesis catalysts. *Adv. Synth. Catal.*, **353** (14–15), 2701–2707.

30 (a) Breitenlechner, S., Fleck, M., Müller, T.E., and Suppan, A. (2004) Solid catalysts on the basis of supported ionic liquids and their use in hydroamination reactions. *J. Mol. Catal. A: Chemical*, **214** (1), 175–179; (b) Sievers, C., Jiménez, O., Knapp, R., Lin, X., Müller, T.E., Türler, A., Wierczinski, B., and Lercher, J.A.

(2008) Palladium catalysts immobilized in thin films of ionic liquid for the direct addition of aniline to styrene. *J. Mol. Catal. A: Chemical*, **279** (1), 187–199.

31 Mehnert, C.P. (2005) Supported ionic liquid catalysis. *Chem. Eur. J.*, **11** (1), 50–56.

32 Joni, J., Haumann, M., and Wasserscheid, P. (2009) Development of a supported ionic liquid phase (SILP) catalyst for slurry-phase friedel–crafts alkylations of cumene. *Adv. Synth. Catal.*, **351** (3), 423–431.

33 Kernchen, U., Etzold, B., Korth, W., and Jess, A. (2007) Solid catalyst with ionic liquid layer (SCILL) – a new concept to improve selectivity illustrated by hydrogenation of cyclooctadiene. *Chem. Eng. Technol.*, **30** (8), 985–994.

34 Arras, J., Steffan, M., Shayeghi, Y., Ruppert, D., and Claus, P. (2009) Regioselective catalytic hydrogenation of citral with ionic liquids as reaction modifiers. *Green Chem.*, **11** (5), 716–723.

35 Arras, J., Paki, E., Roth, C., Radnik, J., Lucas, M., and Claus, P. (2010) How a supported metal is influenced by an ionic liquid: In-Depth characterization of SCILL-Type palladium catalysts and their hydrogen adsorption. *J. Phys. Chem. C*, **114** (23), 10520–10526.

36 Fenske, D., Sonström, P., Stöver, J., Wang, X., Borchert, H., Parisi, J., Kolny-Olesiak, J., Bäumer, M., and Al-Shamery, K. (2010) Colloidally prepared Pt nanoparticles for heterogeneous gas-phase catalysis: Influence of ligand shell and catalyst loading on CO oxidation activity. *ChemCatChem*, **2** (2), 198–205.

37 Jürgens, B., Borchert, H., Ahrenstorf, K., Sonström, P., Pretorius, A., Schowalter, M., Gries, K., Zielasek, V., Rosenauer, A., Weller, H., and Bäumer, M. (2008) Colloidally prepared nanoparticles for the synthesis of structurally well-defined and highly active heterogeneous catalysts. *Angew. Chem. Int. Ed.*, **47** (46), 8946–8949.

38 Sobota, M., Happel, M., Amende, M., Paape, N., Wasserscheid, P., Laurin, M., and Libuda, J. (2011) Ligand effects in SCILL model systems: site-specific interactions with Pt and Pd nanoparticles. *Adv. Mater.*, **23** (22–23), 2617–2621.

39 Crooks, R.M., Zhao, M., Sun, L., Chechnik, V., and Yeung, L.K. (2001) Dendrimer-encapsulated metal nanoparticles: Synthesis, characterization, and applications to catalysis. *Acc. Chem. Res.*, **34** (3), 181–190.

40 Cremer, T., Killian, M., Gottfried, J.M., Paape, N., Wasserscheid, P., Maier, F., and Steinrück, H.-P. (2008) Physical vapor deposition of [EMIM][Tf2N]: A new approach to the modification of surface properties with ultrathin ionic liquid films. *Chem. Phys. Chem.*, **9** (15), 2185–2190.

41 Virtanen, P., Karhu, H., Kordas, K., and Mikkola, J.-P. (2007) The effect of ionic liquid in supported ionic liquid catalysts (SILCA) in the hydrogenation of α,β-unsaturated aldehydes. *Chem. Eng. Sci.*, **62** (14), 3660–3671.

42 Mikkola, J.-P., Virtanen, P.P., Kordás, K., Karhu, H., and Salmi, T.O. (2007) SILCA—Supported ionic liquid catalysts for fine chemicals. *Appl. Catal. A-Gen.*, **328** (1), 68–76.

43 Virtanen, P., Mikkola, J.-P., and Salmi, T.O. (2007) Kinetics of citral hydrogenation by supported ionic liquid catalysts (SILCA) for fine chemicals. *Ind. Eng. Chem. Res.*, **46** (26), 9022–9031.

44 Virtanen, P., Mikkola, J.-P., and Salmi, T.O. (2009) Kinetics of cinnamaldehyde hydrogenation by supported ionic liquid catalysts (SILCA). *Ind. Eng. Chem. Res.*, **48** (23), 10335–10342.

45 Virtanen, P., Karhu, H., Toth, G., Kordás, K., and Mikkola, J.-P. (2009) Towards one-pot synthesis of menthols from citral: Modifying supported ionic liquid catalysts (SILCAs) with Lewis and Brùnsted acids. *J. Catal.*, **263** (2), 209–219.

46 Ruta, M., Laurenczy, G., Dyson, P.J., and Kiwi-Minsker, L. (2008) Pd nanoparticles in a supported ionic liquid phase: highly stable catalysts for selective acetylene hydrogenation under continuous-flow conditions. *J. Phys. Chem. C*, **112** (46), 17814–17819.

47 Kovtunov, K.V., Zhivonitko, V.V., Kiwi-Minsker, L., and Koptyug, I.V. (2010)

Parahydrogen-induced polarization in alkyne hydrogenation catalyzed by Pd nanoparticles embedded in a supported ionic liquid phase. *Chem. Commun.*, **46**, 5764–5766.

48 Gruttadauria, M., Liotta, L.F., Salvo, A.M. P., Giacalone, F., La Parola, V., Aprile, C., and Notoa, R. (2011) Multi-layered, covalently supported ionic liquid phase (mlc-SILP) as highly cross-linked support for recyclable palladium catalysts for the suzuki reaction in aqueous medium. *Adv. Synth. Catal.*, **353** (11–12), 2119–2130.

49 Knapp, R., Wyrzgol, S.A., Reichelt, M., Hammer, T., Morgner, H., Müller, T.E., and Lercher, J.A. (2010) Corrugated ionic liquid surfaces with embedded polymer stabilized platinum nanoparticles. *J. Phys. Chem. C*, **114** (32), 13722–13729.

50 Gelesky, M.A., Chiaro, S.S.X., Pavan, F.A., dos Santos, J.H.Z., and Dupont, J. (2007) Supported ionic liquid phase rhodium nanoparticle hydrogenation catalysts. *Dalton Trans.*, 5549–5553.

51 Brondani, D., Zapp, E., Vieira, I.C., Dupont, J., and Scheeren, C.W. (2011) Gold nanoparticles in an ionic liquid phase supported in a biopolymeric matrix applied in the development of a rosmarinic acid biosensor. *Analyst*, **136**, 2495–2505.

52 Zapp, E., Brondani, D., Vieira, I.C., Dupont, J., and Scheeren, C.W. (2011) Bioelectroanalytical determination of rutin based on bi-enzymatic sensor containing iridium nanoparticles in ionic liquid phase supported in clay. *Electroanalysis*, **23** (3), 764–776.

53 Arras, J., Steffan, M., Shayeghi, Y., and Claus, P. (2008) The promoting effect of a dicyanamide based ionic liquid in the selective hydrogenation of citral. *Chem. Commun.*, 4058–4060.

54 Arras, J., Ruppert, D., and Claus, P. (2009) Supported ruthenium catalysed selective hydrogenation of citral in presence of [NTf_2] (based ionic liquids. *Appl. Catal. A-Gen.*, **371** (1–2), 73–77.

55 Knapp, R., Jentys, A., and Lercher, J.A. (2009) Impact of supported ionic liquids on supported Pt catalysts. *Green Chem.*, 11, 656–661.

56 Knapp, R., Wyrzgol, S.A., Jentys, A., and Lercher, J.A. (2010) Water–gas shift catalysts based on ionic liquid mediated supported Cu nanoparticles. *J. Catal.*, **276** (2), 280–291.

57 Sobota, M., Schmid, M., Happel, M., Amende, M., Maier, F., Steinrück, H.-P., Paape, N., Wasserscheid, P., Laurin, M., Gottfried, J.M., and Libuda, J. (2010) Ionic liquid based model catalysis: interaction of [BMIM][Tf_2N] with Pd nanoparticles supported on an ordered alumina film. *Phys. Chem. Chem. Phys.*, **12**, 10610–10621.

58 Burguete, M.I., Galindo, F., Garcia-Verdugo, E., Karbass, N., and Luis, S.V. (2007) Polymer supported ionic liquid phases (SILPs) versus ionic liquids (ILs): How much do they look alike. *Chem. Commun.*, 3086–3088.

59 Sans, V., Karbass, N., Burguete, M.I., Compañ, V., García-Verdugo, E., Luis, S.V., and Pawlak, M. (2011) Polymer-supported ionic-liquid-like phases (SILLPs): transferring ionic liquid properties to polymeric matrices. *Chem. Eur. J.*, **17** (6), 1894–1906.

60 Wan, Y. and Zhao, D. (2007) On the controllable soft-templating approach to mesoporous silicates. *Chem. Rev.*, **107** (7), 2821–2860.

61 Zhu, H., Lee, B., Dai, S., and Overbury, S.H. (2003) Coassembly synthesis of ordered mesoporous silica materials containing Au nanoparticles. *Langmuir*, **19** (9), 3974–3980.

62 Zanella, R., Sandoval, A., Santiago, P., Basiuk, V.A., and Saniger, J.M. (2006) New preparation method of gold nanoparticles on SiO_2. *J. Phys. Chem. B*, **110** (17), 8559–8565.

63 Karimi, B., Elhamifar, D., Clark, J.H., and Hunt, A.J. (2010) Ordered mesoporous organosilica with ionic-liquid framework: An efficient and reusable support for the Palladium-Catalyzed Suzuki–Miyaura coupling reaction in water. *Chem. Eur. J.*, **16** (27), 8047–8053.

64 Jin, Y., Wang, P., Yin, D., Liu, J., Qiu, H., and Yu, N. (2008) Gold nanoparticles stabilized in a novel periodic mesoporous organosilica of SBA-15 for styrene

epoxidation. *Micropor. Mesopor. Mat.*, **111** (1–3), 569–576.

65 Jin, Y., Zhuang, D., Yu, N., Zhao, H., Ding, Y., Qin, L., Liu, J., Yin, D., Qiu, H., Fu, Z., and Yin, D. (2009) Epoxidation of styrene over gold nanoparticles supported on organic–inorganic hybrid mesoporous silicas with aqueous hydrogen peroxide. *Micropor. Mesopor. Mat.*, **126** (1–2), 159–165.

66 Yu, N., Ding, Y., Lo, A.-Y., Huang, S.-J., Wu, P.-H., Liu, C., Yin, D., Fu, Z., Yin, D., Hung, C.-T., Lei, Z., and Liu, S.-B. (2011) Gold nanoparticles supported on periodic mesoporous organosilicas for epoxidation of olefins: Effects of pore architecture and surface modification method of the supports. *Micropor. Mesopor. Mat.*, **143** (2–3), 426–434.

67 Yin, D., Qin, L., Liu, J., Li, C., and Jin, Y. (2005) Gold nanoparticles deposited on mesoporous alumina for epoxidation of styrene: Effects of the surface basicity of the supports. *J. Mol. Catal. A: Chemical*, **240** (1–2), 40–48.

68 Patil, N.S., Uphade, B.S., Jana, P., Bharagava, S.K., and Choudhary, V.R. (2004) Epoxidation of styrene by anhydrous t-butyl hydroperoxide over reusable gold supported on MgO and other alkaline earth oxides. *J. Catal.*, **223** (1), 236–239.

69 Taylor, B., Lauterbach, J., and Delgass, W. N. (2005) Gas-phase epoxidation of propylene over small gold ensembles on TS-1. *Appl. Catal. A: Gen.*, **291** (1–2), 188–198.

70 Wang, J., Yu, N., Zheng, A., Yang, J., Wu, D., Sun, Y., Ye, C., and Deng, F. (2006) Mesoporous MSU materials functionalized with sulfonic group: A multinuclear NMR and theoretical calculation study. *Micropor. Mesopor. Mat.*, **89** (1–3), 219–226.

71 Beck, J.S., Vartuli, J.C., Roth, W.J., Leonowicz, M.E., Kresge, C.T., Schmitt, K.D., Chu, C.T.W., Olson, D.H., and Sheppard, E.W. (1992) A new family of mesoporous molecular sieves prepared with liquid crystal templates. *J. Am. Chem. Soc.*, **114** (27), 10834–10843.

72 Zhao, D., Feng, J., Huo, Q., Melosh, N., Fredrickson, G.H., Chmelka, B.F., and Stucky, G.D. (1998) Triblock copolymer syntheses of mesoporous silica with periodic 50 to 300 angstrom pores. *Science*, **279** (5350), 548–552.

73 Zhao, D., Huo, Q., Feng, J., Chmelka, B.F., and Stucky, G.D. (1998) Nonionic triblock and star diblock copolymer and oligomeric surfactant syntheses of highly ordered, hydrothermally stable, mesoporous silica structures. *J. Am. Chem. Soc.*, **120** (24), 6024–6036.

74 Choudary, B.M., Mulukutla, R.S., and Klabunde, K.J. (2003) Benzylation of aromatic compounds with different crystallites of MgO. *J. Am. Chem. Soc.*, **125** (8), 2020–2021.

75 Choudary, B.M., Kantam, M.L., Ranganath, K.V.S., Mahendar, K., and Sreedhar, B. (2004) Bifunctional nanocrystalline MgO for chiral epoxy ketones via claisen–schmidt condensation–asymmetric epoxidation reactions. *J. Am. Chem. Soc.*, **126** (11), 3396–3397.

76 Choudary, B.M., Ranganath, K.V.S., Pal, U., Kantam, M.L., and Sreedhar, B. (2005) Nanocrystalline MgO for asymmetric henry and michael reactions. *J. Am. Chem. Soc.*, **127** (38), 13167–13171.

77 Jeevanandam, P. and Klabunde, K.J. (2002) A study on adsorption of surfactant molecules on magnesium oxide nanocrystals prepared by an aerogel route. *Langmuir*, **18** (13), 5309–5313.

78 Utamapanya, S., Klabunde, K.J., and Schlup, J.R. (1991) Nanoscale metal oxide particles/clusters as chemical reagents. Synthesis and properties of ultrahigh surface area magnesium hydroxide and magnesium oxide. *Chem. Mater.*, **3** (1), 175–181.

79 Kantam, M.L., Reddy, R.S., Pal, U., Sreedhar, B., and Bhargava, S. (2008) Transfer hydrogenation of carbonyl compounds catalyzed by ruthenium nanoparticles stabilized on nanocrystalline magnesium oxide by ionic liquids. *Adv. Synth. Catal.*, **350** (14–15), 2231–2235.

80 Kantam, M.L., Pal, U., Sreedhar, B., Bhargava, S., Iwasawa, Y., Tada, M., and

Choudary, B.M. (2008) Aerobic alcohol oxidation by ruthenium species stabilized on nanocrystalline magnesium oxide by basic ionic liquids. *Adv. Synth. Catal.*, **350** (9), 1225–1229.

81 Centeno, M., Hidalgo, M., Dominguez, M., Navío, J., and Odriozola, J. (2008) Titania-supported gold catalysts: comparison between the photochemical phenol oxidation and gaseous CO oxidation performances. *Catal. Lett.*, **123** (3–4), 198–206.

82 Hadjiivanov, K.I. and Klissurski, D.G. (1996) Surface chemistry of titania (anatase) and titania-supported catalysts. *Chem. Soc. Rev.*, **25** (1), 61–69.

83 Busca, G., Saussey, H., Saur, O., Lavalley, J.C., and Lorenzelli, V. (1985) FT-IR characterization of the surface acidity of different titanium dioxide anatase preparations. *Appl. Catal.*, **14**, 245–260.

84 Nakashima, T. and Kimizuka, N. (2003) Interfacial synthesis of hollow TiO_2 microspheres in ionic liquids. *J. Am. Chem. Soc.*, **125** (21), 6386–6387.

85 Avellaneda, R.S., Ivanova, S., Sanz, O., Romero-Sarria, F., Centeno, M.A., and Odriozola, J.A. (2009) Ionic liquid templated TiO_2 nanoparticles as a support in gold environmental catalysis. *Appl. Catal. B: Environ.*, **93** (1–2), 140–148.

86 Turkevich, J. (1985) Colloidal gold. Part II: Colour, coagulation, adhesion, alloying and catalytic properties. *Gold Bull.*, **18** (4), 86–91.

87 Douglas, F., Yañez, R., Ros, J., Marín, S., de la Escosura-Muñiz, A., Alegret, S., and Merkoçi, A. (2008) Silver, gold and the corresponding core shell nanoparticles: synthesis and characterization. *J. Nano Res.*, **10** (1), 97–106.

88 Carriazo, J.G., Martínez, L.M., Odriozola, J.A., Moreno, S., Molina, R., and Centeno, M.A. (2007) Gold supported on Fe, Ce, and Al pillared bentonites for CO oxidation reaction. *Appl. Catal. B: Environ.*, **72** (1–2), 157–165.

89 Roucoux, A., Schulz, J., and Patin, H. (2002) Reduced transition metal colloids: A novel family of reusable catalysts? *Chem. Rev.*, **102** (10), 3757–3778.

90 Zhang, H., Hussain, I., Brust, M., Butler, M.F., Rannard, S.P., and Cooper, A.I. (2005) Aligned two- and three-dimensional structures by directional freezing of polymers and nanoparticles. *Nature Mater.*, **4**, 787–793.

91 Ikegami, S. and Hamamoto, H. (2009) Novel recycling system for organic synthesis via designer polymer-gel catalysts. *Chem. Rev.*, **109** (2), 583–593.

92 Hu, B., Wu, T., Ding, K., Zhou, X., Jiang, T., and Han, B. (2010) Seeding growth of Pd/Au bimetallic nanoparticles on highly cross-linked polymer microspheres with ionic liquid and solvent-free hydrogenation. *J. Phys. Chem. C*, **114** (8), 3396–3400.

93 Miao, S., Liu, Z., Han, B., Huang, J., Sun, Z., Zhang, J., and Jiang, T. (2005) Ru nanoparticles immobilized on montmorillonite by ionic liquids: A highly efficient heterogeneous catalyst for the hydrogenation of Benzene. *Angew. Chem. Int. Ed.*, **45** (2), 266–269.

94 Tao, R., Miao, S., Liu, Z., Xie, Y., Han, B., An, G., and Ding, K. (2009) Pd nanoparticles immobilized on sepiolite by ionic liquids: efficient catalysts for hydrogenation of alkenes and Heck reactions. *Green Chem.*, **11**, 96–101.

95 Miao, S., Liu, Z., Zhang, Z., Han, B., Miao, Z., Ding, K., and An, G. (2007) Ionic liquid-assisted immobilization of Rh on attapulgite and its application in cyclohexene hydrogenation. *J. Phys. Chem. C*, **111** (5), 2185–2190.

96 Ortego, J.D., Kowalska, M., and Cocke, D.L. (1991) Interactions of montmorillonite with organic compounds – adsorptive and catalytic properties. *Chemosphere*, **22** (8), 769–798.

97 Ruiz-Hitzky, E. (2001) Molecular access to intracrystalline tunnels of sepiolite. *J. Mater. Chem.*, **11**, 86–91.

98 (a) Park, M.J., Lee, J.K., Lee, B.S., Lee, Y.-W., Choi, I.S., and Lee, S.-G. (2006) Covalent modification of multiwalled carbon nanotubes with imidazolium-based ionic liquids: Effect of anions on solubility. *Chem. Mater.*, **18** (6), 1546–1551; (b) Chun, Y.S., Shin, J.Y., Song, C.E., and Lee, S.-G. (2008) Palladium

nanoparticles supported onto ionic carbon nanotubes as robust recyclable catalysts in an ionic liquid. *Chem. Commun.*, 942–944.

99 Shin, J.Y., Kim, Y.S., Lee, Y., Shim, J.H., Lee, C., and Lee, S.-G. (2011) Impact of anions on electrocatalytic activity in palladium nanoparticles supported on ionic liquid–carbon nanotube hybrids for the oxygen reduction reaction. *Chem. Asian J.*, **6** (8), 2016–2021.

100 Safavi, A., Maleki, N., Iranpoor, N., Firouzabadi, H., Banazadeh, A.R., Azadi, R., and Sedaghatia, F. (2008) Highly efficient and stable palladium nanocatalysts supported on an ionic liquid-modified xerogel. *Chem. Commun.*, 6155–6157.

101 Abu-Reziq, R., Wang, D., Post, M., and Alper, H. (2007) Platinum nanoparticles supported on ionic liquid-modified magnetic nanoparticles: Selective hydrogenation catalysts. *Adv. Synth. Catal.*, **349** (13), 2145–2150.

102 Sheldon, R.A. (1992) Organic Synthesis – Past, Present and Future. (Advantages of Incorporating Catalysis to Organic Synthesis), *Chemistry & Industry, London*, Dec.7th, 903–906.

103 Weisz, P.B. and Hicks, J.S. (1962) The behaviour of porous catalyst particles in view of internal mass and heat diffusion effects. *Chem. Eng. Sci.*, **17** (4), 265–275.

104 Bischoff, K.B. (1967) An extension of the general criterion for importance of pore diffusion with chemical reactions. *Chem. Eng. Sci.*, **22** (4), 525–530.

105 Hintermair, U., Francio, G., and Leitner, W. (2011) Continuous flow organometallic catalysis: new wind in old sails. *Chem. Commun.*, **47** (13), 3691–3701.

106 Browne, D.L., Deadman, B.J., Ashe, R., Baxendale, I.R., and Ley, S.V. (2011) Continuous flow processing of slurries: Evaluation of an agitated cell. *Reactor Org. Process Res. Dev.*, **15**, 693–697, and references therein.

107 Selected suppliers for lab scale flow systems: http://www.thalesnano.com, http://www.ehrfeld.com (24 November 2011), http://www.amtechuk.com, http://syrris.com, http://www.keyboardchemistry.com, http://www.helgroup.com, http://www.uniqsis.com (6 September 2012).

7
Nanostructured Materials Synthesis in Supercritical Fluids for Catalysis Applications

Samuel Marre and Cyril Aymonier

7.1
Introduction: Properties of Supercritical Fluids

Processes using supercritical fluids (SCFs) have attracted increasing interest over the past 30 years. This was particularly motivated by concerns about environmental aspects, given the capability of some SCFs, such as supercritical carbon dioxide ($scCO_2$), to replace toxic industrial solvents and the possibility of tuning the solvent characteristics for highly specific reactions or separations.

A fluid is said to be 'supercritical' when the temperature and pressure conditions exceed its critical coordinates in a *p*, *T* diagram (critical temperature: T_c and pressure: p_c, see Figure 7.1) [1, 2]. At the critical point (CP), the density of the gas phase equals that of the liquid phase leading to a disappearance of the gas / liquid interface. The critical opalescence phenomenon [3] observed at CP can be interpreted, from a microscopic point of view, by fluctuations of the local density with a correlation length of the same order of magnitude as that of visible light. Fluids close to their CP are characterized by mechanical and thermal instabilities. For instance, at CP, the isothermal compressibility $\left(K_T = -\frac{1}{V}\left(\frac{\partial V}{\partial P}\right)_T\right)$ diverges as evidenced by: $\left(\frac{\partial P}{\partial V}\right)_T = 0$ and $\left(\frac{\partial^2 P}{\partial V^2}\right)_T = 0$ (where *V* is the molar volume). The value of this isothermal compressibility remains important in the current conditions of using supercritical fluids ($1 < \frac{T}{T_c} < 2$ and $1 < \frac{p}{p_c} < 3$).

As reported in the insert in Figure 7.1, the local organization of molecules includes regions of both gas and liquid local densities simultaneously. This local organization is dynamic (with a lifetime of approximately 100 ps) and these local fluctuations of density allow us to explain the increase in the fluid compressibility. In order to discuss this point, we will study the case of a high-pressure vessel with a volume *V* and filled with *n* molecules at *T*.

In this high-pressure vessel, the fluid can be described as a system consisting of microdomains with localized densities of both gaseous and liquid levels which are in dynamic equilibrium. If the volume *V* is reduced with a mechanical action (at constant temperature and number of molecules), the system will react and molecules in gas-like density microdomains will move to liquid-like density

Nanomaterials in Catalysis, First Edition. Edited by Philippe Serp and Karine Philippot.

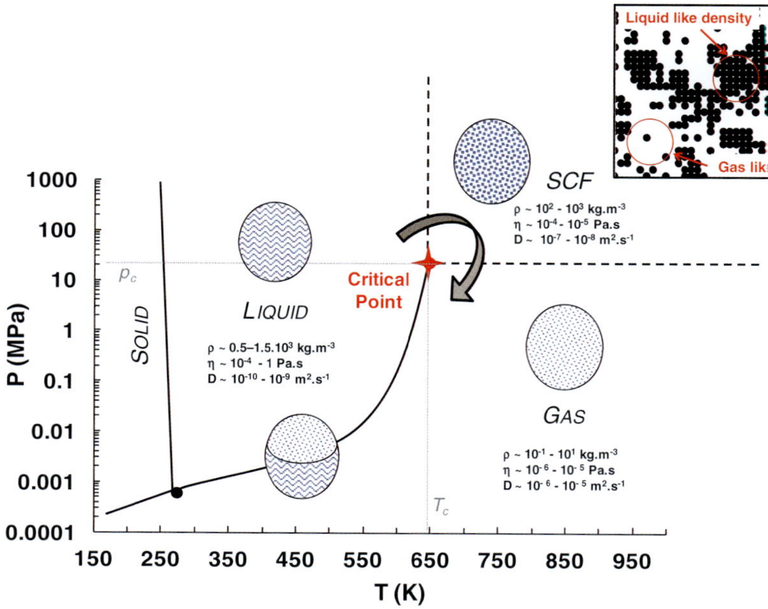

Figure 7.1 *p*, *T* phase diagram of pure water with schematic representation of the molecules distribution in a pure supercritical fluid (insert).

microdomains, resulting in a very weak increase of pressure and explaining the high fluid compressibility.

The classical description of solvation takes into account short-distance interactions, while long-distance interactions – due to the critical behavior of the fluid – are explained by the evolution of the solute partial molar volume, which is a macroscopic value. The large volume contraction is interpreted according to the solvent structure and the intermolecular forces. In the case of an attractive mixture, the addition of a solute causes a change in the solvent organization and generates the aggregation of solvent molecules close to the solute. This phenomenon, which means the aggregation of solvent molecules around the solute, cannot be independent of local density fluctuations.

Probably the most interesting aspect of the supercritical domain resides in the possibility of moving from the liquid to the gaseous state (or the opposite) without experiencing a first order phase transition. This means that there is continuity in the properties of the fluid between the gas and the liquid states (see arrow Figure 7.1). The consequence is that SCFs have 'hybrid' thermophysical properties, intermediate between liquids and gases, which are continuously adjustable with small variations of pressure and temperature.

Supercritical fluids exhibit gas-like viscosities and diffusivities while having liquid-like densities (Figure 7.1). Liquid-like densities enable the dissolution of many molecules, such as metallic precursors for the preparation of nanocatalysts, in noticeable concentration [4].

The low viscosities of SCFs are particularly interesting for continuous flow processes where pressure drop can reach high values (i.e., small scales or high flow rates). For instance, considering cylindrical geometries (pipe), the pressure drop can be expressed as:

$$\Delta P = \frac{8\eta L}{\pi R^4} Q \tag{7.1}$$

where η is the fluid viscosity (Pa s), L the reactor length (m), R the tubing radius (m) and Q the volumetric flow rate ($m^3\,s^{-1}$). Therefore, there is a linear relationship between ΔP and η resulting in much lower pressure drop for SCFs than liquids. Low viscosities also induce high diffusivities, typically encountered in gas phases, which greatly contribute to enhance mixing and to address limitation of mass transfer limited processes.

Furthermore, SCFs exhibit no surface tension, which turns out to be an undeniable advantage for processes involving surface and interface chemistries; these properties make SCFs ideal solvent systems for processing heterogeneous catalysis [5, 6].

Thanks to these unique properties, SCFs have been exploited in a large variety of applications, among which is extraction. Taking advantage of the continuous variations of density with p or T variation, it is possible to selectively tune the solvation power of the SCF versus a molecule, as discussed above. These characteristics have been extensively used for caffeine extraction, for instance [7, 8].

Supercritical antisolvent process is yet another industrial application of SCFs [9]. This process consists in generating high supersaturation by contacting a SCF (generally $scCO_2$) with a solute dissolved in a solvent. The nonsolubility of the solute in the [solvent–antisolvent] mixture leads to the precipitation of the solute. This is mainly used in the pharmaceutical industry to generate micro- or nanosized powder of organic compounds, to increase for instance the bioavailability [10, 11] or to encapsulate particles [12] or active pharmaceutical ingredients (APIs) [13]. The main interest – compared with liquid antisolvent process – is the high supersaturation that can be generated and tuned thanks to the fast diffusion of the SCF within the liquid solvent phase as well as the generated powders, which are solvent-free and do not require further drying steps.

Another long-term application of SCFs is supercritical water oxidation (SCWO), which was initially developed for the degradation of aqueous wastes in a controlled manner at much lower temperature than the conventional combustion process ('cold combustion', $673\,K < T < 873\,K$). The wastewater is pressurized beyond the critical pressure of water (i.e., 22.1 MPa) and further mixed with an oxidant (H_2O_2, O_2 or air) at high temperatures [14]. The reaction products consist mainly of water, CO_2 and other gases and solids depending on the initial composition of the waste. From this initial oxidation process, supercritical water (scH_2O) processes have now been widely extended to biomass valorization [14, 15], materials recycling [16] and chemistry [17, 18].

From these applications, there is a route towards materials synthesis in SCFs, which has been an intensively investigated field for more than 20 years [4, 19, 20]. This will be the scope of this book chapter, discussed in details in the following sections.

Admittedly, the unique thermophysical properties of SCFs are greatly advantageous for chemical reactivity, mainly because the high diffusivity of molecules enables the mass transfer limitations that generally occur in liquid phase processes to be overcome. The miscibility of gases in all proportions within SCFs has been also used for oxidation or reduction of organic compounds with O_2 and H_2 respectively, without experiencing gas-in-liquid dissolution limitations [5, 21–23]. Additionally, the environmental interest of SCFs such as $scCO_2$ or scH_2O brings opportunities for replacing some more harmful solvent systems [5, 6].

Since they are considered to be sustainable solvents, while being nonflammable and exhibiting non- to low toxicity, water and carbon dioxide are the most frequently used SCFs. A particular advantage of $scCO_2$ is its low critical point (see Table 7.1) [24].

Supercritical carbon dioxide can be used as reagent in C—C coupling reactions; however, it is mostly used as an inert solvent. It largely behaves as a nonpolar solvent and its relative dielectric constant exhibits little variations ($\varepsilon = 1.1$–1.5) in the range of p, T conditions conventionally used. Cosolvents are generally added to $scCO_2$ to increase the solubility of polar substances. Carbon dioxide is an ideal reaction medium for oxidation [25] as it is stable in the presence of oxygen, whereas CO can be formed in the case of hydrogenation reactions (depending on the catalyst and reaction conditions) leading to catalyst poisoning [26]. Nevertheless, its high miscibility with hydrogen makes $scCO_2$ a very useful reaction medium in hydrogenation, and more generally in heterogeneously catalyzed reactions. Therefore, hydrogenations of increased selectivity can be carried out. In contrast, the main limitation of scH_2O ($T_c = 647$ K, $p_c = 22.1$ MPa) as compared with $scCO_2$ ($T_c = 304$ K, $p_c = 7.38$ MPa) is

Table 7.1 Critical coordinates and densities of some conventionally used supercritical fluids.

Fluids	T_c (K)	p_c (MPa)	ϱ_c (kg m^{-3})
CO_2	304	7.38	468
H_2O	647	22.10	322
NH_3	405.4	11.10	235
CH_3CH_3	305.4	4.88	207
$CH_3CH_2CH_3$	369.8	4.25	221
$CH_3(CH_2)_4CH_3$	507	2.90	233
CH_3OH	513	8.00	272
CH_3CH_2OH	516	6.39	276
$(CH_3)_2CHOH$	508.6	5.37	274
CH_3COCH_3	508	4.76	278

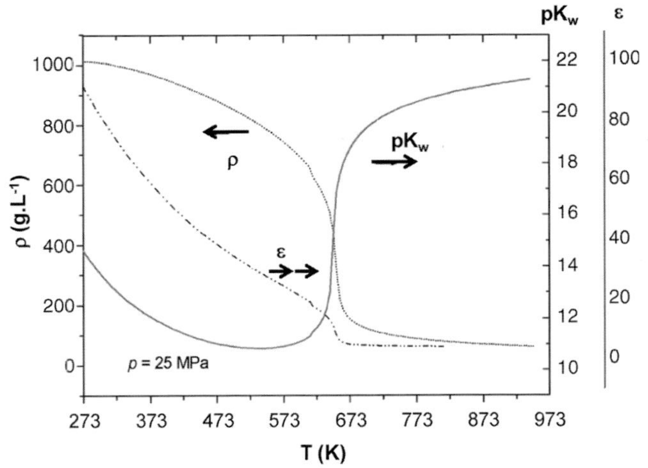

Figure 7.2 Density (ϱ), relative dielectric constant (ε) and negative logarithmic ion product of water (pK_w) at 25 MPa as a function of temperature (from [6]).

its relatively high critical point. However, no other solvent exhibits properties which change more strongly as a function of pressure and temperature (near or above the critical point) than water (see Figure 7.2). In particular, the high relative static dielectric constant ε of liquid water (78.5 at $T=298$ K and $p=0.1$ MPa) drops to a value of about 6 at the critical point. The reasons of this decrease are found in the reduced number of hydrogen bonds (decrease of the density and high temperature) [27]. One of the consequences is the high solubility of organic compounds and gases in scH_2O and a low one for salts. Another key property of hot compressed water is the evolution of the dissociation equilibrium of water $H_2O \leftrightarrows H^+ + OH^-$, which is characterized by the ion product K_w. The evolution obtained with thermodynamic calculation (Figure 7.2) shows an increase of the ion product from 10^{-14} for liquid water ($T=298$ K, $p=0.1$ MPa) up to around 10^{-11} in near critical conditions ($K_w=10^{-11.4}$, $T=593$ K, $p=25$ MPa), then a decrease is observed as soon as the critical point is crossed over ($K_w=10^{-20.9}$ at $T=693$ K, $p=25$ MPa) [28]. Therefore the chemistry in hot compressed water can take benefit of ionic or free radical reaction mechanisms by adjusting pressure and temperature.

Other SCFs have been later used in chemistry (including catalysis) or materials synthesis, such as ammonia (NH_3) [29], various alkanes or hydrocarbons (ethane, pentane, hexane, toluene, . . .) and alcohols (methanol, ethanol, isopropanol). Although much less attractive for their environmental interest, hydrocarbons may nevertheless be advantageous. This is the case when they are not only used as reaction media, but also as reactants or when the reaction medium has special benefits.

In brief, the properties of SCFs bring advantages over conventional processes including:

- No limitation of the reaction rate by a gas-liquid phase boundary.
- Adjustable dissolution behavior.
- Easy separation.
- High total reaction rates due to rapid transfer processes.
- High heat transfer rates.

Aside from these specific properties and advantages, one should also consider the use of SCFs for sustainable development and for the development of green processes [30, 31].

Following this description of the specific properties of SCFs, which are at the root of their applications in different fields of research, we will focus on the potential use of SCFs methods to design nanocatalysts in a controlled way. We propose first to discuss the interest of using SCFs synthesis approaches to design nanocatalysts as nanopowders. Although there are very few examples to date, the potential is great. We will then deal with the added value brought by this approach to form nanostructured catalyst as supported nanoparticles. The last section of this chapter will be dedicated to the presentation of a breakthrough technology, supercritical microfluidics' (SCμF), which has potential interest for processing catalytic reactions, but also to access high precision design of nanocrystals. To conclude, perspectives of using of SCFs synthesis methods for the design of nanocatalysts will be discussed.

7.2
Synthesis of Nanopowders as Nanocatalysts in SCFs

The use of supercritical fluids in the field of materials science starts from the beginning of the 1990s and is closely linked to the rise of nanosciences and nanotechnologies associated with the development of sustainable chemistry and engineering [4, 20, 32, 33].

The design of inorganic materials using methods involving supercritical fluids can be achieved by two distinct processes. Particle formation can thus be based on either a physical transformation or a chemical reaction. Papers exist which present a good overview of the state of the art concerning the first kind of process [13, 34], in which the driving force of particle nucleation and growth is a physical transformation, such as a rapid depressurization or an antisolvent effect.

Over the past decade, the second group of processes has received considerable attention [1]. In such methods, SCFs are used as reaction media to synthesize a number of materials (metals, semiconductors, nitrides, oxides, etc.) from micrometer down to nanometer scale. This approach is more suited for the control of the nucleation and growth of inorganic and hybrid organic/inorganic nanostructures. Different chemical reactions, which take their inspiration from solution chemistry, are performed in supercritical fluids for materials synthesis:

- *Supercritical hydrothermal reactions* with metal nitrates and metal–organic precursors [33],

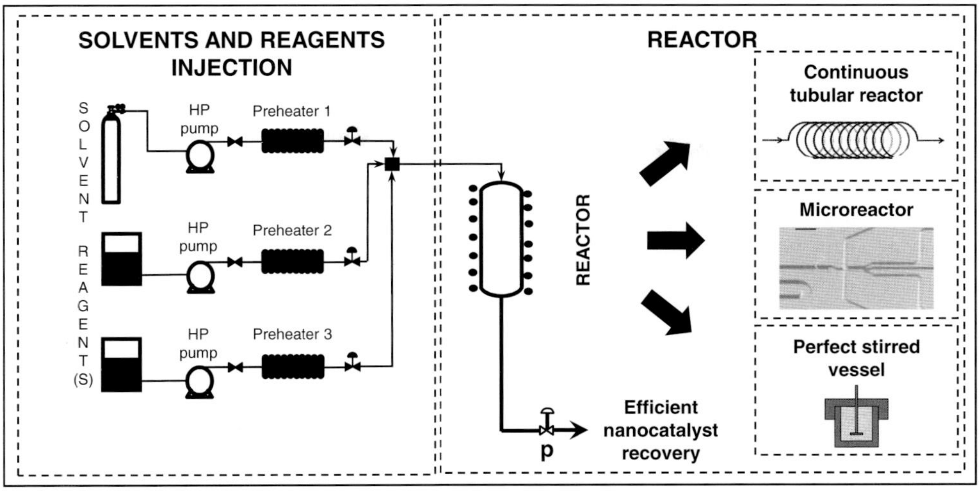

Figure 7.3 Flowsheet of a continuous process of nanomaterials synthesis using supercritical fluids.

- *Thermal decomposition* of metal–organic precursors [35],
- *Red/Ox reactions* of metal salts or metal–organic precursors, generally, in presence of hydrogen [36],
- *Sol/gel reactions* in or assisted by a SCF [37].

These four chemical reactions are mainly used to synthesize materials, especially nanomaterials, in SCFs and can be carried out using continuous micro- or millifluidic reactor. The easy scalability of this technology has recently been demonstrated with the establishment of the first commercial plant (1000 tons per year – Hanwha Chemical in South Korea) for the continuous supercritical hydrothermal synthesis of $LiFePO_4$, launched in November 2010.

Figure 7.3 shows a schematic representation of a continuous facility for processing materials in SCFs. The control of the aforementioned chemical reactions, and thus of materials characteristics (size, morphology, structure, and composition) is achieved by choosing specific operating parameters (Table 7.1): solvent nature, pressure, temperature, reagent nature and concentration, reactor technology, residence time in the reactor, additive agents (co-solvent, catalyst, surfactants . . .).

As in every process, the heart is the reactor. It is possible to switch from stirred vessel reactors to plug flow reactors (micro- up to millifluidic at laboratory scale). The choice of the reactor depends on the desired hydrodynamics, linked to the expected characteristics of the nanomaterials. In this section, we will consider the synthesis of nanocatalysts with the continuous millifluidic reactor. The subject of (stirred) vessel reactors will be discussed in Section 7.3 and that of a new microfluidic device in Section 7.4.

The synthesis of nanomaterials using SCFs has been reviewed several times over the last five years [1, 4, 20, 38, 39]. This method has been used to synthesize various inorganic materials (metals, semiconductors, nitrides, oxides, etc.) with controlled size from micrometer down to nanometer scale and complex shapes and compositions (multi-cation materials): Cu, Ni, Pd, Ag, Ge, Si, CdSe, ZnO, Al_2O_3, AlOOH, NiO, TiO_2, Cr_2O_3, ZrO_2, CeO_2, Fe_2O_3, Fe_3O_4, $CoFe_2O_4$, $NiFe_2O_4$, $ZnFe_2O_4$, $LiCoO_2$, $LiFePO_4$, $Ba_xSr_{1-x}TiO_3$ $(0 < x < 1)$, $Nd_2NiO_{4+\delta}$, La_2CuO_4, ... The control of the chemical reaction, and thus of materials characteristics (size, morphology, structure, composition and architecture) is achieved by choosing specific operating parameters [40, 41]. This approach has been extended in the last 20 years from scH_2O to other SCFs used to synthesize nanostructures, mainly $scCO_2$/alcohols (ethanol, isopropanol ...) [35, 42, 43], sc-MeOH [44], sc-iPrOH [45], alcohols/water (ethanol, isopropanol ...) [46–48], $scNH_3$ [29] or sc-alkanes (hexane, ...) [49]. This variety of solvents opens a route towards the use of numerous precursors and for a very rich chemistry at the origin of the formation of materials with a wide range of nature and composition. Beyond the versatility, other advantages of this synthetic method for material design reside in the high crystallinity of the materials obtained, the high specific surface area (nucleation is favored over growth), and the opportunity for new discoveries. The synthesis of organic-inorganic hybrid nanoparticles has also been investigated to support the development of hybrid materials generally taking advantage of the physico-chemical property of the inorganic nanoparticles and the processability of the polymer matrix. Two main routes have been proposed, adapted from the continuous synthesis of inorganic nanocrystals in SCFs: (i) the separation of nucleation and growth in the supercritical reactor and the functionalization in a vessel operating at room pressure and temperature [50]; and (ii) the *in situ* functionalization in the supercritical reactor [51]. Note that the latter can also be performed by polymer impregnation leading to organic–inorganic hybrid nanoparticles [52].

Today the method of nanomaterial synthesis using SCFs is considered to be a cutting-edge technology with the development of the first industrial applications as previously revealed [33], and the possibility of designing efficient materials in comparison with more conventional synthetic approaches in many fields, such as energy, microelectronics or pharmacy [53]. Surprisingly, although the advantages regarding the material characteristics would fit nicely with catalyst requirements, very few publications refer to the catalytic activity of nanomaterials synthesized in SCFs.

The ability of the method to separately control the growth and functionalization steps was demonstrated with the preparation of palladium nanoparticles (PdNPs) capped with thiol or stabilized with ionic liquid [BMIM][PF_6] (1-*n*-butyl-3 methylimidazolium hexafluorophosphate) as 'heterogeneous soluble catalysts.' The colloidal solution of [BMIM][PF_6], which plays the role of solvent and stabilizing agent, contains Pd NPs of 2.1 ± 0.4 nm. The catalytic efficiency of the [BMIM][PF_6]-functionalized Pd NPs was tested (1% in mass of reactive agents) using Heck reaction between iodobenzene (3 mmol) and styrene (3 mmol), using triethylamine as a base. The turnover number (TON) was found to be 2500, which is in good agreement with the catalytic activity obtained with IL stabilized Pd NPs of average size of 6 nm [54–57]. Since in this case the nanoparticles are smaller (2 nm) and so have a larger surface available for the catalytic process, this result is in good agreement with the J. Dupont *et al.* paper that has demonstrated that the 6 nm nanoparticles act only as reservoirs for smaller catalytical species [50, 54–57].

Very recently, the synthesis of a new generation of 'bulk' catalyst precursors for hydrodesulfurization (HDS) using SCFs was proposed [58, 59]. The hydrodesulfurization problem has gained increasing importance because the adverse effect of sulfur in petrol and diesel fuels on the effectiveness of catalytic exhaust gas after-treatment technologies is well established [60–63]. Therefore, it is crucial to remove +99.99% of sulfur from typical crude oil containing 1.5 wt% sulfur, and the removal process has been termed deep or ultra-deep hydrodesulfurization (Deep-HDS). Given the refining industry's cyclical margin, today, refineries are looking into more innovative ways to perform hydrodesulfurization, in particular through the design of a new generation of catalysts with very high desulfurization activity, particularly for distillate fuels at low to medium pressure, catalysts to be run in the existing process units. In this context an original and innovative method has been proposed to synthesize a new generation of HDS catalysts using SCFs: Nickel Molybdate (NiMoO4). The idea was to take advantage of the specific properties of SCFs in materials science to achieve the synthesis of a contaminant-free 'bulk' HDS catalyst with a defined composition and developing a highly specific surface area.

This bulk catalyst precursor with a controlled composition can be synthesized in a continuous reactor with a reaction time lower than 1 min in a water/alcohol system at 563 K and 23 MPa.

Under these synthetic conditions, contaminant-free catalyst precursors of HDS are obtained. These materials present a high specific surface area (up to 200 $m^2 g^{-1}$), which has never been reported before for bulk $NiMoO_4$. The powders obtained directly after synthesis mainly contained the hydrated isomorph of $NiMoO_4.0.75H_2O$ (Figure 7.4a), which has been reported as one of the most active precursor materials for Deep-HDS [64, 65].

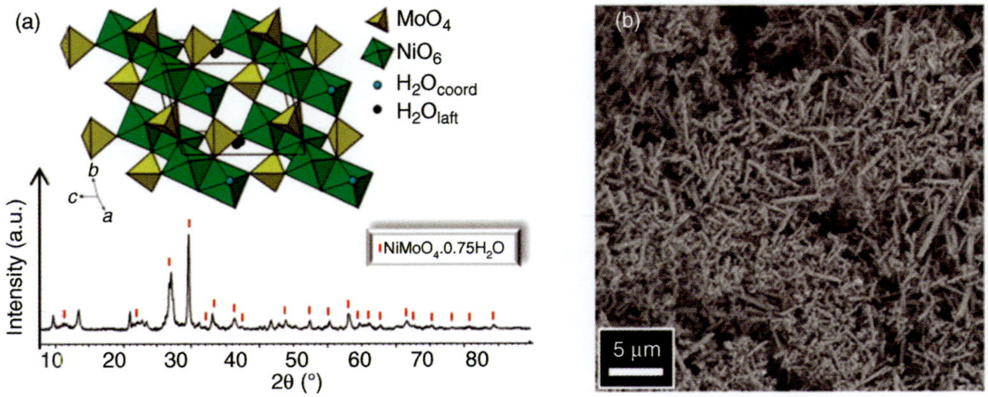

Figure 7.4 (a) XRD pattern of a stoichiometric NiMo based powder synthesized in sc-H_2O/EtOH mixture and structure of $NiMoO_4.0.75H_2O$ and (b) SEM micrograph of the nanostructured powder obtained.

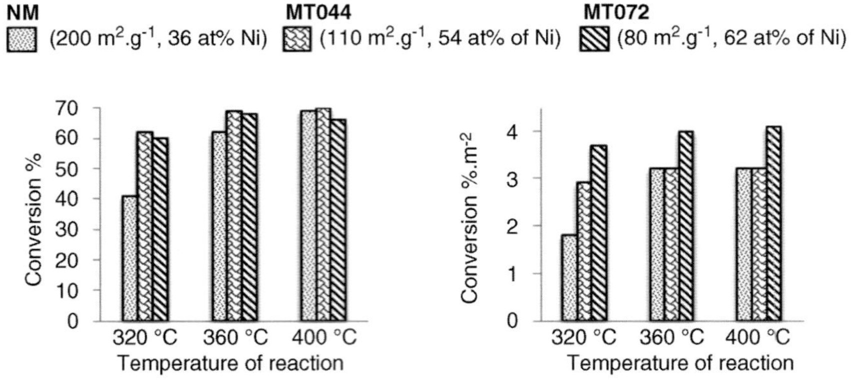

Figure 7.5 Conversion of 4,6-DMDBT in a model feedstock ($P = 3.5$ MPa, LHSV $= 39\,h^{-1}$, $M_G/M_L = 5$). NM: commercial catalyst used as reference, MT044 and MT072: 2 nanostructured catalysts synthesized using SCFs.

Upon thermal treatment of the powders at 673 K, two phases are synthesized, depending on the initial metal concentration ratios. When $[Ni]/[M]o \leq 1$, the α-$NiMoO_4$ isomorph is obtained together with $MoO_3.H_2O$, while in the case of $[Ni]/[Mo] > 1$, the α-$NiMoO_4$ coexists with NiO and β-$NiMoO_4$, the high temperature isomorph [64, 65].

Comparison of catalytic activities of a commercial catalyst and the material synthesized in SCFs for deep-HDS [66] is shown in Figure 7.5. 'Bulk' HDS catalysts, synthesized in SCFs, perform better than the commercial reference materials. For similar conversion under similar operating conditions, a lower mass of material was required compared with the commercial reference material. Alternatively, for similar conversion with similar amount of catalysts, we can work at lower temperature with materials synthesized in SCFs compared with the commercial counterpart.

In brief, from metal precursors dissolved in a water/alcohol supercritical mixture, it is possible to obtain contaminant-free $NiMoO_4$ of high specific surface area containing mainly the hydrate phase $NiMoO_4.0.75H_2O$.

This 'bulk' catalyst exhibits better hydrodesulfurization activity of sulfur compounds than conventional commercial reference materials. It catalyzes both the hydrogenation reaction and the Deep-HDS reaction at lower temperature or with lower amounts of material. Promising results were also obtained on the conversion of a real feedstock, a light cycle oil among the most difficult to treat, with results comparable to the commercial reference material.

In direct connection with the next section, SCF methods for the synthesis of nanomaterials were also applied to the design of ceria, titania or even zirconia nanoparticles [67], which could be used as catalyst support.

Supercritical water has been used for the continuous hydrothermal synthesis of well controlled TiO_2 nanoparticles (about 20 nm) in the anatase structure from inorganic titanium salts such as $TiCl_4$ and $Ti(SO_4)_2$ [68]. TiO_2 nanoparticles were also obtained from the thermal decomposition of titanium tetraisopropoxide

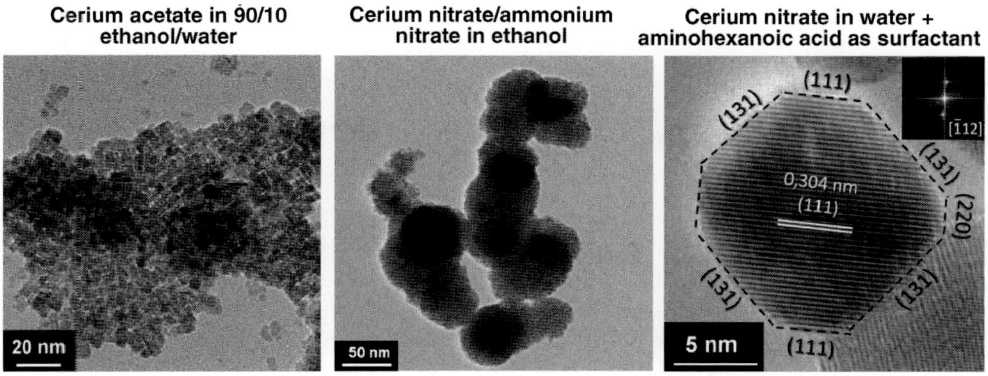

Figure 7.6 Ceria nanocrystals synthesized in SCFs using various solvent systems with or without surfactants (adapted from D. Mesguich PhD thesis [67]).

(TTIP) or diisopropoxititanium bis(acetylacetonate) (DIPBAT) in supercritical ethanol, isopropyl alcohol and CO_2/alcohol [69–71]. Similarly, zirconia has been synthesized by hydrothermal reaction in scH_2O from zirconyl nitrate. The crystallographic ZrO_2 structure depends on the reactive medium: t-ZrO_2 can be obtained in acid medium and the amorphous phase in alkaline medium (with NaOH addition), which transfers into orthorhombic ZrO_2 after thermal treatment at 823 K. The mean size of particles in these conditions was 5–6 nm [72]. Powders containing 78% t-ZrO_2 and 22% m-ZrO_2, with spherical particles 4.2 nm in size has been made from zirconium acetate. The stabilization of t-ZrO_2 in these conditions is attributed to the size effect [73].

An interesting example concerns the synthesis in SCFs of ceria nanoparticles. The first synthesis of ceria have been reported in supercritical water ($Ce(NO_3)_3$ as precursor, 673 K, 30 MPa, residence time of about 1 s) with the formation of nanoparticles of 20 nm with an octahedron shape [74]. More recently, it was shown that the size and the shape of ceria nanoparticles could be tuned by changing the nature of the reaction medium (from water to water/alcohol) and/or of the precursor (cerium acetate, cerium nitrate or cerium ammonium nitrate) and/or adding surface modification agents (Figure 7.6) [67]. The role of surfactants is critical since it was theoretically predicted that some crystal planes are more reactive than others [75]. Organic-ligand-assisted syntheses in supercritical solvents have emerged in order to produce colloidal cerium oxide nanocrystals with tailor-made cerium oxide crystal planes [44, 76, 77]. Usually, fatty acids are used, since the carboxylic group has been proved to easily graft upon CeO_2 surface [76, 78].

Therefore, the mastering of the operating parameters such as the considered solvent systems or the presence of surfactants provides improved control over the nucleation/growth process and the surface chemistry and consequently over the final nanostructures characteristics.

In contrast to the synthesis of nanopowders as nanocatalysts in SCFs mentioned above, the design of nanostructured catalysts with supported nanoparticles has been more studied, and offers real opportunities for the future.

7.3 Synthesis of Supported Nanoparticles as Nanocatalysts in SCFs

Supercritical fluids chemical deposition (SFCD) processes have been used to deposit various metallic nanoparticles onto a wide range of substrates for microelectronic, optical and catalytic applications. Two strategies are commonly employed: (i) a kinetically-controlled surface nanostructure formation, based on a homogeneous nucleation and heterogeneous growth induced by a chemical reaction and simulated by a bi-modal process [20, 79]; and (ii) a thermodynamically-controlled surface nanostructure formation based on the adsorption of metal precursor onto the substrate followed by a wide variety of precursor transformation methods [80, 81]. Both strategies enable the formation of small (1–10 nm) metallic nanoparticles on surfaces and are detailed below.

7.3.1 Kinetically-Controlled SFCD Process (K-SFCD)

The kinetically-controlled deposition is probably the easiest way to deposit particles in supercritical fluid. The mechanism (Figure 7.7a) involves a first mode, generating homogeneous nuclei within the supercritical fluid medium [79, 82]. As mentioned earlier, the properties of such fluids largely favor nucleation, thanks to the

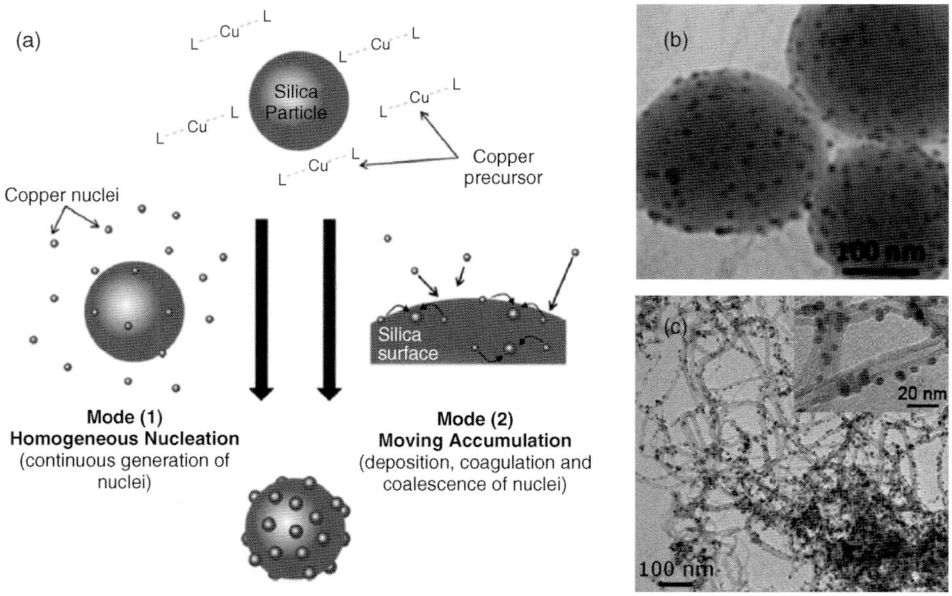

Figure 7.7 (a) Two-step mechanism of the kinetically-controlled decoration of substrates (silica sheres) with nanoparticles (copper) (b) copper nanoparticles deposited on the surface of silica spheres [79]; and (c) Pd nanoparticles deposited on the surface of CNTs through SFCD [83].

high supersaturation that can be reached. A metal precursor is dissolved in a supercritical media (generally $scCO_2$ or a mixture of $scCO_2$ + cosolvent), in which the substrate is suspended. Transformation of the metal precursor can be obtained either by thermal decomposition (for temperatures above 473 K) or by reduction with H_2. A second mode consists in (i) the random deposition of the nuclei onto the substrate surface to minimize their surface energy; followed by (ii) the heterogeneous growth of the nanoparticles by moving coagulation of the deposited nuclei (coagulation and coalescence). This process was recently demonstrated and modeled in the particular case of copper nanoparticles deposited onto the surface of silica spheres (Figure 7.7a and b). Copper hexafluoroacetylacetonate precursor was used as copper source and dissolved in a supercritical H_2/ethanol/CO_2 mixture (5/5/95% molar in composition). It was shown that the deposition can be realized for low operating temperatures (down to 373 K) by using H_2 as reducing agent. This is of particular interest for the nanostructuration of organic materials such as polymer, which cannot withstand high temperatures. Additionally, the size of the synthesized nanoparticles was controlled in a wide range (4 to 20 nm) playing on temperature conditions, concentration and residence time.

The evolution of the particle size distribution was described by a partial differential equation named the 'aerosol general dynamic equation,' which is conventional for simultaneous aerosol nucleation, condensation, coalescence, and coagulation. The size of nanoparticles is therefore primarily dependent on the transformation kinetics of the precursor. A good agreement was obtained between experimental and calculated primary particle size for the model, including both the coalescence and the aggregation mechanisms. The proposed mechanism for the silica sphere decoration process (i.e., homogeneous nucleation followed by heterogeneous growth) was confirmed for temperature conditions below 423 K, where the reduction of the copper precursor to copper metal by H_2 is the favored transformation pathway. However, it might not be suitable anymore for higher temperatures ($T > 448$ K), where there is a competition between the reduction by H_2 and the thermal decomposition. This would indeed change the decomposition kinetics and thus the parameters of the model.

A similar procedure can be applied to various kinds of substrates, as demonstrated in a recent paper for the deposition of Pd nanocrystals on carbon nanotubes (CNTs) [83]. Results show a homogeneous distribution of dispersed metallic PdNPs of about 5 nm (Figure 7.7c). The amount of deposited palladium can be adjusted by modifying the initial precursor quantity and the residence time. Furthermore, successful deposition is realized by using soft conditions and dried CNTs are eventually obtained.

7.3.2 Thermodynamically-Controlled SFCD Process (T-SFCD)

The thermodynamically-controlled approach for depositing nanocrystals from a supercritical fluid media is a two-step process. First, metal precursors are adsorbed on the surface of a given substrate. In a second step, the as-deposited precursor is reduced to its metal form to synthesize supported nanoparticles [81].

Several studies have reported the determination of the adsorption thermodynamics and kinetics of several precursors on various substrates, and very precise amounts of precursors can therefore be deposited. Indeed, the adsorption or sorption kinetics are governed by the mass transfer resistance at the external surface of the substrate, which can be quantified knowing the mass transfer coefficient of the precursor in the supercritical fluid mixture. The kinetics of precursor adsorption on the substrate can be investigated either in a dynamic or a static mode. In the static mode, the precursor and the substrate are placed in a vessel heated up to the desired temperature and charged with the fluid. The determination of the adsorption thermodynamics and kinetics is obtained by monitoring the concentration decay curves isothermally over time, as demonstrated in the case of a ruthenium precursor [Ru(cod)$(tmhd)_2$] on a carbon aerogel from $scCO_2$ [84]. In the dynamic mode, a stream of precursor supercritical solution is passed continuously over the substrate and the effluent concentration is monitored as a function of time. This technique was used to study the adsorption of [Pt(cod)Me_2] from $scCO_2$ on two porous supports, silica gel tablets and a monolithic corderite [85].

In a second step, the transformation of this absorbed precursor by different means (chemical reduction in a supercritical fluid with a reducing agent, chemical reduction with pure hydrogen at elevated temperature, thermal reduction in a supercritical fluid, thermal decomposition in an inert atmosphere, or chemical conversion with hydrogen or air at atmospheric pressure) allows the formation of size-controlled nanoparticles (Figure 7.8) [86, 87]. For the formation of metallic films, the reduction is generally carried out in the supercritical phase to ensure a steady transfer of precursor molecules to the surface for film growth

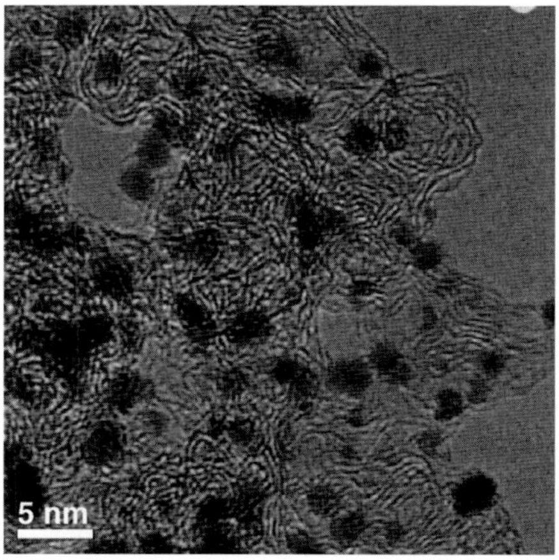

Figure 7.8 Pt nanoparticles deposited on carbon aerogel through thermodynamically-controlled SFCD [86].

and for removal of the reaction by-products. The size and morphology of the nanoparticles obtained during this reduction step depend on several parameters, such as the reduction method, the nature of the substrate, the precursor and the amount of precursor on the surface of the substrate, the reaction temperature and reaction time. This method was thereafter extended to other materials and substrates for decorating the internal or external material surface [80, 81, 88–91]. In a general manner, the supported nanocatalysts synthesized in SCFs exhibit a higher activity compared with commercially available catalysts. Additionally, a better cyclability has also been demonstrated in the case of CNT-supported platinum nanocatalysts, which were reused at least six times for the hydrogenation reaction without losing activity [92].

Both surface nanostructuration strategies using SCFs approaches were used to develop supported nanocatalysts for various applications. Several examples are summarized in Table 7.2 (nonexhaustive list).

We give one particular example dealing with hydrogen storage. Kinetically-controlled SFCD was used to develop advanced nanostructured materials that improve the kinetic sorption of hydrogen [105, 106]. In this field of research, the most promising material is low cost, high weight capacity magnesium, which can store up to 7.6 wt% of H_2 [109]. The characteristics of magnesium make it compatible with most of the required industrial specifications; however, an important challenge concerns the reduction of the sorption temperatures which are, for magnesium, higher than 573 K. To improve sorption kinetics at lower temperatures, the addition of a finely divided catalyst to the magnesium powder is the most investigated solution. In particular, catalysts enabling the dissociation of H_2 (Pd, Ni, Co, Fe) have been largely studied. The dispersion of these finely divided catalysts with magnesium particles is generally performed through the ball milling technique [110]. This helps to reduce the size of the catalyst, while increasing the defects in magnesium crystals. However, this process has several limitations including the weak mechanical link between the Mg and the catalyst, leading to decantation of the dopant during cycling. Kinetically-controlled SFCD was therefore investigated as an alternative process to deposit catalyst nanocrystals (Pd and Ni) on the surface of magnesium powder to improve H_2 sorption kinetics. Supercritical H_2/alcohol/CO_2 mixtures were used to decorate Mg microcrystals with Ni or Pd nanoparticles (size $\sim$2–10 nm), which assembly demonstrates improved hydrogen sorption kinetics and cyclability (Figure 7.9a). Interestingly, in the case of Ni, an interphase (Mg_2Ni) is formed at the interface between the Ni nanocatalyst and the Mg crystal (Figure 7.9b), whereas nothing similar occurs in the case of palladium, since no intermetallic alloys exist between Pd and Mg. The Mg–Ni interphase results in a chemical link between the metallic nanoparticles and the magnesium, which could potentially explain the improvement in the sorption kinetics observed when the samples synthesized through SFCD are used compared with the ball milling process.

SFCD turns out to be a process of choice for the surface and volume nanostructuration of materials. Supported metal and oxide catalysts with small size

Table 7.2 Examples of supported nanocatalysts synthesized in SCFs and their applications.

Materials	Synthesis Process/Fluids	Nanoparticles Size (nm)	Considered Catalytic Reactions	References
Pt on CNTs	**K-SFCD** $MeOH/CO_2$	5–15	Hydrogenation of benzene	[92]
Ni on silica – Al_2O_3 or CNTs	**K-SFCD** MeOH	≈30	Gasification of biomass for H_2 production	[93]
Pt on carbon	**T-SFCD** CO_2 impregnation then thermolysis	1–5	Oxygen reduction for direct methanol fuel cells	[94]
Pd on CNTs	**K-SFCD** H_2/CO_2	/	Suzuki coupling	[95]
Pd, Rh and Ru on CNTs	**K-SFCD** H_2/CO_2	5–10	Hydrogenation of olefine	[96]
Ru on MOF	**K-SFCD** $MeOH/CO_2$	≈2	Hydrogenation of cyclohexene and benzene	[97]
Pd on silica aerogel	**T-SFCD** CO_2 impregnation then H_2 reduction	15–20	Hydrogenation of cyclohexene	[98]
Pd on silica aerogel	**T-SFCD** CO_2 impregnation then thermolysis	50–100	Hydrogenation of 2.5-dihydro-2.5-dimethoxy-furane	[99]
Pt/Ru on CNTs	**K-SFCD** $H_2/MeOH/CO_2$	≈5	Oxygen reduction for direct methanol fuel cells	[100, 101]
Pd in mesoporous silica (SBA-15)	**T-SFCD** CO_2 impregnation then H_2 reduction	6–8	Reduction of 4-nitrophenol to 4-aminophenol	[102]
Rh in mesoporous silica (SBA-15)	**T-SFCD** CO_2 impregnation then thermolysis	5–10	Hydrogenation of phenol	[103]
Ag in mesoporous silica (SBA-15)	**T-SFCD** CO_2 impregnation then thermolysis	3–7	Oxidation of CO	[104]
Ni and Pd on Mg crystals	**K-SFCD** $H_2/iPrOH/CO_2$	2–15	Hydrogen storage in magnesium powder	[105, 106]
NiMo in nanoporous SiO_2-Al_2O_3 (Al-HMS)	**K-SFCD** $MeOH/CO_2$	2–3	Hydrodesulfurization	[107]
Pt/SnO_2 on Al_2O_3	**T-SFCD** CO_2 impregnation then H_2 reduction	<7 nm	Oxidation of CO	[108]

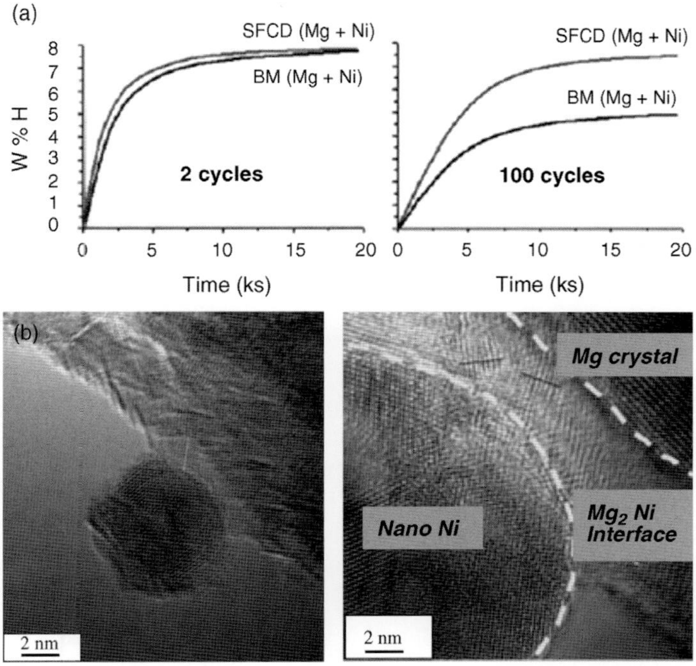

Figure 7.9 (a) Hydrogen sorption in magnesium powder. Comparison of the cyclability at T = 573 K when Ni nanocatalysts are contacted with Mg powder with the ball milling technique and when Ni nanoparticles are deposited through SFCD; (b) HR-TEM snapshot of a Ni nanoparticle deposited on the surface of a magnesium crystal and enlargement of the interface, highlighting the Mg_2Ni interphase [106].

(1–20 nm) and high loading (up to 40 wt%) can be synthesized. This method can be applied to various kinds of substrates both organic and inorganic.

7.4 Supercritical Microfluidic Synthesis of Nanocrystals

Supercritical fluids are mostly used in high pressure/high temperature (HP/HT) blind stainless steel reactors, from milliliter up to liter scale, which are appropriate for studying chemical reactions and nanosized structures synthesis and for developing proof of concept of processes. However, it is difficult to implement characterization techniques, fast screening and optimization of the synthesis conditions to optimize the final characteristics and properties of the desired nanocrystals. To address these limitations, it has become critical to develop new synthesis tools integrated with advanced control of the operating parameters.

Such control of the parameters can be provided by microfluidics, which has advanced biological and chemical research, in particular through the integration of analytic techniques [111] with cell handling [112], biochemical assays, chemical synthesis [113, 114] and micro- and nanostructures synthesis [115–118]. These tools exhibit several advantages over conventional macroscale devices. Indeed, microreactors offer a solution to feedback control of temperature, feed streams, reproducibility, *in situ* reaction monitoring [111] using sensor integration, rapid screening of parameters [119], fast mass and heat transfer [120] and low reagent consumption during optimization. Additionally, such small, controlled reaction volumes enable reactions to be performed under more aggressive conditions (confining for example small amount of hazardous products) [121–123]. However, these tools were only very recently made compatible for HP/HT processes such as SCF processes [124–126]. The development of HP/HT microreactors able to withstand supercritical conditions has offered additional advantages by considerably expanding the synthesis space. This led to a substantial enlargement of the set of solvents, precursors and surfactants available for materials synthesis [127, 128]. In particular, working under supercritical conditions at microscale combines the advantage of size reduction provided by microsystems to the unique properties of SCFs. The so-called 'supercritical microfluidics' (SCμF) therefore address some limitations of macroscale SCFs reactors. There are three main motivations for developing the synthesis of advanced nanocrystals using SCμF: (i) a better understanding of what is going on for a better control through *in situ* characterization opportunities; (ii) an exploitation of the advantages of microfluidics – hydrodynamics control, enhancement of mass and heat transfer, reproducibility, rapid screening of parameters, and low reagent consumption during optimization – for the design and synthesis of high quality nanocrystals; and (iii) a process intensification for the development of cheaper, safer and greener technologies.

The technological breakthrough of developing 'on chip' supercritical microreactors makes it possible to integrate characterization techniques, enables control of the reaction parameters, as well as collection of time- and space-resolved information. Potential characterization techniques include physical measurements (thermocouples and piezoelectric pressure monitoring) and chemical measurements via *in situ* spectroscopy such as ultraviolet-visible (UV-vis) and infrared (IR). Conventional online characterization techniques, well-developed for microfluidic applications, can be used as well, such as microparticle image velocimetry or dynamic light scattering (DLS), allowing measurement of flow profiles and particles size distribution (PSD), respectively [127]. Such an experimental set-up is now operating for the understanding of the chemistry, and nucleation/growth phenomena in SCFs but also for the synthesis of high quality nanocrystals [129]. The use of this well-characterized and controlled set-up with optimization software could provide useful laboratory-scale tools for process optimization and fast screening of reaction conditions (about 100 conditions a week), similarly to what can be done in conventional microfluidic conditions.

In this section, we present, through selected examples, recent developments in SCμF processes for synthesizing nanostructures. It is important to underline that SCμF is a recent area of research with growing interest. However, very few studies have been published so far. There is still no example of applications of nanostructures synthesized in SCμF for catalysis applications. Nevertheless, the aim of this section is to demonstrate that this new approach allows us to achieve highly precise synthesis of nanostructures. This could later be applied to the optimized design of catalysts, for example, favoring multistep, one-pot reactions.

7.4.1 Supercritical Microreactors

Commonly used microfluidic systems made of polymers, such as poly(dimethyl) siloxane (PDMS), cannot be employed for SCF-based processes, given their poor temperature and pressure resistance. However, researchers have developed several other strategies to access SCF-compatible working conditions, by using more robust fabrication materials.

First, metal-based microreactors are particularly advantageous for processes involving high heat load and toxic chemicals because of their good chemical compatibility and thermal resistance, except for strong acids. Metal microreactors are typically fabricated using conventional machining, electroforming, electrodischarge machining (EDM), or laser ablation [130]. They have been demonstrated in scH_2O gasification of organic molecules in the temperature and pressure range as high as 923 K and 25 MPa [131, 132] and in materials synthesis [133]. These microreactors can be built without clean room facilities, but current micromachining tools limit feature sizes to few hundreds of micrometers. Additionally, similarly to macroscale stainless steel reactors, implementation of *in situ* characterization techniques is not easy since metal does not offer any optical access.

Glass/glass microreactors [134] are typically fabricated through HF/BOE (buffered oxide etchant) etching or sandblasting of either borosilicate [135, 136] or soda-lime glasses [137, 138] with further thermal bonding. They have the advantage of easy optical access, providing a mean for a wide range of *in situ* characterization techniques. Although high pressures can be reached using these microreactors [139], the low thermal conductivity of glass limits applications that require good heat transfer. However, such microsystems exhibit acceptable ranges of use for most applications using $scCO_2$, in a low temperature range (293–373 K).

Reactors microfabricated in silicon and Pyrex [124, 140] by silicon etching followed by anodic bonding provide good thermal, chemical and pressure characteristics, while maintaining optical access through the Pyrex side. The good thermal conductivity of silicon devices make them excellent candidates for HT processes, such as fuel conversion [140–143]. These microsystems present a good compromise between metal and glass microreactors. Additionally, glass and silicon-based microfluidic reactors offer the opportunity to modify the wettability properties of the channels, thanks to the wide range of silane chemistry for surface modifications [144–146].

Initiated in 2005, supercritical microfluidics was first applied to organic chemical reactions before being used for nanomaterials synthesis. Some examples are described below.

7.4.2
Nanocrystals Synthesis in SCμF

In term of the synthesis of nanocrystals, the advantages of supercritical microfluidics have recently been demonstrated in a study of the synthesis of CdSe quantum dots (QDs) in a HP/HT microreactor, by comparing the high-boiling-point squalane solvent with supercritical hexane (sc-hexane). The supercritical fluid synthesis produced a narrower full width at half maximum (FWHM) and PSD than liquid phase synthesis under the same conditions [49]. In fact, comparison of the data shows that the size distribution percentage (standard deviation/d_{avg}) for QDs synthesized in sc-hexane, 4–6% (FWHM: 25–27 nm), is significantly smaller than for that for QDs synthesized in liquid squalane, 9–12% (FWHM: 41–49 nm). So the use of a supercritical solvent in a microfluidic reactor results in narrower residence time distribution, producing homogeneous reaction conditions ideal for nanocrystal synthesis. Recently, it was shown that it is possible to develop original microfluidic-based continuous processes to control chemistry, thermodynamics, physico-chemistry and hydrodynamics for the synthesis of well controlled and high quality nanocrystals of indium phosphides [147] and exciton luminescent zinc oxide nanocrystals [129].

Baek *et al.* proposed continuous three-stage silicon-based microfluidic system (operating pressure of 6.5 MPa) consisting of mixing (403–448 K), aging (473–613 K), and sequential injection stages (353–613 K) for the production of high-quality InP nanocrystals within 2 minutes with sizes ranging from 2 to 3.2 nm (Figure 7.10) [147].

Another original approach offered by SCμF was proposed to separate the nucleation and growth of the nanocrystals and the functionalization for the synthesis of ZnO nanocrystals with advanced optical properties [129]. The experiments were carried out in a coaxial flowing microsystem made of two fused silica capillaries (inner diameters: $\phi_1 = 100\,\mu m$, $\phi_2 = 250\,\mu m$) for a total length of 1 m as shown in Figure 7.11a. The precursor solution – $Zn(acac)_2.H_2O$ in an ethanol solution containing H_2O_2 – was injected in an inner silica capillary. Meanwhile, a second solution containing ligands (trioctylphosphine – TOP, oleylamine – OL or oleic acid – OA) in ethanol was injected externally, the overall residence time being fixed at 10 s, the temperature at 523 K and the pressure at 25 MPa. The conventional limitation of clogging of microchannels was overcome by acting on both hydrodynamics (co-flow system) and chemistry (addition of H_2O_2 to accelerate the hydrolysis of Zn $(acac)_2.H_2O$ and to confine the formation of ZnO nanocrystals in the inner flow). The influence of several operating parameters was carefully studied demonstrating the separation of the nucleation/growth steps from the functionalization steps. Typical TEM micrograph of ZnO-TOP nanocrystals is shown in Figure 7.11b (insert). Small (3.7 ± 0.5 nm) spherical nanocrystals are obtained. Electron diffraction pattern corresponds to pure würtzite-type ZnO (Figure 7.11c). This method therefore

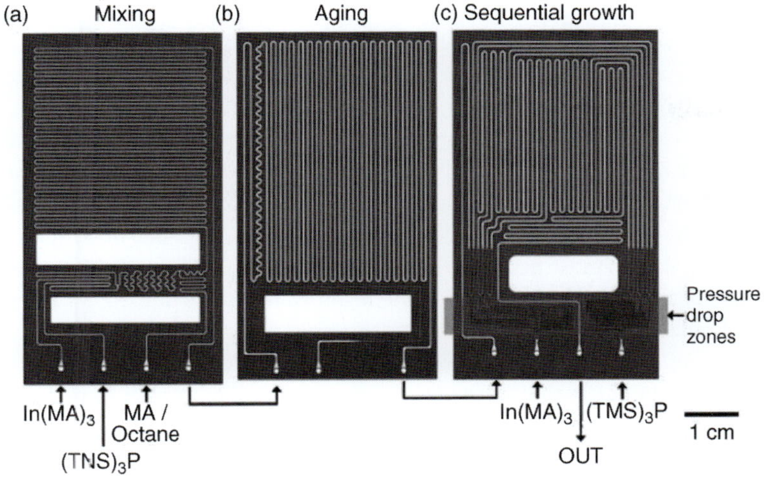

Figure 7.10 Three-stage high-temperature and high-pressure microfluidic system developed by Baek *et al.* with (a) a mixing stage; (b) an aging stage; and (c) a sequential injection microreactor with six additional injection channels.

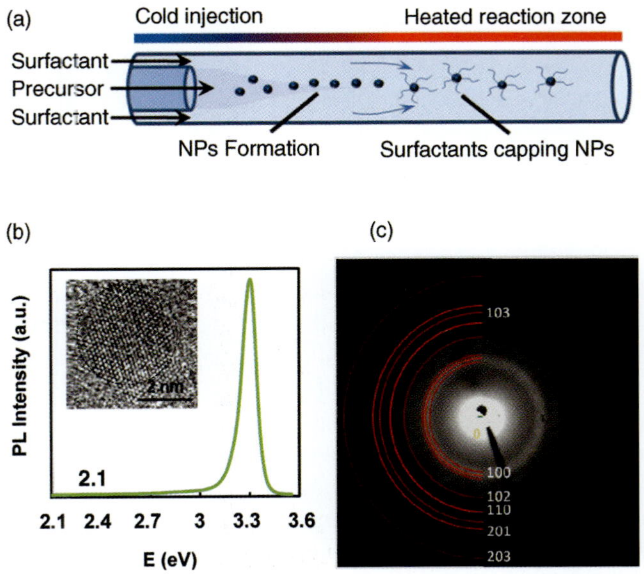

Figure 7.11 (a) Experimental supercritical microfluidics set-up with coaxial injection developed by Roig *et al.* [129]; (b) HRTEM of a single ZnO nanocrystals and photoluminescence spectrum of ZnO nanocrystals obtained through supercritical microfluidics ($\lambda_{excitation} = 320$ nm); (c) electron diffraction snapshot of as-synthesized ZnO nanocrystals.

allows for easily interchanging capping ligands without influencing the ZnO nanocrystals size or structure (würtzite structure in each case). The associated photoluminescence (PL) spectrum exhibits only one narrow peak ($E = 3.30$ eV) attributed to excitonic recombination (Figure 7.11b), while no broad emission related to defect states was observed below the band gap. This is to be compared with (i) size and shape-controlled ZnO QDs synthesized through the conventional low temperature liquid phase method, for which a low intensity excitonic peak is generally observed combined with large defect luminescence centered at 2.20 eV ($\lambda = 600$ nm) and (ii) large crystalline ZnO nanostructures with exciton PL obtained through high-temperature gas-phase synthesis approaches. The result is the access to ZnO UV emitter for development of LEDs and lasers or gas sensors.

From this brief overview, it appears that the supercritical microfluidic approach offers new opportunities in the continuous synthesis of nanomaterials from a better understanding to more efficient advanced nanostructured materials, which could be applied in the near future to the field of catalysis.

7.5
Conclusion

Processes using supercritical fluids have attracted increasing interest over the past 30 years, in particular because some SCFs are able to replace toxic industrial solvent systems. Additionally, SCFs have 'hybrid' thermophysical properties, intermediate between liquids and gases, which are continuously adjustable with small variations of pressure and temperature for highly specific reactions or separations. To date, SCFs have been used in a wide range of applications including separation, extraction, processing of organic materials for pharmaceutical applications, biomass valorization, recycling materials, wastewater treatment, chemistry and materials synthesis.

The use of SCFs in the field of materials science is closely linked to the rise of nanosciences and nanotechnologies associated with the development of sustainable chemistry and engineering. Beyond their process versatility, other advantages of SCF synthetic methods stand in the high crystallinity of the materials obtained, the high specific surface area (nucleation is favored over growth), the possible process separation of nucleation/growth from the functionalization step and the opportunity for new discoveries. However, although the advantages regarding the material characteristics might fit nicely with catalysis requirements, to date very few publications refer to the catalytic activity of nanomaterials synthesized in SCFs.

We have presented in here, through selected examples, the huge potential of such synthetic methods, both for nanopowder nanocatalysts and supported nanoparticles, the latter having been much more studied.

In particular, we highlighted the recent synthesis of a new generation of 'bulk' catalyst precursors for hydrodesulfurization using SCFs. This compound exhibits better activity for hydrodesulfurization of sulfur compounds than the commercial reference studied, by catalyzing the hydrogenation reaction and the Deep-HDS reaction at lower temperature. Promising results were obtained for the conversion of a real feedstock, a light cycle oil among the most difficult to treat and with results comparable to the commercial reference.

Concerning the synthesis of supported nanoparticles, SCFs have been used to deposit various metallic nanoparticles onto a wide range of substrates for micro-electronic, optical, and catalytic applications. Two strategies are commonly employed: (i) a kinetically-controlled surface nanostructure formation based on the homogeneous nucleation and the heterogeneous growth induced by a chemical reaction and (ii) a thermodynamically-controlled surface nanostructure formation based on the adsorption of a metal precursor onto the substrate followed by a wide variety of precursor transformation methods. Both strategies allow the formation of small nanoparticles on surfaces and have been applied to the development of supported nanocatalysts. Supported metal and oxide catalysts with small size (1–20 nm) and high loading (up to 40 wt%) were synthesized and applied to various kind of substrates both organic and inorganic.

Finally, we have introduced the new supercritical microfluidics approach for synthesizing nanomaterials with high precision. Although there is to date no example of applications of nanostructures synthesized through this method for catalysis applications, there is an avenue towards its application to the optimized design of catalysts, for example, favoring multistep reactions in a single pot.

The supercritical microfluidics approach offers new opportunities in the continuous synthesis of well-controlled nanocatalysts from nanopowders to more complex architectures such as supported nanoparticles. To date this promising area of research is still little explored, but it raises many promises for catalysis in a near future.

References

1 Cansell, F., Aymonier, C., and Loppinet-Serani, A. (2003) Review on materials science and supercritical fluids. *Curr. Opin. Solid St. M.*, **7**, 331–340.

2 Desimone, J.M., Maury, E.E., Menceloglu, Y.Z., McClain, J.B., Romack, T.J., and Combes, J.R. (1994) Dispersion polymerizations in supercritical carbon dioxide. *Science*, **265**, 356–359.

3 Abraham, F.F. (1977) Theory of critical opalescence for a classical fluid. *Chem. Phys. Lett.*, **47**, 179–181.

4 Cansell, F. and Aymonier, C. (2009) Design of functional nanostructured materials using supercritical fluids. *J. Supercrit. Fluid*, **47**, 508–516.

5 Kruse, A. and Vogel, H. (2008) Heterogeneous catalysis in supercritical media – I. Carbon dioxide. *Chem. Eng. Technol.*, **31**, 23–32.

6 Kruse, A. and Vogel, H. (2008) Heterogeneous catalysis in supercritical media: 2. Near-critical and supercritical water. *Chem. Eng. Technol.*, **31**, 1241–1245.

7 Peker, H., Srinivasan, M.P., Smith, J.M., and McCoy, B.J. (1992) Caffeine extraction rates from coffee beans with supercritical carbon-dioxide. *AICHE J.*, **38**, 761–770.

8 Mehr, C.B., Biswal, R.N., Collins, J.L., and Cochran, H.D. (1996) Supercritical carbon dioxide extraction of caffeine from Guarana. *J. Supercrit. Fluid*, **9**, 185–191.

9 Reverchon, E. (1999) Supercritical antisolvent precipitation of micro- and nano-particles. *J. Supercrit. Fluid*, **15**, 1–21.

10 Lee, L.Y., Wang, C.H., and Smith, K.A. (2008) Supercritical antisolvent production of biodegradable micro- and nanoparticles for controlled delivery of paclitaxel. *J. Control. Release*, **125**, 96–106.

11 Lopez-Periago, A.M., Vega, A., Subra, P. *et al.* (2008) Supercritical CO_2 processing of polymers for the production of materials with applications in tissue engineering and drug delivery. *J. Mater. Sci.*, **43**, 1939–1947.

12 Marre, S., Cansell, F., and Aymonier, C. (2008) Tailor-made surface properties of particles with a hydrophilic or hydrophobic polymer shell mediated by supercritical CO_2. *Langmuir*, **24**, 252–258.

13 Jung, J. and Perrut, M. (2001) Particle design using supercritical fluids: Literature and patent survey. *J. Supercrit. Fluid*, **20**, 179–219.

14 Loppinet-Sérani, A., Aymonier, C., and Cansell, F. (2010) Supercritical water for environmental technologies. *J. Chem. Technol. Biotechnol.*, **85**, 583–589.

15 Loppinet-Serani, A., Aymonier, C., and Cansell, F. (2008) Current and foreseeable applications of supercritical water for energy and the environment. *ChemSusChem*, **1**, 486–503.

16 Morin, C., Loppinet-Serani, A., Cansell, F., and Aymonier, C. (2012) Near sub and supercritical solvolysis of carbon reinforced polymers (CFRPs) for recycling carbon fibers as a valuable ressource: state of the art. *J. Supercrit. Fluid*, **66**, 232–240.

17 Savage, P.E. (1999) Organic chemical reactions in supercritical water. *Chem. Rev.*, **99**, 603–621.

18 Savage, P.E. (2009) A perspective on catalysis in sub- and supercritical water. *J. Supercrit. Fluid*, **47**, 407–414.

19 Johnston, K.P. and da Rocha, S.R.P. (2009) Colloids in supercritical fluids over the last 20 years and future directions. *J. Supercrit. Fluid*, **47**, 523–530.

20 Aymonier, C., Loppinet-Serani, A., Reveron, H., Garrabos, Y., and Cansell, F. (2006) Review of supercritical fluids in inorganic materials science. *J. Supercrit. Fluid*, **38**, 242–251.

21 Fraga-Dubreuil, J. and Poliakoff, M. (2006) Organic reactions in high-temperature and supercritical water. *Pure Appl. Chem.*, **78**, 1971–1982.

22 Jessop, P.G., Ikariya, T., and Noyori, R. (1995) Homogeneous catalysis in supercritical fluids. *Science*, **269**, 1065–1069.

23 Niessen, H.G., Eichhorn, A., Woelk, K., and Bargon, J. (2002) Homogeneous hydrogenation in supercritical fluids mediated by colloidal catalysts. *J. Mol. Catal. A: Chemical*, **182**, 463–470.

24 Beckman, E.J. (2004) Supercritical and near-critical CO_2 in green chemical synthesis and processing. *J. Supercrit. Fluid*, **28**, 121–191.

25 Kasnevich, A.D. and Beckman, E.J. (2004) Oxidation reactions in CO_2. *Chim. Oggi.*, **22**, 38–40.

26 Burgener, M., Ferri, D., Grunwaldt, J.D., Mallat, T., and Baiker, A. (2005) Supercritical carbon dioxide: An inert solvent for catalytic hydrogenation? *J. Phys. Chem. B*, **109**, 16794–16800.

27 Cochran, H.D., Cummings, P.T., and Karaborni, S. (1992) Solvation in supercritical water. *Fluid Phase Equilibr.*, **71**, 1–16.

28 Park, J.H. and Park, S.D. (2002) Kinetics of cellobiose decomposition under subcritical and supercritical water in continuous flow system. *Korean J. Chem. Eng.*, **19**, 960–966.

29 Desmoulins-Krawiec, S., Aymonier, C., Loppinet-Serani, A. *et al.* (2004) Synthesis of nanostructured materials in supercritical ammonia: nitrides, metals and oxides. *J. Mater. Chem.*, **14**, 228–232.

30 Aymonier, C., Erriguible, A., Marre, S., Serani, A., and Cansell, F. (2007) Processes using supercritical fluids: A sustainable approach for the design of functional nanomaterials. *Int. J. of Chem. Reactor Eng.*, **5**, A77.

31 Dahl, J.A., Maddux, B.L.S., and Hutchison, J.E. (2007) Toward greener

nanosynthesis. *Chem. Rev.*, **107**, 2228–2269.

32 Cansell, F., Aymonier, C., and Loppinet-Serani, A. (2003) Review on materials science and supercritical fluids. *Curr. Opin. Solid St. M.*, **7**, 331.

33 Adschiri, T., Lee, Y.W., Goto, M., and Takami, S. (2011) Green materials synthesis with supercritical water. *Green Chem.*, **13**, 1380–1390.

34 Hakuta, Y., Hayashi, H., and Arai, K. (2003) Fine particle formation using supercritical fluids. *Curr. Opin. Solid St. M.*, **7**, 341–351.

35 Cansell, F., Chevalier, B., and Demourgues, A. (1999) Supercritical fluid processing: A new route for materials synthesis. *J. Mater. Chem.*, **9**, 67–75.

36 Watkins, J.J., Blackburn, J.M., and McCarthy, T.J. (1999) Chemical fluid deposition: Reactive deposition of platinum metal from carbon dioxide solution. *Chem. Mater.*, **11**, 213–215.

37 Moner-Girona, M., Roig, A., Molins, E., and Llibre, J. (2003) Sol-gel route to direct formation of silica aerogel microparticles using supercritical solvents. *J. Sol.-Gel. Sci. Techn.*, **26**, 645–649.

38 Reverchon, E. and Adami, R. (2006) Nanomaterials and supercritical fluids. *J. Supercrit. Fluid*, **37**, 1–22.

39 Holmes, J.D., Lyons, D.M., and Ziegler, K.J. (2003) Supercritical fluid synthesis of metal and semiconductor nanomaterials. *Chem. Eur. J.*, **9**, 2144–2150.

40 Adschiri, T., Kanazawa, K., and Arai, K. (1992) Rapid and continuous hydrothermal crystallization of metal-oxide particles in supercritical water. *J. Am. Ceram. Soc.*, **75**, 1019–1022.

41 Adschiri, T., Kanazawa, K., and Arai, K. (1992) Rapid and continuous hydrothermal synthesis of boehmite particles in subcritical and supercritical water. *J. Am. Ceram. Soc.*, **75**, 2615–2618.

42 Pessey, V., Garriga, R., Weill, F., Chevalier, B., Etourneau, J., and Cansell, F. (2002) Control of particle growth by chemical transformation in supercritical CO_2/ethanol mixtures. *J. Mater. Chem.*, **12**, 958–965.

43 Besnard, M., Tassaing, T., Danten, Y. *et al.* (2006) Bringing together fundamental and applied science: The supercritical fluids route. *J. Mol. Liq.*, **125**, 88–99.

44 Kim, J., Park, Y.S., Veriansyah, B., Kim, J.D., and Lee, Y.W. (2008) Continuous synthesis of surface-modified metal oxide nanoparticles using supercritical methanol for highly stabilized nanofluids. *Chem. Mater.*, **20**, 6301–6303.

45 Bocquet, J.F., Chhor, K., and Pommier, C. (1999) Barium titanate powders synthesis from solvothermal reaction and supercritical treatment. *Mater. Chem. Phys.*, **57**, 273–280.

46 Reveron, H., Aymonier, C., Loppinet-Serani, A., Elissalde, C., Maglione, M., and Cansell, F. (2005) Single-step synthesis of well-crystallized and pure barium titanate nanoparticles in supercritical fluids. *Nanotechnology*, **16**, 1137–1143.

47 Reveron, H., Elissalde, C., Aymonier, C., Bidault, O., Maglione, M., and Cansell, F. (2005) Supercritical fluid route for synthesizing crystalline barium strontium titanate nanoparticles. *J. Nanosci. Nanotech.*, **5**, 1741–1744.

48 Hald, P., Becker, J., Bremholm, M. *et al.* (2006) Supercritical propanol-water synthesis and comprehensive size characterisation of highly crystalline anatase TiO_2 nanoparticles. *J. Solid State Chem.*, **179**, 2674–2680.

49 Marre, S., Park, J., Rempel, J., Guan, J., Bawendi, M.G., and Jensen, K.F. (2008) Supercritical continuous-microflow synthesis of narrow size distribution quantum dots. *Adv. Mater.*, **20**, 4830–4834.

50 Moisan, S., Marty, J.D., Cansell, F., and Aymonier, C. (2008) Preparation of functional hybrid palladium nanoparticles using supercritical fluids: a novel approach to detach the growth and functionalization steps. *Chem. Commun.*, **12**, 1428–1430.

51 Sasaki, T., Ohara, S., Naka, T. *et al.* (2010) Continuous synthesis of fine $MgFe_2O_4$ nanoparticles by supercritical

hydrothermal reaction. *J. Supercrit. Fluid*, **53**, 92–94.

52 Moisan, S., Martinez, V., Weisbecker, P., Cansell, F., Mecking, S., and Aymonier, C. (2007) General approach for the synthesis of organic-inorganic hybrid nanoparticles mediated by supercritical CO_2. *J. Am. Chem. Soc.*, **129**, 10602–10606.

53 Marre, S., Roig, Y., and Aymonier, C. (2012) Supercritical microfluidics: Opportunities in flow-through chemistry and materials science. *J. Supercrit. Fluid*, **66**, 251–264.

54 Cassol, C.C., Umpierre, A.P., Machado, G., Wolke, S.I., and Dupont, J. (2005) The role of Pd nanoparticles in ionic liquid in the Heck reaction. *J. Am. Chem. Soc.*, **127**, 3298–3299.

55 Antonietti, M., Kuang, D.B., Smarsly, B., and Yong, Z. (2004) Ionic liquids for the convenient synthesis of functional nanoparticles and other inorganic nanostructures. *Angew. Chem. Int. Ed.*, **43**, 4988–4992.

56 Scheeren, C.W., Machado, G., Dupont, J., Fichtner, P.F.P., and Texeira, S.R. (2003) Nanoscale Pt(0) particles prepared in imidazolium room temperature ionic liquids: Synthesis from an organometallic precursor, characterization, and catalytic properties in hydrogenation reactions. *Inorg. Chem.*, **42**, 4738–4742.

57 Dupont, J., Fonseca, G.S., Umpierre, A. P., Fichtner, P.F.P., and Teixeira, S.R. (2002) Transition-metal nanoparticles in imidazolium ionic liquids: Recycable catalysts for biphasic hydrogenation reactions. *J. Am. Chem. Soc.*, **124**, 4228–4229.

58 Theodet, M., Martinez, C., Corma, A., and Aymonier, C. *Design of high surface area nanostructured bulk catalysts of controlled composition in supercritical water/alcohol mixtures for Deep-HDS*. In preparation.

59 Aymonier, C., Theodet, M., Sanchez, C., and Corma, A. (2011) Novel bulk catalyst and method for obtaining it. European Patent 10306202.2.

60 Summers, J.C. and Baron, K. (1979) Effects of SO_2 on the performance of noble-metal catalysts in automobile exhaust. *J. Catal.*, **57**, 380–389.

61 Engstrom, P., Amberntsson, A., Skoglundh, M., Fridell, E., and Smedler, G. (1999) Sulfur dioxide interaction with NOx storage catalysts. *Appl. Catal. B: Environ.*, **22**, L241–L248.

62 Sedlmair, C., Seshan, K., Jentys, A., and Lercher, J.A. (2002) Studies on the deactivation of NOx storage-reduction catalysts by sulfur dioxide. *Catal. Today*, **75**, 413–419.

63 Corro, G. (2002) Sulfur impact on diesel emission control – A review. *React. Kinet. Catal. L.*, **75**, 89–106.

64 Brito, J.L., Barbosa, A.L., Albornoz, A., Severino, F., and Laine, J. (1994) Nickel molybdate as precursor of HDS catalysts – effect of phase-composition. *Catal. Lett.*, **26**, 329–337.

65 Brito, J.L. and Barbosa, A.L. (1997) Effect of phase composition of the oxidic precursor on the HDS activity of the sulfided molybdates of Fe(II), Co(II), and Ni(II). *J. Catal.*, **171**, 467–475.

66 Theodet, M. (2010) New generation of 'bulk' catalst precursors for hydrodesulfurization synthesized in supercritical fluids. PhD Université de Bordeaux.

67 Mesguish, D. (2010) Elaboration de matériaux nanostructurés pour piles à combustible SOFC: application à $Nd_2NiO_{4+\delta}$ et $Ce_{1-x}A_xO_{2-y}$. PhD Université de Bordeaux.

68 Adschiri, T., Hakuta, Y., and Arai, K. (2000) Hydrothermal synthesis of metal oxide fine particles at supercritical conditions. *Ind. Eng. Chem. Res.*, **39**, 4901–4907.

69 Courtecuisse, V.G., Chhor, K., Bocquet, J. F., and Pommier, C. (1996) Kinetics of the titanium isopropoxide decomposition in supercritical isopropyl alcohol. *Ind. Eng. Chem. Res.*, **35**, 2539–2545.

70 Chhor, K., Bocquet, J.F., and Pommier, C. (1992) Syntheses of submicron TiO_2 powders in vapor, liquid and supercritical phases, a comparative-study. *Mater. Chem. Phys.*, **32**, 249–254.

71 Alonso, E., Montequi, I., Lucas, S., and Cocero, M.J. (2007) Synthesis of titanium oxide particles in supercritical CO_2: Effect of operational variables in the characteristics of the final product. *J. Supercrit. Fluid*, **39**, 453–461.

72 Aimable, A., Muhr, H., Gentric, C., Bernard, F., Le Cras, F., and Aymes, D. (2009) Continuous hydrothermal synthesis of inorganic nanopowders in supercritical water: Towards a better control of the process. *Powder Technol.*, **190**, 99–106.

73 Becker, J., Hald, P., Bremholm, M. *et al.* (2008) Critical size of crystalline ZrO_2 nanoparticles synthesized in near- and supercritical water and supercritical isopropyl alcohol. *ACS Nano*, **2**, 1058–1068.

74 Hakuta, Y., Onai, S., Terayama, H., Adschiri, T., and Arai, K. (1998) Production of ultra-fine ceria particles by hydrothermal synthesis under supercritical conditions. *J. Mater. Sci. Lett.*, **17**, 1211–1213.

75 Sayle, D.C., Maicaneanu, S.A., and Watson, G.W. (2002) Atomistic models for CeO_2(111), (110), and (100) nanoparticles, supported on yttrium-stabilized zirconia. *J. Am. Chem. Soc.*, **124**, 11429–11439.

76 Zhang, J., Ohara, S., Umetsu, M., Naka, T., Hatakeyama, Y., and Adschiri, T. (2007) Colloidal ceria nanocrystals: A tailor-made crystal morphology in supercritical water. *Adv. Mater.*, **19**, 203.

77 Veriansyah, B., Chun, M.S., and Kim, J. (2011) Surface-modified cerium oxide nanoparticles synthesized continuously in supercritical methanol: Study of dispersion stability in ethylene glycol medium. *Chem. Eng. J.*, **168**, 1346–1351.

78 Aronoff, Y.G., Chen, B., Lu, G., Seto, C., Schwartz, J., and Bernasek, S. (1997) Stabilization of self-assembled monolayers of carboxylic acids on native oxides of metals. *J. Am. Chem. Soc.*, **119**, 259–262.

79 Marre, S., Erriguible, A., Perdomo, A., Cansell, F., Marias, F., and Aymonier, C. (2009) Kinetically controlled formation of supported nanoparticles in low temperature supercritical media for the development of advanced nanostructured materials. *J. Phys. Chem. C*, **113**, 5096–5104.

80 Zhang, Y. and Erkey, C. (2006) Preparation of supported metallic nanoparticles using supercritical fluids: A review. *J. Supercrit. Fluid*, **38**, 252–267.

81 Erkey, C. (2009) Preparation of metallic supported nanoparticles and films using supercritical fluid deposition. *J. Supercrit. Fluid*, **47**, 517–522.

82 Marre, S., Cansell, F., and Aymonier, C. (2006) Design at the nanometer scale of multifunctional materials using supercritical fluid chemical deposition. *Nanotechnology*, **17**, 4594–4599.

83 Majimel, M., Marre, S., Garrido, E., and Aymonier, C. (2011) Supercritical fluid chemical deposition as an alternative process to CVD for the surface modification of materials. *Chem. Vapor Depos.*, **17**, 342–352.

84 Zhang, Y., Cangul, B., Garrabos, Y., and Erkey, C. (2008) Thermodynamics and kinetics of adsorption of bis(2,2,6,6-tetramethyl-3,5-heptanedionato) (1,5-cyclooctadiene) ruthenium(II) on carbon aerogel from supercritical CO_2 solution. *J. Supercrit. Fluid*, **44**, 71–77.

85 Aschenbrenner, O., Dahmen, N., Schaber, K., and Dinjus, E. (2008) Adsorption of dimethyl(1,5-cyclooctadiene)platinum on porous supports in supercritical carbon dioxide. *Ind. Eng. Chem. Res.*, **47**, 3150–3155.

86 Saquing, C.D., Kang, D., Aindow, M., and Erkey, C. (2005) Investigation of the supercritical deposition of platinum nanoparticles into carbon aerogels. *Micropor. Mesopor. Mat.*, **80**, 11–23.

87 Zhang, Y., Kang, D.F., Aindow, M., and Erkey, C. (2005) Preparation and characterization of ruthenium/carbon aerogel nanocomposites via a supercritical fluid route. *J. Phys. Chem. B*, **109**, 2617–2624.

88 Bayrakceken, A., Kitkamthorn, U., Aindow, M., and Erkey, C. (2007) Decoration of multi-wall carbon nanotubes with platinum nanoparticles using supercritical deposition with thermodynamic control of metal loading. *Scripta Mater.*, **56**, 101–103.

89 Cangul, B., Zhang, L.C., Aindow, M., and Erkey, C. (2009) Preparation of carbon black supported Pd, Pt and Pd-Pt nanoparticles using supercritical CO_2 deposition. *J. Supercrit. Fluid*, **50**, 82–90.

90 Miao, S.D., Zhang, C.L., Liu, Z.M. *et al.* (2008) Highly efficient nanocatalysts supported on hollow polymer nanospheres: Synthesis, characterization, and applications. *J. Phys. Chem. C*, **112**, 774–780.

91 Chatterjee, M., Ikushima, Y., Hakuta, Y., and Kawanami, H. (2006) *In situ* synthesis of gold nanoparticles inside the pores of MCM-48 in supercritical carbon dioxide and its catalytic application. *Adv. Synth. Catal.*, **348**, 1580–1590.

92 Yen, C.H., Cui, X.L., Pan, H.B., Wang, S. F., Lin, Y.H., and Wai, C.M. (2005) Deposition of platinum nanoparticles on carbon nanotubes by supercritical fluid method. *J. Nanosci. Nanotech.*, **5**, 1852–1857.

93 Taylor, A.D., Dileo, G.J., and Sun, K. (2009) Hydrogen production and performance of nickel based catalysts synthesized using supercritical fluids for the gasification of biomass. *Appl. Catal. B: Environ.*, **93**, 126–133.

94 Ang, S.Y. and Walsh, D.A. (2010) Highly stable platinum electrocatalysts for oxygen reduction formed using supercritical fluid impregnation. *J. Power Sources*, **195**, 2557–2563.

95 Pan, H.B., Yen, C.H., Yoon, B., Sato, M., and Wai, C.M. (2006) Recyclable and ligandless Suzuki coupling catalyzed by carbon nanotube-supported palladium nanoparticles synthesized in supercritical fluid. *Synthetic Commun.*, **36**, 3473–3478.

96 Ye, X.R., Lin, Y.H., Wang, C.M., Engelhard, M.H., Wang, Y., and Wai, C. M. (2004) Supercritical fluid synthesis and characterization of catalytic metal nanoparticles on carbon nanotubes. *J. Mater. Chem.*, **14**, 908–913.

97 Zhao, Y.J., Zhang, J.L., Song, J.L. *et al.* (2011) Ru nanoparticles immobilized on metal-organic framework nanorods by supercritical CO_2-methanol solution: highly efficient catalyst. *Green Chem.*, **13**, 2078–2082.

98 Morley, K.S., Licence, P., Marr, P.C. *et al.* (2004) Supercritical fluids: A route to palladium-aerogel nanocomposites. *J. Mater. Chem.*, **14**, 1212–1217.

99 Jespersen, H.T., Standeker, S., Novak, Z., Schaumburg, K., Madsen, J., and Knez, Z. (2008) Supercritical fluids applied to the sol-gel process for preparation of AEROMOSILS/palladium particle nanocomposite catalyst. *J. Supercrit. Fluid*, **46**, 178–184.

100 Lin, Y.H., Cui, X.L., Yen, C.H., and Wai, C.M. (2005) PtRu/carbon nanotube nanocomposite synthesized in supercritical fluid: A novel electrocatalyst for direct methanol fuel cells. *Langmuir*, **21**, 11474–11479.

101 An, G.M., Yu, P., Mao, L.Q. *et al.* (2007) Synthesis of PtRu/carbon nanotube composites in supercritical fluid and their application as an electrocatalyst for direct methanol fuel cells. *Carbon*, **45**, 536–542.

102 Morere, J., Tenorio, M.J., Torralvo, M.J., Pando, C., Renuncio, J.A.R., and Cabanas, A. (2011) Deposition of Pd into mesoporous silica SBA-15 using supercritical carbon dioxide. *J. Supercrit. Fluid*, **56**, 213–222.

103 Lee, B.I., Bae, D., Kang, J.K., Kim, H., and Byeon, S.H. (2009) Synthesis of SBA-15 Supported Rh nanoparticles with high loading density and its catalytic hydrogenation of phenol in supercritical carbon dioxide. *B. Kor. Chem. Soc.*, **30**, 1701–1702.

104 Yin, J.Z., Zhang, C.J., Xu, Q.Q., and Wang, A.Q. (2009) Preparation, characterization and catalysis properties of Ag/SBA-15 nanocomposite by supercritical fluid deposition. *J. Inorg. Mater.*, **24**, 129–132.

105 Denis, A., Sellier, E., Aymonier, C., and Bobet, J.L. (2009) Hydrogen sorption properties of magnesium particles decorated with metallic nanoparticles as catalyst. *J. Alloy. Compd.*, **476**, 152–159.

106 Aymonier, C., Denis, A., Roig, Y. *et al.* (2010) Supported metal NPs on magnesium using SCFs for hydrogen storage: Interface and interphase

characterization. *J. Supercrit. Fluid*, **53**, 102–107.

107 Alibouri, M., Ghoreishi, S.M., and Aghabozorg, H.R. (2009) Hydrodesulfurization activity of NiMo/Al-HMS nanocatalyst synthesized by supercritical impregnation. *Ind. Eng. Chem. Res.*, **48**, 4283–4292.

108 Garrido, G.I., Patcas, F.C., Upper, G., Tuerk, M., Yilmaz, S., and Kraushaar-Czarnezki, B. (2008) Supercritical deposition of Pt on SnO_2-coated Al_2O_3 foams: Phase behaviour and catalytic performance. *Appl. Catal. A: Gen.*, **338**, 58–65.

109 Schlapbach, L. and Zuttel, A. (2001) Hydrogen-storage materials for mobile applications. *Nature*, **414**, 353–358.

110 Bobet, J.L., Akiba, E., and Darriet, B. (2001) Study of Mg-M (M=Co, Ni and Fe) mixture elaborated by reactive mechanical alloying: hydrogen sorption properties. *Int. J. Hydrogen Energy*, **26**, 493–501.

111 deMello, A.J. (2006) Control and detection of chemical reactions in microfluidic systems. *Nature*, **422**, 394–402.

112 El-Ali, J., Sorger, P.K., and Jensen, K.F. (2006) Cells on chips. *Nature*, **442**, 403–411.

113 Hartman, R.L. and Jensen, K.F. (2009) Microchemical systems for continuous-flow synthesis. *Lab Chip*, **9**, 2495–2507.

114 Hartman, R.L., McMullen, J.P., and Jensen, K.F. (2011) Deciding whether to go with the flow: Evaluating the merits of flow reactors for synthesis. *Angew. Chem. Int. Ed.*, **50**, 7502–7519.

115 Marre, S. and Jensen, K.F. (2010) Synthesis of micro and nanostructures in microfluidic systems. *Chem. Soc. Rev.*, **39**, 1183–1202.

116 Nightingale, A.M. and de Mello, J.C. (2010) Microscale synthesis of quantum dots. *J. Mater. Chem.*, **20**, 8454–8463.

117 Abou-Hassan, A., Sandre, O., and Cabuil, V. (2010) Microfluidics in inorganic chemistry. *Angew. Chem. Int. Ed.*, **49**, 6268–6286.

118 Il Park, J., Saffari, A., Kumar, S., Gunther, A., and Kumacheva, E. (2010) Microfluidic synthesis of polymer and inorganic particulate materials. *Ann. Rev. Mater. Res.*, **40**, 415–443.

119 Jensen, K.F., Ajmera, S.K., Firebaugh, S. L. *et al.* (2000) Microfabricated chemical systems for product screening and synthesis, in *Automated Synthetic Methods for Speciality Chemicals* (ed. W. Hoyle), Springer, pp. 14–24.

120 Gervais F T. and Jensen, K.F. (2006) Mass transport and surface reactions in microfluidic systems. *Chem. Eng. Sci.*, **61**, 1102–1121.

121 Brandt, J.C. and Wirth, T. (2009) Controlling hazardous chemicals in microreactors: Synthesis with iodine azide. *Beilstein J. Org. Chem.*, **5**, 30.

122 O'Brien, A.G., Levesque, F., Suzuki, Y., and Seeberger, P.H. (2011) Safe use of azides in continuous flow. *Chim. Oggi.*, **29**, 57.

123 Panke, G., Schwalbe, T., Stirner, W., Taghavi-Moghadam, S., and Wille, G. (2003) *A Practical Approach of Continuous Processing to High Energetic Nitration Reactions in Microreactors Synthesis-Stuttgart*, Thieme, pp. 2827–2830.

124 Marre, S., Adamo, A., Basak, S., Aymonier, C., and Jensen, K.F. (2010) Design and packaging of microreactors for high pressure and high temperature applications. *Ind. Eng. Chem. Res.*, **49**, 11310–11320.

125 Oosterbroek, R.E., Hermes, D.C., Kakuta, M. *et al.* (2006) Fabrication and mechanical testing of glass chips for high-pressure synthetic or analytical chemistry. *Microsyst. Technol.*, **12**, 450–454.

126 Trachsel, F., Hutter, C., and von Rohr, P.R. (2008) Transparent silicon/glass microreactor for high-pressure and high-temperature reactions. *Chem. Eng. J.*, **135**, S309–S316.

127 Marre, S., Baek, J., Park, J., Bawendi, M.G., and Jensen, K.F. (2009) High-pressure/high-temperature microreactors for nanostructure synthesis. *JALA*, **14**, 367–373.

128 Benito-Lopez, F., Tiggelaar, R.M., Salbut, K. *et al.* (2007) Substantial rate enhancements of the esterification reaction of phthalic anhydride with methanol at high pressure and using

supercritical CO_2 as a co-solvent in a glass microreactor. *Lab Chip*, **7**, 1345–1351.

129 Roig, Y., Marre, S., Cardinal, T., and Aymonier, C. (2011) Supercritical microfluidics synthesis of exciton luminescent ZnO nanocrystals. *Angew. Chem. Int. Ed.*, **50**, 12071–12074.

130 Lichtenberg, J. and Baltes, H. (2005) Low-cost, chemically resistant microreactors fabricated by laser micromachining in stainless steel. *Micro TAS*, **2**, 351–353.

131 Goodwin, A.K. and Rorrer, G.L. (2008) Conversion of glucose to hydrogen-rich gas by supercritical water in a microchannel reactor. *Ind. Eng. Chem. Res.*, **47**, 4106–4114.

132 Goodwin, A.K. and Rorrer, G.L. (2009) Conversion of xylose and xylose-phenol mixtures to hydrogen-rich gas by supercritical water in an isothermal microtube flow reactor. *Energy & Fuels*, **23**, 3818–3825.

133 de la Iglesia, O., Sebastian, V., Mallada, R. *et al.* (2007) Preparation of Pt/ZSM-5 films on stainless steel microreactors. *Catal. Today*, **125**, 2–10.

134 Kikutani, Y., Hibara, A., Uchiyama, K., Hisamoto, H., Tokeshi, M., and Kitamori, T. (2002) Pile-up glass microreactor. *Lab Chip*, **2**, 193–196.

135 Iliescu, C., Chen, B., and Miao, J. (2008) On the wet etching of Pyrex glass. *Sensor Actuat. A-Phys.*, **143**, 154–161.

136 Tay, F.E.H., Iliescu, C., Jing, J., and Miao, J.M. (2006) Defect-free wet etching through pyrex glass using Cr/Au mask. *Microsyst. Technol.*, **12**, 935–939.

137 Mazurczyk, R., El Khoury, G., Dugas, V. *et al.* (2008) Low-cost, fast prototyping method of fabrication of the microreactor devices in soda-lime glass. *Sensor Actuat. B: Chem.*, **128**, 552–559.

138 Zhu, J.J., Cheng, J., Ang, S.S., and Wang, H. (2006) Microfabrication of microfluidic channels on soda-lime glass. Proceedings of the 8th International Conference on Progress of Machining Technology (ICPMT 2006), pp. 389–392.

139 Tiggelaar, R.M., Benito-Lopez, F., Hermes, D.C. *et al.* (2007) Fabrication, mechanical testing and application of high-pressure glass microreactor chips. *Chem. Eng. J.*, **131**, 163–170.

140 Jensen, K.F. (2006) Silicon-based microchemical systems: Characteristics and applications. *MRS Bull.*, **31**, 101–107.

141 Kelley, S.C., Deluga, G.A., and Smyrl, W. H. (2002) Miniature fuel cells fabricated on silicon substrates. *AICHE J.*, **48**, 1071–1082.

142 Sabate, N., Esquivel, J.P., Santander, J. *et al.* (2008) Passive direct methanol fuel cells in silicon technology. *J. New Mater. Electrochem. Sys.*, **11**, 143–146.

143 Wu, X.H., Guo, H., Ye, F., and Ma, C.F. (2009) Application of silicon in micro fuel cells. *Progress Chem.*, **21**, 1344–1348.

144 Appelhans, D., Ferse, D., Adler, H.J.P. *et al.* (2000) Self-assembled monolayers prepared from omega-thiophene-functionalized n-alkyltrichlorosilane on silicon substrates. *Colloid Surface A*, **161**, 203–212.

145 Kulkarni, S.A. and Vijayamohanan, K.P. (2007) Interfacial behavior of alkyltrichlorosilane monolayers on silicon: Control of flat-band potential and surface state distribution using chain length variation. *Surf. Sci.*, **601**, 2983–2993.

146 Srinivasan, U., Houston, M.R., Howe, R. T., and Maboudian, R. (1998) Alkyltrichlorosilane-based self-assembled monolayer films for stiction reduction in silicon micromachines. *J. Microelectromech. S.*, **7**, 252–260.

147 Baek, J., Allen, P.M., Bawendi, M.G., and Jensen, K.F. (2011) Investigation of indium phosphide nanocrystal synthesis using a high-temperature and high-pressure continuous flow microreactor. *Angew. Chem. Int. Ed.*, **50**, 627–630.

8
Recovery of Metallic Nanoparticles

Inge Geukens and Dirk E. De Vos

8.1
Introduction

For a long time, catalysis was divided into two main domains: heterogeneous and homogeneous catalysis. With the focus on the use of nanoparticles (NPs) in catalysis, the line between homogeneous and heterogeneous catalysis has become less clear. With their high surface area, nanoparticles usually display higher activities compared with purely heterogeneous catalysts, but in contrast to what is often said, recovering them after reaction is not always easy. There are some situations in which recycling is not necessary, for example for the catalytic production of hydrogen, where the gas is easily removed from the liquid reaction mixture during synthesis [1], but such situations are very rare and in a majority of cases, recycling is required. In this chapter, a condensed, but complete overview is given of the different strategies suggested in literature to recycle metallic nanoparticles (MNPs) and to reuse them several times for the same reaction, together with specific examples to elucidate the mechanisms behind these strategies.

8.2
Immobilization on a Solid Support

A first strategy to recycle nanoparticles is by immobilizing them on various supports (Figure 8.1). There are numerous examples of the immobilization of catalytic MNPs on both organic (activated carbon, carbon nanotubes, mesoporous carbon materials, organic polymeric membranes…) [2–5] and inorganic materials (zeolites, mesoporous silicates…) [6, 7]. Recently, hybrid materials such as metal organic frameworks (MOFs) [8, 9] and other composite materials [10, 11] have also been used as supports. Based on all these support materials, catalysts with proven recyclability have been designed. Several methods are available to immobilize the particles on the surface of a support, but they can more or less be divided into two categories (Figure 8.1): the particles can first be formed separately from the support

Nanomaterials in Catalysis, First Edition. Edited by Philippe Serp and Karine Philippot.

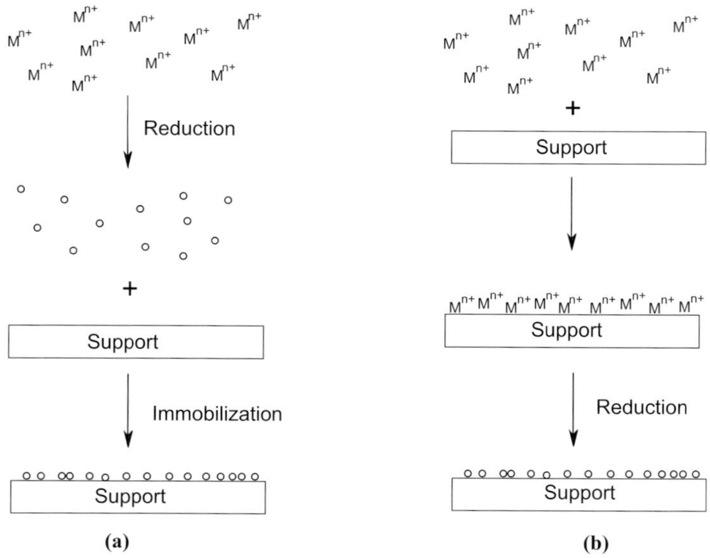

Figure 8.1 Two different strategies to immobilize NPs on a support: (a) metal precursors (M^{n+}) are first reduced to NPs (o) in the presence of a stabilizer and then immobilized on the support; (b) metal precursors are first adsorbed on the support and then reduced to nanoparticles.

via reduction of a metal precursor in the presence of a stabilizer and then be immobilized on the support [12], or the metal precursor can first be contacted with the support and then be reduced. This reduction can be performed using conventional reducing agents like sodium borohydride ($NaBH_4$) [4, 5, 8, 10, 13], hydrazine hydrate [9] or hydrogen gas [7], but can also be accomplished during reaction, like for example during cross-coupling reactions [11, 14, 15].

A third, alternative possibility to immobilize NPs is to produce the support simultaneously in the presence of the NPs and/or metal precursors so that the latter are entrapped within the support. Silver nanoparticles (AgNPs) were formed in this manner inside a poly(vinylalcohol) (PVA) membrane through spin coating of a solution of the polymer and the silver precursor on a supporting structure followed by *in situ* thermal reduction of the silver precursor. The supporting structure was removed later, allowing easy access of the reactants and products to and from the particles during reaction. The catalyst could be recycled up to 30 times for the reduction of 4-nitrophenol with $NaBH_4$ without any significant loss of activity (Figure 8.2) [2]. Another example is the synthesis of the mesoporous silicate, MCM-41, in the presence of preformed palladium nanoparticles (PdNPs), which could be recycled four times without any loss in activity for the hydrogenation of phenylacetylene [6].

An important aspect in all of the immobilization methods that needs to be taken into account is the stabilization of the particles on the support to ensure that the

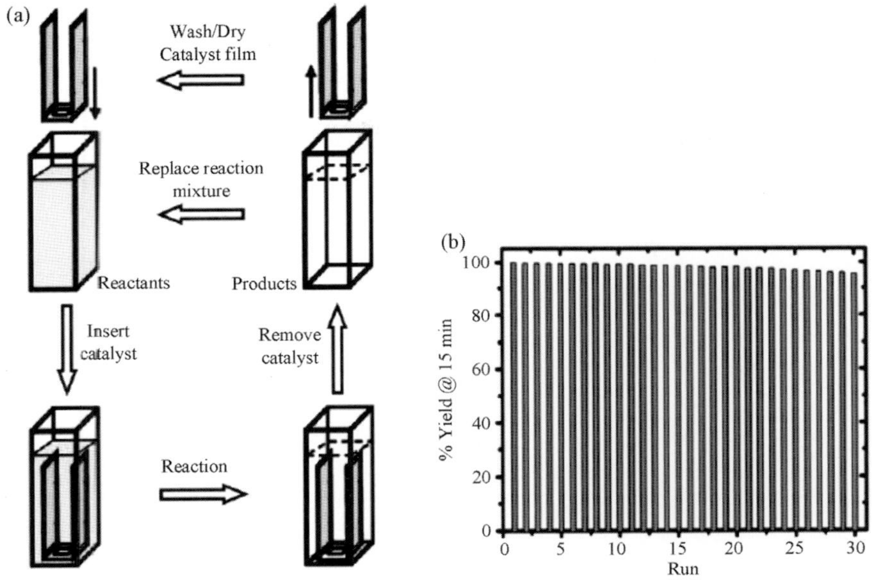

Figure 8.2 (a) Schematic representation of the reuse cycle of the Ag-PVA film. (b) Reaction yield at 15 min for the reduction of 4-nitrophenol by $NaBH_4$ at 298 K in repeated runs catalyzed by the same Ag–PVA film [2].

formation of larger NPs is inhibited and that leaching of particles into the reaction liquor is suppressed. This stabilization can be achieved by using conventional stabilizers, like poly(*N*-vinyl-2-pyrrolidone) (PVP) [5, 8], PVA [4] or dendrimers [16], but can also be provided by the support itself. A first option is to modify the surface of the support with linkers/stabilizers to increase the affinity between the nanoparticles/metal precursors and the support surface. This is usually done when silica-based supports, like MCM materials or amorphous SiO_2, are used, but can also be applied to other supports, for example chitosan, of which the amine groups were transformed into Schiff base ligands to accommodate PdNPs [14]. The linker itself can, in principle, consist of almost any molecule, as long as the NPs are stabilized by it and/or show affinity for it. Mesoporous silica supports containing ethylene bridges, and functionalized with N-heterocyclic carbenes can for example be used to accommodate PdNPs [15] and fluorous compounds can be attached to silica gel to immobilize PdNPs stabilized by fluorous ligands [17]. Even charged compounds, like sulfonate groups, can be placed on the surface of the support to attract and immobilize the metal precursor, which can be reduced afterwards [13]. A second stabilization option is pore confinement: when the particles are situated within a pore or a cage of the support, growth is sterically inhibited. MOFs have frequently been proposed for this reason to stabilize nanoparticles. Gold nanoparticles (AuNPs) were, for example, immobilized inside the cages of the porous Fe-terephthalate

MIL-101 and were used for the aerobic oxidation of 4-methoxybenzyl alcohol with good recyclability [8].

Although immobilizing NPs on a support provides a good way to recycle them, it can be debated whether the resulting materials are still to be considered 'nanoparticle catalysts' as opposed to traditional heterogeneous metal catalysts. Problems often encountered with heterogeneous catalysts, like diffusion limitation or pore blockage are usually also challenging for NPs immobilized on a support.

8.3
Multiple Phases

A second strategy which has been applied many times in homogeneous catalysis is to work with multiphasic systems. Nanoparticles are produced in a phase that does not mix with the phase containing the products and/or reactants at conditions other than the reaction conditions (Figure 8.3), or alternatively, the products can be extracted after reaction with a solvent that does not mix with the NP phase. In the first case, extensive mixing is required during reaction to obtain a good phase transfer between the nanoparticle phase and the reactants/products phase. In some cases, one phase is formed at the reaction temperature.

The easiest and greenest solution is to produce the NPs in an *aqueous phase*. Many organic solvents do not mix with water and hence recycling is easily achieved through extraction of the products after reaction and reusing the residual aqueous catalyst phase for a next run. Several examples exist using such a system, mostly using PdNPs as catalyst for cross-coupling or hydrogenation reactions and using stabilizers like tetrabutylammonium bromide (TBAB) [18–21]. Typically, diethylether or related compounds are used as extracting solvent, but other greener solvents, like supercritical CO_2, can also be used [22, 23]. Recyclability using these types of biphasic aqueous systems is usually quite good, with reuses from four to ten runs without substantial activity losses.

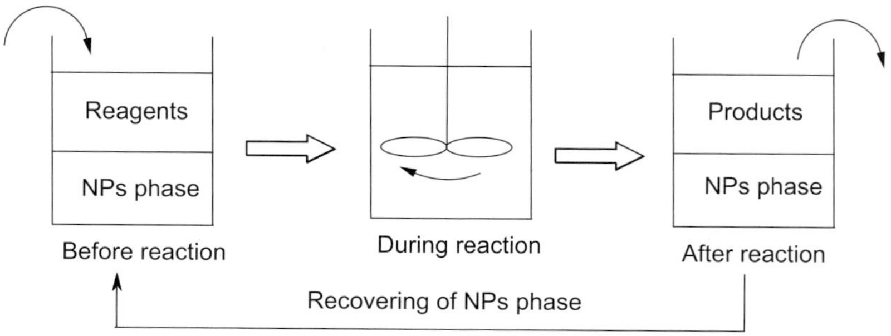

Figure 8.3 Idealized schematic representation of using nanoparticles (NPs) in a biphasic system. During reaction, either a single phase can be formed, or an emulsion is obtained through mixing.

A second phase in which NPs are often produced is *polyethylene glycol* (PEG), which also does not mix with solvents like diethylether or *n*-hexane at room temperature. An extra advantage of using PEG is that it has a reducing capacity: metal salts can be added in the oxidized state and be reduced *in situ* by the polyol groups of PEG. PdNPs and platinum nanoparticles (PtNPs) have been produced this way in PEG of various molecular weights (800–6000 g mol^{-1}) mainly for use in hydrogenation reactions [24–27]. Just as with the aqueous systems, recyclability is quite good, with reuses from six to ten runs with conservation of activity.

Nanoparticles can also be modified so that they show more affinity for a certain phase. One way to do this is to modify the NPs with fluorous compounds so that their affinity for *fluorous solvents* like trifluorotoluene, FC-72 (C_6F_{14}) and perfluoroheptane is strongly enhanced. This adaptation is accomplished by using fluorous phosphine, nitrogen- or sulfur-based ligands as stabilizers for the synthesis of the NPs. The second phase is formed by the reactants and/or products of the reaction or by conventional solvents like *N,N*-dimethylformamide (DMF) or tetrahydrofuran (THF), which do not mix with the fluorous solvents at room temperature [28–30]. Some of these systems show leaching of the NPs into the organic phase and/or incomplete phase separation, but there are examples in which the catalyst could be recycled for 12 runs without extensive leaching into the organic phase [30].

A final class of solvents which is very suitable for working with multiple phases is the group of *ionic liquids* (IL). These solvents have many advantages for use in nanoparticle applications. Not only do most ILs not mix with organic solvents like diethylether or *n*-hexane; but the cation/anion of the ionic liquid can also be used as stabilizer for the NPs, so that no extra stabilizers need to be added. Moreover, the IL can be chemically modified, resulting in task-specific or functionalized ILs [31], enabling their affinity to be adjusted for a certain phase. Several examples exist of performing cross-coupling reactions with NPs in ionic liquids and extracting the products after reaction with organic solvents like diethylether, *n*-hexane or cyclohexane, such as NiNPs in phosphonium-based ILs for aromatic amination reactions [32], copper-bronze-NPs in TBAB and tetrabutylammonium acetate for the Heck reaction [33] and PdNPs for Heck reactions in task-specific alkylammonium based ILs (Figure 8.4). In all cases the IL serves not only as reaction medium, which can be easily recycled, but also as stabilizer/ligand of the NPs, and in some cases it can even be used as base [34].

Extraction with an organic solvent is not always necessary. In some cases the products and/or reactants do not mix with the IL phase at room temperature, but form one phase at reaction temperature, which results in spontaneous phase separation after reaction when cooling back to room temperature. Janiak and coworkers produced Rh, Ir and Co NPs in imidazolium and ammonium based ILs with varying anions [tetrafluoroborate (BF_4^-), trifluoromethanesulfonate (triflate) and bis(trifluoromethylsulfonyl)imide (bistriflimide)] for the hydrogenation of cyclohexene. At room temperature the upper phase consists of cyclohexane and/or cyclohexene and the lower phase consists of the IL

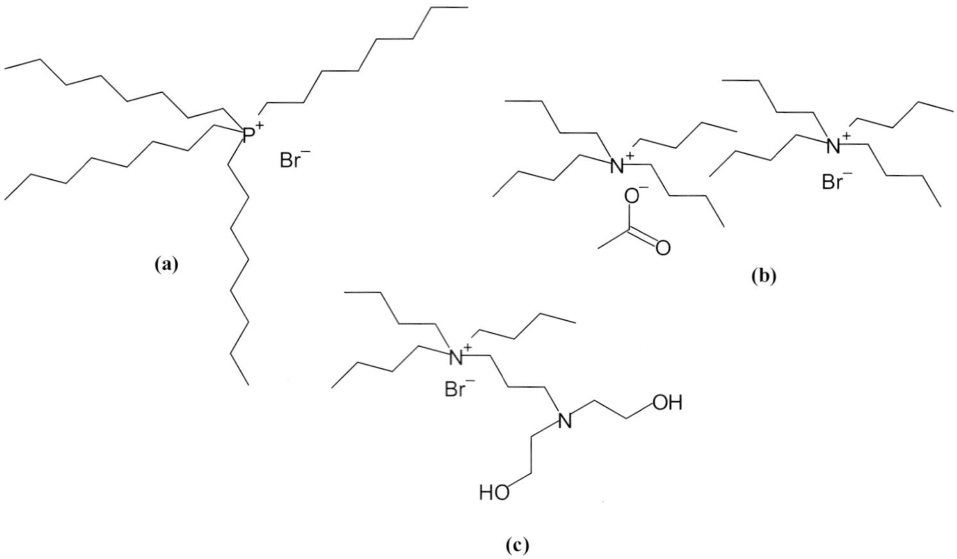

Figure 8.4 Examples of ionic liquids used for synthesis of nanoparticles applied in cross-coupling reactions. The products are extracted after reaction with an organic solvent. (a) tetraoctylphosphonium bromide [32]; (b) tetrabutylammonium acetate/tetrabutylammonium bromide [33]; and (c) 4-di(hydroxyethyl)aminobutyl tributylammonium bromide [34].

containing the NPs, whereas during reaction a single phase is formed [35]. The turnover frequency (TOF) of the NPs even increased after recycling of the catalyst, which the authors ascribed to surface restructuring and/or the formation of more active N-heterocyclic Rh species [36].

Working with multiple phases is not limited to the use of two phases. Canton and coworkers developed a triphasic system using trioctylmethylammonium chloride as IL to produce PdNPs for the hydrodehalogenation reaction of 4-chloropropiophenone. This IL has a density between that of water and isooctane, with neither of which it mixes, resulting in the formation of three phases. The formation of a third, aqueous phase is advantageous because it can capture side products which are formed during the reaction. The system could be recycled four times with only a small loss in activity [37].

There are other ways to recycle NP-containing ILs other than working with multiple phases. First, one can work with the fact that ILs possess a negligible vapor pressure, so that more volatile organic products can simply be removed *in vacuo* from the reaction mixture, though this is not applicable for every reaction or IL. An example is the production of RuNPs in 1-butyl-3-methylimidazolium [BMIM] [bistriflimide] for the hydrogenation of 1,3-cyclohexadiene, which showed less than 4% loss in activity after two runs [38]. Further, the IL can be immobilized on solid substrates, forming a supported ionic liquid phase (SILP, see also Chapter 6) [39].

PdNPs were synthesized in such a silica-supported imidazolium-based IL phase for the hydrogenation of nitrobenzene and successfully recycled 15 times without significant loss of activity [40]. Finally, membrane filtration techniques can be used to separate the IL with the NPs from the products. This will be discussed in Section 8.6.

8.4
Precipitation and Redispersion

Another strategy which is available to recover NPs from a reaction mixture is to precipitate them after reaction and redisperse them for a new cycle. In this section we will first discuss the systems that recycle NPs in this manner by means of centrifugation and look at the possible methods to modulate the NPs to improve their precipitation without decreasing the catalytically-active surface of the particles. Further, we will give an overview of the methods used for stabilizing/destabilizing the NPs so that precipitation and redispersion can be obtained without the need for centrifugation.

8.4.1
Centrifugation

There are several examples in literature in which NPs are separated after reaction through centrifugation. These examples can roughly be divided into two groups. A first group uses pure NPs that are usually somewhat 'larger' (between 20 and 100 nm) than most NPs applied in catalysis (usually below 20 nm) and do not use ligands or surfactants to stabilize them during reaction so that centrifugation is facilitated. When using purely MNPs like NiNPs or CuNPs, a significant part of the activity is usually lost during recycling due to aggregation and oxidation of the NPs [41, 42]. There are some examples though in which recycling is accomplished without considerable loss in activity or aggregation of the NPs, like copper oxide-NPs used for the synthesis of benzoxazoles, which can be used up to five runs [43] and AgNPs used for the synthesis of β-enaminones and β-enamino esters, which were reused up to three times [44]. The second group of examples uses NPs that have been modified to improve their ease of centrifugation, but without enlarging the catalytic NPs themselves. Matsumara and coworkers produced PtNPs of around 2 nm surrounded by a porous carbon shell of 30 to 40 nm for use in hydrogenation reactions (Figure 8.5). The carbon shell not only facilitates the precipitation of the NPs, but also inhibits them from coagulating, whereas the porous structure still allows easy access of the products and reactants to the surface of the Pt [45]. Similar systems exist, using silica instead of carbon [46], and all systems can be well recycled without major loss in activity and/or aggregation of the particles. Nevertheless, the use of centrifugation has its limitations, especially when going to larger scale applications. For this reason, other strategies are necessary to induce

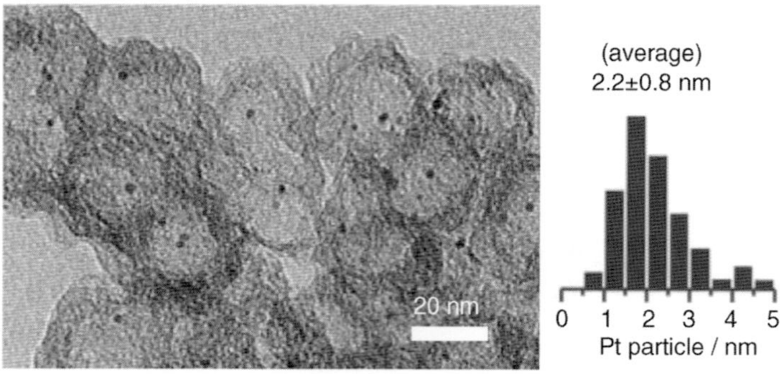

Figure 8.5 TEM picture and Pt particle size distribution of Pt NPs surrounded by a porous carbon shell [45].

precipitation of the NPs without the necessity of performing centrifugation. This will be discussed in the next section.

8.4.2 Adjustment of the Stabilization Conditions

Nanoparticles often need stabilizing agents to ensure that the particles stay in the dispersed state and do not aggregate during reaction. One strategy to induce precipitation of MNPs is therefore to adjust the solution conditions, so that the stabilization ability of the agents is reduced or even completely switched off, which allows the nanoparticles to aggregate and be separated from the reaction mixture. Ideally, this action should be reversible, so that the particles can be redispersed for a new reaction. Several strategies have been developed to obtain such a system.

A first option is to work with photo-destructible surfactants. Wyatt and coworkers [47] synthesized AuNPs in a water-in-oil emulsion using 4-hexylphenylazosulfonate as surfactant. Ultraviolet light breaks down this component into insoluble hydrophobic products, resulting in flocculation of the AuNPs and hence their recovery (Figure 8.6). A limitation of this technique is that the surfactants are irreversibly lost so that redispersion of the AuNPs is not possible.

A second option is to vary the temperature. Certain amphiphilic block copolymers can form reverse micelle systems of which the critical micelle concentration increases with increasing temperature. Using this type of system, nanoparticles can first be formed at a low temperature inside the reverse micelles and be recovered by increasing the temperature of the mixture, which causes the reverse micelles to decompose and the NPs, which were located inside the reverse micelle, to coagulate and precipitate. Zhang and coworkers [48] developed such a reverse micelle system and synthesized AgNPs in a mixture of water and *p*-xylene using poly(ethylene oxide)–poly(propylene oxide)–poly(ethylene oxide) (PEO–PPO–PEO) as surfactant at 288 K. By increasing the temperature to 318 K, precipitation was

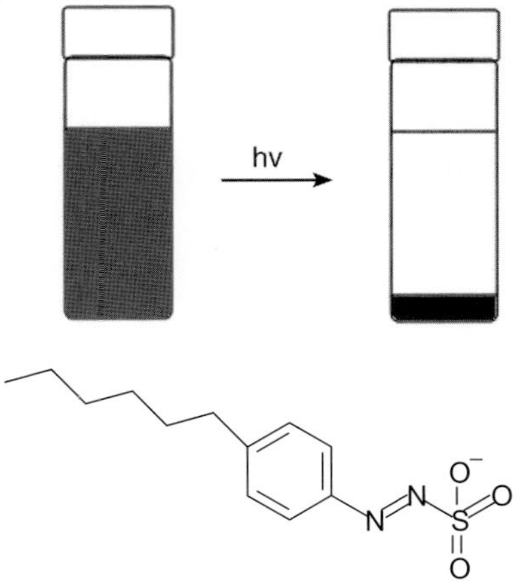

Figure 8.6 Precipitation of NPs using photodestructible surfactants. The surfactant, 4-hexylphenylazosulfonate (structure below), is broken down under influence of ultraviolet light, resulting in precipitation of the NPs [47].

accomplished, though no attempt was made to redisperse the particles after precipitation.

Thirdly, one can vary the solvent of the reaction mixture to increase or decrease the attraction between particles. The easiest way to do this is to add a solvent to the reaction mixture to which the NPs show no affinity, so that the particles tend to cluster together, which results in precipitation. Many examples exist where the NPs are precipitated by simply adding methanol or ethanol to the reaction mixture after reaction, such as PdNPs stabilized by phosphine dendrimer ligands for hydrogenation and cross-coupling reactions [49] or CoNPs used in the Pauson–Khand reaction [50] (Scheme 8.1), which were both recycled up to four times without significant loss in activity using this technique.

Other solvents can also be used to induce precipitation. Zhang *et al.* synthesized AgNPs using an isooctane/water system with sodium bis(2-ethylhexyl)sulfosuccinate (AOT) and tetraethylene glycol dodecyl ether ($C_{12}E_4$) as surfactants [51]. By introducing supercritical CO_2 in the system, the attractive forces between the

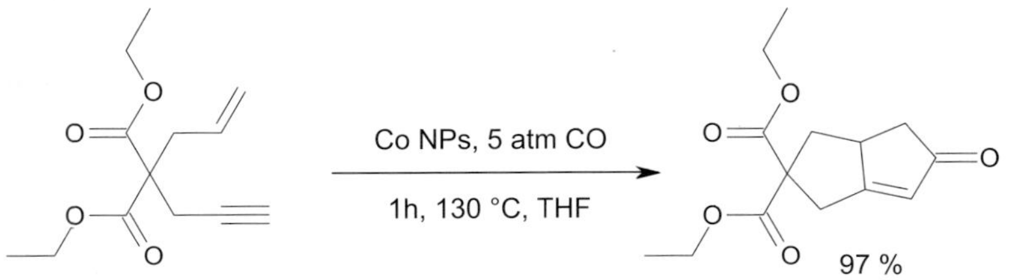

Scheme 8.1 Pauson–Khand reaction using Co NPs as catalyst which were precipitated by adding methanol to the reaction mixture after reaction [50].

Figure 8.7 Photographs of reaction mixtures containing PtNPs modified with pH-responsive stabilizers. (a) at pH 9.5; (b) 10 min after adjusting the pH to 2.5; (c) 60 min after adjusting the pH to 2.5; and (d) 120 min after adjusting the pH to 2.5 (Reprinted from [53], Copyright © (2010), with permission from Elsevier).

particles are increased, which results in coalescence and precipitation of the particles. A similar system was developed for the synthesis of AuNPs by Myakonkaya *et al.* by using mixed reverse micelles consisting of a water/*n*-octane system and using AOT and hexaethylene glycol monododecyl ether ($C_{12}E_6$) as surfactants. This time, squalane was the solvent used for the precipitation of the NPs. The particles were used for the oxidation reaction of alcohols, but some gold was lost during the recycling process, resulting in a lower TOF in the second run [52].

Finally, pH-responsive surfactants can be used to recover the NPs after reaction. Yang and coworkers introduced —COOH groups into octa(aminophenyl)silsesquioxane, forming a pH-sensitive octa(*N*,*N*-diacetic acid phenylamine)silsesquioxane. Modification of PtNPs with these molecules led to pH-responsive NPs. Above pH 4, the carboxylate groups are negatively charged, which results in repulsion between the different particles and leads to a stable NP solution, which can be used for the hydrogenation of aldehydes. Below pH 2.5, the —COOH groups are neutralized, stimulating the aggregation of the particles and allowing them to be easily separated from the reaction mixture (Figure 8.7). The nanoparticles were used five times without significant loss in activity [53]. A similar catalytic system was developed by Zhang and coworkers, which consisted of a polystyrene core surrounded by a porous pH-sensitive polymer shell, which can chelate the PdNPs. The particles were used for Suzuki coupling reactions and by varying the pH, they could be reused four times without loss in activity [54].

8.5
Magnetic Separation

A method that has gained a lot of attention over the last 10 years is the magnetic recovery of NPs. The basic principle consists of separating the NPs from the reaction mixture through use of an external magnet, which facilitates washing and reuse of the catalyst (Figure 8.8).

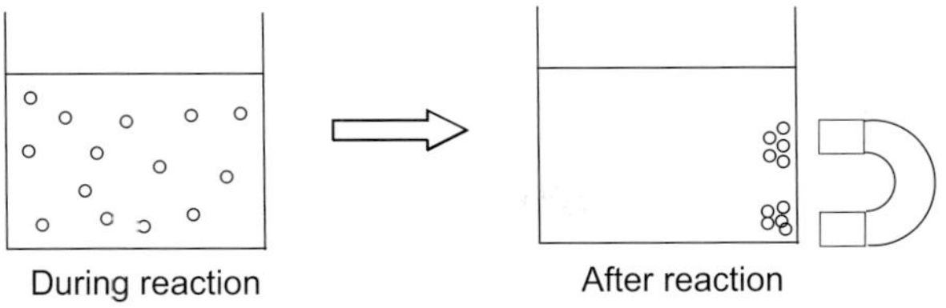

Figure 8.8 Separation of magnetic nanoparticles using an external magnet.

There are different strategies to make a catalyst suitable for magnetic separation. The easiest method is of course when the catalyst itself is magnetic. Examples can be found in applications that use FeNi [55], PdNi [56], CoAu [57] or Cu ferrite ($CuFe_2O_4$) [58] NPs as catalyst. In all the other cases, it is necessary to attach the catalyst to a magnetic nanoparticle. The latter can in principle be made out of any magnetic material, but the most applied are iron based oxides like magnetite (Fe_3O_4), maghemite (γ-Fe_2O_3), Co or Cu ferrite ($CoFe_2O_4$/$CuFe_2O_4$) and Ba ferrite ($BaFe_{12}O_{19}$). Purely metallic Fe, Ni and Co NPs also can be used. The attachment of the magnetic NPs to the catalyst can roughly be divided into three preparation types.

In the first type, organic linkers are connected to the magnetic NPs which in return are connected to the catalytic species (Figure 8.9). This catalyst can either be another nanoparticle [59–61], a homogeneous complex [62–68] or even an enzyme [69, 70] or an organocatalyst [71–74]. The linker can be connected directly to the magnetic particle, or a layer of silica [59, 60, 67, 71, 74] is deposited on the magnetic particle to facilitate the linking process. Sometimes the catalyst is directly attached to the magnetic NP [75, 76] or two linkers [66, 67, 69, 74] are used to hold one catalyst. The linkers are usually covalently bound to the surface, but examples of coordination through π–π interactions [63, 66] also exist.

The second type of synthesis method directly deposits a catalytic layer on the magnetic NP, without the use of any linker, so that a catalytic shell is formed around the magnetic NP (Figure 8.10). Examples of such catalysts are Pt on CoNPs for use in hydrogenation reactions of alkenes [77], TiO_x on barium ferrite-NPs as photocatalysts for the degradation of organic compounds [78], Ag on NiNPs for the production of hydrogen [79], and Au on magnetite-NPs for the reduction of 4-nitrophenol [80].

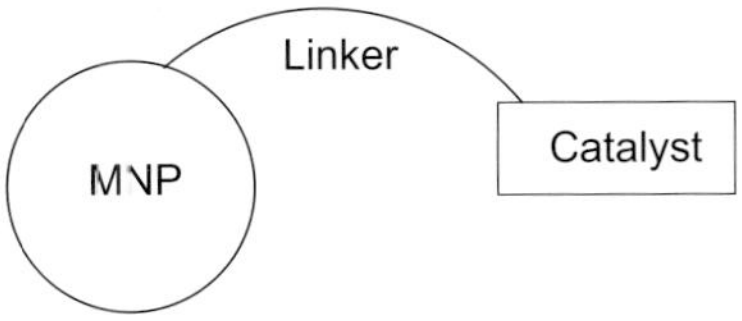

Figure 8.9 First preparation type for the attachment of a catalyst to a magnetic NP: the catalyst is linked to the magnetic nanoparticle using an organic linker.

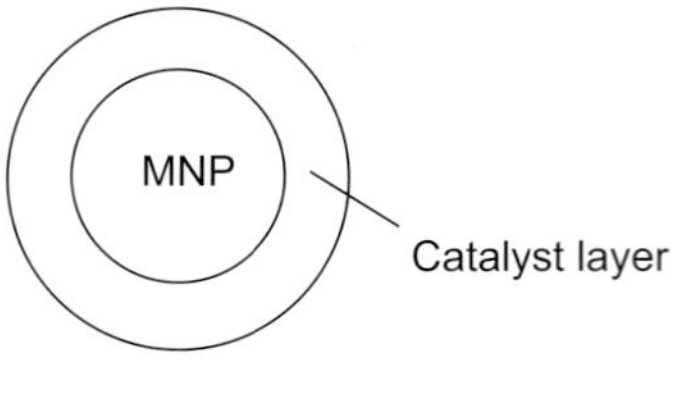

Figure 8.10 Second preparation type for the attachment of a catalyst to a magnetic NP: the catalyst is directly deposited as a shell around the magnetic nanoparticle.

In the third preparation type, a porous matrix layer which contains the catalytic species is formed around the magnetic NPs (Figure 8.11). The porous matrix can consist of polymers [81–84], silica [85], carbon [86, 87], and so on, and again a layer of silica can be deposited on the magnetic NP to facilitate the matrix attachment [83, 85].

Independently of what synthesis type is used for the magnetic nanocatalysts, recyclability is usually very good. Up to ten recycling runs with practically no loss in activity are no exceptions [71, 77, 87], and even examples with up to 20 runs are at hand [68, 73, 88]. Though excellent recyclabilities are attained, the application of strong magnets will probably limit their use in large scale industrial applications. For more extended information on the synthesis and catalytic applications of magnetic NPs, two recent reviews are available [89, 90].

8.6 Filtration

A final strategy for NP recuperation is filtration: the NPs are separated from the reaction mixture by a selective barrier which retains the NPs and allows free passage for the products and/or reactants (Figure 8.12). Depending on the selectivity of the passage, different filtration types can be identified, which can vary from microfiltration (0.05–10 μm) after precipitation of the NPs (see Section 8.4) to nanofiltration (1–100 nm) using very selective membranes [91]. Though this is an interesting technique to use, not many examples can be found in literature that use filtration for the separation of MNPs.

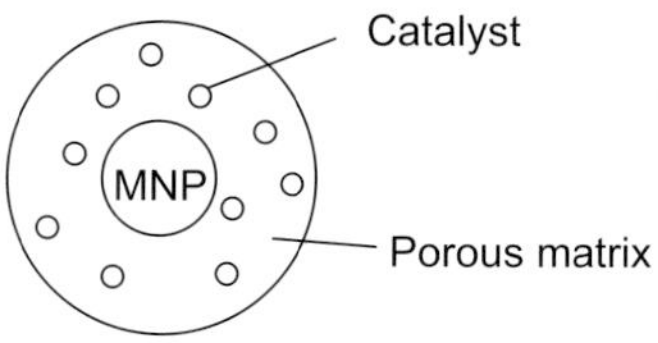

Figure 8.11 Third preparation type for the attachment of a catalyst to a magnetic NP: a porous matrix is formed around the magnetic nanoparticle in which the catalytic species are attached.

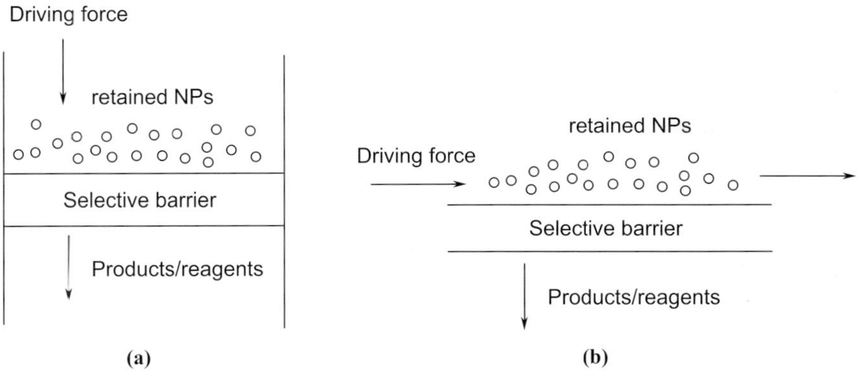

Figure 8.12 Schematic representation of separating nanoparticles via (a) dead-end filtration; (b) cross-flow filtration.

Within our group, solvent-resistant cross-linked polyimide membranes (Figure 8.13) were applied for the separation of Au^0, Ag^0 and Au^0–Ag^0 NPs. They were used in the hydrogenation of α,β-unsaturated aldehydes and ketones to allylic alcohols using *N,N*-dimethylacetamide or *N,N*-dimethylformamide as solvent. The MNPs themselves were only ~6 nm large, but because they were stabilized by PVP, the actual size of the particles was enlarged, making it possible for them to be separated using ultrafiltration. After three runs, >90% of the activity remained, without the selectivity being affected [92, 93]. A similar system was developed earlier for the separation of Au^0 NPs stabilized by PVA used in the oxidation of 1,2-diols to α-hydroxy-carboxylates. A poly(dimethyl)siloxane (PDMS) layer supported on a polyacrylonitrile/polyester carrier was used as membrane using *tert*-butanol as solvent and more than 85% of the activity was retained after three runs [94].

Besides polymers like PVP and PVA, dendrimers are also used to modulate the actual size of the NPs. Mecking and coworkers used hybrids of PdNPs with

Figure 8.13 SEM pictures of an asymmetric membrane consisting of a cross-linked polyimide top layer on a polyethylene/polypropylene carrier used for the separation of Ag^0, Au^0 and Au^0–Ag^0 NPs (Reprinted from [93], Copyright © (2007), with permission from Elsevier).

dendrimers in a continuous ultrafiltration membrane reactor for the hydrogenation of cyclohexene. A commercial solvent-resistant Koch MPF-50 membrane was used in dead-end mode applying a trans-membrane pressure of 1 MPa. The activity lowered gradually and after 30 runs the TOF had dropped from $1200\,h^{-1}$ to $870\,h^{-1}$. The authors ascribed this to fouling occurring during the dead-end operating mode and to the slight increase in size of the NPs [95].

When using ILs as medium, membranes can also be used to separate the products from the IL phase [96]. Though this has already been applied for separating ILs containing homogeneous catalysts [97], there are, to our knowledge, no examples of separating ILs containing pure metallic NPs using membranes.

Finally, the NPs can be modulated to facilitate filtration. Lee and coworkers anchored PdNPs (8–10 nm) on a larger SiO_2 NP (∼200 nm) through use of organic linkers which can coordinate the PdNPs and used them as catalysts for Suzuki–Miyaura coupling reactions. X-ray photoelectron spectroscopy (XPS) revealed that the Pd was present in a mixed oxidation state of both Pd^0 and Pd^{2+}. The particles were recycled after reaction by microfiltration (pore size 0.2 μm) and although 26% Pd was lost after the first recycling, the catalytic activity remained unchanged and the catalyst could be reused for four more times without significant loss in activity or further loss of Pd [98].

8.7 Conclusions

With the increasing attention attracted by the use of nanoparticles in catalysis, the demand for easy-to-apply methods to recycle these particles after reaction has grown equally. We have shown in this chapter that there are roughly five basic methods available to recycle MNPs and in most of the cases, these methods are inspired by similar techniques used for homogeneous catalysts. The introduction of new methods such as the magnetic separation of NPs provides interesting ways to recycle NPs, but the question remains if these methods are easily applicable on a larger scale and allow a larger number of NP types to be used on industrial scale.

References

1 Yan, J.-M., Zhang, X.-B., Shioyama, H., and Xu, Q. (2010) Room temperature hydrolytic dehydrogenation of ammonia borane catalyzed by Co nanoparticles. *J. Power Sources*, **195** (4), 1091–1094.

2 Hariprasad, E. and Radhakrishnan, T.P. (2010) A highly efficient and extensively reusable 'dip catalyst' based on a silver-nanoparticle-embedded polymer thin film. *Chem. Eur. J.*, **16** (48), 14378–14384.

3 Datta, K.K.R., Reddy, B.V.S., Ariga, K., and Vinu, A. (2010) Gold nanoparticles embedded in a mesoporous carbon nitride stabilizer for highly efficient three-component coupling reaction. *Angew. Chem. Int. Ed.*, **49** (34), 5961–5965.

4 Villa, A., Wang, D., Dimitratos, N., Su, D., Trevisan, V., and Prati, L. (2010) Pd on carbon nanotubes for liquid phase

alcohol oxidation. *Catal. Today*, **150** (1–2), 8–15.

5 Mertens, P.G.N., Vandezande, P., Ye, X., Poelman, H., De Vos, D.E., and Vankelecom, I.F.J. (2008) Membrane occluded gold palladium nanoclusters as heterogeneous catalysts for the selective oxidation of alcohols to carbonyl compounds. *Adv. Synth. Catal.*, **350** (9), 1241–1247.

6 Dominguez-Dominguez, S., Berenguer-Murcia, A., Linares-Solano, A., and Cazorla-Amoros, D. (2008) Inorganic materials as supports for palladium nanoparticles: Application in the semi-hydrogenation of phenylacetylene. *J. Catal.*, **257** (1), 87–95.

7 Chatterjee, M., Ikushima, Y., Hakuta, Y., and Kawanami, H. (2006) *In situ* synthesis of gold nanoparticles inside the pores of MCM-48 in supercritical carbon dioxide and its catalytic application. *Adv. Synth. Catal.*, **348** (12–13), 1580–1590.

8 Liu, H., Liu, Y., Li, Y., Tang, Z., and Jiang, H. (2010) Metal-organic framework supported gold nanoparticles as a highly active heterogeneous catalyst for aerobic oxidation of alcohols. *J. Phys. Chem. C*, **114** (31), 13362–13369.

9 Gao, S., Zhao, N., Shu, M., and Che, S. (2010) Palladium nanoparticles supported on MOF-5: A highly active catalyst for a ligand- and copper-free Sonogashira coupling reaction. *Appl. Catal., A*, **388** (1–2), 196–201.

10 Mennecke, K., Cecilia, R., Glasnov, T.N., Gruhl, S., Vogt, C., Feldhoff, A., Vargas, M.A.L., Kappe, C.O., Kunz, U., and Kirschning, A. (2008) Palladium(0) nanoparticles on glass-polymer composite materials as recyclable catalysts: A comparison study on their use in batch and continuous flow processes. *Adv. Synth. Catal.*, **350** (5), 717–730.

11 Tamami, B., Allahyari, H., Ghasemi, S., and Farjadian, F. (2011) Palladium nanoparticles supported on poly(*N*-vinylpyrrolidone)-grafted silica as new recyclable catalyst for Heck cross-coupling reactions. *J. Organomet. Chem.*, **696**, 594–599.

12 Hutchings, G.J., Lopez-Sanchez, J.A., and Dimitratos, N. (2011) A catalyst and a method of making the same. WO 2011/023953, filed on Aug. 27, 2010 and issued March, 3, 2011.

13 Domenech, B., Munoz, M., Muraviev, D., and Macanas, J. (2011) Polymer-stabilized palladium nanoparticles for catalytic membranes: ad hoc polymer fabrication. *Nanoscale Res. Lett.*, **6** (1), 406–411.

14 Makhubela, B.C.E., Jardine, A., and Smith, G.S. (2011) Pd nanosized particles supported on chitosan and 6-deoxy-6-amino chitosan as recyclable catalysts for Suzuki-Miyaura and Heck cross-coupling reactions. *Appl. Catal., A*, **393** (1–2), 231–241.

15 Yang, H., Han, X., Li, G., Ma, Z., and Hao, Y. (2010) Mesoporous ethane-silicas functionalized with a bulky *N*-heterocyclic carbene for Suzuki-Miyaura coupling of aryl chlorides and benzyl chlorides. *J. Phys. Chem. C*, **114** (50), 22221–22229.

16 Sun, L. and Crooks, R.M. (2002) Dendrimer-mediated immobilization of catalytic nanoparticles on flat, solid supports. *Langmuir*, **18** (21), 8231–8236.

17 Bernini, R., Cacchi, S., Fabrizi, G., Forte, G., Niembro, S., Petrucci, F., Pleixats, R., Prastaro, A., Sebastian, R.M., Soler, R. *et al.* (2008) Phosphine-free perfluoro-tagged palladium nanoparticles supported on fluorous silica gel: application to the Heck reaction. *Organic Lett.*, **10** (4), 561–564.

18 Callis, N.M., Thiery, E., Le Bras, J., and Muzart, J. (2007) Palladium nanoparticles-catalyzed chemoselective hydrogenations, a recyclable system in water. *Tetrahedron Lett.*, **48** (46), 8128–8131.

19 Zhang, Z., Zha, Z., Gan, C., Pan, C., Zhou, Y., Wang, Z., and Zhou, M.-M. (2006) Catalysis and regioselectivity of the aqueous heck reaction by Pd(0) nanoparticles under ultrasonic irradiation. *J. Org. Chem.*, **71** (11), 4339–4342.

20 Mejias, N., Pleixats, R., Shafir, A., Medio-Simon, M., and Asensio, G. (2010) Water-soluble palladium nanoparticles: click synthesis and applications as a recyclable catalyst in Suzuki cross-couplings in aqueous media. *Eur. J. Org. Chem.*, (26), 5090–5099.

21 Zhou, J., Li, X., and Sun, H. (2010) An efficient and recyclable water-soluble

cyclopalladated complex for aqueous Suzuki reactions under aerial conditions. *J. Organomet. Chem.*, **695** (2), 297–303.
22 Liu, R., Wu, C., Wang, Q., Ming, J., Hao, Y., Yu, Y., and Zhao, F. (2009) Selective hydrogenation of citral catalyzed with palladium nanoparticles in CO_2-in-water emulsion. *Green Chem.*, **11** (7), 979–985.
23 Ohde, M., Ohde, H., and Wai, C.M. (2005) Recycling nanoparticles stabilized in water-in-CO_2 microemulsions for catalytic hydrogenations. *Langmuir*, **21** (5), 1738–1744.
24 Ma, X., Jiang, T., Han, B., Zhang, J., Mao, S., Ding, K., An, G., Xie, Y., Zhou, Y., and Zhu, A. (2008) Palladium nanoparticles in polyethylene glycols: Efficient and recyclable catalyst system for hydrogenation of olefins. *Catal. Commun.*, **9** (1), 70–74.
25 Pillai, U.R. and Sahle-Demessie, E. (2004) Phenanthroline-stabilized palladium nanoparticles in polyethylene glycol - an active and recyclable catalyst system for the selective hydrogenation of olefins using molecular hydrogen. *J. Mol. Catal. A: Chemical*, **222** (1–2), 153–158.
26 Cheng, H., Xi, C., Meng, X., Hao, Y., Yu, Y., and Zhao, F. (2009) Polyethylene glycol-stabilized platinum nanoparticles: The efficient and recyclable catalysts for selective hydrogenation of o-chloronitrobenzene to o-chloroaniline. *J. Colloid Interface Sci.*, **336** (2), 675–678.
27 Harraz, F.A., El-Hout, S.E., Killa, H.M., and Ibrahim, I.A. (2012) Palladium nanoparticles stabilized by polyethylene glycol: Efficient, recyclable catalyst for hydrogenation of styrene and nitrobenzene. *J. Catal.*, **286**, 184–192.
28 Lantos, D., Contel, M., Larrea, A., Szabo, D., and Horvath, I.T. (2006) Fluorous phosphine-assisted recycling of gold catalysts for hydrosilylation of aldehydes. *QSAR Comb. Sci.*, **25** (8–9), 719–722.
29 Rocaboy, C. and Gladysz, J.A. (2003) Thermomorphic fluorous imine and thioether palladacycles as precursors for highly active Heck and Suzuki catalysts; evidence for palladium nanoparticle pathways. *New J. Chem.*, **27** (1), 39–49.
30 Chechik, V. and Crooks, R.M. (2000) Dendrimer-encapsulated Pd nanoparticles as fluorous phase-soluble catalysts. *J. Am. Chem. Soc.*, **122** (6), 1243–1244.
31 Davis, J.H. (2004) Task-specific ionic liquids. *Chem. Lett.*, **33** (9), 1072–1077.
32 Geukens, I., Fransaer, J., and De Vos, D.E. (2011) Aromatic amination of aryl chlorides catalyzed by Ni nanoparticles in aliphatic phosphonium based ionic liquids. *ChemCatChem*, **3** (9), 1431–1434.
33 Calò, V., Nacci, A., Monopoli, A., Ieva, E., and Cioffi, N. (2005) Copper bronze catalyzed Heck reaction in ionic liquids. *Org. Lett.*, **7** (4), 617–620.
34 Wang, L., Li, H., and Li, P. (2009) Task-specific ionic liquid as base, ligand and reaction medium for the palladium-catalyzed Heck reaction. *Tetrahedron*, **65** (1), 364–368.
35 Redel, E., Krämer, J., Thomann, R., and Janiak, C. (2009) Synthesis of Co, Rh and Ir nanoparticles from metal carbonyls in ionic liquids and their use as biphasic liquid-liquid hydrogenation nanocatalysts for cyclohexene. *J. Organomet. Chem.*, **694** (7–8), 1069–1075.
36 Vollmer, C., Redel, E., Abu Shandi, K., Thomann, R., Manyar, H., Hardacre, C., and Janiak, C. (2010) Microwave irradiation for the facile synthesis of transition metal nanoparticles (NPs) in ionic liquids (ILs) from metal–carbonyl precursors and Ru, Rh, and Ir NP/IL dispersions as biphasic liquid–liquid hydrogenation nanocatalysts for cyclohexene. *Chem. Eur. J.*, **16** (12), 3849–3858.
37 Perosa, A., Tundo, P., Selva, M., and Canton, P. (2006) Triphasic liquid systems: generation and segregation of catalytically active Pd nanoparticles in an ammonium-based catalyst-philic phase. *Chem. Commun.*, (43), 4480–4482.
38 Campbell, P.S., Santini, C.C., Bayard, F., Chauvin, Y., Collière, V., Podgorsek, A., Costa Gomes, M.F., and Sá, J. (2010) Olefin hydrogenation by ruthenium nanoparticles in ionic liquid media: Does size matter? *J. Catal.*, **275** (1), 99–107.
39 Van Doorslaer, C., Wahlen, J., Mertens, P.G.N., Binnemans, K., and De Vos, D.E. (2010) Immobilization of molecular catalysts in supported ionic liquid phases. *Dalton Trans.*, **39** (36), 8377–8390.

40 Li, J., Shi, X., Bi, Y.Y., Wei, J., and Chen, Z.G. (2011) Pd nanoparticles in ionic liquid brush: a highly active and reusable heterogeneous catalytic assembly for solvent-free or on-water hydrogenation of nitroarene under mild conditions. *ACS Catal.*, **1** (6), 657–664.

41 Kidwai, M., Mishra, N.K., Bansal, V., Kumar, A., and Mozumdar, S. (2007) Cu-nanoparticle catalyzed O-arylation of phenols with aryl halides via Ullmann coupling. *Tetrahedron Lett.*, **48** (50), 8883–8887.

42 Kumar, A., Dewan, M., Saxena, A., De, A., and Mozumdar, S. (2010) Knoevenagel condensation catalyzed by chemo-selective Ni-nanoparticles in neutral medium. *Catal. Commun.*, **11** (8), 679–683.

43 Saha, P., Ramana, T., Purkait, N., Ali, M.A., Paul, R., and Punniyamurthy, T. (2009) Ligand-free copper-catalyzed synthesis of substituted benzimidazoles, 2-aminobenzimidazoles, 2-aminobenzothiazoles, and benzoxazoles. *J. Org. Chem.*, **74** (22), 8719–8725.

44 Bhatte, K.D., Tambade, P.J., Dhake, K.P., and Bhanage, B.M. (2010) Silver nanoparticles as an efficient, heterogeneous and recyclable catalyst for synthesis of β-enaminones. *Catal. Commun.*, **11** (15), 1233–1237.

45 Ikeda, S., Ishino, S., Harada, T., Okamoto, N., Sakata, T., Mori, H., Kuwabata, S., Torimoto, T., and Matsumura, M. (2006) Ligand free platinum nanoparticles encapsulated in a hollow porous carbon shell as a highly active heterogeneous hydrogenation catalyst. *Angew. Chem.*, **118** (42), 7221–7224.

46 Wang, Y., Biradar, A.V., Duncan, C.T., and Asefa, T. (2010) Silica nanosphere-supported shaped Pd nanoparticles encapsulated with nanoporous silica shell: Efficient and recyclable nanocatalysts. *J. Mater. Chem.*, **20**, 7834–7841.

47 Vesperinas, A., Eastoe, J., Jackson, S., and Wyatt, P. (2007) Light-induced flocculation of gold nanoparticles. *Chem. Commun.*, (38), 3912–3914.

48 Zhang, R., Liu, J., Han, B.X., He, J., Liu, Z.M., and Zhang, J.L. (2003) Recovery of nanoparticles from $(EO)_8(PO)_{50}(EO)_8$/*p*-xylene/H_2O microemulsions by tuning the temperature. *Langmuir*, **19** (21), 8611–8614.

49 Wu, L., Li, Z.W., Zhang, F., He, Y.M., and Fan, Q.H. (2008) Air stable and highly active dendritic phosphine oxide stabilized palladium nanoparticles: preparation, characterization and applications in the carbon-carbon bond formation and hydrogenation reactions. *Adv. Synth. Catal.*, **350** (6), 846–862.

50 Kim, S.W., Son, S.U., Lee, S.S., Hyeon, T., and Chung, Y.K. (2001) Colloidal cobalt nanoparticles: a highly active and reusable Pauson–Khand catalyst. *Chem. Commun.*, (21), 2212–2213.

51 Zhang, J.L., Han, B.X., Liu, J.C., Zhang, X.G., He, J., Liu, Z.M., Jiang, T., and Yang, G.Y. (2002) Recovery of silver nanoparticles synthesized in AOT/$C_{12}E_4$ mixed reverse micelles by antisolvent CO_2. *Chem. –Eur. J.*, **8** (17), 3879–3883.

52 Myakonkaya, O., Deniau, B., Eastoe, J., Rogers, S.E., Ghigo, A., Hollamby, M., Vesperinas, A., Sankar, M., Taylor, S.H., Bartley, J.K. *et al.* (2010) Recycling nanocatalysts by tuning solvent quality. *J. Colloid Interface Sci.*, **350** (2), 443–446.

53 Zhou, W., Wang, J., Wang, C., Du, Y., Xu, J., and Yang, P. (2010) A novel reusable platinum nanocatalyst. *Mater. Chem. Phys.*, **122** (1), 10–14.

54 Zhang, M. and Zhang, W. (2008) Pd nanoparticles immobilized on pH-responsive and chelating nanospheres as an efficient and recyclable catalyst for Suzuki reaction in water. *J. Phys. Chem. C*, **112** (16), 6245–6252.

55 Yan, J.-M., Zhang, X.-B., Han, S., Shioyama, H., and Xu, Q. (2009) Magnetically recyclable Fe-Ni alloy catalyzed dehydrogenation of ammonia borane in aqueous solution under ambient atmosphere. *J. Power Sources*, **194** (1), 478–481.

56 Wu, Y., Wang, D., Zhao, P., Niu, Z., Peng, Q., and Li, Y. (2011) Monodispersed Pd-Ni nanoparticles: composition control synthesis and catalytic properties in the Miyaura-Suzuki reaction. *Inorg. Chem.*, **50** (6), 2046–2048.

57 Yan, J.-M., Zhang, X.-B., Akita, T., Haruta, M., and Xu, Q. (2010) One-step seeding growth of magnetically recyclable Au@Co

core-shell nanoparticles: highly efficient catalyst for hydrolytic dehydrogenation of ammonia borane. *J. Am. Chem. Soc.*, **132** (15), 5326–5327.

58 Swapna, K., Murthy, S.N., and Nageswar, Y.V.D. (2011) Magnetically separable and reusable copper ferrite nanoparticles for cross-coupling of aryl halides with diphenyl diselenide. *Eur. J. Org. Chem.*, (10), 1940–1946.

59 Jacinto, M.J., Kiyohara, P.K., Masunaga, S.H., Jardim, R.F., and Rossi, L.M. (2008) Recoverable rhodium nanoparticles: Synthesis, characterization and catalytic performance in hydrogenation reactions. *Appl. Catal., A*, **338** (1–2), 52–57.

60 Jacinto, M.J., Landers, R., and Rossi, L.M. (2009) Preparation of supported Pt(0) nanoparticles as efficient recyclable catalysts for hydrogenation of alkenes and ketones. *Catal. Commun.*, **10** (15), 1971–1974.

61 Zhang, F., Jin, J., Zhong, X., Li, S., Niu, J., Li, R., and Ma, J. (2011) Pd immobilized on amine-functionalized magnetite nanoparticles: a novel and highly active catalyst for hydrogenation and Heck reactions. *Green Chem.*, **13**, 1238–1243.

62 Stevens, P.D., Li, G.F., Fan, J.D., Yen, M., and Gao, Y. (2005) Recycling of homogeneous Pd catalysts using superparamagnetic nanoparticles as novel soluble supports for Suzuki, Heck, and Sonogashira cross-coupling reactions. *Chem. Commun.*, (35), 4435–4437.

63 Yoon, T.J., Lee, W., Oh, Y.S., and Lee, J.K. (2003) Magnetic nanoparticles as a catalyst vehicle for simple and easy recycling. *New J. Chem.*, **27** (2), 227–229.

64 Ding, S., Xing, Y., Radosz, M., and Shen, Y. (2006) Magnetic nanoparticle supported catalyst for atom transfer radical polymerization. *Macromolecules*, **39** (19), 6399–6405.

65 Che, C., Li, W., Lin, S., Chen, J., Zheng, J., Wu, J.-C., Zheng, Q., Zhang, G., Yang, Z., and Jiang, B. (2009) Magnetic nanoparticle-supported Hoveyda-Grubbs catalysts for ring-closing metathesis reactions. *Chem. Commun.*, (40), 5990–5992.

66 Wittmann, S., Schaetz, A., Grass, R.N., Stark, W.J., and Reiser, O. (2010) A recyclable nanoparticle-supported palladium catalyst for the hydroxycarbonylation of aryl halides in water. *Angew. Chem. Int. Ed.*, **49** (10), 1867–1870.

67 Shylesh, S., Wang, L., and Thiel, W.R. (2010) Palladium(II)-phosphine complexes supported on magnetic nanoparticles: filtration-free, recyclable catalysts for Suzuki-Miyaura cross-coupling reactions. *Adv. Synth. Catal.*, **352** (2–3), 425–432.

68 Bai, D., Wang, Q., Song, Y., Li, B., and Jing, H. (2011) Synthesis of cyclic carbonate from epoxide and CO_2 catalyzed by magnetic nanoparticle-supported porphyrin. *Catal. Commun.*, **12** (7), 684–688.

69 Wang, W., Xu, Y., Wang, D.I.C., and Li, Z. (2009) Recyclable nanobiocatalyst for enantioselective sulfoxidation: facile fabrication and high performance of chloroperoxidase-coated magnetic nanoparticles with iron oxide core and polymer shell. *J. Am. Chem. Soc.*, **131** (36), 12892–12893.

70 Xin, J., Ju-Fang, L., Ping-Ying, H., Xu-Yan, D., Lu-Lu, G., Liang, Y., Yuan-Cheng, C., Fang, W., Yuan-Di, Z., and Hong, C. (2010) Immobilized protease on the magnetic nanoparticles used for the hydrolysis of rapeseed meals. *J. Magn. Magn. Mater.*, **322** (14), 2031–2037.

71 O' Dalaigh, C., Corr, S.A., Gun'ko, Y., and Connon, S.J. (2007) A magnetic-nanoparticle-supported 4-*N*, N-dialkylaminopyridine catalyst: Excellent reactivity combined with facile catalyst recovery and recyclability. *Angew. Chem.*, **119** (23), 4407–4410.

72 Gleeson, O., Tekoriute, R., Gun'ko, Y.K., and Connon, S.J. (2009) The first magnetic nanoparticle-supported chiral DMAP Analogue: highly enantioselective acylation and excellent recyclability. *Chem. Eur. J.*, **15** (23), 5669–5673.

73 Tucker-Schwartz, A.K. and Garrell, R.L. (2010) Simple preparation and application of TEMPO-coated Fe_3O_4 superparamagnetic nanoparticles for selective oxidation of alcohols. *Chem. Eur. J.*, **16** (42), 12718–12726.

74 Wang, B.G., Ma, B.C., Wang, Q., and Wang, W. (2010) Superparamagnetic nanoparticle supported (S)-diphenylprolinol trimethylsilyl ether as a recyclable catalyst for asymmetric Michael addition in water. *Adv. Synth. Catal.*, **352** (17), 2923–2928.

75 Long, W., Gill, C.S., Choi, S., and Jones, C.W. (2010) Recoverable and recyclable magnetic nanoparticle supported aluminium isopropoxide for ring-opening polymerization of ε-caprolactone. *Dalton Trans.*, **39** (6), 1470–1472.

76 Ranganath, K.V.S., Kloesges, J., Schäfer, A.H., and Glorius, F. (2010) Asymmetric nanocatalysis: *N*-heterocyclic carbenes as chiral modifiers of Fe_3O_4/Pd nanoparticles. *Angew. Chem. Int. Ed.*, **49** (42), 7786–7789.

77 Jun, C.H., Park, Y.J., Yeon, Y.R., Choi, J.R., Lee, W.R., Ko, S.J., and Cheon, J. (2006) Demonstration of a magnetic and catalytic Co@Pt nanoparticle as a dual-function nanoplatform. *Chem. Commun.*, (15), 1619–1621.

78 Fu, W., Yang, H., Li, M., Chang, L., Yu, Q., Xu, J., and Zou, G. (2006) Preparation and photocatalytic characteristics of core-shell structure $TiO_2/BaFe_{12}O_{19}$ nanoparticles. *Mater. Lett.*, **60** (21–22), 2723–2727.

79 Guo, H., Chen, Y., Chen, X., Wen, R., Yue, G.H., and Peng, D.L. (2011) Facile synthesis of near-monodisperse Ag@Ni core–shell nanoparticles and their application for catalytic generation of hydrogen. *Nanotechnology*, **22**, 195604.

80 Wu, Y., Zhang, T., Zheng, Z., Ding, X., and Peng, Y. (2010) A facile approach to Fe_3O_4@Au nanoparticles with magnetic recyclable catalytic properties. *Mater. Res. Bull.*, **45** (4), 513–517.

81 Wang, Y. and Lee, J.-K. (2007) Recyclable nano-size Pd catalyst generated in the multilayer polyelectrolyte films on the magnetic nanoparticle core. *J. Mol. Catal. A: Chemical*, **263** (1–2), 163–168.

82 Amali, A.J. and Rana, R.K. (2009) Stabilisation of Pd(0) on surface functionalised Fe_3O_4 nanoparticles: magnetically recoverable and stable recyclable catalyst for hydrogenation and Suzuki-Miyaura reactions. *Green. Chem.*, **11**, 1781–1786.

83 Zhu, Y., Shen, J., Zhou, K., Chen, C., Yang, X., and Li, C. (2011) Multifunctional magnetic composite microspheres with in situ growth Au nanoparticles: a highly efficient catalyst system. *J. Phys. Chem. C*, **115** (5), 1614–1619.

84 Kong, A.G., Wang, P., Zhang, H.G., Yang, F., Huang, S.P., and Shan, Y.K. (2012) One-pot fabrication of magnetically recoverable acid nanocatalyst, heteropolyacids/chitosan/Fe_3O_4, and its catalytic performance. *Appl. Catal., A*, **417**, 183–189.

85 Ge, J., Zhang, Q., Zhang, T., and Yin, Y. (2008) Core-satellite nanocomposite catalysts protected by a porous silica shell: controllable reactivity, high stability, and magnetic recyclability. *Angew. Chem. Int. Ed.*, **120** (46), 9056–9060.

86 Yoon, H., Ko, S., and Jang, J. (2007) Nitrogen-doped magnetic carbon nanoparticles as catalyst supports for efficient recovery and recycling. *Chem. Commun.*, (14), 1468–1470.

87 Shen, B., Wang, Y., and Wang, Z. (2010) Magnetically recyclable C@(Au@Fe) supported Pd nanoparticles for the Heck cross-coupling reactions. *Reac. Kinet. Mech. Cat.*, **101** (2), 387–396.

88 Shin, K.S., Choi, J.-Y., Park, C.S., Jang, H.J., and Kim, K. (2009) Facile synthesis and catalytic application of silver-deposited magnetic nanoparticles. *Catal. Lett.*, **133** (1–2), 1–7.

89 Shylesh, S., Schünemann, V., and Thiel, W.R. (2010) Magnetically separable nanocatalysts: bridges between homogeneous and heterogeneous catalysis. *Angew. Chem. Int. Ed.*, **49** (20), 3428–3459.

90 Lim, C.W. and Lee, I.S. (2010) Magnetically recyclable nanocatalyst systems for the organic reactions. *Nano Today*, **5**, 412–434.

91 Mulder, M. (1996) *Basic Principles of Membrane Technology*, Kluwer Academic Publishers, Dordrecht.

92 Mertens, P.G.N., Vandezande, P., Ye, X., Poelman, H., Vankelecom, I.F.J., and De Vos, D.E. (2009) Recyclable Au^0, Ag^0 and Au^0-Ag^0 nanocolloids for the

chemoselective hydrogenation of α,β-unsaturated aldehydes and ketones to allylic alcohols. *Appl. Catal., A*, **355** (1–2), 176–183.

93 Mertens, P.G.N., Poelman, H., Ye, X., Vankelecom, I.F.J., Jacobs, P.A., and De Vos, D.E. (2007) Au^0 nanocolloids as recyclable quasihomogeneous metal catalysts in the chemoselective hydrogenation of alpha,beta-unsaturated aldehydes and ketones to allylic alcohols. *Catal. Today*, **122** (3–4), 352–360.

94 Mertens, P.G.N., Bulut, M., Gevers, L.E.M., Vankelecom, I.F.J., Jacobs, P.A., and De Vos, D.E. (2005) Catalytic oxidation of 1,2-diols to α-hydroxy-carboxylates with stabilized gold nanocolloids combined with a membrane-based catalyst separation. *Catal. Lett.*, **102** (1–2), 57–61.

95 Sablong, R., Schlotterbeck, U., Vogt, D., and Mecking, S. (2003) Catalysis with soluble hybrids of highly branched macromolecules with palladium nanoparticles in a continuously operated membrane reactor. *Adv. Synth. Catal.*, **345** (3), 333–336.

96 Van Doorslaer, C., Glas, D., Peeters, A., Odena, A.C., Vankelecom, I.F. J., Binnemans, K., Mertens, P.G.N., and De Vos, D.E. (2010) Product recovery from ionic liquids by solvent-resistant nanofiltration: application to ozonation of acetals and methyl oleate. *Green Chem.*, **12** (10), 1726–1733.

97 Wong, H., Pink, C.J., Ferreira, F.C., and Livingston, A.G. (2006) Recovery and reuse of ionic liquids and palladium catalyst for Suzuki reactions using organic solvent nanofiltration. *Green Chem.*, **8** (4), 373–379.

98 Shin, J.Y., Lee, B.S., Jung, Y., Kim, S.J., and Lee, S. (2007) Palladium nanoparticles captured onto spherical silica particles using a urea cross-linked imidazolium molecular band. *Chem. Commun.*, (48), 5238–5240.

9
Carbon Nanotubes and Related Carbonaceous Structures

Dang Sheng Su

9.1
Introduction

In catalysis, many catalysts are metals, metal oxides or other metallic compounds dispersed on supports. The role of the support is to maintain catalytically-active species in a highly dispersed phase. The support should have a specific surface structure to anchor the active phase, have a high surface area or porosity to support sufficient catalyst, and be inert enough under reaction conditions (high temperature, presence of basic or acidic media, oxidizing or reducing gases) [1, 2]. In addition, the support should present good mechanical stability, and electron conductivity in the case of electrocatalysis or photocatalysis. Among a wide range of possible supports, there are three types that combine these characteristics optimally, and they account for most industrial catalyst supports: alumina, silica, and carbon. In addition, titania plays a special role in photocatalysis, either as catalyst or support.

Various carbon materials such as graphite, carbon black, activated carbon (AC), activated carbon fibers, glassy carbon, pyrolytic carbon and polymer-derived carbon have been used to prepare carbon-supported catalysts. Carbon presents some unique characteristics that alumina or silica does not have. Carbon is inert in acidic or basic environment [2], and can be functionalized by various methods so as to control its surface chemistry. In addition, carbon is abundant in nature and environmentally friendly. However, which support should be used is solely determined by the chemical reactions with the aspects of high conversion rate, high selectivity, long-term stability and acceptable costs. High surface area activated carbon and carbon black are the carbon materials of choice for most carbon-supported catalysts [2]. Nanocarbons, especially carbon nanotubes (CNTs) possess some structural characteristics due to their 'nano' size and unique morphology (Figure 9.1). These characteristics give to CNTs some electronic, physical and chemical properties [3] that distinguish them from classical carbons. This has driven many research activities using CNTs as 'nanosupport' (Figure 9.2) for metal or oxide nanoparticles, a research initiative that dated back to 1991, soon after the report paper by Ijima [4]. Figure 9.2 shows an example of CNT-supported Pd_2Ga catalyst with a narrow

Nanomaterials in Catalysis, First Edition. Edited by Philippe Serp and Karine Philippot.

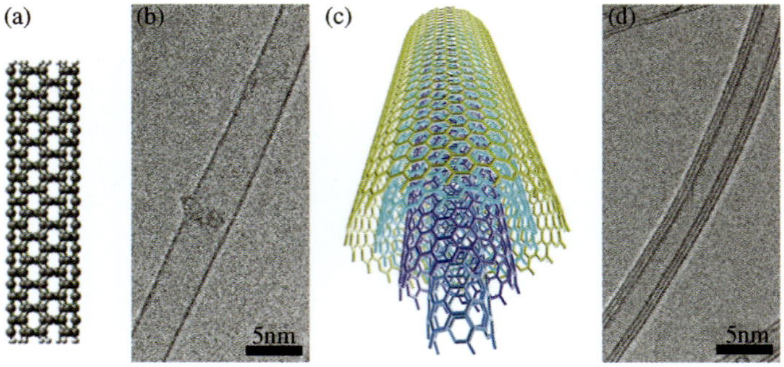

Figure 9.1 Schematic illustration of (a) single-walled and (c) multiwalled CNT. Transmission electron microscopy images of (b) a single-walled and (d) multiwalled CNT. (Courtesy of Dr. Qiang Zhang, Tsinghua University, China. (b) and (d) are reprinted with permission from [8]. Copyright © 2010 Wiley-VCH Verlag GmbH & Co.)

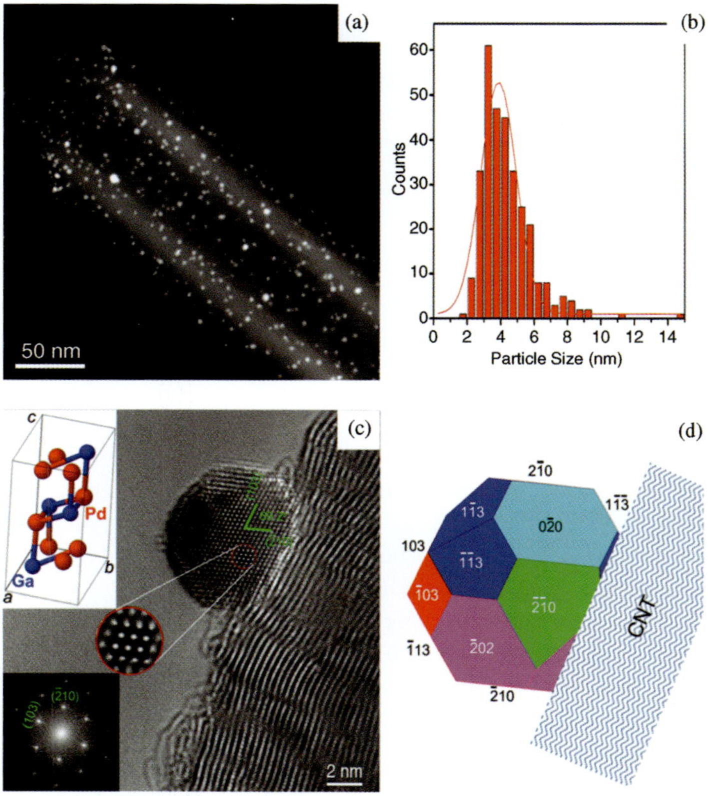

Figure 9.2 (a) Scanning transmission electron microscopy (STEM) image of Pd2Ga catalyst supported on a CNT; (b) size distribution of Pd2Ga nanoparticles calculated from STEM images; (c) high-resolution TEM image of a Pd2Ga nanoparticle on a CNT with insets of (top) crystallographic model of Pd2Ga and (bottom) the Fast Fourier Transform of the local HRTEM image. The circled area represents the image simulation; (d) Wulff construction of the corresponding Pd2Ga nanoparticle in (a). (Reproduced with permission from [9]. Copyright © 2011 Wiley-VCH Verlag GmbH & Co.)

size-distribution and structure of a well-faceted particle. However, the real driving forces for the rapid development of CNTs in catalysis are the increasing demand for new catalysts that should be more efficient, more energy-saving, more resource-saving and environmentally friendly, and for renewable and alternative energies needed for the sustainable development of our society [5–7].

9.2
Carbon Nanotubes as Nanosupport

Why could CNTs be used as nanosupport and what is the advantage of CNTs compared with classical carbon as catalyst support?

Firstly, CNTs differ from graphite in the bonding state of carbon atoms. We know that a CNT is a closed graphene cylinder (Figure 9.1). The rolling-up of the graphene sheet to form the tube causes a rehybridization of carbon bonding orbitals (nonplanar sp^2 configuration), thus leading to a splaying-out of the π density of the graphene sheet, whose magnitude depends on the diameter of the tube [10]. The curvature also affects the charge transfer between supported atoms and carbon, and the values of magnetic moments of the supported atoms on the nanotube wall. The most stable anchoring sites vary sensibly between graphite and CNTs due to the different curvature of the surfaces onto which the active species can be deposited.

Secondly, CNTs differ from activated carbon or carbon black in the atomic ordering degree. Activated carbons do not usually have well-defined long-range ordering, so that the oxidation resistance is reduced [11]. On the other hand, each CNT is a rolled-up graphene with well-defined short-range and long-range ordering, although they may contain some defects. Usually, CNTs are more resistant to oxidation than AC. This property is of special importance when CNTs are used as catalyst support or catalyst for high temperature reactions in the presence of oxygen [12].

Thirdly, CNTs have a different porous structure compared with AC. The specific surface area of CNTs can range from $50\,m^2\,g^{-1}$ [multiwalled carbon nanotubes (MWCNTs) with 50 graphene walls) to $1315\,m^2\,g^{-1}$ (single-walled CNTs) [13]. Theoretically, the porous structure of CNTs is directly linked to their tubular structure and the pore size corresponds to the inner diameters of opened CNTs with related narrow distribution. Activated carbons have micropores, mesopores and macropores. For catalytic applications, the micropores of the AC may hinder the transport of reactant/product molecules, and cause polymerization or coke formation leading to the deactivation of the catalyst [14].

Fourth, CNTs have higher thermo- and electro-conductivity than ACs or carbon black. The high electro-conductivity of CNTs is especially important when they are used to support electro-active or photo-active materials since they permit a facile transport or separation of electrons or electron and hole, respectively.

Finally, due to the tubular structure of CNTs, the cavities in the material can be used in catalysis, differently from other carbon materials or other metal oxide supports. The curvature and the cavity of a CNT may produce the following effects:

(i) to prevent the coalescence of metal nanoparticles spatially confined inside the tube; (ii) to modify the metal–support interface in a way that alters the surface properties of metal nanoparticles and suppresses their sintering; (iii) the confined space may increase the frequency of adsorption of reactant with the active sites and thus may increase turnover frequency (TOF) of particles inside CNTs with respect to those outside. However, since the wall will inevitably block a certain amount of active surface of the metal nanoparticles inside a CNT, and CNTs have a huge aspect ratio, the in-and-out transfer of reactant and product molecules can be significantly lowered by the low opening on the tip [15]. Ball milling is proposed as an efficient method to cut nanotubes [15–18]. Segmented CNTs present typical length-distributions, ranging between 200 and 800 nm. This is still too long compared with the typical diameters of CNTs, at which the confinement effect should be pronounced (less than 10 nm, or even smaller). It is worth noting that residual metallic particles coming from the production process can be found in the inner cavity of CNTs that makes the inner cavity of as-produced CNTs inaccessible [19].

The following achievements or progress have made the wide application of CNTs in catalysis possible: (i) CNTs can be produced on an industrial scale, becoming commercially available at relative low price [20, 21]; (ii) techniques for the functionalization of CNTs have been developed and are well-established [22], allowing the addition of certain groups with heteroatoms onto CNTs as active reaction centers, or as anchoring sites for metal complexes or particles [23, 24]; (iii) the synthetic methods for the preparation of metal nanoparticles on CNTs have become established, [25] and novel methods to selectively localize nanoparticles inside or outside CNTs have been developed [26–28].

9.3 Purification and Functionalization

9.3.1 CNT Purification

Metal (oxide) particles as catalyst residues and soot-like amorphous carbon as pyrolysis byproduct are the common impurities present in CNTs produced by catalytic chemical vapor deposition (CVD). The metal (oxide) residues can reduce the oxidation resistance of CNTs. Amorphous carbon impurity exhibits chemical and physical properties differing from those of the graphitic CNTs. Many methods have been developed to purify CNTs [29]. According to the approach used, these methods can be either chemical or physical. For application in catalysis, chemical methods are mostly used, including selective oxidation using a variety of oxidants such as CO_2 [30, 31], air [32–34], HNO_3 [35–37] and H_2O_2 [38, 39]. The superiority of air oxidation over liquid-phase oxidative treatment has been reported [40]. However, of all the purification methods, treatment with HNO_3 remains that most used. It simultaneously removes the catalyst particles and pyrolytic carbons, and functionalizes CNTs providing oxygen-containing group [41]. Unfortunately, purification of

CNTs may have side-effects, some of which are useful, some detrimental. For instance, oxidation of CNTs can open CNT caps, or make CNTs thinner [42]. Physical purification methods include filtration [43–45], centrifugation [46], and high temperature annealing [47–49]. The first two physical methods are mainly used to obtain high quality single-walled CNTs, while the last one may give rise to very pure CNTs with high graphitization degree but also make them chemically inert.

Recently, it was demonstrated that ultrasonic treatment is an effective and non-destructive method to remove amorphous carbon debris from as-received and HNO_3-NaOH treated CNTs [50]. Ultrasonic treatment causes acoustic streaming and jet pulses in CNT suspensions, allowing the loosely attached amorphous carbon to drop off, and further decompose into fragments via a pyrolytic process inside the microbubbles. Such a method is an easy-to-handle, noncorrosive, eco-friendly and fast process, and competitive with respect to the more frequently used HNO_3 treatment.

9.3.2
CNT Functionalization

The chemical functionalization of CNTs is an essential step in their multiple applications [22]. CNTs have strong graphitic character, and their surface is more inert than AC or carbon black. Introducing surface functionalities to CNTs is important for many applications. The functional groups and the delocalized electrons on the graphitic structure determine the acid/base and/or hydrophilic/hydrophobic character of the carbon surface, thus tailoring the chemical as well as the electronic properties of the surface [51–54]. Thus, such chemical functionalization does affect the physico-chemical properties of CNTs. Indeed, the functional groups on the carbon surface can be used as anchor sites in the synthesis of carbon-based composite materials [55]. Altering the chemical properties of CNTs has become an important topic for their application as catalysts [56] or support materials. For the application of CNTs in catalysis, the most important functional groups are those consisting in O- and N-containing moieties anchored on the edges or defects of graphene sheets, as schematically illustrated in Figure 9.3.

The methods employed to functionalize CNTs can be divided into electrochemical [57, 58], wet chemical, or gas-phase ones. Gas-phase functionalization can occur *in situ* during the CVD process, via a reaction where heteroatoms are added into or onto CNTs [59–63], or *ex situ*, for instance, through high-temperature treatment of CNTs in ammonia [64, 65]. According to the chemical nature of the functional groups, the functionalization can be defined as covalent or noncovalent. Covalent functionalization involves a permanent change of the material surface, giving reactive groups that can further form covalent bonds with other molecules [55, 66]. Noncovalent functionalization does not involve formal chemical bond formation between a molecule and the surface of CNTs. Examples of this type of interaction include van der Waals forces, electrostatic forces, hydrogen bonding, and other attractive forces [67].

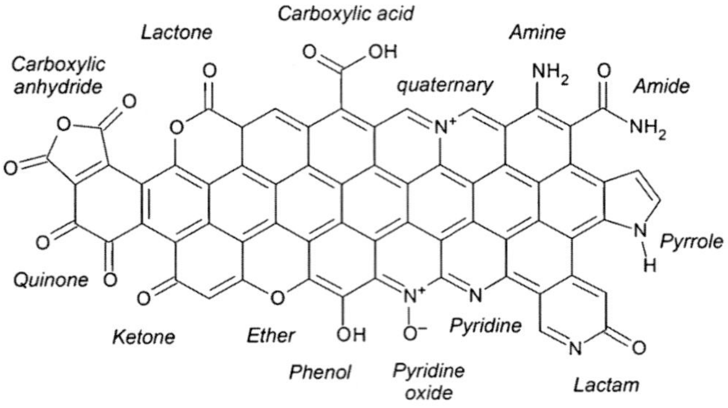

Figure 9.3 Various oxygen and nitrogen species and functional groups on the prismatic sites of sp^2 carbon basal plane.

9.3.2.1 **Functionalization of CNTs by Wet Chemical Methods**

Many efforts have been made during the last two decades to functionalize CNTs by wet chemical methods. The above-mentioned chemical purification process can functionalize CNTs to a certain extent. The most popular and easy covalent functionalization method is the oxidative treatment of CNTs with an acid [68]. Nitric acid is very often used for purification, and at the same time it functionalizes CNTs. Oxygen-containing functional groups are generated in the skeleton of CNTs, but a large number of defects are introduced. This may reduce the electrical conductivity and resistance to corrosion. Further functionalization, for instance, amination can follow to obtain the desired functional groups (Figure 9.4).

Other covalent functionalization of CNTs include sonochemical treatment [70, 71], silane-assisted treatment [72, 73], ionic liquid treatment [74], electrochemical modification [58, 75]. KOH is also used to functionalize MWCNTs as platinum supports for the oxygen reduction reaction [76]. Ultrasonic treatment has the advantage over

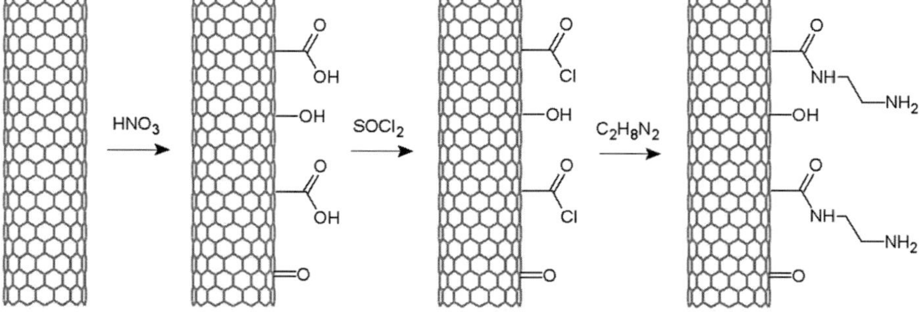

Figure 9.4 Schematic illustration of the functionalization of CNTs by oxidation/amination. (Reproduced with permission from [69]. Copyright © 2009 Wiley-VCH Verlag GmbH & Co.)

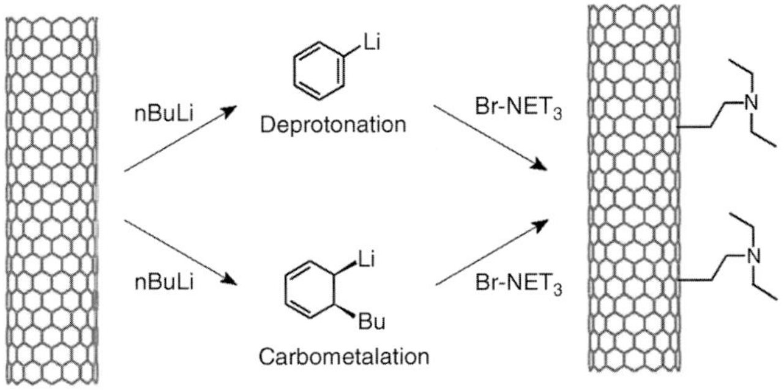

Figure 9.5 Functionalization of CNTs by deprotonation-carbometalation and subsequent electrophilic attack by the bromotriethylamine. (Reproduced with permission from [69]. Copyright © 2009 Wiley-VCH Verlag GmbH & Co.)

refluxing treating methods of CNTs, with regard to the resulting surface structure and electrochemical characteristics [73]. CNTs activated by ultrasonic treatment exhibited a more uniform surface, shorter length, higher separation, openings and modified surfaces with oxygen-bearing functional groups, which favors metal particle loading.

A more elegant and efficient synthetic route to functionalize CNTs with single-site basic groups has recently been reported [69]. This method is based on the direct covalent grafting of the desired basic functional groups onto existing structural defects of MWCNTs (Figure 9.5). CNTs are first reacted with an excess of nBuLi, which is a strong base as well as a good nucleophile. Thus, C—H bonds located near defects can be deprotonated and replaced by C—Li bonds. In parallel, nBuLi can also react with MWCNTs via a carbometalation reaction, which is a common reaction known to proceed through a local breaking of the aromatic system. In the second step, 2-diethylaminoethylbromide is added to the activated MWCNTs in order to perform an electrophilic attack on the C—Li bond (Figure 9.5). A new C—C bond is formed between the ethyl group of the amine and the CNT, thus leading to a covalent functionalization of the nanotube with triethylamine and formation of LiBr as a by-product. This new route can give a very homogeneous functionalization with a high number of easily accessible basic groups. The concentration of such functional groups approaches that of Brùnsted acid sites in zeolites [69].

Noncovalent functionalization of CNTs is particularly attractive since it offers the possibility of grafting the functional groups without affecting the electronic network of the tubes [76, 77]. The noncovalent interaction based on van der Waals forces or π–π-stacking interactions [55, 78, 79] preserves the integrity of CNTs but the addition of organic molecules or polymers may weaken the electron transport and affect the binding between catalytic active species and CNTs.

Liquid-phase functionalization must be completed by separation of the treated CNTs by filtration. This process might be difficult especially for small diameter CNTs. Filtration is always accompanied by a loss of considerable amount of CNTs. The required subsequent drying process frequently leads to the agglomeration of

CNTs, thus impairing the applicability. Gas-phase treatments thus seem to be an attractive alternative to avoid all these problems.

9.3.2.2 **Functionalization of CNTs by Gas-Phase Reactions**

The advantage of gas-phase oxidation or functionalization of CNTs is the simplicity of the process. This includes the exposure of CNTs to ozone [80], carbon dioxide [42], water vapor [81], H_2O_2 [82], or oxygen plasma [83]. Normally, more carbonyls rather than carboxyls are generated by these methods due to the intrinsic lack of hydrogen atoms. Recently, a patented method [84] has been developed for the oxygen functionalization of CNTs in the gas phase using nitric acid vapor [85]. The relevant set-up is shown in Figure 9.6a. CNTs were loaded into the reactor and heated to temperatures between 398 K and 523 K. Concentrated HNO_3 was filled into the round-bottomed flask and heated under magnetic stirring to 398 K. The condenser was connected with an open-end exhaust line to air. The design of the set-up effectively prevents the reflux of liquid HNO_3 collected by the condenser onto the CNT sample. The treatment was fully conducted under gas-phase conditions, and the wetting of CNTs by liquid HNO_3 could be completely avoided. Differing from other gas-phase oxidation processes, this method may lead to selective functionalization by just changing the treatment temperature; and a higher concentration of carboxyl groups was obtained (Figure 9.6b). This method is very efficient: higher O/C ratios than those obtained by treatment with liquid nitric acid were obtained (Figure 9.6c). If compared with the conventional treatment with liquid HNO_3, 5 h HNO_3 vapor treatment at 398 K seems to be sufficient for the usual applications of CNTs, for example, as support for catalysts via impregnation. After the HNO_3 vapor treatment, the CNTs can be directly collected without any additional process such as filtration, washing, or drying. There is no change of the morphology, bulk volume or apparent density of the CNTs due to the gas-phase treatment. It should be pointed out here that, due to the azeotropic boiling point limitation of concentrated HNO_3 (395 K), it is not possible to carry out the conventional HNO_3 treatment at temperatures as high as 423 K or 473 K at atmospheric pressure, thus limiting the creation of certain species within a certain period of treatment.

Amination of CNTs with ammonia at high temperature is an efficient way to functionalize CNTs with N-containing groups after CVD production [64, 65]. Nitrogen can be incorporated into the graphene sheets in form of pyridinic, quaternary, pyrrolic and N-oxides (Figure 9.3). A surface concentration of nitrogen as high as 18% can be reached [65]. Amination with NH_3 and HCN was widely used for the functionalization of AC [86, 87]. One advantage of amination of CNTs is the possibility of tailoring the surface basic/acidic properties by just changing the amination temperature [65] and the homogenous distribution of nitrogen and nitrogen containing functional groups over CNTs.

In situ methods have been developed to incorporate nitrogen into CNTs during the production by a CVD process employing nitrogen-containing precursors [53, 71, 88], such as pyridine (C_6H_5N) [56, 89], triazine ($C_3N_3Cl_2NH_2$) [90],

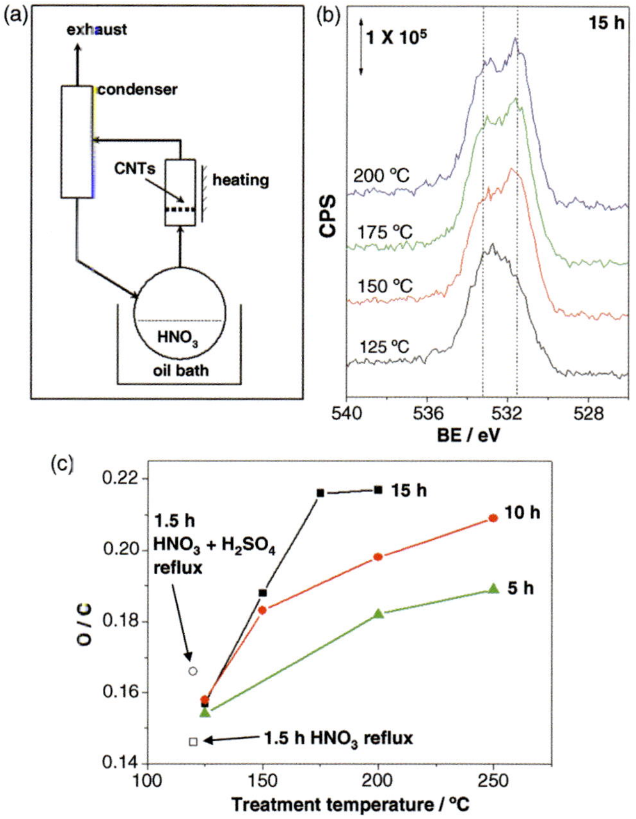

Figure 9.5 (a) Diagram of the set-up used for gas-phase functionalization of CNTs with HNO_3 vapor; (b) XPS spectra of CNTs treated by HNO_3 vapor at different temperatures for 15 h; (c) Oxygen to carbon ratios obtained from XPS surface atomic concentration of CNTs treated by HNO_3 vapor for different periods of time at different temperatures. The oxygen to carbon ratios obtained by conventional acid treatment are included for comparison. (Reproduced with permission from [85]. Copyright © 2009 Elsevier.)

dimethylformamide ($HOCN(CH_3)_2$) [91], Ni phthalocyanine ($NiC_{32}N_8H_{16}$) [92], benzylamine/ferrocene ($PhCH_2NH_2/Fe(C_5H_5)_2$) [93], acetonitrile/tetrahydrofuran (CH_3CN/C_4H_8O) [94]. ferrocene/melamine ($[Fe(C_5H_5)_2]/C_3N_6H_9$) [95], cresol/ferrocene ($C_7H_8O/[Fe(C_5H_5)_2]$) [96] and xylene/ferrocene ($C_8H_{10}/[Fe(C_5H_5)_2]$) [97]. Ammonia added to the hydrocarbon feedstock has also been used as source of nitrogen in a CVD process [98–100]. The amount of N incorporated in CNTs depends on synthetic parameters such as temperature, C/N source and the catalyst used [56, 101]. However, studies reported in the literature have highlighted difficulties in controlling the growth process, which very often lead to different structures of carbon nano materials such as fishbone and bamboo-like structures, or to a mixture of them [102, 103].

9.4 Preparation of CNT-Supported Catalysts

CNT-supported catalysts can be prepared by two main reaction pathways [104]: the direct growth or deposition of catalyst nanoparticles on the CNT surface, or in an alternative approach, the anchoring of preformed nanoparticles onto CNTs using covalent linking through organic fragments or other metal–support interactions.

9.4.1 Growing Metal Nanoparticles Directly on the CNT Surface

The method most often used to prepare CNT-supported catalyst consists in growing the particles directly on CNTs. Ion exchange [105, 106], organometallic grafting [107], incipient wetness impregnation [108] and some other methods are available [109]. Preparing CNT-supported catalyst by reducing metallic salts in the presence of CNTs remains one of the most used methods (Figure 9.7). Noble metals (Pt, Au, Pd, Ag, Rh, Ru) and transition metal catalysts supported on CNTs have been prepared by this method. Various reducing agents such as H_2, $NaBH_4$, and N_2H_4, have been used to chemically reduce the metal salts [110]. The reduced metal atoms anchored to CNTs can develop to particles by controlling the reaction time, salt concentration, and temperature. The resulting metal particles are then anchored on the CNT walls mostly through van der Waals interactions, which in many cases are strong enough to guarantee good adhesion. Heat and light have been used to accelerate the reduction of the metal cation.

The metal-containing species and their catalytic activity depend on the electronic structure and the functional groups on the surface. When PtRu nanoparticles were prepared on nitrogen-functionalized CNTs, more dispersed and uniform metal nanoparticles were obtained compared with the PtRu nanoagglomerates on unfunctionalized CNTs. The anchoring of noble metal particles was enhanced through the activation of nitrogen-neighboring carbon atoms because of the large

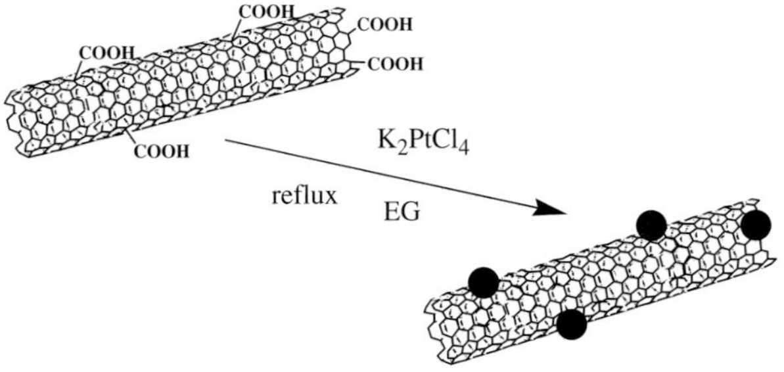

Figure 9.7 Illustration of the preparation of metal (Pt) particles on CNTs via the reduction of the metal salt (K_2PtCl_2) in the presence of CNTs in a reducing agent (ethylene glycol = EG). (Reproduced with permission from [105]. Copyright © 2001 American Chemical Society.)

electron affinity of nitrogen in the N-functionalized CNTs. Treatment in nitric acid can create oxygenated functions on the nanotube walls and thus increase the loading of metal particles [111]. Carboxylic groups [112] or phenol groups [113] on the CNT surface can act as anchors in the formation and deposition of metal particles. The synthesis of metal-containing species on CNTs has been well documented in the literature [104].

9.4.2
Anchoring Pre-Formed Nanoparticles on CNTs

If metallic nanoparticles are pre-formed via, for instance, colloidal chemistry, they can be linked to the CNT surface via covalent or weaker bonds. The linker between nanoparticles and CNTs can be of two types: (i) the functional groups present on nanoparticles form covalent bonds with the functional groups on the CNT surface or (ii) the linker simply sticks to the CNT surface through weak intermolecular interactions such as π–π stacking, or hydrophobic/electrostatic attractions [55].

Oxidized CNTs have plenty of —COOH functionalities on the surface. When they react with aliphatic amines, amides are formed thus leading to soluble CNTs. As shown in Figure 9.8 [114], Au nanoparticles were easily grafted on CNTs via the

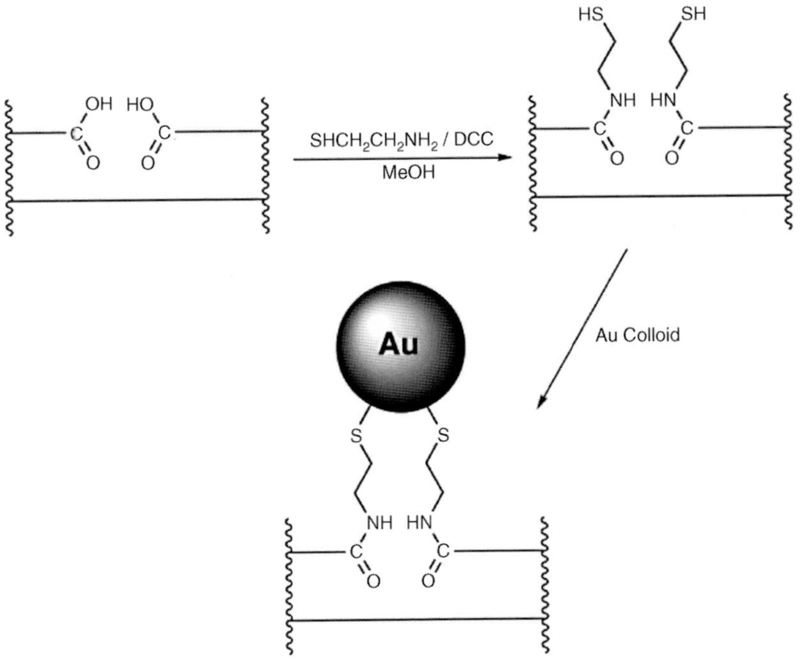

Figure 9.8 Schematic representation of the reaction scheme used to tether gold colloidal nanoparticles to oxidized SWCNT defect sites. (Reproduced with permission from [114]. Copyright © 2002 Royal Society of Chemistry.)

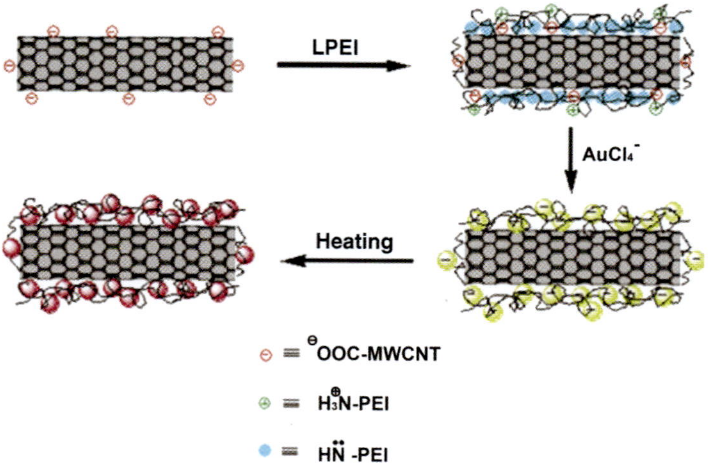

Figure 9.9 Schematic illustration of the experimental procedure by which PEI possibly interacts with acid-functionalized CNTs through electrostatic interaction and physisorption. (Reproduced with permission from [117]. Copyright © 2006 American Chemical Society.)

formation of amiothiols or bifunctional thiols. The number of nanoparticles attached to CNT surface thus depends on the amount of carboxylates, or other functional groups.

The surface of CNTs can include aromatic rings, providing the formation sites for π–π stacking. For instance, the pyrene molecule which bears a long aliphatic chain terminated by a thiol group can act as a linker between Au and CNTs by π–π stacking of the pyrene moiety to the CNT surface [78, 115]. If oxidized CNTs are treated with an ionic polyelectrode, positively or negatively charged CNTs are available to anchor charged nanoparticles [116]. As shown in Figure 9.9, polyethyleneimine (PEI), an amino-rich cationic polyelectrolyte that effectively interacts with CNTs via physisorption on the CNT sidewalls, has been used both as reducing and protecting agent for the preparation of Au nanoparticles [117]. This method needs neither additional steps to oxidize CNTs nor the use of further reducing agents. When electrostatic interactions are responsible for the grafting of metal nanoparticle, the coverage and loading of nanoparticles can be tuned by linkers and appropriate synthesis parameters.

9.4.3
Selective Preparation of Catalysts on CNTs

Preparing nanoparticles selectively localized inside or outside CNTs is the prerequisite to study the confinement effect of CNTs. Although some other techniques have been developed for the selective preparation of metal catalysts on CNTs, wet chemistry appears to be the most simple, versatile and up-scalable method. However, the capillary effect of CNTs seems to be strongly dependent on surface functionalization and their diameters. CNTs with internal diameters close to and less than 10 nm are filled to a lesser extent, or even remain empty by wet impregnation [118]. Ma *et al.* used this capillary effect to place Pt particles on CNTs of different diameters: Pt particles are introduced inside CNTs of large diameter and outside CNTs of smaller diameter [119]. Unfortunately, by varying the tube diameter, both the curvature and the size of the inner cavity are changed and no selective placing of nanoparticles could be achieved. Appropriate methods should be further developed for the selective placing of nanoparticles inside or outside CNTs.

9.4.3.1 Selective Placing of Metal Catalysts Inside CNTs

Two step biphasic impregnation [27] This synthetic procedure relies on the fact that (i) CNTs have a very good affinity for organic solvents; (ii) a solvent with a low surface tension (for instance an ethanolic solution containing nickel nitrate) will easily wet and penetrate inside the CNT; and (iii) a second, metal-free solvent (for instance distilled water) will remain outside the nanotube and protect its outer surface from metal deposition if the liquid/solid interface energy is higher than that of the first solvent (Figure 9.10 left). The precise

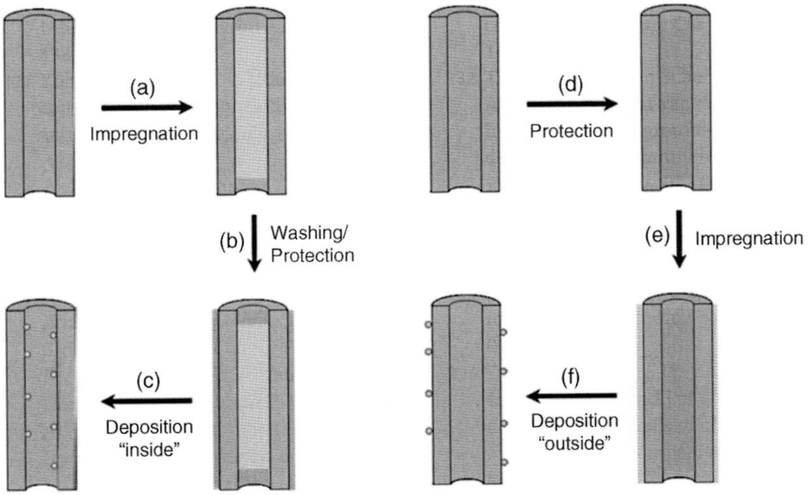

Figure 9.10 Schematic view of a longitudinal cross section of a CNT during the different steps for the selective deposition of nanoparticles *inside* CNTs (left) and *outside* CNTs (right). (a) Impregnation with the ethanolic solution containing the metal precursor. (b) Impregnation with distilled water to wash and protect the outer surface from metal deposition. (c) After drying, calcination and reduction, metal particles decorate the inner surface of the CNT with high selectivity. Right, Scheme for the selective deposition of nanoparticles *outside* CNTs. (d) Impregnation with an organic solvent to protect the inner tubule from metal deposition. (e) Impregnation with an aqueous solution containing the metal precursor. (f) After drying, calcination and reduction, metal particles decorate the outer surface of the CNT with high selectivity. (Reproduced with permission from [27]. Copyright © 2009 American Chemical Society.)

control of the volume of each phase allows the filling of the CNT with the organic phase and the covering of its outer surface with the aqueous one. After drying, calcination and reduction, metal particles decorate the inner surface of the CNT with a high selectivity. In this way, one can decorate the inner wall of CNTs, without the need of discriminating the diameter of the opening CNTs and without any other step of functionalization than a treatment with nitric acid.

Impregnation and selective washing [26] This method is similar to that above, and consists in impregnating CNTs with an *excess* of the precursor solution under ultrasonic conditions to ensure that most of the tubes are filled. A suitable solvent is then used to wash off the solution sticking onto the external walls of CNTs, yet keeping the solution in the internal cavity unchanged. The following calcination leaves the metal particles inside the CNTs.

Molecular recognition [28] Eva Castillejos *et al.* suggest another efficient way to drive nanoparticles into CNTs. They stabilize nanoparticles with a ligand exhibiting an affinity for the CNT graphene layers (interaction) and introduce on the outer surface of CNTs different surface species that induce weak or repulsive interaction between the nanoparticles and CNT external surface. Due to the affinity/anti-affinity of the ligand and surface species, the nanoparticles are driven into the cavity of CNTs.

9.4.3.2 Selective Placing of Metal Catalyst Outside CNTs

Placing or growing metal particles on the outer surface of closed CNTs is an easy task. However, selectively decorating only the outer surface of opened CNTs is far more challenging. It is a troublesome task to prepare metal/oxide particles only on the outer surface of CNTs of large diameters that can easily suck in any solution. One strategy is to block or protect the inner channels from the solution containing the metal precursor. A two-step impregnation is suggested (Figure 9.10, right): an incipient wetness impregnation with an organic solvent having a low surface tension, which consequently easily wets and fills the tubes; then an aqueous solution containing the metal precursor is added [27]. Because of the higher liquid/solid interface energy, this solution cannot penetrate into the CNTs. Thus, the channels remain protected, and decoration only occurs on the outer surface. Several organic solvents have been tested to obtain metal nanoparticles anchored outside CNTs [27].

The procedure illustrated might however not be obvious as its efficiency depends on the balance between all the interface energies, that is, liquid/solid for each solvent employed but also liquid/liquid. Thus, several parameters could influence the choice of the solvent. In particular, its miscibility with water and its boiling point should be considered. Firstly, the aqueous solution ought to penetrate into the tubes by diffusing into the organic solvent. Secondly, if the organic solvent evaporates faster than the aqueous solution, the latter might be able to enter into the tubes. In both cases an unselective decoration would take place.

9.4.4
Localizing the Catalyst Particles Supported on CNTs

The determination of the location of supported nanoparticles is as difficult as the selective preparation of catalyst nanoparticles inside or outside a CNT. Indeed X-ray diffraction (XRD) or other spectroscopic methods could not provide the exact information on the spatial distribution of supported nanoparticles. Nevertheless, information about the exact location of catalyst particles, inside or outside CNTs, is of fundamental importance to evaluate catalytic performances and in particular the confinement effect of CNTs.

Transmission electron microscopy (TEM) remains the only common tool to visualize the nanoparticle location [120]. However, TEM images provide only project 2D information, and multiple images from different directions are required to determine the particle location with certainty [27, 121]. Electron tomography [122–126], a technique that combines a series of TEM images acquired at different tilt angles to give a 3D representation of the object, can provide full 3D morphological information and thus the exact location of the particles (inside or outside CNTs).

The TEM image of a Ru/CNT catalyst [121] in Figure 9.11a is ambiguous with respect to the location of Ru particle: the particles imaged on the channel of the CNT cannot be discriminated as inside or outside CNTs. The electron tomography analysis by numerical cross-sections all along the tomogram clearly reveals the position of the particles: outside the tube (Figure 9.11b), inside the tube (Figure 9.11c and e), and inside the tube wall (Figure 9.11d), respectively. Note that the particle in Figure 9.11d is almost invisible in the TEM image. Figure 9.11b-e

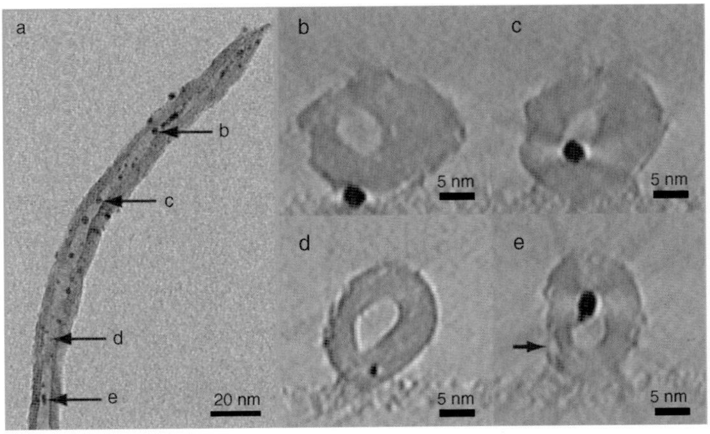

Figure 9.11 Electron tomography analysis of a Ru/CNT catalyst, showing (a) conventional TEM image and (b–e) numerical cross-sections all along the tomogram, perpendicular to the TEM image plane in the direction of the arrows. Only the electron tomogram cross-sections provide certainty about the actual location of the ruthenium nanoparticles. (Reproduced with permission from [121]. Copyright © 2011 Wiley-VCH Verlag GmbH & Co.)

also illustrate that the structure of the nanotube is more complex than the commonly assumed cylinder model.

The tomograms can not only afford a qualitative description of the 3D catalyst morphology, but also give a detailed quantification on the nanometer scale [124, 127]. With the two-step biphasic impregnation method to place Ni particles inside CNTs, J.P. Tessonnier *et al.* found, by means of quantitative electron tomography analysis, that about 75% of the Ni particles are truly located inside the tube [27].

9.5
Applications of CNT-Supported Catalysts

9.5.1
Liquid-Phase Reactions

Liquid-phase reactions are mainly used for the production of fine chemicals. Precious metals supported on carbon are the mostly used catalysts. The processes are conducted in stirred tank batch reactors. About 30% of the catalysts consist in supported palladium, most of them being used for hydrogenation reactions, for example, the saturation of carbon–carbon and carbon-heteroatom multiple bonds [128–132], the synthesis of amines from nitro compounds [133, 134], or the carbon–carbon coupling reactions such as Suzuki [135, 136] or Heck reactions [137]. Activated carbon and carbon black are common supports for precious metal catalysts [2]. One of the shortcomings of the AC or carbon black supported catalysts is the leaching of the active phase during the reaction. CNTs, and in particular the functionalized ones, provide a strong metal–support interaction and therefore strong anchoring of metal nanoparticles. This strong metal–support interaction is also a reason why catalysts supported on CNTs exhibit higher selectivity in some liquid-phase reactions, due to the facile electron transfer from the support to catalyst [138]. Another important reason to use CNTs instead of AC as support is that the unique mesoporous structure of CNTs without micropores permits a facile mass transport and prevents the coking by blocking of micropores by reactants.

9.5.1.1 Hydrogenation

One of the first reported applications of CNTs in heterogeneous catalysis was their use as supports for 3–7 nm Ru nanoparticles in the hydrogenation of cinnamaldehyde (CALD) [139]. CALD contains both a C=C and a C=O bond in an α,β-unsaturated arrangement (Figure 9.12). Depending on which bonds is activated, hydrocinnamaldehyde (HCALD), cinnamyl alcohol (CA) or phenyl propanol (PP) can be obtained via the hydrogenation reaction. The possible reaction pathways are shown in Figure 9.12. In fact, the selective hydrogenation of the carbonyl group of the α,β-unsaturated aldehyde yielding the unsaturated alcohol remains a challenging task. This is of particular interest because of the importance of such alcohols in the fine chemicals industry [140].

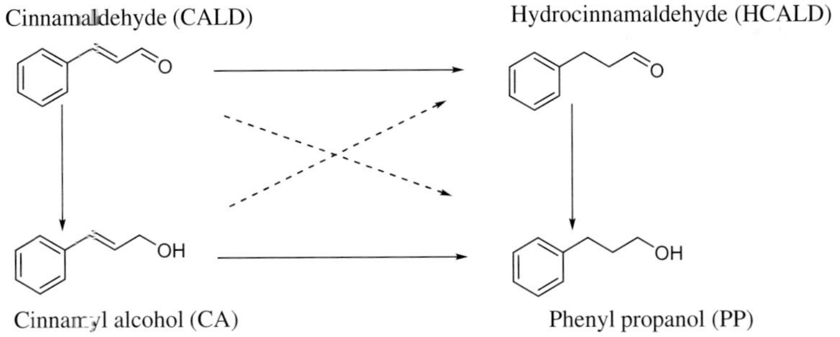

Figure 9.12 Reaction pathways for the hydrogenation of cinnamaldehyde.

Pt, Pd, Rh, Ru metallic nanoparticles and Pt-Co, Pt-Ru bimetallic nanoparticles supported on CNTs show a high catalytic activity in this reaction. Pd and Rh are highly selective for HCALD, but not active for CA formation. For example, when using a 1% (w/w) Rh/CNT to produce HCALD, a selectivity of 100% was obtained. The catalytic activity of Rh/CNT was three times higher than that of a 1% (w/w) Rh/C, despite the fact that the latter has a surface area of $700\,m^2\,g^{-1}$, three times higher than that observed with CNTs as support ($180\,m^2\,g^{-1}$) [107]. Tessonnier *et al.* reported the interesting confinement effect of Pd nanoparticle catalysts deposited on the inner walls of CNTs for the selective hydrogenation of CALD to HCALD [141], explained by the unusual interaction between Pd nanoparticles and the inner walls of the CNTs coupled with the relative lack of oxygenated surface groups on them.

In a comprehensive study on the selective hydrogenation of CALD to CA using monometallic (Pt, Ru) and bimetallic (Pt-Ru) catalysts supported on various carbon supports (MWCNT, SWCNT, CNF, AC), Serp *et al.* found that nanocarbon-supported catalysts present higher CALD conversions than those obtained with AC as support, but poor selectivity toward CA (<32%) [138]. After a heat treatment under nitrogen at 973 K, Pt/CNT and especially PtRu/CNT catalysts exhibit an increase in the conversion rate and the selectivity toward CA (>66%). Heat treatment did not influence the conversion and led to a decrease in selectivity towards CA on AC-supported catalysts. Such activity change is related to the influence of carbon surface chemistry on the adsorption of different aromatic compounds [142, 143]. On heat-treated CNT samples, a π–π interaction between the CA π electron ring and the basic π sites of the CNT surface may increase the CA adsorption capacity. By removal of the electronegative oxygen atoms from the carbon surface, an electron transfer from the CNT to the metal is enhanced leading the increase in CA selectivity. The electron transfer from the nanotube support to metal particles has already been reported [110].

CNT-supported catalysts have also been tested for other hydrogenation reactions. For the hydrogenation of methyl-9-octadecenoate and 2-methyl-2-pentenal, Karaousis *et al.* [144] reported higher activity of CNT supported Pd with respect to a AC supported catalyst. The TOF of hydrogenation of *trans*-stilbene to

1,2-diphenylethane over Pt/CNT catalysts was reported to be six times higher than that over Pt/AC catalysts. The performances of metal catalysts supported on CNTs can be fully or partly interpreted by the structural and chemical characteristics of CNTs compared with other carbons and by the resulting metal–support interactions. It was concluded that for liquid-phase hydrogenation reactions, the use of mesoporous nanostructured carbon supports prevents significant mass transfer limitations thus improving the catalyst activity.

9.5.1.2 **Oxidation**

Metal-catalyzed alcohol oxidation in the liquid phase using oxygen as oxidant represents a well known interesting catalytic process, but its industrial exploitation is limited by the strong deactivation of the catalysts [145]. Ru, Pt, Pd and Au have been used as monometallic catalysts. The main requirement for a high selectivity and long durability has not been achieved yet.

Liquid-phase oxidation of benzyl alcohol to benzaldehyde has been investigated to reveal the differences between CNTs and AC as support for metal particles [146]. Figure 9.13 shows the possible reaction pathways for benzyl alcohol oxidation. Catalytic tests have shown that Pd/CNT behaves differently from Pd/AC. Indeed, Pd/CNT exhibits a lower activity, but a higher selectivity toward benzaldehyde than Pd/AC. The Pd/AC catalyst provides a slightly higher selectivity toward toluene. The catalytic performance is also strongly dependent on the solvent used. A maximum of selectivity (92%) was obtained on Pd/CNT for a 50% volume of cyclohexane; the corresponding value on Pd/AC is 74%. Modification of CNTs with ammonia before catalyst preparation can enhance the activity and increase the TOF of the same reaction significantly [147].

In liquid-phase reactions, the leaching of the active catalytic species remains a big problem for their application. In the above benzyl alcohol oxidation reaction,

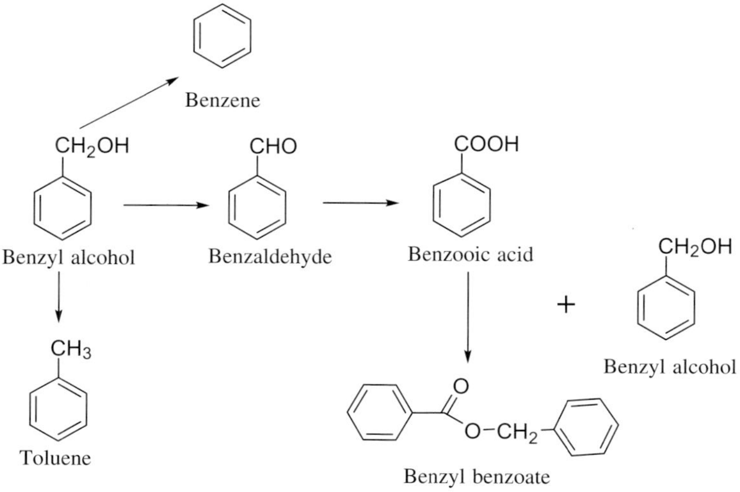

Figure 9.13 Reaction scheme for benzyl alcohol oxidation.

Pd on both AC and CNTs dissolves into the solution (from this the question arises whether the reaction is homogeneously or heterogeneously catalyzed). However, Pd leaching can be limited when CNTs are used as support: Pd/CNT exhibits a slightly more stable cycling (50% activity loss after seven runs) than Pd/AC (70% activity loss after seven runs), even if a structural change in the catalysts is observed at the end of the reaction. A consistent improvement in the long-term use of such catalysts has been observed by modifying the monometallic Pd with gold. On recycling, Pd–Au bimetallic catalyst shows a reduced Pd leaching and exhibits stable catalytic performances over eight runs. Under similar conditions Au–Pd on AC is less stable [146].

9.5.2 Gas-Phase Reactions

Catalytic gas-phase reactions occur mostly at high temperature, and in many cases also under high pressure (for instance, for ammonia synthesis or hydrogenation reactions). Mechanical and thermal stability of the support must be considered when choosing which one ought to be used. In addition, the support should present a high external surface area to facilitate mass transport under robust reaction conditions. A suitable metal–support interaction is needed that hinders the sintering of metal particles at high temperature under oxidative or reductive condition. CNTs have been tested as support for many gas-phase reactions. Two typical examples are discussed below.

9.5.2.1 Fischer–Tropsch Synthesis

Syngas conversion (Fischer–Tropsch (FT) synthesis) is one of the most important reactions to convert coal [148] or biomass to fuel [149]. The reaction scheme is as follows:

$$nCO + (2n + 1)H_2 \rightleftharpoons C_nH_{2n+2} + nH_2O$$

$$nCO + (2n)H_2 \rightleftharpoons C_nH_{2n} + nH_2O$$

$$nCO + (2n)H_2 \rightleftharpoons C_nH_{2n+1}OH + (n - 1)H_2O$$

FT synthesis was already industrialized [150] in the 1930s. The product distribution (alkane, alkene, alcohol) are usually too broad and unselective, extended from C1 to C80 chains! Various supported catalysts such as Fe, Co, Ru on different supports have been tested for this reaction [151, 152]. Developing FT catalyst with a controllable selectivity toward the hydrocarbons produced remains a challenge.

Wang *et al.* have systematically investigated the role of supports in this reaction using Ru nanoparticles of the same size [153]. As it is shown in Table 9.1, the Ru/CNT catalyst provides a lower conversion rate than that of Ru/Al_2O_3, but higher than all the catalysts on other supports, also higher than that Ru/AC and Ru/graphite. Moreover, Ru/CNT exhibits the highest selectivity to diesel fuel (C_{10}-C_{20}). Their study revealed that the surface chemistry of CNTs determines the product selectivity [153]. Indeed, the pretreatment of CNTs with concentrated HNO_3 is necessary to obtain high C_{10}-C_{20} selectivity. The acid groups (—COOH) on the CNT generated

Table 9.1 Effect of the supports used for Ru catalysts on FT synthesis [153].

Catalyst	CO conv./%	Selectivity/C-mol %				
		CH_4	C_2-C_4	C_5-C_9	C_{10}-C_{20}	C_{21}^+
Ru/SiO_2	**32**	6.8	**10**	**20**	**39**	**24**
Ru/Al_2O_3	40	3.1	6.8	17	36	37
Ru/TiO_2	20	14	36	22	22	5.7
Ru/ZrO_2	22	8.4	37	27	15	13
Ru/NaY	23	11	14	37	37	1.9
Ru/HY	25	12	13	41	31	3.0
Ru/H-beta	24	13	21	52	14	0
Ru/AC	11	20	20	22	26	12
Ru/graphite	20	2.9	12	25	55	5.8
Ru/CNT	34	8.3	9.1	21	60	1.9

Reaction conditions: Ru, 3 wt%, W = 0.5 g; $H_2/CO = 1$; P = 2.0 MPa; F = 20 mL min^{-1}; T = 533 K; t = 10 h.
Ru/CNT exhibits a high selectivity to C_{10}-C_{20} (diesel fuel, 60%).

by liquid-phase HNO_3 treatment do enhance the hydrocracking of heavier hydrocarbons, thus affording higher diesel fuel selectivity. This is an important advantage when using CNTs as support since the surface of CNTs is tunable by pretreatment.

Tavasoli *et al.* [154, 155] compared the performance of Co/CNTs with that of Co/γ-Al_2O_3 and found that CNT as support could reduce the metal-support interactions and enhance the Co reducibility. High Co dispersion or small Co particle size is achieved by using CNTs. It seems that the use of CNTs can facilitate both the reducibility and the dispersion of Co species. The small size of Co particles over the CNTs leads to a shift of product to lighter hydrocarbons. An increase in Co loading on CNTs could increase the size of Co particles and thus enhance the C_5+ selectivity. Other aspects such as confinement effect [156–158] were accounted for the different performance of CNTs as support in this reaction.

9.5.2.2 **Ammonia Decomposition**

Ammonia decomposition, the reverse process of ammonia synthesis, is nowadays intensively investigated as a possible way to provide CO_x-free hydrogen for fuel, and to remove ammonia from the reformate of internal gasification combined cycle power plants, which might be widely employed in the future as a CO_2-removal technology cells [159]. One ammonia molecule contains three hydrogen atoms, which makes its hydrogen storage capacity equal to 17.7%, that is, an energy density of 3000 Wh kg^{-1}. This is much higher than that of methanol and other fuels. However, compared with the well-established and industrialized Haber–Bosch process for ammonia synthesis [160], the ammonia decomposition process is underdeveloped and requires substantial improvements before it can be considered as a practical contribution to the energy supply toolbox.

Ru is the most active catalyst for NH_3 decomposition. Xu and Au *et al.* demonstrated that the catalytic performance of Ru catalyst is support-dependent [161–164]. Under similar reaction conditions, the order of NH_3 conversion can be ranked as Ru/CNTs > Ru/MgO > Ru/TiO_2 > Ru/Al_2O_3 > Ru/ZrO_2 > Ru/AC. The excellent catalytic performance of Ru/CNTs is due to the high dispersion of Ru and to the high purity of CNTs. Moreover, the conductivity of the support is considered as an important reason for the high catalytic efficiency. A conductive support is beneficial for the transfer of electrons from promoter and/or support to Ru, a process that would facilitate the recombinative desorption of surface N atoms [161].

The structure–function correlations for Ru/CNT in the catalytic decomposition of ammonia have recently been investigated [165]. Ru particles of various sizes are prepared and located inside or outside CNTs with different structure order (measured by I_D/I_G bands ratio of Raman spectra). Figure 9.14 shows the correlation of the catalytic performance with the structural parameters of the support and of the active mass. The catalytic activity of Ru nanoparticles is inversely related to their size; larger particles exhibit more active sites per particle than small particles suggesting that no relation between exposed surface and active site density exists for this reaction. The size of Ru particles can be

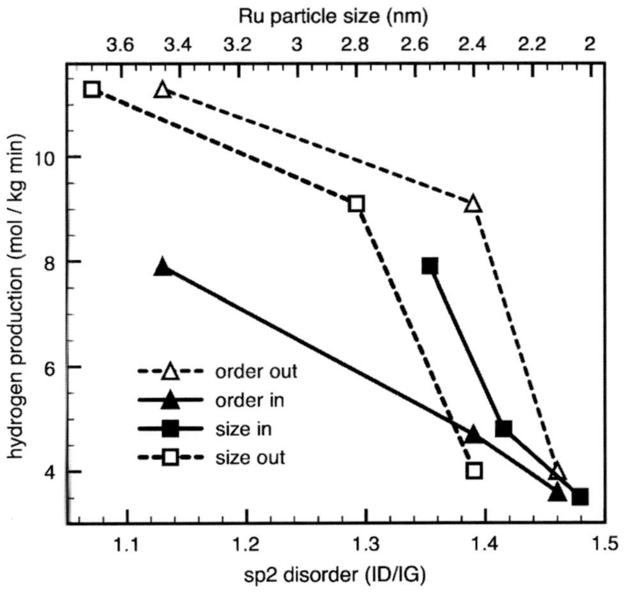

Figure 9.14 Correlations of the catalytic performance in ammonia decomposition with the structural parameters of the support (CNTs) and of the active mass (Ru particles). 'in' and 'out' indicate the Ru particles inside or outside CNTs. (Reproduced with permission from [165]. Copyright © 2010 Wiley-VCH Verlag GmbH & Co.)

effectively controlled by a combination of modifying the site density of defects on the support and by localization of the active metal inside or outside CNTs. A higher degree of graphitization leads to weaker metal–carbon interaction and hence to larger particles. In addition, no positive confinement effect is observed for this reaction [165].

Fresh commercial CNTs containing residual Co or Fe nanoparticles are highly active in NH_3 decomposition [166] Adding cobalt to iron can significantly improve the stability of the catalyst in ammonia decomposition [167]. In addition, a higher reaction rate (6.8 kg NH_3 kg_{cat}^{-1} h^{-1} at 823 K) is obtained on alloyed Fe–Co nanoparticles. The tunable surface properties of CNTs and the bending of their terminal graphene sheets lead to localization of electron density and support the formation of direct metal-carbon interactions, giving stability to nanometer-sized metal clusters.

9.5.3 Electrocatalysis

There are some essential differences between electrocatalysis and conventional catalysis. Electrocatalysis involves proton, oxygen and electron transfer processes at gas/solid, liquid/solid and gas/liquid/solid interfaces. The additional requirements for a good support are optimized transport properties and maximized extension of the various interfaces. Due to this complexity, it is reasonable to check whether CNT-supports could perform better than the standard supports in electrocatalysis.

Up to now the most studied application of CNTs in electrocatalysis is their use in fuel cells as support of the active phase [168–170]. Fuel cells, based on the fuel oxidation reaction (FOR) and oxygen reduction reaction (ORR), are electrochemical energy devices that convert the chemical energy stored in the fuel (typically hydrogen or methanol) and in the oxidizer directly into electricity. The most-used electrocatalyst for fuel cells, both for anode and cathode, is Pt (or Pt alloy) nanoparticles supported on high surface area supports. Currently, carbon black (for instance Vulcan XC-72) is the most-used support. It is commonly available and cheap. However, despite its high surface area, the carbon black-based electrocatalyst support presents two main problems: (i) it induces significant mass transfer limitations due to its dense structure, leading to a very low Pt utilization [171], and (ii) carbon black is known to undergo electrochemical oxidation, forming surface oxides such as —OH, —COOH and —C=O, which further react to give CO_2 at the cathode of the fuel cell [172]. As carbon corrodes, Pt nanoparticles will detach from the carbon black and aggregate into larger particles resulting in the loss of Pt surface area and in the subsequent lowering of the performance of fuel cells [173, 174].

Wang *et al.* investigated the durability of CNTs and carbon black, Vulcan XC-72, as Pt supports for proton exchange membrane fuel cells [175]. They found that CNTs are electrochemically more stable than Vulcan XC-72 with less surface oxide formation and 30% lower corrosion current under the

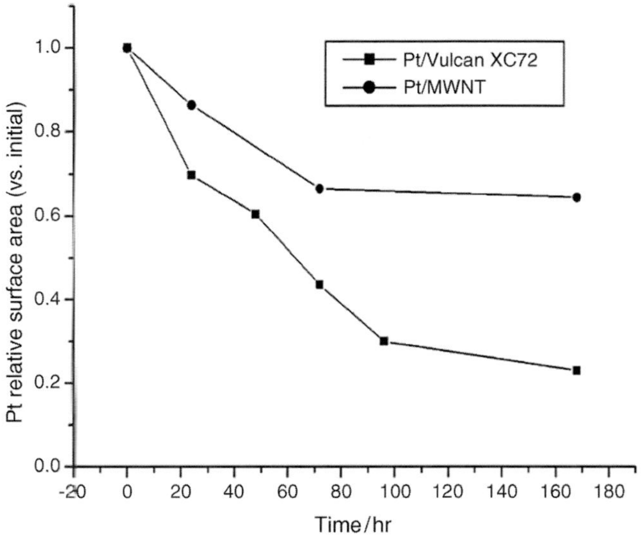

Figure 9.15 Comparison of Pt active surface area loss for Vulcan XC-72 and CNTs at different time intervals during oxidation treatment [175].

investigated condition. As a result of high corrosion resistance, CNTs show lower loss of Pt surface area (Figure 9.15) and higher oxygen reduction reaction activity when used as fuel cell catalyst supports.

Compared with commercial Pt/carbon black catalysts, Pt/SWCNT-based hydrogen fuel cells show improved performance at higher anodic pressures [176]. P. Lamat found that the activation enthalpies for O_2 reduction on Pt/SWCNT and Pt/C exhibits similar values, thus confirming that the rate-determining step involved in the oxygen reduction is the same. Pt/SWCNT film on a rotating disk electrode, compared with a commercial Pt/carbon black catalyst, shows a lower onset potential and a higher electron transfer rate constant for oxygen reduction. Such a good performance is attributed to the high porosity of SWCNT, which facilitates the diffusion of the reactant and thus the interaction with the Pt surface. Improved stability of the SWCNT support is also confirmed by the minimal change in the oxygen reduction current during repeated cycling over a period of 36 h [177].

The nature of the support contributes to the changes in the performance of electrocatalysts. B.Q. Xu *et al.* made a comprehensive study on the question as to whether single- or multiwalled CNTs should be used as catalyst support for methanol oxidation [178]. They used MWCNTs (CVD production with inner diameter 3–10 nm, outer diameter. 6–20 nm, ratio of length to diameter 100–1000, 175 $m^2 g^{-1}$) and single-walled CNTs (arc-discharge production with diameters of 1.25 ± 0.2 nm and lengths between 1 and 10 μm, 129 $m^2 g^{-1}$) to prepare SWCNT/Nafion and MWCNT/Nafion composite film electrodes. Lower onset and peak potentials in CO stripping voltammograms on Pt-SWCNT/Nafion catalyst reveal a higher tolerance to CO poisoning on Pt. In methanol electrooxidation reaction, Pt-SWCNT/Nafion catalyst exhibits a significantly higher current density, lower onset potentials and lower charge transfer resistances [178]. Further investigations have confirmed that Pt supported on SWCNTs exhibits higher performance than Pt/MWCNT and Pt/XC-72 [179].

Recently, carbide-CNT composites have been prepared as supports for fuel cells [180, 181]. Highly active Pt-WC_x/CNT electrocatalysts for the ORR have been developed by the combination of tungsten carbide with CNTs as electrocatalyst supports. The nanostructured WC_x/CNTs do show electrocatalytic activity for oxygen reduction reaction. The deposition of Pt nanoparticles on WC_x/CNTs results in a higher electrocatalytic activity for the oxygen reduction reaction and better immunity to methanol than Pt/CNT catalysts. The unique electrocatalytic properties of the novel Pt-WC_x/CNT electrocatalyst are attributed to a synergistic effect between Pt, WC_x and CNTs. Similar synergistic effect is also found for Pt, Mo_2C and CNTs [181].

9.5.4
Photocatalysis

Photocatalysis consists in the acceleration of a photoreaction in the presence of a catalyst. The photocatalytic activity depends on the ability of the catalyst to create electron–hole pairs, which generate free radicals (hydroxyl radicals $HO^{\cdot}$) ready to undergo secondary reactions. As shown in Figure 9.16, when titanium dioxide is irradiated with ultraviolet light of wavelength corresponding to its band gap energy, the electrons from the valence band can be excited to the conduction band. These photogenerated electron–hole pairs can either recombine or take part in redox reactions. The rate of recombination is fast (a few nanoseconds); the majority of photogenerated electron–hole pairs recombine with dissipation of heat. Some electron–hole pairs can migrate to the surface, where they react with adsorbed electron acceptors and donors [182, 183]. The migration and surface reactions are slow processes occurring from tens of nanoseconds to milliseconds.

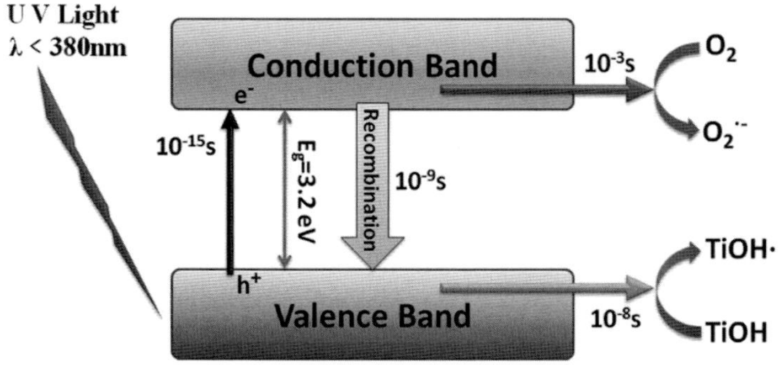

Figure 9.16 Illustration of the basic steps in photochemical mechanisms. Adapted from [183].

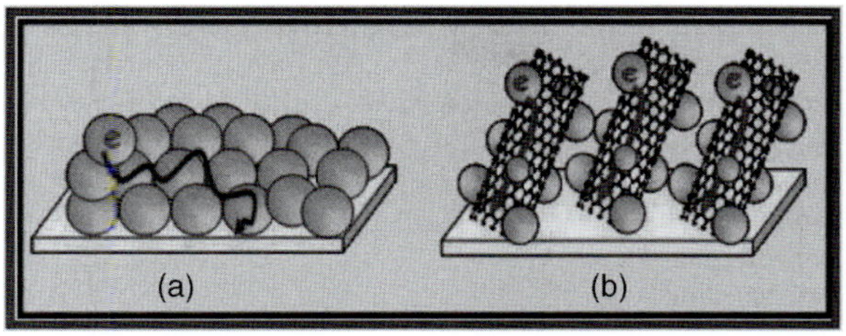

Figure 9.17 Electron transport (a) through nanostructured TiO_2 particles and (b) through CNTs. (Reproduced with permission from [184]. Copyright © 2007 American Chemical Society.)

The current efforts in photocatalysis are focused on (i) increasing the energy efficiency and (ii) improving the activity of the catalysts. The first effort is to engineer the band-gap of photo-active materials to harvest the sunlight with the highest efficiency. Single-walled CNTs with a certain chirality are semiconducting but with a very narrow band-gap that may be not suitable for band-gap engineering. However, a study reports that a shift of ∼100 mV in the apparent Fermi level of the SWCNT-TiO_2 system with respect to the unsupported TiO_2 system indicates the Fermi level equilibration between the two systems [184].

The main purpose of using CNTs in photocatalysis (or photoelectrocatalysis) is to improve the efficiency of the catalysts rather than the band-gap engineering. This can be achieved either by improving the transport pathway of photo-induced electrons thus reducing the probability for recombination of electrons and holes (Figure 9.17), or by stabilizing the active particles through enhanced anchoring. A. Kongkanand *et al.* showed that titanium dioxide nanoparticles dispersed on SWCNT films improve photo-induced charge separation and transport of carriers to the collecting electrode surface [184]. There is also report that the presence of MWCNTs can efficiently inhibit the agglomeration of the TiO_2 nanoparticles and the formation of brookite TiO_2, so that CNT supported TiO_2 composite catalyst possesses more photocatalytically active sites than pure titania [185].

The use of CNTs in photocatalysis is mainly as support for photocatalytically active materials to form composite catalysts. The mostly tested reactions are the photodegradation of organic molecules, mainly dye molecules, using CNT-TiO_2 composites as catalysts. It is reported [186, 187] that anchoring of TiO_2 onto CNTs apparently induces a synergistic effect on phenol removal, which can result in the complete disappearance of phenol in 4 h, whereas the complete removal of phenol (>95%) is observed in 6 h for TiO_2 under UV light. In the case of visible light irradiation, this synergistic effect is more remarkable in that the complete removal of phenol is accomplished in 5 h and more than a double phenol removal rate is observed. Neat TiO_2 can only reach 40.6% of removal within the same reaction time, and the complete elimination of phenol is not attained after 10 h [186, 187]. The enhanced photodegradation of phenol under visible light irradiation was also observed by other groups [188]. This phenomenon is also explained by a kind of band-gap engineering between TiO_2 and CNTs, but it is more likely due to the loss of crystallinity of the TiO_2 nanoparticles, as evidenced by the red-shift of the composite catalyst in UV-Vis spectra compared with Degussa TiO_2 P25 samples.

9.6 Other Related Carbonaceous Materials

9.6.1 Graphene and Graphene Oxide

Graphene is a material consisting in a 2D layer of sp^2 hybridized carbon atoms, the parent material of CNTs. Chemical oxidation of graphite results in exfoliated sheets, called graphene oxide (GO). The use of graphene as a 2D catalyst support with the possibility to harness its redox properties opens up new opportunities for designing next-generation catalysts. Single graphene sheet has a theoretical surface area of $2600\,m^2\,g^{-1}$, making graphene highly desirable for use as a 2D catalyst support. Incorporation of catalyst particles onto an individual graphene or reduced graphene oxide (RGO) sheets with good distribution can provide great versatility in carrying out catalytic processes, and is a new member in carbon-supported catalysts that currently attract special research efforts (Figure 9.18) [189, 190]. The graphene-supported catalysts most often reported consist of noble metal nanoparticles, including Au, Pt, Pd, Ru, Rh, Ir, and of transition metal nanoparticles including Fe,

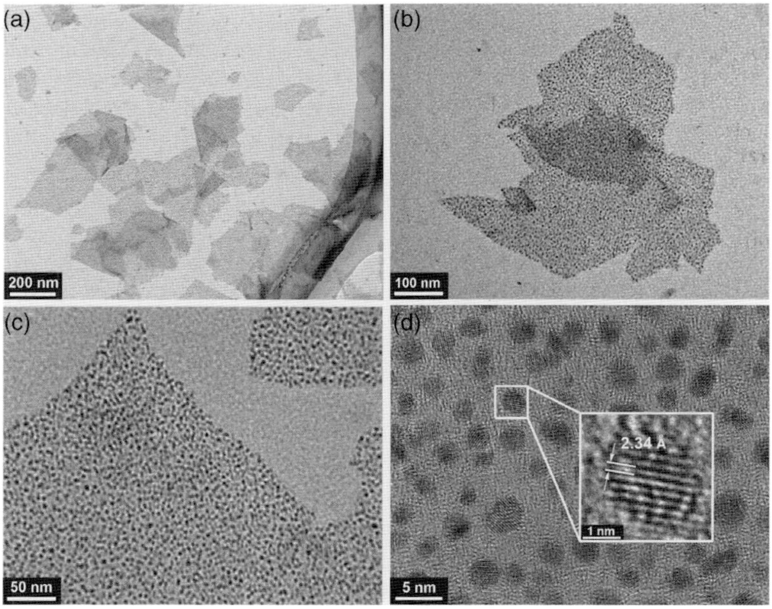

Figure 9.18 Low (a), mid (b, c), and high (d) magnification TEM images of Au nanoparticles uniformly decorated on reduced graphene oxide sheets. (Reproduced with permission from [190]. Copyright © 2011 Royal Society of Chemistry.)

Co Ni. Various inorganic nanomaterials were also synthesized and supported on graphene-based supported including TiO_2, ZnO, SnO_2, MnO_2, Fe_3O_4, Fe_2O_3, Co_3O_4, NiO, ZrO_2, SiO_2,Cu_2O, RuO_2, Al_2O_3, MoO_3, $ZnFeO_4$, $BiWO_6$ and $LiFePO_4$. Other materials like CdS, CdSe, and ZnS have also been used to fabricate graphene-based composites. Details of these graphene-based supported catalysts or composites with corresponding reference for each system are reviewed and summarized in a recent paper by B.F. Machado and P. Serp [191]. The advantages of graphene-based composites in applications such as the Li-ion batteries, supercapacitors, fuel cells, photovoltaic devices, photocatalysis, as well as Raman enhancement are critically reviewed by Hua Zhang and co-workers [192].

Ali R. Siamaki *et al.* demonstrated that both Pd/G and Pd/GO exhibit excellent catalytic activity for the carbon–carbon cross-coupling reactions under ligand-free ambient conditions [193]. The Pd/G containing 7.9 wt.% palladium showed a remarkable turnover frequency (108 000 h^{-1}) in the Suzuki cross-coupling reactions using microwave irradiation heating. The outstanding reactivity and recyclability of the Pd/G catalyst is assigned to the high concentration of Pd(0) nanoparticles very well dispersed on the surface of the graphene sheets.

Homogenous complex $Ru(PPh_3)_3Cl_2$ immobilized on amine-functionalized GO demonstrated comparable selectivity for hydrogenation of cinnamaldehyde to that of the homogenous complex, although with lower activity [194]. The abundant hydroxyl groups on the surface of GO was favorable for covalent anchoring of alkoxide ligand. As a result, the modified GO leads to a better selectivity than unmodified or other carbon supports. More importantly, the Ru-GO catalyst is the most stable one in terms of both selectivity and activity.

Hemin supported on graphene shows enhanced biomimetic oxidative performance of pyrogallol than the free molecules [195]. Graphene could prevent self-dimerization of hemin to form inactive species. The 2D graphene with large accessible surface area provides perfect platform for the reaction turnover rate as well as binding interactions. Meanwhile, the cation-π interaction between iron and graphene takes a great part in the reaction which similar to cysteine or histidine in enzymes.

Pt particles dispersed on RGO were employed as fuel cell electrocatalysts [196]. The partially reduced Pt-RGO-based fuel cell delivered a maximum power of 161 $mW\,cm^{-2}$, compared with 96 $mW\,cm^{-2}$ for an unsupported Pt-based fuel cell. This work demonstrates the usefulness of graphene as a support to anchor electrocatalyst particles. The complexity of incorporating such composites in devices demands adapted strategies to improve the conductive properties of the graphene-based supports.

Zhang *et al.* applied a graphene–P25 TiO_2 catalyst to the photocatalytic degradation of organic compounds and compared against bare P25 TiO_2 and a CNT–P25 TiO_2 composite (Figure 9.19) [197]. The graphene composite was found to have high dye absorption capacity, extended light absorption range, and enhanced charge separation and transportation properties. The authors attributed the enhanced photocatalytic activity in the degradation of methylene blue dye under both UV and visible light to the 2D conjugated structure of

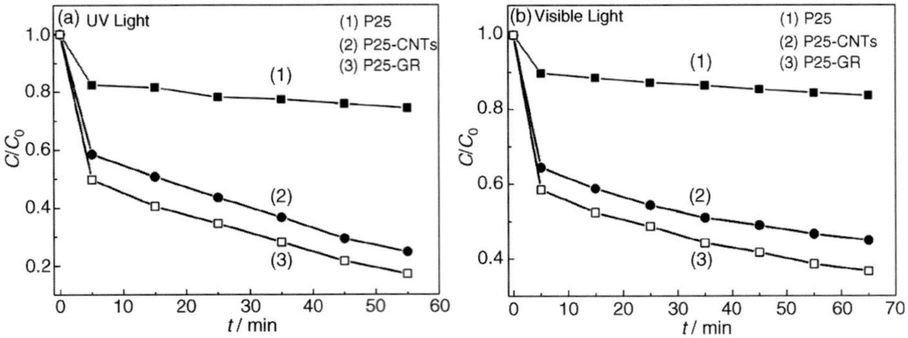

Figure 9.19 Photocatalytic degradation of methylene blue under (a) UV and (b) visible light ($\lambda > 400$ nm) with P25 TiO_2, CNT–P25 TiO_2 and graphene–P25 TiO_2. (Reproduced with permission from [197]. Copyright © 2010 American Chemical Society.)

graphene, which facilitated a better platform for dye adsorption and charge transportation. An effective reduction in charge recombination because of improved contact between graphene and TiO_2 nanoparticles was reported, increasing the photocatalytic activity [198]. All these reported results are promising and could open important perspectives for improving the photocatalytic activity of graphene–TiO_2 composites by optimizing the morphology and distribution of TiO_2 nanoparticles on graphene sheets [191].

9.6.2
Carbon Nanofibers

Carbon nanofibers (CNFs) are nanostructures with graphene layers arranged as stacked cones, cups or plates (Figure 9.20). CNFs with graphene layers wrapped into perfect cylinders with a hollow cavity are CNTs.

CNF were already reported 100 years ago [200–202]. Their structural details were revealed when TEM became available [203–206]. CNFs can be produced in CVD process with transition metal as catalyst. Apart from the different structure order of graphene layers in CNTs and CNTs (Figure 9.20), the main difference between CNTs and CNFs is the lack of a hollow capacity for the latter. The diameters of CNFs are generally higher than those of CNTs and can reach 500 nm. CNFs have high electrical resistivity, but otherwise exhibit quite similar properties of CNTs, and usually have higher graphitization degree and less amorphous carbon impurity [19]. There are as many efforts to use CNFs as that to use CNTs in catalysis as support [207, 208]. In fact, the above-mentioned methods for purification, functionalization of CNTs as well as the method for placing catalyst nanoparticles on CNTs are applicable for CNFs.

Due to their peculiar structure and morphology, CNFs allow for a better dispersion than other supports [209, 210]. The alignment of graphene sheet in CNFs determines their surface chemistry and reactivity. The edge of graphene sheets is chemically more active for anchoring functional groups and catalyst particles. This structural difference between CNFs and CNTs may have some consequences on the catalyst supported on them. For instance, hydrous RuO_2 nanoparticles with an average size of 1.35 nm can be easily prepared on herringbone CNFs, but the nanoparticles tend to agglomerate if cylindrical CNTs are employed as support.

Park *et al.* found that Ni catalyst supported on CNFs yield higher conversion values in ethylene and buta-1,3-diene hydrogenation than Ni catalysts supported on γ-alumina and AC [209]. The strong metal-support interaction between Ni and CNFs causes spreading of the metal onto the support surface and thus a good dispersion. The Ni crystallites deposited on CNFs were found to adopt very thin hexagonal structures. CNF-supported Pd catalysts (<7 nm) have high activity for Heck reactions of olefins with aryl halides, are stable at relative high temperatures, and do not react with oxygen [211]. De Jong's group showed the promising performance of Co/CNF catalysts for the FT reaction [212, 213]. Hydrogenation of phenol in a medium of supercritical CO_2 at a rather low temperature of 323 K over Rh/CNFs has been tested showing promising results [214]. Other investigated reactions include selective propane oxidation (Mo/CNFs) [215], catalytic wet-air oxidation of phenol aqueous solution and thermal decarboxylation of formic acid [216]. Pt/CNF catalysts were reported to be about 1.2 times more active than Pt/AC or Pt/Al_2O_3 for the hydrogenation of toluene [217]. CNF supports allow also here a high Pt dispersion and higher concentration of Pt(0). However, due to their specific morphology with more exposure of prismatic planes shown in Figure 9.20, CNF-supports may suffer gasification under net oxidization conditions at high temperature causing a significant textural change and thus the decrease in catalytic activity [216].

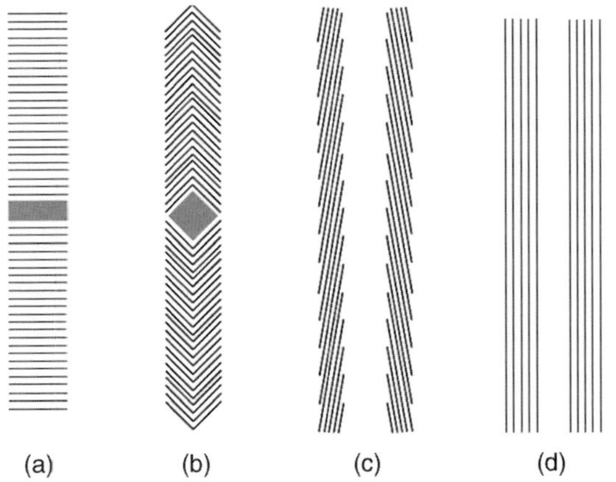

Figure 9.20 Schematic representation of (a) platelet CNF, (b) fishbone CNF, (c) hollow CNF and (d) MWCNT. In the morphology based definition, both (c) and (d) can be called multiwalled CNT although they do not present the same structure. (Reproduced with permission from [199]. Copyright © 2009 Elsevier.)

For the applications of CNF-supported catalysts in electrocatalysis, Steigerwall *et al.* [218] studied the different behavior of systems in which Pt-Ru was supported on graphitic CNFs, SWCNTs, MWCNTs, and herringbone CNFs, and found that Pt-Ru supported on graphitic CNFs shows better performance as an anode for direct methanol fuel cells than Pt-Ru supported on the other three support. This has been related to the different electrical conductivity of these four carbon materials. Using vapor grown CNFs as electrode filler and catalyst support for PEM fuel cell, Pt loading could reduce 15%, and the corresponding cell performance could be improved, especially at the high current density range. More details about the application of CNFs in fuel cell can be found in literature [19, 219].

9.6.3
Mesoporous Carbon

Ordered mesoporous carbon (OMC) materials are ideally suitable as support materials for advanced catalysts due to the well-developed regular mesopores, high surface area, thermal and chemical stability, and conductive framework. In particular, the confined geometry of uniform nanometer-sized (2–10 nm) mesopores in the OMC materials favors the formation of catalytically active species on nanometer scales. Recent studies on the development of catalysts exploiting the inherent or controlled properties of OMC materials have demonstrated their significant potential in various catalytic and environmental applications [220].

Gaurav Gupta *et al.* showed that pre-synthesized Pt nanocrystals (< 4 nm) on graphitic mesoporous carbon could constitute a stable fuel cell catalyst [221]. This catalyst shows negligible loss in mass activity and active surface area after an accelerated durability test (1000 cycles, 0.5–1.2 V), whereas commercial Pt on amorphous carbon loses 70% of activity and area. Strong Pt-graphite interactions coupled with partial charge transfer facilitate high dispersion and loadings up to 20 wt%. The high resistance to oxidation of the graphitized carbon, along with the strong Pt-C interactions, help to maintain electrical contact between the metal and carbon while mitigating Pt dissolution, ripening, and coalescence. The enhanced performance of Pt/OMC with respect to the commercial catalyst has also been reported for direct methanol fuel cells [222]. Sulfur-containing OMC can enhance the stability of the Pt catalyst [223]. Recently, it is reported that Ni and Co nanoparticles within OMC supports have significant effect on methanol oxidation reaction and oxygen reduction reaction of fuel cells [224].

The catalytic activity of Ru nanoparticles supported on OMC was evaluated for the hydrogenation of cinnamaldehyde [225]. With respect to the traditional Ru/AC catalyst, the Ru-OMC samples exhibited much higher activity (2–14-fold) and selectivity up to 60% toward cinnamyl alcohol. OMC containing molecular-level dispersed Pd clusters in the carbon walls exhibited high selectivity for the oxidation of alcohols to aldehydes [226].

9.7 Summary

The use of CNTs and related nanocarbon materials in catalysis has been a cutting-edge research area of chemistry and materials science. CNTs, together with CNFs, are typical 1D nanomaterials presenting novel structures and tunable electronic density, mediated surface chemistry, extraordinary mechanical and thermal properties with fascinating chemical reactivity.

Soon after Ijima's high impact paper on CNTs, these materials were used as nanosupports in catalysis. CNT supported catalysts exhibit, in most cases under given reaction conditions, novel catalytic activity and high selectivity. The use of CNTs and graphene as catalysts and catalyst supports brings new insight into nanocatalysis and provides general understanding of the chemistry and physics of carbon materials, which is a powerful impetus for sustainable green chemistry.

However, the surface chemistry and dynamics of catalytic reactions on CNT/CNF/graphene supported catalysts is far from well understood. Even with 20 years research on CNTs, the delicate structure control of CNTs with certain chirality, diameter, length, sites, and defect density is still an open question [227]. The number and type of defects on CNTs surface, as anchoring sites for functional groups and for nanoparticles, cannot be controlled during the production of CNTs [199]. Suitable methods to control the functionalization of CNTs with desired groups at the kilogram scale are still missing. There is no well-established method to give a quantitative description of CNTs with respect to defects per square nanometer or to functional groups per gram. Thus, it is still not straightforward to correlate the electronic state and micro/nanostructure of CNTs with their effect as catalyst or nanosupport.

The reactions using CNTs as nanosupport or catalyst are normally performed at laboratory scale. It is far away from an industrial application and still demands further efforts, especially from chemical engineering and process engineering. In this sense, catalysis requires not only large amounts of CNTs or CNFs with well-defined surface chemical and mechanical properties, but also nanocarbons formed into larger objects to optimize the reaction processes and allow effective contact with reacting matrices. Loose CNTs/CNFs are not suitable for this use as their supra-structural properties cannot be controlled [228]. The use of nanoscopic catalysts is hampered in large-scale fixed-bed reactors because of the large pressure drop across the catalyst bed and because of the problems associated with the handling and transportation of these materials. Macroscopic shaping of nanocarbons without damaging their intimate structure remains the only choice, and promising progress has been reported in this field [229].

Not discussed in this chapter are the modeling and calculation on CNTs as nanosupport, the efforts to bring a general understanding of the CNT-catalyst support and its correlation with catalytic performance. Numerous papers on these topics can be found in the literature [230–233].

References

1 Arai, H. and Machida, M. (1996) Thermal stabilization of catalyst supports and their application to high-temperature catalytic combustion. *Appl. Catal. A: General*, **138** (2), 161–176.

2 Auer, E., Freund, A., Pietsch, J., and Tacke, T. (1998) Carbons as supports for industrial precious metal catalysts. *Appl. Catal. A: General*, **173** (2), 259–271.

3 Harris, P.J.F. (ed.) (2009) *Carbon Nanotube Science*, Cambridge University Press.

4 Iijima, S. (1991) Helical microtubules of graphitic carbon. *Nature*, **354** (6348), 56–58.

5 Su, D.S. and Thomas, A. (2010) Nanochemical concepts for a sustainable energy supply. *ChemSusChem*, **3** (2), 120–121.

6 Su, D.S. and Schloegl, R. (2010) Nanostructured carbon and carbon nanocomposites for electrochemical energy storage applications. *ChemSusChem*, **3** (2), 136–168.

7 Su, D.S. (2012) Editorial: Chemistry of energy conversion and storage. *ChemSusChem*, **5** (3), 443–445.

8 Zhao, M.-Q., Zhang, Q., Jia, X.-L., Huang, J.-Q., Zhang, Y.-H., and Wei, F. (2010) Hierarchical composites of single/double-walled carbon nanotubes interlinked flakes from direct carbon deposition on layered double hydroxides. *Adv. Funct. Mater.*, **20** (4), 677–685.

9 Shao, L.D., Zhang, W., Armbruster, M., Teschner, D., Girgsdies, F., Zhang, B.S., Timpe, O., Friedrich, M., Schlogl, R., and Su, D.S. (2011) Nanosizing intermetallic compounds onto carbon nanotubes: active and selective hydrogenation catalysts. *Angew. Chem. Int. Ed.*, **50** (43), 10231–10235.

10 Nunzi, F., Mercuri, F., Sgamellotti, A., and Re, N. (2002) The coordination chemistry of carbon nanotubes: a density functional study through a cluster model approach. *J. Phys. Chem. B*, **106** (41), 10622–10633.

11 Serp, P., Corrias, M., and Kalck, P. (2003) Carbon nanotubes and nanofibers in catalysis. *Appl. Catal. A: General*, **253** (2), 337–358.

12 Frank, B., Rinaldi, A., Blume, R., Schlogl, R., and Su, D.S. (2010) Oxidation stability of multiwalled carbon nanotubes for catalytic applications. *Chem. Mater.*, **22** (15), 4462–4470.

13 Peigney, A., Laurent, C., Flahaut, E., Bacsa, R.R., and Rousset, A. (2001) Specific surface area of carbon nanotubes and bundles of carbon nanotubes. *Carbon*, **39** (4), 507–514.

14 Guerreroruiz, A. and Rodriguezramos, I. (1994) Oxydehydrogenation of ethylbenzene to styrene catalyzed by graphites and activated carbons. *Carbon*, **32** (1), 23–29.

15 Khlobystov, A.N. (2011) Carbon nanotubes: from nano test tube to nano-reactor. *ACS Nano*, **5** (12), 9306–9312.

16 Maurin, G., Stepanek, I., Bernier, P., Colomer, J.F., Nagy, J.B., and Henn, F. (2001) Segmented and opened multiwalled carbon nanotubes. *Carbon*, **39** (8), 1273–1278.

17 Konya, Z., Zhu, J., Niesz, K., Mehn, D., and Kiricsi, I. (2004) End morphology of ball milled carbon nanotubes. *Carbon*, **42** (10), 2001–2008.

18 Kukovecz, A., Kanyo, T., Konya, Z., and Kiricsi, I. (2005) Long-time low-impact ball milling of multi-wall carbon nanotubes. *Carbon*, **43** (5), 994–1000.

19 Serp, P. and Figueiredo, J. (2009) *Carbon Materials for Catalysis*, John Wiley & Sons, Inc., New York, USA.

20 Tessonnier, J.P. and Su, D.S. (2011) Recent progress on the growth mechanism of carbon nanotubes: A review. *ChemSusChem*, **4** (7), 824–847.

21 Zhang, Q., Huang, J.-Q., Zhao, M.-Q., Qian, W.-Z., and Wei, F. (2011) Carbon nanotube mass production: principles and processes. *ChemSusChem*, **4** (7), 864–889.

22 Karousis, N., Tagmatarchis, N., and Tasis, D. (2010) Current progress on the chemical modification of carbon nanotubes. *Chem. Rev.*, **110** (9), 5366–5397.

23 Banerjee, S., Hemraj-Benny, T., and Wong, S.S. (2005) Covalent surface

chemistry of single-walled carbon nanotubes. *Adv. Mater.*, **17** (1), 17–29.

24 Peng, X.H. and Wong, S.S. (2009) Functional covalent chemistry of carbon nanotube surfaces. *Adv. Mater.*, **21** (6), 625–642.

25 Peng, X.H., Chen, J.Y., Misewich, J.A., and Wong, S.S. (2009) Carbon nanotube-nanocrystal heterostructures. *Chem. Soc. Rev.*, **38** (4), 1076–1098.

26 Fu, Q., Weinberg, G., and Su, D.S. (2008) Selective filling of carbon nanotubes with metals by selective washing. *New Carbon Mater.*, **23** (1), 17–20.

27 Tessonnier, J.P., Ersen, O., Weinberg, G., Pham-Huu, C., Su, D.S., and Schlogl, R. (2009) Selective deposition of metal nanoparticles inside or outside multiwalled carbon nanotubes. *ACS Nano*, **3** (8), 2081–2089.

28 Castillejos, E., Debouttiere, P.J., Roiban, L., Solhy, A., Martinez, V., Kihn, Y., Ersen, O., Philippot, K., Chaudret, B., and Serp, P. (2009) An efficient strategy to drive nanoparticles into carbon nanotubes and the remarkable effect of confinement on their catalytic performance. *Angew. Chem. Int. Ed.*, **48** (14), 2529–2533.

29 Hou, P.X., Liu, C., and Cheng, H.M. (2008) Purification of carbon nanotubes. *Carbon*, **46** (15), 2003–2025.

30 Smith, M.R., Hedges, S.W., LaCount, R., Kern, D., Shah, N., Huffman, G.P., and Bockrath, B. (2003) Selective oxidation of single-walled carbon nanotubes using carbon dioxide. *Carbon*, **41** (6), 1221–1230.

31 Delpeux, S., Szostak, K., Frackowiak, E., and Beguin, F. (2005) An efficient two-step process for producing opened multiwalled carbon nanotubes of high-purity. *Chem. Phys. Lett.*, **404** (4–6), 374–378.

32 Shaffer, M.S.P., Fan, X., and Windle, A. H. (1998) Dispersion and packing of carbon nanotubes. *Carbon*, **36** (11), 1603–1612.

33 Young Soo, P., Young Chul, C., Keun Soo, K., Dong-Chul, C., Dong Jae, B., Kay Hyeok, A., Seong Chu, L., Xiao Yan, Z., and Young Hee, L. (2001) High yield purification of multiwalled carbon nanotubes by selective oxidation during thermal annealing. *Carbon*, **39** (5), 655–661.

34 Ando, Y., Zhao, X., and Shimoyama, H. (2001) Structure analysis of purified multiwalled carbon nanotubes. *Carbon*, **39** (4), 569–574.

35 Rinzler, A.G., Liu, J., Dai, H., Nikolaev, P., Huffman, C.B., Rodriguez-Macias, F.J., Boul, P.J., Lu, A.H., Heymann, D., Colbert, D.T., Lee, R.S., Fischer, J.E., Rao, A.M., Eklund, P.C., and Smalley, R.E. (1998) Large-scale purification of single-wall carbon nanotubes: process, product, and characterization. *Appl. Phys. A-Mater. Sci. Process*, **67** (1), 29–37.

36 Hu, H., Zhao, B., Itkis, M.E., and Haddon, R.C. (2003) Nitric acid purification of single-walled carbon nanotubes. *J. Phys. Chem. B*, **107** (50), 13838–13842.

37 Shao, L., Tobias, G., Salzmann, C.G., Ballesteros, B., Hong, S.Y., Crossley, A., Davis, B.G., and Green, M.L.H. (2007) Removal of amorphous carbon for the efficient sidewall functionalisation of single-walled carbon nanotubes. *Chem. Commun.*, (47), 5090–5092.

38 Hernadi, K., Siska, A., Thien-Nga, L., Forro, L., and Kiricsi, I. (2001) Reactivity of different kinds of carbon during oxidative purification of catalytically prepared carbon nanotubes. *Solid State Ion.*, **141**, 203–209.

39 Suzuki, T., Suhama, K., Zhao, X.L., Noue, S., Nishikawa, N., and Ando, Y. (2007) Purification of single-wall carbon nanotubes produced by arc plasma jet method. *Diam. Relat. Mater.*, **16** (4–7), 1116–1120.

40 Suri, A. and Coleman, K.S. (2011) The superiority of air oxidation over liquid-phase oxidative treatment in the purification of carbon nanotubes. *Carbon*, **49** (9), 3031–3038.

41 Ismail, A.F., Goh, P.S., Tee, J.C., Sanip, S.M., and Aziz, M. (2008) A review of purification techniques for carbon nanotubes. *Nano*, **3** (3), 127–143.

42 Tsang, S.C., Harris, P.J.F., and Green, M.L.H. (1993) Thinning and opening of carbon nanotubes by oxidation using carbon-dioxide. *Nature*, **362** (6420), 520–522.

43 Bonard, J.M., Stora, T., Salvetat, J.P., Maier, F., Stockli, T., Duschl, C., Forro, L., deHeer, W.A., and Chatelain, A. (1997) Purification and size-selection of carbon nanotubes. *Adv. Mater.*, **9** (10), 827–831.

44 Shelimov, K.B., Esenaliev, R.O., Rinzler, A.G., Huffman, C.B., and Smalley, R.E. (1998) Purification of single-wall carbon nanotubes by ultrasonically assisted filtration. *Chem. Phys. Lett.*, **282** (5–6), 429–434.

45 Bandow, S., Rao, A.M., Williams, K.A., Thess, A., Smalley, R.E., and Eklund, P. C. (1997) Purification of single-wall carbon nanotubes by microfiltration. *J. Phys. Chem. B*, **101** (44), 8839–8842.

46 Yu, A.P., Bekyarova, E., Itkis, M.E., Fakhrutdinov, D., Webster, R., and Haddon, R.C. (2006) Application of centrifugation to the large-scale purification of electric arc-produced single-walled carbon nanotubes. *J. Am. Chem. Soc.*, **128** (30), 9902–9908.

47 Andrews, R., Jacques, D., Qian, D., and Dickey, E.C. (2001) Purification and structural annealing of multiwalled carbon nanotubes at graphitization temperatures. *Carbon*, **39** (11), 1681–1687.

48 Huang, W., Wang, Y., Luo, G.H., and Wei, F. (2003) 99.9% Purity multiwalled carbon nanotubes by vacuum high-temperature annealing. *Carbon*, **41** (13), 2585–2590.

49 Wang, Y., Wu, J., and Wei, F. (2003) A treatment method to give separated multiwalled carbon nanotubes with high purity, high crystallization and a large aspect ratio. *Carbon*, **41** (15), 2939–2948.

50 Rinaldi, A., Frank, B., Su, D.S., Hamid, S.B.A., and Schlogl, R. (2011) Facile removal of amorphous carbon from carbon nanotubes by sonication. *Chem. Mat.*, **23** (4), 926–928.

51 Terrones, M., Ajayan, P.M., Banhart, F., Blase, X., Carroll, D.L., Charlier, J.C., Czerw, R., Foley, B., Grobert, N., Kamalakaran, R., Kohler-Redlich, P., Ruhle, M., Seeger, T., and Terrones, H. (2002) N-doping and coalescence of carbon nanotubes: synthesis and electronic properties. *Appl. Phys. A: Mater. Sci. Process*, **74** (3), 355–361.

52 Balasubramanian, K. and Burghard, M. (2005) Chemically functionalized carbon nanotubes. *Small*, **1** (2), 180–192.

53 Ewels, C.P. and Glerup, M. (2005) Nitrogen doping in carbon nanotubes. *J. Nanosci. Nanotechnol.*, **5** (9), 1345–1363.

54 Chen, J., Rao, A.M., Lyuksyutov, S., Itkis, M.E., Hamon, M.A., Hu, H., Cohn, R.W., Eklund, P.C., Colbert, D.T., Smalley, R.E., and Haddon, R.C. (2001) Dissolution of full-length single-walled carbon nanotubes. *J. Phys. Chem. B*, **105** (13), 2525–2528.

55 Tasis, D., Tagmatarchis, N., Bianco, A., and Prato, M. (2006) Chemistry of carbon nanotubes. *Chem. Rev.*, **106** (3), 1105–1136.

56 van Dommele, S., de Jong, K.P., and Bitter, J.H. (2006) Nitrogen-containing carbon nanotubes as solid base catalysts. *Chem. Commun.*, (46), 4859–4861.

57 Nikolaev, P., Bronikowski, M.J., Bradley, R.K., Rohmund, F., Colbert, D.T., Smith, K.A., and Smalley, R.E. (1999) Gas-phase catalytic growth of single-walled carbon nanotubes from carbon monoxide. *Chem. Phys. Lett.*, **313** (1–2), 91–97.

58 Balasubramanian, K., Friedrich, M., Jiang, C.Y., Fan, Y.W., Mews, A., Burghard, M., and Kern, K. (2003) Electrical transport and confocal Raman studies of electrochemically modified individual carbon nanotubes. *Adv. Mater.*, **15** (18), 1515–1518.

59 Grobert, N., Terrones, M., Trasobares, S., Kordatos, K., Terrones, H., Olivares, J., Zhang, J.P., Redlich, P., Hsu, W.K., Reeves, C.L., Wallis, D.J., Zhu, Y.Q., Hare, J.P., Pidduck, A.J., Kroto, H.W., and Walton, D.R.M. (2000) A novel route to aligned nanotubes and nanofibres using laser-patterned catalytic substrates. *Appl. Phys. A: Mater. Sci. Process*, **70** (2), 175–183.

60 Liu, J., Czerw, R., Webster, S., Carroll, D. L., Park, J.H., Park, Y.W., and Terrones, M. (2003) Advances in CNx nanotube growth, nanotube-based devices. (ed. Bernier, Carroll, Kim, Roth), *MRS Symp. Proc.*, **772**, 105–111.

61 Liu, J., Czerw, R., and Carroll, D.L. (2005) Large-scale synthesis of highly aligned nitrogen doped carbon nanotubes by injection chemical vapor deposition methods. *J. Mater. Res.*, **20** (2), 538–543.

62 Lee, C.J., Lyu, S.C., Kim, H.W., Lee, J.H., and Cho, K.I. (2002) Synthesis of bamboo-shaped carbon-nitrogen nanotubes using C_2H_2-NH_3-$Fe(CO)_5$ system. *Chem. Phys. Lett.*, **359** (1–2), 115–120.

63 Maldonado, S., Morin, S., and Stevenson, K.J. (2006) Structure, composition, and chemical reactivity of carbon nanotubes by selective nitrogen doping. *Carbon*, **44** (8), 1429–1437.

64 Arrigo, R., Haevecker, M., Schloegl, R., and Su, D.S. (2008) Dynamic surface rearrangement and thermal stability of nitrogen functional groups on carbon nanotubes. *Chem. Commun.*, (40), 4891–4893.

65 Arrigo, R., Haevecker, M., Wrabetz, S., Blume, R., Lerch, M., McGregor, J., Parrott, E.P.J., Zeitler, J.A., Gladden, L.F., Knop-Gericke, A., Schloegl, R., and Su, D.S. (2010) Tuning the acid/base properties of nanocarbons by functionalization via amination. *J. Am. Chem. Soc.*, **132** (28), 9616–9630.

66 Hirsch, A. (2002) Functionalization of single-walled carbon nanotubes. *Angew. Chem. Int. Ed.*, **41** (11), 1853–1859.

67 Lin, Y., Taylor, S., Li, H.P., Fernando, K. A.S., Qu, L.W., Wang, W., Gu, L.R., Zhou, B., and Sun, Y.P. (2004) Advances toward bioapplications of carbon nanotubes. *J. Mater. Chem.*, **14** (4), 527–541.

68 Ebbesen, T.W., Hiura, H., Bisher, M.E., Treacy, M.M.J., ShreeveKeyer, J.L., and Haushalter, R.C. (1996) Decoration of carbon nanotubes. *Adv. Mater.*, **8** (2), 155–157.

69 Tessonnier, J.P., Villa, A., Majoulet, O., Su, D.S., and Schlogl, R. (2009) Defect-mediated functionalization of conversion. *Angew. Chem. Int. Ed.*, **48** (35), 6543–6546.

70 Xing, Y.C. (2004) Synthesis and electrochemical characterization of uniformly-dispersed high loading Pt nanoparticles on sonochemically-treated carbon nanotubes. *J. Phys. Chem. B*, **108** (50), 19255–19259.

71 Yang, C.W., Hu, X.G., Wang, D.L., Dai, C.S., Zhang, L., Jin, H.B., and Agathopoulos, S. (2006) Ultrasonically treated multiwalled carbon nanotubes (Mwcnts) as PtRu catalyst supports for methanol electrooxidation. *J. Power Sources*, **160** (1), 187–193.

72 Velasco-Santos, C., Martinez-Hernandez, A.L., Lozada-Cassou, M., Alvarez-Castillo, A., and Castano, V.M. (2002) Chemical functionalization of carbon nanotubes through an organosilane. *Nanotechnology*, **13** (4), 495–498.

73 Sun, X., Li, R., Villers, D., Dodelet, J.P., and Desilets, S. (2003) Composite electrodes made of Pt nanoparticles deposited on carbon nanotubes grown on fuel cell backings. *Chem. Phys. Lett.*, **379** (1–2), 99–104.

74 Zhao, Z.W., Guo, Z.P., Ding, J., Wexler, D., Ma, Z.F., Zhang, D.Y., and Liu, H.K. (2006) Novel ionic liquid supported synthesis of platinum-based electrocatalysts on multiwalled carbon nanotubes. *Electrochem. Commun.*, **8** (2), 245–250.

75 Bahr, J.L., Yang, J.P., Kosynkin, D.V., Bronikowski, M.J., Smalley, R.E., and Tour, J.M. (2001) Functionalization of carbon nanotubes by electrochemical reduction of aryl diazonium salts: a bucky paper electrode. *J. Am. Chem. Soc.*, **123** (27), 6536–6542.

76 Chaoxiong, H., Shuqin, S., Jinchao, L., Maragou, V., and Tsiakaras, P. (2010) Koh-activated multiwalled carbon nanotubes as platinum supports for oxygen reduction reaction. *J. Power Sources*, **195** (21), 7409–7414.

77 Moghaddam, M.J., Taylor, S., Gao, M., Huang, S.M., Dai, L.M., and McCall, M.J. (2004) Highly efficient binding of DNA on the sidewalls and tips of carbon nanotubes using photochemistry. *Nano Lett.*, **4** (1), 89–93.

78 Liu, L., Wang, T.X., Li, J.X., Guo, Z.X., Dai, L.M., Zhang, D.Q., and Zhu, D.B. (2003) Self-assembly of gold nanoparticles to carbon nanotubes using a thiol-terminated pyrene as interlinker. *Chem. Phys. Lett.*, **367** (5–6), 747–752.

79 Yang, D.Q., Hennequin, B., and Sacher, E. (2006) XPS demonstration of pi-pi interaction between benzyl mercaptan and multiwalled carbon nanotubes and their use in the adhesion of Pt nanoparticles. *Chem. Mater.*, **18** (21), 5033–5038.

80 Byl, O., Liu, J., and Yates, J.T. (2005) Etching of carbon nanotubes by ozone - a surface area study. *Langmuir*, **21** (9), 4200–4204.

81 Xia, W., Hagen, V., Kundu, S., Wang, Y. M., Somsen, C., Eggeler, G., Sun, G.G., Grundmeier, G., Stratmann, M., and Muhler, M. (2007) Controlled etching of carbon nanotubes by iron-catalyzed steam gasification. *Adv. Mater.*, **19** (21), 3648–3652.

82 Cailu, X., Junfeng, C., Yan, C., Qingyou, H., Hahn, C., Liaw, P.K., and Dehai, W. (2006) Influence of the surface treatment on the deposition of platinum nanoparticles on the carbon nanotubes. *Adv. Eng. Mater.*, **8** (1–2), 73–77.

83 Bubert, H., Brandl, W., Kittel, S., Marginean, G., and Toma, D. (2002) Analytical investigation of plasma-treated carbon fibres. *Anal. Bioanal. Chem.*, **374** (7–8), 1237–1241.

84 Muhler, M. and Xia, W. (2007) DE Patent 102 006 007 208.

85 Xia, W., Jin, C., Kundu, S., and Muhler, M. (2009) A highly efficient gas-phase route for the oxygen functionalization of carbon nanotubes based on nitric acid vapor. *Carbon*, **47** (3), 919–922.

86 Stohr, B., Boehm, H.P., and Schlogl, R. (1991) Enhancement of the catalytic activity of activated carbons in oxidation reactions by thermal-treatment with ammonia or hydrogen-cyanide and observation of a superoxide species as a possible intermediate. *Carbon*, **29** (6), 707–720.

87 Shao, Y.Y., Sui, J.H., Yin, G.P., and Gao, Y.Z. (2008) Nitrogen-doped carbon nanostructures and their composites as catalytic materials for proton exchange membrane fuel cell. *Appl. Catal. B: Environ.*, **79** (1–2), 89–99.

88 Belz, T., Bauer, A., Find, J., Gunter, M., Herein, D., Mockel, H., Pfander, N., Sauer, H., Schulz, G., Schutze, J., Timpe, O., Wild, U., and Schlogl, R. (1998) Structural and chemical characterization of N-doped nanocarbons. *Carbon*, **36** (5–6), 731–741.

89 Sen, R., Satishkumar, B.C., Govindaraj, A., Harikumar, K.R., Raina, G., Zhang, J. P., Cheetham, A.K., and Rao, C.N.R. (1998) B-C-N, C-N and B-N nanotubes produced by the pyrolysis of precursor molecules over Co catalysts. *Chem. Phys. Lett.*, **287** (5–6), 671–676.

90 Terrones, M., Grobert, N., Olivares, J., Zhang, J.P., Terrones, H., Kordatos, K., Hsu, W.K., Hare, J.P., Townsend, P.D., Prassides, K., Cheetham, A.K., Kroto, H. W., and Walton, D.R.M. (1997) Controlled production of aligned-nanotube bundles. *Nature*, **388** (6637), 52–55.

91 Tang, C.C., Bando, Y., Golberg, D., and Xu, F.F. (2004) Structure and nitrogen incorporation of carbon nanotubes synthesized by catalytic pyrolysis of dimethylformamide. *Carbon*, **42** (12–13), 2625–2633.

92 Suenaga, K., Yudasaka, M., Colliex, C., and Iijima, S. (2000) Radially modulated nitrogen distribution in CNx nanotubular structures prepared by CVD using Ni phthalocyanine. *Chem. Phys. Lett.*, **316** (5–6), 365–372.

93 Terrones, M., Kamalakaran, R., Seeger, T., and Ruhle, M. (2000) Novel nanoscale gas containers: encapsulation of N(2) in CN(X) nanotubes. *Chem. Commun.*, **23**, 2335–2336.

94 Glerup, M., Castignolles, M., Holzinger, M., Hug, G., Loiseau, A., and Bernier, P. (2003) Synthesis of highly nitrogen-doped multiwalled carbon nanotubes. *Chem. Commun.*, **20**, 2542–2543.

95 Terrones, M., Terrones, H., Grobert, N., Hsu, W.K., Zhu, Y.Q., Hare, J.P., Kroto, H.W., Walton, D.R.M., Kohler-Redlich, P., Ruhle, M., Zhang, J.P., and Cheetham, A.K. (1999) Efficient route to large arrays of CN(X) nanofibers by pyrolysis of ferrocene/melamine mixtures. *Appl. Phys. Lett.*, **75** (25), 3932–3934.

96 Zou, L., Lv, R., Kang, F., Gan, L., and Shen, W. (2008) Preparation and application of bamboo-like carbon

nanotubes in lithium ion batteries. *J. Power Sources*, **184** (2), 566–569.

97 Gan, L., Lv, R., Du, H., Li, B., and Kang, F. (2009) Highly dispersed Pt nanoparticles by pentagon defects introduced in bamboo-shaped carbon nanotube support and their enhanced catalytic activity on methanol oxidation. *Carbon*, **47** (7), 1833–1840.

98 Lee, Y.T., Kim, N.S., Bae, S.Y., Park, J., Yu, S.C., Ryu, H., and Lee, H.J. (2003) Growth of vertically aligned nitrogen-doped carbon nanotubes: control of the nitrogen content over the temperature range 900–1100 degrees C. *J. Phys. Chem. B*, **107** (47), 12958–12963.

99 Trasobares, S., Stephan, O., Colliex, C., Hug, G., Hsu, W.K., Kroto, H.W., and Walton, D.R.M. (2001) Electron beam puncturing of carbon nanotube containers for release of stored N_2 gas. *Eur. Phys. J. B*, **22** (1), 117–122.

100 Wang, X.B., Hu, W.P., Liu, Y.Q., Long, C. F., Xu, Y., Zhou, S.Q., Zhu, D.B., and Dai, L.M. (2001) Bamboo-like carbon nanotubes produced by pyrolysis of iron (Ii) phthalocyanine. *Carbon*, **39** (10), 1533–1536.

101 Louis, B., Gulino, G., Vieira, R., Amadou, J., Dintzer, T., Galvagno, S., Centi, G., Ledoux, M.J., and Pham-Huu, C. (2005) High yield synthesis of multiwalled carbon nanotubes by catalytic decomposition of ethane over iron supported on alumina catalyst. *Catal. Today*, **102**, 23–28.

102 Deck, C.P. and Vecchio, K. (2006) Prediction of carbon nanotube growth success by the analysis of carbon-catalyst binary phase diagrams. *Carbon*, **44** (2), 267–275.

103 Teo, K.B.K., Singh, C., Chhowalla, M., and Milne, W.I. (2004) Catalytic synthesis of carbon nanotubes and nanofibers in *Encyclopedia of Nanoscience and Nanotechnology*. (ed. H.S. Nalwa) American Scientific Publishers, California

104 Georgakilas, V., Gournis, D., Tzitzios, V., Pasquato, L., Guldi, D.M., and Prato, M. (2007) Decorating carbon nanotubes with metal or semiconductor nanoparticles. *J. Mater. Chem.*, **17** (26), 2679–2694.

105 Lordi, V., Yao, N., and Wei, J. (2001) Method for supporting platinum on single-walled carbon nanotubes for a selective hydrogenation catalyst. *Chem. Mater.*, **13** (3), 733–737.

106 Ros, T.G., Keller, D.E., van Dillen, A.J., Geus, J.W., and Koningsberger, D.C. (2002) Preparation and activity of small rhodium metal particles on fishbone carbon nanofibres. *J. Catal.*, **211** (1), 85–102.

107 Giordano, R., Serp, P., Kalck, P., Kihn, Y., Schreiber, J., Marhic, C., and Duvail, J.L. (2003) Preparation of rhodium catalysts supported on carbon nanotubes by a surface mediated organometallic reaction. *Eur. J. Inorg. Chem.*, **4**, 610–617.

108 Pham-Huu, C., Keller, N., Charbonniere, L.J., Ziessle, R., and Ledoux, M.J. (2000) Carbon nanofiber supported palladium catalyst for liquid-phase reactions. an active and selective catalyst for hydrogenation of C=C bonds. *Chem. Commun.*, **19**, 1871–1872.

109 Benoit, L., Begin, D., Ledoux, M.J., and Phamhuu, C. (2008) *Ordered Porous Solids: Recent Advances and Prospects*, Elsevier Science.

110 Choi, H.C., Shim, M., Bangsaruntip, S., and Dai, H.J. (2002) Spontaneous reduction of metal ions on the sidewalls of carbon nanotubes. *J. Am. Chem. Soc.*, **124** (31), 9058–9059.

111 Toebes, M.L., van der Lee, M.K., Tang, L. M.in 't Veld, M.H.H., Bitter, J.H., van Dillen, A.J., and de Jong, K.P. (2004) Preparation of carbon nanofiber supported platinum and ruthenium catalysts: comparison of ion adsorption and homogeneous deposition precipitation. *J. Phys. Chem. B*, **108** (31), 11611–11619.

112 van der Lee, M.K., van Dillen, A.J., Bitter, J.H., and de Jong, K.P. (2005) Deposition precipitation for the preparation of carbon nanofiber supported nickel catalysts. *J. Am. Chem. Soc.*, **127** (39), 13573–13582.

113 Plomp, A.J., Su, D.S., de Jong, K.P., and Bitter, J.H. (2009) On the nature of oxygen-containing surface groups on carbon nanofibers and their role for platinum deposition-an XPS and titration

study. *J. Phys. Chem. C*, **113** (22), 9865–9869.

114 Azamian, B.R., Coleman, K.S., Davis, J. J., Hanson, N., and Green, M.L.H. (2002) Directly observed covalent coupling of quantum dots to single-wall carbon nanotubes. *Chem. Commun.*, **4**, 366–367.

115 Detriche, S., Devillers, S., Seffer, J.F., Nagy, J.B., Mekhalif, Z., and Delhalle, J. (2011) The use of water-soluble pyrene derivatives to probe the surface of carbon nanotubes. *Carbon*, **49** (9), 2935–2943.

116 Jiang, K.Y., Eitan, A., Schadler, L.S., Ajayan, P.M., Siegel, R.W., Grobert, N., Mayne, M., Reyes-Reyes, M., Terrones, H., and Terrones, M. (2003) Selective attachment of gold nanoparticles to nitrogen-doped carbon nanotubes. *Nano Lett.*, **3** (3), 275–277.

117 Hu, X.G., Wang, T., Qu, X.H., and Dong, S.J. (2006) *In situ* synthesis and characterization of multiwalled carbon nanotube/Au nanoparticle composite materials. *J. Phys. Chem. B*, **110** (2), 853–857.

118 Wang, C.F., Guo, S.J., Pan, X.L., Chen, W., and Bao, X.H. (2008) Tailored cutting of carbon nanotubes and controlled dispersion of metal nanoparticles inside their channels. *J. Mater. Chem.*, **18** (47), 5782–5786.

119 Ma, H.X., Wang, L.C., Chen, L.Y., Dong, C., Yu, W.C., Huang, T., and Qian, Y.T. (2007) Pt nanoparticles deposited over carbon nanotubes for selective hydrogenation of cinnamaldehyde. *Catal. Commun.*, **8** (3), 452–456.

120 Su, D.S. (2011) Special issue: Advanced electron microscopy for catalysis. *ChemCatChem*, **3** (6), 919–920.

121 Friedrich, H., Guo, S.J., de Jongh, P.E., Pan, X.L., Bao, X.H., and de Jong, K.P. (2011) A quantitative electron tomography study of ruthenium particles on the interior and exterior surfaces of carbon nanotubes. *ChemSusChem*, **4** (7), 957–963.

122 Weyland, M. (2002) Electron tomography of catalysts. *Top. Catal.*, **21** (4), 175–183.

123 Ziese, U., de Jong, K.P., and Koster, A. J. (2004) Electron tomography: A tool for 3D structural probing of heterogeneous catalysts at the nanometer scale. *Appl. Catal. A: General*, **260** (1), 71–74.

124 Friedrich, H., de Jongh, P.E., Verkleij, A. J., and de Jong, K.P. (2009) Electron tomography for heterogeneous catalysts and related nanostructured materials. *Chem. Rev.*, **109** (5), 1613–1629.

125 Thomas, J.M. and Midgley, P.A. (2010) The merits of static and dynamic high-resolution electron microscopy (HREM) for the study of solid catalysts. *ChemCatChem*, **2** (7), 783–798.

126 Su, D.S. (2010) Electron tomography: From 3D statics to 4D dynamics. *Angew. Chem. Int. Ed.*, **49** (50), 9569–9571.

127 Ersen, O., Werckmann, J., Houlle, M., Ledoux, M.J., and Pham-Huu, C. (2007) 3D electron microscopy study of metal particles inside multiwalled carbon nanotubes. *Nano Lett.*, **7** (7), 1898–1907.

128 Pinna, F., Menegazzo, F., Signoretto, M., Canton, P., Fagherazzi, G., and Pernicone, N. (2001) Consecutive hydrogenation of benzaldehyde over Pd catalysts – influence of supports and sulfur poisoning. *Appl. Catal. A: General*, **219** (1–2), 195–200.

129 Jen, P.-H., Hsu, Y.-H., and Lin, S.D. (2007) The activity and stability of Pd/C catalysts in benzene hydrogenation. *Catal. Today*, **123** (1–4), 133–141.

130 Cabiac, A., Delahay, G., Durand, R., Trens, P., Plee, D., Medevielle, A., and Coq, B. (2007) The influence of textural and structural properties of Pd/Carbon on the hydrogenation of cis,trans,trans-1,5,9-cyclododecatriene. *Appl. Catal. A: General*, **318**, 17–21.

131 Watanabe, S. and Arunajatesan, V. (2010) Influence of acid modification on selective phenol hydrogenation over Pd/Activated carbon catalysts. *Top. Catal.*, **53** (15–18), 1150–1152.

132 Bulushev, D.A. and Ross, J.R.H. (2011) Vapour phase hydrogenation of olefins by formic acid over a Pd/C catalyst. *Catal. Today*, **163** (1), 42–46.

133 Neri, G., Musolino, M.G., Milone, C., Pietropaolo, D., and Galvagno, S. (2001) Particle size effect in the catalytic hydrogenation of 2,4-dinitrotoluene over Pd/C catalysts. *Appl. Catal. A: General*, **208** (1–2), 307–316.

134 Bouchenafa-Saib, N., Grange, P., Verhasselt, P., Addoun, F., and Dubois, V. (2005) Effect of oxidant treatment of date pit active carbons used as Pd supports in catalytic hydrogenation of nitrobenzene. *Appl. Catal. A: General*, **286** (2), 167–174.

135 Felpin, F.X., Ayad, T., and Mitra, S. (2006) Pd/C: an old catalyst for new applications - its use for the Suzuki-Miyaura reaction. *Eur. J. Org. Chem.*, **12**, 2679–2690.

136 Mori, Y. and Seki, M. (2002) Pd(OH)(2)/C (Pearlman's catalyst): a highly active catalyst for Fukuyama, Sonogashira, and Suzuki coupling reactions. *J. Org. Chem.*, **68** (4), 1571–1574.

137 Kohler, K., Heidenreich, R.G., Krauter, J. G.E., and Pietsch, M. (2002) Highly active palladium/activated carbon catalysts for Heck reactions: Correlation of activity, catalyst properties, and Pd leaching. *Chem. Eur. J.*, **8** (3), 622–631.

138 Vu, H., Goncalves, F., Philippe, R., Lamouroux, E., Corrias, M., Kihn, Y., Plee, D., Kalck, P., and Serp, P. (2006) Bimetallic catalysis on carbon nanotubes for the selective hydrogenation of cinnamaldehyde. *J. Catal.*, **240** (1), 18–22.

139 Planeix, J.M., Coustel, N., Coq, B., Brotons, V., Kumbhar, P.S., Dutartre, R., Geneste, P., Bernier, P., and Ajayan, P.M. (1994) Application of carbon nanotubes as supports in heterogeneous catalysis. *J. Am. Chem. Soc.*, **116** (17), 7935–7936.

140 Gallezot, P. and Richard, D. (1998) Selective hydrogenation of alpha,beta-unsaturated aldehydes. *Catal. Rev. Sci. Eng.*, **40** (1–2), 81–126.

141 Tessonnier, J.P., Pesant, L., Ehret, G., Ledoux, M.J., and Pham-Huu, C. (2005) Pd nanoparticles introduced inside multiwalled carbon nanotubes for selective hydrogenation of cinnamaldehyde into hydrocinnamaldehyde. *Appl. Catal. A: General*, **288** (1–2), 203–210.

142 Garcia, T., Murillo, R., Cazorla-Amoros, D., Mastral, A.M., and Linares-Solano, A. (2004) Role of the activated carbon surface chemistry in the adsorption of phenanthrene. *Carbon*, **42** (8–9), 1683–1689.

143 Nevskaia, D.M., Castillejos-Lopez, E., Guerrero-Ruiz, A., and Munoz, V. (2004) Effects of the surface chemistry of carbon materials on the adsorption of phenol-aniline mixtures from water. *Carbon*, **42** (3), 653–665.

144 Karousis, N., Tsotsou, G.E., Evangelista, F., Rudolf, P., Ragoussis, N., and Tagmatarchis, N. (2008) Carbon nanotubes decorated with palladium nanoparticles: synthesis, characterization, and catalytic activity. *J. Phys. Chem. C*, **112** (35), 13463–13469.

145 Mallat, T. and Baiker, A. (2004) Oxidation of alcohols with molecular oxygen on solid catalysts. *Chem. Rev.*, **104** (6), 3037–3058.

146 Villa, A., Wang, D., Dimitratos, N., Su, D. S., Trevisan, V., and Prati, L. (2010) Pd on carbon nanotubes for liquid phase alcohol oxidation. *Catal. Today*, **150** (1–2), 8–15.

147 Villa, A., Wang, D., Spontoni, P., Arrigo, R., Su, D., and Prati, L. (2010) Nitrogen functionalized carbon nanostructures supported Pd and Au-Pd Nps as catalyst for alcohols oxidation. *Catal. Today*, **157** (1–4), 89–93.

148 Liu, G.J., Larson, E.D., Williams, R.H., Kreutz, T.G., and Guo, X.B. (2011) Making Fischer-Tropsch fuels and electricity from coal and biomass: performance and cost analysis. *Energy Fuels*, **25**, 415–437.

149 Huber, G.W., Iborra, S., and Corma, A. (2006) Synthesis of transportation fuels from biomass: chemistry, catalysts, and engineering. *Chem. Rev.*, **106** (9), 4044–4098.

150 Leckel, D. (2009) Diesel production from Fischer-Tropsch: The past, the present, and new concepts. *Energy Fuels*, **23**, 2342–2358.

151 van Niekerk, M.J., Fletcher, J.C.Q., and Oconnor, C.T. (1996) Effect of catalyst modification on the conversion of methanol to light olefins over Sapo-34. *Appl. Catal. A: General*, **138** (1), 135–145.

152 Iglesia, E. (1997) Design, synthesis, and use of cobalt-based Fischer-Tropsch synthesis catalysts. *Appl. Catal. A: General*, **161** (1–2), 59–78.

153 Kang, J.C., Zhang, S.L., Zhang, Q.H., and Wang, Y. (2009) Ruthenium nanoparticles supported on carbon nanotubes as efficient catalysts for selective conversion of synthesis gas to diesel fuel. *Angew. Chem. Int. Ed.*, **48** (14), 2565–2568.

154 Tavasoli, A., Sadagiani, K., Khorashe, F., Seifkordi, A.A., Rohaniab, A.A., and Nakhaeipour, A. (2008) Cobalt supported on carbon nanotubes - a promising novel Fischer-Tropsch synthesis catalyst. *Fuel Process Technol.*, **89** (5), 491–498.

155 Tavasoli, A., Abbaslou, R.M.M., Trepanier, M., and Dalai, A.K. (2008) Fischer-Tropsch synthesis over cobalt catalyst supported on carbon nanotubes in a slurry reactor. *Appl. Catal. A: General*, **345** (2), 134–142.

156 Chen, W., Fan, Z.L., Pan, X.L., and Bao, X.H. (2008) Effect of confinement in carbon nanotubes on the activity of Fischer-Tropsch iron catalyst. *J. Am. Chem. Soc.*, **130** (29), 9414–9419.

157 Abbaslou, R.M.M., Tavassoli, A., Soltan, J., and Dalai, A.K. (2009) Iron catalysts supported on carbon nanotubes for Fischer-Tropsch synthesis: effect of catalytic site position. *Appl. Catal. A: General*, **367** (1–2), 47–52.

158 Pan, X.L., Fan, Z.L., Chen, W., Ding, Y.J., Luo, H.Y., and Bao, X.H. (2007) Enhanced ethanol production inside carbon-nanotube reactors containing catalytic particles. *Nat. Mater.*, **6** (7), 507–511.

159 Schüth, F., Palkovits, R., Schlögl, R., and Su, D.S. (2012) Ammonia as a possible element in an energy infrastructure: catalysts for ammonia decomposition. *Energy Environ. Sci.*, **5**, 6278–6289.

160 Erisman, J.W., Sutton, M.A., Galloway, J., Klimont, Z., and Winiwarter, W. (2008) How a century of ammonia synthesis changed the world. *Nature Geosci.*, **1** (10), 636–639.

161 Yin, S.F., Xu, B.Q., Ng, C.F., and Au, C.T. (2004) Nano Ru/CNTs: A highly active and stable catalyst for the generation of COx-free hydrogen in ammonia decomposition. *Appl. Catal. B: Environ.*, **48** (4), 237–241.

162 Yin, S.F., Zhang, Q.H., Xu, B.Q., Zhu, W.X., Ng, C.F., and Au, C.T. (2004) Investigation on the catalysis of COx-free hydrogen generation from ammonia. *J. Catal.*, **224** (2), 384–396.

163 Yin, S.F., Xu, B.Q., Wang, S.J., Ng, C.F., and Au, C.T. (2004) Magnesia-carbon nanotubes (MgO-CNTs) nanocomposite: novel support of Ru catalyst for the generation of COx-free hydrogen from ammonia. *Catal. Lett.*, **96** (3–4), 113–116.

164 Yin, S.F., Xu, B.Q., Zhu, W.X., Ng, C.F., Zhou, X.P., and Au, C.T. (2004) Carbon nanotubes-supported Ru catalyst for the generation of COx-free hydrogen from ammonia. *Catal. Today*, **93-5**, 27–38.

165 Zheng, W.Q., Zhang, J., Zhu, B., Blume, R., Zhang, Y.L., Schlichte, K., Schlogl, R., Schuth, F., and Su, D.S. (2010) Structure-function correlations for Ru/CNT in the catalytic decomposition of ammonia. *ChemSusChem*, **3** (2), 226–230.

166 Zhang, J., Comotti, M., Schuth, F., Schlogl, R., and Su, D.S. (2007) Commercial Fe- or Co-containing carbon nanotubes as catalysts for NH_3 decomposition. *Chem. Commun.*, **19**, 1916–1918.

167 Zhang, J., Muller, J.O., Zheng, W.Q., Wang, D., Su, D.S., and Schlogl, R. (2008) Individual Fe-Co alloy nanoparticles on carbon nanotubes: structural and catalytic properties. *Nano Lett.*, **8** (9), 2738–2743.

168 Li, W.Z., Liang, C.H., Qiu, J.S., Zhou, W. J., Han, H.M., Wei, Z.B., Sun, G.Q., and Xin, Q. (2002) Carbon nanotubes as support for cathode catalyst of a direct methanol fuel cell. *Carbon*, **40** (5), 791–794.

169 Li, W.Z., Liang, C.H., Zhou, W.J., Qiu, J. S., Zhou, Z.H., Sun, G.Q., and Xin, Q. (2003) Preparation and characterization of multiwalled carbon nanotube-supported platinum for cathode catalysts of direct methanol fuel cells. *J. Phys. Chem. B*, **107** (26), 6292–6299.

170 Li, W.Z., Liang, C.H., Zhou, W.J., Qiu, J. S., Li, H.Q., Sun, G.Q., and Xin, Q. (2004) Homogeneous and controllable Pt particles deposited on multi-wall carbon

nanotubes as cathode catalyst for direct methanol fuel cells. *Carbon*, **42** (2), 436–439.

171 Ralph, T.R. and Hogarth, M.P. (2002) Catalysis for low temperature fuel cells Part I: The cathode challenges. *Platinum Met. Rev.*, **46** (1), 3–14.

172 Kangasniemi, K.H., Condit, D.A., and Jarvi, T.D. (2004) Characterization of Vulcan electrochemically oxidized under simulated PEM fuel cell conditions. *J. Electrochem. Soc.*, **151** (4), E125–E132.

173 Liu, J.G., Zhou, Z.H., Zhao, X.X., Xin, Q., Sun, G.Q., and Yi, B.L. (2004) Studies on performance degradation of a direct methanol fuel cell (DMFC) in life test. *Phys. Chem. Chem. Phys.*, **6** (1), 134–137.

174 Shao, Y.Y., Yin, G.P., and Gao, Y.Z. (2007) Understanding and approaches for the durability issues of Pt-based catalysts for PEM fuel cell. *J. Power Sources*, **171** (2), 558–566.

175 Wang, X., Li, W.Z., Chen, Z.W., Waje, M., and Yan, Y.S. (2006) Durability investigation of carbon nanotube as catalyst support for proton exchange membrane fuel cell. *J. Power Sources*, **158** (1), 154–159.

176 Girishkumar, G., Rettker, M., Underhile, R., Binz, D., Vinodgopal, K., McGinn, P., and Kamat, P. (2005) Single-wall carbon nanotube-based proton exchange membrane assembly for hydrogen fuel cells. *Langmuir*, **21** (18), 8487–8494.

177 Kongkanand, A., Kuwabata, S., Girishkumar, G., and Kamat, P. (2006) Single-wall carbon nanotubes supported platinum nanoparticles with improved electrocatalytic activity for oxygen reduction reaction. *Langmuir*, **22** (5), 2392–2396.

178 Wu, G. and Xu, B.Q. (2007) Carbon nanotube supported Pt electrodes for methanol oxidation: a comparison between multi- and single-walled carbon nanotubes. *J. Power Sources*, **174** (1), 148–158.

179 Chen, Y., Zhang, G.J., Ma, J.A., Zhou, Y.M., Tang, Y.W., and Lu, T.H. (2010) Electro-oxidation of methanol at the different carbon materials supported Pt nano-particles. *Int. J. Hydrogen Energy*, **35** (19), 10109–10117.

180 Liang, C.H., Ding, L., Li, C.A., Pang, M., Su, D.S., Li, W.Z., and Wang, Y.M. (2010) Nanostructured WC(X)/CNTs as highly efficient support of electrocatalysts with low Pt loading for oxygen reduction reaction. *Energy Environ. Sci.*, **3** (8), 1121–1127.

181 Pang, M., Li, C., Ding, L., Zhang, J., Su, D.S., Li, W.Z., and Liang, C.H. (2010) Microwave-assisted preparation of Mo(2) C/CNTs nanocomposites as efficient electrocatalyst supports for oxygen reduction reaction. *Ind. Eng. Chem. Res.*, **49** (9), 4169–4174.

182 Hoffmann, M.R., Martin, S.T., Choi, W. Y., and Bahnemann, D.W. (1995) Environmental applications of semiconductor photocatalysis. *Chem. Rev.*, **95** (1), 69–96.

183 Krishna, V., Pumprueg, S., Lee, S.H., Zhao, J., Sigmund, W., Koopman, B., and Moudgil, B.M. (2005) Photocatalytic disinfection with titanium dioxide coated multi-wall carbon nanotubes. *Process. Saf. Environ. Protect.*, **83** (B4), 393–397.

184 Kongkanand, A., Dominguez, R.M., and Kamat, P.V. (2007) Single wall carbon nanotube scaffolds for photoelectrochemical solar cells. Capture and transport of photogenerated electrons. *Nano Lett.*, **7** (3), 676–680.

185 Wang, Q., Yang, D., Chen, D.M., Wang, Y.B., and Jiang, Z.Y. (2007) Synthesis of anatase titania-carbon nanotubes nanocomposites with enhanced photocatalytic activity through a nanocoating-hydrothermal process. *J. Nanopart. Res.*, **9** (6), 1087–1096.

186 Wang, W.D., Serp, P., Kalck, P., Silva, C. G., and Faria, J.L. (2008) Preparation and characterization of nanostructured MWCNT-TiO_2 composite materials for photocatalytic water treatment applications. *Mater. Res. Bull.*, **43** (4), 958–967.

187 Wang, W.D., Serp, P., Kalck, P., and Faria, J.L. (2005) Photocatalytic degradation of phenol on MWNT and titania composite catalysts prepared by a modified sol-gel method. *Appl. Catal. B: Environ.*, **56** (4), 305–312.

188 An, G.M., Ma, W.H., Sun, Z.Y., Liu, Z. M., Han, B.X., Miao, S.D., Miao, Z.J., and Ding, K.L. (2007) Preparation of titania/carbon nanotube composites using supercritical ethanol and their photocatalytic activity for phenol degradation under visible light irradiation. *Carbon*, **45** (9), 1795–1801.

189 Kamat, P.V. (2010) Graphene-based nanoarchitectures. anchoring semiconductor and metal nanoparticles on a two-dimensional carbon support. *J. Phys. Chem. Lett.*, **1** (2), 520–527.

190 Yang, X., Xu, M.S., Qiu, W.M., Chen, X. Q., Deng, M., Zhang, J.L., Iwai, H., Watanabe, E., and Chen, H.Z. (2011) Graphene uniformly decorated with gold nanodots: In situ synthesis, enhanced dispersibility andapplications. *J. Mater. Chem.*, **21** (22), 8096–8103.

191 Machado, B.F. and Serp, P. (2012) Graphene-based materials for catalysis. *Catal. Sci. & Technol.*, **2** (1), 54–75.

192 Huang, X., Qi, X.Y., Boey, F., and Zhang, H. (2012) Graphene-based composites. *Chem. Soc. Rev.*, **41** (2), 666–686.

193 Siamaki, A.R., Khder, A.E.S., Abdelsayed, V., El-Shall, M.S., and Gupton, B.F. (2011) Microwave-assisted synthesis of palladium nanoparticles supported on graphene: A highly active and recyclable catalyst for carbon-carbon cross-coupling reactions. *J. Catal.*, **279** (1), 1–11.

194 Dongil, A.B., Bachiller-Baeza, B., Guerrero-Ruiz, A., and Rodriguez-Ramos, I. (2011) chemoselective hydrogenation of cinnamaldehyde: a Comparison of the immobilization of Ru-phosphine complex on graphite oxide and on graphitic surfaces. *J. Catal.*, **282** (2), 299–309.

195 Xue, T., Jiang, S., Qu, Y., Su, Q., Cheng, R., Dubin, S., Chiu, C.-Y., Kaner, R., Huang, Y., and Duan, X. (2012) Graphene-supported hemin as a highly active biomimetic oxidation catalyst. *Angew. Chem. Int. Ed.*, **51** (16), 3822–3825.

196 Seger, B. and Kamat, P.V. (2009) Electrocatalytically active graphene-platinum nanocomposites. Role of 2-D carbon support in PEM fuel cells. *J. Phys. Chem. C*, **113** (19), 7990–7995.

197 Zhang, H., Lv, X.J., Li, Y.M., Wang, Y., and Li, J.H. (2010) P25-graphene composite as a high performance photocatalyst. *ACS Nano*, **4** (1), 380–386.

198 Liu, J., Bai, H., Wang, Y., Liu, Z., Zhang, X., and Sun, D.D. (2010) Self-assembling TiO_2 nanorods on large graphene oxide sheets at a two-phase interface and their anti-recombination in photocatalytic applications. *Adv. Funct. Mater.*, **20** (23), 4175–4181.

199 Tessonnier, J.P., Rosenthal, D., Hansen, T.W., Hess, C., Schuster, M.E., Blume, R., Girgsdies, F., Pfander, N., Timpe, O., Su, D.S., and Schlogl, R. (2009) Analysis of the structure and chemical properties of some commercial carbon nanostructures. *Carbon*, **47** (7), 1779–1798.

200 Hughes, T. and Chambers, C. (1889) Manufacture of carbon filaments, US patent no 405 480.

201 Schuetzenberger, P. and Schuetzenberger, L. (1890) Sur quelques faits relatifs à l'histoire du carbone. *C. R. Acad. Sci. Paris*, **111**, 774–778.

202 Pelabon, C. and Pelabon, H. (1903) Sur une variete de carbone filamenteux. *Acad. Sci. Paris*, **137**, 706–708.

203 Radushkevich, L. and Lukyanovich, V. (1952) Structure of the carbon produced in the thermal decomposition of carbon monoxide on an iron catalyst. *Zh Fiz Khim*, **26**, 88–95, in Russian.

204 Hillert, M. and Lange, N. (1958) The structure of graphite filaments. *Z. Kristallogr.*, **111**, 24–34.

205 Oberlin, A., Endo, M., and Koyama, T. (1976) Filamentous growth of carbon through benzene decomposition. *J. Cryst. Growth*, **32** (3), 335–349.

206 Wiles, P.G. and Abrahamson, J. (1978) Carbon-fiber layers on arc electrodes 1. Their properties and cool-down behavior. *Carbon*, **16** (5), 341–349.

207 De Jong, K.P. and Geus, J.W. (2000) Carbon nanofibers: catalytic synthesis and applications. *Catal. Rev. Sci. Eng.*, **42** (4), 481–510.

208 Bessel, C.A., Laubernds, K., Rodriguez, N.M., and Baker, R.T.K. (2001) Graphite nanofibers as an alectrode for fuel cell applications. *J. Phys. Chem. B*, **105** (6), 1115–1118.

209 Park, C. and Baker, R.T.K. (1998) Catalytic behavior of graphite nanofiber supported nickel particles. 2. The influence of the nanofiber structure. *J. Phys. Chem. B*, **102** (26), 5168–5177.

210 Chambers, A., Nemes, T., Rodriguez, N.M., and Baker, R.T.K. (1998) Catalytic behavior of graphite nanofiber supported nickel particles. 1. Comparison with other support media. *J. Phys. Chem. B*, **102** (12), 2251–2258.

211 Zhu, J., Zhou, J.H., Zhao, T.J., Zhou, X.G., Chen, D., and Yuan, W.K. (2009) Carbon nanofiber-supported palladium nanoparticles as potential recyclable catalysts for the Heck reaction. *Appl. Catal. A: General*, **352** (1–2), 243–250.

212 Bezemer, G.L., van Laak, A., van Dillen, A.J., and de Jong, K.P. (2004) Cobalt supported on carbon nanofibers – a promising novel Fischer-Tropsch catalyst, in *Natural Gas Conversion Vii* (eds X. Bao and Y. Xu), Elsevier Science Bv, Amsterdam, pp. 259–264.

213 Bezemer, G.L., Falke, U., van Dillen, A.J., and de Jong, K.P. (2005) Cobalt on carbon nanofiber catalysts: auspicious system for study of manganese promotion in Fischer-Tropsch catalysis. *Chem. Commun.*, **6**, 731–733.

214 Wang, H.J., Zhao, F.Y., Fujita, S.I., and Arai, M. (2008) Hydrogenation of phenol in $scCO_2$ over carbon nanofiber supported Rh catalyst. *Catal. Commun.*, **9** (3), 362–368.

215 Ressler, T., Walter, A., Scholz, J., Tessonnier, J.P., and Su, D.S. (2010) Structure and properties of a Mo oxide catalyst supported on hollow carbon nanofibers in selective propene oxidation. *J. Catal.*, **271** (2), 305–314.

216 Taboada, C.D., Batista, J., Pintar, A., and Levec, J. (2009) Preparation, characterization and catalytic properties of carbon nanofiber-supported Pt, Pd, Ru monometallic particles in aqueous-phase reactions. *Appl. Catal. B: Environ.*, **89** (3–4), 375–382.

217 Zhou, M., Lin, G., and Zhang, H. (2007) Pt catalyst supported on multiwalled carbon nanotubes for hydrogenation-dearomatization of toluene. *Chin. J. Catal.*, **28** (3), 210–216.

218 Steigerwalt, E.S., Deluga, G.A., and Lukehart, C.M. (2002) Pt-Ru/carbon fiber nanocomposites: synthesis, characterization, and performance as anode catalysts of direct methanol fuel cells. A search for exceptional performance. *J. Phys. Chem. B*, **106** (4), 760–766.

219 Lee, K., Zhang, J.J., Wang, H.J., and Wilkinson, D.P. (2006) Progress in the synthesis of carbon nanotube- and nanofiber-supported Pt electrocatalysts for PEM fuel cell catalysis. *J. Appl. Electrochem.*, **36** (5), 507–522.

220 Taguchi, A. and Schuth, F. (2005) Ordered mesoporous materials in catalysis. *Microp. Mesop. Mater.*, **77** (1), 1–45.

221 Gupta, G., Slanac, D.A., Kumar, P., Wiggins-Camacho, J.D., Kim, J., Ryoo, R., Stevenson, K.J., and Johnston, K.P. (2010) Highly stable Pt/ordered graphitic mesoporous carbon electrocatalysts for oxygen reduction. *J. Phys. Chem. C*, **114** (24), 10796–10805.

222 Joo, S.H., Pak, C., You, D.J., Lee, S.A., Lee, H.I., Kim, J.M., Chang, H., and Seung, D. (2006) Ordered mesoporous marbons (OMC) as supports of electrocatalysts for direct methanol fuel cells (DMFC): effect of carbon precursors of OMC on DMFC performances. *Electrochim. Acta*, **52** (4), 1618–1626.

223 Hyung Ik, L., Sang Hoon, J., Jin Hoe, K., Dae Jong, Y., Ji Man, K., Jung-Nam, P., Hyuk, C., and Chanho, P. (2009) Ultrastable Pt nanoparticles supported on sulfur-containing ordered mesoporous carbon via strong metal-support interaction. *J. Mater. Chem.*, **19** (33), 5934–5939.

224 Su, F., Poh, C.K., Zeng, J., Zhong, Z., Liu, Z., and Lin, J. (2012) Pt nanoparticles supported on mesoporous carbon nanocomposites incorporated with Ni or Co nanoparticles for fuel cells. *J. Power Sources*, **205**, 136–144.

225 Gao, P., Wang, A.W., Wang, X.D., and Zhang, T. (2008) Synthesis and catalytic performance of highly ordered Ru-containing mesoporous carbons for hydrogenation of cinnamaldehyde. *Catal. Lett.*, **125** (3–4), 289–295.

226 Lu, A.H., Li, W.C., Hou, Z.S., and Schuth, F. (2007) Molecular level dispersed Pd clusters in the carbon walls of ordered mesoporous carbon as a highly selective alcohol oxidation catalyst. *Chem. Commun.*, (10), 1038–1040.

227 Su, D.S. (2011) 20 years of carbon nanotubes. *ChemSusChem*, **4** (7), 811–813.

228 Su, D.S., Chen, X.W., Weinberg, G., Klein-Hofmann, A., Timpe, O., Hamid, S.B.A., and Schlogl, R. (2005) Hierarchically structured carbon: synthesis of carbon nanofibers nested inside or immobilized onto modified activated carbon. *Angew. Chem. Int. Ed.*, **44** (34), 5488–5492.

229 Chizari, K., Deneuve, A., Ersen, O., Florea, I., Liu, Y., Edouard, D., Janowska, I., Begin, D., and Cuong, P.H. (2012) nitrogen-doped carbon nanotubes as a highly active metal-free catalyst for selective oxidation. *ChemSusChem*, **5** (1), 102–108.

230 Menon, M., Andriotis, A.N., and Froudakis, G.E. (2000) Curvature dependence of the metal catalyst atom interaction with carbon nanotubes walls. *Chem. Phys. Lett.*, **320** (5–6), 425–434.

231 Giannozzi, P., Car, R., and Scoles, G. (2003) Oxygen adsorption on graphite and nanotubes. *J. Chem. Phys.*, **118** (3), 1003–1006.

232 Pannopard, P., Khongpracha, P., Warakulwit, C., Namuangruk, S., Probst, M., and Limtrakul, J. (2012) Structures, energetics and reaction mechanisms of nitrous oxide on transition-metal-doped and -undoped single-wall carbon nanotubes. *ChemPhysChem*, **13** (2), 583–587.

233 Hu, X., Wu, Y., Li, H., and Zhang, Z. (2010) Adsorption and activation of O_2 on nitrogen-doped carbon nanotubes. *J. Phys. Chem. C*, **114** (21), 9603–9607.

10
Nano-oxides

Vasile Hulea and Emil Dumitriu

10.1
Introduction

Metal oxides are a very important class of materials, displaying a large range of properties and applications [1]. Due to their low toxicity, high chemical/thermal stability and availability, metal oxides play a major role in many areas of technological application, being used in the fabrication of microelectronic circuits, sensors, piezoelectric devices, fuel cells, coatings for the passivation of surfaces against corrosion, and as catalysts. Almost all catalysts used in the chemical and petrochemical industries and to control of environmental pollution involve an oxide as active phase, promoter or support [2, 3]. Though metal oxides are structurally organized according to various crystalline patterns, and bonding of elements can vary from ionic to metallic, their physical and chemical properties are mainly determined by defects, including edges, corners, and vacancies, and other surface imperfections. For instance, these imperfections are crucial for determining the catalytic abilities of materials. As such defects are mainly related to the surface atoms, their proportion is greatly enhanced by a large surface area, which can be obtained by decreasing the size of oxide particles down to a nanometer range. On the nanometer scale, oxide particles exhibit unusual and size-dependent physical and chemical properties that differ from those of bulk materials. Therefore, these materials have attracted great scientific interest in recent years. Moreover, the nano-oxides have been introduced as commercial products in a number of applications in areas including electronics, magnetics, optics, cosmetics, biomedicine, energy, environmental detection/control and catalysis.

The catalytic performances of nano-oxides have been explored in a large number of fundamental studies, but only a few review articles have been published on this subject [4, 5]. The aim of the present review is to provide an overview of the current knowledge available about the catalytic properties and applications of nano-oxides. Through representative catalytic reactions we wish to show how the particle size and morphology of nano-oxides affect their behavior. In order to better understand the interdependence of the catalyst properties and catalytic results, we include in

Nanomaterials in Catalysis, First Edition. Edited by Philippe Serp and Karine Philippot.

the first section of this chapter some information relating to nano-oxide synthesis and characterization.

10.2 Synthesis and Characterization of Nano-oxides

10.2.1 Design of Metal Oxide Nanoparticles

Optimization of the catalytic performances of nano-oxides needs careful control of their characteristics, such as composition, size, shape, morphology, and so on, and this control can be achieved by choosing the appropriate preparation method and specific operating parameters. The development of systematic methods for the synthesis of metal oxide nanostructures is still a challenge for academic and industrial research, but considerable efforts have been devoted during the last decade to improving the usual methods of synthesis. Generally, the syntheses of nano-oxides, as well as of other nanoscale materials, can be classified according to the following two strategies: (i) *bottom-up* strategy – which is based on building up nanostructures from atomic or molecular precursors, which come together to form clusters, and subsequently nanoparticles, and (ii) *top-down* strategy – which are based on physically tearing down larger building blocks. Between the two approaches, the *bottom-up* is preferred because the fundamental building blocks are atoms and molecules, which can be easily controlled during the nanomaterial preparation. Different *bottom-up* methods have been developed to synthesize a variety of nano-oxides, such as sol–gel process, reverse micelles, homogeneous precipitation, microwave hydrothermal methods, spray pyrolysis, and combustion. Because these strategies have been periodically described in detailed review papers and books [4, 6–14], we will not describe these synthetic methods, but just provide some fundamental information, as well as useful approaches for controlling the size and morphology of the oxide nanoparticles.

In an excellent paper, Jun *et al.* [8] have reviewed the progress in the shape control of metal oxide nanocrystals with various geometrical motifs, obtained by non-hydrolytic synthetic methods. In this work, three mechanisms for the shape-controlled synthesis of nanocrystals were proposed and widely discussed: (i) seed-mediated solution–liquid–solid growth; (ii) shape transformation through oriented attachment; and (iii) kinetically induced anisotropic growth. In a more recent paper, Jolivet *et al.* [12] have reviewed their results in the synthesis of metal (Fe, Al, Ti and Mn) oxide nanoparticles based on the chemistry of cations in aqueous solution (*chimie douce*). They showed that for these chemical systems, the variations in size and/or shape are tightly related to the variation of the electrostatic surface charge density of the particles, which induces a variation of the oxide–solution interfacial tension, and, consequently, a decrease of the surface energy. Strict control of the physico-chemical conditions of the precipitation, essentially the acidity and ionic strength in the absence of complexing species, enables the tailoring of

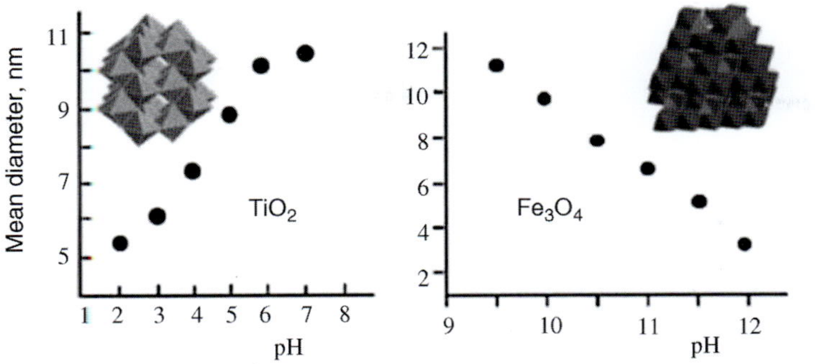

Figure 10.1 Variation of particle diameter with the pH of precipitation for anatase (TiO_2) and maganetite (Fe_2O_3) (Adapted with permission from [12]. Copyright © (2010) Elsevier).

the particle size in the range 2–15 nm (Figure 10.1) and, in some cases, of their morphology. Additionally, modification of particle shape and texture can be obtain through different effects such as adsorption, control of growth kinetic steps and oriented attachment (controlled aggregation).

The same research group reported that the complexing agents, such as polyols, might be used as size and shape controllers of oxide nanoparticles synthesized by soft chemistry in aqueous solution [15]. Thus, the use of polyols during the synthesis of nano-alumina caused a modification of the usual morphology, specifically resulting in an increase of (101) faces area.

Similar phenomena have been previously reported for the TiO_2 syntheses, where anatase nanoparticles exhibiting only the (101) faces have been obtained using glutamic acid as starting material [16], whereas ellipsoidal morphology were found in the presence of triethanolamine [17]. He *et al.* [18] introduced the concept 'solubility-controlled process,' where the thermodynamic and kinetics of the nucleation and growth of nanocrystals can be separated and controlled by changing the solubility of reagents. Through such approach, the controlled syntheses of monodispersed 2-nm (spherical), 2.5-nm (cubelike) and 4.7-nm (cubelike) Co_3O_4 crystals have been suitably achieved.

Evidence for the great influence of the nature of starting reagents and solvents on the control of the shapes of nanoparticle was often seen. For example, CeO_2 nanorods and nanocubes were selectively synthesized using cerium(III) chloride and cerium(III) nitrate as precursor, respectively [19]. Using cobalt acetate and ammonia as starting materials, and adjusting the ratio ethanol/water in the solvent, Dong *et al.* [20] have obtained Co_3O_4 nanoparticles with average sizes of 3.5, 6, 11, 19 and 70 nm. Reviewing the syntheses of indium oxide nanoparticles, Niederberger *et al.* [21] pointed out the crucial effect of the metal oxide precursors and the solvents on the particle morphology control (Figure 10.2).

It has also been reported that the size and the morphology of the metal nano-oxides can be easily tailored through proper choice of the parameters and conditions for the preparation reaction. Polshettiwar *et al.* [22] developed a synthetic protocol for metal (Fe, Co, Mn, Cr, Mo) oxides with three-dimensional (3D)

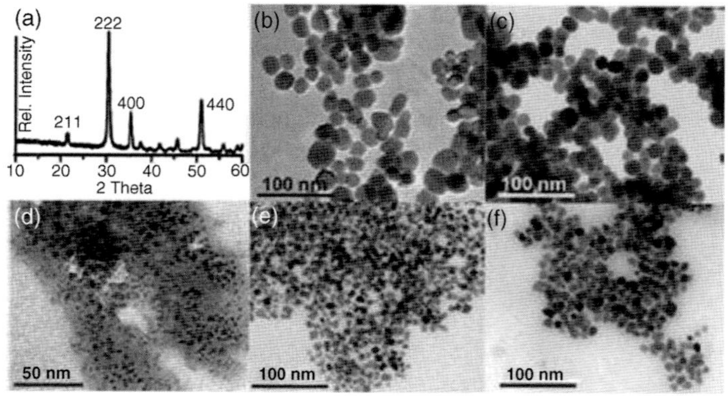

Figure 10.2 (a) XRD pattern of indium oxide nanoparticles; TEM images of the indium nanoparticles obtained in (b) indium isopropoxide-benzyl alcohol; (c) indium isopropoxide-acetophenone; (d) indium acetylacetonate-2-butanone; (e) indium acetylacetonate-benzylamanine and (f) indium acetylacetonate-acetophenone (Adapted with permission from [21]. Copyright © (2006) Springer, Netherlands).

nanostructures under microwave irradiation conditions, in pure water without using any reducing or capping reagent. They found that the size and morphology (octahedra, spheres, triangular rods, pine, hexagonal snowflake-like) of the metal oxide was properly tailored by varying substrate concentration and reaction temperature. In the case of reverse microemulsion syntheses [7, 23–26], it has been observed that the morphology of the product may be controlled by varying different parameters, such as concentration of water and surfactant, nature of surfactant and nonpolar phase (oil). Sun *et al.* [27] have reported the solvothermal preparation of CeO_2 nanorods. The nanoparticle morphology was controlled by adjusting cerium source, surfactant, solvent compositions, reaction temperature and duration.

Cr_2O_3 nanopowders were successfully synthesized using EDTA as a chelating agent [28]. It was found that the crystallite size decreased from 35 to 23 nm when the pH increased from 3 to 10.

Very recently, it has been reported that unconventional materials, such as natural polymers and metal organic frameworks (MOFs) can be used as templates for preparing nano-oxides with controlled size and morphology. Thus, CeO_2 [29], MgO [30] or NiO, CuO, Cr_3O_4 [31] with very small uniform particles have been prepared by a method based on metal–biopolymer (alginate, chitosan) complexes. For example, ceria with particle size of 5 nm has been obtained by means of electrostatic binding of Ce^{4+} to alginate gel, subsequent supercritical CO_2 drying, and calcination. Yang *et al.* [32] reported that MOFs can be used as templates for the synthesis of ZnO particles with diameter size in the range of 20–35 nm. Using Zn-terephthalic acid MOF-5 as a precursor, after combustion of the organic species, ZnO nanoparticles with 3D cubic morphologies have been obtained. Similarly, Chen *et al.* [33] prepared CuO nanosheets from $Cu_3(btc)_2$ type MOF (btc = benzene-1,3,5-tricarboxylate). Using Co_3O_4 as an example, Wang *et al.* [34] showed that the direct pyrolysis of cobalt nitrate accommodated in the pores of ZIF-8 MOF

[Zn(MeIM)$_2$, MeIM = 2-methylimidazole] yields oxide nanoparticles with a mean size of 18 nm. A comparable method, based on pyrolysis of organic ligands, was previously proposed by Jana *et al.* [35]. In this case, fatty acids were used for controlling the size, shape, and size distribution of oxide (Cr, Mn, Fe, Co, Ni) nanocrystals. The metal oxides were obtained after the pyrolysis of metal fatty acid salts.

An effective approach to control and stabilize the oxide nanoparticles consists in confining these nanoparticles inside of porous rigid materials. Zeolite, mesostructured silica, and MOF are the most used micro/mesoporous hosts for different nano-oxides. There are two usual strategies to prepare these catalytic systems. In the first, vapors of molecular metal compounds are absorbed inside pores, and in a second step they are converted into metal nano-oxides by hydrolysis/oligomerization [36] or oxidation/annealing [37]. Alternatively, metal oxide clusters can be formed inside pores by calcinations of impregnated supports obtained either by 'two-solvents' impregnation method [38, 39], by 'incipient wetness impregnation' method and gentle drying at room temperature [40], or by 'solvent-free molten-salt infiltration' [41]. Following these approaches, a variety of confined nano-oxides such as monometallic MnO_2 [38], Co_3O_4 [39, 41], CeO_2, NiO [41], ZnO [42, 43] and bimetallic NiO·CuO [40] can be obtained.

Generally, the nanoparticles confined in pores and cages exhibited uniform sizes, high thermal and mechanical stability, as well as remarkable catalytic performance, which will be detailed later, in Section 10.3. For example, high and thermostable dispersions of the mixed oxide CuNi nanoparticles confined in the channels of ordered mesoporous SBA-15 silica were reported by Ungureanu *et al.* [40].

The thermal stability of these mixed oxides both under calcination and reduction conditions was explained by a geometric stabilization of nanoparticles in the mesopores of SBA-15 by confinement effect and by a strong interaction between NiO and silica walls via nickel phyllosilicate phases. Figure 10.3 shows some representative transmission electron microscopy (TEM) images for the calcined NiO·CuO/SBA-15 samples which illustrate that the rod-like oxide nanoparticles are well dispersed (i.e., diameters of 8.5 nm and length/diameter ratio of 3–4), uniformly distributed and with a low degree of agglomeration.

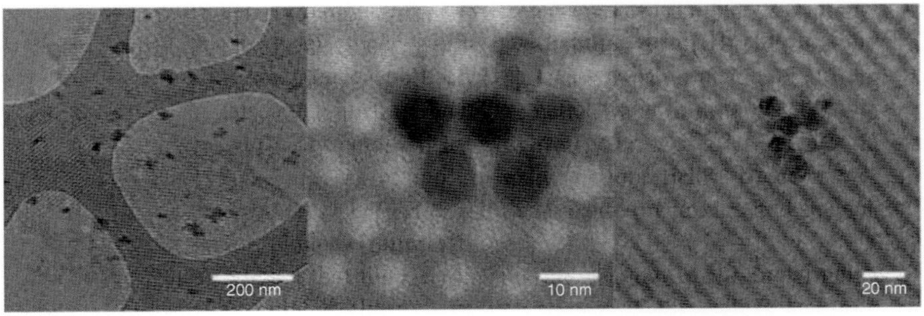

Figure 10.3 TEM images at different resolutions for NiO·CuO/SBA-15 calcined at 773 K (Reprinted with permission from [40]. Copyright © (2011) The Royal Society of Chemistry).

10.2.2
Size-Dependent Oxide Properties

Understanding the size- and shape-dependence of physical and chemical properties in nanoscale particles is an essential step towards the design, manufacture and use of nanomaterials. Therefore, intense efforts have been made to characterize the structure and the reactivity of nano-oxides. To characterize and explore possible correlations among their structural, electronic, and chemical properties various methods have been employed [3, 44, 45]. Usually, the structural characteristics are explored individually; sometimes even *in situ*, using techniques such as X-ray diffraction (XRD), extended X-ray absorption fine structure (EXAFS), X-ray absorption near edge structure (XANES), Raman spectroscopy, advanced TEM, and atomic force microscopy. To explore the surface properties and chemical reactivity, nanoparticles are studied at the ensemble level, mainly by spectral measurements of adsorbed gas molecules. To investigate the electronic characteristics of oxide nanoparticles, XANES, optical absorption, valence and core-level photoemission measurements are commonly employed.

Based on these experimental techniques it has been shown that the nanoscaled oxides offer a variety of novel features that are different compared with their bulk counterparts. The most important factors causing the unusual properties of nano-oxides are the limited size, the very high surface area, the high percentage of under-coordinated atoms on the surface, as well as the quantum effects. These factors can change or enhance properties such as structural characteristics, electronic properties and physico-chemical surface behavior [5, 44]. Indeed, as the particle size decreases, the increasing number of surface atoms generates structural perturbations associated with changes in cell parameters and lattice symmetry [46–48]. On the other hand, it was found that at the nanoscale range, the oxides frequently adopt the most stable morphologies and shapes. For example, prepared MgO, CaO and BaO nanoparticles showed a nearly perfect distorted cubic shape, exhibiting mainly the nonpolar (100) faces [44]. Additionally, for these nano-oxides, the density of the surface functions, such as hydroxyl groups, was higher than that of bulk oxides.

The nanostructure produces the so-called quantum size or confinement effects, which essentially arise from the presence of discrete, atom-like electronic states. For example, although the bulk anatase is an indirect band gap semiconductor, the nano-TiO_2 is likely a direct semiconductor [49, 50]. In the case of zirconium oxide nanoparticles, an increase in the band gap energy was observed, and its increase was attributed to quantum confinement effects (when the particle diameters range between 10 and 100 nm) or to crystalline-amorphous transition (when the diameter is lower than 10 nm) [51]. It should be noted that the modification in the band gap energy causes a major modification of the optical absorption properties. Additionally, the unique electronic properties exhibited by the nanoscale metal oxide particles are to a large extent responsible for the particular catalytic behavior exhibited by these materials.

The physico-chemical features are mainly affected by changes in structural and electronic properties. For example, the enhanced concentration of under-coordinated atoms (e.g., from corners or edges) or O vacancies on the surface of oxide nanoparticle strongly modified the sorption, acid/base or redox characteristics with respect to bulk oxides [52]. These defects have a positive electron affinity and often serve as centers of adsorption in catalytic reactions. From an electronic viewpoint, modifications in the chemical reactivity can arise from changes in the size of the band gap when going from bulk oxides to nano-oxides. The nanostructure of metal oxides also affects mechanical and thermochemical properties. For example, the Young's modulus of ZnO nanowires was found to increase when the wire diameter is reduced [53]. Similarly, both tensile modulus and bending modulus were found to increase as the wire diameter decreased from 80 to 20 nm [54]. Sayle and Sayle [55] have pointed out that ceria nanorods with exposed (110) facets exhibit higher elastic deformation than the brittle bulk phase. Experimental results indicated that it is much easier to reduce CeO_2 and $Ce_{1-x}Zr_xO_2$ nanoparticles than bulk like powder of these oxides [56]. According to the authors, there was an expansion in the unit cell of the reduced particles probably as a consequence of a partial $Ce^{4+} \rightarrow Ce^{3+}$ transformation and the sorption of hydrogen into the bulk of the material. Consequently, the $Ce_{1-x}Zr_xO_2$ nanoparticles interact with H_2 and reduce at lower temperatures than bulk $Ce_{1-x}Zr_xO_2$ systems. Similarly, Han *et al.* [57] have found that the reduction of CeO_2 nanorods in H_2 effective at lower temperature (623 K) with respect to both spherical nanoparticles and bulk ceria (763 K). Consequently, ceria nanorods displayed higher catalytic activity for the water-gas shift reaction compared to the bulk ceria (see also the Section 10.3.1.1).

10.3 Catalytic Applications of Nano-oxides

Metal oxide nanoparticles as heterogeneous catalysts have received much attention in the recent years. They can be used either as an active phase or support for developing various catalytic systems, which show real potential in major applications such as organic transformations, destruction of hazardous chemicals, green chemistry, and so on. Some review articles [4, 5] dealt with the catalytic behavior of nano-oxides. However, aspects such as catalytic behavior vs. particle size and morphology or nano-oxides as support for catalysts have not been examined systematically. In this section, we wish to address these particular aspects by reviewing articles mainly published during the last 10 years.

10.3.1 Nano-oxides as Active Phases for Catalytic Applications

10.3.1.1 Catalytic Behavior Related to the Characteristics of Nanoparticles

10.3.1.1.1 Bulk Materials Compared with Nanoparticles Unsupported nanoparticles of metal oxides such as MgO [24, 58], Mn_3O_4, Co_3O_4 [59], Ce_2O [60], ZnO [61–64], TiO_2 [65], RuO_2 [66], Fe_2O_3 [67, 68] have successfully been used as catalysts for various chemical applications. In order to underline the favorable effect of the particle size, the catalytic behaviors of these nanomaterials were compared with those of the corresponding bulk oxides. Generally, the nanosized oxides showed superior performances. For example, in the oxidation of water at 295 K, the oxygen yield for nanostructured Co_3O_4 clusters (diameter of 7.6 nm) confined in SBA-15 exceeded that of the bare micrometer-sized particles by a factor of 1550 [59]. Furthermore, the liquid-phase oxidation of benzyl alcohol to benzaldehyde proceeded with excellent activity (85% conversion) and a selectivity of 35% in the presence of nano-γ-Fe_2O_3 particles with diameters of 3–5 nm [68]. This activity was higher by a factor of eight relative to the corresponding bulk Fe_2O_3. The effectiveness of nano-Fe_2O_3 (particle size of 3–5 nm) over the micro-α-Fe_2O_3 (powder of 5 μm) was also shown for carbon monoxide oxidation into CO_2 [67]. Thus, at 623 K, the nanocatalysts can catalyze oxidation of almost 100% of CO, whereas the α-Fe_2O_3 powder can only catalyze oxidation of less than 5% of CO.

Huang *et al.* [63] used three different ZnO particle sizes in a catalytic ozonation process for removing 2,4,6-trichlorophenol (TCP) from water. The global rate constants obtained at steady state conditions for average particle size of 31 nm, 0.46 μm, and 27.6 μm were 0.24, 0.13 and 0.09 min^{-1}, respectively (Figure 10.4).

Jang *et al.* [64] compared the photocatalytic activities of nano- and microscale ZnO crystals. ZnO nanoparticles having a higher population of polar Zn (0001) faces showed the highest photocatalytic activity for H_2O_2 generation.

In most publications, the high activity of the nano-oxide systems has been simply attributed either to the high specific surface area of nanoparticles [59, 60, 62, 63, 66] or the change of surface property (charge, defects, low-coordinated sites), that are not common for the corresponding bulk oxides [24, 63, 64, 67]. In a more scientific way, Choudary *et al.* explained the remarkable performances of MgO nanoparticles

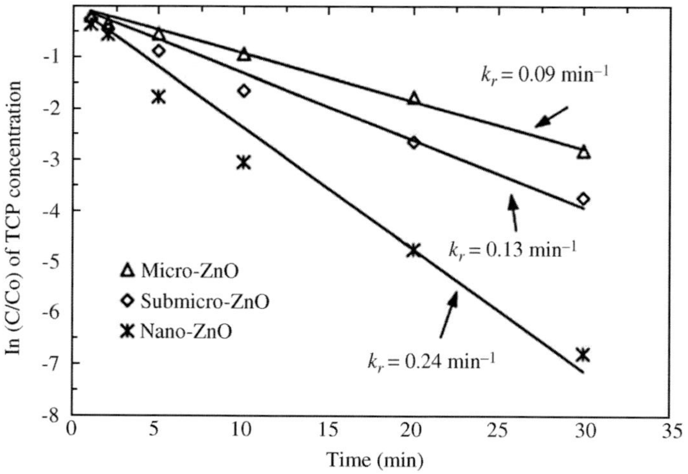

Figure 10.4 Rate constant of total removal of TCP by ozonolization over ZnO with different particle size (Reprinted with permission from [63]. Copyright (2005) Elsevier).

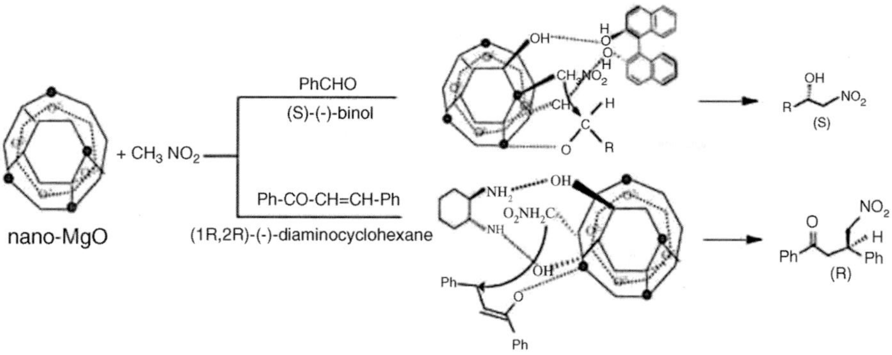

Figure 10.5 Proposed mechanism for asymmetric Henry and Michael reactions catalyzed by nano-MgO (Adapted with permission from [69]. Copyright © (2005) American Chemical Society).

for asymmetric Claisen–Schmidt, Michael and Henry reactions, taking into account both the texture and the particular morphology of nanocrystallites [58, 69]. For these reactions, aerogel prepared MgO nanoparticles (very small, irregular stacks of square plates exhibiting numerous crystal faces, edges and corners, SSA: $590\,m^2\,g^{-1}$) showed higher activity and especially higher enantioselectivity (ee) than the conventional prepared MgO nanoparticles (thin hexagonal platelets, SSA: $250\,m^2\,g^{-1}$) or commercial MgO (large cubic crystals, SSA: $30\,m^2\,g^{-1}$). Concerning the activity, the aerogel-MgO having 3D polyhedral structure and various exposed crystal planes, leads to inherently high surface reactivity per unit area. On the other hand, it was found that Brønsted hydroxyls are the sole contributors for the ee. According to the authors, the hydrogen bond interactions between the —OH or $—NH_2$ groups of reactants and the —OH groups of nano-MgO (Figure 10.5) are essential for the induction of enantioselectivity.

The prepared aerogel- and conventional-MgO samples possess the same average concentrations of surface —OH groups. A possible rationale for the display of higher ee by the aerogel-MgO is that the —OH groups present on edge and corner sites on this materials, which are stretched in 3D space, are more isolated and accessible for the chiral ligand for greater alignment, whereas the hindered —OHs situated on flat planes in closer proximity with each other present in relatively large numbers on conventional-MgO disable proper alignment for the chiral ligand.

10.3.1.1.2 **Effect of the Size of the Nanoparticles** In some fundamental studies, metal oxide nanoparticles of different sizes were prepared, characterized and tested as heterogeneous catalysts. When the catalytic behaviors of nanoparticles with variable sizes are compared, two main trends emerge. First, the catalytic performance increases with the decrease of the average size of particles [20, 70–72]. This trend is related to the fact that nanoparticles with smaller size have better dispersion and higher surface area, which are factors important for catalysis efficiency. Such behavior has been clearly pointed out by Dong *et al.* [20] for the ozonation degradation of phenol catalyzed by Co_3O_4 nanoparticles with average sizes of 3.5, 6, 11, 19 and 70 nm. Similarly, for water oxidation in alkaline solution, the activities of cubic Co_3O_4 nanoparticles with average diameters of 5.9, 21.1 and 46.9 nm properly correlate with their Brunauer-Emmett-Teller (BET) surface areas [72]. Analogous correlations were found in two very recent studies, which explored the effect of the calcination temperature on ZnO [70] and Ce_2O [71] nanoparticle size and their catalytic activity. It was found that the phenol degradation efficiency under ultraviolet (UV) laser irradiation decreases with the increase in calcination temperature of ZnO due to the agglomeration of particles and the decrease in specific surface area. Also, the CeO_2 nanocubes calcined at 823 K, which possessed smaller crystallite sizes and larger BET surface area, exhibited better *o*-xylene oxidation performance than those calcined above this temperature.

In a second trend, the catalytic activity cannot be directly related to the decrease of the nanoparticle size. In this case, the highest activity is related to an optimal parameter, such as particle size [73], concentration of surface vacancies [74] or catalytic sites. For example, Dodd *et al.* [73] using ZnO nanoparticles with diameters between 28 and 57 nm, reported that there exists an optimum particle size of approximately 33 nm for which the photocatalytic activity is maximized. The existence of this optimal particle size is attributable to an increase in the charge carrier recombination rate, which counteracts the increased activity arising from the higher specific surface area for a sufficiently small particle size. An atypical behavior was found in the case of copper ceria doped nanoparticles, for which the decrease of crystallite size resulted in lower catalytic activity in the methanol and ethylene total oxidation at low temperature [74]. According to the authors, this effect can be related to an optimal concentration of oxygen vacancies and an appropriate defect structure formed during substitution of copper into ceria lattice. Such vacancies and defect structure are necessary for an effective transport of oxygen ions as well as electrons within nanocrystallites.

Martin Alonso *et al.* [75] prepared a series of CaO nanoparticles by thermal treatment of different precursors. Particles with size of 37, 47, 67 and 116 nm were obtained from calcium acetate, oxalate, carbonate and nitrate, respectively. CaO samples (excepting that of 116 nm) displayed comparable surface area values, ranging from 22 to 26 $m^2\,g^{-1}$. These solids showed distinct reaction rates in the catalytic transesterification of triglycerides with methanol: 780 (carbonate), 670 (acetate), 450 (oxalate) and 20 mmol ester$\cdot h^{-1}\,g_{catal.}^{-1}$ (nitrate). CaO obtained by decomposition of calcium carbonate, which catalyzes triglycerides transesterification with higher rates, showed the highest amount and surface density of very strong base sites. As a result, the authors suggest that the key parameter to explain the measured reaction rates consists in the surface density of these basic sites and not in the particle size and surface area.

10.3.1.1.3 **Effect of the Morphology of the Nanoparticles** As discussed above, oxide nanoparticles with different morphologies but comparable particle sizes can be prepared using appropriate procedures. Catalytic data obtained in the presence of such materials have been reported in a series of publications and, generally, they pointed out the dependencies between the surface reactivity and the morphological properties. Some applications and results involving metal oxide nanoparticles are reviewed in this section.

TiO_2 and ZnO are usually used as materials in photocatalytic reactions. For both oxides it has been reported that the morphology and crystal structure have major effect on the photocatalytic activity. Kim *et al.* [76] have prepared titania nanoparticles with diameters between 10 and 36 nm in nonionic water:oil microemulsions. The amorphous as-prepared particles were transformed into the anatase phase at 733 K, and further into the rutile phase at 1123 K. It was found that the anatase phase showed considerable photocatalytic activity for the degradation of phenol, whereas both the amorphous phase and the rutile phase were inactive for this reaction. The same oxide, with nanotube and nanowire shapes has been tested as photocatalysts for the degradation of rhodamine B (RhB) and methyl orange in aqueous solutions, under simulated solar light irradiation [65]. Both TiO_2 nanocatalysts displayed higher degradation activity compared to commercial P25 TiO_2 precursor. Additionally, the nanotubes exhibited activities 10% higher than those exhibited by the nanowire shape titania. Lu *et al.* [77] have evaluated the photovoltaic properties and the photocatalytic activity in the photodegradation of RhB of ZnO samples with two-dimensional (2D) plate-like and 3D flowerlike morphologies. Despite their comparable crystallite sizes and shapes, the corresponding photovoltaic properties and photocatalytic performances are quite different. Thus, the RhB conversion was 42 and 67% for plate-like ZnO and flower-like ZnO, respectively. Compared to the plate-like ZnO, the flower-like ZnO exhibited more efficient separation of photogenerated electron-hole pairs in the UV region, so that the surface redox process by photogenerated electrons and holes may occur more easily, as indicated by the photocatalytic degradation of RhB. The photodegradation of RhB has been also performed over uniform single-crystalline ZnO nanodisks and nanowires having well-developed facets [78]. ZnO nanodisks with a high population of (0001) facet showed better catalytic activity by comparison to that of the ZnO nanowires, indicating that the photocatalytic activity depends on specific crystal planes. Xi *et al.* [79] have prepared Cu_xO ($x = 1, 2$) nanocrystals with variable CuO/Cu_2O ratio and variable shapes, decreasing from sphere to flower and wire. The photocatalytic ability of CuO-dominated nanostructures (sphere and flower morphology) was much higher than that of Cu_2O-dominated nanowires, owing to the absorption of broader wavelength energy due to the narrower band gap of CuO (1.27 eV) than that of Cu_2O (2.0 eV).

In order to identify active sites on ZnO powdered catalysts in methanol synthesis, Wilmer *et al.* [80] prepared and characterized five ZnO samples with different degrees of crystallinity. With respect to catalytic behavior, the experimental results demonstrated that there exist no overall linear correlation between methanol synthesis activity and the BET surface area, indicating that methanol synthesis is a structure-sensitive reaction over ZnO. It was found that only part of the exposed surface is active, and surface imperfection has a negative effect on the activity. According to the authors, the polar ZnO faces are assumed to be highly relevant for the catalytic activity under methanol synthesis conditions.

Recently it has been reported that Co_3O_4 nanocrystals were able to activate the oxidation of CO at low temperature with performances comparable to those of noble metal catalysts [81–83]. The catalytic activity of cobalt oxides strongly depended on the nanocrystal morphology, and more exactly on the crystal planes which allow preferential exposure of catalytically active sites. For example, for CO oxidation under a normal gas stream containing more than 3 ppm H_2O the catalytic activity of the (110) crystal planes of Co_3O_4 nanocrystals was higher than that of either the (100) or (111) planes [81]. Co_3O_4 has a spinel structure containing Co^{3+} in an octahedral coordination and Co^{2+} in a tetrahedral coordination. The former is regarded as the active site for CO oxidation, whereas the latter is almost inactive. Surface differential diffraction studies showed that the (001) and (111) planes contain only Co^{2+} cations, while the (110) plane is composed mainly of Co^{3+} cations. Consequently, at 200 K rod-shaped Co_3O_4 nanocrystals, which mainly consisted of (110) exposed planes, showed much higher catalytic activity than did conventional nanoparticles of this material. When Co_3O_4 nanocrystals having different shapes – plate-like, rod-like, cubical, and roughly spherical – were tested in CO oxidation in a dry gas stream, the plate-like nanocrystals (P-Co_3O_4) mainly exposing (111) planes showed the highest catalytic activity [82]. Indeed, at 200 K their catalytic activity was 9.5×10^{-2} μmol m^{-2} s^{-1}, which is an order of magnitude higher than that of the other nanocrystals. The apparent activation energies (E_a) also depended on the nanoparticle shape, as illustrated in Figure 10.6. P-Co_3O_4 had the lowest E_a among the four samples.

Zhou *et al.* [84] demonstrated that the exposed crystal plane of ceria plays an essential role in determining its catalytic oxidation properties. They prepared single-crystalline CeO_2 nanorods with well-defined crystal planes ((001) and (110)) by a facile solution-based hydrothermal method. These nanorods showed higher CO oxidation activity than the conventional CeO_2 nanoparticles because of their more reactive planes.

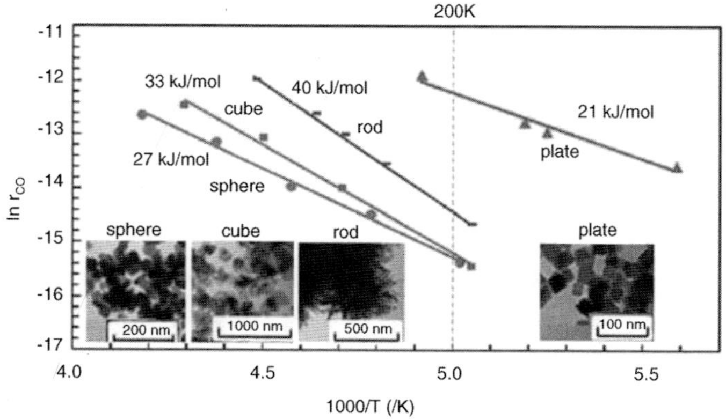

Figure 10.6 TEM images and Arrhenius plots for the rate of CO oxidation over differently shaped Co_3O_4 (Adapted with permission from [82]. Copyright © (2011) The Royal Society of Chemistry).

Co_3O_4 crystals with nanorods and nanopolyhedral morphology were used as catalyst precursors for COS hydrodesulfurization [85]. It has been shown that the catalytic properties of Co_3O_4 nanocrystals were also dependent on the nature of their surface structure and the crystal planes. Cobalt-based nanorods exhibited high activity in presulfurization process and the resultant sulfide is superior to nanopolyhedra in the hydrodesulfurization (HDS) of carbonylsulfide. The significant enhancement in sulfidation activity of the nanorod-shape catalyst is originally attributed to higher reactivity of Co_3O_4 nanorods, which bring on deeper sulfidation and more active sulfur species to the sulfide catalyst than the nanopolyhedra. Rhadfi *et al.* [86] explored the catalytic behaviors of Mn_2O_3 single crystals with oval shape (mean size of 7.9 nm) and rhombohedron shape (mean size of 12.6 nm) for the decomposition H_2O_2 and the degradation of methylene blue. Different kinetic parameters were clearly observed, depending on the particle size and shape. Thus for rhombohedron-like particles, a linear relationship is found when plotting k_{obs} as a function of the nanoparticles concentration, indicating a first-order with respect to the catalyst concentration, that is, $v = k\ [Mn_3O_4][H_2O_2]$. On the contrary, for oval-like sample, the k_{obs} constant did not vary linearly with the Mn_3O_4 concentration, indicating a more complex kinetic rate law. According to the authors, these results could be explained by differences in the nature and/or quantity of surface active sites due to the different morphologies.

MnO_2 nanotubes, nanorods, and nanoparticles prepared using a hydrothermal method, were compared for selective catalytic reduction (SCR) of nitrogen oxides (NO_x) [87]. MnO_2 nanorods exhibited the highest activity for reduction of NO_x at low temperatures (373–573 K). The superior catalytic behavior of this sample was attributed to its low crystallinity, more lattice oxygen, high reducibility, and a large number of strong acid sites. Perez-Ramirez *et al.* [88] investigated the role of (110) and (101) facets of nano-RuO_2 catalyst in NH_3 oxidation. The binding energies of intermediates and products were lower on (101) surface than those on the (110) surface. Additionally, reaction barriers on RuO_2(101) were lowered by about 0.3 eV compared to the RuO_2(110). These differences affected the catalytic activity, but they were not likely to explain the selectivity differences, which were attributed rather to the pressure effect.

Using pure ceria and Pd-loaded ceria nanotubes and nanorods, one-dimensional (1D)-ceria, as catalysts for the water-gas shift (WGS) reaction, Han *et al.* [57] showed that the catalytic activity can be correlated with the crystal faces present in the different nano crystalline morphologies. While bulk ceria is very poor as WGS catalysts, pure 1D-ceria and Pd-loaded 1D-ceria displayed catalytic activity starting at 473 K. This behavior has been attributed to high amount of oxygen vacancies present on CeO_2 nanorods, mainly due to their preferential exposed (110) and (100) facets. Liang *et al.* [89] have prepared MnO_2 nanorods of α, β, γ and δ crystal phases via a hydrothermal method, which were tested as catalysts for CO oxidation. The higher activity of the δ- and α-MnO_2 nanorods was found to be strongly influenced by their crystalline phase. Thus, the catalytic activity decreased in the order: $\alpha \sim \delta > \gamma > \beta$.

Table 10.1 Effect of the nano-oxide morphology over the catalytic activity.

Nano-oxide	Morphology/ shape (facet)	Catalyzed reaction	Activity	Ref.
TiO_2	amorphous	photocatalysis	0% (conversion)	76
	anatase		10%	
	rutile		100%	
ZnO	2D-plate	RhB photocatalysis	42% (conversion)	77
	3D-flower		67%	
ZnO	disk (0001)	RhB photocatalysis	disk > wire	78
	wire			
CuO-Cu_2O	sphere, wire	photocatalysis	sphere > wire	79
Co_3O_4	rod (110)	CO oxidation	4–6 $\mu mol \cdot g^{-1} \cdot s^{-1}$	81
	sphere (100) (111)		very low	
Co_3O_4	plate (111)	CO oxidation	$9.5 \cdot 10^{-2}\ \mu mol \cdot m^{-2} \cdot s^{-1}$	82
	rod		$0.8 \cdot 10^{-2}\ \mu mol \cdot m^{-2} \cdot s^{-1}$	
	cube		$0.7 \cdot 10^{-2}\ \mu mol \cdot m^{-2} \cdot s^{-1}$	
	sphere		$0.7 \cdot 10^{-2}\ \mu mol \cdot m^{-2} \cdot s^{-1}$	
Co_3O_4	belt (011)	CO oxidation	$7.4 \cdot 10^{-3} \cdot s^{-1}$	83
	cube (001)		$2.7 \cdot 10^{-3} \cdot s^{-1}$	
CeO_2	rod	CO oxidation	2 $\mu mol \cdot g^{-1} \cdot s^{-1}$	84
	sphere		0.66 $\mu mol \cdot g^{-1} \cdot s^{-1}$	
Co_3O_4	rod	COS HDS	$T_{50} = 443$ K[a]	85
	polyhedra		$T_{50} = 503$ K[a]	
Mn_2O_3	oval crystal	H_2O_2 decomposition	$4.6 \cdot 10^{-2}\ m^{-2} \cdot min^{-1}$	86
	rhomboedron		$8.9 \cdot 10^{-2}\ m^{-2} \cdot min^{-1}$	
MnO_2	tube	NOx-SCR	3.3 $mL \cdot g^{-1} \cdot s^{-1}$	87
	rod		13.4 $mL \cdot g^{-1} \cdot s^{-1}$	
	particle		5.1 $mL \cdot g^{-1} \cdot s^{-1}$	
MnO_2	rod: α-, β-, δ- and γ-crystal phase	CO oxidation	$\alpha \sim \delta > \gamma > \beta$	89

a) Temperature corresponding to 50% conversion.

To better reflect the role of the nano-oxide morphology on its catalytic performance, we have summarized in Table 10.1 the most important data contained in the publications discussed in this section.

10.3.1.2 Catalysis by Unsupported Oxide Nanoparticles

10.3.1.2.1 **Organic Reactions** As shown above, the catalytic potential of unsupported nanoparticles of metal oxides has been proved in a number of useful organic reactions. Some additional catalytic systems and their applications in both fine and bulk chemical synthesis will be discussed as examples in this section.

Nano-MgO is widely used as a catalyst for a number of organic reactions because of its stability in the unsupported form, as well as the presence of different sites on the surface, including Lewis acid (Mg^{2+}), Lewis base (O^{2-}), isolated hydroxyls groups and cationic and anionic vacancies [90]. Nanocrystalline MgO has been used as a recyclable catalyst in asymmetric Henry and Michael reactions [69], Strecker reaction [91], Claisen–Schmidt condensation [24, 58], the asymmetric aldol reaction [92], the organic carbonates synthesis [93], Meerwein–Ponndorf–Verley reduction [94], and so on. MgO is also a very efficient heterogeneous basic catalyst for synthesis of biodiesel [30, 95]. Thus, in supercritical or subcritical conditions, the complete transesterification of soybean oil was realized in only 10 min.

Bulk oxides, such as Co_3O_4, CeO_2, Fe_2O_3 and Mn_3O_4 are known to be effective catalysts for partial oxidation reactions of organic molecules. Several reports showed that the nanoparticles of these oxides are also able to catalyze the oxidation reaction. For example, Co_3O_4 nanocrystals with average particle size of 30 and 50 nm were successfully employed as catalysts for cyclohexane oxidation to cyclohexanol and cyclohexanone, with molecular oxygen as oxidant [96]. Cyclohexane oxidation, which is a very important industrial application, can be also catalyzed by other nanostructured catalysts, such as Fe_2O_3 and mixed Fe–Co oxide [97, 98]. Recently, Ghiaci *et al.* [99] have reported the efficient allylic oxidation of cyclohexene to the corresponding α,β-unsaturated ketone catalyzed by trimetallic $RuO_2/Co_3O_4/CeO_2$ nanoparticles. This catalytic system may be easily prepared by a reverse micelle approach. Finally, Akhlaghinia *et al.* [100] found that ceria nanoparticles efficiently catalyzed the oxidation of benzylic C—H bonds to carbonyl compounds.

The epoxidation of olefins is an extremely important class of catalytic reactions because of the epoxides versatility in preparing many chemical intermediates. Usually, oxide matrices containing certain metal atoms, including Ti, Mo, V, or W [101, 102], are suitable for use as catalysts for this reaction. In recent reports, nanosize oxide particles, such as Co_3O_4, Mn_3O_4 and Nb_2O_5–SiO_2, have been investigated for olefin epoxidation activation. Uniform sphere-like Co_3O_4 and Mn_3O_4 nanocrystals, prepared by a simple sonochemical method exhibited remarkable performances for the epoxidation of styrene and cyclooctene by anhydrous *tert*-butyl hydroperoxide [103]. Nb_2O_5–SiO_2 nanocomposites containing up to 23.0 wt.% Nb were active and stable catalysts for the epoxidation of cyclooctene [104]. The catalytic properties were related to acid sites and to the presence of NbO_x species with different coordination of niobium cations.

10.3.1.2.2 **Hazardous Compound Abatement** Heterogeneous catalysis plays an innovative role in the development of new technologies to prevent and control the polluting emissions. Catalysis, especially that based on noble metals and metal oxides, is one of the most powerful tools used in gas and liquid effluent treatments, especially when it is based on noble metals and metal oxides. In this section, representative catalytic results exhibited by nano-oxides in catalytic combustion, photocatalysis, pollutant degradation by Fenton-like reaction and ozonation will be briefly discussed.

Catalytic combustion technology is suitable to control the emissions of volatile organic compounds and carbon monoxide. Although the supported Pt and Pd nanoparticles are usually employed in this application, many research groups reported the interesting catalytic behavior of the less expensive catalysts, such as nanoscale metal oxides. Chen *et al.* [105] used monodisperse MnO_2 hollow nanospheres as effective catalysts for the complete oxidative conversion of formaldehyde to CO_2 and H_2O at low temperatures. A very promising catalyst for the low-temperature oxidation of carbon monoxide and methane, consisting of ultra-fine particles of ferric oxide (average size of 3 nm and specific surface area of $250\,m^2\,g^{-1}$) was proposed by Kwon *et al.* [106]. Thus, over nano-Fe_2O_3 CO is completely converted into CO_2 at 623 K. Under similar conditions, the α-Fe_2O_3 powder (3 μm) can only catalyze oxidation of less than 40% of CO.

Nanoscale CeO_2 doped with different metal oxides were successfully used for CO oxidation. For example, Ce-Pr-oxide solid solutions with high specific surface area exhibited high oxygen storage capacity and CO oxidation activity [107]. Nanosized ceria–zirconia, ceria–hafnia and ceria–terbia solid solutions were also used as catalysts for CO oxidation [108]. Among the investigated samples, the ceria–hafnia catalyst exhibited the best catalytic potential. Ce-Cu-Pr mixed oxides prepared by the sol–gel method showed higher activity in CO conversion than the pure nano-CeO_2 [109]. It was found that the enhanced reactivity was closely related to the higher concentrations of the oxygen vacancies and the chemisorbed CO in the catalysts. Mai *et al.* [110] prepared a series of $Ce_{1-x}Cu_xO_2$ nanocomposite catalysts with various copper contents by a simple hydrothermal method at low temperature without any surfactants. The $Ce_{1-x}Cu_xO_2$ nanocomposites exhibited higher catalytic activity in the CO oxidation compared to pure CeO_2 and CuO catalysts. The excellent catalytic performance was associated with the characteristics of nanocomposites and synergic effect between the redox couples of Cu^+/Cu^{2+} and Ce^{3+}/Ce^{4+}. The preferential CO oxidation (PROX) is a process by which the residual CO in hydrogen destined to fuel-cells is catalytically oxidized to CO_2. Besides noble metals, metal oxides are usually employed as catalysts in this process. One of the most active and selective metal oxides in this process is CuO, whose activity is significantly increased by the inclusion of cerium oxide in the catalyst composition [111].

Photocatalysis is an effective way for removing organic pollutants in both gas and liquid phases. TiO_2 is by far the most known and investigated photocatalyst, due to its chemical stability, nontoxicity and well positioned valence and conduction bands. To implement real life application of this photocatalyst, considerable research effort has been invested into increasing the photocatalytic activity of TiO_2, particularly by enhancing its ability to be activated by visible light. Two general approaches have been developed: the modification of the chemical composition and the control of the particle size. By reducing the particle size at the nanometric length scale, the band gap and the position of the valence and conducting bands in TiO_2 are strongly modified, with beneficial effect on the photocatalytic activity. For example, the titanium oxide species with very small particles exhibit absorption bands in the wavelength regions shifted to shorter wavelength regions as compared with those of the bulk TiO_2 catalyst [112]. Such a shift to the shorter wavelength (attributed to the size quantization effect) strongly improves the photocatalytic efficiency and the reactivity of nano-Ti-oxide particles. Similar spectroscopic changes were observed for ZnO nanoparticles ($d < 5$ nm) compared to the bulk material [113]. The photocatalysis basic principles and the applications of nano-oxides in the photocatalytic process have been extensively discussed in some excellent reviews in the literature. The reader is referred in particular to those by Aprile *et al.* [36], Wells [114], Colon-Ibanez *et al.* [115], and Chaturvedi *et al.* [116].

In order to destroy the refractory organic compounds from contaminated waters, advanced chemical oxidation processes based on the Fenton-like reaction or the ozonation reaction are usually used. Oxide-based nanoparticles were tested as efficient catalysts for both applications. For example, ethylene glycol and phenol were efficiently destroyed at room temperature by the Fenton-like reaction using iron (III) oxide-based nanocatalysts in the presence of hydrogen peroxide without UV or visible radiation sources [117]. Mn_3O_4 nanocrystals were highly active in the degradation of methylene blue by the Fenton-like reaction. The particles were true heterogeneous catalysts, which can be recycled without loss of activity and without noticeable morphologic modification. [86]. Recently, Polzer *et al.* [118] reported the oxidative degradation of organic dyes with H_2O_2 catalyzed by ultrathin manganese oxide nanoparticles immobilized on spherical polyelectrolytes brushes.

Nano-oxides also showed remarkable performance in the degradation of organic pollutants by ozonation. Cerium oxides with distinct morphologies and particle size were efficiently used as catalysts in the conversion of oxalic acid, aniline and a reactive dye [119]. Co_3O_4 nanoparticles with average sizes of 3.5, 6, 11, 19 and 70 nm exhibited remarkable catalytic behavior for the ozonation degradation of phenol [20]. Jung and Choi [120] found that nanosized ZnO is able to catalyze the degradation of *p*-chlorobenzoic acid.

10.3.1.3 Catalysis by Supported Oxide Nanoparticles

During the catalytic applications, the very small size of the nanoparticles can involve serious problems, including mass transport limitations, excessive pressure drops, difficulties in separation and reuse, and so on. An efficient mode to overcome these drawbacks is to support the nanoparticles on materials with larger size particles. Various supports have been used to stabilize metal or oxide nanoparticles: synthetic and natural polymers, sand, nonporous oxides, porous oxides, or carbon-based materials. Among them, ordered porous oxides and nanotube-type materials appeared to be the most appropriate supports for oxide nanoparticles.

10.3.1.3.1 Nano-oxides Supported on Ordered Mesoporous/Microporous Materials

Ordered microporous and mesoporous materials have been used extensively as catalysts and catalyst supports in various chemical applications [121, 122]. The advantage of using these solids in catalysis is their high surface area, large pore volume, and especially their well-defined pore architecture. It was found that these unusual porous systems are well suited to stabilize nanoparticles of different metals [123–125] or oxides [122]. Therefore, the deposition of catalytically active nanoparticles on ordered porous materials became an important approach for the design of useful catalysts. Although most publications focused on metal nanoparticles confined in ordered porous materials, interesting results have been reported on solid catalysts involving oxides nanoparticles. According to these studies, the oxide nanoparticles are mainly confined inside the pores of mesoporous materials and zeolites possessing cavities (ex. faujasite), while for most of zeolites these particles are located rather on the external surface of crystals. Usually the particle size and dispersion, as well as the thermal stability of oxides supported on ordered porous materials have been superior to those of non-ordered porous supports. Table 10.2 summarizes catalytic systems and their applications reported in recent publications.

These data illustrate the diversity of catalysts prepared from different nano-oxides and ordered porous supports. In addition, these materials have proved interesting performances for several catalytic reactions, including partial or total oxidation of organic compounds, Friedel–Crafts reactions, biodiesel production, CO, SO_2 and NO_x abatement, photocatalytic degradation of pollutants, and so on. We will briefly discuss only selected examples, in order to emphasize the favorable effect of the support properties (surface area, pore system and volume) on the oxide nanoparticles and their catalytic behavior. Martinez *et al.* [134] prepared iron-containing catalysts following different synthetic routes and silica supports (amorphous, zeolitic and mesostructured materials). Activity and stability of these materials were assessed on the photo-Fenton degradation of phenolic aqueous solutions using near UV irradiation at room temperature and initial neutral pH. Activity and stability depended on the environment of iron species and features of silica support. A nanocomposite material of crystalline iron oxides supported over mesostructured SBA-15 material was shown to be the most active catalyst for the degradation of phenolic aqueous solutions, achieving an outstanding overall catalytic performance accompanied with a noteworthy stability. A highly active heterogeneous Fenton catalyst was also prepared by impregnating iron oxide nanoparticles in alumina coated mesoporous SBA-15 silica [131]. The alumina layer supported on silica played a crucial role in the high activity, by increasing the dispersion of the iron oxide nanoparticles and facilitating the redox cycle of iron species. Using a low-temperature route, Srinivasan and Bandyopadhyaya [132] prepared very small SnO_2 nanoparticles (3.5 nm) inside of 6.3 nm pore size SBA-15 silica. The fast photodegradation rate of RhB achieved with this catalyst indicated the good accessibility of SnO_2 nanoparticles by Rh B molecule.

Nanosized NiO-doped cage cubic Pm3n mesoporous silica materials with high Ni contents (Ni/Si$\sim$1) were used as catalysts in model oxidation reactions of organic amines [133]. Among all the catalysts used, the ordered cage catalyst with open, uniform pore architectures, high surface area and large pore volumes allowed efficient adsorption and diffusion of organic amines to the active site of NiO clusters, leading to a high degree of conversion and a high reaction rate. Nanostructured Mn oxide clusters supported on mesoporous silica KIT-6 are very

Table 10.2 Nano-oxides supported on porous materials: catalytic systems and applications.

Nano-oxides	Porous supports	Catalytic applications	Ref.
Co_3O_4	SBA-15	Water dissociation	59, 126
Co_3O_4, NiO, CeO_2	SBA-15	Hydrodechlorination of chlorobenzene	41
Mn_3O_4, Co_3O_4	SBA-15	Methane combustion	127
Fe_2O_3, Cr_2O_3	SBA-15	Methanol decomposition	128
TiO_2	SBA-15	Photocatalytic degradation of rhodamine 6G	129
MoO_x	SBA-15	Partial ethane oxidation	130
Iron oxides	SBA-15	Photo-Fenton oxidation	131
SnO_2	SBA-15	Photocatalytic degradation of rhodamine B	132
NiO	Cubic Pm3n mesoporous silica	Oxidation of organic amines	133
Iron oxides	SBA-15 silica, zeolites amorphous silica	Photo-Fenton oxidation of phenol	134
In_2O_3, Ga_2O_3, ZnO	MCM-41	Benzoylation of benzene by benzoyl chloride	135
CeO_2	MCM-41	Oxidative dehydrogenation of alcanes	136
Fe_2O_3	MCM-41	Oxidation of benzyl alcohol to benzaldehyde	137
Iron oxides	Fe-MCM-41	Sulfur dioxide oxidation	138
In_2O_3, Ga_2O_3	MCM-41, zeolites, silica-alumina, alumina, zirconia	Benzylation of benzene with benzyl chloride	139
Iron oxides	HMS (hexagonal mesoporous silica)	N-Alkylations of aniline with benzyl alcohol	140
Manganese oxides	Mesoporous silica KIT-6	Water oxidation	141
γ-Fe_2O_3	Y faujasite	Olefin epoxidation	142
RuO_2	Y faujasite	Aerobically oxidize of alcohols to aldehydes and ketones	66
TiO_2	Y faujasite	Photocatalytic degradation of aquatic pollutants	143
CaO, MgO	X faujasite	Transesterification of sunflower oil	144
RuO_2	ZSM-5	Olefine hydrogenation	145
RhOx	Beta, ZSM-5	NOx reduction	146
TiO_2	Beta	Photocatalytic oxidation of methanol	147

efficient water oxidation catalysts under mild conditions [141]. The high surface area silica support seams to be critical for the integrity of the catalytic system by offering a perfect, stable dispersion of the nanostructured Mn oxide clusters. In addition, the silica environment plays an important role for maintaining activity by protecting the active Mn centers of the catalyst from deactivation by surface restructuring.

Iron oxide supported on mesoporous MCM-41 silica was very active in the oxidation of SO_2 [138]. This catalyst showed higher activity compared with the iron oxide supported on a commercial high surface area amorphous silica support. Also, MCM-41 was the best support for In_2O_3 in the acylation of benzene [139]. Indeed, under similar conditions, the use of supports such as zeolites, silica-alumina, alumina, or zirconia leads to lower activity. Gonzalez-Arellano *et al.* [140] explored the catalytic performances in N-alkylation of aniline of supported iron oxide nanoparticles on a variety of porous materials. Among the reported catalysts, iron oxide supported on hexagonal mesoporous silica was found to be highly active. In addition, this catalyst preserved >90% of their initial activity and selectivity after three reuses. No Fe-leaching was detected in the final reaction mixture upon the recycling of the catalyst.

Superparamagnetic nanosized crystals of γ-Fe_2O_3 were synthesized onto the surface of the NaY-zeolite by immobilizing a polynuclear iron complex and subsequent calcination in oxygen [142]. This material showed a remarkable catalytic efficiency in epoxidation reactions with various olefins using *tert*-BuOOH as oxidant. A synergetic Fe–Al effect in Fe_2O_3 nanoparticles supported on MCM-41 mesoporous aluminosilicates compared with pure siliceous silicates has been demonstrated by a remarkably superior catalytic activity of the former in the microwave-assisted selective oxidation of benzyl alcohol to benzaldehyde [137]. Nakatsuji *et al.* [146] showed that the morphology of Rh and RhO_x nanoparticles supported on zeolites depended on the type of zeolite and synthesis conditions. Thus, the nanoparticles loaded on the Na-beta zeolite were highly dispersed particles, whereas those dispersed on Na-ZSM-5 were agglomerated particles. The nanoparticles supported on Na-ZSM-5 catalyzed NO*x* reduction more efficiently in lean conditions using hydrogen and CO than those supported on the Na-beta zeolite.

10.3.1.3.2 **Oxide Nanoparticles Supported on Nanotubes** Materials with nanotubular morphology are widely studied due to their properties, including the mesopore-range internal diameter (about 4–10 nm), the unique combination of physico-chemical performances, and their potential applications in various fields [148, 149]. Carbon and titania nanotubes have also showed interesting performances as supports for metal and oxide nanoparticles (see also Chapters 9 and 11 of this book). The few examples about the metal nano-oxides which are described below confirm this assertion. RuO_2 nanoparticles (1.35 nm) supported by herringbone carbon nanotubes (CNTs) have been synthesized by a homogenous-oxidation precipitation method with H_2O_2 [150]. CNTs exhibited better ability to induce the formation of highly dispersed nanoparticles, compared to common used supports, such as γ-Al_2O_3 and activated carbon. Additionally, RuO_2/CNTs exhibited high activity and excellent selectivity in the aerobic oxidation of alcohols to aldehydes or ketones. Catalysts consisting of CeO_2 nanoparticles located on the outer surface or in the tubular channel of CNTs were used for the SCR of NO with ammonia [151]. The NO conversion was related to the specific surface area, the crystal size of CeO_2 and the dispersion of CeO_2. Qin and Li [152] investigated the catalytic behavior of CuO nanoparticles confined in CNTs using both molecular dynamic simulations and ozone decomposition probe reactions. They found that the effects of the confined environment of CNTs induced more reactants into the channel. Interface interactions between reactants and CuO nanoparticles in the channel and charge transfer accelerated the decomposition of ozone into oxygen molecule and atomic oxygen species.

Bavykin *et al.* [153] deposited Ru(III) hydrated oxide (1–2 nm) onto the surface of TiO_2 nanotubes via ion exchange followed by alkali treatment. High-resolution transmission electron microscopy and XRD data showed that ruthenium is highly dispersed on both the inner and the outer surfaces of the mesopores. The catalytic activity of Ru(III)/TiO_2 catalyst for the selective oxidation of alcohols by oxygen was studied in a continuous multichannel, compact reactor. The turnover frequency (TOF) reached $450\,h^{-1}$, greater than that of the Ru(III)/Al_2O_3 catalyst (TOF = $335\,h^{-1}$). Titanium nanotubes-confined ceria were used for SCR of NO with ammonia. In comparison with the catalysts supported by TiO_2 nanoparticles, the confined ceria showed a superiority in this reaction due to the improved redox potential and special adsorption of NH_3 [154].

10.3.1.4 Oxide Nanocatalysts for Green Chemistry

Green chemistry is a concept that promotes and encourages chemical applications that minimize the use and generation of hazardous substances and maximize the raw material/energy efficiency. Heterogeneous catalysis has proven to be one of the most important tools for the accomplishment of such principles. Recently, Polshettiwar and Varma [155] reviewing a number of publications demonstrated that the nanomaterials represent an important group of catalysts for green chemistry development. This review paper mainly focused on the metal nanoparticles and their behavior for processes performed in aqueous medium, and under nonconventional microwave heating. We selected here some representative applications, in order to highlight the catalytic potential of metal oxide nanoparticles for cleaner synthesis.

Zhan *et al.* [66] have reported one of the earliest examples in this field. They found that nano-RuO_2 particles confined in Y zeolite cages showed much more activity than the bulk RuO_2 catalyst for the oxidation of alcohols to the corresponding aldehydes and ketones. In processes performed under aerobic conditions, without using either a cocatalyst or a sacrificial reducing agent, the nano-RuO_2 particles displayed extraordinarily high activity and selectivity.

Beller and co-workers [68] reported the performance of magnetically recyclable γ-nano-Fe_2O_3 catalysts in oxidation reactions with hydrogen peroxide used as a clean oxidizing agent. Alcohols and aromatic olefins were oxidized to the corresponding aldehydes with activities closed to that of the homogeneous Fe^{3+} catalyst, in liquid-phase processes, under very mild conditions. They also proposed an original, environmentally benign method for the direct coupling of sulfonamides and alcohols in the presence of magnetite-immobilized nano-Ru catalysts [156]. Thus, a domino dehydrogenation-condensation-hydrogenation sequence of reagents has been realized with excellent yields and selectivity in a process giving only water as side product. The catalyst was easily isolated and reused because of its magnetic property.

In recent publications it was shown that a number of multistep organic syntheses can be easily performed through one-pot protocols thanks to the use of metal nano-oxide catalysts. For example, two different solvent-free green methods for α-amino-phosphonates synthesis, based on recyclable nano-oxide catalysts, have been developed by Redy *et al.* [157] and Agawane and Nagarkar [60], respectively. In the first case, the one-pot synthesis of α-aminophosphonates from aldehyde, amine, and diethyl phosphate has been efficiently performed with nano Fe_3O_4, a magnetically recyclable and economically viable catalyst. In the second one, nano CeO_2 catalyzed with high yield and selectivity the ultrasound irradiation assisted condensation of aldehyde, amine and triethyl phosphite.

Rahman *et al.* [158] reported the application of nano In_2O_3 as a nontransition metal oxide catalyst for the synthesis of propargylamines. With this catalyst, a simple and efficient one-pot synthesis, involving terminal alkyne, dichloromethane, and secondary amine could be carried out under mild conditions. It was also possible to elaborate robust protocols for organic synthesis by using native silica nanoparticles as catalysts. Thus, the anti-Markovnikov addition of thiols to inactivated alkenes and alkynes, at room temperature, under solvent-free conditions, was proposed as new green route to linear and vinyl thioethers [159]. The same group reported that nanosized silica was an excellent catalyst for the bis-Michael addition of α,β-unsaturated ketones with active methylene compounds performed in one step, at room temperature [160]. Using ZnO nanoparticles as catalyst, Dharma *et al.* [161] described an efficient and environmentally benign one-pot solvent-free procedure for the synthesis of oxazinone derivatives.

10.3.2
Nano-oxides as Supports for Active Phases

Metal oxides with porous structure, including Al_2O_3, TiO_2, Fe_2O_3, ZrO_2, CeO_2, and so on, are extensively used as supports to manufacture robust heterogeneous catalysts for organic and inorganic reactions. Their particular properties are the high surface area combined with unusually chemical, mechanical and thermal stabilities. In most cases, the physico-chemical properties of the support strongly influence the catalyst performance. As these properties frequently depend on the particle morphology and size, it is expected that the metal oxide with nanosized particles exhibit particular catalytic properties, when used as supports. Indeed, in a number of recent studies it has been reported that the oxide nanosupports offer new opportunities for the design of supported metals or oxides with potential use in catalysis. Part of these results is summarized in Table 10.3.

Table 10.3 Overview of the reactions catalyzed by active phases on various nano-oxide supports.

Catalyst	Preparation Method of Nanosupport	Catalytic Application	Ref.
Au/CeO_2	Controlled aging in acid aqueous media of $Ce(NO_3)_3$	Aerobic oxidation of alcohols	162
Au/CeO_2	Alginate template procedure and supercritical CO_2 drying	Water photocatalysis/visible light	29
Au/Co_3O_4	Solvothermal synthesis from cobalt oxalate and hydrothermal synthesis from cobalt acetate	Complete oxidation of ethylene traces	163
Fe_2O_3/ZrO_2	Sol–gel	Selective catalytic reduction of NOx	164
Au/TiO_2	Sol–gel from Ti(IV) isopropoxide	Hydrogen production from ethanol under photo-irradiation	165
Pd/TiO_2	Solvothermal sol–gel of from Ti *n*-butoxide	Liquid-phase selective hydrogenation of phenylacetylene to styrene	166
Au/Fe_2O_3	Nucleation-oxidation	H_2O_2 reduction	167
NiO/CeO_2, ZrO_2 and $Ce_{0.8}Zr_{0.2}O_2$	Co-precipitation	Carbon dioxide reforming of methane	168
Co_3O_4/ZrO_2, CeO_2 and $Ce_{0.8}Zr_{0.2}O_2$	Reverse microemulsion	CO preferential oxidation	169
Fe/MgO	Sol–gel	Removal of H_2S in a biogas	170
Pt/Al_2O_3	Flame spray pyrolysis	Enantioselective hydrogenation	171
Pd/Al_2O_3	Flame spray pyrolysis	Enantioselective hydrogenation	172
Pt-Ba/Al_2O_3	Flame spray pyrolysis	Lean-NOx storage-reduction	173
Multi-noble metal particles/Al_2O_3	Flame spray pyrolysis	Partial oxidation of methane	174
Pt-Sn/Al_2O_3	Flame spray pyrolysis	Dehydrogenation of propane	175
Co/ZrO_2	Flame spray pyrolysis	CO hydrogenation	176
Ag/ZnO	Flame spray pyrolysis	UV-photodegradation of methylene blue	177
RuO_2/SnO_2, ZrO_2, TiO_2, Al_2O_3, SiO_2	Chloride hydrolysis or commercial materials	Selective oxidation of methanol and ethanol	178
Ru(III) hydrated oxide/TiO_2 Nanotubes	Alkali hydrothermal transformation of TiO_2	Selective oxidation of alcohols by oxygen	153
Au/TiO_2 nanotubes	Alkali hydrothermal transformation of TiO_2	CO oxidation	179
Au/TiO_2 or Ag/SnO_2	Colloidal particles	Photocatalysis	180

(continued)

Table 10.3 (*Continued*)

Catalyst	Preparation Method of Nanosupport	Catalytic Application	Ref.
Li/CaO	Commercial CaO	Biodiesel production from karanja and jatropha oils transesterification	181
KF/CaO	Commercial CaO	Biodiesel production from Chinese tallow seed oil transesterification	182
WO_3/TiO_2	Commercial TiO_2	UV-photocatalytic decomposition of 1,4-dichlorobenzene	183
Fe/TiO_2	Commercial TiO_2	Oxidation of cyclohexane with O_2	98
$Co/Al_2O_3–ZrO_2$	Flame spray pyrolysis (ZrO_2); commercial Al_2O_3	CO hydrogenation	184
$Pt/ZrO_2/SiO_2$	Deposition–precipitation	Reforming of methane with CO_2	185
CeO_2/ZnO	Homogeneous precipitation + micro-emulsion	Oxidative coupling of methane with CO_2	186
Ce or La/γ-Al_2O_3 doped with CuO	Chemical precipitation combined with supercritical CO_2 drying	CO oxidation	187
CuO/CeO_2	Alcohothermal synthesis	CO oxidation	188
Au/CeO_2 nanorods	Surfactant-assisted hydrothermal route	Low-temperature water-gas shift reaction	189
Au/TiO_2 nanotubes	TiO_2 reaction with NaOH	Low temperature water-gas shift reaction	190
Pd/CeO_2, ZnO, TiO_2	Organometallics chemical vapor deposition	Selective butadiene hydrogenation	191

Most of publications pointed up the main characteristics of these catalytic systems and their attractive performances in various chemical applications. Aspects, such as the effect of the particle size of the support on the catalytic activity have not been studied in a systematic way. For instance, the beneficial effect of the nanometer-sized supports as compared to bulk supports was revealed only in few publications. According to these studies the remarkable catalytic effect exhibited by the nanosupports can be attributed to their textural (i.e., high surface area), morphological or electronic properties. Here we discuss some representative results.

Nadeem *et al.* [165] investigated the photoconversion of ethanol into hydrogen on Au nanoparticles supported on TiO_2 with micro and nanoparticle size. The nanosized catalyst shows a ca. one order of magnitude higher hydrogen production per unit mass compared with the micro-sized system. Moreover, the induction period of the nanosized particles is less than that corresponding to the micro-sized particles (Figure 10.7).

The activities of the two catalysts were equivalent for a normalization based on surface area, indicating that while nanoparticles of TiO_2 enhance the reaction, this effect is merely geometrical and does not result from any changes to intrinsic

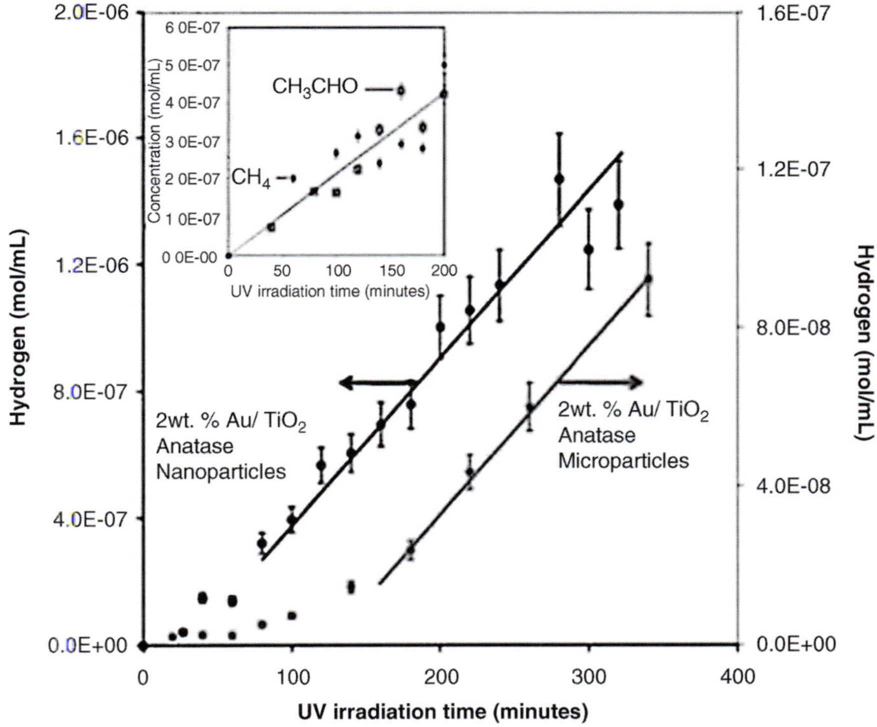

Figure 10.7 Hydrogen formation as a function of time during the photocatalytic reaction of ethanol over 2 wt.% Au/TiO_2 anatase nanoparticles and 2 wt.% Au/TiO_2 anatase microparticles at 300 K (Reprinted with permission from [165]. Copyright © (2010) Elsevier).

electronic properties. Similarly, the nanosized ZrO_2 support in Fe_2O_3/ZrO_2 catalyst showed better activity compared to bulk ZrO_2 in the SCR of NO*x* [164]. For instance at 523 K, the apparent TOF was $6.0\,10^{-4}\,s^{-1}$ for the nanoscaled catalyst and only $0.6\,10^{-4}\,s^{-1}$ for the bulk material. In this case, the higher SCR activity of the nano-ranged Fe_2O_3/ZrO_2 catalyst as compared to the bulk sample was also evidenced when the SCR rate was referred to the surface area of the substrate. More recently, Primo *et al.* [29] have reported the unprecedented photocatalytic activity of 1 wt% Au-supported CeO_2 nanoparticles for oxygen generation from water under visible light irradiation. The photocatalytic activity strongly depended on the ceria particle size. Thus, the ceria sample prepared by a biopolymer template procedure and having a particle size of 5 nm was a far more efficient support than the commercial CeO_2 with a larger particle size.

Weerachawanasak *et al.* [166] showed that in the case of Pd/TiO_2 catalyst, the particle size of the support influenced the strength of the metal-support interaction. Thus, after reduction at 773 K, strong interactions were found for the nanosized TiO_2 supported Pd catalyst, but not for the micron-sized TiO_2 supported one. The strong interaction produced a great beneficial effect on the catalyst performance in the liquid-phase selective hydrogenation of phenylacetylene to styrene. It was suggested that during high-temperature reduction, the inner Ti^{3+} (responsible for promoting strong metal–support interaction) in the large crystallite size TiO_2 was more difficult to diffuse to the Pd^0 surface than the surface Ti^{3+} in the smaller crystallite size ones. Moreover, sintering of Pd^0 metal species was observed over the micro-particles.

Gold nanoparticles (2–5 nm) supported on nanocrystalline CeO_2 (about 5 nm) were particularly active, selective, and recyclable catalyst for the oxidation of alcohols using molecular oxygen at atmospheric pressure under solvent-free and base-free conditions [162]. According to the authors, the key point concerning the excellent activity is the presence of the surface oxygen vacancies, a structural feature that arises from the nanometric size of ceria, and the presence of positive gold species in a cluster that contains many Au^0 atoms. These special properties reveal the synergetic metal-support effect in the oxidation reaction activation, as suggested in Figure 10.8.

Similar effect has been pointed out by Lee *et al.* [167] for a catalyst consisting of 6 nm Au particles supported on 17 nm Fe_3O_4 particles (Figure 10.9). By studying the H_2O_2 reduction catalyzed by the individual Au and Fe_3O_4 nanoparticles, they demonstrated that the enhanced catalysis of Au/Fe_3O_4 arises from the polarization effect at the Au/Fe_3O_4 interface, where Fe_3O_4 becomes more active.

Not only the size but also the morphology of the nanoparticle can influence its catalytic activity. Xue *et al.* [163] tested Au/Co_3O_4 catalysts for the complete oxidation of ethylene. The Au/Co_3O_4-nanorods catalyst exhibited the highest ethylene conversion (93.7% at 273 K), whereas gold dispersed on Co_3O_4 nanopolyhedra and nanocubes showed lower catalytic activities. This difference in catalytic activities has been attributed to the nature of exposed crystal planes of Co_3O_4 supports. HRTEM revealed that the Co_3O_4-nanorods predominantly exposes (110) planes, while the dominant exposed planes of Co_3O_4–nanopolyhedra and –nanocubes are

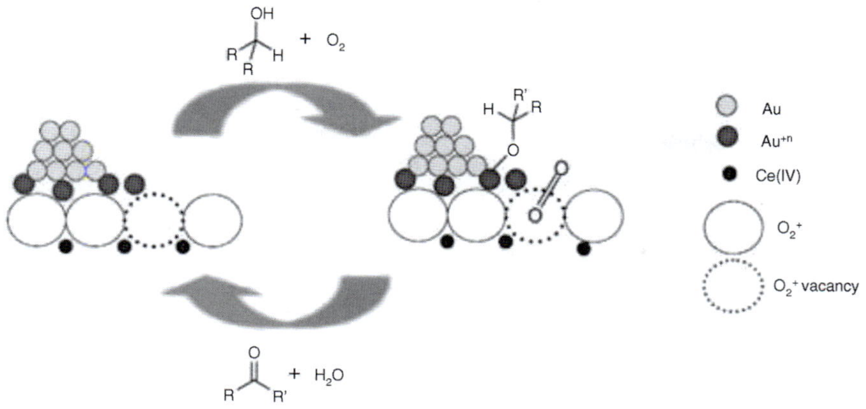

Figure 10.8 Proposed mechanism for aerobic alcohol oxidation on Au/CeO_2 catalyst (Adapted with permission from [162]. Copyright © (2006) Elsevier).

(011) and (001) planes, respectively. Compared to the (011) and (001) planes, the (110) planes exhibit the maximum amount of oxygen vacancies, which play a major role in ethylene oxidation. Si and Flytzani-Stephanopoulos [192] have prepared gold–ceria catalysts by depositing gold on the different facets of ceria nanorods, nanocubes, and nanopolyhedra in a two-step process that involves controlled hydrothermal synthesis of the cerium oxide shapes followed by deposition/precipitation of gold. A strong shape/crystal plane effect of CeO_2 on the gold–ceria activity for the WGS reaction reaction has been identified. The rodlike ceria was most active than other samples for gold stabilization/activation. This was attributed to their exposed (110) and (100) facets, which appeared to be energetically more favorable for introducing oxygen vacancies that stabilize the metallic phase during catalytic reaction.

Liu and Iglesias [178] reported the high catalytic performance of RuO_2 clusters supported on nano-SnO_2, ZrO_2, TiO_2, Al_2O_3, and SiO_2 in the selective alcohol

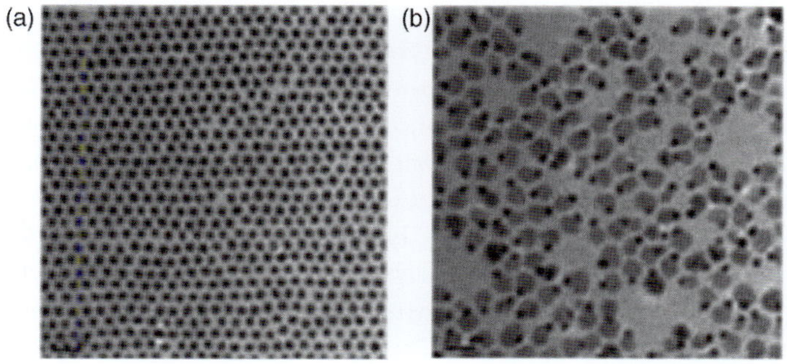

Figure 10.9 TEM images of the as-synthesized (a) 6 nm Au nanoparticles, (b) 6–17 nm Au-Fe_3O_4 nanoparticles [167]. (Reproduced with permission from [167]. Copyright © (2010) Wiley-VCH Verlag Gmbh).

oxidation. The methanol can be converted into formaldehyde, methylformate, and dimethoxymethane with high rates and selectivity (>99%) at low temperatures (300–400 K). Ethanol also reacts to form predominately acetaldehyde and diethoxyethane. It was found that the acid-base properties of the support influence the RuO_2 reducibility, and thus the type of intermediates, the turnover rates and the selectivities. For example, acid sites on Al_2O_3 and SiO_2 favor dimethoxymethane formation from methanol, while redox and amphoteric sites on SnO_2, ZrO_2, and TiO_2 preferentially form methylformate.

A special class of nanosupports consists of magnetic nanoparticles, which are key components for the magnetic nanocatalyst design and preparation (see also Chapter 8 of this book). In a typical preparation protocol, the magnetic nanoparticles (commonly based on iron oxides) are modified either with organic functions or covered with polymers and oxides. In order to generate the catalytic sites, the resulted hybrid or composite nanomaterials are functionalized with desirable chemical species (metals, metal complexes, oxides, enzymes). These catalysts, which can be easily separated from the reaction mixture with a simple external magnet, are promising candidates for cleaner chemical applications. Therefore, although 'magnetically recoverable nanocatalysts' is a recent topic, a large number of studies devoted to their preparation and application have already been reported. Moreover, these publications have been overviewed in a series of extensive review articles [155, 193–197] that we recommend to readers.

10.4
Conclusions and Perspectives

This chapter, focusing on the catalytic behavior of metal nano-oxides, shows through a number of examples the high potential of these materials in various chemical applications. The 'size' and 'shape' effects on the catalytic properties of nano-oxides have been discussed in this paper, and we can reach some conclusions: (i) in recent years, enormous progress has been made in both the synthesis of metal oxide nanostructures and the development of new applications catalyzed by nano-oxides; (ii) compared with its bulk counterpart, the oxide nanoparticles exhibit new and unique physical and chemical properties; (iii) generally, the nanosized oxide showed superior catalytic behavior compared with that exhibited by the bulk oxide; (iv) in most cases, the catalytic performance has been found to be dependent upon particle size and morphology; and (v) much more research is required to produce the knowledge necessary to better understand the unusual properties of nano-oxides and to use them in order to design the 'ideal' catalyst (the most active and selective).

Until now, most of the catalytic applications of nano-oxides which have been reported relate to the domain of fine chemical syntheses, but there are favorable perspectives for their extension into bulk chemicals, where an important contribution will have the nano-oxides supported on ordered micro- and mesoporous materials. At the same time, very interesting perspectives can be noted for nano-oxides in the field of hazardous compound abatement, ranging from small scale applications (small plants and devices) up to those at large scale (e.g., waste water treatment). Additional challenges include scale-up and safety of large-scale nano-oxide production processes.

References

1 Fierro, J.L.G. (2006) *Metal Oxides: Chemistry and Applications*, CRC Press, Boca Raton, FL.

2 Ertl, G., Knözinger, H., Schüth, F., and Weitkamp, J. (eds) (2008) *Handbook of Heterogeneous Catalysis*, Wiley-VCH Verlag GmbH, Weinheim, Germany.

3 Jackson, S.D. and Hargreaves, S.J. (eds) (2009) *Metal Oxide Catalysts*, vol. **1**, Wiley-VCH Verlag GmbH, Weinheim, Germany.

4 Hu, J., Chen, L., and Richards, R. (2009) Properties, synthesis and applications of highly dispersed metal oxide catalysts, in *Metal Oxide Catalysis* (eds S.D. Jackson and S.J. Hargreaves), Wiley-VCH Verlag GmbH, Weinheim, Germany, pp. 613–664.

5 Rodriguez, J.A. and Fernández-García, M. (eds) (2007) *Synthesis, Properties, and Applications of Oxide Nanomaterials*, John Wiley & Sons, Inc., Hoboken, USA.

6 Cushing, B.L., Kolesnichenko, V.L., and O'Connor, C.J. (2004) Recent advances in the liquid-phase syntheses of inorganic nanoparticles. *Chem. Rev.*, **104**, 3893–3946.

7 Eriksson, S., Nylen, U., Rojas, S., and Boutonnet, M. (2004) Preparation of catalysts from microemulsios and their applications in heterogeneous catalysis. *Appl. Catal. A*, **265**, 207–219.

8 Jun, Y.W., Choi, J.S., and Cheon, J. (2006) Shape control of semiconductor and metal oxide nanocrystals through nonhydrolytic colloidal routes. *Angew. Chem. Int. Ed.*, **45**, 3414–3439.

9 Niederberger, M. (2007) Nonaqueous sol-gel routes to metal oxide nanoparticles. *Acc. Chem. Res.*, **40**, 793–800.

10 D'Souza, L. and Richards, R. (2007) Chapter 3. Synthesis of nano-oxide nanoparticles: Liquid-solid transformations, in *Synthesis, Properties, and Applications of Oxide Nanomaterials* (eds J.A. Rodriguez and M. Fernández-García), John Wiley & Sons, Inc., Hoboken, USA.

11 Byrappa, K. and Adschiri, T. (2007) Hydrothermal technology for nanotechnology. *Prog. Cryst. Growth Charact. Mater.*, **53**, 117–166.

12 Jolivet, J.P., Cassaignon, S., Chanéac, C., Chiche, D., Durupthy, O., and Portehault, D. (2010) Design of metal oxide nanoparticles: Control of size, shape, crystalline structure and functionalization by aqueous chemistry. *C. R. Chimie*, **13**, 40–51.

13 Tiginyanu, I.M., Lupan, O., Ursaki, V.V., Chow, L., and Enachi, M. (2011) Nanostructures of metal oxides, in *Comprehensive Semiconductor Science and Technology*, vol. **3** (eds P. Bhattacharya, R. Fornari, and H. Kamimura), Elsevier, Amsterdam, pp. 396–479.

14 Rao, C.N.R., Ramakrishna Matte, H.S.S., Voggu, R., and Govindaraj, A. (2012) Recent progress in the synthesis of inorganic nanoparticles. *Dalton Trans.*, **41**, 5089–5120.

15 Chiche, D., Chaneac, C., Revel, R., and Jolivet, J.P. (2011) Use of polyols as particle size and shape controllers: Application to boehmite synthesis from sol-gel routes. *Phys. Chem. Chem. Phys.*, **13**, 6241–6248.

16 Durupthy, O., Bill, J., and Aldinger, F. (2007) Bio-inspired synthesis of crystalline TiO_2: Effect of amino acids on nanoparticles structure and shape. *Cryst. Growth Des.*, **7**, 2696–2704.

17 Sugimoto, T., Zhou, X., and Muramatsu, A. (2003) Synthesis of uniform anatase TiO_2 nanoparticles by gel–sol method: 4. Shape control. *J. Colloid Interface Sci.*, **259**, 53–61.

18 He, T., Chen., D., Jiao, X., Wang, Y., and Duan, Y. (2005) Solubility-controlled synthesis of high-quality Co_3O_4 nanocristals. *Chem. Mater.*, **17**, 4023–4030.

19 Wu, Q., Zhang, F., Xiao, P., Tao, H., Wang, X., Hu, Z., and Lu, Y. (2008) Great influence of anions for controllable synthesis of CeO_2 nanostructures: From nanorods to nanocubes. *J. Phys. Chem. C*, **112**, 17076–17080.

20 Dong, Y., He, K., Yin, L., and Zhang, A. (2007) A facile route to controlled synthesis of Co_3O_4 nanoparticles and their environmental catalytic properties. *Nanotechnology*, **18**, 435602.

21 Niederberger, M., Garnweitner, G., Buha, J., Polleux, J., Ba, J., and Pinna, N. (2006) Nonaqueous synthesis of metal oxide nanoparticles: Review and indium oxide as case study for the dependence of particle morphology on precursors and solvents. *J. Sol.-Gel. Sci. Techn.*, **40**, 259–266.

22 Polshettiwar, V., Baruwati, B., and Varma, R.S. (2009) Self-assembly of metal oxides into three-dimensional nanostructures: Synthesis and application in catalysis. *ACS Nano*, **3**, 728–736.

23 Boutonnet, M., Lögdberg, S., and Svensson, E.E. (2008) Recent developments in the application of nanoparticles prepared from w/o microemulsions in heterogeneous catalysis. *Curr. Opin. Colloid Interface Sci.*, **13**, 270–286.

24 Ganguly, A.A., Trinh, P., Ramanujachary, K.V., Ahmad, T., Mugweru, A., and Ganguli, A.K. (2011) Reverse micellar based synthesis of ultrafine MgO nanoparticles (8–10 nm): Characterization and catalytic properties. *J. Colloid Interface Sci.*, **353**, 137–142.

25 Ahmad, T., Ganguly, A., Ahmed, J., Ganguli, A.K., and Alhartomy, O.A.A. (2011) Nanorods of transition metal oxalates: A versatile route to the oxide nanoparticles. *Arabian J. Chem.*, **4**, 125–134.

26 Supakanapitak, S., Boonamnuayvitaya, V., and Jarudilokkul, S. (2012) Synthesis of nanocrystalline CeO_2 particles by different emulsion methods. *Mater. Characterization*, **67**, 83–82.

27 Sun, C., Li, H., Zhang, H., Wang, Z., and Chen, L. (2005) Controlled synthesis of CeO_2 nanorods by a solvothermal method. *Nanotechnology*, **16**, 1454–1463.

28 Aghaie-Khafri, M. and Kakaei Lafdani, M.H. (2012) A novel method to synthesis Cr_2O_3 nanopowders using EDTA as a chelating agent. *Powder Technol.*, **222**, 152–159.

29 Primo, A., Marino, T., Corma, A., Molinari, R., and García, H. (2011) Efficient visible-light photocatalytic water splitting by minute amounts of gold supported on nanoparticulate CeO_2 obtained by a biopolymer templating method. *J. Am. Chem. Soc.*, **133**, 6930–6933.

30 Almerindo, G.I., Probst, L.F.D., Campos, E.M., de Almeida, R.M., Meneghetti, S.M.P., Meneghetti, M.R., Clacens, J.M., and Fajardo, H.V. (2011) Magnesium oxide prepared via metal-chitosan complexation method: Application as catalyst for transesterification of soybean oil and catalyst deactivation studies. *J. Power Sources*, **196**, 8057–8063.

31 Horga, R., Di Renzo, F., and Quignard, F. (2007) Ionotropic alginate aerogels as precursors of dispersed oxide phases. *Appl. Catal. A*, **325**, 251–255.

32 Yang, S.J., Im, J.H., Kim, T., Lee, K., and Park, C.R. (2011) MOF-derived ZnO and ZnO@C composites with high photocatalytic activity and adsorption capacity. *J. Hazard. Mater.*, **186**, 376–382.

33 Chen, L., Zhao, C., Wei, Z., Wang, S., and Gu, Y. (2011) Direct synthesis and characterization of spongy CuO with nanosheets from $Cu_3(btc)_2$ microporous metal-organic framework. *Mater. Lett.*, **65**, 446–449.

34 Wang, W., Li, Y., Zhang, Y., He, D., Liu, H., and Liao, S. (2011) Metal-organic framework as a host for synthesis of nanoscale Co_3O_4 as an active, catalyst for CO oxidation. *Catal. Commun.*, **12**, 875–879.

35 Jana, N.R., Chen, Y., and Peng, X. (2004) Size- and shape-controlled magnetic (Cr, Mn, Fe, Co, Ni) oxide nanocrystals via a simple and general approach. *Chem. Mater.*, **16**, 3931–3935.

36 Aprile, C., Corma, A., and Garcia, H. (2008) Enhancement of the photocatalytic activity of TiO_2 trough spatial structuring and particle size control: From subnanometric to submilimetric length scale. *Phys. Chem. Chem. Phys.*, **10**, 769–783.

37 Müler, M., Zhang, X., Wang, Y., and Fischer, R.A. (2009) Nanometer-sized titania hosted inside MOF-5. *Chem. Commoun.*, 119–121.

38 Imperor-Clerc, M., Bazin, D., Appay, M. D., Beaunier, P., and Davidson, A. (2004) Crystallization of β-MnO_2 nanowires in the pores of SBA-15 silicas: *In situ* investigation using synchrotron radiation. *Chem. Mater.*, **16**, 1813–1821.

39 Lopes, I., El Hassan, N., Guerba, H., Wallez, G., and Davidson, A. (2006) Size-induced structural modifications affecting Co_3O_4 nanoparticles patterned in SBA-15 silicas. *Chem. Mater.*, **18**, 5826–5828.

40 Ungureanu, A., Dragoi, B., Chirieac, A., Royer, R., Duprez, D., and Dumitriu, E. (2011) Synthesis of highly thermostable copper-nickel nanoparticles confined in the channels of ordered mesoporous SBA-15 silica. *J. Mater. Chem.*, **21**, 12529–12541.

41 Tang, C., Zhang, H., Sun, C., Li, J., Qi, L., Quan, Y., Gao, F., and Dong, L. (2011) An efficient strategy for highly loaded, well dispersed and thermally stable metal oxide catalysts. *Catal. Commun.*, **12**, 1075–1078.

42 Cheng, X.W., Meng, Q.Y., Chen, J.Y., and Long, Y.C. (2012) A facile route to synthesize mesoporous ZSM-5 zeolite incorporating high ZnO loading in mesopores. *Micropor. Mesopor. Mater.*, **135**, 198–203.

43 Lihitkar, P.B., Violet, S., Shirolkar, M., Singh, J., Srivastava, O.N., Naik, R.H., and Kulkarni, S.K. (2012) Confinement of zinc oxide nanoparticles in ordered mesoporous silica MCM-41. *Mater. Chem. Phys.*, **133**, 850–856.

44 Fernandez-Garcia, M., Martinez-Arias, A., Hanson, J.C., and Rodriguez, J.A. (2004) Nanostructured oxides in chemistry: Characterization and properties. *Chem. Rev.*, **104**, 4063–4104.

45 Wang, Y. and Völl, C. (2009) Chemical reactions on metal oxide surfaces investigated by vibrational spectroscopy. *Surf. Sci.*, **603**, 1589–1599.

46 Ayyub, P., Multani, M., Barma, M., Palkar, V.R., and Vijayaraghavan, R. (1988) Size-induces structural phase transition and hyperfine properties of microcrystalline F_2O_3. *J. Phys. C: Solid State Phys.*, **21**, 2229–2245.

47 Garvie, R.C. and Goss, M.F. (1986) Intrinsic size dependence of the phase transformation temperature in zirconia micro-crystals. *J. Mater. Sci.*, **21**, 1253–1257.

48 Paun, C., Safonova, O.V., Szlachetko, J., Abdala, P.M., Nachtegaal, M., Sa, J., Kleymenov, E., Cervellino, A., Krumeich, F., and van Bokhoven, J.A. (2012) Polyhedral CeO_2 nanoparticles: Size-dependent geometrical and electronic structure. *J. Phys. Chem. C*, **116**, 7312–7317.

49 Serpone, N., Lawless, D., and Khairutdinov, R. (1995) Size effects on the photophysical properties of colloidal anatase TiO_2 particles: Size quantization versus direct transitions in this indirect semiconductor? *J. Phys. Chem.*, **98**, 16646–16654.

50 Reddy, K.M., Manorama, S.V., and Reddy, A.R. (2002) Bandgap studies on anatase titanium dioxide nanoparticles. *Mater. Chem. Phys.*, **78**, 239–245.

51 Kosacki, I., Petrovsky, V., and Anderson, H.U. (1999) Band gap energy in nanocrystalline ZrO_2: 16%Y thin films. *Appl. Phys. Lett.*, **74**, 341–343.

52 Lucas, E., Decker, S., Khaleel, A., Seitz, A., Futlz, S., Ponce, A., Li, W., Carnes, C., and Klabunde, K.J. (2001) Nanocrystalline metal oxides as unique chemical reagents/sorbents. *Chem. Eur. J.*, **7**, 2505–2510.

53 Chen, C.Q., Shi, Y., Zhang, Y.S., Zhu, J., and Yan, Y.J. (2006) Size dependence of Young's modulus in ZnO nanowires. *Phys. Rev. Lett.*, **96**, 075505.

54 Xu, F., Qin, Q., Mishra, A., Gu, Y., and Zhu, Y. (2010) Mechanical properties of ZnO nanowires under different loading modes. *Nano Res.*, **3**, 271–280.

55 Sayle, T.X.T. and Sayle, D.C. (2010) Elastic deformation in ceria nanorods *via*

a fluorite-to-rutile phase transition. *ACS Nano*, **4**, 879–886.

56 Rodriguez, J.A., Hanson, J.C., Kim, J.-Y., Liu, G., Iglesias-Juez, A., and Fernandez-Garcia, M. (2003) Properties of CeO_2 and $Ce_{1-x}Zr_xO_2$ Nanoparticles: X-ray absorption near-edge spectroscopy, density functional, and time-resolved X-ray diffraction studies. *J. Phys. Chem. B*, **107**, 3535–3543.

57 Han, W.Q., Wen, W., Hanson, J.C., Teng, X., Marinkovic, N., and Rodriguez, J.A. (2009) One-dimensional ceria as catalyst for the low-temperature water-gas shift reaction. *J. Phys. Chem. C*, **113**, 21949–21955.

58 Choudary, B.M., Kantam, M.L., Ranganath, K.V.S., Mahendar, K., and Sreedhar, B. (2004) Bifunctional nanocrystalline MgO for chiral epoxy ketones via Claisen-Schmidt condensation-asymmetric epoxidation reactions. *J. Am. Chem. Soc.*, **126**, 3396–3397.

59 Jiao, F. and Frei, H. (2009) Nanostructured cobalt oxide clusters in mesoporous silica as efficient oxygen-evolving catalysts. *Angew. Chem. Int. Ed.*, **48**, 1841–1844.

60 Agawane, S.M. and Nagarkar, J.M. (2011) Nano ceria catalyzed synthesis of α-aminophosphonates under ultrasonication. *Tetrahedron Lett.*, **52**, 3499–3504.

61 Velmurugan, R., Selvam, K., Krishnakumar, B., and Swaminathan, M. (2011) An efficient reusable and antiphotocorrosive nano ZnO for the mineralization of Reactive Orange 4 under UV- a light. *Sep. Purif. Tech.*, **80**, 119–124.

62 Kansal, S.K., Ali, A.H., Kapoor, S., and Bahnemann, D.W. (2011) Synthesis of flower like zinc oxide nanostructure and its application as a photocatalyst. *Sep. Purif. Tech.*, **80**, 125–130.

63 Huang, W.J., Fang, G.C., and Wang, L.C. (2005) A nanometer-ZnO catalyst to enhance the ozonation of 2,4,6-trichlorophenol in water. *Colloid. Surfaces A*, **260**, 45–51.

64 Jang, E.S., Won, J.H., Hwang, S.J., and Choy, J.H. (2006) Fine tuning of the face orientation of ZnO crystals to optimize their photocatalytic activity. *Adv. Mater.*, **18**, 3309–3312.

65 Guo, C., Xu, J., He, Y., Zhang, Y., and Wang, Y. (2011) Photodegradation of rhodamine B and methyl orange over one-dimensional TiO_2 catalysts under simulated solar irradiation. *Appl. Surf. Sci.*, **257**, 3798–3803.

66 Zhan, B.Z., White, M.A., Sham, T.K., Pincock, J.A., Doucet, R.J., Ramana Rao, K.V., Robertson, K.N., and Cameron, T.S. (2003) Zeolite-confined nano-RuO_2: A green, selective, and efficient catalyst for aerobic alcohol oxidation. *J. Am. Chem. Soc.*, **125**, 2195–2199.

67 Li, P., Miser, D.E., Rabiei, S., Yadav, R.T., and Hajaligol, M.R. (2003) The removal of carbon monoxide by iron oxide nanoparticles. *Appl. Catal. B*, **43**, 151–162.

68 Shi, F., Tse, M.K., Pohl, M.M., Bruckner, A., Zhang, S., and Beller, M. (2007) Catalysis: Improved activity and selectivity of free nano-Fe_2O_3 in selective oxidations. *Angew. Chem. Int. Ed.*, **46**, 8866–8868.

69 Choudary, B.M., Ranganath, K.V.S., Pal, U., Kantam, M.L., and Sreedhar, B. (2005) Nanocrystalline MgO for asymmetric Henry and Michael reactions. *J. Am. Chem. Soc.*, **127**, 13167–13171.

70 Hayat, K., Gondal, M.A., Khaleda, M.M., Ahmed, S., and Shemsi, A.M. (2011) Nano ZnO synthesis by modified sol gel method and its application in heterogeneous photocatalytic removal of phenol from water. *Appl. Catal. A*, **393**, 122–129.

71 He, L., Yu, Y., Zhang, C., and He, H. (2011) Complete catalytic oxidation of o-xylene over CeO_2 nanocubes. *J. Environ. Sci.*, **23**, 160–165.

72 Esswein, A.J., McMurdo, M.J., Ross, P.N., Bell, A.T., and Tilley, T.D. (2009) Size-dependent activity of Co_3O_4 nanoparticle anodes for alkaline water electrolysis. *J. Phys. Chem. C*, **113**, 15068–15072.

73 Dodd, A.C., McKinley, A.J., Saunders, M., and Tsuzuki, T. (2006) Effect of particle size on the photocatalytic activity of nanoparticulate zinc oxide. *J. Nanoparticle Res.*, **8**, 43–51.

74 Dziembaj, R., Molenda, M., Chmielarz, L., Zaitz, M.M., Piwowarska, Z., and Rafalska-Łasocha, A. (2011) Optimization of Cu doped ceria nanoparticles as catalysts for low-temperature methanol and ethylene total oxidation. *Catal. Today*, **169**, 112–117.

75 Martín Alonso, D., Vila, F., Mariscal, R., Ojeda, M., López Granados, M., and Santamaría-González, J. (2010) Relevance of the physico-chemical properties of CaO catalysts for the methanolysis of triglycerides to obtain biodiesel. *Catal. Today*, **158**, 114–120.

76 Kim, E.J. and Hahn, S.H. (2001) Microstructure and photoactivity of titania nanoparticles prepared in nonionic w:o microemulsions. *Mater. Sci. Eng. A*, **303**, 24–29.

77 Lu, Y., Wang, L., Wang, D., Xie, T., Chen, L., and Lin, Y. (2011) A comparative study on plate-like and flower-like ZnO nanocrystals surface photovoltage property and photocatalytic activity. *Mater. Chem. Phys.*, **129**, 281–287.

78 Zeng, J.H., Jin, B.B., and Wang, Y.F. (2009) Facet enhanced photocatalytic effect with uniform single-crystalline zinc oxide nanodisks. *Chem. Phys. Lett.*, **472**, 90–95.

79 Xi, Y., Hu, C., Gao, P., Yang, R., He, X., Wang, X., and Wan, B. (2010) Morphology and phase selective synthesis of CuxO (x = 1, 2) nanostructures and their catalytic degradation activity. *Mater. Sci. Eng. B*, **166**, 113–117.

80 Wilmer, H., Kurtz, M., Klementiev, K. V., Tkachenko, O.P., Grünert, W., Hinrichsen, O., Birkner, A., Rabe, S., Merz, K., Driess, M., Wöll, C., and Muhler, M. (2003) Methanol synthesis over ZnO: A structure-sensitive reaction? *Phys. Chem. Chem. Phys.*, **5**, 4736–4742.

81 Xie, X., Li, Y., Liu, Z.Q., Haruta, M., and Shen, W. (2009) Low-temperature oxidation of CO catalysed by Co_3O_4 nanorods. *Nature*, **458**, 746–749.

82 Teng, Y., Kusano, Y., Azuma, M., Haruta, M., and Shimakawa, Y. (2011) Morphology effects of Co_3O_4 nanocrystals catalyzing CO oxidation in a dry reactant gas stream. *Catal. Sci. Tech.*, **1**, 920–922.

83 Hu, L., Sun, K., Peng, Q., Xu, B., and Li, Y. (2010) Surface active sites on Co_3O_4 nanobelt and nanocube model catalysts for CO oxidation. *Nano Res.*, **3**, 363–368.

84 Zhou, K., Wang, X., Sun, X., Peng, Q., and Li, Y. (2005) Enhanced catalytic activity of ceria nanorods from well-defined reactive crystal planes. *J. Catal.*, **229**, 206–212.

85 Wang, X., Ding, L., Zhao, Z., Xu, W., Meng, B., and Qiu, J. (2011) Novel hydrodesulfurization nano-catalysts derived from Co_3O_4 nanocrystals with different shapes. *Catal. Today*, **175**, 509–514.

86 Rhadfi, T., Piquemal, J.Y., Sicard, L., Herbst, F., Briot, E., Benedetti, M., and Atlamsani, A. (2010) Polyol-made Mn_3O_4 nanocrystals as efficient Fenton-like catalysts. *Appl. Catal. A*, **386**, 132–139.

87 Tian, W., Yang, H., Fan, X., and Zhang, X. (2011) Catalytic reduction of NOx with NH_3 over different-shaped MnO_2 at low temperature. *J. Hazard. Mater.*, **188**, 105–109.

88 Perez-Ramirez, W.J., Lopez, N., and Kondratenko, E.V. (2010) Pressure and materials effects on the selectivity of RuO_2 in NH_3 oxidation. *J. Phys. Chem. C*, **114**, 16660–16668.

89 Liang, S.H., Bulgan, F.T.G., Zong, R.L., and Zhu, Y.F. (2008) Effect of phase structure of MnO_2 nanorod catalyst on the activity for CO oxidation. *J. Phys. Chem. C*, **112**, 5307–5315.

90 Reddy, C.V. and Kantam, M.L. (2011) Recent developments on catalytic applications of nano-crystalline magnesium oxide. *Catal. Surveys Asia*, **5**, 89–110.

91 Kantam, M.L., Mahendar, K., Sreedhar, B., and Choudary, B.M. (2008) Synthesis of α-amino nitriles through Strecker reaction of aldimines and ketoimines by using nanocrystalline magnesium oxide. *Tetrahedron*, **64**, 3351–3360.

92 Choudary, B.M., Chakrapani, L.T., Kumar, K.V.R., and Kantum, M.L. (2006) Direct asymmetric aldol reaction catalyzed by nanocrystallinemagnesium oxide. *Tetrahedron*, **62**, 9571–9576.

93 Kantam, M.L., Pal, U., Sreedhar, B., and Choudary, B.M. (2007) An efficient synthesis of organic carbonates using nanocrystalline magnesium oxide. *Adv. Synth. Catal.*, **349**, 1671–1675.

94 Heidari, H., Abedini, M., Nemati, A., and Amini, M.M. (2009) Nanocrystalline magnesium oxide as a versatile heterogeneous catalyst for the Meerwein-Ponndorf-Verley reduction of cyclohexanone into cyclohexanol: Effect of preparation method of magnesium oxide on yield. *Catal. Lett.*, **130**, 266–270.

95 Wang, L. and Yang, J. (2007) Transesterification of soybean oil with nano-MgO or not in supercritical and subcritical methanol. *Fuel*, **86**, 328–333.

96 Zhou, L., Xu, J., Miao, H., Wang, F., and Li, X. (2005) Catalytic oxidation of cyclohexane to cyclohexanol and cyclohexanone over Co_3O_4 nanocrystals with molecular oxygen. *Appl. Catal. A*, **292**, 223–228.

97 Kesavan, V., Sivanand, P.S., Chandrasekaran, S., Koltypin, Y., and Gedanken, A. (1999) Catalytic aerobic oxidation of cycloalkanes with nanostructured amorphous metals and alloys. *Angew. Chem. Int. Ed.*, **38**, 3521–3523.

98 Perkas, N., Koltypin, Y., Palchik, O., Gedanken, A., and Chandrasekaran, S. (2001) Oxidation of cyclohexane with nanostructured amorphous catalysts under mild conditions. *Appl. Catal. A*, **209**, 125–130.

99 Ghiaci, M., Aghabararia, B., Botelho do Rego, A.M., Ferraria, A.M., and Habibollahi, S. (2011) Efficient allylic oxidation of cyclohexene catalyzed by trimetallic hybrid nano-mixed oxide (Ru/Co/Ce). *Appl. Catal. A*, **393**, 225–230.

100 Akhlaghinia, B., Ebrahimabadi, H., Goharshadi, E.K., Samiee, S., and Rezazadeh, S. (2012) Ceria nanoparticles as an efficient catalyst for oxidation of benzylic C—H bonds. *J. Mol. Catal. A: Chemical*, **357**, 67–72.

101 Cojocariu, A.M., Mutin, P.H., Dumitriu, E., Fajula, F., Vioux, A., and Hulea, V. (2010) Mild oxidation of bulky organic compounds with hydrogen peroxide over mesoporous TiO_2-SiO_2 xerogels prepared by nonhydrolytic sol-gel. *Appl. Catal. B*, **97**, 407–413.

102 Maciuca, A.L., Ciocan, C.E., Dumitriu, E., Fajula, F., and Hulea, V. (2008) V-, Mo- and W-containing layered double hydroxides as effective catalysts for mild oxidation of thioethers and thiophenes with H_2O_2. *Catal. Today*, **138**, 33–37.

103 Askarinejad, A., Bagherzadeh, M., and Morsali, A. (2010) Catalytic performance of Mn_3O_4 and Co_3O_4 nanocrystals prepared by sonochemical method in epoxidation of styrene and cyclooctene. *Appl. Surf. Sci.*, **256**, 6678–6682.

104 Aronne, A., Turco, M., Bagnasco, G., Ramis, G., Santacesaria, E., Di Serio, M., Marenna, E., Bevilacqua, M., Cammarano, C., and Fanelli, E. (2008) Gel derived niobium–silicon mixed oxides: Characterization and catalytic activity for cyclooctene epoxidation. *Appl. Catal. A*, **347**, 179–185.

105 Chen, H., He, J., Zhang, C., and He, H. (2007) Self-assembly of novel mesoporous manganese oxide nanostructures and their application in oxidative decomposition of formaldehyde. *J. Phys. Chem. C*, **111**, 18033–18038.

106 Kwon, S.C., Fana, M., Wheelock, T.D., and Saha, B. (2007) Nano- and micro-iron oxide catalysts for controlling the emission of carbon monoxide and methane. *Sep. Purif. Tech.*, **58**, 40–48.

107 Reddy, B.M., Thrimurthulu, G., Katta, L., Yamada, Y., and Park, S.E. (2009) Structural characteristics and catalytic activity of nanocrystalline ceria-praseodymia solid solutions. *J. Phys. Chem. C*, **113**, 15882–15890.

108 Reddy, B.M., Katta, L., and Thrimurthulu, G. (2001) Novel nanosized $Ce_xZr_{1-x}O_2$, $Ce_xHf_{1-x}O_2$ and $Ce_xTb_{1-x}O_{2-\delta}$ solid solutions: Structural characteristics and catalytic performance. *Catal. Today*, **175**, 585–592.

109 Pu, Z.Y., Liu, X.S., Jia, A.P., Xie, Y.L., Lu, J.Q., and Luo, M.F. (2008) Enhanced activity for CO oxidation over Pr- and Cu-doped CeO_2 catalysts: Effect of oxygen vacancies. *J. Phys. Chem. C*, **112**, 15045–15051.

110 Mai, H., Zhang, D., Shi, L., Yan, T., and Li, H. (2011) Highly active $Ce_{1-x}Cu_xO_2$ nanocomposite catalysts for the low temperature oxidation of CO. *Appl. Surf. Sci.*, **257**, 7551–7559.

111 Lopez, I., Valdes-Solis, T., and Marban, G. (2008) An attempt to rank copper-based catalysts used in the CO-PROX reaction. *Int. J. Hydrogen Energy*, **33**, 197–205.

112 Yamashita, H., Takeuchi, M., and Anpo, M. (2004) Visible-light-sensitive photocatalysts, in *Encyclopedia of Nanoscience and Nanotechnology*, vol. **10**, American Scientific Publishers, pp. 639–654.

113 Bahnemann, D.W., Kormann, C., and Hoffmann, M.R. (1987) Preparation and characterization of quantum size zinc oxide: A detailed spectroscopic study. *J. Phys. Chem.*, **91**, 3789–3798.

114 Wells, R.P.K. (2009) Oxide materials in photocatalytic processes, in *Metal Oxide Catalysis* (eds S.D. Jackson and S.J. Hargreaves), Wiley-VCH Verlag GmbH, Weinheim, Germany, pp. 755–770.

115 Colón-Ibáñez, G., Belver-Coldeira, C., and Fernández-García, M. (2007) Chapter 17. Nanostructured oxides in photo-catalysis, in *Synthesis, Properties, and Applications of Oxide Nanomaterials* (eds M. Fernández-García and J.A. Rodriguez), John Wiley & Sons, Inc., Hoboken, USA.

116 Chaturvedi, S., Dave, P.N., and Shah, N.K. (2011) Applications of nano-catalyst in new era. *J. Saudi Chem. Soc*, 015. doi: 10.1016/j.jscs.2011.01

117 Zelmanov, G. and Semiat, R. (2008) Iron (3) oxide-based nanoparticles as catalysts in advanced organic aqueous oxidation. *Water Res.*, **42**, 492–498.

118 Polzer, F., Wunder, S., Lu, Y., and Ballauff, M. (2012) Oxidation of an organic dye catalyzed by MnO_x nanoparticles. *J. Catal.*, **289**, 80–87.

119 Orge, C.A., Órfão, J.J.M., Pereira, M.F. R., Duarte de Farias, A.M., Rabelo Neto, R.C., and Fraga, M.A. (2011) Ozonation of model organic compounds catalysed by nanostructured cerium oxides. *Appl. Catal. B*, **103**, 190–199.

120 Jung, H. and Choi, H. (2006) Catalytic decomposition of ozone and para-chlorobenzoic acid (pCBA) in the presence of nanosized ZnO. *Appl. Catal. B*, **66**, 288–294.

121 Corma, A. (1997) From microporous to mesoporous molecular sieve materials and their use in catalysis. *Chem. Rev.*, **97**, 2373–2419.

122 Tahuchi, A. and Schüth, F. (2005) Ordered mesoporous materials in catalysis. *Micropor. Mesopor. Mater.*, **77**, 1–45.

123 Hulea, V., Brunel, D., Galarneau, A., Philippot, K., Chaudret, B., Kooyman, P.J., and Fajula, F. (2005) Synthesis of well-dispersed ruthenium nanoparticles inside mesostructured porous silica under mild conditions. *Micropor. Mesopor. Mater.*, **79**, 185–194.

124 Roucous, A., Schulz, J., and Patin, H. (2002) Reduced transition metal colloids: A novel family of reusable catalysts? *Chem. Rev.*, **102**, 3757–3778.

125 Ichikawa, M. (2000) 'Ship-in-bottle' catalyst technology. Novel templating fabrication of platinum group metals nanoparticles and wires in micro/mesopores. *Platinum Metals Rev.*, **44**, 3–14.

126 Somorjai, G.A., Frei, H., and Park, J.Y. (2009) Advancing the frontiers in nanocatalysis, biointerfaces, and renewable energy conversion by innovations of surface techniques. *J. Am. Chem. Soc.*, **131**, 16589–16605.

127 Laugel, G., Arichi, J., Molière, M., Kiennemann, A., Garin, F., and Louis, B. (2008) Metal oxides nanoparticles on SBA-15: Efficient catalyst for methane combustion. *Catal. Today*, **138**, 38–42.

128 Tsoncheva, T., Roggenbuck, J., Paneva, D., Dimitrov, M., Mitov, I., and Fröba, M. (2010) Nanosized iron and chromium oxides supported on mesoporous CeO_2 and SBA-15 silica: Physico-chemical and catalytic study. *Appl. Surf. Sci.*, **257**, 523–530.

129 Busuioc, A.M., Meynen, V., Beyers, E., Cool, P., Bilba, N., and Vansant, E.F. (2007) Growth of anatase nanoparticles inside the mesopores of SBA-15 for photocatalytic applications. *Catal. Commun.*, **8**, 527–530.

130 Lou, Y., Wang, H., Zhang, Q., and Wang, Y. (2007) SBA-15-supported molybdenum

oxides as efficient catalysts for selective oxidation of ethane to formaldehyde and acetaldehyde by oxygen. *J. Catal.*, **247**, 245–255.

131 Lim, H., Lee, J., Jin, S., Kim, J., Yoon, J., and Hyeo, T. (2006) Highly active heterogeneous Fenton catalyst using iron oxide nanoparticles immobilized in alumina coated mesoporous silica. *Chem. Commun.*, 463–465.

132 Srinivasan, N.R. and Bandyopadhyaya, R. (2012) Highly accessible SnO_2 nanoparticle embedded SBA-15 mesoporous silica as a superior photocatalyst. *Micropor. Mesopor. Mater.*, **149**, 166–171.

133 El-Safty, S.A., Kiyozumi, Y., Hanaoka, T., and Mizukami, F. (2008) Nanosized NiO particles wrapped into uniformly mesocaged silica frameworks as effective catalysts of organic amines. *Appl. Catal. A*, **337**, 121–129.

134 Martinez, F., Calleja, G., Melero, J.A., and Molina, R. (2007) Iron species incorporated over different silica supports for the heterogeneous photo-Fenton oxidation of phenol. *Appl. Catal. B*, **70**, 452–460.

135 Choudhary, V.R. and Jana, S.K. (2002) Benzoylation of benzene and substituted benzenes by benzoyl chloride over In_2O_3/Si-MCM-41 catalyst. *J. Mol. Catal. A: Chemical*, **184**, 247–255.

136 Corberan, V.C. (2005) Novel approaches for the improvement of selectivity in the oxidative activation of light alkanes. *Catal. Today*, **99**, 33–41.

137 Balu, A.M., Pineda, A., Yoshida, K., Campelo, J.M., Gai, P.L., Luque, R., and Angel Romero, A. (2010) Fe/Al synergy in Fe_2O_3 nanoparticles supported on porous aluminosilicate materials: Excelling activities in oxidation reactions. *Chem. Commun.*, **46**, 7825–7827.

138 Wingen, A., Anastasievic, N., Hollnagel, A., Werner, D., and Schüth, F. (2000) Fe–MCM-41 as a catalyst for sulfur dioxide oxidation in highly concentrated gases. *J. Catal.*, **193**, 248–254.

139 Choudhary, V.R., Jana, S.K., and Kiran, B.P. (2000) Highly active Si-MCM-41-supported Ga_2O_3 and In_2O_3 catalysts for Friedel-Crafts-type benzylation and acylation reactions in the presence or absence of moisture. *J. Catal.*, **192**, 257–261.

140 Gonzalez-Arellano, C., Yoshida, K., Luque, R., and Gai, P.L. (2010) Highly active and selective supported iron oxide nanoparticles in microwave-assisted N-alkylations of amines with alcohols. *Green Chem.*, **12**, 1281–1287.

141 Jiao, F. and Frei, H. (2010) Nanostructured manganese oxide clusters supported on mesoporous silica as efficient oxygen-evolving catalysts. *Chem. Commun.*, **46**, 2920–2922.

142 Dutta, B., Jana, S., Bhattacharjee, A., Gütlich, P., Iijima, S.I., and Koner, S. (2010) γ-Fe_2O_3 nanoparticle in NaY-zeolite matrix: Preparation, characterization, and heterogeneous catalytic epoxidation of olefins. *Inorg. Chim. Acta*, **363**, 696–704.

143 Zhang, G., Choi, W., Kim, S.H., and Hong, S.B. (2011) Selective photocatalytic degradation of aquatic pollutants by titania encapsulated into FAU-type zeolites. *J. Hazard. Mater.*, **188**, 198–205.

144 Luz Martínez, S., Romero, R., Lopez, J.C., Romero, A., Sanchez Mendieta, V., and Natividad, R. (2011) Preparation and characterization of CaO nanoparticles/NaX zeolite catalysts for the transesterification of sunflower oil. *Ind. Eng. Chem. Res.*, **50**, 2665–2670.

145 Altwasser, S., Glaser, R., Lo, A.S., Liu, P. H., Chao, K.J., and Weitkamp, J. (2006) Incorporation of RuO_2 nanoparticles into MFI-type zeolites. *Micropor. Mesopor. Mater.*, **89**, 109122.

146 Nakatsuji, T., Yamaguchi, T., Li, J., Sato, N., and Matsuzono, Y. (2009) Synthesis and characterization of Rh and RhOx particles supported on zeolites with high activity of lean NOx–CO–H_2 reaction. *Catal. Commun.*, **10**, 763–767.

147 Lafjaha, M., Djafri, F., Bengueddach, A., Keller, N., and Keller, V. (2011) Beta zeolite supported sol-gel TiO_2 materials for gas phase photocatalytic applications. *J. Hazard. Mater.*, **186**, 1218–1225.

148 Iijima, G.C. (1991) Helical microtubules of graphitic carbon. *Nature*, **354**, 56–58.

149 Pyrzynska, K. (2011) Carbon nanotubes as sorbents in the analysis of pesticides. *Chemosphere*, **83**, 1407–1413.

150 Fu, X., Yu, H., Peng, F., Wang, H., and Qian, Y. (2007) Facile preparation of RuO_2/CNT catalyst by a homogenous oxidation precipitation method and its catalytic performance. *Appl. Catal. A*, **321**, 190–197.

151 Chen, X., Gao, S., Wang, H., Liu, Y., and Wu, Z. (2011) Selective catalytic reduction of NO over carbon nanotubes supported CeO_2. *Catal. Commun.*, **14**, 1–5.

152 Qin, W. and Li, X. (2011) A theoretical study on the catalytic effect of nanoparticle confined in carbon nanotube. *Chem. Phys. Lett.*, **502**, 96–100.

153 Bavykin, D.V., Lapkin, A.A., Plucinski, P.K., Friedrich, J.M., and Walsh, F.C. (2005) TiO_2 nanotube-supported ruthenium(III) hydrated oxide: A highly active catalyst for selective oxidation of alcohols by oxygen. *J. Catal.*, **235**, 10–17.

154 Wang, H.Q., Chen, X.B., Weng, X.L., Liu, Y., Gao, S., and Wu, Z.B. (2011) Enhanced catalytic activity for selective catalytic reduction of NO over titanium nanotube-confined CeO_2 catalyst. *Catal. Commun.*, **12**, 1042–1045.

155 Polshettiwar, V. and Varma, R.S. (2010) Green chemistry by nano-catalysis. *Green Chem.*, **12**, 743–754.

156 Shi, F., Tse, M.K., Zhou, S., Pohl, M.M., Radnik, J., Hübner, S., Jähnisch, K., Brückner, A., and Beller, M. (2009) Green and efficient synthesis of sulfonamides catalyzed by nano-Ru/Fe_3O_4. *J. Am. Chem. Soc.*, **131**, 1775–1779.

157 Reddy, B.V.S., Krishna, A.S., Ganesh, A. V., and Kumar, G.G.K.S.N. (2011) Nano Fe_3O_4 as magnetically recyclable catalyst for the synthesis of α-aminophosphonates in solvent-free conditions. *Tetrahedron Lett.*, **52**, 1359–1362.

158 Rahman, M., Bagdi, A.K., Majee, A., and Hajra, A. (2011) Nano indium oxide catalyzed efficient synthesis of propargylamines via C—H and C—Cl bond activations. *Tetrahedron Lett.*, **52**, 4437–4439.

159 Banerjee, S., Das, J., and Santra, S. (2009) Native silica nanoparticle catalyzed anti-Markovnikov addition of thiols to inactivated alkenes and alkynes: A new route to linear and vinyl thioethers. *Tetrahedron Lett.*, **50**, 124–127.

160 Banerjee, S. and Santra, S. (2009) Remarkable catalytic activity of silica nanoparticle in the bis-Michael addition of active methylene compounds to conjugated alkenes. *Tetrahedron Lett.*, **50**, 2037–2040.

161 Dhrama Rao, G.B., Kaushik, M.P., and Halve, A.K. (2012) An efficient synthesis of naphtha[1,2-*e*]oxazinone and 14-substituted-14*H*-dibenzo[*a,j*]xanthenes derivatives promoted by zinc oxide nanoparticle under thermal and solvent-free conditions. *Tetrahedron Lett.*, **53**, 2741–2744.

162 Abad, A., Almela, C., Corma, A., and Garcia, H. (2006) Efficient chemoselective alcohol oxidation using oxygen as oxidant. Superior performance of gold over palladium catalysts. *Tetrahedron*, **62**, 6666–6672.

163 Xue, W.J., Wang, Y.F., Li, P., Liu, Z.T., Hao, Z.P., and Ma, C.Y. (2011) Morphology effects of Co_3O_4 on the catalytic activity of Au/Co_3O_4 catalysts for complete oxidation of trace ethylene. *Catal. Commun.*, **12**, 1265–1268.

164 Djerad, S., Geiger, B., Schott, F.J.P., and Kureti, S. (2009) Synthesis of nanosized ZrO_2 and its use as catalyst support in SCR. *Catal. Commun.*, **10**, 1103–1106.

165 Nadeem, M.A., Murdoch, M., Waterhouse, G.I.N., Metson, J.B., Keane, M.A., Llorca, J., and Idriss, H. (2010) Photoreaction of ethanol on Au/TiO_2 anatase: Comparing the micro to nanoparticle size activities of the support for hydrogen production. *J. Photochem. Photobiol. A*, **216**, 250–255.

166 Weerachawanasak, P., Praserthdam, P., Arai, M., and Pan, J. (2008) A comparative study of strong metal–support interaction and catalytic behavior of Pd catalysts supported on micron- and nanosized TiO_2 in liquid-phase selective hydrogenation of phenylacetylene. *J. Mol. Chem. A*, **279**, 133–139.

167 Lee, Y., Garcia, M.A., Huls, N.A.F., and Sun, S.H. (2010) Synthetic tuning of catalytic properties of Au-Fe_3O_4

nanoparticles. *Angew. Chem. Int. Ed.*, **49**, 1271–1274.

168 Roh, H.S., Potdar, H.S., and Jun, K.W. (2004) Carbon dioxide reforming of methane over co-precipitated Ni–CeO_2, Ni–ZrO_2 and Ni–Ce–ZrO_2 catalysts. *Catal. Today*, **93–95**, 39–44.

169 Zhao, Z., Lin, X., Jin, R., Dai, Y., and Wang, G. (2011) High catalytic activity in CO PROX reaction of low cobalt-oxide loading catalysts supported on nano-particulate CeO_2-ZrO_2 oxides. *Catal. Commun.*, **12**, 1448–1451.

170 Rakmak, N., Wiyaratn, W., Bunyakan, C., and Chungsiriporn, J. (2010) Synthesis of Fe/MgO nano-crystal catalysts by sol–gel method for hydrogen sulfide removal. *Chem. Eng. J.*, **162**, 84–90.

171 Strobel, R., Stark, W.J., Mädler, L., Pratsinis, S.E., and Baiker, A. (2003) Flame-made platinum/alumina: Structural properties and catalytic behaviour in enantioselective hydrogenation. *J. Catal.*, **213**, 296–304.

172 Strobel, R., Krumeich, F., Stark, W.J., Pratsinis, S.E., and Baiker, A. (2004) Flame spray synthesis of Pd/Al_2O_3 catalysts and their behavior in enantioselective hydrogenation. *J. Catal.*, **222**, 307–314.

173 Piacentini, M., Strobe, R., Maciejewski, M., Pratsinis, S.E., and Baiker, A. (2006) Flame-made Pt-Ba/Al_2O_3 catalysts: Structural properties and behavior in lean-NOx storage-reduction. *J. Catal.*, **243**, 43–56.

174 Hannemann, S., Grunwaldt, J.D., Lienemann, P., Günther, D., Krumeich, F., Pratsinis, S.E., and Baiker, A. (2007) Combination of flame synthesis and high-throughput experimentation: The preparation of alumina-supported noble metal particles and their application in the partial oxidation of methane. *Appl. Catal. A*, **316**, 226–239.

175 Pisduangdaw, S., Panpranot, J., Methastidsook, C., Chaisuk, C., Faungnawakij, K., Praserthdam, P., and Mekasuwandumrong, O. (2009) Characteristics and catalytic properties of Pt–Sn/Al_2O_3 nanoparticles synthesized by one-step flame spray pyrolysis in the dehydrogenation of propane. *Appl. Catal. A*, **370**, 1–6.

176 Chaisuk, C., Boonpitak, P., Panpranot, J., and Mekasuwandumrong, O. (2011) Effects of Co dopants and flame conditions on the formation of Co/ZrO_2 nanoparticles by flame spray pyrolysis and their catalytic properties in CO hydrogenation. *Catal. Commun.*, **12**, 917–922.

177 Height, M.J., Pratsinis, S.E., Mekasuwandumrong, O., and Praserthdam, P. (2006) Ag-ZnO catalysts for UV-photodegradation of methylene blue. *Appl. Catal. B*, **63**, 305–312.

178 Liu, H. and Iglesia, E. (2005) Selective oxidation of methanol and ethanol on supported ruthenium oxide clusters at low temperatures. *J. Phys. Chem. B*, **109**, 2155–2163.

179 Méndez-Cruz, M., Ramírez-Solís, J., and Zanella, R. (2011) CO oxidation on gold nanoparticles supported over titanium oxide nanotubes. *Catal. Today*, **166**, 172–179.

180 Kamat, P.V. and Meisel, D. (2002) Nanoparticles in advanced oxidation processes. *Curr. Opin. Colloid Interface Sci.*, **7**, 282–287.

181 Kaur, M. and Ali, A. (2011) Lithium ion impregnated calcium oxide as nano catalyst for the biodiesel production from karanja and jatropha oils. *Renewable Energy*, **36**, 2866–2871.

182 Wen, L., Wang, Y., Lu, D., Hu, S., and Han, H. (2010) Preparation of KF/CaO nanocatalyst and its application in biodiesel production from Chinese tallow seed oil. *Fuel*, **89**, 2267–2271.

183 Kwon, Y.T., Song, K.Y., Lee, W.I., Choi, G.J., and Do, Y.R. (2000) Photocatalytic behavior of WO_3-loaded TiO_2 in an oxidation reaction. *J. Catal.*, **191**, 192–199.

184 Burakorn, T., Panpranot, J., Mekasuwandumrong, O., Chaisuk, C., Praserthdam, P., and Jongsomjit, B. (2008) Characterization of cobalt dispersed on the mixed nanoscale alumina and zirconia supports. *J. Mater. Process Tech.*, **206**, 352–358.

185 Reddy, G.K., Loridant, S., Takahashi, A., Delichère, P., and Reddy, B.M. (2010)

Reforming of methane with carbon dioxide over $Pt/ZrO_2/SiO_2$ catalysts-effect of zirconia to silica ratio. *Appl. Catal. A*, **389**, 92–100.

186 He, Y., Yang, B., and Cheng, G. (2004) On the oxidative coupling of methane with carbon dioxide over CeO_2/ZnO nanocatalysts. *Catal. Today*, **98**, 595–600.

187 Sun, G., Mu, X., Zhang, Y., Cui, Y., Xia, G., and Chen, Z. (2011) Rare earth metal modified $CuO/\gamma\text{-}Al_2O_3$ catalysts in the CO oxidation. *Catal. Commun.*, **12**, 349–352.

188 Zheng, X., Wang, S., Wang, S., Zhang, S., Huang, W., and Wu, S. (2004) Copper oxide catalysts supported on ceria for low-temperature CO oxidation. *Catal. Commun.*, **5**, 729–732.

189 Yuan, Z.Y., Idakiev, V., Vantomme, A., Tabakova, T., Ren, T.Z., and Su, B.L. (2008) Mesoporous and nanostructured CeO_2 as supports of nanosized gold catalysts for low-temperature water-gas shift reaction. *Catal. Today*, **131**, 203–210.

190 Idakiev, V., Yuan, Z.Y., Tabakova, T., and Su, B.L. (2005) Titanium oxide nanotubes as supports of nanosized gold catalysts for low temperature water-gas shift reaction. *Appl. Catal. A*, **281**, 149–155.

191 Dal Santo, V., Gallo, A., Naldoni, A., and Sordelli, L. (2012) Selective butadiene hydrogenation by Pd nanoparticles deposed onto nanosized oxide supports by CVD of Pd-hexafluoroacetylacetonate. *Inorg. Chim. Acta*, **380**, 216–222.

192 Si, R. and Flytzani-Stephanopoulos, M. (2008) Shape and crystal-plane effects of nanoscale ceria on the activity of Au-CeO_2 catalysts for the water–gas shift reaction. *Angew. Chem. Int. Ed.*, **47**, 2884–2887.

193 Lu, A.H., Salabas, E.L., and Schüth, F. (2007) Magnetic nanoparticles: Synthesis, protection, functionalization, and application. *Angew. Chem. Int. Ed.*, **46**, 1222–1244.

194 Teja, A.S. and Koh, P.Y. (2009) Synthesis, properties, and applications of magnetic iron oxide nanoparticles. *Prog. Cryst. Growth Charact. Mater.*, **55**, 22–45.

195 Shylesh, S., Schünemann, V., and Thiel, W.R. (2010) Magnetically separable nanocatalysts: Bridges between homogeneous and heterogeneous catalysis. *Angew. Chem. Int. Ed.*, **49**, 3428–3459.

196 Polshettiwar, V., Luque, R., Fihri, A., Zhu, H., Bouhrara, M., and Basset, J.M. (2011) Magnetically recoverable nanocatalysts. *Chem. Rev.*, **111**, 3036–3075.

197 Xu, P., Zeng, G.M., Huang, D.L., Feng, C.L., Hu, S., Zhao, M.H., Lai, C., Wei, Z., Huang, C., Xie, G.X. and Liu, Z.F. (2012) Use of iron oxide nanomaterials in wastewater treatment: A review. *Sci. Total Environ*, **424**, 1–10.

11
Confinement Effects in Nanosupports

Xiulian Pan and Xinhe Bao

11.1
Introduction

Catalysis is widely involved in our modern life because the production of most industrially important chemicals and environmental protection rely on catalytic processes. The role of catalysts is to change the reaction rate and direct the product selectivity. Porous materials have been used for a long time in heterogeneous catalysis either directly as catalysts or catalyst supports. The use of supports with different physical and chemical properties can tailor catalytic performance by maintaining the active phase in a highly dispersed state and modulating the relevant chemical properties through interactions. In particular, the pore structure remains one of most important parameters.

Confined within pores, molecules frequently exhibit different behavior from that in the bulk. For example, the freezing and melting behavior of confined liquids has long been recognized to be different from that in the bulk [1]. Derouane and Chang reported that heat adsorption of ammonia and several simple amines can be modified in zeolites MOR and MFI [2]. Guegan *et al.* showed by molecular simulation that the density of confined methanol was higher than that of bulk methanol and varies regularly across the diameter of the silica cylindrical pore [3]. Baiker and coworkers demonstrated by *in situ* Fourier-transform infrared (FTIR) that the CO_2 density was always higher in the mesopores of silica than in the bulk, even under supercritical conditions [4]. Modified behavior of the confined phase including adsorption, phase transitions, diffusion, and structure might directly affect chemical reactions [5]. For example, Lesthaeghe reviewed how zeolite confinement effects influenced the reactivity of bulky molecules in the methanol-to-olefin process [6].

Confinement in zeolite pores made it possible to tune the selectivity of a chemical reaction, which is well-known as shape selectivity [7, 8], due to the molecular-size dimensions of the pores and to the interaction with the pore walls governing the behavior and properties of reactants, intermediates and products. The confined nanospace provided by mesoporous silicas with relatively well-defined pore structures has also been widely explored for asymmetric catalysis particularly with heterogeneous chiral catalysts inserted within the pores [9, 10]. Thomas *et al.*

Nanomaterials in Catalysis, First Edition. Edited by Philippe Serp and Karine Philippot.

commented that the spatial restrictions imposed by the nanocavities within which the reactions occur opened up the prospect of performing heterogeneous enantio-selective conversions in a novel manner including hydrogenations, oxidations, and aminations [9]. Within well-controlled surfaces and pores, some of the reaction channels might be hindered or even blocked, which consequently modified the selectivity of the reaction [10].

In addition, metallic nanoparticles can be entrapped and stabilized within pores, thus preventing their possible aggregation due to high surface energy. The orientation and configuration of metal nanoparticles can also be modified due to the spatial restriction, which could directly influence their catalytic activity through interaction with reactants, intermediates and/or products. For example, Zhan showed that RuO_2 nanoparticles of around 1.3 nm confined within faujasite zeolite cages exhibited a two-dimensional (2D) chain structure, in which RuO_6 units were connected by two shared oxygen atoms. This composite was much more active than bulk hydrous RuO_2 in the aerobic oxidation of benzyl alcohol under the same conditions [11].

In addition to zeolites, mesoporous silicates and aluminosilicates, carbon nanotubes (CNTs) represented a new type of material with a well-defined pore structure, which has aroused wide interest since their discovery in 1991 [12]. CNTs are distinguished by their nearly 1D pore structure and the graphitic state of the pore walls, which provide an intriguing confinement environment for catalysis.

11.2 Confinement Effects in Carbon Nanotubes

Carbon nanotubes can be considered as graphene layers rolling up and forming a tubular structure with graphitic walls aligned along the tube axis. This unique structure provides potential applications in a vast range of fields, which have stimulated wide efforts to develop controllable synthetic techniques [13]. Depending on the techniques, reaction conditions and metal catalysts, single-walled carbon nanotubes (SWCNTs) with a single graphene layer, double-walled carbon nanotubes (DWCNTs) with two graphene layers, and multiwalled carbon nanotubes (MWCNTs) with more than two coaxial graphene layers can be produced. The diameters of SWCNTs typically range from less than 1 nm to around 2 nm while those of DWCNTs and MWCNTs can be in the range of 2–100 nm or even larger. Figure 11.1 shows the arrangement of the graphene layers in SWCNTs, DWCNTs, and MWCNTs. Note that noncoaxial CNTs (having an herringbone structure) have also been produced, showing the graphitic shells positioned at a certain angle with respect to the tube axis [14]. In addition, nanotubes with internal closures resulting in closed compartments along the tube axis, that is, the so called 'bamboo-like' morphology have also been synthesized [15]. Herewith we will focus only on the co-axial nanotubes.

The unique tubular morphology of CNTs in the nanometer scale attracts wide attention for studying chemistry in their channels [16–22]. Modified adsorption

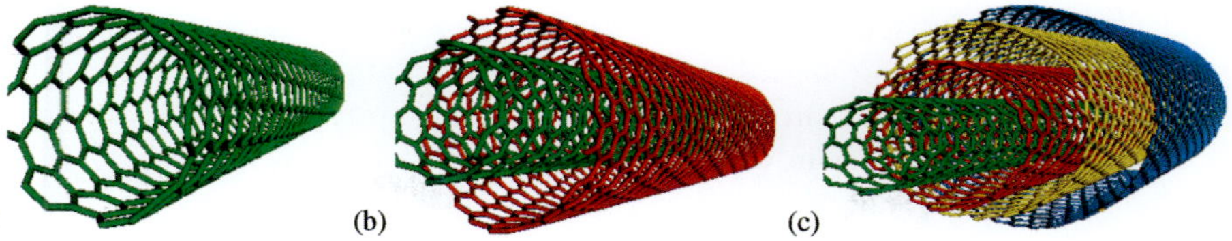

Figure 11.1 Representation of (a) single-walled carbon nanotube; (b) double-walled carbon nanotube; (c) multiwalled carbon nanotube.

and diffusion, and structural properties have been reported for a variety of molecules and nanomaterials inserted in the channels compared with their counterparts either on the outer walls of CNTs or in the bulk. The increasing numbers of studies demonstrate that confinement inside CNTs may affect chemical reactions with respect to those occurring on the outer surface of CNTs [22–24]. Such different behavior originates from both the physical (spatial restriction of the channels without blocking diffusion pathway of molecules) and chemical (interaction of confined species with the CNT walls) factors inside CNTs.

11.2.1
Spatial Restriction of the Carbon Nanotube Channels

The most obvious confinement effect is represented by the spatial restriction in the CNT channels. For example, Sloan *et al.* inserted KI into SWCNT channels with a diameter of ~1.4 nm and observed that KI crystals were three atomic layers thick by high resolution transmission electron microscope (HRTEM) [25]. These crystals exhibited a significant lattice distortion with respect to bulk crystals. The crystal grew with [001] (relative to bulk KI) parallel to the tube axes and consisted of continuous tetragonally distorted bilayer crystals composed of alternating columns of K-I and I-K pairs when viewed along [100]. This lattice distortion arose as a consequence of an interaction between the KI crystal and the tube wall. Similar effects were also observed for singular and binary halide crystals, and other materials encapsulated within SWCNTs [26].

The relatively rigid nanoscale channels can thus be used as templates or restricted spaces for the synthesis of nanomaterials, or to create novel composite materials with new structures and properties [27–35]. For example, Zhang *et al.* demonstrated that subnanometer sized titanium oxide particles can be obtained inside DWCNT channels with inner diameters around 1.0–1.5 nm [27]. The bright dots in the high-angle annular-dark field (HAADF) electron microscopy image (Figure 11.2b, which provides sub-angstrom resolution) corresponding to the heavy element titanium, were neatly aligned in the resulting sample, and the size of the majority of them was around 0.2 nm, even though some were probably bigger under the electron microscope due to overlapping of individual dots. Larger particles were only observed inside a few bigger nanotubes. These particles exhibited a rather good stability since no obvious aggregation was observed under the electron microscope when the sample was *in situ* heated up to about 773 K. On the other

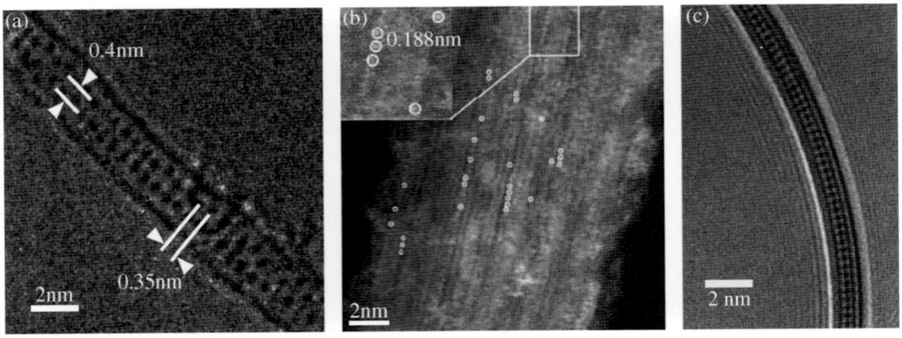

Figure 11.2 (a) HRTEM image of a two atom thick KI crystal formed within a discrete 1.4 nm diameter SWCNT. Each dark spot corresponds to a two atom K-I (or I-K) column [25]; (b) HAADF image of DWCNT bundles encapsulating titania particles with the inset showing an enlarged section marked with a grey frame and some bright dots marked with grey circles [27]; (c) HRTEM image of a single-atomic chain of Eu encapsulated in DWCNTs [30].

hand, the outside titania particles were larger because of the lack of physical restriction, and tended to aggregate. This method is expected to enable the synthesis of other sub-nanometer-sized metal and metal oxide clusters [27]. Correa-Duarte *et al.* reported linear assemblies of silica-coated gold nanoparticles using CNTs as templates [28]. Kim showed that γ-Fe particles with an fcc crystal structure (known to be stable in the temperature range 1185–1667 K in the bulk) retained their stability at room temperature inside MWCNTs [29].

Kitaura *et al.* synthesized ultrathin Eu nanowires in CNT channels [30]. By changing their types (SWCNT, DWCNT, or MWCNT) and the corresponding diameter of the CNTs, nanowires with different number of chains can be prepared (Figure 11.2c). Such composites showed a 10 times larger magnetic susceptibility than bulk Eu crystals at 4.25 K. Rao and coworkers reported interesting magnetic properties of iron nanowires encapsulated in aligned CNT bundles [31]. Elias *et al.* synthesized FeCo nanowires, which always exhibited the FeCo (110) plane parallel to the CNT axis. Such composites showed large coercive fields (e.g., 900 Oe) at room temperature [32]. Liu *et al.* synthesized long continuous cobalt nanorods with a few micrometers length and 20 nm in diameter within MWCNTs with average diameter about 40 nm [33]. X-ray diffraction (XRD) analysis showed that these cobalt nanorods possessed an fcc instead of the stable hexagonal structure. Pham-Huu *et al.* synthesized cobalt ferrite nanowires inside CNTs at atmospheric pressure and low temperatures by using the confinement effect, a process that would otherwise require high pressure hydrothermal conditions [34].

Thus, using the spatial restriction effect of CNT channels of varying sizes, a variety of nanomaterials, including ultrathin nanowires and particles of even subnanometer size, can be synthesized, which would be usually intrinsically unstable particularly under elevated temperature and pressure or difficult to obtain under mild conditions. In addition, the thin carbon shell can provide effective protection to the encapsulated materials against oxidation under ambient conditions. For example, CNT-encapsulated Eu nanowires were found to retain their structure upon exposure to air for one month [30]. Pelletier *et al.* showed that copper azide, an environmentally acceptable explosive but sensitive to electrostatic charges, could be stabilized via encapsulation into anodic aluminum oxide template CNTs [35]. At the same time, the well-ordered open structure of the tubes still allow diffusion of molecules in and out of the channels. These CNT-encapsulate composites with novel structures and properties may find application as catalysts, sensors, and magnetic materials.

11.2.2
Adsorption Inside Carbon Nanotubes

The CNT walls are essentially curved graphene layers, in which the sp^2 hybridization electronic structure is deformed. As a result, there is charge redistribution and polarization on the CNT surface [36]. This could cause molecules interacting differently with CNT surfaces with respect to planar graphene surface even for physisorption because of electrostatic forces and van der Waals interactions [37–42].

Adsorption sites with different adsorption energies have thus been identified on CNTs by theoretical simulations, including the interior sites (interior surface), groove and exterior sites (exterior surface of CNTs) [39–42]. The groove sites are narrow troughs on the outside of the bundles where a pair of nanotubes meets. It was demonstrated that a series of molecules, for example, H_2, CF_4 and 1-heptene, bind most strongly on the interior graphene surfaces of CNTs, followed by groove sites and exterior sites [39–42]. The difference depended mainly on diameters but was independent of the chirality [40]. Henwood *et al.* showed that the binding energy can be larger (by a factor of 2) for hydrogen inside nanotubes (9, 0) and (10, 0) than for adsorption outside or onto the graphene layer [41]. This was attributed to the curvature and the hydrogen–carbon electronic interactions, as analysis of the electron density evidenced the importance of charge redistribution.

Yates and coworkers employed temperature programmed desorption (TPD) and infrared (IR) spectroscopy and resolved these different adsorption sites experimentally for water, alkane, and alkene [39]. TPD spectra were obtained in a rapidly pumped ultrahigh vacuum system. The adsorbed molecules with different binding energies desorbed at different temperatures during the programmed heating process. Figure 11.3a displays the TPD spectra of *n*-pentane with increasing exposures. In combination with Monte Carlo simulation of the desorption process from three types of adsorption sites (Figure 11.3b), the highest desorption temperature peak in Figure 11.3a was assigned to the interior sites of SWCNTs, the next highest to the groove sites, and the one at lower temperature regime, to the nanotube exterior surface [39].

A first principles calculation indicated that both CO and H_2 were bound more strongly onto the interior surface of SWCNTs than onto the planar graphene surface, followed by those on the exterior surface [42]. In addition, CO interacted more

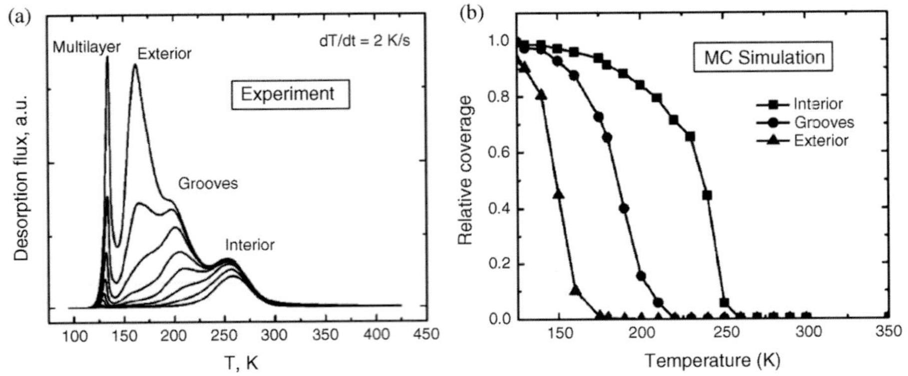

Figure 11.3 (a) Temperature programmed desorption profiles of *n*-pentane in SWCNTs with increasing exposure; (b) Monte Carlo simulated desorption of *n*-pentane from three types of adsorption sites [39].

strongly with the CNT surface than H_2. In combination with Monte Carlo simulations, it was found that CO and H_2 molecules were present in a higher density inside CNT channels with respect to the outside (Figure 11.4) when CNTs were placed in a mixture of CO and H_2 (volume ratio $CO/H_2 = 0.5$). This enrichment process generally became greater inside the smaller nanotubes at lower temperatures and higher pressures. Furthermore, the ratio of CO/H_2 inside CNTs was significantly higher than that in the bulk [42], as shown in Figure 11.4. This finding is significant for chemical reactions because a density increase could lead to an increased yield for reactions in which there is a decrease in the total number of moles according to Le Chatelier's principle. Furthermore, an altered ratio of one

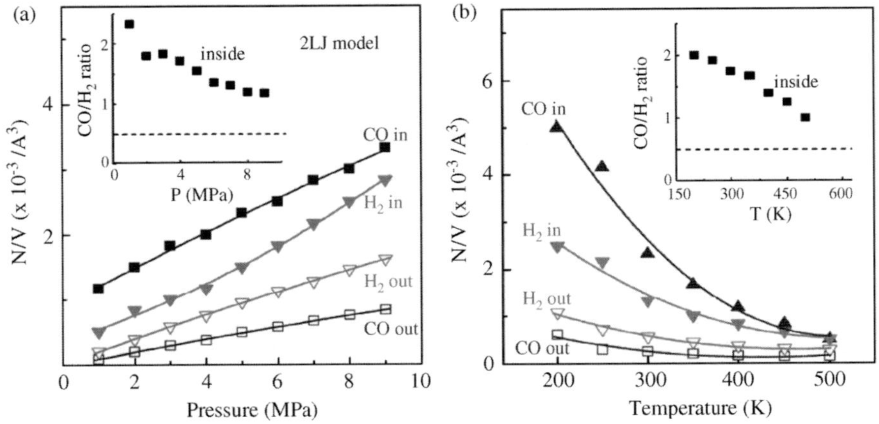

Figure 11.4 Theoretical prediction of molecular density of CO and H_2 in and outside SWCNTs (10, 10) (a) as a function of pressure; (b) as a function of temperature with the inset showing the ratio of CO/H_2 in comparison to the value 0.5 outside [42].

molecule over the other inside CNTs would modify product selectivity. However, direct experimental evidence of locally increased concentrations is still lacking.

11.2.3
Diffusion Inside Carbon Nanotubes

Diffusion in CNTs has been the subject of intense theoretical and experimental studies during recent years because of their potential applications in separation, chemical reactions, drug delivery, and sensing. In mesoporous materials diffusion is mainly governed by viscous or Poiseuille flow, Knudsen diffusion and surface diffusion. When the mean free molecular path is larger than the pore diameter and the molecular concentration is low, diffusion is usually controlled by the interaction with the pore walls, following the Knudsen diffusion mechanism. For most gases, transport in CNTs of a few nanometers in diameter would have been expected to follow Knudsen diffusion, which is characterized by a diffusion rate two to three orders of magnitude lower than that in the gas phase [43]. However, a faster transport inside CNTs than in the bulk is frequently reported.

The diffusivity of N_2 was theoretically predicted to be faster in SWCNTs with diameters ranging from 0.86 to 1.6 nm than that in the bulk [44]. Localization of N_2 into the core region, either by direct confinement in narrow tubes or by bilayer formation at higher loadings in larger nanotubes, could significantly enhance molecular diffusion in SWCNTs. The formation of this core region was a consequence of the steric confinement of the adsorbate by the CNT [44]. Molecular dynamic simulation showed that the diffusion of *n*-heptane inside individual SWCNTs was more than 30 times faster than the bulk diffusion at the same temperature even when the interior cavity was fully loaded [45]. This was attributed to the molecule ordering induced by confinement inside CNTs. Indeed, *n*-heptane molecules tended to be oriented parallel to the nanotube axis, which made the translational motion along the axis less hindered, leading to a faster diffusion. Experimental measurements were also carried out on the long-range self-diffusivity of *n*-heptane adsorbed on SWCNTs at 275 K [45]. This was performed by observing the displacement kinetics of heptane using labeled 1-deuteroheptane. The results showed that the measured diffusion coefficient was eight orders of magnitude lower than its simulated value inside an individual nanotube. This was likely complicated by the exchange of heptane molecules between different nanotubes.

Faster diffusion rates of adsorbed gases were also theoretically predicted in SWCNTs than in zeolites with similar pore sizes. For example, using atomistic simulations, Sholl and coworkers reported that light gases such as H_2 and CH_4 transported much faster in SWCNT (10, 10) than in silicalite or ZSM-12 [46]. Comparison with another SWCNT (12, 8) indicated that chirality was not important to the diffusion of light gases, since these two SWCNTs with different chirality exhibited almost the same diffusivities. The enhanced diffusion rate was frequently attributed to the smooth surface of CNTs, so that the presence of defects might slow down the diffusion process [45].

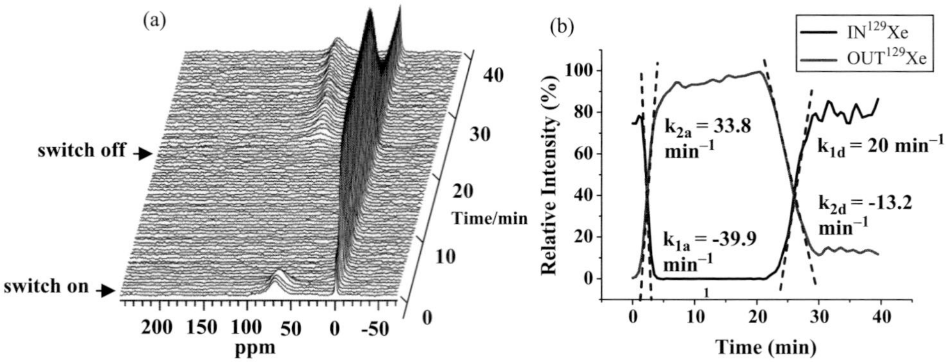

Figure 11.5 (a) *In situ* HP ^{129}Xe MAS NMR spectra recorded as a function of time during adsorption and desorption of methanol in SWCNTs at 253 K; (b) Change in the peak intensities of IN-Xe and OUT-Xe versus time. The solid lines represent normalized experimental data, and the dashed lines represent the best fit to the data of ^{129}Xe in nanochannels [47].

However, a recent experimental study on the diffusion rate of methanol using technical SWCNTs (inner diameter 2–2.5 nm), which are expected to contain defects, still showed a faster diffusion rate than in MCM-41 (average pore size 2 nm) [47]. This study employed solid state nuclear magnetic resonance (NMR) with highly sensitive hyperpolarized (HP) ^{129}Xe as the probe molecule. As shown in Figure 11.5, the HP ^{129}Xe magic angle spinning (MAS) NMR spectra taken at 253 K exhibited a signal at 0 ppm, which is attributed to Xe located outside of the channels (OUT-Xe), and another at lower field originating from Xe inside SWCNTs (IN-Xe). When the SWCNTs pre-adsorbed with ^{129}Xe were exposed to flowing methanol vapor, the intensity of IN-Xe dropped rapidly indicating that methanol drove ^{129}Xe out of the channels. When methanol was switched off, the IN-Xe signal did appear again, becoming more intense with time and indicating desorption of methanol and re-entering of ^{129}Xe in the SWCNT channels.

From the relative peak intensity change of IN-Xe and OUT-Xe (Figure 11.5b), where the signals were normalized to their highest intensities, one can estimate the methanol diffusion rate associated with the adsorption process to be $-39.9\,\text{min}^{-1}$. Compared with the initial adsorption process in empty MCM-41 pores, this rate was eight times faster. Monte Carlo and molecular dynamic simulations confirmed the faster diffusion coefficient of methanol in a SWCNT (10, 10) than in the pores of Si-VFI zeolite (1.5 nm size), in which diffusion falls in the Knudsen regime [47].

Fast diffusion of molecules encourages wide research efforts to design separation membranes based on CNTs [48]. Majumder *et al.* fabricated a membrane consisting of aligned MWCNTs with an inner diameter of about 7 nm crossing a solid polystyrene film [49]. The flow rates of water, ethanol, isopropanol, and of nonpolar liquids such as hexane and decane were four to five orders of magnitude larger than those predicted by the hydrodynamic equations. The permeability of gases such as N_2, CO_2, Ar, H_2 and CH_4 was inversely related with their molecular weight by an exponent of 0.42, close to predictions of Knudsen diffusion (0.5) [50]. However, the magnitude of the fluxes was 15- to 30-fold higher than that predicted from Knudsen diffusion kinetics and consistent with specular momentum reflection inside smooth pores. Holt *et al.* assembled a membrane with aligned DWCNTs with a diameter <2 nm [51]. It was found that the air flux exceeded the predictions of the Knudsen diffusion model by more than one order of magnitude and the water flow exceeded the values calculated from continuum hydrodynamics models by more than three orders of magnitude. Furthermore, both of the permeabilities of air and water through these membranes were several orders of magnitude higher than those of commercial polycarbonate membranes despite that they have pore sizes of one order of magnitude smaller [51]. Interestingly, the single-component selectivity for most of the gases exhibited the expected inverse-square-root scaling of molecular mass with the exception of hydrocarbons, the selectivity of which was higher. This was attributed to the preferential interaction of hydrocarbons with CNT sidewalls. Izadi-Najafabadi *et al.* recently used vertically aligned and densified SWCNTs (average diameter 2.6 nm) forest as electrode material [52]. They reported that the ion diffusion coefficient (KSCN in acetonitrile) in such an electrode was close to the theoretical limit of perfectly straight pores, assuming Fick's first law of diffusion. Indeed, it was 20 times larger than that of activated carbon electrodes [52]. The above experimental studies using technical CNTs confirmed fast transport inside CNT pores, which demonstrate the feasibility of creating a practical membrane separator or other device offering a high flux without degrading the selectivity.

11.2.4
Interaction of Confined Materials with the Graphene Layers of Carbon Nanotubes

In addition to the physical effects of confinement inside CNTs, the chemical nature of pore walls can influence the structure and properties of the confined materials through interactions with the graphene layers of CNTs. This has been observed for fullerenes, organic molecules, metallocenes, metal atoms and clusters, and ionic compounds.

The most frequently studied molecule inside CNTs is fullerene and its derivatives, which are well-known as peapods [53]. Khlobystov *et al.* reviewed the impact of the confinement on the molecular packing, orientation, translation, rotation, and reactivity for fullerenes and some metallocenes [54]. The interaction between fullerene molecules and nanotubes was mainly governed by van der Waals forces whereas endohedral metallofullerenes containing one or more metal atoms inside the carbon cage had an asymmetric charge distribution as a result of interaction of the incarcerated metal atom with the cage. For example, the effect of dipolar interactions on molecular orientations in CNTs was observed for $Ce@C_{82}$ molecules [54]. They tended to align their dipolar moments along the nanotube axis so as to maximize the electrostatic interactions between neighboring units, which was not observed in the bulk crystal or in solutions of Ce@C_{82}. With the help of low-temperature scanning tunneling microscopy (STM), Hornbaker *et al.* reported modification of the local electronic structure of SWCNTs in which C_{60} molecules were encapsulated [55]. Combined with theoretical calculations they showed that a periodic array of C_{60} molecules can give rise to new electronic bands, due to the interaction between the SWCNT band structure and the C_{60} molecular orbitals. Lee *et al.* also observed a strongly localized bandgap change in CNTs upon encapsulation of Gd metallofullerenes by STM, as shown in Figure 11.6a [56]. A bandgap of ~0.5 eV was narrowed down to ~0.1 eV at sites into which endohedral metallofullerenes were inserted. This change in the bandgap was explained by local elastic strain and charge transfer at metallofullerene sites.

Another type of molecule that has been frequently studied is the metallocenes, such as bis(cyclopentadienyl) iron [$FeCp_2$] and [$CoCp_2$]. They are organometallic molecules with a Fe or Co atom sandwiched between two cyclopentadienyl rings. Khlobystov and coworkers investigated the interaction of [$CoCp_2$] and bis(ethylcyclopentadienyl) cobalt molecules [$Co(EtCp)_2$] with SWCNTs by photoluminescence and ultraviolet-visible (UV-vis) absorption spectroscopy (Figure 11.6b) [57]. A red shift in the spectra was observed for the SWCNT-encapsulated [$CoCp_2$] and [$Co(EtCp)_2$] (Figure 11.6b), and was attributed to a change in the charge state of cobalt from +2 to +3 due to electron transfer to the nanotube. Charge transfer from metallocene molecules was also evidenced by the shift of the G band of CNT in Raman spectroscopy, which was attributed to carbon atom vibrations along the nanotube axis [58]. Shiozawa *et al.* analyzed the valence-band electronic structure of the composite formed with ferrocene inserted into the cavity of SWCNTs and compared it with the blank SWCNTs (i.e., without the encapsulates) using ultraviolet photoemission spectroscopy (Figure 11.6c) [59]. Taking potassium-intercalated SWCNTs as a reference, they concluded that the charge transfer density along the tube axis was 0.0067 e $Å^{-1}$ from [$FeCp_2$] to SWCNT. This charge transfer direction was reversed upon vacuum annealing at 873 K, which led to iron particle decorated DWCNTs [59]. By combination of nonequilibrium Green function techniques and density functional theory, Garcia-Suarez *et al.* predicted that 0.4–0.6 electrons were transferred in the CNT-encapsulated [$CoCp_2$] per [$CoCp_2$] molecule [60]. It was predicted that any other molecule with large electron affinity or small ionization potential would produce p-doped or n-doped SWCNTs, respectively, through the electron transfer between the molecules and the nanotubes [61].

Theoretical simulations showed that the interaction of transition metal atoms with the inner CNT walls differs from that with planar graphite layers with respect to bonding sites, magnetic moments and charge-transfer direction [62]. The work function was lowered upon decoration of CNTs with Ru nanoparticles, suggesting electron transfer from Ru to the CNT [63]. The first principles calculations on models of Ru_6 clusters located inside and outside CNT (10, 10) also showed electron transfer from the Ru_6 cluster to carbon [64]. Mulliken population analysis indicated that the inside Ru_6 cluster donated 2.41 electrons to the CNT while the outside one transferred 1.66 electrons to the CNT. In the model consisting of an icosahedral Ru_{13} cluster and a (12, 12) CNT, the inside Ru_{13} cluster donated 0.8 electrons more

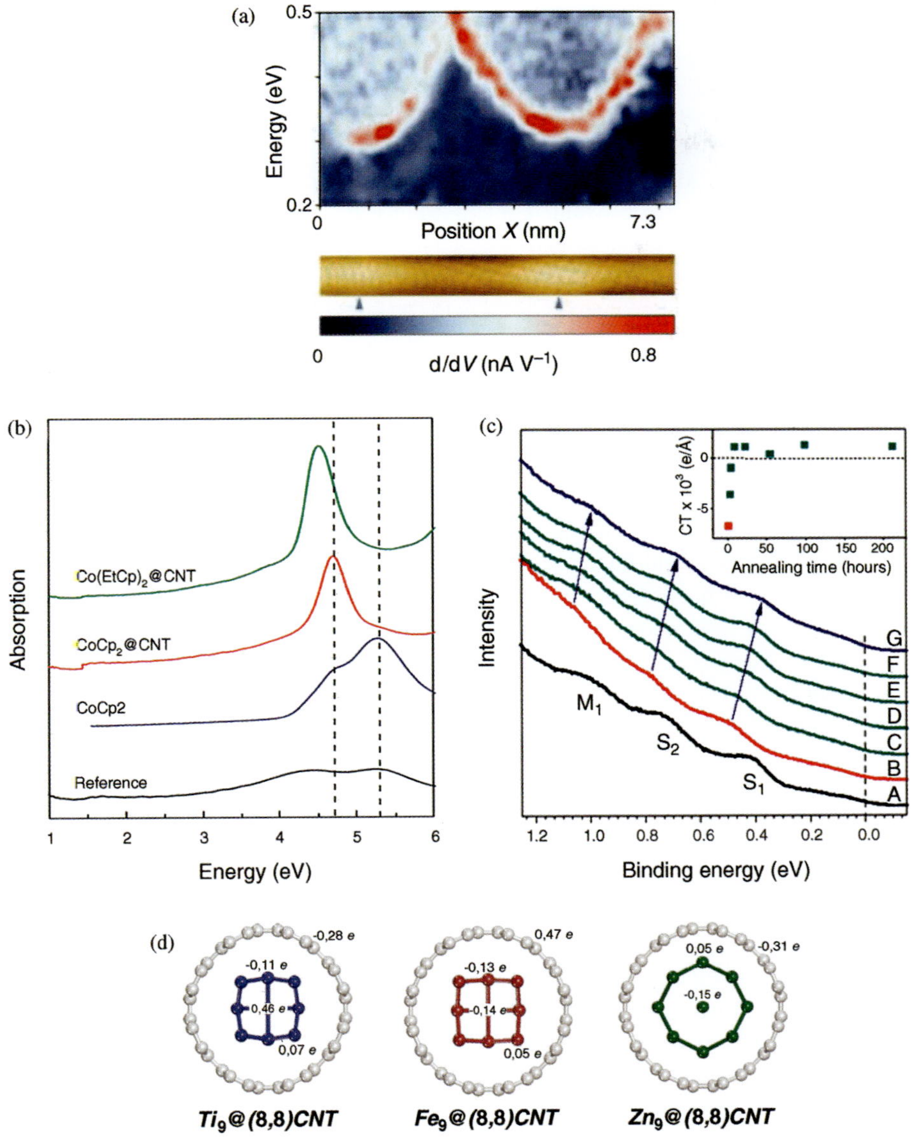

Figure 11.6 (a) Topographic image of a 7.3-nm-long (11,9) Gd metallofullerene@SWNT at a sample bias voltage of 0.5 V. The sites of inserted Gd metallofullerenes are indicated by arrows [56]; (b) UV-vis absorption spectra of SWCNT-confined [$CoCp_2$] indicating electron transfer from cobalt ions to the nanotube [57]; (c) He I photoemission spectra for A: pristine SWCNT, B: [$FeCp_2$]@NT, C-F: Fe doped DWCNT or [$FeCp_2$]@NT annealed at 873 K for 2, 8, 54, and 212 h, respectively; G: [$FeCp_2$]@NT annealed at 1423 K for 1 h. Inset: annealing time dependence of charge transfer density (CT) for the outer tube in e/Å [59]; (d) Calculated charges on atoms for medium-sized metal wires inside the armchair (8,8) CNT [65].

to the CNT surface than the outside Ru_{13} cluster. When a Ru_6 cluster was placed inside a DWCNT (8, 8)@(13, 13), it transferred 1.41 electrons more to carbon than the outside Ru_6 cluster. As a result, the inside and outside Ru clusters exhibited a slightly different electron density. This offers a fine tuning of the electronic structure of nanomaterials and CNTs.

The density functional theory (DFT) study on the structures of the Ti, Fe and Zn nanowires and clusters inside CNTs presented by Ivanovskaya *et al.* indicated that the interaction between the encapsulated metals and CNTs varied with specific metals [65]. Analysis of the electronic structure showed that the partially occupied Ti3*d* states were considerably hybridized with the C2*p* states for a Ti nanowire inside a CNT (Ti_n@CNT). For Fe_n@CNT composites, the bands in the energy region between −4 and +2 eV consisted predominantly of Fe3*d* states. In contrast, carbon and zinc valence states were well separated and their hybridization was very weak. A similar trend was observed for clusters of these metals. In addition, there was a charge transfer q of 0.47 e^- per unit cell from carbon to iron atoms for a Fe_9 nanowire in a CNT (8, 8). The transferred charge was not homogeneously distributed over the iron atoms. An opposite charge transfer direction was observed for Ti_9@CNT (8, 8) and Zn_9@CNT (8, 8) systems [65].

Warner *et al.* observed direct interaction of metal clusters with CNT walls using *in situ* low-voltage aberration-corrected HRTEM with an ultrafast temporal resolution of 80 ms [66]. Small metal clusters were formed on the nanotube under electron beam irradiation starting from La@C_{82} metallofullerene encapsulates. These metal clusters escaped from the interior of CNTs by first opening the CNT sidewalls, moving from inner to the outer surface, sealing the hole back up, detaching from the nanotube, and then rolling away. All this pointed to interaction of metal clusters with the graphene wall, which had catalyzed cleavage and formation of C–C bonds. Chamberlain *et al.* also demonstrated direct experimental evidence of the interaction of catalytically active rhenium atoms with SWCNTs at ambient temperature from aberration-corrected HRTEM [67]. These atoms readily reacted with C_2 biradicals deriving from fullerenes forming ReC_2 species, which bonded with the interior surface of SWCNTs upon irradiation by low energy electron beam. Thus, catalyzed by Re atoms, the structurally perfect inner surface of SWCNT was deformed or ruptured forming a metastable asymmetric nanoprotrusion with an open edge, and finally the nanoprotrusion was closed up via a slow symmetrization process.

Chen *et al.* introduced iron oxide nanoparticles into CNT channels, and their reduction was found to be facilitated with respect to the particles located on the outer walls of CNTs, as evidenced by *in situ* TEM, XRD and Raman spectroscopy [68, 69]. Dalai and coworkers deposited iron oxide particles of similar sizes inside and outside CNTs, and a temperature programmed reduction study showed facilitated reduction for the encapsulated iron oxide [70]. Facilitated reduction was also observed for Fe_3O_4 nanowires inside CNTs, which took place at 843 K in contrast to 973 K or more for a mechanical mixture of Fe_3O_4 powder with CNTs [71]. Even in the presence of an external reducing atmosphere, such as H_2 and CO, the reduction of CNT-confined iron oxide particles also took place at a temperature

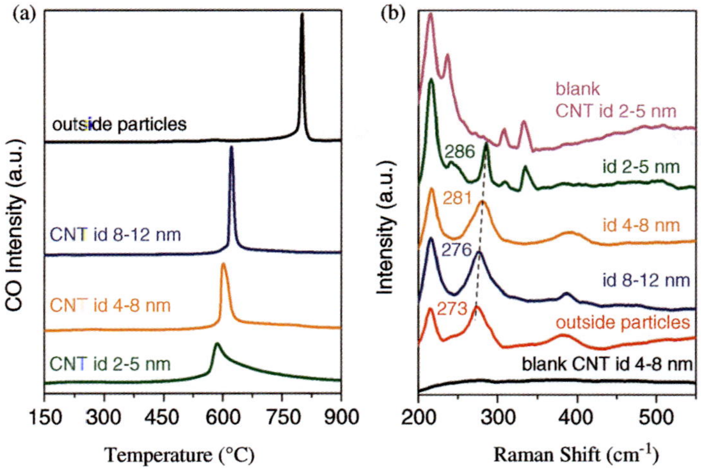

Figure 11.7 (a) Temperature-programmed desorption profiles in He for the CNT-confined Fe_2O_3 particles with different inner diameters; (b) their Raman spectra [69].

much lower than that for the outside iron oxide particles [69]. This is quite remarkable considering that the diffusion of H_2 or CO inside the CNT channels might slow down the reaction to a certain extent with respect to that occurring on the freely accessible outer walls. Such a facilitated reduction was also observed for nanoparticles of other metal oxides, such as ruthenium [64], rhodium, cobalt and nickel oxides.

Raman spectroscopy (Figure 11.7b) suggested that the Eg Fe—O band red shifted for Fe_2O_3 nanoparticles on the outer CNT walls with respect to bulk Fe_2O_3, in agreement with the general observation for Fe_2O_3 nanoparticles in earlier reports [69]. However, this frequency shift was reversed when the Fe_2O_3 particles were located on the interior of 8–12 nm inner diameter CNTs. It further blue-shifted within smaller inner diameter CNTs. Although the particle size decreased when CNT channels became narrower, the stepwise blue-shift of the Fe—O band does not line up with the red shift generally reported for nano-sized Fe_2O_3 implying the presence of an interaction of the Fe_2O_3 particles with the inner CNT surface. Furthermore, this interaction became stronger within smaller CNTs. Recently, Chen *et al.* also observed blue-shift of Mn-O frequency by 6 cm^{-1} with respect to the MnO_2 particles located on the outer walls of CNTs, which they ascribed to the partial reduction of MnO_2 to Mn_2O_3 by the CNT walls [72].

X-ray photoelectron spectroscopy (XPS), X-ray absorption near-edge spectroscopy (XANES), scanning transmission X-ray microscopy and Raman spectroscopy suggested that there was also an electronic interaction between the confined titania clusters and the DWCNT interior wall [27, 73]. This electron transfer was temperature dependent and was strong enough to be observed above 573 K. For example, the Ti 2p XPS peak was shifted 0.5 eV to a higher binding energy for TiO_x confined

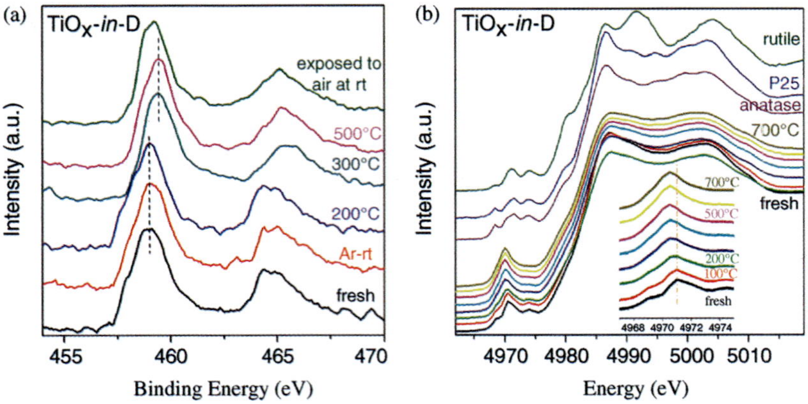

Figure 11.8 (a) Ti 2p XPS spectra; (b) XANES features of TiO_x-in-D upon thermal treatment in He [27].

inside DWCNTs (TiO_x-in-D) (Figure 11.8a), whereas that of TiO_x on the outside (TiO_x-out-D) did not exhibit obvious changes. Correspondingly, the 4970.5 eV peak in the pre-edge region was downshifted to 4970.0 eV for TiO_x-in-D at 573 K (Figure 11.8b).

Ilie *et al.* studied the effects of the encapsulation of the ionic compound KI in SWCNTs with a diameter distribution that peaks around 0.9–1.0 nm by Raman and optical absorption spectroscopy [74]. The encapsulation appeared to involve charge transfer and strain, which shrank both the C—C bonds and the overall nanotube along the axial direction. Estimation of the values of strain and charge transfer suggested a low level of electron depletion from the nanotube wall in a larger tube, while it was stronger in the narrower tubes, manifested through bleached regions in the optical absorption spectrum and the quenching of the Breit-Wigner-Fano (BWF) metallic feature of the Raman G-band [74].

11.3
Metal Catalyst-Free Chemical Reactions inside Carbon Nanotubes

The modified adsorption, diffusion of molecules and electronic structure of confined materials are expected to result in different reactivity. Theoretical studies predict that chemical reactions can be influenced by confinement inside CNT channels due to the significantly reduced volume and the interaction of molecules with CNT walls compared with those in the bulk [22, 75–77]. Santiso *et al.* identified different effects including changes in the thermodynamic state of the system due to interactions with the pore walls, selective adsorption, and geometrical constraints that can affect chemical equilibrium, kinetic rates and reaction mechanism [22]. The relative importance of each factor could be very different for different reactions.

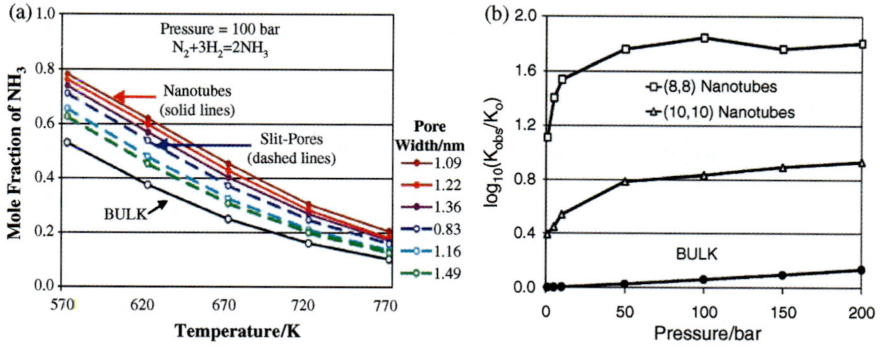

Figure 11.9 Effects of confinement on (a) ammonia synthesis; (b) HI decomposition in slit-shaped pores and carbon nanotubes of various sizes [77].

For example, the D and H_2 exchange reaction to form HD and H was predicted to be considerably enhanced inside CNTs compared with the gas-phase reaction [75]. Furthermore the reaction probabilities were considerably higher inside smaller CNTs. This enhancement was attributed to be steric in nature, having to do with the alignment of H_2 along the tube axis. Another theoretical study suggested that the isomerization of *n*-butane, 1-butene and 1,3-butadiene could be significantly modified by steric hindrance inside CNTs with a pore size comparable to the molecular dimensions [76].

Figure 11.9 shows simulated results on ammonia synthesis and bimolecular hydrogen iodide decomposition over models of carbon slit-pores and nanotubes using Reactive Monte Carlo method [77]. These results indicate that the equilibrium yield and kinetics of chemical reactions could be enhanced by confinement inside pores, and that the confinement effects could be different according to the structure and surface chemistry of the catalyst support material. For example, confinement within slit-shaped pores increased the conversion by almost 40%, and within nanotubes conversion can be increased by an additional 10–20%. The rate was enhanced within narrower pores, as the density was often greater in confined spaces. The enhancement in ammonia synthesis largely resulted from selective adsorption of the ammonia product. If some —COOH groups were introduced on the surface, the reaction could be further shifted toward formation of more NH_3 molecules due to favorable interaction of NH_3 with the surface, and its dipolar character [77].

Halls *et al.* showed by hybrid density functional theory that the polarizability of CNTs was responsible for a reaction involving dipolar species [78]. The enthalpy and activation energy of the Menshutkin SN_2 reaction

$$H_3N + H_3CCl \rightarrow H_3NCH_3{}^+ + Cl^-$$

in which the charged product $[H_3NCH_3]^+[Cl]^-$ was formed, were modified. Dipolar species confined within nanotubes interacted with CNTs via an induced image dipole on the tubes. Thus, the produced species were stabilized inside CNTs with a significantly reduced reaction volume relative to the gas phase, making the overall reaction more favorable. For example, the reaction endothermicity was reduced by more than 23 kcal mol^{-1} within the (9, 0) nanotubes. The activation energy was also significantly reduced compared with that in the gas phase. The effect of CNT confinement on reaction enthalpies was proposed to closely resemble that of solvation in low-dielectric solvents [78].

In addition to these theoretical simulations on metal-catalyst free chemical reactions, which show the effects of confinement inside CNTs, Kondratyuk *et al.* carried out an interesting experimental study on the reaction between 1-heptene and hydrogen atoms [79]. A lower reactivity of physisorbed 1-heptene toward atomic hydrogen was observed inside SWCNTs, compared with 1-heptene molecules adsorbed on external SWCNT sites. This was attributed to the shielding effect of nanotube walls, which prevented the reactive atomic H species from entering and undergoing reaction with the double bond of the confined reactant [79].

11.4 Catalytic Reactions over Metal Particles Confined Inside Carbon Nanotubes

Although metal and metal oxide species could be successfully introduced into CNT channels shortly after the discovery of CNTs [16–20], catalysis over confined metal species has been studied only recently. This is largely due to the difficulty in obtaining bulk-scale filled nanotubes for practical catalytic applications, particularly inside nanotubes with inner diameters smaller than 10 nm. Nevertheless, there is growing interest in this topic. A number of studies have shown that activity, selectivity and even stability can be modified via confinement with respect to the reactions over metal catalysts attached to the outer walls of CNTs and other carbon materials.

11.4.1 Liquid-Phase Catalytic Reactions

The $[HRh(CO)(PPh_3)_3]$ complex was deposited on CNTs with open tips for the liquid-phase hydroformylation of propylene [80]. This catalyst gave a turnover frequency (TOF) of 0.10 s^{-1} and a molar ratio of the linear/branched products $n/i = 9$ in comparison with a TOF of 0.06 s^{-1} and $n/i = 6$ for the same complex on CNTs without previous acid treatment and hence considered to have closed tips. Although no direct evidence was provided for the location of Rh, the higher activity and regioselectivity for the former catalyst could be attributed to Rh species inside CNTs and to their surface consisting of six-membered C-rings [80].

Pd particles were introduced into MWCNTs with inner diameters of 5–10 nm for benzene hydrogenation [81]. The resulting material exhibited a TOF twice as high as that on zeolite Y and AC-supported Pd catalysts, although zeolite Y and AC had much higher surface areas than CNTs.

Pd particles with a size of 4–6 nm were also introduced inside MWCNTs with an average inner diameter as large as 40 nm [82]. This catalyst showed good performance in the selective hydrogenation of cinnamaldehyde, which contains both C=O and C=C bonds. 100% of conversion was obtained after 25 h in comparison to 27 h for a high surface area activated charcoal supported Pd catalyst. More significantly, a much higher selectivity (90% versus around 50%) to hydrocinnamaldehyde (due to selective C=C bond hydrogenation) was obtained. Peculiar metal–support interactions, the absence of micropores and oxygenated surface groups on the CNT support were proposed to be responsible for these results.

Serp and co-workers reported a much better catalytic performance for PtRu particles inside MWCNTs with an average inner diameter of 40 nm for the same reaction, but with opposite selectivity [83]. The CNT-confined PtRu catalyst had 80% particles (with a size of 2 nm) located inside MWCNTs and the rest on their outer walls (with a size of 2.5 nm). The TOF on this CNT-confined PtRu catalyst was almost three times higher and the selectivity toward cinnamyl alcohol (resulting from C=O bond hydrogenation) was 95%, more than twice that on an unsupported PtRu catalyst with similar particle size (2.2 nm). Moreover, the selectivity was linearly correlated to the percentage of nanoparticles located inside CNTs (Figure 11.10). These results could be attributed to the confinement of both the active phase and the reactants in the inner cavity of CNTs [83]. Ma *et al.* compared two Pt catalysts: Pt particles on the outer surface of CNT-1 (with inner diameters of less than 10 nm) and Pt particles inside CNT-2 (with inner diameters between 60 and 100 nm) for the selective hydrogenation of cinnamaldehyde [84]. On Pt/CNTs-1 the product of complete hydrogenation of both C=C and C=O bonds was mainly obtained, while Pt/CNTs-2 gave a higher selectivity to cinnamyl alcohol (about 60%).

Chen *et al.* reported that a Pt nanocatalyst modified by cinchonidine was encapsulated within CNTs was very active for asymmetric hydrogenation of α-ketoesters [85]. The TOF was more than six times higher than that of Pt nanoparticles dispersed on the outside of CNTs, one order of magnitude higher than that of Pt/Al_2O_3 and more than one order of magnitude higher than that of Pt/AC. The enantioselectivity was also significantly superior to Pt outside, Pd/Al_2O_3 and Pt/AC. The superior catalytic performance of the CNT-encapsulated Pt catalyst was attributed mainly to the unique properties of the nanochannels of the CNTs as they can readily enrich both cinchonidine and the reactants.

So far, most of the experimental studies on reactions conducted on confined catalysts have been limited to MWCNTs with relatively larger inner diameters. Recently, Zhang *et al.* showed that subnanometer-sized titania particles confined in DWCNTs were active in catalytic propylene epoxidation [27], which was usually achieved by single titanium sites embedded in silica matrices, such as Ti-containing silicalite-1 (TS-1). Compared with titania particles located outside DWCNTs, and to titania inside MWCNTs (i.d. 4–8 nm), DWCNT-confined titania showed a formation rate of propylene oxide eight times higher than that on the outer surface of DWCNTs and twice as high as that in MWCNTs under the same reaction conditions. Note that no conversion was detected over blank DWCNTs, and that commercial P25 gave a 20 times lower activity. This enhanced activity is probably due to small titania clusters and to the interaction of confined titania with the DWCNT surface.

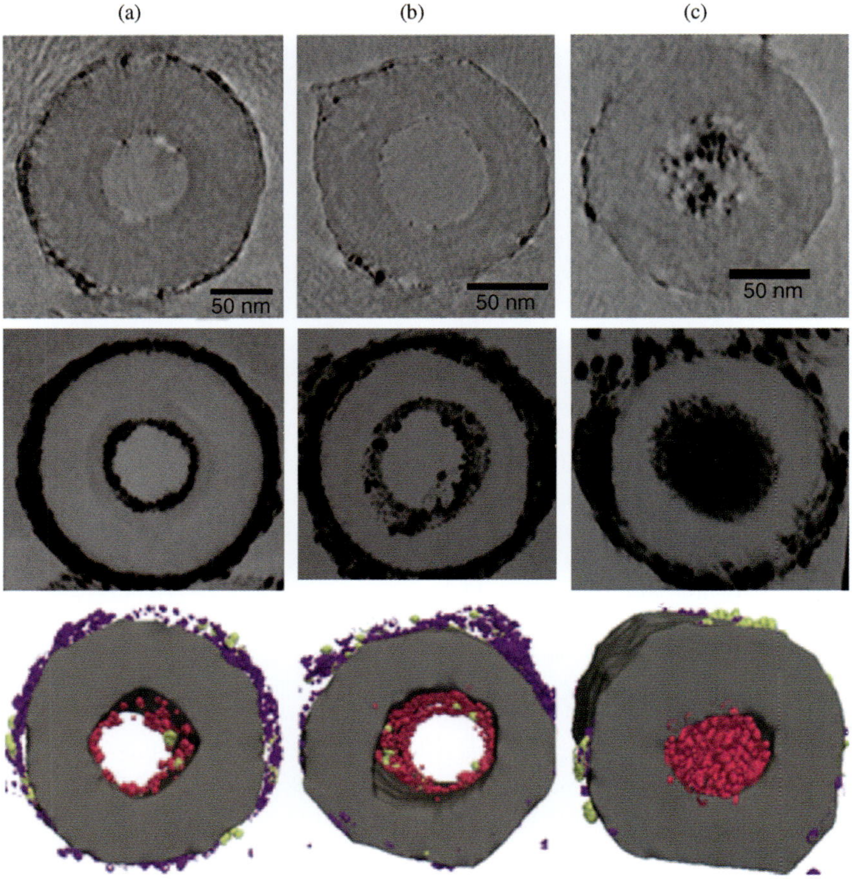

Figure 11.10 (a) 5 wt% PtRu/CNT2 with 90% of the particles located on the external CNT walls; (b) 11 wt% PtRu@CNT1 with about 30% of the particles inside CNTs; (c) 23 wt% PtRu@CNT3 with 80% of particles located inside CNTs. Violet: PtRu NPs outside; red: PtRu NPs inside; yellow: Au NPs [83].

11.4.2
Gas-Phase Catalytic Reactions

Iron nanoparticles were confined inside MWCNTs (i.d. 4–8 nm) (Fe-in) for Fischer–Tropsch synthesis [86]. For comparison, iron was also deposited on the outside of CNTs (Fe-out). TEM analysis indicated that over 70% of the iron particles of Fe-in were distributed inside CNT channels while almost all iron particles of Fe-out were on their outer surface. The confinement benefited syngas conversion since CO conversion was almost 1.5 times, and the yield in C_{5+} hydrocarbons (hydrocarbons containing more than 5 carbon atoms) was twice as high as that on Fe-out (Figure 11.11a). In addition, the yield of C_{5+} hydrocarbons was six times

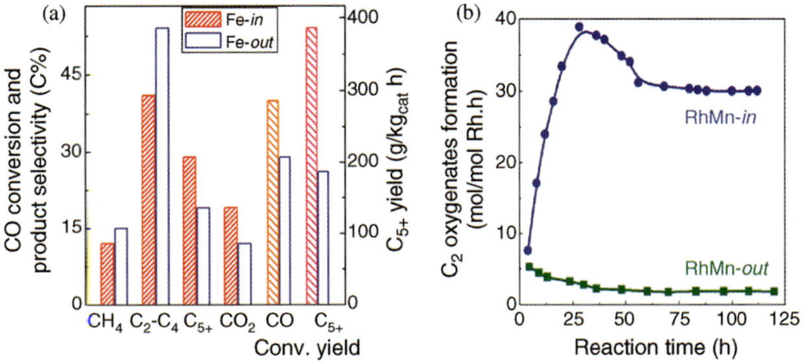

Figure 11.11 (a) Fischer–Tropsch synthesis activity of iron catalyst confined in CNTs compared with the outside iron catalyst [24]; (b) Syngas conversion to C_2 oxygenates on CNT-confined RhMn and RhMn catalysts deposited outside [24].

higher than that on an iron catalyst supported on the XC-72 carbon black, which had a surface area similar to that of the MWCNTs.

Characterization with *in situ* HRTEM, XRD, Raman spectroscopy, and TPD revealed that the reduction of Fe_2O_3 particles was facilitated within MWCNTs with respect to the outside Fe_2O_3 particles [86]. Consequently, a relatively higher content of iron carbide was formed over Fe-in than that on Fe-out. Iron carbide is generally considered to be an active phase for Fischer–Tropsch synthesis. Better reducibility and enhanced catalytic performance of CNT-confined iron (Fe-in) were also observed by Dalai and coworkers [70]. The Fe-in catalyst had 70–80% of Fe particles located inside CNTs with a particle size of 6–11 nm, whist the Fe-out particles were slightly smaller (5–9 nm). The results demonstrated that confinement inside CNTs gave a better long term stability although the initial CO conversion was similar at 2 MPa and a space velocity of $2\,L\,g^{-1}\,h^{-1}$). Fe-in was much more stable since CO conversion was 6–10% higher than that over Fe-out in the temperature range 538–558 K, and selectivity to C_{5+} hydrocarbons was higher by 13% after 125 h on stream [70]. The above results indicate that the FTS activity difference between inside and outside catalysts is sensitive to pressure and space velocity.

Confinement was also found to significantly enhance the catalytic activity of a bicomponent RhMn catalyst (RhMn-in) in syngas conversion to C_2 oxygenates with respect to the external catalyst (RhMn-out) [87], as shown in Figure 11.11b. TEM indicated that the particle size was not the only factor affecting the activity because the two fresh catalysts had similar size distributions. For comparison, a catalyst was supported on SBA-15, with a pore diameter of 6–8 nm similar to that of the CNTs employed. This catalyst exhibited a much lower activity and a higher selectivity to methane (byproduct) under the same reaction conditions, suggesting that the spatial restriction may not be the only effect. In addition to the probable increased local density of reactants within small channels, Raman spectroscopy indicated that the activation of CO could have been modified to provide a facilitated CO dissociation inside CNTs with respect to the catalyst outside [87].

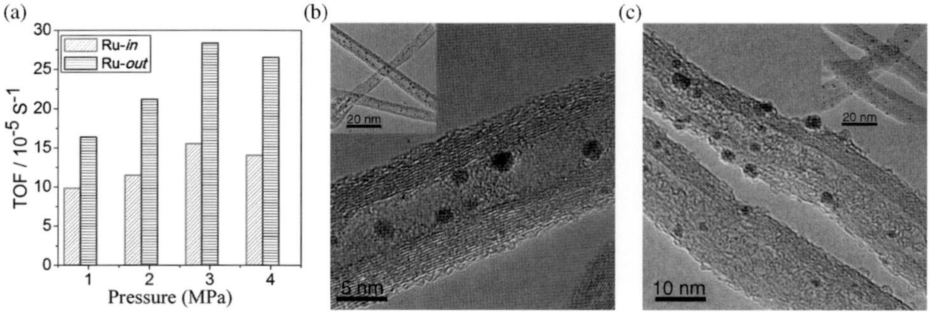

Figure 11.12 (a) Activity of the CNT-confined Ru and outside Ru in ammonia synthesis [24]; (b) TEM image for Ru nanoparticles confined inside MWCNTs; (c) TEM image for Ru nanoparticles dispersed on the outer walls of MWCNTs.

Li *et al.* prepared 5 wt% Ru nanoparticles confined in the channels (Ru/CNTs-in), dispersed on the outer surface of CNTs (Ru/CNTs-out) and also on other supports [88]. HRTEM, XRD and H_2-temperature programmed reduction revealed highly dispersed Ru nanoparticles with a mean size of 1.8 nm for both catalysts. In the preferential oxidation of CO in the presence of hydrogen (CO-PROX) (1.0% CO, 1.0% O_2 and 98% H_2), the Ru/CNTs-in catalyst showed the best activity with 100% CO conversion at 323–393 K. However, lower CO conversions and narrower temperature window were observed over the 5 wt% Ru/CNTs-out catalyst and other supported catalysts under identical conditions.

By contrast, confinement of Ru inside CNTs (Ru-in) was found to have a negative impact on ammonia synthesis [64]. The TOF on Ru-in was much lower than that on Ru-out (the outside Ru catalyst) (Figure 11.12a). TEM (Figure 11.12b and c) and chemisorption experiments confirmed that the particle sizes of Ru-in and Ru-out were similar, with ~90% of particles falling in the range 2–5 nm. Even after reaction, the particle size distribution changed only slightly, indicating that both catalysts were rather stable under reaction conditions. Adsorption microcalorimetry showed that the initial differential heat for CO adsorption on Ru-out was 12 kJ mol^{-1} higher than that over Ru-in, indicating a stronger adsorption on Ru-out. The different strength of adsorption sites was most likely related to the electron density on metal surfaces since the employed Ru-in and Ru-out presented a similar particle size. First principles calculations indicated a lower electron density on the inside Ru with respect to the outside Ru clusters [64]. This slightly different electron density could possibly cause the different activity in the electrophilic process of N_2 dissociative adsorption. Interestingly, our recent study indicated that the inside and outside difference in activity can be reversed if iron was used as a catalyst, that is, the inside Fe catalyst exhibiting a higher activity than the outside Fe under the same conditions [89].

Ammonia decomposition, however, benefits from confinement of the bimetallic $CoFe_5$ inside CNTs with an average inner diameter of 40 nm ($CoFe_5$-in) [90]. 96 wt% of the particles of $CoFe_5$-in were located inside the channels and that

$CoFe_5$-out had 73 wt% of particles located on the outer surface, as revealed by TEM. The two fresh catalysts exhibited a similar particle size distribution and the same $CoFe_2O_4$(311) planes. NH_3 conversion on $CoFe_5$-out was roughly half that on $CoFe_5$-in after ~16 h on stream. The higher conversion on $CoFe_5$-in was attributed to the better thermal stability of the inside particles [90]. Electron transfer within the graphene walls was expected to be much smaller in large CNTs because of their weaker curvature [35]. Thus it likely had little influence on the electronic structure of metal particles inside wide CNTs. The above results show that the activity difference between the inside and outside catalysts differs in ammonia synthesis and decomposition. This could be due to confinement effect on specific metals in CNTs of different diameters.

Centi and Perathoner reported gas-phase photoelectrocatalytic conversion of CO_2 to isopropanol and other alcohols/hydrocarbons using nanoconfined carbon electrodes based on Fe [91]. The results showed that the localization of iron inside the tubes led to enhanced productivity and also influenced the product distribution.

Nhut *et al.* deposited NiS_2 inside MWCNTs with an inner diameter of 50–80 nm and used it as a catalyst for oxidation of H_2S [92]. The confined NiS_2 exhibited an enhanced activity and resistance to the solid sulfur deposition compared with NiS_2 supported on macroscopic SiC grains. The better performance of the confined NiS_2 was attributed to the absence of bottled pores with minimum diffusional phenomena, a confinement effect inside CNTs, the highly faceted morphology of NiS_2 particles, as well as the improved storage capacity of solid sulfur due to the higher void volume outside the MWCNTs compared with SiC [92].

Reactions on confined catalysts are studied by theoretical simulations to a lesser extent. A recent DFT calculation has indicated that the confinement effects depend not only on the specific reaction, but also on the chosen metal catalyst and its size [93]. This study examined confinement of small Au_n ($n = 1$–5) clusters within SWCNTs, and the effects in H_2 dissociation. It was found that the encapsulated Au clusters interacted more strongly with metallic CNTs than with semi-conducting ones, for which charge transfer from CNTs to Au clusters played an important role because negative charged Au clusters were unfavorable for H_2 adsorption compared with free standing Au clusters. Therefore, encapsulation of gold clusters into CNTs did not favor H_2 dissociation though thermodynamic dissociation was enhanced inside CNTs except for encapsulated Au dimer, which provided higher activity than the outside one [93].

Qin *et al.* investigated the catalytic effect of CuO nanoparticles confined in CNTs using molecular dynamic simulations and DFT calculations [94]. Ozone decomposition and hydroxyl radical generation were used as probe reactions. MD simulation showed that there was increased density of O_3 in the nanotube. In addition, confinement could change the electronic properties of CuO particles. Interface interactions between reactants and CuO nanoparticles in the channel and charge transfer were proposed to be responsible for the accelerated activity.

11.5 Summary

The tubular morphology of CNTs with diameters in the nanometer scale has created great opportunities for studying chemistry in confined spaces not only because the thin carbon walls provide protection against environmental degradation, but also because the tubes exhibit intriguing confinement effects. Encapsulation of molecules and guest materials has been demonstrated with potential to generate unique 1D molecular nanostructures that may bring forth novel electronic, magnetic and catalytic properties.

An increasing number of studies show that confinement could allow the fine tuning of catalytic performances without much changing the catalyst composition. Looking into the nature of the confinement effects, they can be either physical or chemical. The physical nature is straightforward, which is related to the geometry of the CNT channels, that is, to the space restriction on the orientation and conformation of confined molecules and nanocatalysts. The chemical factors are due to the chemical nature of the pores, which influences the interaction of the confined materials with the pore wall. The chemical nature of the pores can be tuned by introducing heteroatoms into the graphene network, such as boron and nitrogen, which is often concomitant with modified electronic structures [95, 96]. This can further affect the adsorption, diffusion and selective stabilization of certain reactants, intermediates or products through interactions, and thus can have a profound effect on reaction equilibria and rates. Interaction can also modify the structures of catalysts, and may completely alter the reaction mechanisms.

It appears that all the above effects can influence reactions to different extents depending on the selected metals, the diameter of CNTs, and the specific reactions. It is frequently a collective result of more than just one of these effects. Although much effort has been made to assess the relative importance of each effect in order to gain fundamental insights, they are still rather difficult to distinguish in most cases. It would certainly require CNTs with better defined size, structure and properties, and efficient techniques for filling small nanotubes at a bulk scale, as well as advanced characterization techniques, to examine the catalytic role of local structures under working conditions. At the same time, theoretical simulations on models closer to realistic systems could offer further insight into confinement effects.

Potentials as catalytic supports have been widely demonstrated for CNTs since their discovery. Their commercial availability on the large scale has also brought CNTs one step closer to application. However, there are still a lot of challenges, for example, fabrication of technical CNT-based catalysts mainly because of shaping difficulties. In this respect, successful growth of CNTs on silica spheres [97], alumina particles [98] and inert SiC substrates [99] recently offers a possible solution since silica, alumina and SiC are compactable, and are already used as catalyst supports in many industrial processes.

Acknowledgment

We acknowledge the financial support from the Ministry of Science and Technology of China (No. 2011CBA00503) and the National Science Foundation of China (No. 11079005 and 21033009).

References

1 Christenson, H.K. (2001) Confinement effects on freezing and melting. *J. Phys.: Condens. Matter*, **13**, R95–R133.
2 Derouane, E.G. and Chang, C.D. (2000) Confinement effects in the adsorption of simple bases by zeolites. *Microporous Mesoporous Mater.*, **35–36**, 425–433.
3 Guegan, R., Morineau, D., and Alba-Simionesco, C. (2005) Interfacial structure of an H-bonding liquid confined into silica nanopores with surface silanols. *Chem. Phys.*, **317**, 236–244.
4 Schneider, M.S., Grunwaldt, J.-D., and Baiker, A. (2004) Near-critical CO_2 in mesoporous silica studied by in situ FTIR spectroscopy. *Langmuir*, **20**, 2890–2899.
5 Goettmann, F. and Sanchez, C. (2007) How does confinement affect the catalytic activity of mesoporous materials? *J. Mater. Chem.*, **17**, 24–30.
6 Lesthaeghe, D., Speybroeck, V.V., and Waroquier, M. (2009) Theoretical evaluation of zeolite confinement effects on the reactivity of bulky intermediates. *Phys. Chem. Chem. Phys.*, **11**, 5222–5226.
7 Smit, B. and Maesen, T.L.M. (2008) Towards a molecular understanding of shape selectivity. *Nature*, **451**, 671–678.
8 Derouane, E.G. (1986) Shape selectivity in catalysis by zeolites: the nest effect. *J. Catal.*, **100**, 541–544.
9 Thomas, J.M. and Raja, R. (2008) Exploiting nanospace for asymmetric catalysis: confinement of immobilized, single-site chiral catalysts enhances enantioselectivity. *Acc. Chem. Res.*, **41**, 708–720.
10 Fraile, J.M., García, J.I., Herrerías, C.I., Mayoral, J.A., and Pires, E. (2009) Enantioselective catalysis with chiral complexes immobilized on nanostructured supports. *Chem. Soc. Rev.*, **38**, 695–706.
11 Zhan, B.-Z., White, M.A., Sham, T.-K., Pincock, J.A., Doucet, R.J., Rao, K.V.R., Robertson, K.N., and Cameron, T.S. (2003) Zeolite-confined nano-RuO_2: A green, selective, and efficient catalyst for aerobic alcohol oxidation. *J. Am. Chem. Soc.*, **125**, 2195–2199.
12 Iijima, S. (1991) Helical microtubules of graphitic carbon. *Nature*, **354** (6348), 56–58.
13 Dai, H. (2002) Carbon nanotubes: Synthesis, integration, and properties. *Acc. Chem. Res.*, **35**, 1035–1044.
14 de Jong, K.P. and Geus, J.W. (2000) Carbon nanofibers: catalytic synthesis and applications. *Catal. Rev. Sci. Eng.*, **42** (4), 481–510.
15 Lee, C.J., Park, J.H., and Park, J. (2000) Synthesis of bamboo-shaped multiwalled carbon nanotubes using thermal chemical vapor deposition. *Chem. Phys. Lett.*, **323**, 560–565.
16 Ajayan, P.M. and Iijima, S. (1993) Capillarity-induced filling of carbon nanotubes. *Nature*, **361**, 333–334.
17 Ajayan, P.M., Ebbesen, T.W., Ichihashi, T., Iijima, S., Tanigaki, K., and Hiura, H. (1993) Opening carbon nanotubes with oxygen and implications for filling. *Nature*, **362**, 522–525.
18 Guerretpiecourt, C., Le Bouar, Y., Loiseau, A., and Pascard, H. (1994) Relation between metal electronic structure and morphology of metal compounds inside carbon nanotubes. *Nature*, **372**, 761–765.
19 Dujardin, E., Ebbesen, T.W., Hiura, H., and Tanigaki, K. (1994) Capillarity and wetting of carbon nanotubes. *Science*, **265**, 1850–1852.

20 Ugarte, D., Chatelain, A., and de Heer, W.A. (1996) Nanocapillarity and chemistry in carbon nanotubes. *Science*, **274**, 1897–1899.

21 Gimenez-Lopez, M.D.C., Moro, F., Torre, A.L., Gómez-García, C.J., Brown, P.D., van Slageren, J., and Khlobystov, A.N. (2011) Encapsulation of single-molecule magnets in carbon nanotubes. *Nat. Commun.* doi: 10.1038/ncomms1415

22 Santiso, E.E., George, A.M., Turner, C.H., Kostov, M.K., Gubbins, K.E., Buongiorno-Nardelli, M., and Sliwinska-Bartkowiak, M. (2006) Adsorption and catalysis: The effect of confinement on chemical reactions. *Appl. Surf. Sci.*, **252**, 766–777.

23 Serp, P. and Castillejos, E. (2010) Catalysis in carbon nanotubes. *ChemCatChem*, **2**, 41–47.

24 Pan, X. and Bao, X. (2011) The effects of confinement inside carbon nanotubes on catalysis. *Acc. Chem. Res.*, **44**, 553–562.

25 Sloan, J., Novotny, M.C., Bailey, S.R., Brown, G., Xu, C., Williams, V.C., Friedrichs, S., Flahaut, E., Callender, R.L., York, A.P.E., Coleman, K.S., Green, M.L. H., Dunin-Borkowski, R.E., and Hutchison, J.L. (2000) Two layer 4:4 co-ordinated KI crystals grown within single walled carbon nanotubes. *Chem. Phys. Lett.*, **329**, 61–65.

26 Sloan, J., Kirkland, A.I., Hutchison, J.L., and Green, M.L.H. (2002) Integral atomic layer architectures of 1D crystals inserted into single walled carbon nanotubes. *Chem. Commun.*, 1319–1332.

27 Zhang, H., Pan, X., Liu, J., Qian, W., Wei, F., Huang, Y., and Bao, X. (2011) Enhanced catalytic activity of subnanometer titania clusters confined inside double-wall carbon nanotubes. *ChemSusChem*, **4**, 975–980.

28 Correa-Duarte, M.A., Sobal, N., Liz-Marzan, L.M., and Giersig, M. (2004) Linear assemblies of silica-coated gold nanoparticles using carbon nanotubes as templates. *Adv. Mater*, **16**, 2179–2184.

29 Kim, H. and Sigmund, W. (2005) Iron nanoparticles in carbon nanotubes at various temperatures. *J. Crystal Growth*, **276**, 594–605.

30 Kitaura, R., Nakanishi, R., Saito, T., Yoshikawa, H., Awaga, K., and Shinohara, H. (2009) High-yield synthesis of ultrathin metal nanowires in carbon nanotubes. *Angew. Chem. Int. Ed.*, **48**, 8298–8302.

31 Satishkumar, B.C., Govindaraj, A., Vanitha, P.V., Raychaudhuri, A.K., and Rao, C.N.R. (2002) Barkhausen jumps and related magnetic properties of iron nanowires encapsulated in aligned carbon nanotube bundles. *Chem. Phys. Lett.*, **362**, 301–306.

32 Elias, A.L., Rodriguez-Manzo, J.A., McCartney, M.R., Golberg, D., Zamudio, A., Baltazar, S.E., Lopez-Urias, F., Munoz-Sandoval, E., Gu, L., Tang, C.C., Smith, D.J., Bando, Y., Terrones, H., and Terrones, M. (2005) Production and characterization of single-crystal FeCo nanowires inside carbon nanotubes. *Nano Lett.*, **5**, 467–472.

33 Liu, S., Zhu, J., Mastai, Y., Felner, I., and Gedanken, A. (2000) Preparation and charastristics of carbon nanotubes filled with cobalt. *Chem. Mater.*, **12**, 2205–2211.

34 Pham-Huu, C., Keller, N., Estournes, C., Ehret, G., and Ledoux, M.J. (2002) Synthesis of $CoFe_2O_4$ nanowires in carbon nanotubes. A new use of the confinement effect. *Chem. Commun.*, 1882–1883.

35 Pelletier, V., Bhattacharyya, S., Knoke, I., Forohar, F., Bichay, M., and Gogotsi, Y. (2010) Copper azide confined inside templated carbon nanotubes. *Adv. Funct. Mater.*, **20**, 3168–3174.

36 Haddon, R.C. (1993) Chemistry of the fullerenes: the manifestation of strain in a class of continuous aromatic molecules. *Science*, **261**, 1545–1550.

37 Kostov, M.K., Cheng, H., Cooper, A.C., and Pez, G.P. (2002) Influence of carbon curvature on molecular adsorptions in carbon-based materials: a force field approach. *Phys. Rev. Lett.*, **89**, 146105.

38 Britz, D.A. and Khlobystov, A.N. (2006) Noncovalent interactions of molecules with single walled carbon nanotubes. *Chem. Soc. Rev.*, **35**, 637–659.

39 Kondratyuk, P., Wang, Y., Johnson, J.K., and Yates, J.T. (2005) Observation of a one-dimensional adsorption site on carbon nanotubes: adsorption of alkanes of different molecular lengths. *J. Phys. Chem. B*, **109**, 20999–21005.

40 Ferre-Vilaplana, A. (2005) Ab initio computational investigation of

physisorption of molecular hydrogen on achiral single-walled carbon nanotubes. *J. Chem. Phys.*, **122**, 214724.

41 Henwood, D. and Carey, J.D. (2007) Ab initio investigation of molecular hydrogen physisorption on graphene and carbon nanotubes. *Phys. Rev. B*, **75**, 245413.

42 Guan, J., Pan, X., Liu, X., and Bao, X. (2009) Syngas segregation induced by confinement in carbon nanotubes: A combined first-principle and Monte-Carlo study. *J. Phys. Chem. C*, **113**, 21687–21692.

43 Malek, K. and Coppens, M.O. (2003) Knudsen self- and Fickian diffusion in rough nanoporous media. *J. Chem. Phys.*, **119**, 2801–2811.

44 Arora, G., Wagner, N.J., and Sandler, S.I. (2004) Adsorption and diffusion of molecular nitrogen in single wall carbon nanotubes. *Langmuir*, **20**, 6268–6277.

45 Kondratyuk, P., Wang, Y., Liu, J., Johnson, J.K., and Yates, J.T. (2007) Inter- and intratube self-diffusion in n-heptane adsorbed on carbon nanotubes. *J. Phys. Chem. C*, **111**, 4578–4584.

46 Skoulidas, A.I., Ackerman, D.M., Johnson, J.K., and Sholl, D.S. (2002) Rapid transport of gases in carbon nanotubes. *Phys. Rev. Lett.*, **89**, 185901.

47 Xu, S., Zhang, W., Li, X., Han, X., Liu, X., and Bao, X., *Enhanced Diffusion of Methanol in Single-walled Carbon Nanotubes: An Evidence by in situ Hyperpolarized* ^{129}*Xe* NMR. Submitted for publication.

48 Hinds, B.J., Chopra, N., Rantell, T., Andrews, R., Gavalas, V., and Bachas, L.G. (2003) Aligned multiwalled carbon nanotubes membranes. *Science*, **303**, 62–65.

49 Majumder, M., Chopra, N., Andrews, R., and Hinds, B.J. (2005) Enhanced flow in carbon nanotubes. *Nature*, **438**, 44.

50 Majumder, M., Chopra, N., and Hinds, B.J. (2011) Mass transport through carbon nanotubes membranes in three different regimes: ionic diffusion and gas and liquid flow. *ACS Nano*, **5**, 3867–3877.

51 Holt, J.K., Park, H.G., Wang, Y., Stadermann, M., Artyukhin, A.B., Grigoropoulos, C.P., Noy, A., and Bakajin, O. (2006) Fast mass transport through sub-2-nanometer carbon nanotubes. *Science*, **312**, 1034–1037.

52 Izadi-Najafabadi, A., Futaba, D.N., Iijima, S., and Hata, K. (2010) Ion diffusion and electrochemical capacitance in aligned and packed single-walled carbon nanotubes. *J. Am. Chem. Soc.*, **132**, 18017–18019.

53 Smith, B.W., Monthioux, M., and Luzzi, D.E. (1998) Encapsulated C_{60} in carbon nanotubes. *Nature*, **396**, 323–324.

54 Khlobystov, A.N., Britz, D.A., and Briggs, G.A.D. (2005) Molecules in carbon nanotubes. *Acc. Chem. Res.*, **38**, 901–909.

55 Hornbaker, D.J., Kahng, S.-J., Misra, S., Smith, B.W., Johnson, A.T., Mele, E.J., Luzzi, D.E., and Yazdani, A. (2002) Mapping the one-dimensional electronic states of nanotube peapod structures. *Science*, **295**, 828–831.

56 Lee, J., Kim, H., Kahng, S.-J., Kim, G., Son, Y.-W., Ihm, J., Kato, H., Wang, Z.W., Okazaki, T., Shinohara, H., and Kuk, Y. (2002) Bandgap modulation of carbon nanotubes by encapsulated metallofullerenes. *Nature*, **415**, 1005–1008.

57 Li, L.J., Khlobystov, A.N., Wiltshire, J.G., Briggs, G.A.D., and Nicholas, R.J. (2005) Diameter-selective encapsulation of metallocenes in single-walled carbon nanotubes. *Nat. Mater.*, **4**, 481–485.

58 Dresselhaus, M.S., Dresselhausb, G., Saito, R., and Jorio, A. (2005) Raman spectroscopy of carbon nanotubes. *Phys. Reports*, **409**, 47–99.

59 Shiozawa, H., Pichler, T., Kramberger, C., Grüneis, A., Knupfer, M., Büchner, B., Zólyomi, V., Koltai, J., Kürti, J., Batchelor, D., and Kataura, H. (2008) Fine tuning the charge transfer in carbon nanotubes via the interconversion of encapsulated molecules. *Phys. Rev. B*, **77**, 153402.

60 Garcia-Suarez, V.M., Ferrer, J., and Lambert, C.J. (2006) Tuning the electrical conductivity of nanotubes-encapsulated metallocene wires. *Phys. Rev. Lett.*, **96**, 106804.

61 Takenobu, T., Takano, T., Shiraishi, M., Murakami, Y., Ata, M., Kataura, H., Achiba, Y., and Iwasa, Y. (2003) Stable and controlled amphoteric doping by encapsulation of organic molecules inside carbon nanotubes. *Nat. Mater.*, **2**, 683–688.

62 Menon, M., Andriotis, A.N., and Froudakis, G.E. (2000) Curvature dependence of the metal catalyst atom interaction with carbon nanotubes walls. *Chem. Phys. Lett.*, **320**, 425–434.

63 Liu, C., Kim, K.S., Baek, J., Cho, Y., Han, S., Kim, S.-W., Min, N.-K., Choi, Y., Kim, J.-U., and Lee, C.J. (2009) Improved field emission properties of double-walled carbon nanotubes decorated with Ru nanoparticles. *Carbon*, **47**, 1158–1164.

64 Guo, S., Pan, X., Gao, H., Yang, Z., Zhao, J., and Bao, X. (2010) Probing the electronic effect of CNTs in catalysis: NH_3 synthesis with Ru nanoparticles. *Chem. Eur. J.*, **16**, 5379–5384.

65 Ivanovskaya, V.V., Köhler, C., and Seifert, G. (2007) 3d metal nanowires and clusters inside carbon nanotubes: structural, elecronic, magnetic properties. *Phys. Rev. B*, **75**, 075410.

66 Warner, J.H., Ito, Y., Ruemmeli, M.H., Buechner, B., Shinohara, H., and Briggs, G.A.D. (2009) Capturing the motion of molecular nanomaterials encapsulated within carbon nanotubes with ultrahigh temporal resolution. *ACS Nano*, **3**, 3037–3044.

67 Chamberlain, T.W., Meyer, J.C., Biskupek, J., Leschner, J., Santana, A., Besley, N.A., Bichoutskaia, E., Kaiser, U., and Khlobystov, A.N. (2011) Reactions of the inner surface of carbon nanotubes and nanoprotrusion processes imaged at the atomic scale. *Nat. Chem.*, **3**, 732–737.

68 Chen, W., Pan, X., Willinger, M.G., Su, D.S., and Bao, X. (2006) Facile autoreduction of iron oxide/carbon nanotube encapsulates. *J. Am. Chem. Soc.*, **128**, 3136–3137.

69 Chen, W., Pan, X., and Bao, X. (2007) Tuning of redox properties of iron and iron oxides via encapsulation within carbon nanotubes. *J. Am. Chem. Soc.*, **129**, 7421–7426.

70 Abbaslou, R.M.M., Tavassoli, A., Soltan, J., and Dalai, A.K. (2009) Iron catalysts supported on carbon nanotubes for Fischer–Tropsch synthesis: effect of catalytic site position. *Appl. Catal. A: General*, **367**, 47–52.

71 Cao, F., Zhong, K., Gao, A., Chen, C., Li, Q., and Chen, Q. (2007) Reducing reaction of Fe_3O_4 in nanoscopic reactors of a-CNTs. *J. Phys. Chem. B*, **111**, 1724–1728.

72 Chen, W., Fan, Z., Gu, L., Bao, X., and Wang, C. (2010) Enhanced capacitance of manganese oxide via confinement inside carbon nanotubes. *Chem. Commun.*, **46**, 3905–3907.

73 Zhang, H., Wang, J., Pan, X., Hu, Y., and Bao, X. (2012) Local structure of titania decorated double-walled carbon nanotube characterized by scanning transmission X-ray microscopy. *J. Chem. Phys.*, **136**, 174701.

74 Ilie, A., Bendall, J.S., Roy, D., Philp, E., and Green, M.L.H. (2006) Effects of KI encapsulation in single-walled carbon nanotubes by Raman and optical absorption spectroscopy. *J. Phys. Chem. B*, **110**, 13848–13857.

75 Lu, T., Goldfield, E.M., and Gray, S.K. (2008) Chemical reactivity within carbon nanotubes: a quantum mechanical study of the $D + H_2 \rightarrow HD + H$ reaction. *J. Phys. Chem. C*, **112**, 2654–2659.

76 Santiso, E.E., Nardelli, M.B., and Gubbins, K.E. (2008) A remarkable shape-catalytic effect of confinement on the rotational isomerization of small hydrocarbons. *J. Chem. Phys.*, **128**, 034704.

77 Turner, C.H., Brennan, J.K., Pikunic, J., and Gubbins, K.E. (2002) Simulation of chemical reaction equilibria and kinetics in heterogeneous carbon micropores. *Appl. Surf. Sci.*, **196**, 366–374.

78 Halls, M.D. and Schlegel, H.B. (2002) Chemistry inside carbon nanotubes: the Menshutkin S_N2 reaction. *J. Phys. Chem. B.*, **106**, 1921–1925.

79 Kondratyuk, P. and Yates, J.T. (2007) Effects of molecular confinement inside single walled carbon nanotubes on chemical reactivity – atomic H + 1-heptene. *J. Am. Chem. Soc.*, **129**, 8736–8739.

80 Zhang, Y., Zhang, H., Lin, G., Chen, P., Yuan, Y., and Tsai, K.R. (1999) Preparation, characterization and catalytic hydroformylation properties of carbon nanotubes-supported Rh-phosphine catalyst. *Appl. Catal. A: General*, **187**, 213–224.

81 Zhang, A.M., Dong, J.L., Xu, Q.H., Rhee, H.K., and Li, X.L. (2004) Palladium cluster filled in inner of carbon nanotubes and their catalytic properties in liquid phase benzene hydrogenation. *Catal. Today*, **93–95**, 347–352.
82 Tessonnier, J., Pesant, L., Ehret, G., Ledoux, M.J., and Pham-Huu, C. (2005) Pd nanoparticles introduced inside multi-walled carbon nanotubes for selective hydrogenation of cinnamaldehyde into hydrocinnamaldehyde. *Appl. Catal. A: General*, **288**, 203–210.
83 Castillejos, E., Debouttiere, P.J., Roiban, L., Solhy, A., Martinez, V., Kihn, Y., Ersen, O., Philippot, K., Chaudret, B., and Serp, P. (2009) An efficient strategy to drive nanoparticles into carbon nanotubes and the remarkable effect of confinement on their catalytic performance. *Angew. Chem. Int. Ed.*, **48**, 2529–2533.
84 Ma, H.X., Wang, L.C., Chen, L.Y., Dong, C., Yu, W.C., Huang, T., and Qian, Y.T. (2007) Pt nanoparticles deposited over carbon nanotubes for selective hydrogenation of cinnamaldehyde. *Catal. Commun.*, **8** (3), 452–456.
85 Chen, Z., Guan, Z., Li, M., Yang, Q., and Li, C. (2011) Enhancement of the performance of a platinum nanocatalyst confined within carbon nanotubes for asymmetric hydrogenation. *Angew. Chem. Int. Ed.*, **50**, 4913–4917.
86 Chen, W., Fan, Z.L., Pan, X.L., and Bao, X.H. (2008) Effect of confinement in carbon nanotubes on the activity of Fischer–Tropsch iron catalyst. *J. Am. Chem. Soc.*, **130** (29), 9414–9419.
87 Pan, X.L., Fan, Z.L., Chen, W., Ding, Y.J., Luo, H.Y., and Bao, X.H. (2007) Enhanced ethanol production inside carbon-nanotube reactors containing catalytic particles. *Nat. Mater.*, **6** (7), 507–511.
88 Li, B., Wang, C., Yi, G., Lin, H., and Yuan, Y. (2011) Enhanced performance of Ru nanoparticles confined in carbon nanotubes for CO preferential oxidation in a H_2 – rich stream. *Catal. Today*, **164**, 74–79.
89 Guo, S., Pan, X., Gao, D., and Bao, X., *Confinement effects on iron and ruthenium catalysts for Fischer–Tropsch and ammonia synthesis.* to be published.
90 Zhang, J., Muller, J.O., Zheng, W.Q., Wang, D., Su, D.S., and Schlogl, R. (2008) Individual Fe-Co alloy nanoparticles on carbon nanotubes: structural and catalytic properties. *Nano Lett.*, **8** (9), 2738–2743.
91 Centi, G. and Perathoner, S. (2010) Problems and perspectives in nanostructured carbon-based electrodes for clean and sustainable energy. *Catal. Today*, **150**, 151–162.
92 Nhut, J.-M., Nguyen, P., Pham-Huu, C., Keller, N., and Ledoux, M.J. (2004) Carbon nanotubes as nanosized reactor for the selective oxidation of H_2S into elemental sulfur. *Catal. Today*, **91–92**, 91–97.
93 Li, Z., Chen, Z.-X., Kang, G.-J., and He, X. (2011) Does confinement effect always enhance catalytic activity? A theoretical study of H_2 dissociation on CNT supported gold clusters. *Catal. Today*, **165**, 25–31.
94 Qin, W. and Li, X. (2011) A theoretical study on the catalytic effect of nanoparticle confined in carbon nanotube. *Chem. Phys. Lett.*, **502**, 96–100.
95 Blase, X., Charlier, J.C., de Vita, A., Car, R., Redlich, P., Terrones, M., Hsu, W.K., Terrones, H., Carroll, D.L., and Ajayan, P.M. (1999) Boron-mediated growth of long helicity-selected carbon nanotubes. *Phys. Rev. Lett.*, **83**, 5078.
96 Czerw, R., Terrones, M., Charlier, J.C., Blase, X., Foley, B., Kamalakaran, R., Grobert, N., Terrones, H., Tekleab, D., Ajayan, P.M., Blau, W., Rühle, M., and Carroll, D.L. (2001) Identification of electron donor states in N-doped carbon nanotubes. *Nano Lett.*, **1**, 457–460.
97 Zhang, Q., Huang, J., Zhao, M., Qian, W., Wang, Y., and Wei, F. (2008) Radial growth of vertically aligned carbon nanotube arrays from ethylene on ceramic spheres. *Carbon*, **46**, 1152–1158.
98 Philippe, R., Caussat, B., Falqui, A., Kihn, Y., Kalck, P., Bordère, S., Plee, D., Gaillard, P., Bernard, D., and Serp, P. (2009) An original growth mode of MWCNTs on alumina supported iron catalysts. *J. Catal.*, **263**, 345–358.
99 Cambaz, Z.G., Yushin, G., Osswald, S., Mochalin, V., and Gogotsi, Y. (2008) Noncatalytic synthesis of carbon nanotubes, graphene and graphite on SiC. *Carbon*, **46**, 841–849.

12
In Silico Nanocatalysis with Transition Metal Particles: Where Are We Now?

Iann C. Gerber and Romuald Poteau

12.1
Introduction

Materials at the 1–20 nanometer length scale exhibit physical and chemical properties different from those of the bulk or of molecular compounds. We will essentially focus in this chapter on nanoparticle (NP) catalysts synthesized via colloid chemistry, such as **a**, Figure 12.1. Chemical methods [1, 2] allow control of the size, crystal structure, and surface chemistry of monodisperse NPs and their resulting electronic properties [3, 4]. Monodispersity is also a prerequisite for self-assembly of NPs into superlattices. Metallic NPs are particularly appealing owing to their structural, magnetic, electric, optical, or catalytic properties [5]. Nanocatalysis takes place at the frontier between homogeneous and heterogeneous catalysis [6–8]. It is a field of intense research, owing to the opportunity of designing new-generation catalysts with high selectivity. The question addressed in the present chapter is the following: *is theoretical chemistry able to provide clues about the structure and reactivity of colloidal NPs, despite the complexity of such a problem?*

Quantum chemists have for a while now provided valuable molecular insights in inorganic, surface chemistry, and in the solid state. A long period was dominated by qualitative analysis provided by extended Hückel theory (EHT), symmetry, and molecular orbital (MO) theory [10–16], as well as by tight-binding methods [17, 18]. A new era began some 20 years ago, aiming at providing quantitative theoretical data. Thanks to density functional theory (DFT) methods [19–21], state-of-the-art organometallic studies, and more specifically in homogeneous catalysis, often associate both experimental and theoretical investigations with success [22–27]. DFT is also involved in heterogeneous catalysis both to provide theoretical insights and in the rational design of efficient catalysts [28–30]. But, until these early 2010s, quantum chemistry (QC) has been scarcely involved in the field of the chemistry of NPs, because it is difficult to accurately describe such complex systems; they exhibit a nanoscale metallic core with ligands on its surface, often partially mobile, and not so easy to characterize experimentally. Beyond semi-empirical approaches [31], theoretical chemists face the dilemma on the one hand of considering scale models, that is, small organometallic clusters (schemes **b** and **c**, Figure 12.1), on the other

Nanomaterials in Catalysis, First Edition. Edited by Philippe Serp and Karine Philippot.

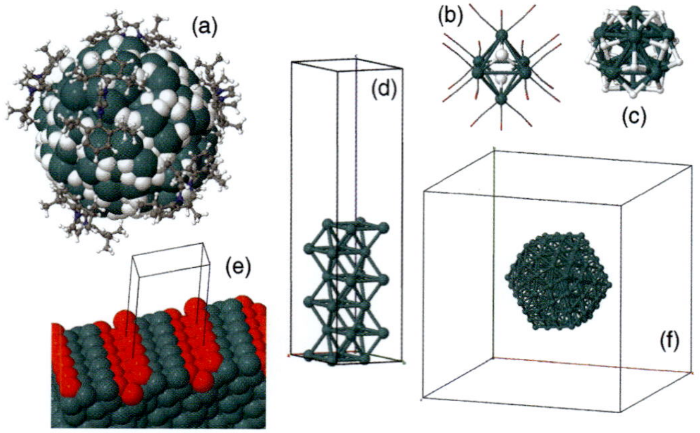

Figure 12.1 Ruthenium models for nanoscience and catalysis studies: space filling model of a 1.8 nm hcp Ru nanoparticle stabilized by 8 NHC ligands and accommodating 1.5 hydrides per surface Ru. **a:** small $[Ru_6H_2(CO)_{18}]$ organometallic cluster, considered as a scale model: **b**: hydrogenated $[Ru_{13}]$ cluster; **c**: 6-layer periodic slab model of a Ru(0001) surface; **d**: 7-layer periodic slab model of a stepped Ru(10$\bar{1}$5) surface, with B_5 sites (*vide infra*) highlighted in red; **e**: naked Ru_{147} nanocluster in a periodic supercell, **f**. (Adapted from ref. [9].)

hand taking advantage of periodic-boundary methods, that is, considering the facets of NPs as infinite surfaces or steps (schemes **d** and **e**, Figure 12.1). However, although such models may shed light on the properties of organometallic NPs, such species are expected to have physical and chemical properties intermediate between small molecular compounds and the bulk material, resulting from surface or quantum size effects [32].

Saying that nanocatalysis is based on the properties of nanoparticles sounds like tautology, but it is important to put the recent results into the historical perspective of cluster science – the forerunner of nanoscience. In the 1980s, naked NPs were the subject of intense research activity, motivated by the need to investigate an almost continuous range from the atom to an upper limit of several hundred thousand atoms. Using molecular-beam techniques, physicists produced metal clusters in the gas phase and measured their abundance, polarizability, ionization potential, optical and photoelectron spectra [33]. A key paper from Knight's group appeared in 1984 [34], published independently and simultaneously with an analysis of the electronic properties of alkali clusters in terms of a self-consistent jellium approach [35, 36]. Knight and co-workers produced and detected alkali Na_n clusters ($n = 4$–100), extensively studied from then on until the mid-1990s as prototypes of metal clusters. The mass spectra showed a striking feature: the remarkable abundance of clusters at given sizes, the so-called magic numbers ($n = 8$, 20, 40, 58, and 92). They were rationalized in the same paper in a simple and elegant way, thanks to a spherically symmetric rounded well, describing sodium clusters as effective atoms: clusters in which the valence electrons filled the electronic shells of the effective atom (1s, 1p, 1d, 2s, 1f, 2p, 1g, . . .) were produced more abundantly.

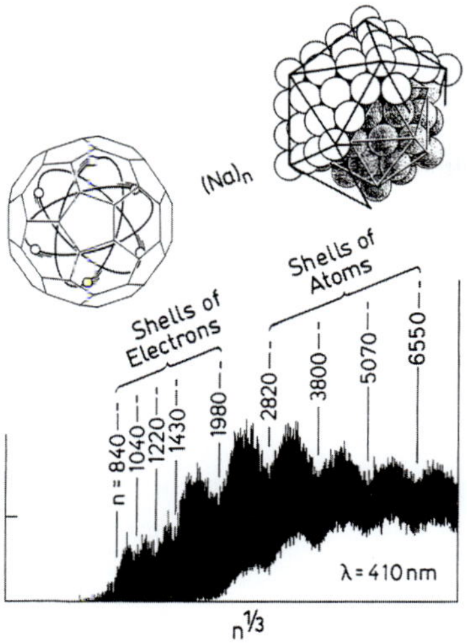

Figure 12.2 Merged mass spectra of sodium clusters Na*n* photoionized with 400 nm and 410 nm light and plotted as a function of $n^{1/3}$. (Reproduced with permission from [38]. Copyright © (2000) Elsevier.)

Mass spectra have been recorded for many metals, and have shown magic numbers (see ref. [33] and references therein). An example of mass spectrum, recorded in the early 1990s by the group of T.P. Martin [37, 38], is shown in Figure 12.2. The periodicity of the abundance peaks is analyzed in terms of a transition from electronic shell effects below about 2000 sodium atoms and geometric shell effects beyond. As a matter of fact, the magic numbers observed for large sizes follow the finite power series $\frac{1}{3}(10K^3 - 15K^2 + 11K - 3)$ which corresponds either to cuboctahedra or MacKay icosahedra with K shells of atoms. It was suggested that metal atoms of the smallest clusters are mobile, so that they behave as metal droplets. This collective mobility of the ionic cores is not possible in the largest clusters, which must be considered as small crystallites [39].

Alkali M_n clusters must of course only be considered as prototypes of transition metal (TM) clusters, since they have a completely different electronic structure, with valence *s* electrons delocalized in an ionic M_n^{n+} field. These different length scale regimes are however very interesting to notice and should probably apply to TM particles, as underlined for example by van Santen [40]. In that context, the filling of atomic shells in superatom complexes has recently been applied to understand the relation between the number of surface ligands with respect to the size of the metallic core in the case of colloidal gold clusters [41, 42].

In the following, we will highlight some recent successful theoretical work which is considered as representative of the potential contribution of computational chemistry in the field of nanocatalysis. After a brief resumé of some striking experimental results in balance with theoretical results aiming at showing that size and shape play a role on the activity and selectivity of chemical reactions over metallic NPs (Section 12.2.1), some molecular and theoretical insights will be given on the structure and electronic structure of TM NPs (Sections 12.2.3 and 12.3.6). Reasoning based on electronic band theory closely related to molecular frontier orbital theory will also be described (Section 12.3). Finally, Section 12.4 will be devoted to recently published pioneering first-principle calculations of multistep pathways and energy profiles. Only a few research groups are involved in such studies. However, owing to the continuing increase of computer power and the development of powerful QC codes, such studies should be more and more directed to experimental work so that joint experimental/theoretical papers are expected to become the new state-of-the-art in nanocatalysis.

We also underline in Section 12.3.5 some methodological and computational bottlenecks that the computational chemist currently encounters. Although DFT methods are currently theoretical tools of choice both for studying the shape, electronic structure, thermodynamic and kinetic parameters, as well as spectroscopic properties of organometallic NPs, they are not free of artifacts. This is mainly related to the fact that the exact exchange-correlation functional is still unknown, whereas the most commonly used approximate functionals lack long-range effects, or are not free from self-interaction error. The reader interested in this topic should refer to the dedicated literature [19, 21, 43, 44]. An article of J. Hafner [45] is also a good *point d'entrée* to applications, successes and issues of plane-wave periodic DFT methods, which are most often used to describe extended metallic systems or large metal clusters, such as the Ru_{147} one, scheme **f**, shown in Figure 12.1. Going back to catalysis and to the determination of thermodynamic and kinetic data, from a pragmatic and operational point of view, let us remind that the expected DFT accuracy on energy differences is roughly $\pm 5\,kcal\,mol^{-1}$ [46, 47], with pathological cases which resist this simple rule of thumb. From our knowledge of theoretical homogeneous catalysis, we suggest that the most relevant way to use DFT is to consider the relative activity and selectivity of catalysts belonging to a same family. It is then preferable to take advantage of knowing catalysts in their molecular intimacy to yield models, trends, and insights, rather than expecting a number in perfect agreement with experiments – which nevertheless often happens to be the case.

12.2
Surface Chemistry and Chemistry on Facets of Nanoparticles: Is it the Same ?

12.2.1
The Experimental Evidence: Size and Shape Matter

Colloidal chemistry provides several strategies for synthesizing mono-disperse NPs of various sizes and shapes [5]. Several authors have demonstrated that, owing to high surface-to-volume ratios of nanometer sized NPs, size affects their catalytic properties [6, 48–51]. For example, pyrrole hydrogenation catalyzed by well-defined platinum NPs between 0.8 and 5.0 nm supported onto mesoporous silica was demonstrated to be size sensitive [52]. While ring hydrogenation was structure insensitive, ring opening occurred more easily over large particles compared with

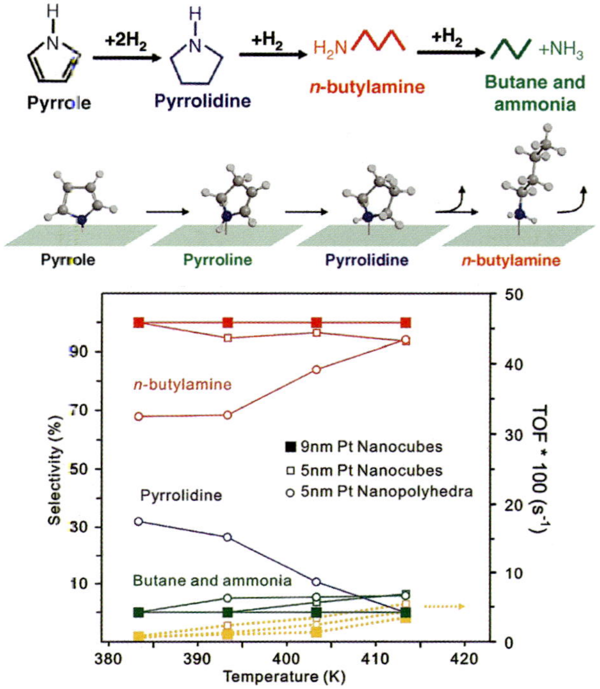

Figure 12.3 Hydrogenation of pyrrole catalyzed on Pt NPs. Upper part: proposed molecular adsorption over a Pt surface; lower part: selectivity for Pt nanocrystals with different size and shape as a function of temperature. (Adapted with permission from [53]. Copyright 2009 American Chemical Society.)

smaller ones. It was suggested that the phenomenon was caused by surface roughness or electronic effects that change with particle size, with more unsaturated surface bonds on small NPs [53]. The selectivity of pyrrole hydrogenation over Pt NPs also exhibits shape dependence: nanocubes enhanced ring-opening ability and thus showed a higher selectivity to *n*-butylamine as compared with nanopolyhedra (Figure 12.3).

More recently, Kliewer *et al.* [54] have compared the adsorption and hydrogenation of furan over Pt(111) and Pt(100) single-crystal surfaces and size-controlled Pt NPs from 1 to 7 nm. The specificity of such a study is the joint use of gas chromatography and sum frequency generation vibrational spectroscopy carried out *in situ* to elucidate the surface species. In this case, the three NP sizes exhibited similar selectivity trends with temperature. It was however significantly different, compared with the single crystals, with the dominant production of propylene – a ring cracking product – at higher temperature (above 100°C) and selectivities clearly function of the NP size at 80°C.

Let us now consider the Fischer–Tropsch (FT) reaction, of prime scientific and industrial importance [55, 56]. In this reaction a mixture of CO and H_2, (syngas) is converted into hydrocarbons, which can be used for the production of fuels, thus being an alternative to crude oil. The effect of particle size has been studied in the case of cobalt, which is well-known for its activity and selectivity [57–59]. Iglesia and co-workers have shown that the turnover frequency (TOF), which is a measure of the surface-specific activity, is not a function of the NP size in the range 9 to 200 nm, whereas it decreases for smaller Co NPs (<10 nm) [60]. Similar conclusions were later found by de Jong and co-workers [61], who proposed an explanation of the origin of the Co NP size effect based on the analysis of steady-state isotropic transient kinetic analysis (SSITKA). The lower TOF for Co NPs <6 nm was proposed to be caused by both blocking of edge/corner sites and a lower intrinsic activity at the small terraces.

Ammonia synthesis, summarized by $N_2 + 3H_2 = 2NH_3$, is also sensitive to NP size. In such context, it has been known since the seminal work of van Hardeveld and van Montfoort in 1966, that nitrogen adsorption on TM crystallites depends on their diameter [48]. It was rationalized in terms of the presence of the so-called B_5 sites on the surface. This reaction is of utmost importance in the chemical industry, since it is a preliminary step of catalytic ammonia synthesis. On Fe/MgO NPs with iron cores between 1.5 and 30 nm, the ammonia synthesis at atmospheric pressure and between 570 and 680 K was found to be a structure-sensitive reaction, with the TOF of the small particles being an order of magnitude smaller than that for larger particles [62].

12.2.2
Can this Diversity of Observations be Rationalized by Theoretical Insights?

The catalytic behavior of NPs can at least be classified, following the early suggestions of Che and Bennett [49]. They distinguished three types of reactivity scheme in the particle size regime between 2 and 20 nm. In an attempt to assist in the rational design of new catalytic systems and on the basis of computed activation energies of elementary reactions on various TM surfaces, van Santen [40] provided a molecular picture explaining the three classes of structure sensitivity (Figure 12.4). In class I type reactions, the rate may show a maximum (Ib) or may increase (Ia) as a function of increasing particle size. This is observed for reactions in which the rate-limiting step involves the formation or the breaking of a π chemical bond. van Santen has underlined with the associated molecular inset in Figure 12.4 that such π-bond activation requires step sites on the surface, a geometrical feature hardly possible below a certain NP size. The other molecular inset in Figure 12.4 illustrates that methane activation, that is, the activation of a σ-bond, is typical of class II type reactions, which show a decreasing TOF when the particle size increases. The insensitivity of some reactions to the particle size, such as hydrogen recombining with adsorbed alkyl species, is of class III type.

12.2.3
Structural and Chemical Bonding Knowledge: A Mandatory Prerequisite

Experimentalists are able to synthesize colloidal NPs with various shapes (cubes, spheroidal structures, rods, filaments, platelets, multipodal buildings . . .) by

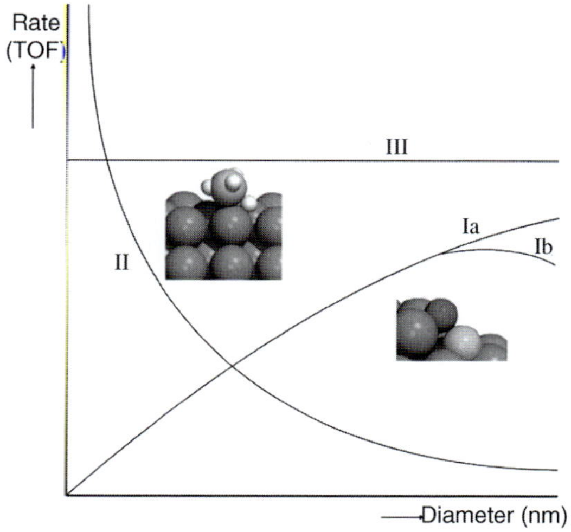

Figure 12.4 The three classes of structure sensitivity-particle size relationship adapted by van Santen [40] from the schemes proposed by Che and Bennett [49]. (Reproduced with permission from [40]. Copyright © (2009) American Chemical Society.)

varying the precursors, the amount of ancillary ligands used as stabilizing agents and the surrounding atmosphere (see refs. [63] and [64] in the case of platinum). While little is known on the mechanism of growth nor on the relationship between the final structure and the experimental conditions, a large amount of information gathered by experimental groups has drawn a picture of the role of ancillary ligands (acid, amine, . . .) and of the surrounding gas (H_2, Ar, O_2 . . .) on the final shape of NPs [65].

The metallic core of particles exhibits several shapes and happens to have a very versatile structure depending on the parameters of their synthesis. Some of the polyhedra adopted by metallic NPs are shown in Figure 12.5. Reviews of the shape and structure of actual nanometer-sized particles can be found in refs. [2], [66], [67] and [68], whereas a systematic enumeration of structural characteristics of some clusters is given in ref. [69]. Silver (from an experimental point of view) and iron or platinum NPs (theoretically speaking) are three examples of great interest.

12.2.3.1 **Silver**

The metallic NPs often exhibit multiply twinned morphology [67], that is, NPs with fivefold symmetry such as decahedra or icosahedra [71]. Transition electron microscopy (TEM) revealed that ~7 nm diameter silver NPs, enclosed by thiolate ligands bound on the surface, adopt a truncated octahedron shape or a tetrahedral shape, depending on experimental conditions [72, 73]. Starting the synthesis process with air-stable silver phosphine complexes as silver precursor and amine molecules serving as reducing agent and NP-capping ligand, leads to multiply twinned structures [74]. It is noteworthy that silver NPs prepared by $NaBH_4$ reduction exhibit

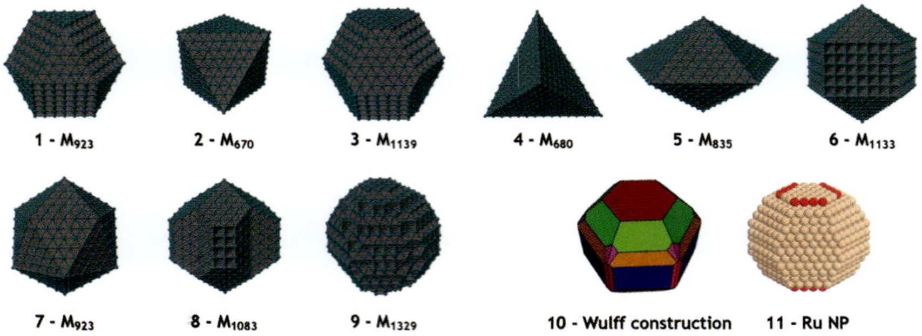

Figure 12.5 Some possible shapes for a ~3 nm metallic core (interatomic distance taken as 2.6 Å). **1:** Cubooctahedron; **2:** octahedron; **3:** truncated octahedron (or Wulff polyhedron); **4:** tetrahedron; **5:** decahedron; **6:** Ino decahedron; **7:** icosahedron; **8:** Marks decahedron; **9:** spherical piece on an hcp crystal; **10:** Wulff construction for a large Ru particle: **11:** DFT-optimized Ru NP with an average diameter of 2.9 nm (atoms that belong to B_5 sites are shown in red on the marble representation). (**10** and **11** are reproduced with permission from [70]. Copyright © (2005) AAAS.)

truncated octahedron morphologies, whereas TEM images of silver NPs prepared by polyol synthesis show icosahedra [75]. A polyol process led to NPs with shapes ruled by the length of the reaction [76]. More spectacular, simple chemical etchants can be used to shape NPs by selectively etching their crystallographic planes. Octahedral silver NPs have recently been etched by a mixture of ammonium hydroxide and hydrogen peroxide [77]. Such ability to shape tailor-made NPs opens very interesting perspectives in nanocatalysis.

12.2.3.2 **Iron**

Until very recently, the use of a finite size approach combined with DFT to describe 'real' nanoclusters has been very limited: not more than 15 atoms until 2004. However access to supercomputers and parallelization/optimization of plane-wave codes made possible the study of structures of clusters up to 641 atoms in the case of free iron clusters [78–80] and to 1415 atoms in the case of ruthenium nanoclusters (ref. [81], see also Section 12.3.6). In the case of iron, it provides an answer to the puzzling question of the structural evolution of iron clusters as a function of the particle size. It suggests that a competition of structures is taking place [78, 79]: icosahedral isomers for small clusters (few dozen of atoms) and body-centered cubic (bcc) lattice structure for large NPs, with a crossover from icosahedral to cubooctahedral geometries expected with a decrease of the surface to volume ratio. This transformation is gradual along a Mackay path between the two geometries in the intermediates shells of metals, from a close-packed core to icosahedral surfaces. Since colloidal iron NPs may exhibit very diverse shapes [82], such theoretical studies partly solved the question of their morphology.

12.2.3.3 **Platinum**

The specific reactivity of metallic NPs depends on the interplay between their size, geometry and electronic structure. In order to theoretically shed light on the role of surface ligands in the control of the structure of NPs, DFT calculations have been undertaken on Pt_{55} nanoclusters capped by amine and thiol ligands [83]. This study, although restricted to icosahedral and cuboctahedral geometries, shows that the structure of Pt_{55} changes with the number and species of adsorbed ligands. According to the metal/ligand ratio, methylamines involve a Mackay transformation from cuboctahedron to isosahedron or the reverse transformation. It was discussed in terms of electronic and geometric effects and showed that surface Pt atoms may rearrange to match the directionality between the Pt d orbitals and the N p orbital in order to allow strong bonding. Methylthiols tend to form two single bonds with the surface and would not give rise to a driving force strong enough for structural transformation.

12.3
Electronic and Geometric Factors that Determine the Reactivity of Metal Surfaces

12.3.1
Introduction

The French chemist Paul Sabatier undertook pioneering work in heterogeneous catalysis at the dawn of the twentieth century, some of them being still in use in the laboratory or as the basis of actual industrial processes. He explained his experimental observations by the intervention of special chemical affinities of reactive intermediates operating at the surface of these catalysts [84, 85]. His idea on the mechanism of catalysis, later formulated as the Sabatier principle, has never left him and was at the basis of his scientific approach. The modern formulation of the Sabatier principle states that an optimum exists for the interaction of a catalytically active surface and the adsorbate, resulting in a maximum for the reaction rate. In other words, if the interaction is too weak, no reaction will take place, whereas if it is too strong, the catalyst will be poisoned. This is clearly illustrated by Balandin's volcano relations between reaction rates and adsorption energies [86, 87]. Some of the theoretical and experimental works that will be commented on later are grounded on Sabatier principle.

12.3.2
Special Sites

It is historically interesting to recall that this concept of specific active sites associated with low-coordinated surface atoms was introduced as early as 1925 by Taylor in the context of catalytic surfaces [88].

Usually the surfaces of NPs, as drawn in Figure 12.5, exhibit corners, edges and terraces, whereas step sites can also form. This is, for example, the case in structure **11** (Figure 12.5) a DFT-optimized Ru NP based on a Wulff construction (**10**) with addition of B_5 step sites [70].

In the same spirit, an overview of DFT modeling of particularly reactive sites at noble and late transition metal surfaces are given in ref. [89]. Several special sites

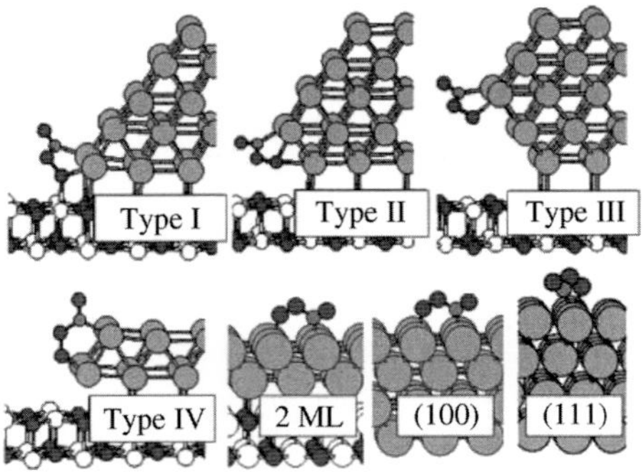

Figure 12.6 Au structures, some of them being supported at MgO(100), with the adsorbed [CO·O_2]a intermediate of the model reaction $CO_g + O_{2,g} \rightarrow O_a + CO_{2,g}$. (Reproduced with permission from [89]. Copyright © (2006) Springer Science and Business Media.)

have been listed, such as sites at the flat surfaces of thin metal films, sites at stepped surfaces, sites at the metal/oxide interface boundary for oxide-supported metal clusters (Figure 12.6), and sites at the perimeter of oxide islands grown on metal surfaces. Such study has been achieved in the context of heterogeneous catalysts exposing transition metal or noble metal surfaces. It may also generate interest in the nanocatalysis field since polyhedral NPs also exhibit terraces, edges, ridges and other defects. Two properties were highlighted: (i) local change in the electronic structure at the reaction sites without geometrical modification; (ii) local changes in the atomic structure of the catalyst – when they are favorable for the reaction, the active sites are called special sites. Reactivity on edge sites, E_n, will also be analyzed in Section 12.4.3 in the case of oxidation on gold NPs.

Disregarding the computational times, obtaining an efficient screening of all possible conformations between a molecular adsorbate and a simple vicinal surface, with very few specific sites, is usually humanly tractable. However when dealing with a nanoparticle presenting rough surfaces, that is, multiplying the 'interesting sites', the human task increases significantly to reach a limit where computational assistance for systematic search is mandatory. There is clearly a need for automatic search of the lowest ground-state, based on geometry's generation by efficient algorithms, probably thanks to data-mining and machine learning process; such a project is already in use in material science [90, 91]. For instance could we imagine using genetic algorithms as has already been done to create decorated nanoparticle structures [92]?

12.3.3
The Electronic Structure Effect in Heterogeneous Catalysis: The *d*-Band Model

As already stated in heterogeneous catalysis, it is well known that the adsorbate-surface bond strength determines the catalytic activity of a given surface. Besides, variations from one transition metal to another one can be easily understood from certain characteristics of the electronic structure of the metallic surface [93]. First theoretical investigations in the 1980s pointed the existing correlations between the reactivity of a metallic surface and the local density of one-electron states at the Fermi level [94–97] and the number of holes in the *d*-bands [98]. One solution to examine the local site reactivity is the use of Fukui functions [99]. The adsorption of chemical species on metal surfaces can be also understood in terms of Hammer and Nørskov's model [93, 100], which, as far as we understand, is not so different from Hoffmann's approach [101]. In this model, the interaction between an adsorbate and a metal surface is based on only a few parameters including the filling, average energy and width of the localized *d*-band of the metal surface, the adsorbate energy levels and the geometric overlap between the surface and adsorbate orbitals. A schematic illustration of the formation of a chemical bond between an adsorbate valence level and the *sp*- and *d*-states of a transition metal surface is presented in Figure 12.7a. When coupling to the *s* and *p* metallic states, adsorbate states are

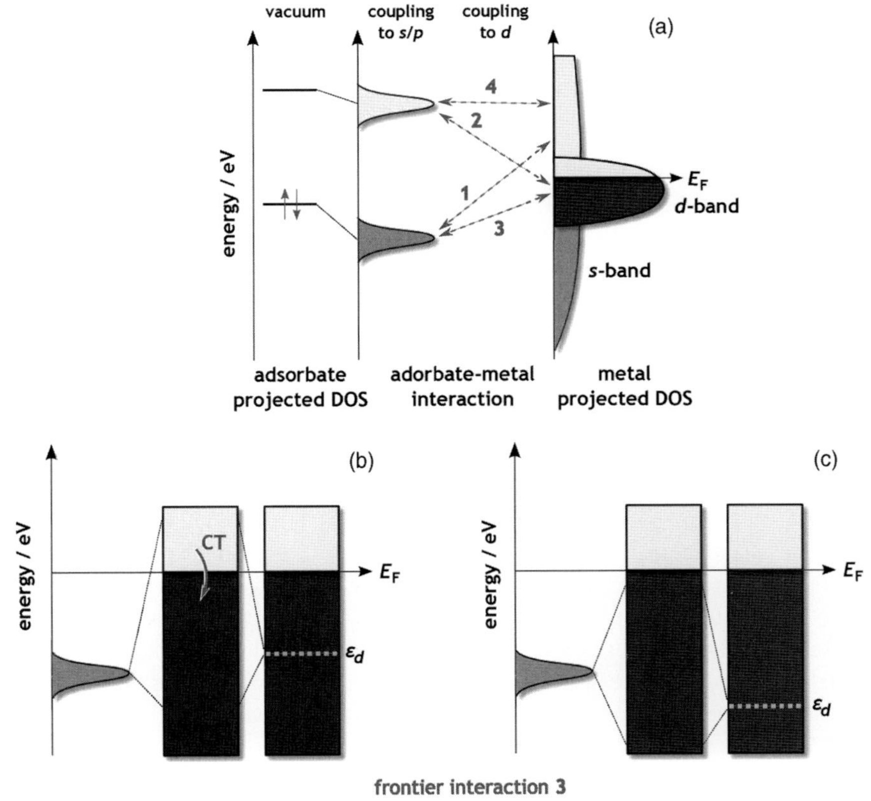

Figure 12.7 Bond formation at a transition-metal surface between an adsorbate valence level and the s- and *d*-states (see text for details). (a) interaction schemes between the adsorbate and metal electronic states; (b) and (c) typical cases for interaction **3**, depending on the position of the d-band center ε_d. (Adapted from ref. [102] thanks to the chemist's view of bonding on surfaces proposed by Hoffmann [15, 101].)

shifted and broadened, but this effect contributes almost equally to the bond energy for all transition metals. This is where the difference with Hoffmann's approach lies: rather than using the free molecular states, Hammer and Nørskov consider the ones that have been renormalized to stronger binding energies through the interactions with the metal *sp* states. The key parameter that controls the bond strength between the adsorbate and the surface is then the *d*-band center ε_d. If the *d*-band center is far below the Fermi level, the adsorbate binding is weak. When the *d*-band center approaches the Fermi level, the interaction with the adsorbate increases and it binds more strongly, due to the filling of anti-bonding states. As these states are always above the *d*-states, it explains why the center of *d*-band is a good indicator of the bond strength [103]. This assertion may seem obscure to molecular chemists familiar with frontier orbital theory and deserves an explanation and a comparison between molecular-surface and intermolecular bonding. This is illustrated in Figure 12.7, by mixing the conceptions of Hoffmann, Hammer and Nørskov, all based on the interactions between the frontier states. The stabilization's energy is given by a second-order perturbation theory term:

$$\Delta E = \frac{V^2}{\varepsilon_{ads} - \varepsilon_d} \tag{12.1}$$

where V is the coupling element between the adsorbate state of energy ε_{ads} and the metal *d* states centered at ε_d.

Among the four frontier state interactions summarized in Figure 12.7a, **1** and **2** are the so-called two-state 2e-stabilizing interactions; **4** denotes the interaction between empty states; **3**, the two-state 4e-interaction, which is unfavorable in the molecular case, deserves comment. Since the corresponding antibonding state may rise above the Fermi level in the adsorbate-surface case, it will dump its electrons at the Fermi level. This charge transfer (CT) process (Figure 12.7b) no longer destabilizes the system and can even enhance the strength of the adsorbate-surface bond, provided that the *d*-band center ε_d is not too low in energy (Figure 12.7c).

This model is particularly useful for predicting binding energies of an adsorbate in the same binding geometry, on different metal surfaces. For instance it predicts a linear relationship between the *d*-band center and the binding energy of the adsorbate to the surface [104]. A large number of calculated and experimental data have been accounted for in this way, as shown in the recent perspective of Nørskov and in several references therein [103]. Since the *d* filling does not change for any transition metal and fixes the *d*-band to the Fermi level, variations in the width of the *d*-band, due to changes in metal coordination number, will affect the position of the *d*-band center. Then each metal system has to compensate for variations in the width by shifting the *d* states up or down in energy, depending on the nature of the change. As a consequence, when scanning the transition metal block in the periodic table from the left to right, the metal becomes more reactive and variations in the adsorption geometry of an adsorbate can appear. Geometrical changes will affect the position and the number of the adsorbate states that couple to the metal states, inducing more complexity.

More interestingly the *d*-band center position also correlates with transition-state energies of reactions that may occur on the surface. This is actually what is expected from frontier orbital theory applied to chemical reactions: we do not have a direct access of transition states, but we can get an estimate of the slope of the path along the reaction coordinate leading to the transition state. Steeper paths are likely to lead to high-energy transition states [11, 105]. As a consequence, linear correlations exist between different adsorption energies on different oriented surfaces, the so-called scaling relations defined in ref. [106]. Correlations between adsorption energies and transition-state energies, the so-called Brønsted-Evans-Polanyi (BEP) relations are also found [107, 108]. These scaling relations can serve firstly as indicators, but more importantly can be used to predict catalytic activity, as we will see later in the case of methane steam reforming (Section 12.4.4.2).

12.3.4
Descriptors and Predictive Studies

The kinetics for a given catalytic reaction is the key property for a catalyst. If one wants to understand deeply a catalytic reaction or to design new reactors, activity and selectivity for different products are two important parameters. These properties relate to the intimate nature of the catalyst, through atomic-level properties such as active sites, adsorption energies, activation barriers of elementary reaction steps. Theoretical work can definitely help to describe these microscopic properties and even in some cases predict macroscopic results thanks to simplified kinetic models. In principle, the kinetics of a given catalytic reaction is obtained by calculating all possible reaction paths and determining activation barrier energies. This huge task starts to be highly prohibitive when including environment effects (surface orientation, ligands coverage . . .) and varying the metals, even with slab models of the surface. However in the recent years, several successes have been obtained where macroscopic kinetics of a simple heterogeneous catalytic reaction have been modeled thanks to DFT calculations [70], [106], [109–111]. Unfortunately, such investigations have to be repeated thousands of times in the search for new catalysts, a task still too prohibitive humanly and computationally speaking even with large supercomputer facilities. Instead one can use simplified surface reactivity models, and focus on only key atomic-level properties (1 or 2 at most), thanks to the correlation relationship described above. These limited descriptors are the basis for an extensive study throughout the transition metal jungle. Taken together they establish some design concepts that have to be fulfilled by experimentalists and catalyst developers. For instance, theoretical volcano plot can be produced, in relation with the fundamental concept of Sabatier in heterogeneous catalysis [84, 85]. TOF are plotted as a function of the selected descriptors, usually the binding energies of atomic species on the surface. A volcano-shaped relationship between the catalytic rate and the adsorption energies is obtained. It allows for the identification of the best catalytic materials for a given reaction as a function of the selected descriptors [112]. In these plots, best catalysts will be found in the vicinity of the volcano top, a consequence of the Sabatier principle.

However, the critical point is to determine active sites. As discussed previously, several sites have to be tested for adsorption, bond breaking and bond formation, and every site defines its own structure and in return defines its own new set of scaling relations. As a consequence, one has to determine which sites are present during the reaction, and to identify which of the resulting relations leads to the highest rate. This procedure has been successfully applied to several simple cases such as the design of a Ni-Fe alloy for methanation [113], the search of better catalysts for ethylene epoxidation [114], the study of alloys for hydrogen-reactions related [115–118], steam reforming [119], oxygen reduction electrocatalysts [120] or fuel cell anode catalysts [121].

12.3.5
Density Functional Theory in Surface Chemistry and Nanocatalysis: Limitations and Challenges

Despite its enormous number of successes in homogeneous catalysis, and in surface and material science, several limitations are still present in DFT [122] calculations based on Kohn–Sham orbitals [123], which can be viewed as the most powerful workhorse in computational molecular scale studies of catalytic processes. Among these drawbacks we can cite the discrimination of stable states in mixed valence compounds, the determination of accurate surface and adsorption energies, a correct description of long-range correlation effects, and a correct description of excited states.

These drawbacks are directly related to the cornerstone of DFT, that is, the approximate exchange–correlation density functional. This functional of the electronic density takes care of the exchange-correlation effects between electrons. To handle this energy term, several degrees of approximations have been proposed during the last 30 years, defined by the inclusion of density derivatives (semi-local or gradient corrected approximation (GGA)), and/or portion of exact-exchange term (hybrid functional), and more recently with the inclusion of explicit many-electron terms. Searching for the ultimate functional describing electronic interactions [124], to achieve chemical accuracy, is probably one of the most active fields in methodological development today. Achieving chemical accuracy starts to be a crucial issue as soon as one searches for very accurate surface and adsorption energies. For instance the emblematic CO on Pt (111) surface test case is very instructive in this respect [125]. Until very recently, none of the available functionals were able to predict the correct adsorption site for CO on Pt (111), due to a systematic overestimation of the interaction between the empty π^* frontier orbital of CO with the substrate [126, 127]. Indeed the general trend of semi-local functionals is to underestimate surface energies and to overestimate chemisorption energies. Only the inclusion of an explicit many-body terms corrects this flaw [128], ideally by placing the frontier orbitals not too close to the Fermi level and keeping the *d*-bandwidth unchanged. When dealing with catalysis involving large molecules, their physisorption and intermolecular interactions will be of great importance [129]. Unfortunately, any standard functionals missed long-range correlation effects by construction [130]: pure nonlocal effects cannot be handled by any local term. In this direction several tracks are followed to take into account for van der Waals interaction with different successes: empiric or semi-empiric corrections [131, 132], DFT calculations combining wave function approach (WFT) in a range separation formalism [133], and explicit nonlocal term, the so-called vdW-DF [134]. Until now, theoretical studies refer most of the time to well-defined active sites on vicinal surfaces of simple NPs structures of pure or bi-alloys of transition metals only. However one of the last well-known drawbacks of DFT is the inability of standard approximations to describe electronic structure (excited states and band-gaps) of inorganic compounds and more specifically on oxide based structures, where strong correlations are present – see for instance the recent work on CeO_2 nanoparticles [135, 136].

12.3.6
Difference between Bulk, Surface and Nanoparticles from a Theoretical Point of View

We shall first consider ruthenium which, together with its derivatives, constitutes an important class of catalysts, involved in hydrogenolysis reactions [137], olefin metathesis [138] and Fischer–Tropsch synthesis [139, 140]. There are a variety of ways to synthesize Ru NPs. Among them the organometallic route is an efficient way to produce monodisperse NPs under mild conditions, starting from an organometallic precursor. At the end of the chemical synthesis, which is based on the decomposition of the precursor under dihydrogen, the metallic core is dressed by mobile hydrides [141] and ligands [142]. The presence of hydrides is related to the fact that hydrogen molecules take part in the reduction of the precursor [143]. A typical organometallic NP, **a**, is sketched in Figure 12.1.

The difference in the electronic structure of bulk, surface and large clusters is illustrated in Figure 12.8 by plotting density of states (DOS) and projected DOS for naked and hydrogenated systems. As expected, the DOS of the Ru bulk and Ru (0001) surface look similar and are characteristic of metallic solids. The *d*-band center for the slab, $-1.22\,\text{eV}$ ($-28\,\text{kcal mol}^{-1}$) is close to the value given in ref. [100] by Hammer and Nørskov. Slight variations are observed for Ru_{13} and Ru_{55}, which may however involve increased adsorption energies for Ru_{13}. Pay attention that ε_d values for surfaces and clusters must be compared carefully (unpublished results). After hydrogenation, the bulk keeps its metallic character, with a stabilization of the lower part of the *d*-band. Ru_{55} and Ru_{13} clusters exhibit a discrete electronic structure with a small band gap, almost tiny in the Ru_{55} case. Hydrogenation increases the band width, but does not really alter the metallic character of the cluster, at least for Ru_{55}.

The results reported in Figure 12.9 are related to the notion of special sites, previously introduced in Section 12.3.2. The hydrogen adsorption energies for top, bridge and hollow sites calculated for the Ru(0001) surface show that H

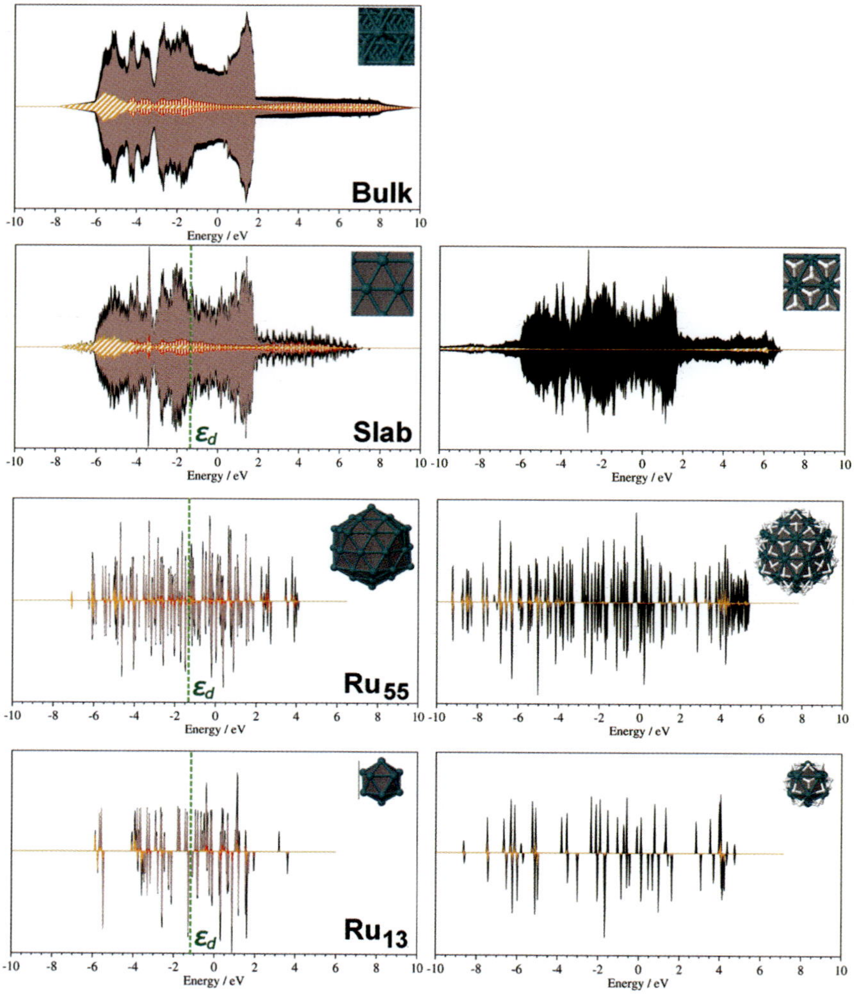

Figure 12.8 α-(upper part) and β-(lower part) states DOS of naked (left) and hydrogenated (right) ruthenium in various phases: bulk, Ru (0001), icosahedral Ru_{13} and Ru_{55}. Black: total DOS; brown: *d*-projected DOS; orange and diagonal-shaded: *s*-projected DOS; red and vertical-shaded: *p*-projected DOS. For hydrogenated systems, only the total DOS and the *s*(H)-projected DOS are shown. The bulk and the slabs being diamagnetic, α- and β-DOS are the same for these systems. The *d*-band center, ε_d, is also shown for pristine species.

are more likely to be adsorbed on hollow fcc or hcp sites [144–146]. However, by a combination of theoretical and experimental ^{2}H NMR, we have expressed the possibility of having, against conventional wisdom, a coexistence of top, bridge, and threefold H atoms on the surface of Ru NPs [147, 144]. It has been confirmed by DFT calculations on a Ru_{55} cluster chosen for its variety of surface sites, both reminiscent of the Ru(0001) surface and exhibiting defects.

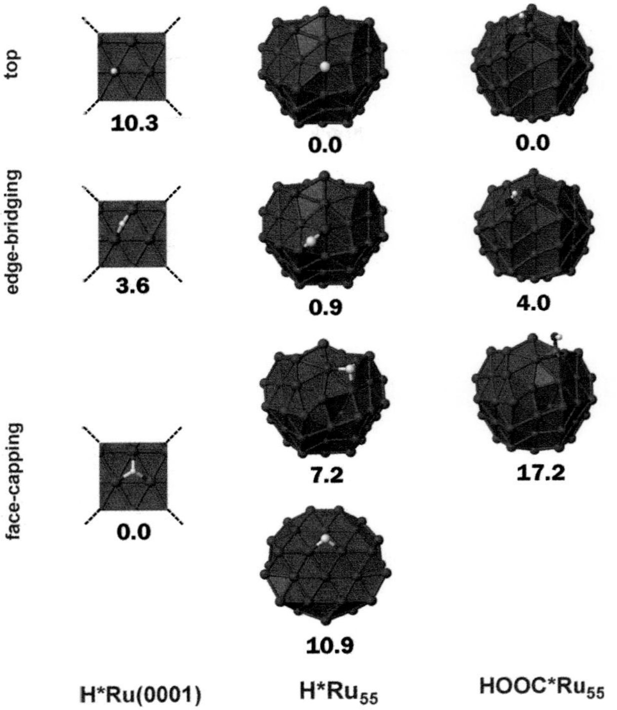

Figure 12.9 Relative stability of various coordination types of H- on a Ru(0001) slab [144] and on a Ru_{55} nanocluster. Adsorption of HOOC- on Ru_{55} is also given (same coordination sites as H-, but with a different viewpoint). The energies (in kcal mol^{-1}) relative to the lowest isomer are given below each structure.

The results reported on Figure 12.9 show that some special sites, are characterized by a low coordination number of the involved Ru atoms due to a particular local subsurface geometry, favor by about 10 kcal mol^{-1} on-top or bridging sites rather than hollow sites. The same observation is made for the —COOH functional group.

How large does a metallic nanoparticle need to be before its surface chemical properties can be considered to be those of a solid, rather than those of a large molecule ? This fundamental question has been addressed in the case of gold in a spectacular paper in terms of performance computing [81]. Calculations of adsorption energies for carbon monoxide and oxygen on a series of gold nanoparticles ranging from 13 to 1415 atoms (i.e., of diameter 0.8–3.7 nm), have been made possible by using up to 32 768 cores on a high performance computer (HPC). It was shown that bulk surface properties are obtained for clusters larger than about 560 atoms (2.7 nm, see Figure 12.10). Pay attention that such mobilization of computing power is quite far beyond the HPC resources usually available for such studies.

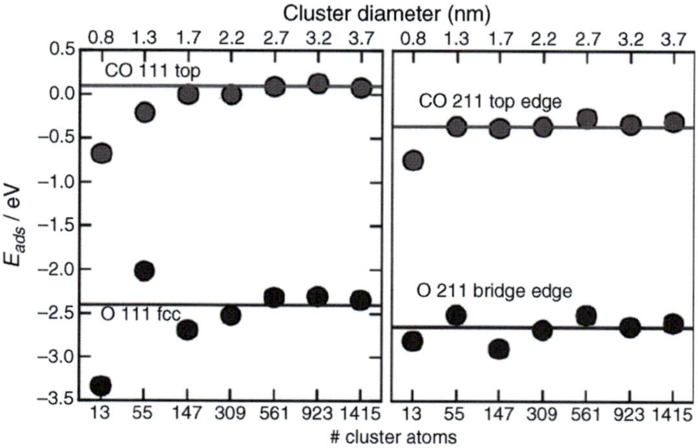

Figure 12.10 Adsorption energies of CO and O on (111) facet (left) and at the edge (right) of Au clusters as a function of cluster size (red and blue disks), compared with adsorption energies on (111) and (211) slab surfaces (red and blue lines). (Reproduced with permission from [81]. Copyright © (2011) Springer Science and Business Media.)

12.4
Theoretical Studies of Multistep Pathways

12.4.1
Methods

One of the major achievements of computational chemistry is its ability to address multistep chemical reactions by the location of reactants, products, intermediates and transition states (TSs) on the potential energy surface (PES) of the system. A typical two-step reaction is modeled in Figure 12.11a. The transition state is the

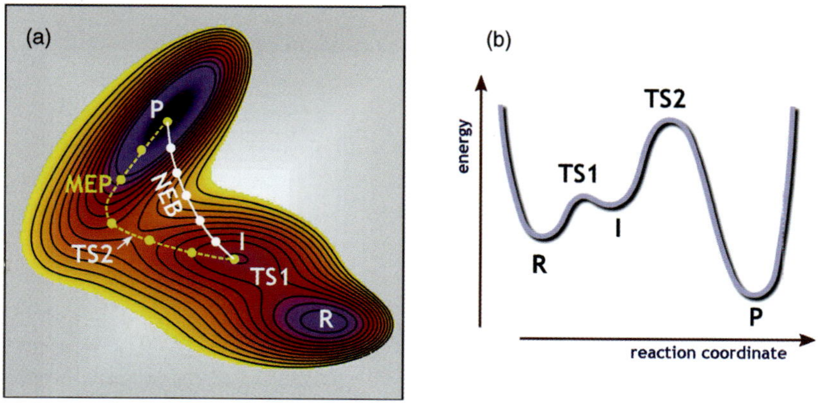

Figure 12.11 (a) Contour plot of the Müller–Brown PES model for a two-step reaction [148]. The MEP is shown as well as an illustration of an early step of a 5-image NEB method; (b) schematic energy profile as a function of the reaction coordinate.

ridge between reactants and products and the calculation of its energy allows the evaluation of the corresponding energy barrier ($\Delta E^{\ddagger}$ or $\Delta H^{\ddagger}$) or the activation barrier ($\Delta G^{\ddagger}$), which can be compared with experimental kinetic data. However, the energy profile drawn in Figure 12.11b contains only thermodynamic and kinetic information about the chemical reaction at 0 K. Let us remind that in transition state theory, the rate constant for step i is given by:

$$k_i = \kappa \frac{k_B T}{h} \exp\left(-\frac{\Delta G^{0,\ddagger}_{i,T}}{k_B T}\right) \tag{12.2}$$

The κ factor, often taken as 1, includes the TS ridge recrossing, as well as quantum and non equilibrium effects.

The energy of products, reactants and intermediates also provides the thermodynamics of elementary reactions. It is related to the equilibrium constant K_i of elementary step i:

$$K_i = \exp\left(-\frac{\Delta G^{0}_{i,T}}{k_B T}\right). \tag{12.3}$$

In addition to ΔG^0_T and $\Delta G^{0,\ddagger}_T$, a careful analysis of the wavefunction and of the geometrical features of all species highlights the key parameters of a given reaction. These kinetic and thermodynamic parameters are usually evaluated in the context of statistical mechanics applied within the harmonic approximation [149]. It requires to compute the harmonic frequencies of the full system. They are now analytically calculated, that is, efficiently, in most molecular-QC packages. Periodic-QC software only offers the possibility to calculate second-order derivatives of the energy with finite-difference methods. This severely limits the practical determination of normal modes of vibration. Mind that such harmonic approximation is very good in many cases, but it may dramatically fail when weak interactions become prominent. An alternative is to achieve *ab initio* molecular dynamics (AIMD) simulations to determine free energies activation barriers, but with a very high computational cost [150, 151].

Nowadays, quantitative approaches based on the harmonic approximation are routinely applied in the context of homogeneous catalysis [22, 27], most of the time within DFT methods [24]. Computational chemistry has also proven its ability to shed light on surface organometallic chemistry [30], or catalysis on transition metal surfaces [28, 103, 112, 152]. The methods used to find local minima on the PES are similar both in homogeneous and heterogeneous catalysis. Most of them use some variants of the quasi-Newton algorithm employing energies and gradients [153]. The tools used to find transition states are somewhat different. They share the evaluation of TS free energies $\Delta G^{0,\ddagger}_T$ within the harmonic approximation at the saddle point of the PES, whereas a rigorous approach would consist in locating the TS on the free energy surface. While equations and methods are grounded on the same basis, from our own experience, we would say that those usually available in molecular QC packages such as Gaussian [154] are more efficient and robust than the nudged elastic band (NEB)-derived methods implemented in periodic-DFT packages such as VASP [155, 156]. The algorithms [157] generally available for optimizing minima and TSs in molecular-QC packages are based on the natural internal coordinate system defined by Pulay and co-workers [158, 159]. Developing the optimization algorithm for crystalline structures is also an important issue [160]. In addition, a standard reaction path following an algorithm, the so-called IRC algorithm [161], allows the minimum energy path (MEP) to be explored from the transition state toward reactants and products. In summary, such a toolbox allows intensive explorations of the PES of large size molecules, a non trivial task regarding their large degrees of freedom and their flexibility.

The algorithms of choice in the periodic-QC domain are all based on the chain method principle [162–165], that is they intend to find reaction pathways when both the initial and final states are known. A set of images is built by a linear interpolation between them. The energy of this string of images is then minimized in order to obtain the MEP, along which the TS is found close to an image. In the NEB variant [166, 167], these images are connected by spring forces that ensure equal spacing of the images along the reaction path. An illustration is given in Figure 12.11a.

12.4.2
Ammonia Synthesis

Based on experimental evidence as well as on a marble model of an incomplete cubooctahedron, van Hardeveld and van Moonfort were the first to postulate that the so-called B_5 step sites are necessary for N_2 activation [48, 168], which is well-known to be the rate-determining step of ammonia synthesis. They introduced the B_n terminology for adsorption sites, where n denotes the number of contact points of an adatom on the site. Given the dependence of the number of such sites as a function of particle diameter, they underlined the necessity to consider the diameter of metal catalyst particles as an important parameter in catalysis in order to maximize the number of active sites (see also the work of van Hardeveld and Hartog, ref. [169]). Using adsorption experiments and DFT calculations [170], it has more recently been shown that N_2 dissociation on Ru(0001) is totally dominated by steps: the experimental adsorption rate at the steps is nine orders of magnitude higher than on terraces at 500 K [170, 171], and the corresponding DFT-GGA energy difference is 35 kcal mol^{-1}.

Later, the multistep pathways for ammonia synthesis on perfect Ru(0001) and stepped Ru surfaces have been calculated thanks to first-principle calculations [70]. The activity of NP ruthenium catalysts has been discussed, based on the assumption that the chemical reactivity of NPs can be inferred from results obtained on model extended surfaces. Beyond the interest of confirming that the dissociative adsorption of N_2 is the rate-determining step and that steps are more reactive than terraces (Figure 12.12), the authors have confirmed the early assumption of van Hardeveld and van Moonfort that the (001)/(101) edges lower their energy by a reconstruction in which the edge row of atoms is removed. It has been obtained by optimizing the basic shape of a 2.9 nm particle provided by a Wulff construction [172–174]. In its standard

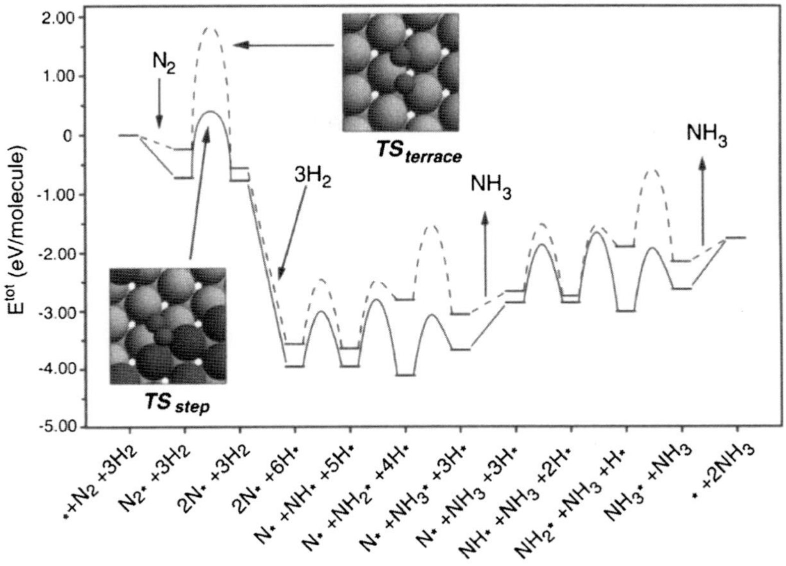

Figure 12.12 DFT energy diagram for NH_3 synthesis over Ru(0001) and stepped Ru surfaces. (Reproduced with permission from [70]. Copyright © (2005) AAAS.)

formulation, this construction consists in determining the equilibrium shape of a crystal with fixed volume V so that its Gibbs free energy is minimized by assuming a configuration of low surface energy. This Wulff construction/DFT optimization strategy gives rise to a hcp Ru NP with steps that contain B_5 sites along the edge (Figure 12.5, compounds **9** and **10**).

12.4.3
Oxidation

12.4.3.1 **Styrene**

Zeng and co-workers have achieved a DFT investigation of structural, electronic, catalytic, and selective properties of the ligand-covered gold nanoparticle $Au_{55}(PPh_3)_{12}C_{16}$ and associated model clusters [175]. Such ligand-protected gold cluster is known as a superatom complex [41, 42]. Interestingly, in the present case the electron count of the metallic core does not correspond to a closed shell superatom complex ($n^* = 48$, i.e., it occurs between the 34 and 58 closed shell values). The investigation of 49 potential isomers by means of DFT-GGA calculations has shown that the lowest isomer exhibits a quasi-icosahedral Au_{55} core, which entails 20 ligand-encompassed Au_6 faces as the active sites (Figure 12.13). After identification of the shape of the Au_{55} core in the lowest isomer and of the organization of the ligands on its surface, transition states have been searched on the $Au_{55}(PPh_3)_3(PH3)_9C_{16}$ model compound in order to examine the dissociative adsorption of O_2 on the surface, followed by the oxidation of styrene by dioxygen.

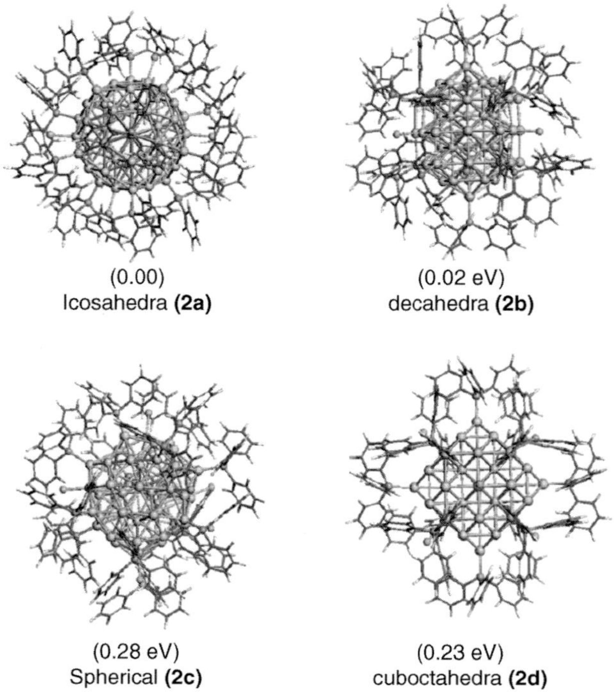

Figure 12.13 Optimized structures and relative energies of four $Au_{55}(PPh_3)_{12}Cl_6$ isomers, among 49 different structures. (Reproduced with permission from [175]. Copyright © (2010) American Chemical Society.)

The authors showed that the -PPh_3 encompassed triangle Au_6 face is the primary active site on $Au_{55}(PPh_3)_{12}C_{16}$, owing to the low coordination of Au atoms in the Au6 'triangle-socket' and to the electron back-donation from phosphine ligands which makes the Au_{55} core negatively charged. Four probable pathways of styrene oxidation under the Eley–Rideal mechanism have been explored, leading to benzaldehyde, acetophenone and styrene epoxide. The reactive site favors the formation of an oxametallacycle intermediate that leads to benzaldehyde, in agreement with the experimental high yield of this product in styrene oxidation.

Although the total number of electrons and nuclei of the cluster is large, such calculation is feasible thanks to modern computational facilities. Such work is undoubtedly a paradigm of the future standard papers in the field, which will combine experiments and theory.

12.4.3.2 **Propylene**

Propylene oxide is an important precursor produced on a large scale industrially, the traditional processes being environmentally harmful [176]. There is accordingly a need to develop new synthetic strategies. In that context, a joint experimental and theoretical study [177] aimed at evaluating the size-dependent selectivity and activity of supported silver nanoclusters between 9 and 23 nm in the partial oxidation of

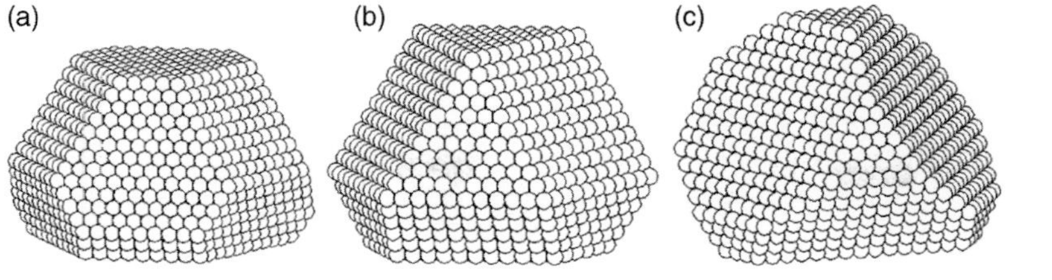

Figure 12.14 Reshaping of supported silver NPs in case of (a) non-oxidized surfaces and (b, c) oxidized surfaces at $p(O_2) = 5 \times 10^{-3}$ atm and $p(O_2) = 1$ atm respectively. (Reproduced with permission from [177]. Copyright © (2011) Elsevier.)

propylene to propylene oxide and acrolein. Smaller clusters preferentially yield acrolein whereas the 23 nm particles are more selective towards the synthesis of propene oxide. The theoretical counterpart to the experimental approach consisted in the determination of the supported particle shapes as a function of silver oxidation – thanks to a Wulff–Kaischew construction [178, 179] – and on the calculation of activation barriers for propylene oxide and acrolein formation on various facets and on the edges of the NPs. The theoretical size- and morphology-dependent activity and selectivity were found to be in good agreement with experimental observations. In such multiscale approach, also followed in the ammonia synthesis study previously mentioned [70], silver NPs are considered as crystals that will arrange themselves such that their Gibbs free energy is minimized by assuming a configuration of low surface energy. DFT calculations of silver (i,j,k) bare and oxidized slabs provide the surface energies that will be considered in the Wulff–Kaischew construction. Figure 12.14 shows the morphology of about 6 nm supported silvers NPs in the case of non-oxidized and oxidized surfaces. However, on the contrary of the 2.9 nm ruthenium particle considered in ref. [70], the giant size of the Ag NPs does not allow a full relaxation of the geometry by means of first-principle calculations. As a consequence, the possible presence of defects on the surface and the nature of such defects cannot be proved. In addition to the possibility to compare the theoretical morphology with HRTEM images, such construction highlights the terraces available for catalysis. The next part of the study consisted in a standard transition states research on given (i,j,k) surfaces and on a model of oxidized NP edge sites.

12.4.3.3 **Aerobic Phenylethanol Oxidation in Aqueous Solution**

Solvent-phase calculations with an implicit consideration of the solvent effect is now a standard for the calculation of energies, structures and properties of moderate-size molecules in solution, thanks to the so-called polarizable continuum model (PCM) [180–182]. Yet, we have already alerted the reader about the difficulty of handling transition state searches for catalytic reactions on metallic NPs. One can easily imagine that taking into account solvent effects in such context is very challenging. Such implementation has been achieved, using a periodic continuum solvation model based on the modified Poisson–Boltzmann equation (CM-MPB) [183] and successfully applied on the controversial aerobic oxidation

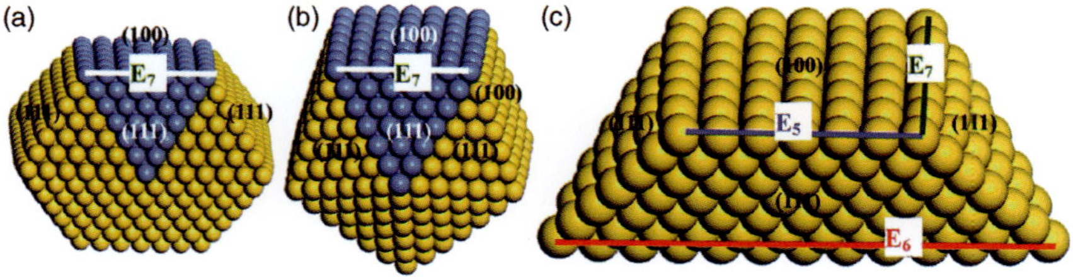

Figure 12.15 Model of gold NPs. (a) decahedral NP; (b) penta-star NP; (c) Au_{252} model cluster, highlighted in blue as a common building block in the Wulff particle models a and b. Edge sites E_n are labeled, where n is the coordination number of gold atoms. (Adapted with permission from [184]. Copyright © (2011) American Chemical Society.)

mechanism on gold NPs [184]. The authors have investigated the phenylethanol oxidation on Au NPs in aqueous solution by means of a combined periodic DFT/CM-MPB approach and have demonstrated that gold NPs are capable to activate O_2 at the solid–liquid interface. They have shown that the presence of water can enhance the adsorption energy of O_2, $E_{ad}(O_2)$, on the edges of ~2 to ~4 nm (Au_{203}, Au_{429}, Au_{781} and Au_{932}, see Figure 12.15) by about 9 kcal mol^{-1}.

The Au_{252} cluster model in Figure 12.15c has been validated by showing that $E_{ad}(O_2)$ at the E_7 sites is within the range of values found for the 2–4 nm Wulff particle models. The mechanism for phenylethanol $Ar(CH_3)CHOH$ oxidation to $Ar(CH_3)C{=}O$ on Au has then been investigated on this large gold cluster. The reader may observe that cluster calculations (such as **c**, Figure 12.1) are an alternative to slab calculations (**d** and **e**, Figure 12.1). Nevertheless, in the present case the cluster model is very large and is to be considered as a common building block for decahedral and penta-star clusters with the same reactive E_7 sites. The O_2 reduction pathways by surface H are reproduced in Figure 12.16, together with the complete mechanism for the aerobic oxidation of 1-phenylethanol investigated in ref. [184]. In addition to the impact of solvent on O_2 adsorption and on oxidation pathways in aqueous solution, the important outcome of this work is the demonstration that (i) the hydroxyl bond can break only with the help of an external base (**4**, Figure 12.16); (ii) the surface H from the alpha-C—H bond cleavage can easily be removed via a H_2O_2 pathway. Such study mixes DFT calculations of structures and energies, multistep reaction investigations with TS searches, implicit solvent treatment using a periodic PCM model, and a good knowledge in chemistry and catalysis. The pioneering work of Shang and Liu is representative of the studies that will be more and more achieved in the near future.

12.4.4 Dissociation

12.4.4.1 Carbon Monoxide

The dissociation of CO on a surface is believed to be the rate-limiting step in the methanation reaction as well as in Fischer–Tropsch synthesis. Carbon monoxide

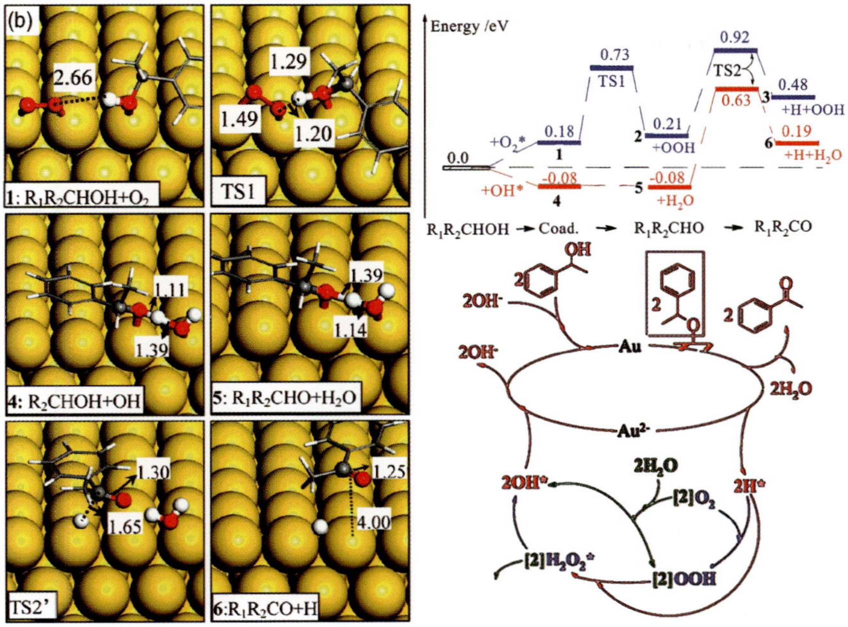

Figure 12.16 ZPE-corrected energy diagram (in eV) of the reaction pathways for 1-phenylethanol oxidation without (blue) and with (red) assistance of the OH base, completed with the mechanism for the aerobic oxidation of 1-phenylethanol and the snapshots of key reaction intermediates and TSs. (Adapted with permission from [184]. Copyright © (2011) American Chemical Society.)

disproportionation is also a reaction needed for the production of carbon nanotubes. Understanding the topologic and electronic role of metal surfaces and NPs in this process is consequently of prime importance.

Kinetic investigations, X-ray photoelectron spectroscopy (XPS) analysis and grid-based plane augmented wave (GPAW) DFT calculations have been undertaken to shed some light on the chemical vapor deposition (CVD) catalysis of single-wall carbon nanotubes on Fe NPs [185]. The detailed chemistry is largely unknown, but the dissociative adsorption of CO followed by the reaction of $O_{(ads)}$ with an undissociated $CO_{(ads)}$ to form $CO_{2(ads)}$ are believed to occur. The theoretical counterpart of the studies published in ref. [185] focused on these steps occurring on an icosahedral Fe_{55} NPs. The barrier for CO dissociation on a Fe_{55} cluster was found to be lower than on surfaces (14.5 kcal mol^{-1} versus 18 kcal mol^{-1}), suggesting that Fe_{55} is more reactive than stepped iron surfaces. The energetic for CO_2 formation on the surface and for the migration of C atoms on and below the surface have been studied as well, showing that subsurface sites can easily accommodate C atoms. Unfortunately, CO_2 formation was found to be the rate-limiting step, in contradiction to experimental kinetic data (activation energy E_a: 24 kcal mol^{-1} versus 7.6 kcal mol^{-1}). The authors concluded that the real clusters, which are bigger in size, most probably present highly reactive defects that cannot be modeled on a small nanocluster.

Actually, van Santen and co-workers were able to describe such defects on a small rhodium particle, by considering a semicubooctahedron of 57 atoms constructed from two rhodium atoms adsorbed on a regular Rh_{55} cubooctahedron and creating B_5 sites [186]. The interesting outcome of this study is the comparison of the size and topological effects on the dissociation of CO on Rh surfaces (flat Rh(100) and Rh(111), Rh(211) with B_5 sites), on a Rh_{13} cluster and on a Rh_{57} nanocluster. It had already been theoretically shown by Mavrikakis *et al.* that CO dissociation is significantly more favored on B_5 sites than on terraces [187]. It has experimentally been shown(see Section 12.2.1) that the rate of CO dissociation on Co particles decreases when they become small [60, 61]. Several pathways were considered by Filot *et al.* [186] Although the Rh_{57} particle shows a DOS structure more similar to that of extended surfaces than Rh_{13}, the energy profiles of CO dissociation somewhat differ. Beyond the calculation of numbers, some rules and trends are inferred from this work: in all cases, CO dissociation is most favorable on B_5 sites; the activation energies decrease with an increasing particle size; with the exception of Rh_{13}, CO interaction with Rh atoms with lower coordination numbers results in stronger bonds – because the metal surface exhibits an upward shift of the *d*-band center, resulting in a better overlap with the CO 5σ and $2\pi^*$ MOs.

12.4.4.2 **Methane Steam Reforming**

It is an industrially important process, allowing the conversion of natural gas or liquid hydrocarbons into synthesis gas or hydrogen, usually performed over nickel-based catalysts [188–190]. A typical reaction starts from methane, CH_4, and ends up with H_2 and CO. This subsection is devoted to a combination of first-principle calculations supported by insight from experiments achieved with metallic NPs, published in 2008 [191]. It aimed at rationalizing the steam reforming process thanks to scaling relationships applied to adsorption and barrier energies and a kinetic model, leading to a so-called volcano plot. The authors considered several late transition metal catalysts. This plot, reported in Figure 12.17, provides a synthetic view of the theoretical overall rate as a function of two key parameters, in the present case the adsorption energies of atomic oxygen and carbon, $\Delta E(O^*)$ and $\Delta E(C^*)$.

Before giving some details about such calculations, let us first analyze the plot, obtained from the kinetic model. Given that the reactivity is probably not governed by a single rate-determining step, but by several, it is easy to figure out that the analysis of energy profiles is not straightforward. As a consequence, there is a need for such synthetic plot for simplifying the understanding of the delicate kinetic and thermodynamic balance factors in such multistep reactions (nine elementary steps are considered in this process, see Section 4.3 in ref. [191]). The rate is calculated to be optimal when the descriptors $\Delta E(O^*)$ and $\Delta E(C^*)$ lie close to 0.0 and 1.2 eV respectively. This optimum lies close to the adsorption energies on Ni, Rh and Ru. The following trend

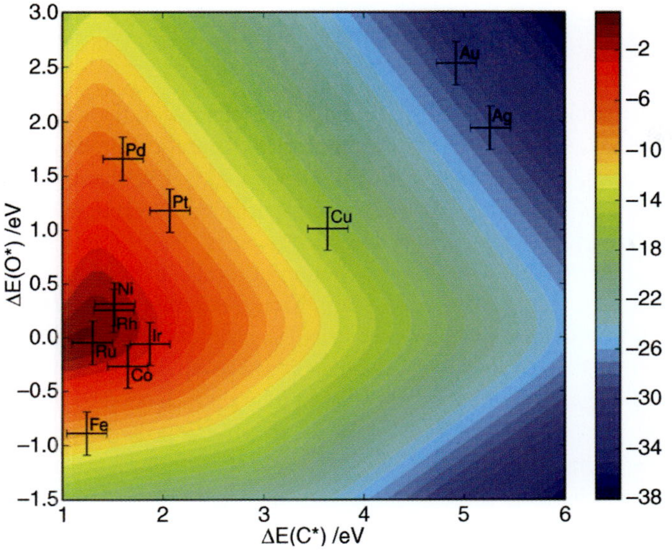

Figure 12.17 Volcano plot of the logarithmic TOF (log10), calculated as a function of atomic oxygen and carbon adsorption energies and in the conditions $T = 773$ K, $P = 1$ bar, 10% conversion. (Adapted with permission from [191]. Copyright © (2008) Elsevier.)

from theory for catalysts activity is thus:

$$\text{Ru} > \text{Rh} > \sim \text{Ni} > \text{Ir} > \text{Pd} \sim \text{Pt}$$

whereas the trend from the experimental part of the study is:

$$\text{Ru} \sim \text{Rh} > \text{Ni} \sim \text{Ir} \sim \text{Pd} \sim \text{Pt}.$$

Some of the discrepancy could originate from the accuracy and sources of error both from experiments and theory as well as from the shape of the experimental NPs, but overall the general trends are similar. The theoretical methodology behind such volcano plot has been briefly summarized in Section 12.3.4. It requires equilibrium and rate constants to be incorporated in the kinetic model, all related to energy differences (Equations 12.2 and 12.3), themselves linearly scaled to atomic C and O adsorption energies. As an example, according to the BEP relationship [192, 193], the CH_4 dissociation barrier height scales with the atomic C adsorption energy, whereas the CO formation barrier scales with both $\Delta E(\text{O}^*)$ and $\Delta E(\text{C}^*)$. Such analysis does not explicitly depend on the metal, since it is the adsorption energies that are related to a given metal. Such abacus surfaces can then be used by superimposing calculated $\Delta E(\text{O}^*)$ and $\Delta E(\text{C}^*)$ values. This theoretical strategy can be applied to any process where the elementary steps can be described by a scaling relationship. At this stage, the reader is probably wondering where the NPs are explicitly modeled in the theoretical part of this work. Actually only models were built, namely stepped surfaces, taking into account the highest TOF for steps compared with terraces, such steps being considered as the important active sites for a number a catalytic processes.

12.5
Conclusion

In a perspective paper published in 2010 [8], Li and Somorjai wrote in their conclusion: 'Because of device complexity in nanotechnologies, theoretical modeling is now playing an increasingly important role in device design. (. . .) Rational design that combines theoretical modeling with advanced experimental techniques will significantly shorten the R&D cycle of new nanotechnologies'. Behind this statement, which we completely agree with, the key notion of local activity of catalytic NPs is implicitly acknowledged. We have seen throughout this chapter that (i) terraces, edges, defects and special sites specificity is a characteristic property of adsorption (see also ref. [194] for surface chemistry and catalysis); (ii) the geometry, electronic structure and reactivity of active sites can be fully characterized by means of quantum chemistry calculations, at least on model systems; (iii) the permanent enhancement of computing power should allow to consider more realistic models of colloidal NPs in their environment in the nearest future.

The utmost importance of the molecular point of view is also the underlying concept of a recent work of Tsang and co-workers, who aimed at locally probing active sites on nanocatalysts by means of ^{13}C NMR guides established from the resonance of non direct ^{13}C atoms of the adsorbate [195]. This remarkable work achieved in the case of formic acid decomposition is grounded in several topics developed in the present chapter: (i) the Sabatier principle of maximizing the catalytic activity by optimizing the adsorption energy of the substrate; (ii) the relationship between adsorption energy and *d*-band center – also shown to correlate with ^{13}C chemical shift values of the adsorbate. The authors have also mixed ^{13}C NMR measurements and DFT calculations. They nevertheless underlined that computational chemistry modeling of NPs remains to be validated under realistic reaction conditions, that is in liquid phase or under pressure, rather than in ultrahigh vacuum (UHV). Their approach undoubtedly avoids such issues since the diagnostic chemical shift is *de facto* determined under realistic conditions.

The apparent distance between theoretical models and the real life of molecular systems is often a debate in the context of theoretical homogeneous catalysis. We shall suggest that gas-phase calculations are relevant so far as solvent is a spectator of chemical reactions. When it is not the case, there are several ways to take solvent effects into account, even now in the NP case [184]. But chasing a perfect quantitative agreement with experimental data should not be the only goal of computational studies. Theoreticians have to provide analysis, models, areas to think about and trends. Besides, important issues may arise when doing DFT calculations, which are related to the empirical character of the functionals. They have an incorrect asymptotic behavior of the exchange–correlation potential, that is a non-physically diffuse density in outer regions. Several developments are under consideration, which shall one day solve this problem. It is an intense field of research in the molecular context, but the theoretical community probably lacks the necessary distance to fairly judge what the most methodologically relevant and computationally economical solutions are. We obviously neglect in the present chapter the potential applicability of wavefunction theory, but for the time being *ab initio* methods do not seem able to handle such complex systems.

This book deals with catalysis, where one key activity on the theoretical side is the ability of computational packages to efficiently explore the PES of nanosystems. Although periodic-DFT methods are often impressive for calculating electronic properties, these types of software do not offer the same efficient algorithms for following reaction paths as those available in molecular-DFT packages. Technology transfer is thus needed between the two quantum-chemistry/-physics communities.

It is also surprising to notice that most of the theoretical work dealing with nanocatalysis only focus on geometries, thermodynamics and kinetics, and sometimes molecular vibrations. Yet, spectroscopy is a tool of choice for a fruitful dialog between theoreticians and experimentalists. In our opinion – based on recent successful studies [144, 147, 196–199] – it should be more often used in theoretical work aiming at interpreting the experimental outcome as well as guiding the rational design of efficient nanocatalysts.

In summary, several recent results obtained by a small number of theoretical groups should be considered as paragons for future state-of-the-art studies in the field of nanocatalysis. If not for the design of new catalysts, this is of interest at least to give interpretations about their catalytic behavior.

Acknowledgments

We gratefully thank Dr. Iker del Rosal, Université Paul Sabatier, Toulouse (France), for his valuable help and for stimulating discussions. The CALMIP and CINES computer centers are acknowledged for providing generous computational time. This work was supported by the ANR (project ANR-08-BLAN-0010-02). R. P. thanks the Service des Relations Internationales of the Université Paul Sabatier (APC actions) for supporting a one-month sabbatical leave in the group of Pr. G. A. Somorjai, LBNL, Berkeley, CA (USA). Discussions with faculty members especially Pr. G. A. Somorjai have been very useful.

References

1 Thomas, J.R. (1966) Preparation and magnetic properties of colloidal cobalt particles. *J. Appl. Phys.*, **37**, 2914–2915.
2 Schmid, G. (1992) Large clusters and colloids. Metals in the embryonic state. *Chem. Rev.*, **92**, 1709–1727.
3 Rodriguez, A., Amiens, C., Chaudret, B., Casanove, M.J., Lecante, P., and Bradley, J.S. (1996) Synthesis and isolation of cuboctahedral and icosahedral platinum nanoparticles. ligand-dependent structures. *Chem. Mater.*, **8** (8), 1978–1986.
4 Peng, X., Manna, L., Yang, W., Wickham, J., Scher, E., Kadavanich, A., and Alivisatos, A. (2000) Shape control of CdSe nanocrystals. *Nature*, **404** (6773), 59–61.
5 Schmid, G. (ed.) (2004) *Nanoparticles. From Theory to Application*, Wiley-VCH Verlag GmbH, Weinheim, Germany.

6 Somorjai, G.A., Tao, F., and Park, J.Y. (2008) The nanoscience revolution: Merging of colloid science, catalysis and nanoelectronics. *Top. Catal.*, **47** (1–2), 1–14.
7 Nørskov, J.K., Bligaard, T., Hvolbaek, B., Abild-Pedersen, F., Chorkendorff, I., and Christensen, C.H. (2008) The nature of the active site in heterogeneous metal catalysis. *Chem. Soc. Rev.*, **37**, 2163–2171.
8 Li, Y. and Somorjai, G.A. (2010) Nanoscale advances in catalysis and energy applications. *Nano Lett.*, **10** (7), 2289–2295.
9 Lara, P., Rivada-Wheelaghan, O., Conejero, S., Poteau, R., Philippot, K., and Chaudret, B. (2011) N-heterocyclic carbene-stabilized ruthenium nanoparticles: a case study for ligands location and their influence on reactivity. *Angew. Chem., Int. Ed. Eng.*, **50**, 12080–12084.
10 Woodward, R. and Hoffmann, R. (1970) *The Conservation of Orbital Symmetry*, Verlag Chemie and Academic Press Inc., Weinheim and New York.
11 Fukui, K. (1975) *Theory of Orientation and Stereoselection*, Springer-Verlag, Berlin.
12 Hoffmann, R. (1982) Building bridges between inorganic and organic chemistry (Nobel Lecture). *Angew. Chem., Int. Ed. Eng.*, **21**, 711–800.
13 Fukui, K. (1982) The role of frontier orbitals in chemical reactions (Nobel Lecture). *Angew. Chem., Int. Ed. Eng.*, **21**, 801–876.
14 Albright, T., Burdett, J., and Whangbo, M.H. (1985) *Orbital Interactions in Chemistry*, John Wiley & Sons, Inc., New York, USA.
15 Hoffmann, R. (1987) How chemistry and physics meet in the solid state. *Angew. Chem., Int. Ed. Eng.*, **26**, 846–878.
16 Burdett, J.K. (1995) *Chemical Bonding in Solids*, Oxford University Press, New York.
17 Slater, J.C. and Koster, G.F. (1954) Simplified LCAO method for the periodic potential problem. *Phys. Rev.*, **94** (6), 1498–1524.
18 Goringe, C., Bowler, D., and Hernández, E. (1997) Tight-binding modelling of materials. *Rep. Prog. Phys.*, **60**, 1447–1512.
19 Parr, R.G. and Yang, W. (1994) *Density-Functional Theory of Atoms and Molecules*, Oxford University Press.
20 Kohn, W. (1999) Nobel lecture: Electronic structure of matter–wave functions and density functionals. *Rev. Mod. Phys.*, **71**, 1253–1266.
21 Koch, W. and Holthausen, M.C. (2001) *A Chemist's Guide to Density Functional Theory*, 2nd edn, Wiley-VCH Verlag GmbH, Weinheim, Germany.
22 Niu, S. and Hall, M.B. (2000) Theoretical studies on reactions of transition-metal complexes. *Chem. Rev.*, **100** (2), 353–406.
23 Eisenstein, O. and Maron, L. (2002) DFT studies of some structures and reactions of lanthanides complexes. *J. Organomet. Chem.*, **6471**–2 (SI), 190–197.
24 Ziegler, T. and Autschbach, J. (2005) Theoretical methods of potential use for studies of inorganic reaction mechanisms. *Chem. Rev.*, **105** (6), 2695–2722.
25 Hunt, P.A. (2007) Organolanthanide mediated catalytic cycles: a computational perspective. *Dalton Trans*, (18), 1743–1754.
26 Boutadla, Y., Davies, D.L., Macgregor, S.A., and Poblador-Bahamonde, A.I. (2009) Mechanisms of C-H bond activation: rich synergy between computation and experiment. *Dalton Trans*, (30), 5820–5831.
27 Balcells, D., Clot, E., and Eisenstein, O. (2010) C-H bond activation in transition metal species from a computational perspective. *Chem. Rev.*, **110** (2), 749–823.
28 Neurock, M. (2003) Perspectives on the first principles elucidation and the design of active sites. *J. Catal.*, **216**, 73–88.
29 Christensen, C.H. and Nørskov, J.K. (2008) A molecular view of heterogeneous catalysis. *J. Chem. Phys.*, **128** (18), 182503.
30 Sautet, P. and Delbecq, F. (2010) Catalysis and surface organometallic chemistry: A view from theory and simulations. *Chem. Rev.*, **110** (3), 1788–1806.
31 Calvo, F. and Carré, A. (2006) Structural transitions and stabilization of palladium

nanoparticles upon hydrogenation. *Nanotechnol.*, **17**, 1292–1299.
32 Halperin, W.P. (1986) Quantum size effects in metal particles. *Rev. Mod. Phys.*, **58**, 533–606.
33 de Heer, W.A. (1993) The physics of metal clusters: experimental aspects and simple models. *Rev. Mod. Phys.*, **65**, 611–676.
34 Knight, W., Clemenger, K., de Heer, W., Saunders, W., Chou, M., and Cohen, M. (1984) Electronic shell structure and abundances of sodium clusters. *Phys. Rev. Lett.*, **52**, 2141–2143.
35 Ekardt, W. (1984) Dynamical polarizability of small metal particles: Self-consistent spherical jellium background model. *Phys. Rev. Lett.*, **52**, 1925–1928.
36 Ekardt, W. (1984) Work function of small metal particles: Self-consistent spherical jellium-background model. *Phys. Rev. B*, **29**, 1558–1564.
37 Martin, T., Bergmann, T., Göhlich, H., and Lange, T. (1990) Observation of electronic shells and shells of atoms in large Na clusters. *Chem. Phys. Lett.*, **172**, 209–213.
38 Martin, T. (2000) From atoms to solids. *Solid State Ionics*, **131** (1–2), 3–12.
39 Martin, T.P., Bergmann, T., Göhlich, H., and Lange, T. (1991) Electronic shells and shells of atoms in metallic clusters. *Z. Phys., D At. Mol. Clust.*, **19** (1–4), 25–29.
40 Van Santen, R.A. (2009) Complementary structure sensitive and insensitive catalytic relationships. *Acc. Chem. Res.*, **42** (1), 57–66.
41 Walter, M., Akola, J., Lopez-Acevedo, O., Jadzinsky, P.D., Calero, G., Ackerson, C.J., Whetten, R.L., Groenbeck, H., and Hakkinen, H. (2008) A unified view of ligand-protected gold clusters as superatom complexes. *Proc. Natl. Acad. Sci. U. S. A.*, **105** (27), 9157–9162.
42 Aikens, C.M. (2011) Electronic structure of ligand-passivated gold and silver nanoclusters. *J. Phys. Chem. Lett.*, **2** (2), 99–104.
43 Burke, K. and Gross, E. (1998) Chap. A guided tour of time-dependent density functional theory, in *Density Functional: Theory and Applications, Lecture Notes in Physics*, vol. **500**, Springer, Berlin, pp. 117–146.
44 Gerber, I.C., Ángyán, J.G., Marsman, M., and Kresse, G. (2007) Range separated hybrid density functional with long-range Hartree-Fock exchange applied to solids. *J. Chem. Phys.*, **127** (5), 054 101–1–9.
45 Hafner, J. (2008) Ab-initio simulations of materials using VASP: Density-functional theory and beyond. *J. Comp. Chem.*, **29** (13), 2044–2078.
46 Schmider, H.L. and Becke, A.D. (1998) Optimized density functionals from the extended G2 test set. *J. Chem. Phys.*, **108** (23), 9624–9631.
47 Zhao, Y. and Truhlar, D.G. (2008) Density functionals with broad applicability in chemistry. *Acc. Chem. Res.*, **41** (2), 157–167.
48 van Hardeveld, R. and van Montfoort, A. (1966) The influence of crystallite size on the adsorption of molecular nitrogen on nickel, palladium and platinum: An infrared and electron-microscopic study. *Surf. Sci.*, **4** (4), 396–430.
49 Che, M. and Bennett, C.O. (1989) The influence of particle-size on the catalytic properties of supported metals. *Adv. Catal.*, **36**, 55–172.
50 Henry, C.R. (1998) Surface studies of supported model catalysts. *Surf. Sci. Rep.*, **31** (7–8), 235–325.
51 Bezemer, G., Bitter, J., Kuipers, H., Oosterbeek, H., Holewijn, J., Xu, X., Kapteijn, F., van Dillen, A., and de Jong, K. (2006) Cobalt particle size effects in the Fischer-Tropsch reaction studied with carbon nanofiber supported catalysts. *J. Am. Chem. Soc.*, **128** (12), 3956–3964.
52 Kuhn, J.N., Huang, W., Tsung, C.K., Zhang, Y., and Somorjai, G.A. (2008) Structure sensitivity of carbon-nitrogen ring opening: Impact of platinum particle size from below 1 to 5nm upon pyrrole hydrogenation product selectivity over monodisperse platinum nanoparticles loaded onto mesoporous silica. *J. Am. Chem. Soc.*, **130** (43), 14026–14027.
53 Tsung, C.K., Kuhn, J.N., Huang, W., Aliaga, C., Hung, L.I., Somorjai, G.A., and Yang, P. (2009) Sub-10nm platinum nanocrystals with size and shape control:

Catalytic study for ethylene and pyrrole hydrogenation. *J. Am. Chem. Soc.*, **131** (16), 5816–5822.

54 Kliewer, C.J., Aliaga, C., Bieri, M., Huang, W., Tsung, C.K., Wood, J.B., Komvopoulos, K., and Somorjai, G.A. (2010) Furan hydrogenation over Pt(111) and Pt(100) single-crystal surfaces and Pt nanoparticles from 1 to 7nm: A kinetic and sum frequency generation vibrational spectroscopy study. *J. Am. Chem. Soc.*, **132** (37), 13088–13095.

55 Fischer, F. and Tropsch, H. (1923) Über die herstellung synthetischer ölgemische (syn-thol) durch aufbau aus kohlenoxyd und wasserstoff. *Brennst. Chem.*, **4**, 276–285.

56 Anderson, R.B. (1984) *The Fischer-Tropsch Synthesis*, Academic Press, New York.

57 Reuel, R.C. and Bartholomew, C.H. (1984) The stoichiometries of H_2 and CO adsorptions on cobalt - effects of support and preparation. *J. Catal.*, **85** (1), 63–77.

58 Reuel, R.C. and Bartholomew, C.H. (1984) Effects of support and dispersion on the CO hydrogenation activity selectivity properties of cobalt. *J. Catal.*, **85** (1), 78–88.

59 Lisitsyn, A.S., Golovin, A.V., and Kuznetsov, V.l., and Yermakov, Y.I. (1985) Properties of catalysts prepared by pyrolysis of $Co_2(CO)_8$ on silica containing surface Ti ions. *J. Catal.*, **95** (2), 527–538.

60 Iglesia, E. (1997) Design, synthesis, and use of cobalt-based Fischer-Tropsch synthesis catalysts. *Appl. Catal., A*, **161** (1–2), 59–78.

61 den Breejen, J.P., Radstake, P.B., Bezemer, G.L., Bitter, J.H., Froseth, V., Holmen, A., and de Jong, K.P. (2009) On the origin of the cobalt particle size effects in Fischer-Tropsch catalysis. *J. Am. Chem. Soc.*, **131** (20), 7197–7203.

62 Dumesic, J., Topsœ, H., Khammouma, S., and Bourdart, M. (1975) Surface, catalytic and magnetic-properties of small iron particles .2. Structure sensitivity of ammonia-synthesis. *J. Catal.*, **37** (3), 503–512.

63 Cabié, M., Giorgio, S., Henry, C.R., Axet, M.R., Philippot, K., and Chaudret, B. (2010) Direct observation of the reversible changes of the morphology of Pt nanoparticles under gas environment. *J. Phys. Chem. C*, **114** (5), 2160–2163.

64 Axet, M.R., Philippot, K., Chaudret, B., Cabié, M., Giorgio, S., and Henry, C.R. (2011) TEM and HRTEM evidence for the role of ligands in the formation of shape-controlled platinum nanoparticles. *Small*, **7** (2), 235–241.

65 Xia, Y., Xiong, Y., Lim, B., and Skrabalak, S.E. (2009) Shape-controlled synthesis of metal nanocrystals: Simple chemistry meets complex physics? *Angew. Chem., Int. Ed. Eng.*, **48** (1), 60–103.

66 Marks, L.D. (1994) Experimental studies of small-particle structures. *Rep. Prog. Phys.*, **57** (6), 603–649.

67 Hofmeister, H. (1998) Forty years study of fivefold twinned structures in small particles and thin films. *Cryst. Res. Technol.*, **33** (1), 3–25.

68 Yacamán, M.J., Ascencio, J.A., Liu, H.B., and Gardea-Torresdey, J. (2001) Structure shape and stability of nanometric sized particles. *J. Vac. Sci. Technol., B*, **19** (4), 1091–1103.

69 Montejano-Carrizales, J.M., Aguilera-Granja, F., and Morán-López, J.L. (1997) Direct enumeration of the geometrical characteristics of clusters. *Nanostruct. Mater.*, **8**, 269–287.

70 Honkala, K., Hellman, A., Remediakis, I.N., Logadottir, A., Carlsson, A., Dahl, S., Christensen, C.H., and Nørskov, J.K. (2005) Ammonia synthesis from first-principles calculations. *Science*, **307** (5709), 555–558.

71 Baletto, F. and Ferrando, R. (2005) Structural properties of nanoclusters: Energetic, thermodynamic, and kinetic effects. *Rev. Mod. Phys.*, **77** (1), 371–423.

72 Wang, Z., Harfenist, S., Vezmar, I., Whetten, R., Bentley, J., Evans, N., and Alexander, K. (1998) Superlattices of self-assembled tetrahedral Ag nanocrystals. *Adv. Mater.*, **10** (10), 808–812.

73 Wang, Z.L. (2000) Transmission electron microscopy of shape-controlled nanocrystals and their assemblies. *J. Phys. Chem. B*, **104** (6), 1153–1175.

74 Tang, Y. and Ouyang, M. (2007) Tailoring properties and functionalities of metal

nanoparticles through crystallinity engineering. *Nat. Mater.*, **6** (10), 754–759.

75 Zhang, Q., Xie, J., Liang, J., and Lee, J.Y. (2009) Synthesis of monodisperse Ag-Au alloy nanoparticles with independently tunable morphology, composition, size, and surface chemistry and their 3-D superlattices. *Adv. Funct. Mater.*, **19** (9), 1387–1398.

76 Mulvihill, M., Tao, A., Benjauthrit, K., Arnold, J., and Yang, P. (2008) Surface-enhanced raman spectroscopy for trace arsenic detection in contaminated water. *Angew. Chem., Int. Ed. Eng.*, **47** (34), 6456–6460.

77 Mulvihill, M.J., Ling, X.Y., Henzie, J., and Yang, P. (2010) Anisotropic etching of silver nanoparticles for plasmonic structures capable of single-particle sers. *J. Am. Chem. Soc.*, **132** (1), 268–274.

78 Tiago, M.L., Zhou, Y., Alemany, M.M.G., Saad, Y., and Chelikowsky, J.R. (2006) Evolution of magnetism in iron from the atom to the bulk. *Phys. Rev. Lett.*, **97** (14), 147201.

79 Rollmann, G., Gruner, M.E., Hucht, A., Meyer, R., Entel, P., Tiago, M.L., and Chelikowsky, J.R. (2007) Shellwise Mackay transformation in iron nanoclusters. *Phys. Rev. Lett.*, **99** (8), 083402.

80 Gruner, M.E. and Entel, P. (2009) Simulating functional magnetic materials on supercomputers. *J. Phys.: Condens. Matter*, **21** (29), 293201.

81 Kleis, J., Greeley, J., Romero, N.A., Morozov, V.A., Falsig, H., Larsen, A.H., Lu, J., Mortensen, J.J., Dulak, M., Thygesen, K.S., Nørskov, J.K., and Jacobsen, K.W. (2011) Finite size effects in chemical bonding: From small clusters to solids. *Catal. Lett.*, **141** (8), 1067–1071.

82 Lacroix, L.M., Lachaize, S., Falqui, A., Respaud, M., and Chaudret, B. (2009) Iron nanoparticle growth in organic superstructures. *J. Am. Chem. Soc.*, **131** (2), 549–557.

83 Ryu, J.H., Han, S.S., Kim, D.H., Henkelman, G., and Lee, H.M. (2011) Ligand-induced structural evolution of Pt_{55} nanoparticles: Amine versus thiol. *ACS Nano*, **5** (11), 8515–8522.

84 Sabatier, P. (1911) Hydrogénations et déshydrogénations par catalyse. *Ber. Dtsch. Chem. Ges.*, **44** (3), 1984–2001.

85 Sabatier, P. (1920) *La Catalyse en Chimie Organique*, 2nd edn, Librairie Polytechnique, Paris, Liège.

86 Balandin, A. (1969) *Modern State of the Multiplet Theory of Heterogeneous Catalysis*, Academic Press, pp. 1–210.

87 Swiegers, G. (2008) *Mechanical Catalysis: Methods of Enzymatic, Homogeneous, and Heterogeneous Catalysis*, John Wiley & Sons, Inc., Hoboken, USA.

88 Taylor, H.S. (1925) A theory of the catalytic surface. *Proc. R. Soc. London, Ser. A*, **108**, 105–111.

89 Hammer, B. (2006) Special sites at noble and late transition metal catalysts. *Top. Catal.*, **37** (1), 3–16.

90 Curtarolo, S., Morgan, D., Persson, K., Rodgers, J., and Ceder, G. (2003) Predicting crystal structures with data mining of quantum calculations. *Phys. Rev. Lett.*, **91**, 135–503.

91 Fischer, C.C., Tibbetts, K.J., Dane, M., and Ceder, G. (2006) Predicting crystal structure by merging data mining with quantum mechanics. *Nat. Mater.*, **5**, 641–646.

92 Froemming, N.S. and Henkelman, G. (2009) Optimizing core-shell nanoparticle catalysts with a genetic algorithm. *J. Chem. Phys.*, **131** (23), 234103.

93 Hammer, B. and Nørskov, J.K. (1995) Electronic factors determining the reactivity of metal surfaces. *Surf. Sci.*, **343** (3), 211–220.

94 Feibelman, P.J. and Hamann, D.R. (1984) Electronic structure of a 'poisoned' transition-metal surface. *Phys. Rev. Lett.*, **52**, 61–64.

95 Feibelman, P.J. and Hamann, D. (1985) Modification of transition metal electronic structure by p, s, cl, and li adatoms. *Surface Science*, **149** (1), 48–66.

96 Falicov, L.M. and Somorjai, G.A. (1985) Correlation between catalytic activity and bonding and coordination number of atoms and molecules on transition metal surfaces: Theory and experimental evidence. *Proc. Natl. Acad. Sci. U. S. A.*, **82** (8), 2207–2211.

97 Yang, W. and Parr, R.G. (1985) Hardness, softness, and the Fukui function in the electronic theory of metals and catalysis. *Proc. Natl. Acad. Sci. U. S. A.*, **82** (20), 6723–6726.

98 Harris, J. and Andersson, S. (1985) h_2 dissociation at metal surfaces. *Phys. Rev. Lett.*, **55**, 1583–1586.

99 Parr, R.G. and Yang, W. (1984) Density functional approach to the frontier-electron theory of chemical reactivity. *J. Am. Chem. Soc.*, **106** (14), 4049–4050.

100 Hammer, B. and Nørskov, J. (2000) Theoretical surface science and catalysis–calculations and concepts, in *Impact of Surface Science on Catalysis, Adv. Catal*, vol. **45** (eds H.K. Bruce and C. Gates), Academic Press, pp. 71–129.

101 Hoffmann, R. (1988) *Solids and Surfaces. A Chemist's View of Bonding in Extended Structures*, Wiley-VCH Verlag GmbH, Weinheim, Germany.

102 Nilsson, A., Pettersson, L.G.M., Hammer, B., Bligaard, T., Christensen, C.H., and Nørskov, J.K. (2005) The electronic structure effect in heterogeneous catalysis. *Catal. Lett.*, **100**, 111–114.

103 Nørskov, J.K., Abild-Pedersen, F., Studt, F., and Bligaard, T. (2011) Density functional theory in surface chemistry and catalysis. *Proc. Natl. Acad. Sci. U. S. A.*, **108** (3), 937–943.

104 Ruban, A., Hammer, B., Stoltze, P., Skriver, H., and Nørskov, J. (1997) Surface electronic structure and reactivity of transition and noble metals. *J. Mol. Catal. A: Chemical*, **115** (3), 421–429.

105 Anh, N.T. (2007) *Frontier Orbitals: A Practical Manual*, Wiley-VCH Verlag GmbH, Weinheim, Germany.

106 Abild-Pedersen, F., Greeley, J., Studt, F., Rossmeisl, J., Munter, T.R., Moses, P.G., Skúlason, E., Bligaard, T., and Nørskov, J.K. (2007) Scaling properties of adsorption energies for hydrogen-containing molecules on transition-metal surfaces. *Phys. Rev. Lett.*, **99**, 016105.

107 Nørskov, J.K., Bligaard, T., and Kleis, J. (2009) Rate control and reaction engineering. *Science*, **324** (5935), 1655–1656.

108 Jiang, T., Mowbray, D.J., Dobrin, S., Falsig, H., Hølbaek, B., Bligaard, T., and Nørskov, J.K. (2009) Trends in co oxidation rates for metal nanoparticles and close-packed, stepped, and kinked surfaces. *J. Phys. Chem. C*, **113** (24), 10548–10553.

109 Hansen, E.W. and Neurock, M. (2000) First-principles-based Monte Carlo simulation of ethylene hydrogenation kinetics on Pd. *J. Catal.*, **196** (2), 241–252.

110 Linic, S. and Barteau, M.A. (2003) Construction of a reaction coordinate and a microkinetic model for ethylene epoxidation on silver from DFT calculations and surface science experiments. *J. Catal.*, **214** (2), 200–212.

111 Reuter, K., Frenkel, D., and Scheffler, M. (2004) The steady state of heterogeneous catalysis, studied by first-principles statistical mechanics. *Phys. Rev. Lett.*, **93**, 116105.

112 Nørskov, J.K., Bligaard, T., Rossmeisl, J., and Christensen, C.H. (2009) Towards the computational design of solid catalysts. *Nat. Chem.*, **1** (1), 37–46.

113 Andersson, M.P., Bligaard, T., Kustov, A., Larsen, K.E., Greeley, J., Johannessen, T., Christensen, C.H., and Nørskov, J.K. (2006) Toward computational screening in heterogeneous catalysis: Pareto-optimal methanation catalysts. *J. Catal.*, **239** (2), 501–506.

114 Linic, S., Jankowiak, J., and Barteau, M.A. (2004) Selectivity driven design of bimetallic ethylene epoxidation catalysts from first principles. *J. Catal.*, **224** (2), 489–493.

115 Greeley, J. and Mavrikakis, M. (2006) Alloy catalysts designed from first principles. *Nat. Mater.*, **3**, 810.

116 Toulhoat, H. and Raybaud, P. (2003) Kinetic interpretation of catalytic activity patterns based on theoretical chemical descriptors. *J. Catal.*, **216** (1–2), 63–72.

117 Studt, F., Abild-Pedersen, F., Bligaard, T., Sørensen, R.Z., Christensen, C.H., and Nørskov, J.K. (2008) Identification of non-precious metal alloy catalysts for selective hydrogenation of acetylene. *Science*, **320** (5881), 1320–1322.

118 Greeley, J., Jaramillo, T.F., Bonde, J., Chorkendorff, I., and Nørskov, J.K. (2006) Computational high-throughput screening

of electrocatalytic materials for hydrogen evolution. *Nat. Mater.*, **5**, 909–913.

119 Besenbacher, F., Chorkendorff, I., Clausen, B.S., Hammer, B., Molenbroek, A.M., Nørskov, J.K., and Stensgaard, I. (1998) Design of a surface alloy catalyst for steam reforming. *Science*, **279** (5358), 1913–1915.

120 Greeley, J., Stephens, I.E.L., Bondarenko, A.S., Johansson, T.P., Hansen, H.A., Jaramillo, T.F., Rossmeis, l.J., Chorkendorff, I., and Nørskov, J.K. (2009) Alloys of platinum and early transition metals as oxygen reduction electrocatalysts. *Nat. Chem.*, **1**, 552–556.

121 Strasser, P., Fan, Q., Devenney, M., Weinberg, W.H., Liu, P., and Nørskov, J.K. (2003) High throughput experimental and theoretical predictive screening of materials - a comparative study of search strategies for new fuel cell anode catalysts. *J. Phys. Chem. B*, **107** (40), 11013–11021.

122 Hohenberg, P. and Kohn, W. (1964) Inhomogeneous electron gas. *Phys. Rev.*, **136**, B864–B871.

123 Kohn, W. and Sham, L. (1965) Self-consistent equations including exchange and correlation effects. *Phys. Rev.*, **140**, A1133–A1138.

124 Perdew, J.P. and Schmidt, K. (2001) Jacob's ladder of density functional approximations for the exchange-correlation energy, in *Density functional theory and its application to materials, American Institute of Physics Conference Series, American Institute of Physics Conference Series*, vol. **577** (eds V.V. Doren, C.V. Alsenoy, and P. Geerlings), AIP, pp. 1–20.

125 Feibelman, P.J., Hammer, B., Nørskov, J.K., Wagner, F., Scheffler, M., Stumpf, R., Watwe, R., and Dumesic, J. (2001) The CO/Pt(111) puzzle. *J. Phys. Chem. B*, **105** (18), 4018–4025.

126 Gil, A., Clotet, A., Ricart, J.M., Kresse, G., Garcia-Hernandez, M., Rösch, N., and Sautet, P. (2003) Site preference of CO chemisorbed on Pt(111) from density functional calculations. *Surf. Sci.*, **530** (1–2), 71–87.

127 Stroppa, A. and Kresse, G. (2008) The shortcomings of semi-local and hybrid functionals: what we can learn from surface science studies. *New J. Phys.*, **10** (6), 063 020.

128 Shimka, L., Harl, J., Stroppa, A., Grüneis, A., Marsman, M., Mittendorfer, F., and Kresse, G. (2010) Accurate surface and adsorption energies from many-body perturbation theory. *Nat. Mater.*, **9**, 741–744.

129 Tkatchenko, A., Romaner, L., Hofmann, O., Zojer, E., Ambrosch-Draxl, C., and Scheffler, M. (2010) van der Waals interactions between organic adsorbates and at organic/inorganic interfaces. *MRS Bull.*, **35**, 435–442.

130 Gerber, I.C. and Ángyán, J.G. (2007) London dispersion forces by range-separated hybrid density functional with second order perturbational corrections: The case of rare gas complexes. *J. Chem. Phys.*, **126** (4), 044103–1–14.

131 Grimme, S. (2004) Accurate description of van der Waals complexes by density functional theory including empirical corrections. *J. Comp. Chem.*, **25**, 1463–1473.

132 Tkatchenko, A. and Scheffler, M. (2009) Accurate molecular van der Waals interactions from ground-state electron density and free-atom reference data. *Phys. Rev. Lett.*, **102**, 073 005.

133 Ángyán, J.G., Gerber, I.C., Savin, A., and Toulouse, J. (2005) van der Waals forces in density functional theory: perturbational long-range electron interaction corrections. *Phys. Rev. A*, **72**, 012 510.

134 Dion, M., Rydberg, H., Schröder, E., Langreth, D.C., and Lundqvist, B.I. (2004) Van der Waals density functional for general geometries. *Phys. Rev. Lett.*, **92**, 246–401.

135 Loschen, C., Bromley, S.T., Neyman, K.M., and Illas, F. (2007) Understanding ceria nanoparticles from first-principles calculations. *J. Phys. Chem. C*, **111** (28), 10142–10145.

136 Vayssilov, G.N., Migani, A., and Neyman, K. (2011) Density functional modeling of the interactions of platinum clusters with CeO_2 nanoparticles of different size. *J. Phys. Chem. C*, **115** (32), 16081–16086.

137 Bond, G.C., Coq, B., Dutartre, R., Ruiz, J.G., Hooper, A.D., Proietti, M.G., Sierra, M.C.S., and Slaa, J.C. (1996) Effect of various pretreatments on the structure and properties of ruthenium catalysts. *J. Catal.*, **161**, 480–494.

138 Six, C., Beck, K., Wegner, A., and Leitner, W. (2000) Ligand effects in novel ruthenium-based ROMP catalysts bearing bidentate phosphines. *Organometallics*, **19**, 4639–4642.

139 McQuire, M.W. and Rochester, C.H. (1995) FTIR study of CO/H_2 reactions over Ru/TiO_2 and $Ru\text{-}Rh/TiO_2$ catalysts at high temperature and pressure. *J. Catal.*, **157**, 396–402.

140 Maitlis, P.M. (2004) Fischer-tropsch, organometallics, and other friends. *J. Organomet. Chem.*, **689**, 4366–4374.

141 Pery, T., Pelzer, K., Buntkowsky, G., Philippot, K., Limbach, H.H., and Chaudret, B. (2005) Direct NMR evidence for the presence of mobile surface hydrides on ruthenium nanoparticles. *ChemPhysChem*, **6**, 605–607.

142 Jansat, S., Picurelli, D., Pelzer, K., Philippot, K., Gómez, M., Muller, G., Lecante, P., and Chaudret, B. (2006) Synthesis, characterization and catalytic reactivity of ruthenium nanoparticles stabilized by chiral N-donor ligands. *New J. Chem.*, **30**, 115–122.

143 Holzwarth, N.A.W. and Chelikowsky, J.R. (1985) Bulk and surface electron states in ruthenium. *Solid State Commun.*, **53** (2), 171–174.

144 del Rosal, I., Truflandier, L., Poteau, R., and Gerber, I.C. (2011) A DFT study of spectroscopic and thermodynamic properties of surfacic hydrides on Ru (0001) model surface: the influence of the coordination modes and the coverage. *J. Phys. Chem. C*, **115**, 2169–2178.

145 Ciobîcă, I.M., Kleyn, A.W., and Santen, R.A.V. (2003) Adsorption and coadsorption of CO and H on ruthenium surfaces. *J. Phys. Chem. B*, **107**, 164–172.

146 Xu, L., Xiao, H., and Zu, X. (2005) Hydrogen adsorption on Ru(0 0 1) surface from density-functional periodic calculations. *Chem. Phys.*, **315**, 155–160.

147 Truflandier, L.A., del Rosal, I., Chaudret, B., Poteau, R., and Gerber, I.C. (2009) Where does hydrogen adsorb on Ru nanoparticles ? A powerful joint ^2H MAS-NMR/DFT study. *ChemPhysChem*, **10**, 2939–2942.

148 Müller, K. and Brown, L.D. (1979) Location of saddle points and minimum energy paths by a constrained simplex optimization procedure. *Theor. Chem. Acc.*, **53** (1), 75–93.

149 McQuarrie, D.A. and Simon, J.D. (1999) *Molecular Thermodynamics*, University Science Book, Sausalito, California.

150 Beveridge, D.L. and Dicapua, F.M. (1989) Free-energy via molecular simulation - applications to chemical and biomolecular systems. *Annu. Rev. Biophys. Bio.*, **18**, 431–492.

151 Raynaud, C., Daudey, J.P., Jolibois, F., and Maron, L. (2006) *Ab Initio* dynamic study of the reaction of $Cl_2$2LaR (R=H, CH_3) with H_2. *J. Phys. Chem. A*, **110**, 101–105.

152 van Santen, R.A., Neurock, M., and Shetty, S.G. (2010) Reactivity theory of transition-metal surfaces: A Brønsted-Evans-Polanyi linear activation energy-free-energy analysis. *Chem. Rev.*, **110** (4), 2005–2048.

153 Schlegel, H.B. (2011) Geometry optimization. *WIRES. Comput. Mol. Sci.*, **1** (5), 790–809.

154 Frisch, M.J., Trucks, G.W., Schlegel, H.B., Scuseria, G.E., Robb, M.A., Cheeseman, J.R., Scalmani, G., Barone, V., Mennucci, B., Petersson, G.A., Nakatsuji, H., Caricato, M., Li, X., Hratchian, H.P., Izmaylov, A.F., Bloino, J., Zheng, G., Sonnenberg, J.L., Hada, M., Ehara, M., Toyota, K., Fukuda, R., Hasegawa, J., Ishida, M., Nakajima, T., Honda, Y., Kitao, O., Nakai, H., Vreven, T., Montgomery J Jr., J.A., Peralta, J.E., Ogliaro, F., Bearpark, M., Heyd, J.J., Brothers, E., Kudin, K.N., Staroverov, V.N., Kobayashi, R., Normand, J., Raghavachari, K., Rendell, A., Burant, J.C., Iyengar, S.S., Tomasi, J., Cossi, M., Rega, N., Millam, J.M., Klene, M., Knox, J.E., Cross, J.B., Bakken, V., Adamo, C., Jaramillo, J., Gomperts, R., Stratmann, R.E., Yazyev, O., Austin, A.J., Cammi, R.,

Pomelli, C., Ochterski, J.W., Martin, R.L., Morokuma, K., Zakrzewski, V.G., Voth, G.A., Salvador, P., Dannenberg, J.J., Dapprich, S., Daniels, A.D., Farkas, Ö., Foresman, J.B., Ortiz, J.V., Cioslowski, J., and Fox, D.J. (2009) *Gaussian 09 Revision A.1*, Gaussian Inc, Wallingford CT.

155 Kresse, G. and Fürthmuller, J. (1996) Efficient iterative schemes for ab initio total-energy calculations using a plane-wave basis set. *Phys. Rev. B*, **54** (16), 11169–11186.

156 Kresse, G. and Fürthmuller, J. (1996) Efficiency of ab-initio total energy calculations for metals and semiconductors using a plane-wave basis set. *Comput. Mater. Sci.*, **6** (1), 15–50.

157 Peng, C., Ayala, P.Y., Schlegel, H.B., and Frisch, M.J. (1996) Using redundant internal coordinates to optimize equilibrium geometries and transition states. *J. Comp. Chem.*, **17**, 49–56.

158 Pulay, P., Fogarasi, G., Pang, F., and Boggs, J.E. (1979) Systematic ab initio gradient calculation of molecular geometries, force constants, and dipole moment derivatives. *J. Am. Chem. Soc.*, **101** (10), 2550–2560.

159 Fogarasi, G., Zhou, X., Taylor, P.W., and Pulay, P. (1992) The calculation of ab initio molecular geometries: efficient optimization by natural internal coordinates and empirical correction by offset forces. *J. Am. Chem. Soc.*, **114** (21), 8191–8201.

160 Bučko, T., Hafner, J., and Ángyán, J.G. (2005) Geometry optimization of periodic systems using internal coordinates. *J. Chem. Phys.*, **122** (12), 124508.

161 Gonzalez, C. and Schlegel, H.B. (1990) Reaction path following in mass-weighted internal coordinates. *J. Phys. Chem.*, **94**, 5523–5527.

162 Liotard, D. and Penot, J.P. (1981) *Numerical Methods in the Study of Critical Phenomena*, Springer, Berlin, Heidelberg, New York, p. 213. Proceedings of a colloquium, Carry-le-Rouet, France, June 2–4, 1980.

163 Pratt, L.R. (1986) A statistical method for identifying transition states in high dimensional problems. *J. Chem. Phys.*, **85** (9), 5045–5048.

164 Elber, R. and Karplus, M. (1987) A method for determining reaction paths in large molecules: Application to myoglobin. *Chem. Phys. Lett.*, **139** (5), 375–380.

165 Liotard, D.A. (1992) Algorithmic tools in the study of semiempirical potential surfaces. *Int. J. Quant. Chem.*, **44** (5), 723–741.

166 Jónsson, H., Mills, G., and Jacobsen, K.W. (1998) Chap. Nudged elastic band method for finding minimum energy paths of transitions, in *Classical and Quantum Dynamics in Condensed Phase Simulations*, World Scientific Publishing Co., Singapore, pp. 385–404.

167 Sheppard, D., Terrell, R., and Henkelman, G. (2008) Optimization methods for finding minimum energy paths. *J. Chem. Phys.*, **128**, 134 106–1–10.

168 van Hardeveld, R. and van Montfoort, A. (1969) Infrared spectra of nitrogen adsorbed on nickel-on-aerosil catalysts: Effects of intermolecular interaction and isotopic substitution. *Surf. Sci.*, **17** (1), 90–124.

169 van Hardeveld, R. and Hartog, F. (1972) Influence of metal particle size in nickel-on-aerosil catalysts on surface size, distribution, catalytic activity, and selectivity. *Adv. Catal.*, **22**, 75–113.

170 Dahl, S., Logadottir, A., Egeberg, R.C., Larsen, J.H., Chorkendorff, I., Törnqvist, E., and Nørskov, J.K. (1999) Role of steps in n_2 activation on Ru(0001). *Phys. Rev. Lett.*, **83**, 1814–1817.

171 Dahl, S., Törnqvist, E., and Chorkendorff, I. (2000) Dissociative adsorption of N_2 on Ru(0001): A surface reaction totally dominated by steps. *J. Catal.*, **192** (2), 381–390.

172 Wulff, G. (1901) Zur frage der geschwindigkeit des wachstums und der auflösung der krystallflachen (on the question of speed of growth and dissolution of crystal surfaces). *Z. Kristallogr. Mineral.*, **34** (5/6), 449–530.

173 von Laue, M. (1943) Wulff's theorem for the balanced form of crystals. *Z. Kristallogr.*, **105** (2), 124–133.

174 Herring, C. (1951) Some theorems on the free energies of crystal surfaces. *Phys. Rev.*, **82** (1), 87–93.
175 Pei, Y., Shao, N., Gao, Y., and Zeng, X. C. (2010) Investigating active site of gold nanoparticle $Au_{55}(PPh_3)_{12}Cl_6$ in selective oxidation. *ACS Nano*, **4** (4), 2009–2020.
176 Kahlich, D., Wiechern, U., and Lindner, J. (2000) *Propylene Oxide*, Wiley-VCH Verlag GmbH, Weinheim, Germany.
177 Molina, L.M., Lee, S., Sell, K., Barcaro, G., Fortunelli, A., Lee, B., Seifert, S., Winans, R.E., Elam, J.W., Pellin, M.J., Barke, I., von Oeynhausen, V., Lei, Y., Meyer, R.J., Alonso, J.A., Rodriguez, A. F., Kleibert, A., Giorgio, S., Henry, C.R., Meiwes-Broer, K.H., and Vajda, S. (2011) Size-dependent selectivity and activity of silver nanoclusters in the partial oxidation of propylene to propylene oxide and acrolein: A joint experimental and theoretical study. *Catal. Today*, **160** (1), 116–130.
178 Kaischew, R. (1950) Equilibrium shape and nucleation work of crystals on substrates. *Commun. Bulg. Acad. Sci.*, **1**, 100–139.
179 Müller, P. and Kern, R. (2000) Equilibrium nano-shape change induced by epitaxial stress: effect of surface stress. *Appl. Surf. Sci.*, **164** (1–4), 68–71.
180 Miertus, S., Scrocco, E., and Tomasi, J. (1981) Electrostatic interaction of a solute with a continuum. A direct utilization of ab initio molecular potentials for the prevision of solvent effects. *Chem. Phys.*, **55**, 117–129.
181 Cossi, M. and Barone, V. (2000) Solvent effect on vertical electronic transitions by the polarizable continuum model. *J. Chem. Phys.*, **112**, 2427–2435.
182 Cossi, M., Scalmani, G., Rega, N., and Barone, V. (2002) New developments in the polarizable continuum model for quantum mechanical and classical calculations on molecules in solution. *J. Chem. Phys.*, **117**, 43–54.
183 Wang, H.F. and Liu, Z.P. (2009) Formic acid oxidation at Pt/H2O interface from periodic DFT calculations integrated with a continuum solvation model. *J. Phys. Chem. C*, **113** (40), 17502–17508.
184 Shang, C. and Liu, Z.P. (2011) Origin and activity of gold nanoparticles as aerobic oxidation catalysts in aqueous solution. *J. Am. Chem. Soc.*, **133** (25), 9938–9947.
185 Lanzani, G., Susi, T., Ayala, P., Jiang, T., Nasibulin, A.G., Bligaard, T., Pichler, T., Laasonen, K., and Kauppinen, E.I. (2010) Mechanism study of floating catalyst CVD synthesis of SWCNTs. *Phys. Status Solidi B*, **247** (11–12), 2708–2712.
186 Filot, I.A.W., Shetty, S.G., Hensen, E.J. M., and van Santen, R.A. (2011) Size and topological effects of rhodium surfaces, clusters and nanoparticles on the dissociation of CO. _*J. Phys. Chem. C*, **115** (0), 14204–14212.
187 Mavrikakis, M., Bümer, M., Freund, H.J., and Nørskov, J.K. (2002) Structure sensitivity of CO dissociation on Rh surfaces. *Catal. Lett.*, **81**, 153–156.
188 Crabtree, G.W., Dresselhaus, M.S., and Buchanan, M.V. (2004) The hydrogen economy. *Phys. Today*, **57** (12), 39–44.
189 Crabtree, G.W., Dresselhaus, M.S., and Buchanan, M.V. (2005) Thoughts on starting the hydrogen economy - reply. *Phys. Today*, **58** (6), 15.
190 Ritter, S.K. (2011) Still chasing the hydrogen economy. *Chem. Eng. News*, **89** (45), 26–29.
191 Jones, G., Jakobsen, J.G., Shim, S.S., Kleis, J., Andersson, M.P., Rossmeisl, J., Abild-Pedersen, F., Bligaard, T., Helveg, S., Hinnemann, B., Rostrup-Nielsen, J. R., Chorkendorff, I., Sehested, J., and Nørskov, J.K. (2008) First principles calculations and experimental insight into methane steam reforming over transition metal catalysts. *J. Catal.*, **259** (1), 147–160.
192 Brönsted, J.N. (1928) Acid and basic catalysis. *Chem. Rev.*, **5** (3), 231–338.
193 Evans, M.G. and Polanyi, M. (1938) Inertia and driving force of chemical reactions. *Trans. Faraday Soc.*, **34**, 11–24.
194 Somorjai, G.A. and Li, Y. (2010) *Surface Chemistry and Catalysis*, 2nd edn, John Wiley & Sons, Inc., Hoboken, USA.
195 Tedsree, K., Chan, C.W.A., Jones, S., Cuan, Q., Li, W.K., Gong, X.Q., and Tsang, S.C.E. (2011) ^{13}C NMR guides rational design of nanocatalysts via

chemisorption evaluation in liquid phase. *Science*, **332** (6026), 224–228.
196 del Rosal, I., Maron, L., Poteau, R., and Jolibois, F. (2008) DFT calculations of ^{1}H and ^{13}C NMR chemical shifts in transition metal complexes. *Dalton Trans.*, 3959–3970.
197 del Rosal, I., Jolibois, F., Maron, L., Philippot, K., Chaudret, B., and Poteau, R. (2009) Ligands effects on the NMR, vibrational and structural properties of tetra- and hexanuclear ruthenium hydrido clusters: a theoretical investigation. *Dalton Trans.*, 2142–2156.
198 Gutmann, T., Walaszek, B., Yeping, X., Wächtler, M., del Rosal, I., Grünberg, A., Poteau, R., Axet, R., Lavigne, G., Chaudret, B., Limbach, H.H., and Buntkowsky, G. (2010) Hydrido-ruthenium cluster complexes as models for reactive surface hydrogen species of ruthenium nanoparticles. ^{2}H solid-state NMR and quantum chemical calculations. *J. Am. Chem. Soc.*, **132** (33), 11759–11767.
199 del Rosal, I., Gutmann, T., Walaszek, B., Gerber, I.C., Chaudret, B., Limbach, H. H., Buntkowsky, G., and Poteau, R. (2011) 2H NMR calculations on polynuclear transition metal complexes: on the influence of local symmetry and other factors. *Phys. Chem. Chem. Phys.*, **13**, 20199–20207.

Index

Nanomaterials in Catalysis, First Edition. Edited by Philippe Serp and Karine Philippot.

d

q

r

s